Springer Proceedings in Mathematics & Statistics

Volume 152

Springer Proceedings in Mathematics & Statistics

This book series features volumes composed of selected contributions from workshops and conferences in all areas of current research in mathematics and statistics, including operation research and optimization. In addition to an overall evaluation of the interest, scientific quality, and timeliness of each proposal at the hands of the publisher, individual contributions are all refereed to the high quality standards of leading journals in the field. Thus, this series provides the research community with well-edited, authoritative reports on developments in the most exciting areas of mathematical and statistical research today.

More information about this series at http://www.springer.com/series/10533

Jean-Yves Beziau · Mihir Chakraborty
Soma Dutta
Editors

New Directions in Paraconsistent Logic

5th WCP, Kolkata, India, February 2014

Springer

Editors
Jean-Yves Beziau
Federal University of Rio de Janeiro
Rio de Janeiro
Brazil

Mihir Chakraborty
Indian Statistical Institute Kolkata
Kolkata
India

Soma Dutta
C.I.T. Campus
Institute for Mathematical Sciences
Chennai, Tamil Nadu
India

ISSN 2194-1009 ISSN 2194-1017 (electronic)
Springer Proceedings in Mathematics & Statistics
ISBN 978-81-322-2717-5 ISBN 978-81-322-2719-9 (eBook)
DOI 10.1007/978-81-322-2719-9

Library of Congress Control Number: 2015955853

Mathematics Subject Classification (2010): 03B53, 03B45, 03B50, 03B52, 03A05

Springer New Delhi Heidelberg New York Dordrecht London

Printed on acid-free paper

Springer (India) Pvt. Ltd. is part of Springer Science+Business Media (www.springer.com)

Foreword

This book constitutes the proceedings of the 5th World Congress of Paraconsistent Logic, a kind of logic whose systematic development effectively began in the middle of the twentieth century.

Loosely speaking, a paraconsistent system of logic can be defined as a system that may be the underlying logic of inconsistent but nontrivial theories, i.e., inconsistent theories, in which there are sentences of their languages that are not provable.

The systems of paraconsistent logic may be envisaged from two basic perspectives: (a) as rivals of classical logic, for instance, when inconsistent though apparently nontrivial set theories are built as rivals of classical set theories and (b) as systems complementary to classical logic, when, for example, a paraconsistent negation is present together with classical negation.

An important characteristic of paraconsistent logic is that it has found numerous applications in philosophy, quantum mechanics, artificial intelligence, traffic control, medicine, economics, finances, and computing. So, we are in the presence of an area of logic which opened up new research directions in philosophy, science, and technology.

This volume is valuable not only due to the technical works it includes, but also because it contributes to show the meaning of paraconsistent logic and its connection with other domains of knowledge. It also testifies to how paraconsistent logic is spreading around the world, the 5th edition of this congress having been organized in India.

Florianópolis
September 2015

Newton C.A. da Costa

Preface

This book is a collection of papers presented at the 5th World Congress on Paraconsistency, which was organized at the Indian Statistical Institute (ISI), Kolkata, India, February 13–17, 2014.

A paraconsistent logic is a logic where there is a nonexplosive negation, i.e., from a proposition and its paraconsistent negation it is not necessarily possible to deduce anything. The expression "paraconsistent logic" was coined in a discussion between Newton da Costa and the Peruvian philosopher Francisco Miró Quesada. This expression had a booming effect as recalled by da Costa:

> Several years ago, I needed a convenient and meaningful denomination for a logic that did not eliminate contradictions from the outset as being false, i.e., as absolutely unacceptable. Miró Quesada helped me. On the one hand, it should be recalled that, by that time, all logics unavoidably condemned contradictions. The new logic in which I worked faced too much resistance, it was badly divulged, and those that got to know it were in general sceptics. By that time I wrote to Miró Quesada, who saw the new logic with great enthusiasm, requesting a name for it. I remember as it was today that he answered with three proposals: it could be called metaconsistent, ultraconsistent or paraconsistent. After commenting on these possible denominations, he stated that, from his viewpoint, he preferred the latter. The term paraconsistent sounded splendid and I began to use it, suggesting that people interested on this logic did the same. Two or three months later, the miracle took place; the term spread through the world, all the centres directly or indirectly related to logic, from northern to southern hemisphere, began to employ it. I believe that few times in the history of science (definitely in the history of logic) something similar has happened, for not only the word run the whole world, but the very logic called by Miró Quesada "paraconsistent" received a formidable push. It became one of the most discussed theories of logic of our time. (da Costa, "La Filosofia dc la Logica de Francisco Miró Quesada Cantuarias," in *Logica, Razon y Humanismo,* Lima, 1992, pp. 69–78.)

Previous world congresses on paraconsistency were organized in the following locations:

- 1st Word Congress on Paraconsistency: Ghent, Belgium (1997)
- 2nd Word Congress on Paraconsistency: Juquehy, Brazil (2000)
- 3rd Word Congress on Paraconsistency: Toulouse, France (2003)
- 4th Word Congress on Paraconsistency: Melbourne, Australia (2008)

In India, paraconsistent logic is still not very well known, but people do have interest in the subject, and a few researchers have taken it quite seriously. In Indian ancient methodology, there was "chatuskoti," which had four corners of which one was both "yes" and "no." This implies that contradiction was not altogether rejected. That is why it was decided to organize the 5th edition of the world congress on paraconsistency in this country.

And to make paraconsistent logic better known in India, we decided to organize tutorials during this event. Three tutorials were given, and they are included in the first part of this book. The other parts of the books contain papers presented during the event, and a few others are by people who were not able to come.

The event was nice and relaxing. The ISI is a charming place surrounded by nature and with a convenient guest house. The people from ISI were enthusiastic and animated to organize this event. The members of the local organizing ISI team included Sisir Roy, Rana Barua, Probal Dasgupta, Kuntal Ghosh, and Guruprasad Kar. Kuntal led the team in an efficient manner, supported by some local students, who helped to make this event a success. One evening a beautiful cruise was organized on the Ganga.

Jean-Yves Beziau
Mihir Chakraborty
Soma Dutta

Contents

Part I Tutorials

1 Tutorial on Inconsistency-Adaptive Logics 3
Diderik Batens

2 Round Squares Are No Contradictions (Tutorial on Negation Contradiction and Opposition) 39
Jean-Yves Beziau

3 On the Philosophy and Mathematics of the Logics of Formal Inconsistency 57
Walter Carnielli and Abilio Rodrigues

Part II Many-valued Systems of Paraconsistent Logic

4 Three-Valued Paraconsistent Propositional Logics 91
Ofer Arieli and Arnon Avron

5 Strong Three-Valued Paraconsistent Logics 131
Jean-Yves Beziau and Anna Franceschetto

6 An Approach to Paraconsistent Multivalued Logic: Evaluation by Complex Truth Values 147
J. Nescolarde-Selva, J.L. Usó-Doménech and K. Alonso-Stenberg

7 A Paraconsistent Logic Obtained from an Algebra-Valued Model of Set Theory 165
Sourav Tarafder and Mihir Kr. Chakraborty

8 Two Consistent Many-Valued Logics for Paraconsistent Phenomena 185
Esko Turunen and J. Tinguaro Rodríguez

Part III Paraconsitency and Modality

9 **On Modal Logics Defining Jaśkowski-Like Discussive Logics** 213
Marek Nasieniewski and Andrzej Pietruszczak

10 **From Possibility Theory to Paraconsistency** 229
Davide Ciucci and Didier Dubois

11 **Modality, Potentiality, and Contradiction in Quantum Mechanics** 249
Christian de Ronde

Part IV Tools and Framework

12 **Consequence–Inconsistency Interrelation: In the Framework of Paraconsistent Logics** 269
Soma Dutta and Mihir K. Chakraborty

13 **Univalent Foundations of Mathematics and Paraconsistency** 285
Vladimir L. Vasyukov

14 **A Method of Defining Paraconsistent Tableaus** 295
Tomasz Jarmużek and Marcin Tkaczyk

15 **Some Adaptive Contributions to Logics of Formal Inconsistency** 309
Diderik Batens

16 **Stipulation and Symmetrical Consequence** 335
Bryson Brown

17 **Logic—The Big Picture** 353
Ross T. Brady

18 **The Evil Twin: The Basics of Complement-Toposes** 375
Luis Estrada-González

19 **Topological Semantics for da Costa Paraconsistent Logics C_ω and C_ω^*** 427
Can Başkent

Part V Philosophical Aspects and Applications of Paraconsistent Logic

20 **Perceiving and Modelling Brightness Contradictions Through the Study of Brightness Illusions** 447
Ashish Bakshi and Kuntal Ghosh

21 **Truth, Trivialism, and Perceptual Illusions** 465
Otávio Bueno

22 Being Permitted, Inconsistencies, and Question Raising 477
Andrzej Wiśniewski

23 On the Type-Free Paracoherent Foundation of Mathematics with the Sedate Extension of Classical Logic by the Librationist Set Theory £, and Specifically on Why £ Is Neither Inconsistent nor Contradictory nor Paraconsistent 509
Frode Bjørdal

24 None of the Above: The Catuṣkoṭi in Indian Buddhist Logic 517
Graham Priest

25 Eastern Proto-Logics 529
Fabien Schang

Part I
Tutorials

Chapter 1
Tutorial on Inconsistency-Adaptive Logics

Diderik Batens

Abstract This paper contains a concise introduction to a few central features of inconsistency-adaptive logics. The focus is on the aim of the program, on logics that may be useful with respect to applications, and on insights that are central for judging the importance of the research goals and the adequacy of results. Given the nature of adaptive logics, the paper may be read as a peculiar introduction to defeasible reasoning.

Keywords Paraconsistent logic · Inconsistency-adaptive logic

Mathematics Subject Classication (2000) 03-01, 03B53, 03B60, 03A05

1.1 Introduction

By a *logic* I shall mean a function that assigns a consequence set to any premise set. So where $\mathcal{L}$ is a language schema, with $\mathcal{F}$ as its set of formulas and $\mathcal{W}$ as its set of closed formulas, a logic is a function $\wp(\mathcal{W}) \to \wp(\mathcal{W})$. The standard predicative language schema, viz. that of **CL** (classical logic), will be called $\mathcal{L}_s$; $\mathcal{F}_s$ its set of formulas and $\mathcal{W}_s$ its set of closed formulas.

Adaptive logics are formal logics but are not deductive logics. They do not define the meaning of logical symbols and are certainly not in the competition for the title 'standard of deduction'—that is, for delineating deductively correct inferences from incorrect inferences and from non-deductive inferences. To the contrary, adaptive logics explicate reasoning processes that are typically not deductive, viz. defeasible reasoning processes.

I am indebted to Mathieu Beirlaen for careful comments on a previous draft.

D. Batens (✉)
Centre for Logic and Philosophy of Science, Ghent University,
Blandijnberg 2, 9000 Gent, Belgium
e-mail: Diderik.Batens@UGent.be

J.-Y. Beziau et al. (eds.), *New Directions in Paraconsistent Logic*,
Springer Proceedings in Mathematics & Statistics 152,
DOI 10.1007/978-81-322-2719-9_1

Sometimes deductive logics are opposed to inductive logics. The expression "inductive logic" may refer to constructions that proceed, for example in terms of probabilities, as in Carnap's work [31]. Where the expression refers to a logic in the sense of the previous paragraph, inductive logics are a specific form of defeasible reasoning, next to many others. Handling inconsistency as described in the present paper is just one of them.[1]

A logic is *formal* iff its consequence relation is defined in terms of logical form. Some people identify this with the Uniform Substitution rule,[2] but that is a mistake because Uniform Substitution defines just one way in which a logic may be formal. Let me quickly spell out a different one. A language or language schema $\mathcal{L}$ will comprise one or more sets of non-logical symbols, for example sentential letters, predicative letters, letters for individual constants, etc. Consider all total functions f that map every such set to itself. Extend f to formulas, $f(A)$ being the result of replacing every non-logical symbol ξ in A by $f(\xi)$. A logic **L** is clearly formal iff the following holds: $A_1, \ldots, A_n \vdash_{\mathbf{L}} B$ iff, for every such f, $f(A_1), \ldots, f(A_n) \vdash_{\mathbf{L}} f(B)$.

Logics may obviously be presented in very different ways. Formal logics are usually presented as sets of rules, possibly combined with the somewhat special rules that are called axioms (and axiom schemata). Apart from many types of 'axiomatizations', logics are standardly characterized by a semantics, which has a rather different function. Deductive logics are typically Tarski logics. This means that they are reflexive ($\Gamma \subseteq Cn_{\mathbf{L}}(\Gamma)$),[3] transitive (if $\Delta \subseteq Cn_{\mathbf{L}}(\Gamma)$, then $Cn_{\mathbf{L}}(\Delta) \subseteq Cn_{\mathbf{L}}(\Gamma)$), and monotonic ($Cn_{\mathbf{L}}(\Gamma) \subseteq Cn_{\mathbf{L}}(\Gamma \cup \Gamma')$ for all Γ'). Another interesting property, which is required if a logic has to have static proofs,[4] is compactness (if $A \in Cn_{\mathbf{L}}(\Gamma)$ then there is a finite $\Gamma' \subseteq \Gamma$ such that $A \in Cn_{\mathbf{L}}(\Gamma')$).

This paper follows several conventions that I better spell out from the start. Classical logic, **CL**, will be taken as the standard of deduction. This is a purely pragmatic decision, not a principled one. Next, all metalinguistic statements are meant in their classical sense. More specifically, the metalinguistic negation will always be classical. So where I say that A is not a **L**-consequence of Γ, I rule out that A is a **L**-consequence of Γ. Similarly, I shall use "false" in its classical sense; no statement can be true as well as false in this sense. An inconsistent situation will be one in which both A and $\neg A$ are true, not one in which A is both true and false. There is a

---

[1] See, for example, [17] for many other real-life examples of reasoning forms for which there is no positive test. The import of a positive test is discussed further in the text.

[2] Uniform Substitution is rule of propositional logic. Predicative classical logic is traditionally axiomatized in terms of a finite set of rules and axiom schemata, rather than axioms. So no substitution rule is then required. Substitution rules in predicate logic have been studied [56] and the outcome is very instructive.

[3] The **L**-consequence set of Γ is defined as $Cn_{\mathbf{L}}(\Gamma) =_{df} \{A \mid \Gamma \vdash_{\mathbf{L}} A\}$.

[4] Just think about usual proofs. Every formula in the proof is a consequence of the premise set and every proof may be extended into a longer proof by applications of the rules.

rather deep divide between paraconsistent logicians on these matters. There are those who claim that 'the true logic' is paraconsistent and that it should always be used, in particular in its own metalanguage. Some of these even take it that classical negation is not coherent, lacks sense and the like. Other paraconsistent logicians, with whom I side, have no objections against the classical negation or against its occurrence in the same language as a paraconsistent negation. This is related to the fact that they are pluralists, either in general or with respect to contexts. They might argue, for example, that consistent domains, like most paraconsistent logics themselves, are more adequately described by **CL** than by a paraconsistent logic.

A warning of a different kind is that the materials discussed in the subsequent pages have been studied at the predicative level. That I shall offer mainly propositional toy examples has a pedagogical rationale.

The last general survey paper that I wrote on adaptive logics was [20]. Meanwhile new results were and are being obtained, some of them are still unpublished. This may be as expected, but one aspect needs to be mentioned from the start. Quite a group of people have contributed to adaptive logics and have published in the field, many more than I shall mention below. While I was always eager to retain the unity of the domain, not everyone attached the same value to unification. Such a situation was obviously very useful to prevent that interesting things are left out of the picture—in principle the aim is to integrate directly or under a translation all potentially realistic first-order defeasible reasoning forms. As we shall see, this integrating frame is the standard format. Little changes were introduced over the years in an attempt to make it as embracing as possible. While most were improvements or clarifications, there was one development that I now consider as misguided. In the end it resulted in the systematic introduction of a set of new symbols to any language. These new symbols had their **CL**-meaning, whence they were called classical. They were added even if they duplicated existing symbols. In the second half of Sect. 1.11, I shall discuss the idea of adding classical symbols and the reasons for not adding them any more today.

The present paper is by no means a summary of all available results on adaptive logics. It merely provides an introduction to the central highlights. Moreover, this paper is explicitly intended as an introduction to inconsistency-adaptive logics, viz. adaptive logics that handle inconsistency. They concern compatibility, inductive generalization, abduction, prioritized reasoning, the dynamics of discussions, belief revision, abstract argumentation theory, deontic logic and so on. Most adaptive logics in standard format are not inconsistency-adaptive and have no connection to paraconsistency. Nevertheless, the present paper can also be read as an introduction to adaptive logics in general, with special attention to handling inconsistency and with illustrations from that domain. The reference section is not a bibliography of inconsistency-adaptive logics.

1.2 The Original Problem

Consider a theory T that was intended as consistent and was given **CL** as its underlying logic: $T = \langle \Gamma, \mathbf{CL} \rangle$, in which Γ is the set of non-logical axioms of T and $Cn_{\mathbf{CL}}(\Gamma)$ is the set of theorems of T, often simply called T. Suppose, however, that T turns out to be inconsistent. There are several well-documented examples of such situation, both in mathematics (Newton's infinitesimal calculus, Cantor's set theory, Frege's set theory, ...) and in the empirical sciences [30, 43, 44, 47, 51–53, 62]. Actually, it is not difficult to find more examples, especially in creative episodes, for example in scientists' notes.

What scientists do in such situations, is look for a consistent replacement for T. As history teaches, however, they do not look for a consistent replacement from scratch. To the contrary, they reason from T, trying to locate the problems in it. This reasoning obviously cannot proceed in terms of **CL** because **CL** validates Ex Falso Quodlibet: $A, \neg A \vdash_{\mathbf{CL}} B$. So the theory T, viz. its set of theorems $Cn_{\mathbf{CL}}(\Gamma)$ is trivial; it contains each and every sentence of the language. If **CL** is the criterion, all one can do is give up the theory and restart from scratch; but scientists do not do so. The upshot is that one should reason about T in terms of a paraconsistent logic, a logic that allows for non-trivial inconsistent theories. Note that any such logic has a semantics that contains inconsistent models—models that verify inconsistent sets of sentences.

It is useful to make a little excursion at this point because many people underestimate the difficulties arising in inconsistent situations. Time and again, people argue that one should figure out where the inconsistency resides and next modify the theory in such a way that the inconsistency disappears. They apparently think that it is easy to separate the consistent parts of a theory from the inconsistencies. Next, if they are very uninformed, they will think that one may choose one half of the inconsistency (or inconsistencies) and add that to the consistent part. If they are a bit better informed, they will realize that a conceptual shift may very well be required, that the new consistent theory should only contain the important statements from the consistent parts, or even a good approximation of them, and should only contain an approximation of one of the 'halves' of the inconsistencies. What is wrong with this reasoning, even with the sophisticated version, is that it is in general impossible to identify the consistent parts of a predicative theory. There is no general positive test for consistency. Being a consistent set of predicative statements is *not* semi-decidable. The set of consistent subsets of a set of predicative statements is *not* semi-recursive. So there is no systematic method, no Turing machine, that is able to identify an arbitrary consistent set as consistent, independent of the number of steps that one allows the Turing machine (or the person who applies the method) to take. So the reasoning from an inconsistent theory can only be explicated in terms of a paraconsistent logic.

Moving from **CL** to a paraconsistent logic has some drastic consequences. Not only Ex Falso Quodlibet, but many other rules are invalidated. Which rules will be invalidated will depend on the chosen paraconsistent logic. If one chooses a compact Tarski logic in which negation is paraconsistent but in which all other logical symbols have the same meaning as in **CL**, then Disjunctive Syllogism and several other rules are definitely invalidated. Incidentally, the *weakest* compact Tarski logic in which negation is paraconsistent but not paracomplete[5] and in which all other logical symbols have their **CL**-meaning is **CLuN**, to which I return in Sect. 1.3.

Let us first have a look at Disjunctive Syllogism (or rather at one of its forms), for example $A \vee B, \neg A / B$. Reasoning about the classical semantics one shows: if $A \vee B$ and $\neg A$ are true, then B is true. Here is one version of the reasoning.

1	$A \vee B$ and $\neg A$ are true	supposition
2	$A \vee B$ is true	from 1
3	$\neg A$ is true	from 1
4	A is true or B is true	from 2
5	A is false	from 3
6	B is true	from 4 and 5

Reasoning about the paraconsistent semantic leads to a very different result because 5 is not derivable from 3. Indeed, both A and $\neg A$ may be true in a paraconsistent model. If that is the case, however, then both $A \vee B$ and $\neg A$ are true even if B is false. So there are models in which $A \vee B$ and $\neg A$ are true and B is false.

Remember that we were considering **CLuN** and paraconsistent extensions of it. We have seen that Disjunctive Syllogism is invalid in **CLuN**. Moreover, as Addition (in particular the variant $A / A \vee B$) is valid, extending **CLuN** with Disjunctive Syllogism would make Ex Falso Quodlibet derivable, whence we would be back at **CL**. Other **CL**-rules are also invalid in **CLuN**, but **CLuN** may be extended with them. Double Negation is among those rules, for example the axiom $\neg\neg A \supset A$ and also its converse. If A is false, $\neg A$ is bound to be true, but $\neg\neg A$ may still be true also. So some paraconsistent models verify $\neg\neg A$ and falsify A. Although $\neg\neg A \supset A$ is invalid in **CLuN**, extending **CLuN** with it results in a paraconsistent logic. This holds for many **CL**-theorems, for example $\neg(\neg A \wedge \neg B) \supset (A \vee B)$. However, extending **CLuN** with several such **CL**-theorems may again result in **CL**.

1.3 Paraconsistent Tarski Logics

The basic paraconsistent logic **CLuN** was already mentioned in the previous section. It is obtained in two steps. First, full positive logic $\mathbf{CL}^+$ is retained. Next, for the negation, Excluded Middle ($\vdash A \vee \neg A$, which is contextually equivalent to $\vdash (A \supset \neg A) \supset \neg A$) is retained, but Ex Falso Quodlibet is not.[6] To avoid confusion, let

---

[5]A logic **L** is paracomplete (with respect to a negation $\neg$) iff some A may false together with its negation $\neg A$; syntactically: iff there are Γ, A and B such that $\Gamma, A \vdash_{\mathbf{L}} B$ and $\Gamma, \neg A \vdash_{\mathbf{L}} B$, but $\Gamma \nvdash_{\mathbf{L}} B$.

[6]In the context of $\mathbf{CL}^+$, Excluded Middle together with Ex Falso Quodlibet define the classical negation.

me characterize **CLuN** semantically. It is obtained from the **CL**-semantics by first removing the clause for negation—the result of this removal is $\mathbf{CL}^+$—and next adding "If $v_M(A) = 0$, then $v_M(\neg A) = \ldots$"[7]

That **CLuN** contains $\mathbf{CL}^+$ warrants that, for example, $\neg p \vdash_{\mathbf{CLuN}} q \supset (\neg p \wedge q)$ because $A \vdash_{\mathbf{CL}^+} B \supset (A \wedge B)$. This is because $\mathbf{CL}^+$ theorem schemata hold for all formulas, formulas of the form $\neg A$ included. However, $\mathbf{CL}^+$ does not have any effect *within* such formulas, in other words within the scope of a negation symbol. As a result of this, Replacement of Equivalents is invalid: $\vdash_{\mathbf{CLuN}} p \equiv (p \wedge p)$ and $\vdash_{\mathbf{CLuN}} \neg p \equiv \neg p$ but $\nvdash_{\mathbf{CLuN}} \neg p \equiv \neg(p \wedge p)$. For the same reason, Replacement of Identicals is invalid: $a = b \vdash_{\mathbf{CLuN}} Pa \equiv Pb$ but $a = b \nvdash_{\mathbf{CLuN}} \neg Pa \equiv \neg Pb$. However, it is easy to extend **CLuN** with Replacement of Identicals.

In the previous section, I referred several times to **CLuN**-models. The reader may wonder what these models precisely look like. For all that was said until now, the **CLuN**-semantics is *indeterministic*. Excluded Middle is retained, $v_M(\neg A) = 1$ whenever $v_M(A) = 0$, but the converse obviously cannot hold because, if it did, Ex Falso Quodlibet would be valid. It is not difficult to restore determinism and the method is interesting because it can be applied rather generally. Two functions play an important role in connection with models. The assignment v is part of the model itself: $M = \langle D, v\rangle$.[8] The assignment fixes the 'meaning' of non-logical symbols. Next, the valuation v_M fixes the 'meaning' of logical symbols. A decent semantics presupposes a complexity ordering $<$ which is such that if $A < B$, then all non-logical symbols that occur in A also occur in B. If the semantics is deterministic, the valuation function defines the valuation value $v_M(A)$ in terms of the assignment function and in terms of valuation values $v_M(B_1)$, ..., $v_M(B_n)$ such that $B_1 < A$, ..., $B_n < A$. So every valuation value $v_M(A)$ is a function of assignment values of formulas B such that $B < A$ and of non-logical symbols that occur in those B. Actually, a deterministic semantics is the standard. If two models are identical $M = \langle D, v\rangle = \langle D', v'\rangle = M'$, whence $D = D'$ and $v = v'$, then they better verify the same formulas. If they do not, then we should describe a semantics in terms of model variants rather than models. Nevertheless, indeterministic semantic systems have been around for more than 30 years, never caused any confusion and were the subject of several interesting systematic studies [3–6].

The official deterministic semantics for **CLuN** is obtained from the indeterministic one by replacing the clause "if $v_M(A) = 0$, then $v_M(\neg A) = 1$" by

$$v_M(\neg A) = 1 \text{ iff } v_M(A) = 0 \text{ or } v(\neg A) = 1\,.$$

Obviously, for this to work, v needs to assign a value to formulas of the form $\neg A$. Note that $v_M(\neg A)$ is still not a function of $v_M(A)$ in the deterministic **CLuN**-semantics. Determinism does not entail truth-functionality.

---

[7]So $p \wedge \neg q \vDash_{\mathbf{CL}^+} \neg q$, $\forall x \neg Px \vDash_{\mathbf{CL}^+} \neg Pa$, and $a = b, Px \vDash_{\mathbf{CL}^+} Pb$, but $a = b, \neg Px \nvDash_{\mathbf{CL}^+} \neg Pb$.

[8]Names and notation may obviously be different and the model may be more complex.

A useful observation is the following. Precisely because, in the two-valued semantics of paraconsistent logics, $v_M(\neg A)$ is not a function of $v_M(A)$, the truth-value of $\neg A$ depends on information not contained in the truth-value of A. Information of this type must naturally be conveyed by the assignment v. Indeed, a model itself, viz. $M = \langle D, v\rangle$, represents a possible situation (or possible state of the world, etc.), whereas the valuation describes the conventions by which we define logical symbols in order to build complex statements—formulas at the schematic level—that enable us to describe the situation. So all information should obviously come from the model itself—the situation, the world, or however you prefer to call it. Moreover, in order to handle not only negation gluts, viz. inconsistencies, but gluts and gaps with respect to any logical symbol, one better lets the assignment map every formula of the language to the set of truth values $\{0, 1\}$.[9]

Incidentally, the view on models presented in the previous paragraph throws some doubt on claims to the effect that classical negation is not a sensible logical operator, among other things because it would be tonk-like. Unless a different approach to logic and models is elaborated, such claims seem not to refer to the situation or world, but to the way in which we handle language. If that is so, one wonders why a modification to our logical operators (for example banning classical negation) is more legitimate than modifying the way in which we handle language.[10]

As already suggested in the previous section, several **CL**-theorems (as well as the corresponding rules) are lost in **CLuN**. Moreover, some of these are such that if **CLuN** is extended with them, even separately, then Ex Falso Quodlibet is derivable, whence we are back to **CL**, or Ex Falso Quodlibet Falsum ($A, \neg A \vdash \neg B$) is derivable, whence we are back to something almost as explosive as **CL**. Disjunctive Syllogism is such a rule. Other examples of such rules are (full) Contraposition, Modus Tollens, Reductio ad Absurdum and Replacement of Equivalents. Let me illustrate the matter for Modus Tollens. In view of $A \vdash_{\mathbf{CLuN}} B \supset A$ and reflexivity, $B \supset A, \neg A \in Cn_{\mathbf{CLuN}}(\{A, \neg A\})$. So extending **CLuN** with Modus Tollens results in $A, \neg A \vdash_{\mathbf{CLuN}} \neg B$ in view of transitivity.

As was also suggested in the preceding section, some **CL**-theorems and **CL**-rules are invalid in **CLuN**, but adding them (separately) to **CLuN** results in a richer paraconsistent logic. Among the striking examples are $\neg\neg A/A$; de Morgan properties; $A, \neg A \vdash B$ for non-atomic A; Replacement of Identicals; and so on. Note that some combinations of such **CL**-theorems and **CL**-rules still result in the validity of Ex Falso Quodlibet or of Ex Falso Quodlibet Falsum.

[9]Take conjunction as an example. The clause allowing for gluts: $v_M(A \wedge B) = 1$ iff ($v_M(A) = 1$ and $v_M(B) = 1$) or $v(A \wedge B) = 1$; the one allowing for gaps: $v_M(A \wedge B) = 1$ iff ($v_M(A) = 1$ and $v_M(B) = 1$) and $v(A \wedge B) = 1$; the one allowing for both: $v_M(A \wedge B) = v(A \wedge B)$.

[10]I heard the claim that restricting the formation rules of natural language so as to classify "this sentence is false" as non-grammatical is illegitimate because the sentence is 'perfect English'. I also heard the claim that invalidating Disjunctive Syllogism is illegitimate because this reasoning form is 'perfectly sound'.

It still seems useful to mention a result from an almost 35-year-old publication [8]. There is an infinity of logics between the propositional fragments of **CLuN** and **CL**. These logics form a mesh. Some of them are maximally paraconsistent in that every extension of them is either propositional **CL** or the trivial logic **Tr**, characterized by $\Gamma \vdash_{\mathbf{Tr}} A$, in other words $Cn_{\mathbf{Tr}}(\Gamma) = \mathcal{W}$. Many propositional paraconsistent logics have a place in this mesh—exceptions are extensions of **CLuN** that validate non-**CL**-theorems like $\neg(A \supset \neg A)$.[11] Other paraconsistent logics are fragments of logics in this mesh, for example Priest's **LP**, which has no detachable implication. Other paraconsistent propositional logics are obviously not within the mesh, for example relevant logics, modal paraconsistent logics, logics that display other gluts or gaps and so on.

An example of a maximal paraconsistent logic is the propositional fragment of a logic which is called **CLuNs** in Ghent because Schütte [59] was the first to describe that propositional fragment. **CLuNs**, fragments of it and slight variants of it were heavily studied and are known under many names [1, 2, 8, 25, 33, 35–40, 57, 61]. **CLuNs** is obtained by extending **CLuN** with axiom schemas to 'drive negations inwards' as well as with an axiom schema that restores Replacement of Identicals: $\neg\neg A \equiv A$, $\neg(A \supset B) \equiv (A \wedge \neg B)$, $\neg(A \wedge B) \equiv (\neg A \vee \neg B)$, $\neg(A \vee B) \equiv (\neg A \wedge \neg B)$, $\neg(A \equiv B) \equiv ((A \vee B) \wedge (\neg A \vee \neg B))$, $\neg(\forall\alpha)A \equiv (\exists\alpha)\neg A$, $\neg(\exists\alpha)A \equiv (\forall\alpha)\neg A$, and $\alpha = \beta \supset (A \supset B)$, in which B is obtained by replacing in A an occurrence of α by β. **CLuNs** has a nice two-valued semantics and several other semantic systems, among which a three-valued one, are adequate for it. I refer the reader elsewhere [25] for this. Priest's **LP** is obtained from **CLuNs** by removing the axioms and semantic clauses for implication and equivalence and defining the symbols in a non-detachable way: $A \supset B =_{df} \neg A \vee B$ and $A \equiv B =_{df} (A \supset B) \wedge (B \supset A)$.

Several paraconsistent logics having been described, we may now return to the original problem and phrase things in a more precise way.

1.4 The Original Problem Revisited

We considered a $T = \langle \Gamma, \mathbf{CL} \rangle$ that turned out inconsistent. T itself is obviously too strong, viz. trivial, to offer a sensible view on 'what T was intended to be'. But we know a way to avoid triviality: replace **CL** by a paraconsistent logic. So let us pick **CLuN** or any other paraconsistent Tarski logic. For nearly all sensible Γ, $T' = \langle \Gamma, \mathbf{CLuN} \rangle$ offers a non-trivial interpretation of 'what T was intended to be'. A little reflection reveals, however, that this T' is too weak.

A toy example will be helpful. Specify the Γ in T to be $\Gamma_1 = \{p, q, \neg p \vee r, \neg q \vee s, \neg q\}$. Note that $\Gamma \nvdash_{\mathbf{CLuN}} s$ and $\Gamma \nvdash_{\mathbf{CLuN}} r$. However, there seems to be a clear difference between p and q. Intuitively speaking, Γ_1 obviously requires that q behaves

[11] This formula is **CL**-equivalent to A but not **CLuN**-equivalent to it.

inconsistently but does not require that p behaves inconsistently. However, and this is interesting, **CLuN** leads to exactly the same insight. Indeed, $\Gamma_1 \vdash_{\mathbf{CLuN}} q \wedge \neg q$ whereas $\Gamma_1 \vdash_{\mathbf{CLuN}} p$ but $\Gamma_1 \nvdash_{\mathbf{CLuN}} \neg p$. Let us see whether something interesting can be done with the help of this apparently interesting distinction.

As p and $\neg p \vee r$ are T-theorems, r was intended as a T-theorem. Similarly, as q and $\neg q \vee s$ are T-theorems, s was *intended* as a T-theorem. However, s better be not a T-theorem. Indeed, intuitively and by **CLuN**, q and $\neg q \vee A$ are T-theorems for every A. So if, relying q, we obtain the conclusion s from $\neg q \vee s$, then, by exactly the same move we obtain the conclusion A from $\neg q \vee A$. The justification for deriving s justifies deriving every formula A because $\neg q \vee A$ is just as much a **CLuN** consequence of Γ_1 as is $\neg q \vee s$. In other words, this kind of reasoning leads to triviality. The matter is very different in the case of r. Indeed, r can be a T-theorem. Relying on p one obtains the conclusion r from$\neg p \vee r$ and there is no other formula of the form $\neg p \vee A$ to which the same move might sensibly be applied.[12] A different way to phrase the matter is by saying that applications of Disjunctive Syllogism of which q is the minor result in triviality, but that applications of Disjunctive Syllogism of which p is the minor do not result in triviality. The reason for the difference is clear: Γ_1 requires q to behave inconsistently, but does not require p to behave inconsistently.

One might take that the preceding paragraphs led to the following insight: what was intended as a T-theorem and can be retained as a T-theorem, should be retained as a T-theorem. Alas, this will not do. Consider another toy example for the non-logical axioms: $\Gamma_2 = \{\neg p, \neg q, p \vee r, q \vee s, \neg t, u \vee t, p \vee q\}$. Clearly, r was intended as a theorem and indeed it can be retained. However, then q, which was also intended as a theorem, should by the same reasoning also be retained. Moreover, if q is retained, then so is $q \vee A$ for every formula A. So, although s was also intended as a theorem, it cannot be retained because, relying on $\neg q$ we cannot only obtain s from $q \vee s$, but we can obtain every formula A from $q \vee A$.

That may seem all right at first sight, but it is not. If you take a closer look at Γ_2, you will see that p and q are strictly on a par. The reasoning in the preceding paragraph relied on the consistent behaviour of p to derive s and q and hence to find out that q behaves inconsistently. However, one may just as well start off by relying on the consistent behaviour of q to obtain s as well as p and hence to find out that p behaves inconsistently. So the insight mentioned at the outset of the previous paragraph should be corrected. Here is the correct version: what was intended as a T-theorem and can be retained as a T-theorem *in view of a systematic and formal account,* should be retained as a T-theorem. A little reflection on the part of the reader will readily reveal that neither r nor s can be retained as consequences of Γ_2, but that u can be so retained.

What is the upshot? We want to replace T by a consistent theory. Obviously, there is no point in pursuing a consistent replacement for a trivial theory—every

[12]As q is **CLuN**-derivable from the premises, so is $\neg p \vee q$. However, relying on p to repeat the move described in the text delivers a formula that was already derivable, viz. q. The same story may be retold for every **CLuN**-consequence of Γ_1 and each time the move will be harmless because nothing new will come out of it.

consistent theory is equally qualified. Moreover, T', in which **CL** is replaced by **CLuN** will be non-trivial for most Γ, but is clearly too weak. However, for most Γ one may strengthen T' by adding certain instances of applications of **CL**-rules that are **CLuN**-invalid. These instances of applications may be added to T' in view of the fact that a systematic distinction can be made between formulas that behave consistently with respect to Γ and others that do not. In this way one obtains T "in its full richness, except for the pernicious consequences of its inconsistency"; one obtains an 'interpretation' of T that is as consistent as possible, and also as much as possible in agreement with the intention behind T.

Of course the matter should still be made precise. This will be done in the next section, but a central clue is the following:

$$\neg A, A \vee B \nvdash_{\mathbf{CLuN}} B \text{ but } \neg A, A \vee B \vdash_{\mathbf{CLuN}} B \vee (A \wedge \neg A)\,.$$

In view of this, one may consider formulas of the form $A \wedge \neg A$ as false, unless and until proven otherwise—unless it turns out that the premises do not permit to consider them as false on systematic grounds. In the first toy example Γ_1 requires that $q \wedge \neg q$ is true, but not that $p \wedge \neg p$ is true: $\Gamma_1 \vdash_{\mathbf{CLuN}} q \wedge \neg q$ whereas $\Gamma_1 \nvdash_{\mathbf{CLuN}} p \wedge \neg p$. Relying on the presumed falsehood of $p \wedge \neg p$, we may take r to be true. The second toy example shows that the matter is slightly more complicated: $\Gamma_2 \vdash_{\mathbf{CLuN}} (p \wedge \neg p) \vee (p \wedge \neg p)$ whereas neither $\Gamma_2 \vdash_{\mathbf{CLuN}} p \wedge \neg p$ nor $\Gamma_2 \vdash_{\mathbf{CLuN}} p \wedge \neg p$. We shall deal with this in the next section.

In order to avoid circularity, it is essential to distinguish between **CLuN**-consequences of a premise set and defeasible consequences derived in view of **CLuN**-consequences. Which formulas behave consistently with respect to a given premise set, will typically be decided in terms of the **CLuN**-consequences of Γ.

1.5 Dynamic Proofs

Dynamic proofs are a typical feature of adaptive logics. The logics were 'discovered' in terms of the proofs. In the first paper written on the topic [10], not the first published, only a rather clumsy semantics was available. The semantics for what became later known as the Minimal Abnormality strategy was described in an article [9] that was written 6 years later but published earlier. A decent semantics for the Reliability strategy appears only in [12]. Dynamic proofs are also typical for adaptive logics because nearly no other approaches to defeasible reasoning present proofs and certainly not proofs that resemble Hilbert proofs. A theoretic account of static proofs as well as dynamic proofs, which turn out to be a generalization of the former, is published [21]; a more extensive account is available on the web [24, Sect. 4.7].

Let us, very naively, have a look at some examples of dynamic proofs. More precise definitions follow in Sect. 1.7, but obtaining a clear and intuitive insight may be more important for the reader. Let us start with a dynamic proof from Γ_1. First have a look at stage 7 of the proof—a stage is a sequence of lines; think about stage 0 as the empty sequence and let the addition of a line to stage n result in stage $n + 1$.

1	p	Prem	$\emptyset$
2	q	Prem	$\emptyset$
3	$\neg p \vee r$	Prem	$\emptyset$
4	$\neg q \vee s$	Prem	$\emptyset$
5	$\neg q$	Prem	$\emptyset$
6	r	1, 3; RC	$\{p \wedge \neg p\}$
7	s	2, 4; RC	$\{q \wedge \neg q\}$

So the premises were introduced and next two conditional steps were taken. Line 6 informs us that r is derivable on the condition that $p \wedge \neg p$ is false and line 7 that s is derivable on the condition that $q \wedge \neg q$ is false. Incidentally, a line with a non-empty condition corresponds nicely and directly with a line from a static proof—in the present case a Hilbert-style **CLuN**-proof. The condition, Δ, of a line is always a finite set of contradictions. Where a line of the dynamic proof contains a line at which A is derived on the condition Δ, the corresponding static **CLuN**-proof contains a line at which $A \vee \bigvee(\Delta)$ is derived—as expected, $\bigvee(\Delta)$ is the disjunction of the members of Δ. So in a sense stage 7 of this dynamic proof is nothing but a static proof in disguise. Note that the rule applied at lines 6 and 7 is called RC (conditional rule) because, as explained, a formula $A \vee \bigvee(\Delta)$ is **CLuN**-derivable from previous members of the proof, but Δ is pushed into the condition.

The way in which dynamics is introduced appears from the continuation of the proof. I do not repeat 1–5, which merely introduce the premises.

6	r	1, 3; RC	$\{p \wedge \neg p\}$	
7	s	2, 4; RC	$\{q \wedge \neg q\}$	✓
8	$q \wedge \neg q$	2, 5; RU	$\emptyset$	

At stage 8 of the proof, $q \wedge \neg q$ is unconditionally derived, viz. at line 8. So the supposition of line 7, viz. that $\{q \wedge \neg q\}$ is false, cannot be upheld. As a result, line 7 is marked, which means that its formula is considered as not derived from the premise set Γ_1.[13] Incidentally, the rule applied at line 8 is called RU (unconditional rule) because (the formula of) 8 is a **CLuN**-consequence of (the formulas of) 2 and 5.

So the dynamics is controlled by marks. Which lines are marked or unmarked is decided by a marking *definition*, which is typical for a strategy. More information on this follows in Sect. 1.7. For now, it is important that the reader understands why line 7 is marked and other lines are unmarked. As far as this specific proof stage is concerned, nothing interesting happens when the proof is continued. No mark will be removed or added to any of these 8 lines.[14] Incidentally, the only line that might

---

[13]Do not read the "not derived" as "not derivable". Indeed, a formula may be derivable in several ways from the same premise set.

[14]A more accurate wording requires that one adds: in a proof from Γ_1 that extends the present stage 8. Indeed, the logic we are considering is non-monotonic. So extending the premise set may result in line 6 being marked.

become marked is line 6. The formulas derived on lines with an empty condition are **CLuN**-consequences of the premises. These are the stable consequences of the premise set. The marks pertain to the supplementary, defeasible consequences of the premise set.

How can I be so sure that the marks of lines 1–8 will not be changed in an extension of the proof from Γ_1? The example is propositional and propositional **CLuN** is decidable in the same sense as propositional **CL**. It is easy enough to prove that $q \wedge \neg q$ is the only contradiction that is **CLuN**-derivable from Γ_1.[15] Beware. As is the case for **CL**, only some fragments of **CLuN** are decidable. So arguing that a predicative proof is stable with respect to certain lines will often be much more complicated than in the present case.

Before we proceed, allow me to summarize that the two components governing dynamic proofs are rules (of inference) and the marking definition. The rules are applied at will by the people who devise the proof—if they are smart, they will follow a certain heuristics. As we shall see, the marking definition operates independently of any human intervention. In view of the stage of the proof, the marking definition determines which lines are marked.

When we consider more examples, a little complication will catch our attention. Here is a dynamic proof from $\Gamma_2 = \{\neg p, \neg q, p \vee r, q \vee s, \neg t, u \vee t, p \vee q\}$.

1	$\neg p$	PREM	$\emptyset$	
2	$\neg q$	PREM	$\emptyset$	
3	$p \vee r$	PREM	$\emptyset$	
4	$q \vee s$	PREM	$\emptyset$	
5	$\neg t$	PREM	$\emptyset$	
6	$u \vee t$	PREM	$\emptyset$	
7	$p \vee q$	PREM	$\emptyset$	
8	r	1, 3; RC	$\{p \wedge \neg p\}$	✓
9	s	2, 4; RC	$\{q \wedge \neg q\}$	✓
10	u	5, 6; RC	$\{t \wedge \neg t\}$	
11	$(p \wedge \neg p) \vee (q \wedge \neg q)$	1, 2, 7; RC	$\emptyset$	

At stage 10 of the proof—when the proof consists of lines 1–10 only—no line is marked. At stage 11, however, lines 8 and 9 are both marked. Why is that? Line 11 gives us the information that either p or q behaves inconsistently on Γ_2, but does not inform us which of both behaves inconsistently. So a natural reaction is to consider both $p \wedge \neg p$ and $q \wedge \neg q$ as unreliable. This is the reaction that agrees with the Reliability strategy—we shall come across other strategies later. According to the Reliability strategy a line is marked if one of the members of its condition is unreliable. At this point in the paper, consider the unreliable formulas as the

[15] The reader might think that, as p is also a **CLuN**-consequence of Γ_1, $(p \wedge q) \wedge \neg(p \wedge q)$ is also a **CLuN**-consequence of Γ_1. This however is mistaken. $\neg q \nvdash_{\mathbf{CLuN}} \neg(p \wedge q)$.

disjuncts of the *minimal* disjunctions of contradictions. If the “minimal” was not there, Addition would cause every contradiction to be unreliable as soon as one contradiction is unreliable.

In both example proofs, some lines were unmarked at a stage and marked at a later stage. The converse move is also possible, as is illustrated by a proof from $\Gamma_3 = \{(p \wedge q) \wedge t, \neg p \vee r, \neg q \vee s, \neg p \vee \neg q, t \supset \neg p\}$.

1	$(p \wedge q) \wedge t$	PREM	$\emptyset$	
2	$\neg p \vee r$	PREM	$\emptyset$	
3	$\neg q \vee s$	PREM	$\emptyset$	
4	$\neg p \vee \neg q$	PREM	$\emptyset$	
5	$t \supset \neg p$	PREM	$\emptyset$	
6	r	1, 2; RC	$\{p \wedge \neg p\}$	✓
7	s	1, 3; RC	$\{q \wedge \neg q\}$	✓
8	$(p \wedge \neg p) \vee (q \wedge \neg q)$	1, 4; RU	$\emptyset$	

Both lines 6 and 7 are marked at stage 8 because $(p \wedge \neg p) \vee (q \wedge \neg q)$ is a minimal disjunction of contradictions that is derived at the stage. However, look what happens if stage 9 looks as follows—I do not repeat 1–5.

6	r	1, 2; RC	$\{p \wedge \neg p\}$	✓
7	s	1, 3; RC	$\{q \wedge \neg q\}$	
8	$(p \wedge \neg p) \vee (q \wedge \neg q)$	1, 4; RU	$\emptyset$	
9	$p \wedge \neg p$	1, 5; RU	$\emptyset$	

At stage 9 of this proof, $(p \wedge \neg p) \vee (q \wedge \neg q)$ is not a minimal disjunction of abnormalities because (the ‘one disjunct disjunction’) $p \wedge \neg p$ was derived. We knew already that either $p \wedge \neg p$ or $q \wedge \neg q$ was unreliable and now obtain the more specific information that it is actually $p \wedge \neg p$ that is unreliable. So $q \wedge \neg q$ is off the hook, whence line 7 is unmarked. Stage 9 of this proof is stable: no mark will be removed or added to lines 1–9 if the stage is extended. Actually, nothing interesting happens in any such extension.

It is time to make the marking more precise. Dynamic proofs need to explicate the dynamic reasoning. So, at the level of the proofs, the dynamics needs to be controlled. The central features for this control are the conditions and the marking definition. The way in which conditions are introduced should be clear by now—precise generic rules follow in Sect. 1.7. However, how does one precisely figure out which lines are marked?

Only some adaptive logics are inconsistency-adaptive. So allow me to use a slightly more general terminology. The formulas that occur in conditions of lines—in the previous examples these were contradictions—are called abnormalities and Ω is the usual name for the set of abnormalities.

A classical disjunction of abnormalities will be called a *Dab-formula*—it goes without saying that a disjunction of formulas is always a disjunction of finitely many formulas. I shall often write $Dab(\Delta)$ to refer to the classical disjunction of the members of a finite $\Delta \subset \Omega$. A *Dab*-formula that is derived in a proof stage by RU at a line with condition $\emptyset$ will be called a *inferred Dab-formula* of the proof stage. Note that a *Dab*-formula introduced by Prem is not an inferred *Dab*-formula in the sense of this definition. $Dab(\Delta)$ is a *minimal inferred Dab-formula* of a proof stage if it is an inferred *Dab*-formula of the proof stage and there is no $\Theta \subset \Delta$ such that $Dab(\Theta)$ is an inferred *Dab*-formula of the proof stage. Where $Dab(\Delta_1), \ldots, Dab(\Delta_n)$ are the minimal inferred *Dab*-formulas of stage s, the set of *unreliable formulas of stage* s is $U_s(\Gamma) = \Delta_1 \cup \ldots \cup \Delta_n$. Where Θ is the condition of line i, line i is *marked* iff $\Theta \cap U_s(\Gamma) \neq \emptyset$. This is the marking definition for the Reliability strategy—every strategy has its own marking definition.

Marks come and go. As they determine which formulas are considered as derived, derivability seems to be unstable; it changes from stage to stage. Let this unstable derivability be called *derivability at a stage*. Apart from it, we want a stable form of derivability, which is called *final derivability* and is noted as $\Gamma \vdash_{\mathbf{CLuN}^r} A$. There are several ways to define final derivability. At this point in my story, the following seems most handy. If A is derived at an unmarked line i of a stage of a proof from Γ and the stage is *stable* with respect to i—line i is not marked in any extension of the stage—then A is finally derived from Γ.

Just as we wanted the stable entity called final derivability, we also want to have some further entities that refer to what is **CLuN**-derivable from the premise set Γ rather than referring to a stage of a proof from Γ.

Definition 1.1 $Dab(\Delta)$ is a *minimal Dab-consequence of* Γ iff $\Gamma \vdash_{\mathbf{CLuN}} Dab(\Delta)$ and, for all $\Delta' \subset \Delta$, $\Gamma \nvdash_{\mathbf{CLuN}} Dab(\Delta')$.

Definition 1.2 Where $Dab(\Delta_1), \ldots, Dab(\Delta_n)$ are the minimal *Dab*-consequences of Γ, $U(\Gamma) = \Delta_1 \cup \ldots \cup \Delta_n$.

The set $U(\Gamma)$ is defined in view of the Reliability strategy. A very different set will be introduced later in view of Minimal Abnormality.

The reader may expect a section on semantics at this point, but I shall only deal with the semantics as defined by the standard format.

1.6 The Standard Format SF

There is a large diversity of adaptive logics. Every new adaptive logic requires that one delineates its syntax (proof theory), its semantics (models), and, what is the hard bit, its metatheory (study of properties of the system). This suggested the search for a common structure for a large set of adaptive logics, if possible for all of them. The idea was that the structure would take care of most of the work beforehand, that the proof theory and semantics would be defined in terms of the common structure and that the metatheoretic properties would be provable from the structure. The common structure would be a function of certain parameters and specifying these would result in a specific adaptive logic with all required features available. This common structure is called the *standard format*.

An adaptive logic **AL** in Standard Format is defined as a triple comprising[16]:

- a *lower limit logic* **LLL**: a logic that has static proofs and contains classical disjunction,
- a *set of abnormalities* Ω, a set of formulas that share a (possibly restricted) logical form or a union of such sets,
- a *strategy* (Reliability, Minimal Abnormality, ...).

That the lower limit logic contains a classical disjunction means that one of the logical symbols is implicitly or explicitly defined in such a way that it has the meaning of the **CL**-disjunction. Explaining the notion of static proofs goes beyond the scope of the present paper, but the reader may for all useful purposes replace the requirement by: a formal and compact Tarski logic.

"Abnormality" is a technical term, different adaptive logics require that different formulas are seen as abnormalities. Only the abnormalities of corrective adaptive logics—those with **LLL** weaker than **CL**—are **CL**-falsehoods. In nearly all inconsistency-adaptive logics, existentially closed contradictions are abnormalities. Also other formulas may belong to the Ω, for example Universally closed contradictions or formulas of the form $A \wedge \neg(A \vee B)$. Some examples of restricted and unrestricted logical forms will be presented below.

Adaptive strategies will be discussed at some length later in this section.

If the lower limit logic **LLL** is extended with a set of rules or axioms that trivialize abnormalities (and no other formulas), then one obtains a logic called the *upper limit logic* **ULL**. Examples follow but it should be clear by now that, for all $A \in \Omega$ and for all $B \in \mathcal{W}$, A/B should be a derivable rule in **ULL**. As Ω is characterized by a logical form, it is in possible to obtain **ULL** by extending **LLL** with a set of rules.

I shall suppose that a characteristic semantics of **LLL** is available. This will enable me to define the semantics of **AL** in terms of the standard format. The **LLL**-models that verify no member of Ω form a semantics for **ULL**.[17] A premise set that has **ULL**-models is often called a *normal premise set*; it does not require that any abnormality is true.

It is instructive to have a closer look at the difference between **ULL** and **AL**. **ULL** extends **LLL** by validating some further rules of inference. **AL** extends **LLL** by validating certain *applications* of **ULL**-rules. The point is easily illustrated in connection to Disjunctive Syllogism. **CL** validates this rule, while in the (not yet precise) toy examples of proofs from Sect. 1.5, some but not all applications of Disjunctive Syllogism were sanctioned as correct. As those examples clarify, it depends on the premises—or should one say on the content of the premises—which applications

---

[16]Names like **LLL**, **AL**, $\mathbf{AL}^r$, and **ULL** are used as generic names to define the standard format and to study its features. The names refer to arbitrary logics that stand in a certain relation to each other.

[17]Similarly for those models together with the trivial model—the model that verifies all formulas.

turn out valid. In other words, adaptive logics display a form of *content guidance*.[18] A different way of phrasing the matter is that $Cn_{\mathbf{AL}}(\Gamma)$ comes to $Cn_{\mathbf{LLL}}(\Gamma)$ extended with what is derivable if *as many* abnormalities are false *as* the premises permit. This phrase is obviously ambiguous, but strategies disambiguate it, as we shall see.

An important supposition on the language $\mathcal{L}$ of **AL** is that it contains a classical disjunction. It may of course contain several disjunctions, but one of them should be classical. In the sequel of this paper, the symbol $\check{\vee}$ will always refer to this disjunction.[19] Similarly, $\sim$ will always refer to a classical negation. This is *not* supposed to occur in every considered language schema.

As we already have seen in Sect. 1.5, we need $\check{\vee}$ for *Dab*-formulas—but see Sect. 1.11 for an alternative. In Sect. 1.5, I also introduced inferred *Dab*-formulas and minimal inferred *Dab*-formulas of a proof stage as well as the notation $Dab(\Delta)$.

Let us consider some examples of adaptive logics. Expressions $\exists A$ will denote the existential closure of A, viz. A preceded by an existential quantifier over every variable free in A.

The adaptive logic $\mathbf{CLuN}^m$ is defined by the following triple:

- lower limit logic: **CLuN**,
- set of abnormalities $\Omega = \{\exists(A \wedge \neg A) \mid A \in \mathcal{F}_s\}$
- strategy: Minimal Abnormality.

The upper limit logic is **CL**, obtained by extending **CLuN** with, for example, the axiom schema $(A \wedge \neg A) \supset B$.[20] It is not difficult to prove that the **CLuN**-models that verify no abnormality form a semantics of **CL**.

The logic $\mathbf{CLuNs}^m$ is defined by:

- lower limit logic: **CLuNs**,
- set of abnormalities $\Omega = \{\exists(A \wedge \neg A) \mid A \in \mathcal{F}_s^a\}$
- strategy: Minimal Abnormality,

in which $\mathcal{F}_s^a$ is the set of atomic (open and closed) formulas of $\mathcal{L}_s$—atomic formulas are those in which no logical symbols occur except possibly for identity $=$. The upper limit logic is **CL**, obtained by extending **CLuNs** with, for example, the axiom schema $(A \wedge \neg A) \supset B$.[21] Semantically: the **CLuNs**-models that verify no abnormality form a **CL**-semantics.

[18] The notion played a rather central role in discussions on scientific heuristics. A very clear and argued position was for example proposed by Dudley Shapere [60].

[19] This obviously does not mean that $\check{\vee}$ is a symbol of the language. It is a conventional name to refer to a symbol of the language that has the meaning of classical disjunction. It may even refer ambiguously: if there are several classical disjunctions, $\check{\vee}$ need not always refer to the same one.

[20] Axioms are suppose to be closed formulas. So $A \in \mathcal{W}_s$. The idea is that **CLuN**-valid rules are fully retained in the extension. One of these rules is: from $\vdash A(a) \supset B$ to derive $\vdash \exists x A(x) \supset B$ provided a does not occur in B.

[21] The axiom schema may be restricted to $A \in \mathcal{W}_s^a$, but there is no need to do so.

Some further examples are easy variants. **CLuN**^{r} is like **CLuN**^{m}, except that Minimal Abnormality is replaced by Reliability. **LP**^{m} is like **CLuNs**^{m} except that **CLuNs** is replaced by Priest's **LP**—see Sect. 1.3 for the relation between **CLuNs** and **LP**.

In these examples **LLL** or the strategy are varied. What about the difference between the set of abnormalities of **CLuN**^{m} as opposed to **CLuNs**^{m}? In a sense this is just a variation. Yet, if the Ωs are exchanged, the resulting variant of **CLuN**^{m} is still an inconsistency-adaptive logic, but its **ULL** is weaker than **CL**—a feature that is difficult to justify with respect to applications. If the Ω are exchanged, the resulting variant of **CLuNs**^{m} is also still an inconsistency-adaptive logic, but it is a flip-flop logic—see Sect. 1.12, where also more variation will be considered.

If an adaptive logic is in standard format, this fact (not specific properties of the logic) provides it with:

- its proof theory,
- its semantics (models),
- most of its metatheory (*including* soundness and completeness).

So the standard format provides guidance in devising new adaptive logics. Moreover, once a new adaptive logic is phrased in standard format, most of the hard work is over.

1.7 SF: Proof Theory

As we already know, every adaptive logic requires a set of rules of inference and a marking definition. The rules of inference are determined by **LLL** and Ω; the marking definition is determined by Ω and by the strategy. We also know that the dynamics of the proofs is controlled by attaching conditions (finite subsets of Ω) to derived formulas, or, if you prefer, to lines at which formulas are derived. We also have seen what is special about annotated dynamic proofs: their lines consist of four rather than three elements: a number, a formula, a justification and a condition. The rules govern the addition of lines, the marking definition determines for every line i at every stage s of a proof whether i is unmarked or marked— this means that it is respectively IN or OUT—in view of (i) the condition of i and (ii) the minimal inferred *Dab*-formulas of stage s.

The rules of inference can be presented as three generic rules. Let Γ be the premise set and let

$$A \quad \Delta$$

abbreviate that A occurs in the proof on the condition Δ.

Prem If $A \in \Gamma$: $\dfrac{\ldots \quad \ldots}{A \quad \emptyset}$

RU If $A_1, \ldots, A_n \vdash_{\mathbf{LLL}} B$: $\dfrac{\begin{array}{ll} A_1 & \Delta_1 \\ \ldots & \ldots \\ A_n & \Delta_n \end{array}}{B \quad \Delta_1 \cup \ldots \cup \Delta_n}$

RC If $A_1, \ldots, A_n \vdash_{\mathbf{LLL}} B \mathbin{\hat{\vee}} Dab(\Theta)$: $\dfrac{\begin{array}{ll} A_1 & \Delta_1 \\ \ldots & \ldots \\ A_n & \Delta_n \end{array}}{B \quad \Delta_1 \cup \ldots \cup \Delta_n \cup \Theta}$

Only RC *introduces* new non-empty conditions (adds a non-empty set to the conditions of the local premises). Prem introduces empty conditions and RU merely carries conditions over and adds them up in a union.

Easy illustrations: RU may be applied in view of $p, p \supset q \vdash_{\mathbf{CLuN}} q$; RC may be applied in view of $p, \neg p \vee q \vdash_{\mathbf{CLuN}} q \mathbin{\hat{\vee}} (p \wedge \neg p)$. In view of the formulation of the antecedent of RU and RC, all rules are *finitary*—have a finite number of local premises. This formulation does not in any way affect the adaptive logic **AL** because **LLL** is a compact logic anyway. Incidentally, it is instructive to review the toy examples in terms of the precise formulation of the rules.

Marking definitions proceed in terms of the minimal inferred *Dab*-formulas at the proof stage. Where $Dab(\Delta_1), \ldots, Dab(\Delta_n)$ are the minimal inferred *Dab*-formulas at stage s, $U_s(\Gamma) = \Delta_1 \cup \ldots \cup \Delta_n$.

Definition 1.3 Marking for Reliability: where Δ is the condition of line i, line i is marked at stage s iff $\Delta \cap U_s(\Gamma) \neq \emptyset$.

The idea behind the definition consists of two steps. First, the minimal inferred *Dab*-formulas of stage s of a proof from Γ provide, at stage s, the best available estimate of the minimal *Dab*-consequences of Γ. So their disjuncts, which are abnormalities, cannot be safely considered as false. Next, the formula of a line can only be considered as derived (by present insights) if the abnormalities in the condition of the line can be considered as false. If they cannot, the line is marked.

However sensible this may sound, Minimal Abnormality offers a more refined approach. A *choice set* of $\Sigma = \{\Delta_1, \Delta_2, \ldots\}$ is a set that contains one element out of each member of Σ. A *minimal choice set* of Σ is a choice set of Σ of which no proper subset is a choice set of Σ. Where $Dab(\Delta_1), \ldots, Dab(\Delta_n)$ are the minimal inferred *Dab*-formulas of stage s, $\Phi_s(\Gamma)$ is the set of the minimal choice sets of $\{\Delta_1, \ldots, \Delta_n\}$.

Definition 1.4 Marking for Minimal Abnormality: where A is the formula and Δ is the condition of line i, line i is marked at stage s iff (i) there is no $\varphi \in \Phi_s(\Gamma)$ such that $\varphi \cap \Delta = \emptyset$, *or* (ii) for some $\varphi \in \Phi_s(\Gamma)$, there is no line at which A is derived on a condition Θ for which $\varphi \cap \Theta = \emptyset$.

The set $\Phi_s(\Gamma)$ is the best estimate, at stage s, of $\Phi(\Gamma)$, which is the set of minimal choice sets of the minimal *Dab*-consequences of Γ. The $\varphi \in \Phi(\Gamma)$ are the minimal sets of abnormalities that are true if Γ is true. On the Minimal Abnormality strategy, a formula A is an adaptive consequence of Γ iff A is a consequence for every $\varphi \in \Phi(\Gamma)$. So, for every $\varphi \in \Phi(\Gamma)$, there should be a Θ such that $A \check{\vee} Dab(\Theta)$ is a **LLL**-consequence of Γ and all members of Θ can be false, viz. none of them is a member of φ.

The difference between Minimal Abnormality and Reliability can be nicely illustrated by means of a toy proof. Considering again $\Gamma_2 = \{\neg p, \neg q, p \vee r, q \vee s, \neg t, u \vee t, p \vee q\}$, let us continue the second proof from Sect. 1.5. The premise lines 1–7 are not repeated.

8	r	1, 3; RC	$\{p \wedge \neg p\}$	$\checkmark$
9	s	1, 4; RC	$\{q \wedge \neg q\}$	$\checkmark$
10	u	5, 6; RC	$\{t \wedge \neg t\}$	
11	$(p \wedge \neg p) \vee (q \wedge \neg q)$	1, 2, 7; RC	$\emptyset$	
12	$r \vee s$	8; RC	$\{p \wedge \neg p\}$	
13	$r \vee s$	9; RC	$\{q \wedge \neg q\}$	

Obviously $\Phi_{13}(\Gamma) = \Phi_{11}(\Gamma) = \{\{p \wedge \neg p\}, \{q \wedge \neg q\}\}$. So, on the Minimal Abnormality strategy, lines 12 and 13 are unmarked. Indeed, if $p \wedge \neg p$ is the case and $q \wedge \neg q$ is not, then $r \vee s$ is in view of line 13. If $q \wedge \neg q$ is the case and $p \wedge \neg p$ is not, then $r \vee s$ is in view of line 12. It follows that, on the Minimal Abnormality strategy, $r \vee s$ is an adaptive consequence of Γ_2. The matter is very different for Reliability. Indeed, $U_{13}(\Gamma) = \{p \wedge \neg p, q \wedge \neg q\}$, whence lines 12 and 13 are marked. As the displayed proof stage is stable for both strategies and $r \vee s$ is not **CLuN**-derivable from Γ_2 on any other condition, $\Gamma_2 \vdash_{\mathbf{CLuN}^m} r \vee s$ but $\Gamma \nvdash_{\mathbf{CLuN}^r} r \vee s$.

In Sect. 1.5, I delineated final derivability in terms of a stable proof stage. This is not very handy as a general definition. Indeed, for some adaptive logics **AL**, premise sets Γ, and formulas A, only infinite **AL**-proofs of A from Γ are stable [12, Sect. 7]. But one obviously cannot write down infinite proofs. For this reason, the official definition of final derivability goes as follows.

Definition 1.5 A is *finally derived* from Γ at line i of a finite proof stage s iff (i) A is the second element of line i, (ii) line i is not marked at stage s, and (iii) every extension of the proof in which line i is marked may be further extended in such a way that line i is unmarked.

Definition 1.6 $\Gamma \vdash_{\mathbf{AL}} A$ (A is *finally* **AL***-derivable* from Γ) iff A is finally derived at a line of a proof stage from Γ.

Establishing final derivability requires (i) a finite proof stage and (ii) a metatheoretic reasoning about extensions of the stage and extensions of these. Some comments on these definitions follow in Sect. 1.10.

1.8 SF: Semantics

The syntactic definition of minimal *Dab*-consequences of Γ was presented in Definition 1.1. As this proceeds in terms of **LLL** and an adequate semantics of this logic is supposed to be known, $Dab(\Delta)$ is a minimal *Dab*-consequence of Γ iff $\Gamma \vDash_{\mathbf{LLL}} Dab(\Delta)$ and, for all $\Delta' \subset \Delta$, $\Gamma \nvDash_{\mathbf{LLL}} Dab(\Delta')$.

Definition 1.7 Where M is a **LLL**-model, $Ab(M) = \{A \in \Omega \mid M \Vdash A\}$.

Consider first adaptive logics $\mathbf{AL}^r$ that follow the Reliability strategy. Let $\mathcal{M}_\Gamma^{\mathbf{LLL}}$ be the set of **LLL**-models of Γ.

Definition 1.8 $M \in \mathcal{M}_\Gamma^r$ (M is a *reliable* model of Γ) iff $M \in \mathcal{M}_\Gamma^{\mathbf{LLL}}$ and $Ab(M) \subseteq U(\Gamma)$.

So the reliable models of Γ are the models of Γ that verify at most reliable abnormalities. Note that there are no reliable models, but only reliable models *of* a set of formulas Γ. The same holds for adaptive models in general.

Definition 1.9 $\Gamma \vDash_{\mathbf{AL}^r} A$ (A is an $\mathbf{AL}^r$-consequence of Γ) iff $M \Vdash A$ for all $M \in \mathcal{M}_\Gamma^r$.

So the $\mathbf{AL}^r$-semantics selects some **LLL**-models of Γ as $\mathbf{AL}^r$-models *of* Γ. The selection depends on Ω and on the strategy.

For adaptive logics $\mathbf{AL}^m$ that follow the Minimal Abnormality strategy, one may proceed in a very different way.

Definition 1.10 $M \in \mathcal{M}_\Gamma^m$ (M is a *minimally abnormal* model of Γ) iff $M \in \mathcal{M}_\Gamma^{\mathbf{LLL}}$ and no $M' \in \mathcal{M}_\Gamma^{\mathbf{LLL}}$ is such that $Ab(M') \subset Ab(M)$.

Definition 1.11 $\Gamma \vDash_{\mathbf{AL}^m} A$ (A is an $\mathbf{AL}^m$-consequence of Γ) iff $M \Vdash A$ for all $M \in \mathcal{M}_\Gamma^m$.

Lemma 1.14 below greatly clarifies the relation between the minimal abnormal models and the marking definition for Minimal Abnormality.

Have a look at Fig. 1.1. For a normal premise set Γ, an adaptive logic simply selects the upper limit models of Γ, and hence delivers the same consequence set as the upper limit logic. Abnormal Γ have no **ULL**-models. Still, some exceptions aside,[22] adaptive logics select a proper subset of the set of **LLL**-models and hence deliver a larger consequence set than **LLL**.

---

[22]The exception may be caused by the logic, which is then called a flip-flop, or by the premise set—for example if the premise set comprises the formulas verified by a **LLL**-model.

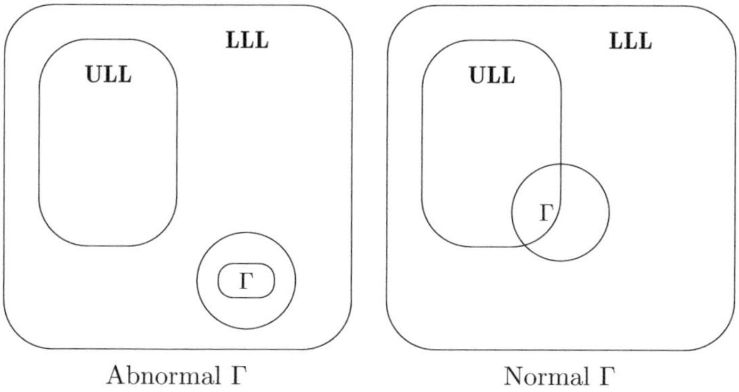

Fig. 1.1 Comparison of Models

1.9 SF: Metatheory

What follows is a selection of theorems. They are selected in view of their importance or in view of the insights they reveal in the context of the present introduction. They are all provable from the standard format [20, 24]. This means that they are provable from the common structure of all adaptive logics in standard format, independent of further specific properties.

Theorem 1.12 *$\Gamma \vDash_{\mathbf{AL}^r} A$ iff $\Gamma \vDash_{\mathbf{LLL}} A \,\check{\vee}\, Dab(\Delta)$ and $\Delta \cap U(\Gamma) = \emptyset$ for a finite $\Delta \subset \Omega$.*

Corollary 1.13 *$\Gamma \vdash_{\mathbf{AL}^r} A$ iff $\Gamma \vDash_{\mathbf{AL}^r} A$. (Soundness and Completeness for Reliability)*

Lemma 1.14 *$M \in \mathcal{M}^m_\Gamma$ iff $M \in \mathcal{M}^{\mathbf{LLL}}_\Gamma$ and $Ab(M) \in \Phi(\Gamma)$.*

Theorem 1.15 *$\Gamma \vdash_{\mathbf{AL}^m} A$ iff $\Gamma \vDash_{\mathbf{AL}^m} A$. (Soundness and Completeness for Minimal Abnormality)*

Strong Reassurance, also called Stopperedness or Smoothness, refers to the following property: if a model of the premises is not selected, this is justified by the fact that a selected model of the premises is less abnormal. If Strong Reassurance is absent, there are infinite sequences of models of a certain Γ in which each member of the sequence is less abnormal than its predecessor. This absence sometimes results in very odd consequence sets [13].

Theorem 1.16 *If $M \in \mathcal{M}^{\mathbf{LLL}}_\Gamma - \mathcal{M}^m_\Gamma$, then there is a $M' \in \mathcal{M}^m_\Gamma$ such that $Ab(M') \subset Ab(M)$. (Strong Reassurance for Minimal Abnormality.)*

Theorem 1.17 *If $M \in \mathcal{M}^{\mathbf{LLL}}_\Gamma - \mathcal{M}^r_\Gamma$, then there is a $M' \in \mathcal{M}^r_\Gamma$ such that $Ab(M') \subset Ab(M)$. (Strong Reassurance for Reliability.)*

All of the following theorems highlight important features of adaptive logics. The reader may find some more fascinating than others. This will depend on the reader's familiarity with certain aspects of non-monotonic reasoning and of defeasible reasoning in general.

Theorem 1.18 *Each of the following obtains:*

1. $\mathcal{M}^m_\Gamma \subseteq \mathcal{M}^r_\Gamma$. *Hence* $Cn_{\mathbf{AL}^r}(\Gamma) \subseteq Cn_{\mathbf{AL}^m}(\Gamma)$.
2. *If* $A \in \Omega - U(\Gamma)$, *then* $M \nVdash A$ *for all* $M \in \mathcal{M}^r_\Gamma$, *whence* $\sim A \in Cn_{\mathbf{AL}^r}(\Gamma)$ *if* $\sim$ *is in* $\mathcal{L}$.
3. *If* $Dab(\Delta)$ *is a minimal Dab-consequence of* Γ *and* $A \in \Delta$, *then some* $M \in \mathcal{M}^m_\Gamma$ *verifies* A *and falsifies all members (if any) of* $\Delta - \{A\}$.
4. $\mathcal{M}^m_\Gamma = \mathcal{M}^m_{Cn_{\mathbf{AL}^m}(\Gamma)}$ *whence* $Cn_{\mathbf{AL}^m}(\Gamma) = Cn_{\mathbf{AL}^m}(Cn_{\mathbf{AL}^m}(\Gamma))$. *(Fixed Point for Minimal Abnormality.)*
5. $\mathcal{M}^r_\Gamma = \mathcal{M}^r_{Cn_{\mathbf{AL}^r}(\Gamma)}$ *whence* $Cn_{\mathbf{AL}^r}(\Gamma) = Cn_{\mathbf{AL}^r}(Cn_{\mathbf{AL}^r}(\Gamma))$. *(Fixed Point for Reliability.)*
6. *For all* $\Delta \subseteq \Omega$, $Dab(\Delta) \in Cn_{\mathbf{AL}}(\Gamma)$ *iff* $Dab(\Delta) \in Cn_{\mathbf{LLL}}(\Gamma)$. *(Immunity.)*
7. *If* $\Gamma' \subseteq Cn_{\mathbf{AL}}(\Gamma)$ *then* $Cn_{\mathbf{AL}}(\Gamma \cup \Gamma') \subseteq Cn_{\mathbf{AL}}(\Gamma)$. *(Cautious Cut.)*
8. *If* $\Gamma' \subseteq Cn_{\mathbf{AL}}(\Gamma)$, *and* $Cn_{\mathbf{AL}}(\Gamma) \subseteq Cn_{\mathbf{AL}}(\Gamma \cup \Gamma')$. *(Cautious Monotonicity.)*

Theorem 1.19 *Each of the following obtains:*

1. *If* Γ *is normal, then* $\mathcal{M}^{\mathbf{ULL}}_\Gamma = \mathcal{M}^m_\Gamma = \mathcal{M}^r_\Gamma$ *whence* $Cn_{\mathbf{AL}^r}(\Gamma) = Cn_{\mathbf{AL}^m}(\Gamma) = Cn_{\mathbf{ULL}}(\Gamma)$.
2. *If* Γ *is abnormal and* $\mathcal{M}^{\mathbf{LLL}}_\Gamma \neq \emptyset$, *then* $\mathcal{M}^{\mathbf{ULL}}_\Gamma \subset \mathcal{M}^m_\Gamma$ *and hence* $Cn_{\mathbf{AL}^r}(\Gamma) \subseteq Cn_{\mathbf{AL}^m}(\Gamma) \subset Cn_{\mathbf{ULL}}(\Gamma)$.
3. $\mathcal{M}^{\mathbf{ULL}}_\Gamma \subseteq \mathcal{M}^m_\Gamma \subseteq \mathcal{M}^r_\Gamma \subseteq \mathcal{M}^{\mathbf{LLL}}_\Gamma$ *whence* $Cn_{\mathbf{LLL}}(\Gamma) \subseteq Cn_{\mathbf{AL}^r}(\Gamma) \subseteq Cn_{\mathbf{AL}^m}(\Gamma) \subseteq Cn_{\mathbf{ULL}}(\Gamma)$.
4. $\mathcal{M}^r_\Gamma \subset \mathcal{M}^{\mathbf{LLL}}_\Gamma$ *iff* $\Gamma \cup \{A\}$ *is* **LLL**-*satisfiable for some* $A \in \Omega - U(\Gamma)$.
5. $Cn_{\mathbf{LLL}}(\Gamma) \subset Cn_{\mathbf{AL}^r}(\Gamma)$ *iff* $\mathcal{M}^r_\Gamma \subset \mathcal{M}^{\mathbf{LLL}}_\Gamma$.
6. $\mathcal{M}^m_\Gamma \subset \mathcal{M}^{\mathbf{LLL}}_\Gamma$ *iff there is a (possibly infinite)* $\Delta \subseteq \Omega$ *such that* $\Gamma \cup \Delta$ *is* **LLL**-*satisfiable and there is no* $\varphi \in \Phi_\Gamma$ *for which* $\Delta \subseteq \varphi$.
7. *If there are* $A_1, \ldots, A_n \in \Omega$ $(n \geq 1)$ *such that* $\Gamma \cup \{A_1, \ldots, A_n\}$ *is* **LLL**-*satisfiable and, for every* $\varphi \in \Phi_\Gamma$, $\{A_1, \ldots, A_n\} \nsubseteq \varphi$, *then* $Cn_{\mathbf{LLL}}(\Gamma) \subset Cn_{\mathbf{AL}^m}(\Gamma)$.
8. $Cn_{\mathbf{AL}^m}(\Gamma)$ *and* $Cn_{\mathbf{AL}^r}(\Gamma)$ *are non-trivial iff* $Cn_{\mathbf{LLL}}(\Gamma)$ *is non-trivial. (Reassurance)*

Theorem 1.20 *If* $\Gamma' \subseteq Cn_{\mathbf{AL}}(\Gamma)$, *then* $Cn_{\mathbf{AL}}(\Gamma \cup \Gamma') = Cn_{\mathbf{AL}}(\Gamma)$. *(Cumulative Indifference.)*

Theorem 1.21 *If* $\Gamma \vdash_{\mathbf{AL}} A$, *then every* **AL**-*proof from* Γ *can be extended in such a way that* A *is finally derived in it. (Proof Invariance)*

Theorem 1.22 *If* $\Gamma' \in Cn_{\mathbf{AL}}(\Gamma)$ *and* $\Gamma \in Cn_{\mathbf{AL}}(\Gamma')$, *then* $Cn_{\mathbf{AL}}(\Gamma) = Cn_{\mathbf{AL}}(\Gamma')$. *(Equivalent Premise Sets)*

1.10 SF: Decidability Matters and a Philosophical Comment

We have seen in Sect. 1.7 that final derivability is established by a finite proof stage and a metatheoretic reasoning about extensions of the stage and extensions of these. It is provable that, if $\Gamma \vdash_{\mathbf{AL}} A$, then A is derived on an unmarked line i of an **AL**-proof stage from Γ that is stable with respect to line i. The inconvenience is that the stage may be infinite,[23] whence Definition 1.5 is superior.

The need for a metatheoretic argument reveals an ambiguity in the notion of a proof. On the one hand, there are proofs in the sense of constructions obtained by correct applications of the rules of inference. On the other hand, a proof in the strong sense establishes *by itself* that a certain formula is derivable from a certain premise set. For compact Tarski logics, there are metatheoretic arguments that show that the existence of a proof in the weak sense warrants the existence of a proof in the strong sense—or that a proof in the weak sense constitutes a proof in the strong sense. For adaptive logics that matter is more sophisticated, as we shall see.

Definition 1.5 has a nice game-theoretic interpretation, actually several related such interpretations. As one might expect, the Proponent's task is to establish the proof, the Opponent's task to defeat it. In the simplest variant, the first move is for the Proponent who should produce a finite proof stage in which A is derived from Γ, say at line i. The next move is for the Opponent, who should extend the proof stage from Γ in such a way that i is marked. In the third move, the Proponent has to further extend the result in such a way that line i is unmarked. The Proponent has a winning strategy if, whatever the second move of the Opponent, the Proponent is able to carry out the third move successfully. Please check that this literally follows Definition 1.5.

For the propositional fragment (and for other decidable fragments of **LLL**), final derivability from finite premise sets is *decidable*. For the full predicative logics, however, there is not even a positive test. Nevertheless, even at the predicative level, there are *criteria* for final derivability. Such criteria were developed by several means, for example a 'block analysis' of proofs [11], specific tableau methods [27, 28], and a specific prospective dynamics [16, 18, 71]. Some of these need some reworking in view of the present standard format. The third approach results in the formulation of proof procedures that provide a criterion. If the procedure stops, the state of the proof reveals whether a certain formula is or is not finally derivable from the premise set; however, it is also possible that the procedure does not stop.

What if no criterion applies? All one can do is act on present insights as revealed by a proof at a stage. This leads to two questions. The first is whether the dynamics of the proofs goes anywhere. In view of the block analysis of proofs (and of the connected block semantics), the following can be established. A stage of a proof provides an insight in the premises and every step of the proof can be either informative or non-informative—this is defined in a precise way. If the step is informative, more insight

[23] Infinite stages can be extended by inserting lines in the sequence.

in the premises is gained; if the step is non-informative, no insight is gained but no insight is lost either.

Sensible proofs contain only informative steps and it is not difficult to avoid uninformative steps. There is, however, no guarantee on convergence because the computational complexity of some adaptive consequence sets, viz. where the logic follows the Minimal Abnormality strategy, is Π^1_1.[24] Let me be more explicit on convergence. There is convergence with respect to the set of *Dab*-consequences of the premise set. There is also convergence with respect to the set of minimal *Dab*-consequences of the premise set Γ. Both sets are recursively enumerable. However, there is no convergence with respect to final derivability from Γ. Suppose that A is derived on a condition, respectively a set of conditions, that warrants its final derivability with respect to $U(\Gamma)$, respectively $\Phi(\Gamma)$. As long as not all minimal *Dab*-consequences of Γ are derived, it is possible that the derivation of a non-minimal *Dab*-consequence of Γ causes A not to be derived at the stage. Needless to say, there is convergence with respect to final derivability whenever the set of minimal *Dab*-consequences of Γ is finite.

If no criterion applies, there is, as announced, a second question: Does the application context require final derivability? Not always. Reconsider the role of inconsistency-adaptive logics with respect to (what I called) the original problem. After certain abnormalities are located and perhaps some abnormalities are narrowed down in view of personal constraints and the like—see Sect. 1.12—one may have a clear idea for replacement and this may be sufficient to launch a hypothesis for a replacement of the inconsistent theory. Several people may launch several hypotheses, but the located problems will usually be common. Even if these are far from complete, some of the launched hypotheses may be successful, for a while or forever. A good example is Frege's set theory. The Russell paradox was known and led to proposals for replacements. Several of these were not shown to be inconsistent until now. So, as far as we can tell, they are worthwhile proposals for consistent set theories. Only after most of these proposals were formulated, the Curry paradox was discovered. So the proposals were made without a full analysis of the inconsistencies in Frege's theory. A similar story may be told, although perhaps less convincingly, about Clausius' removal of an inconsistency from thermodynamics. The aim of applications with respect to creative processes is to arrive at sensible hypothetical proposals for consistent replacements. The means to reach this end is the analysis provided by the inconsistency-adaptive logic(s). In that respect $Cn_{\mathbf{AL}}(\Gamma)$ is merely an ideal. This ideal is studied in order to show that the applied mechanism is coherent and conceptually sound. To the extent that our estimate of $Cn_{\mathbf{AL}}(\Gamma)$ is better, we may arrive at better proposals. We know that, for some **AL** and Γ, the set $Cn_{\mathbf{AL}}(\Gamma)$ is beyond our reach. All we can do is go by present insights and hope that they are not too bad an estimate of the final consequence set. That's life. The only alternatives are dogmatic belief and gardening.

[24] It is ironic that the study of the computational complexity of adaptive logics started with a paper arguing that they are too complex [41]. The philosophical complaints and misunderstandings in that paper were answered in [26]; a mistaken theorem was corrected in [68]. Extremely interesting and more detailed studies followed [54, 55].

1.11 Variants to the Standard Format

The first versions of the standard format were published in [15, 17]. It soon became clear that especially a universal formulation of the proof theory required the presence of a classical disjunction. Other classical logical symbols also proved very useful. If the abnormalities are contradictions or existentially closed contradictions, one better has a classical conjunction around. Having classical negation around also turned out attractive.

Let me illustrate the attractiveness of classical negation in terms of **CLuN**^{r}—the subsequent illustration may be adjusted to any inconsistency-adaptive logic mentioned so far. If $p \wedge \neg p \notin U(\Gamma)$, then each of the following obtain: (i) if $\neg p, p \vee q \in Cn_{\mathbf{CLuN}^r}(\Gamma)$, then $q \in Cn_{\mathbf{CLuN}^r}(\Gamma)$, (ii) if $\neg p, q \supset p \in Cn_{\mathbf{CLuN}^r}(\Gamma)$, then $\neg q \in Cn_{\mathbf{CLuN}^r}(\Gamma)$, (iii) if $\neg p \in Cn_{\mathbf{CLuN}^r}(\Gamma)$, then $\neg(p \wedge q) \in Cn_{\mathbf{CLuN}^r}(\Gamma)$, and so forth and so on. Suppose, however, that **CLuN** is extended with the classical negation $\sim$.[25] As $p \wedge \neg p \notin U(\Gamma)$, we now obtain: if $\neg p \in Cn_{\mathbf{CLuN}^r}(\Gamma)$, then $\sim p \in Cn_{\mathbf{CLuN}^r}(\Gamma)$. Note, however, that this is a very basic step. Once we have derived $\sim p$ by the rule RC, all other steps follow by the rule RU. Indeed, in the version of **CLuN** that contains a classical negation, (i) $\sim p, p \vee q \vdash_{\mathbf{CLuN}} q$, (ii) $\sim p, q \supset p \vdash_{\mathbf{CLuN}} \neg q$, (iii) $\sim p \vdash_{\mathbf{CLuN}} \neg(p \wedge q)$, and so forth and so on. So once the classical negation of p is derived, there is no further need to apply RC. This made classical negation quite interesting.

The situation became even more attractive when it turned out that, in certain combinations of adaptive logics—like in $Cn_{\mathbf{AL2}}(Cn_{\mathbf{AL1}}(\Gamma))$—not all information is carried over to the second logic unless $Cn_{\mathbf{AL1}}(\Gamma)$ contains a classical negation. Moreover, the formulation of the standard format turned out more elegant if classical connectives were around. I tried to avoid $\sim$ in Sect. 1.9—actually, $\sim$ only occurs in Item 2 of Theorem 1.18. However, many transparent and clarifying statements may be phrased as soon as classical negation is around. Just to mention one example: $Cn_{\mathbf{AL}^r}(\Gamma) = Cn_{\mathbf{LLL}}(\Gamma \cup \{\sim A \mid A \in \Omega - U(\Gamma)\})$. Note that, thanks to the presence of $\sim$, this defines the **AL**^{r}-consequences of Γ in terms of its **LLL**-consequences—even $U(\Gamma)$ is so defined. All this, and actually more, suggested the usefulness of classical symbols in general and of classical negation in particular. Moreover, adding the classical logical symbols (in a specific way) turned out to be easy and seemed philosophically unobjectionable. Over the years, this led to the view that, given a premise set $\Gamma \subseteq \mathcal{W}$, it is advisable to formulate adaptive logics handling Γ in terms of the extension of the native $\mathcal{L}$ with the classical symbols that do not belong to $\mathcal{L}$. In the interest of the elegance of the standard format, this was modified to: add classical symbols, even when they duplicate symbols of $\mathcal{L}$, and refer to them by specific 'checked' logical symbols $\check{\neg}$, $\check{\vee}$, etc.[26]

[25] Stepwise: the language $\mathcal{L}_s$ of **CLuN** is extended with the symbol $\sim$ and **CLuN** is extended with axioms or rules that give $\sim$ its classical meaning—for example the schemas $A \supset (\sim A \supset B)$ and $(A \supset \sim A) \supset \sim A$.

[26] The classical symbols were actually *superimposed* on $\mathcal{L}$: in the extended language, they never occur within the scope of the original logical symbols of $\mathcal{L}$.

It later turned out that it was important to distinguish, with respect to proofs, between (what is now called) *Dab*-formulas and inferred *Dab*-formulas.[27] As the added symbols were around anyway, the distinction was originally introduced in terms of the checked disjunction $\check{\vee}$.

There are mainly three reasons why I described a standard format without 'checked' logical symbols. First, the introduction of those symbols is rather tiresome. It requires a motivation and a lengthy and careful formulation. A standard format with checked symbols is definitely more complicated than one without, and one wonders whether the advantages of extending the language outweighs the complication. Next, the addition of classical negation will definitely raise suspicion from the side of dialetheists. So, as the addition is avoidable, it better is avoided—the formulation of a logic should refrain from taking a philosophical stance. Finally, the checked symbols led to confusion, for example to the mistaken claim that adaptive logics are in a sense incomplete because not all semantic consequences would be derivable from premise sets in which occur checked symbols [63, 64].[28]

All that we really need in the standard format is a classical disjunction, to which I *refer* by $\hat{\vee}$. The classical disjunction will occur in *Dab*-formulas and in disjunctions like $B \mathbin{\hat{\vee}} Dab(\Theta)$ in applications of RC. And even the requirement that a classical disjunction should occur in $\mathcal{L}$ may be dropped, as we shall see after the next paragraph.

Do all adaptive logics that fit in the version of the standard format with added classical symbols also fit in the version without such added symbols? Not quite. However, the adaptive logics that do not belong to the standard format in the present (actually restored original)[29] version can be integrated by a single and simple strike. We shall see so in Sect. 1.13.

The requirement that classical disjunction should be a symbol of $\mathcal{L}$ may be dropped by moving to a *multiple-conclusion standard format*. This fact was first seen and used by Sergei Odintsov and Stanislav Speranski [55]; they formulated this version of the standard format for propositional logics, but the generalization to predicative logics is straightforward.

Where **L** is a logic, I shall write $\Gamma \vdash^{mc}_{\mathbf{L}} \Delta$ to express that, according to **L**, one of the members of Δ is true if all members of Γ are true. **LLL** should be specified to be left compact as well as right compact; so if $\Gamma \vdash^{mc}_{\mathbf{L}} \Delta$, then there is a finite $\Gamma' \subseteq \Gamma$ and a finite $\Delta' \subseteq \Delta$ such that $\Gamma' \vdash^{mc}_{\mathbf{L}} \Delta'$. Next, the condition of the rule RC can now be phrased as "If $A_1, \ldots, A_n \vdash^{mc}_{\mathbf{LLL}} \{B\} \cup \Theta$", in which Θ is a finite set as in the original RC. The multiple-conclusion standard format is also handy and interesting from a metatheoretic point of view. Remember the characterization of $\mathbf{AL}^r$ in terms of **LLL** phrased with the help of $\sim$: $Cn_{\mathbf{AL}^r}(\Gamma) = Cn_{\mathbf{LLL}}(\Gamma \cup \{\sim A \mid A \in \Omega - U(\Gamma)\})$. This

[27]The distinction warrants that the reference to a *finite* proof stage in Definition 1.5 is all right.

[28]The mistake is caused by a confusion between symbols and concepts. If $\check{\vee}$ occurs in a premise, and so in $\mathcal{L}$, then $\check{\vee}$ is not a new symbol of the extended language. So one needs to extend the language with another symbol, say $\tilde{\vee}$, and call that the checked disjunction.

[29]All that is new in the restored version is the notion of an inferred *Dab*-formula.

can be phrased without classical negation in multiple-conclusion terms: $\Gamma \vdash^{mc}_{\mathbf{AL}^r} \Delta$ iff $\Gamma \vdash^{mc}_{\mathbf{LLL}} \Delta \cup (\Omega - U(\Gamma))$. The multiple conclusion version of Theorem 1.12 follows from this by right compactness.

1.12 Variation

As adaptive logics are not deductive logics but formal characterizations of methods, a multiplicity of adaptive logics is required for every purpose. It is not up to the logician to decree which methods a scientist should use. This choice is up to the user, viz. the scientist, and perhaps to some extent to philosophers of science. The choice cannot be justified in terms of logical features. It depends on what one learned about how to learn (Shapere), and more precisely about learning within a specific domain. So the logician should provide a multiplicity of adaptive logics. Variation may have two sources. On the one hand, the logician should look at the facts, historical facts most of the time. As the saying justly goes, the facts often outdo our phantasy. On the other hand, the logician is well placed to devise a set of variations in terms of features of the formal machinery.

Let us first have a look at **LLL**-variation. In principle, the lower limit logic can be every formal paraconsistent logic that is reflexive, transitive, monotonic and compact, for which there is a positive test, and that contains a classical disjunction—the latter is not even required in view of the multiple-conclusion standard format. So a multitude of potential lower limit logics is available. Logics between **CLuN** and **CL** (**CLuNs**, da Costa's $\mathbf{C}_n$, ...), fragments of the former, such as **LP**, all **LFI** that have a classical disjunction, Jaśkowski's **D2**,[30] practically all relevant logics, etc. Each of these can be combined with several Ω and with several strategies. Some **LLL** behave in an unexpected way if they are combined with an unsuitable Ω. However, a suitable Ω is usually easily located.

The set of abnormalities Ω may also be varied. We have already seen $\{\exists(A \wedge \neg A) \mid A \in \mathcal{W}_s\}$ as well as a restricted version $\{\exists(A \wedge \neg A) \mid A \in \mathcal{W}^a_s\}$, which is adequate for **CLuNs**, **LP**, and similar logics. At first sight, not much room seems to be left as the lower limit logic **CLuN** combined with $\Omega = \{\exists(A \wedge \neg A) \mid A \in \mathcal{W}^a_s\}$ results in adaptive logics of which **CL** is not the upper limit, whereas the lower limit logic **CLuNs** combined with $\Omega = \{\exists(A \wedge \neg A) \mid A \in \mathcal{W}_s\}$ results in a flip-flop logic—see below.

And yet, some variation is known. One example is that the set of abnormalities is extended as follows: $\Omega = \{\exists(A \wedge \neg A) \mid A \in \mathcal{F}_s\} \cup \{\forall(A \wedge \neg A) \mid A \in \mathcal{F}_s\}$. The effect is rather transparent. Although $\forall(A \wedge \neg A) \vdash_{\mathbf{CLuN}} \exists(A \wedge \neg A)$, it makes a difference whether, next to minimizing $\exists(A \wedge \neg A)$ one also minimizes $\forall(A \wedge \neg A)$. Again, this Ω is suitable for **CLuN**; for **CLuNs** one needs to replace $\mathcal{F}_s$ by $\mathcal{F}^a_s$. Other variations require symbols not in $\mathcal{L}_s$—but **CL**-definable in $\mathcal{L}_s$. A nice example is the consistency operator from logics of formal inconsistency [32]. If **LLL** is a

[30] Adaptive versions of **D2** and other Jaśkowski logics were extensively studied [48–50].

compact such logic (and $\check{\vee}$ is present in its language schema), it may be combined with $\{\neg \circ A \mid A \in \mathcal{W}\}$, possibly restricted to, for example, $\{\neg \circ A \mid A \in \mathcal{W}^a\}$. A few more suitable sets of abnormalities for inconsistency-adaptive logics are known, but it seems wiser to postpone their introduction for a few paragraphs.

So let us turn to variations to the strategy. Reliability and Minimal Abnormality are the oldest and still central strategies. A few others are worth being mentioned. The first strategy that comes to the mind of people new in the domain is the Simple strategy.

Definition 1.23 Marking for Simple: where Δ is the condition of line i, line i is marked at stage s iff some $A \in \Delta$ is an inferred *Dab*-formula of s.

This strategy is suitable iff, in view of properties of **LLL** or of the specific premise set Γ, every minimal *Dab*-consequence of Γ has only one disjunct and so is just an abnormality. It is easily seen that, if this is the case, Reliability, Minimal Abnormality, and Simple define the same adaptive logic. Where Simple is suitable, its semantics is like that of Reliability or Minimal Abnormality—the semantics for those coincide whenever Simple is suitable.

The Normal Selections Strategy was mainly developed in order to characterize some non-monotonic logics known from the literature in terms of an adaptive logic—see Sect. 1.13. The relation with Minimal Abnormality is obvious in view of Sect. 1.8.

Definition 1.24 Marking for Normal Selections: where Δ is the condition of line i, line i is marked at stage s iff $\varphi \cap \Delta = \emptyset$ for all $\varphi \in \Phi_s(\Gamma)$.

The following theorem shows that the computational complexity of adaptive logics that follow the Normal Selections strategy is less complex than the definition suggests.

Theorem 1.25 *Where* $\mathbf{AL}^n$ *is an adaptive logic following the Normal Selections strategy,* $\mathbf{AL}^n$*-final consequence sets are identical to the final consequence sets assigned by an adaptive logic* **AL1** *that is exactly like* $\mathbf{AL}^n$ *except that marking is defined as follows:*
where Δ *is the condition of line i, line i is marked at stage s iff, for a* $\Theta \subseteq \Delta$, *Dab*(Θ) *is an inferred Dab-formula of stage s.*

Definition 1.26 $\Gamma \vDash_{\mathbf{AL}^n} A$ iff, for some $\varphi \in \Phi(\Gamma)$, $M \Vdash A$ for all $M \in \mathcal{M}^m_\Gamma$ with $Ab(M) = \varphi$.

Some adaptive logics **AL** are called flip-flops. For normal premise sets Γ, $Cn_{\mathbf{AL}}(\Gamma) = Cn_{\mathbf{ULL}}(\Gamma)$, which is as desired and holds for all adaptive logics. For abnormal Γ—those that have no **ULL**-models—$Cn_{\mathbf{AL}}(\Gamma) = Cn_{\mathbf{LLL}}(\Gamma)$, which is usually not what one wants. As was explained in Sect. 1.4, a central aim of adaptive logics is to isolate the abnormalities in abnormal Γ and to validate applications of **ULL**-rules whenever no abnormality is involved. Flip-flops do this only in the crudest possible way. In the case of inconsistency-adaptive logics, for example, flip-flops deliver the full **CL**-consequence set of normal Γ and nevertheless avoid triviality in the case of abnormal Γ. Unlikely as it may appear, there are application contexts in which a flip-flop is precisely what one wants. For such cases, it is useful to have a strategy around to define flip-flops.

Definition 1.27 Marking for Flip-Flops: where Δ is the condition of line i, line i is marked at stage s iff $\Delta \neq \emptyset$ and there is at least one inferred *Dab*-formula of s.

The Blindness strategy handles abnormal premise sets as if they were normal. Replacing the strategy of any of the aforementioned inconsistency-adaptive logics by Blindness results in **CL**.

Definition 1.28 Marking for Blindness: mark no lines.

By varying the strategy, one may also define some logic-like entities. A first example is the Single Selection Strategy. It consists in *choosing* a $\varphi \in \Phi_s(\Gamma)$ and in marking lines with condition Δ iff $\varphi \cap \Delta = \emptyset$. The result is not a logic because there is an element of choice that is not specified in the premise set. There are several ways in which the consequence set may be characterized in terms of an adaptive logic. I mention the most obvious one. Let $\mathbf{AL}^s$ have the same lower limit and set of abnormalities as the logic-like object but the Simple strategy instead. The intended consequence set is provably identical to $Cn_{\mathbf{AL}^s}(\Gamma \cup \varphi)$.[31]

Another logic-like entity is defined by the All Selections Strategy. The entity is at best logic-like because it maps premise sets to sets of consequence sets, rather than to consequence sets: $\wp(\mathcal{W}) \to \wp(\wp(\mathcal{W}))$. Each of the consequence sets is associated with a $\varphi \in \Phi(\Gamma)$. One also needs to associate a mark to each $\varphi \in \Phi(\Gamma)$. A line with condition Δ is φ-marked iff $\Delta \cap \varphi \neq \emptyset$.[32]

Leaving strategy variations, let us have a look at some more drastic 'variants'. A first variant comes in a sense to digging deeper in abnormalities. The point is that an inconsistency like $(p \vee q) \wedge \neg(p \vee q)$ may have several 'causes' and that the causes themselves may be considered as abnormalities. The inconsistency $(p \vee q) \wedge \neg(p \vee q)$ may be derivable from the premises because $p \wedge \neg(p \vee q)$ is derivable, or because $q \wedge \neg(p \vee q)$ is derivable. It is also possible that neither of the two is derivable, but that $(p \vee q) \wedge \neg(p \vee q)$ still is. So this leaves us with three different sorts of (non-independent) abnormalities rather than one. What is fascinating in this approach? Let me explain in terms of Reliability. Even if $(p \vee q) \wedge \neg(p \vee q) \in U(\Gamma)$, it is possible that r is derivable on the condition $\{p \wedge \neg(p \vee q)\}$ and that $p \wedge \neg(p \vee q) \notin U(\Gamma)$. On the one hand this approach forms an Ω-variant. On the other hand, a net gain is obtained if one applies this approach to, for example, $Cn_{\mathbf{CLuN}^m}(\Gamma)$ rather than to Γ itself. I refer to a published paper [22] for the precise (but rather lengthy) definition of the new set of abnormalities. It is instructive to compare the new combined logic—call it $\mathbf{CLuN}^m_c$—with the well studied $\mathbf{CLuN}^m$, $\mathbf{CLuNs}^m$, and $\mathbf{LP}^m$. I present one example of a premise set in Table 1.1. The consequence set of the combined logic

[31]The low computational complexity of the consequence set is rather artificial. We suppose that at least one $\varphi \cap \Delta = \emptyset$ is given, but precisely locating a φ may be a very complex task.

[32]The logic-like entity has a rather limited application field. For some Γ, $\Phi(\Gamma)$ is not only infinite but also uncountable.

Table 1.1 Comparison for $\Gamma = \{p, \neg p \vee q, \neg(p \vee r), \neg\neg p \supset s\}$

CLuN^m	**CLuN**^m_c	**CLuNs**^{m}	**LP**^{m}
p	p	p	p
		$\neg p$	$\neg p$
$\neg\neg p$	$\neg\neg p$	$\neg\neg p$	$\neg\neg p$
	$\neg r$	$\neg r$	$\neg r$
q	q		
s	s	s	

is rather fascinating. On the one hand, it extends the **CLuN**^{m}-consequence set. On the other hand, where a member of the **CLuNs**^{m}-consequence set is absent ($\neg p$ in the example), this results in a more interesting consequence (q in the example); an inconsistency is avoided in order to obtain a different consequence.

A very different variant concerns the reduction of abnormalities in terms of plausibilities or preferences. Suppose that $A_1, \ldots, A_n \in \Omega$ and that $A_1 \hat{\vee} \cdots \hat{\vee} A_n$ is a minimal inferred *Dab*-formula at stage s of a proof from Γ. One may have reasons not to consider the n abnormalities A_i as equally affected, but to opt for or against a specific abnormality A_i. Of course, the choice should be made defeasibly to avoid triviality on the one hand and superfluous inconsistency on the other. So one will add the premise $\Diamond A_i$ or $\Diamond\neg A_i$, in which $\Diamond$ functions as a plausibility operator. Abnormalities may be the formulas of the form $\Diamond A \wedge {\sim}A$ and those of the form $\Diamond{\sim}A \wedge A$. So, for example, from $\Diamond{\sim}A$ one may derive ${\sim}A$ on the condition $\{\Diamond{\sim}A \wedge A\}$. The upshot will be that plausible statements will be defeasibly turned into full premises and that *Dab*-formulas from the inconsistency-adaptive logic will be reduced. If $A_1 \hat{\vee} \cdots \hat{\vee} A_n$ is a minimal inferred *Dab*-formula at stage s and A_1 came out of the plausibility logic, then $A_2, \ldots, A_n$ are off the hook. If, to the contrary, ${\sim}A_1$ came out of the plausibility logic, then $A_2 \hat{\vee} \cdots \hat{\vee} A_n$ is **LLL**-derivable and hence is a minimal inferred *Dab*-formula. It is often more appropriate to have different degrees of plausibility available: $\Diamond A$ for very plausible, $\Diamond\Diamond B$ for a bit less plausible, and so on. Technically speaking, one first adds the layers of plausibility statements—as much as possible of the most plausible statements, next as much as possible of the second-most plausible statements, and so on, and finally one applies the inconsistency-adaptive logic. This approach to weeding out abnormalities was studied along with several variants for expressing and handling plausibilities or preferences [19].

And now to a third type of variant, and again a completely different one: other gluts, gaps, and ambiguities. Remember that, in the original problem, the aim was to obtain minimally inconsistent theories that may serve as a starting point to devise a consistent theory. Until now, I have followed the official line of thought: as the theory under consideration is inconsistent, one has to replace **CL** by a paraconsistent logic. This, however, is not the only way out. Inconsistencies may be seen as negation gluts: the classical condition for $\neg A$ to be false is present (in that A is true), but nevertheless $\neg A$ is true. Negation gaps may be understood in a similar way. Moreover, gluts as well as gaps with respect to other logical symbols may also be understood along the same line. We are for example confronted with an existential gap if $\exists x Px$ is false although Pa is true. Furthermore, non-logical symbols may be ambiguous in that different occurrences of the same symbol may have a different meaning, whence different occurrences of the same formula may have different truth values. Sundry gluts or gaps may be allowed, possibly along with ambiguities, in order to avoid triviality; next, the gluts and gaps and ambiguities may be minimized in order to interpret the premise set as much as possible in the way **CL** interprets it—the first ambiguity-adaptive logics were devised by Guido Vanackere [65–67].

The premise set $\Gamma_4 = \{p, r, (p \vee q) \supset s, (p \vee t) \supset \neg r, (p \wedge r) \supset \neg s, (p \wedge s) \supset t\}$ may serve as an illustration. Γ_4 has models (i) of logics that allow for negation gluts, (ii) of logics that allow for negation gaps, (iii) of logics that allow for conjunction gaps as well as disjunction gaps, (iv) of logics that allow for implication gluts, (v) of logics that allow for ambiguities in the non-logical symbols, and of course of logics that allow for several of the mentioned gluts and gaps and ambiguities. Each of these possibilities defines a different adaptive theory. Each of these theories is a sensible solution of the original problem. So, again, a multiplicity of approaches is available and this is as it should be. All those abnormalities *surface* as inconsistencies when one applies **CL** to premise sets, but this does not mean that paraconsistency is the only possible answer. The combinations lead up to adaptive zero logic $\mathbf{CL\emptyset}^m$. In this logic, all meaning is contextual. According to **CL∅** nothing is derivable from any premise set, not even the premises. Nevertheless, the adaptive $\mathbf{CL\emptyset}^m$ assigns to normal premise sets the same consequence set as **CL**. Apart from its own interest, $\mathbf{CL\emptyset}^m$ was shown to have an important heuristic value for determining which combinations of gaps or gluts or ambiguities lead to maximally normal interpretations of a given premise set. A detailed study is available [7].

1.13 Integration

Once the standard format was described, it was not difficult to devise many new logics and this pragmatic attitude led to useful work. However, it is also important to unify the domain of 'defeasible logics'. It is important to find out whether all defeasible logics can be subsumed under the same schema or, if that turns out impossible, whether the number of schemas can be reduced. Needless to say, it cannot be settled today which schemes have most unifying power. However, studying the unifying power of adaptive logics seems sensible because there is a clear underlying concept. This is why a lot of attention was given to integrating existing mechanisms into adaptive logics. There is a book [63] that contains many relevant results and a list of papers that I shall not add to the references.

As I see it, the aim should be to integrate the realistic and potentially realistic defeasible reasoning forms. It goes without saying that truckloads of defeasible mechanisms may be defined, especially in semantic terms. It goes equally without saying that many of them cannot be integrated in any finite set of unifying schemas.

This is as unimportant as it is obvious. Among the possible sources for potentially realistic reasoning forms are (i) defeasible reasoning forms described by different approaches, (ii) old and 'unusual' adaptive logics that are not in standard format, (iii) new defeasible reasoning forms that are useful in view of the philosophy of science, the philosophy of mathematics, and everyday reasoning.

Two examples of integration follow, one 'external' and one 'internal'. The external one concerns the Strong Consequence Relation devised by Nicholas Rescher [58]. Consider a version of **CLuN** with classical negation $\sim$—the variant will not be given a different name. Let Γ' comprise the members of Γ with $\neg$ replaced by $\sim$ and let $\Gamma^{\neg\sim} = \{\neg\sim A \mid A \in \Gamma'\}$. It was proven [14] that $\Gamma \vdash_{\text{Strong}} A$ iff $\Gamma^{\neg\sim} \vDash_{\mathbf{CLuN}^m} A$. So the corrective consequence relation Strong is characterized by (the variant of) the adaptive logic $\mathbf{CLuN}^m$ under a translation. The characterization in adaptive terms reveals at once a whole set of properties of the Strong consequence relation. It also enables one to devise so-called direct proofs: adequate dynamic proofs that proceed in the original language (with one negation symbol) [29].

By internal integration I mean that adaptive logics that are not in standard format are characterized in terms of an adaptive logic in standard format. It may be shown, for example, that adaptive logics following the Normal Selections strategy can be characterized in terms of adaptive logics that follow the Minimal Abnormality strategy. The example I shall use as an illustration here is the one promised in Sect. 1.11: adaptive logics that fall under the standard format with checked logical symbols but not under the standard format without, may (all and in one sweep) be characterized in terms of adaptive logics that fall under the new standard format.

Let **AL1** be the adaptive logic that requires integration because it requires the presence of checked symbols whereas some (or even all) classical symbols are absent from its native language. One simply proceeds as follows. First, the native language $\mathcal{L}$ of **AL1** is extended to $\mathcal{L}^+$ by superimposing $\check{\vee}$[33] as well as all other classical symbols. Next, define **AL2** like **AL1** except that **AL2** is defined over $\mathcal{L}^+$. So, whatever classical symbols were required for defining **AL1** are available in the native language of **AL2**, which is in the present standard format. Finally, define $Cn_{\mathbf{AL}}(\Gamma) = Cn_{\mathbf{AL}^+}(\Gamma) \cap \mathcal{W}$—obviously no translation function is required, or rather, the translation function is such that $\text{tr}(A) = A$. The reader should not be misled by this example. Here integration is nearly obvious. In other cases, however, integration may require quite some ingenuity.

1.14 In Conclusion: Applications

From the very first ideas on, my motivation for developing adaptive logics was always guided by the aim to handle sensible applications in a sensible way. Moreover, this aim was to understand and explicate the actual defeasible reasoning. Attention for models and for formal properties came only afterwards, as a means rather than as an end.

[33] That is (i) $\mathcal{W} \subseteq \mathcal{W}^+$ and (ii) if $A, B \in \mathcal{W}^+$, then $(A \check{\vee} B) \in \mathcal{W}^+$.

We have seen that the 'original problem' was to construct minimally abnormal interpretations of mathematical or empirical theories that were intended as consistent but turned out to be inconsistent. This was the central application context for inconsistency-adaptive logics as well as for combinations of inconsistency-adaptive logics with other adaptive logics.

In the previous paragraph, "theory" should not be taken too literally. There are many cases in which one deals with inadvertently inconsistent premise sets the content of which is much more disparate than are the theorems of a theory. A nice example is that inconsistency-adaptive logics allow one to incorporate the inconsistent case in belief revision [34]. This broadens an existing approach, making room for inconsistency. A similar move may be made with respect to many other approaches, for example question evocation [45]. A different matter is that existing mechanisms that are able to handle inconsistency have more attractive adaptive versions [46].

Graham Priest, who edited my oldest paper on the topic, was fascinated by the application of adaptive logics to a very different problem. Inconsistency-adaptive logics offer the possibility to understand most of classical reasoning and actually to understand it as correct. Not as correct by logical standards, but as correct by logical standards extended with the presumption that inconsistencies are false. For dialetheists the presumption is justified by the low frequency of true inconsistencies. That a person with so different a view on logic saw a use in inconsistency-adaptive logics has been a great source of encouragement.

Recently, a very different type of application turned out to be fascinating. In view of the limitative theorems in mathematics, (i) the axiomatic method is known to have a rather limited scope and (ii) some of our present mathematical theories may very well turn out to be inconsistent and hence, as their underlying logic is **CL**, trivial. In view of each of these facts, it became attractive to phrase theories that have an adaptive logic as their underlying logic. These theories, viz. their set of theorems, are obviously not semi-recursive. That is precisely one of the advantages. Notwithstanding their finitary rules and notwithstanding the simplicity of dynamic proofs-at-a-stage, adaptive logics enable one to axiomatize Π^1_1-complex theories. So although it is too complex, for either humans or Turing machines, to figure out whether some formula is or is not a theorem of the theory, the theory at least defines correctly a certain complex consequence set.[34]

With respect to the possible triviality of classical mathematical theories, the advantage of adaptive theories is similar. Well-wrought inconsistency-adaptive theories display the following feature: if the classical theory is consistent, then the adaptive theory defines exactly the same set of theorems; if the classical theory is inconsistent,

[34]Classical theories, which have **CL** as underlying logic, fail to define such a theory. Their consequence relation is much less complex. If *A* is not a theorem of a classical theory, humans or Turing machines may never find this out. However, if *A* is a theorem of the classical theory, humans or Turing machines will find that out at a finite point in time. As this point may be two million years from here, the point is slightly theoretical.

it is trivial and so pointless, but the adaptive theory, which we may phrase today, will still define a non-trivial consequence set that is 'as close to' the intended consequence set 'as is possible'.

Until now only a few adaptive theories have been formulated and studied [23, 69, 70], but the results seem fascinating.

References

1. Avron, A.: On an implication connective of RM. Notre Dame J. Formal Logic **27**, 201–209 (1986)
2. Avron, A.: Natural 3-valued logics–characterization and proof theory. J. Symbolic Logic **56**, 276–294 (1991)
3. Avron, A.: Non-deterministic matrices and modular semantics of rules. In: J.-Y. Béziau (ed.) Logica Universalis, pp. 149–167. Birkhäuser Verlag, Basel–Boston–Berlin (2005)
4. Avron, A., Konikowska, B.: Multi-valued calculi for logics based on non-determinism. J. Interest Group Pure Appl. Logic **10**, 365–387 (2005)
5. Avron, A., Lev, I.: Non-deterministic multiple-valued structures. J. Logic Comput. **15**, 241–261 (2005)
6. Avron, A., Zamansky, A.: Non-deterministic semantics for logical systems. In: Gabbay, D., Guenthner, F. (ed.) Handbook of Philosophical Logic, vol. 16, pp. 227–304. Springer (2011)
7. Batens, D.: Spoiled for choice? J. Logic Comput. doi:10.1093/logcom/ext019 (1913)
8. Batens, D.: Paraconsistent extensional propositional logics. Logique et Analyse **90–91**, 195–234 (1980)
9. Batens, D.: Dialectical dynamics within formal logics. Logique et Analyse **114**, 161–173 (1986)
10. Batens, D.: Dynamic dialectical logics. In: Priest, G., Routley, R., Norman, J. (eds.) Paraconsistent Logic. Essays on the Inconsistent, pp. 187–217. Philosophia Verlag, München (1989)
11. Batens, D.: Blocks. The clue to dynamic aspects of logic. Logique et Analyse, **150–152**, pp. 285–328 (1995). Appeared 1997
12. Batens, D.: Inconsistency-adaptive logics. In: Orłowska, E. (ed.) Logic at Work. Essays Dedicated to the Memory of Helena Rasiowa, pp. 445–472. Physica Verlag (Springer), Heidelberg, New York (1999)
13. Batens, D.: Minimally abnormal models in some adaptive logics. Synthese **125**, 5–18 (2000)
14. Batens, D.: Towards the unification of inconsistency handling mechanisms. Logic Logical Philos. **8**, 5–31 (2000). Appeared 2002
15. Batens, D.: A general characterization of adaptive logics. Logique et Analyse, **173–175**, 45–68 (2001). Appeared 2003
16. Batens, D.: On a partial decision method for dynamic proofs. In: Decker, H., Villadsen, J., Waragai, T. (eds.) PCL 2002. Paraconsistent Computational Logic, pp. 91–108. (=*Datalogiske Skrifter* vol. 95) (2002). Also available as cs.LO/0207090 at http://arxiv.org/archive/cs/intro.html
17. Batens, D.: The need for adaptive logics in epistemology. In: Gabbay, D., Rahman, S., Symons, J., Van Bendegem, J.P. (eds.) Logic. Epistemology and the Unity of Science, pp. 459–485. Kluwer Academic Publishers, Dordrecht (2004)
18. Batens, D.: A procedural criterion for final derivability in inconsistency-adaptive logics. J. Appl. Logic **3**, 221–250 (2005)
19. Batens, D.: Narrowing down suspicion in inconsistent premise sets. In Malinowski and Pietruszczak [42], pp. 185–209
20. Batens, D.: A universal logic approach to adaptive logics. Log. Univers. **1**, 221–242 (2007)
21. Batens, D.: Towards a dialogic interpretation of dynamic proofs. In Dégremont, C., Keiff, L., Rückert, H. (eds.) Dialogues, Logics and Other Strange Things. Essays in Honour of Shahid Rahman, pp. 27–51. College Publications, London (2009). 558 pp

22. Batens, D.: New arguments for adaptive logics as unifying frame for the defeasible handling of inconsistency. In: Tanaka, K., Berto, F., Mares, E., Paoli, F. (eds.) Paraconsistency: Logic and Applications. Logic, Epistemology, and the Unity of Science, vol. 26, pp. 101–122. Springer, Dordrecht (2013)
23. Batens, D.: The consistency of Peano Arithmetic. A defeasible perspective. In: Allo, P., Van Kerkhove, B. (eds.) Modestly Radical or Radically Modest. Festschrift for Jean Paul van Bendegem on the Occasion of His 60th Birthday, volume 24 (sometimes 22) of Tributes, pp. 11–59. College Publications, London (2014)
24. Batens, D.: Adaptive Logics and Dynamic Proofs. Mastering the Dynamics of Reasoning. 201x. Forthcoming
25. Batens, D., De Clercq, K.: A rich paraconsistent extension of full positive logic. Logique et Analyse. **185–188**, 227–257 (2004). Appeared 2005
26. Batens, D., De Clercq, K., Verdée, P., Meheus, J.: Yes fellows, most human reasoning is complex. Synthese **166**, 113–131 (2009)
27. Batens, D., Meheus, H.: A tableau method for inconsistency-adaptive logics. In Dyckhoff, R. (ed.) Automated Reasoning with Analytic Tableaux and Related Methods, Lecture Notes in Artificial Intelligence, vol. 1847, pp. 127–142. Springer (2000)
28. Batens, D., Meheus, Joke: Shortcuts and dynamic marking in the tableau method for adaptive logics. Stud. Logica **69**, 221–248 (2001)
29. Batens, D., Vermeir, T.: Direct dynamic proofs for the Rescher-Manor consequence relations: the flat case. J. Appl. Non-Class. Logics **12**, 63–84 (2002)
30. Brown, B.: How to be realistic about inconsistency in science. Stud. Hist. Philos. Sci. **21**, 281–294 (1990)
31. Carnap, R.: Logical Foundations of Probability. University of Chicago Press, Chicago (1950)
32. Carnielli, W.A., Coniglio, M.E., Marcos, J.: Logics of formal inconsistency. In: Gabbay, D., Guenthner, F. (eds.) Handbook of Philosophical Logic, vol. 14, pp. 1–93. Springer (2007)
33. Carnielli, W.A., Marcos, J., de Amo, S.: Formal inconsistency and evolutionary databases. Logic Logical Philos. **8**, 115–152 (2001). Appeared 2002
34. De Clercq, K.: Logica in communicatie. Cahiers du Centre de Logique, vol. 14. Academia-Bruylant, Louvain-la-Neuve (2005)
35. D'Ottaviano, I.M.L.: Sobre uma Teoria de Modelos Trivalente (in Portuguese). PhD thesis, State University of Campinas (Brazil), 1982
36. D'Ottaviano, I.M.L.: The completeness and compactness of a three-valued first-order logic. In: Proceedings of the 5th Latin American Symposium on Mathematical Logic, pp. 77–94. Revista Colombiana de Matemáticas, 1–2, 1985
37. D'Ottaviano, I.M.L.: The model extension theorems for $\mathbf{J}_3$-theories. In: Di Prisco C.A. (ed.) Methods in Mathematical Logic: Proceedings of the 6th Latin American Symposium on Mathematical Logic, Lecture Notes in Mathematics, vol. 1130, pp. 157–173. Springer (1985)
38. D'Ottaviano, I.M.L.: Definability and quantifier elimination for J_3-theories. Stud. Logica **46**(1), 37–54 (1987)
39. D'Ottaviano, I.M.L., Epstein, R.L.: A paraconsistent many-valued propositional logic: J_3. Rep. Math. Logic **22**, 89–103 (1988)
40. Esser, O.: A strong model of paraconsistent logic. Notre Dame J. Formal Logic **44**, 149–156 (2003)
41. Horsten, L., Welch, P.: The undecidability of propositional adaptive logic. Synthese **158**, 41–60 (2007)
42. Malinowski, J., Pietruszczak, A. (eds.): Essays in Logic and Ontology. Poznań Studies in the Philosophy of the Sciences and the Humanities, vol. 91. Rodopi, Amsterdam/New York (2006)
43. Meheus, J.: Adaptive logic in scientific discovery: the case of Clausius. Logique et Analyse **143–144**, 359–389 (1993). Appeared 1996
44. Meheus, J.: Clausius' discovery of the first two laws of thermodynamics. A paradigm of reasoning from inconsistencies. Philosophica **63**, 89–117 (1999). Appeared 2001
45. Meheus, J.: Adaptive logics for question evocation. Logique et Analyse **173–175**, 135–164 (2001). Appeared 2003

46. Meheus, J.: An adaptive logic for pragmatic truth. In: Carnielli, W.A., Coniglio, M.E., D'Ottaviano, I.M.L. (eds.) Paraconsistency. The Logical Way to the Inconsistent, pp. 167–185. Marcel Dekker, New York (2002)
47. Meheus, J.: Inconsistencies in scientific discovery. Clausius's remarkable derivation of Carnot's theorem. In: Krach, H., Vanpaemel, G., Marage, P. (eds.) History of Modern Physics. Acta of the XXth International Congress of History of Science, pp. 143–154. Brepols, Turnhout (Belgium) (2002)
48. Meheus, J.: Discussive adaptive logics: handling internal and external inconsistencies. In: Malinowski and Pietruszczak [42], pp. 211–223
49. Nasieniewski, M.: An adaptive logic based on Jaśkowski's logic $\mathbf{D}_2$. Logique et Analyse **185–188**, 287–304 (2004). Appeared 2005
50. Nasieniewski, M.: Wprowadzenie do logik adaptywnych. Universytetu Mikołaja Kopernika, Toruń, Wydawnictwo Naukowe (2008)
51. Nersessian, N.: Inconsistency, generic modeling, and conceptual change in science. In: Meheus, J. (ed.) Inconsistency in Science, pp. 197–211. Kluwer, Dordrecht (2002)
52. Norton, J.: The logical inconsistency of the old quantum theory of black body radiation. Philos. Sci. **54**, 327–350 (1987)
53. Norton, J.: A paradox in Newtonian gravitation theory. PSA **1992**(2), 420–421 (1993)
54. Odintsov, S.P., Speranski, S.O.: On algorithmic properties of propositional inconsistency-adaptive logics. Logic Logical Philos. **21**, 209–228 (2012)
55. Odintsov, S.P., Speranski, S.O.: Computability issues for adaptive logics in multi-consequence standard format. Stud. Logica **101**(6), 1237–1262 (2013). doi:10.1007/s11225-013-9531-2
56. Pogorzelski, W.A., Prucnal, T.: The substitution rule for predicate letters in the first-order predicate calculus. Rep. Math. Logic **5**, 77–90 (1975)
57. Priest, G.: In Contradiction. A Study of the Transconsistent. Nijhoff, Dordrecht (1987)
58. Rescher, N., Manor, R.: On inference from inconsistent premises. Theory Decis. **1**, 179–217 (1970)
59. Schütte, K.: Beweistheorie. Springer, Berlin (1960)
60. Shapere, D.: Logic and the philosophical interpretation of science. In: Weingartner, P. (ed.) Alternative Logics. Do sciences need them?, pp. 41–54. Springer, Berlin, Heidelberg (2004)
61. Smirnova, E.D.: An approach to the justification of semantics of paraconsistent logics. In: Batens, D., Mortensen, C., Priest, G., Van Bendegem, J.P. (eds.) Frontiers of Paraconsistent Logic, pp. 255–262. Research Studies Press, Baldock, UK (2000)
62. Smith, J.: Inconsistency and scientific reasoning. Stud. History Philos. Sci. **19**, 429–445 (1988)
63. Straßer, C.: Adaptive Logic and Defeasible Reasoning. Applications in Argumentation, Normative Reasoning and Default Reasoning, Trends in Logic, vol. 38. Springer (2014)
64. Van De Putte, F., Straßer, C.: Extending the standard format of adaptive logics to the prioritized case. Logique et Analyse **220**, 601–641 (2012)
65. Van Kerckhove, B., Vanackere, G.: Vagueness-adaptive logic: a pragmatical approach to Sorites paradoxes. Stud. Logica **75**, 383–411 (2003)
66. Vanackere, G.: Ambiguity-adaptive logic. Logique et Analyse **159**, 261–280 (1997). Appeared 1999
67. Vanackere, G.: The role of ambiguities in the construction of collective theories. Logique et Analyse, **173–174–175**, 189–214 (2001). Appeared 2003
68. Verdée, P.: Adaptive logics using the minimal abnormality strategy are Π^1_1-complex. Synthese **167**, 93–104 (2009)
69. Verdée, P.: Strong, universal and provably non-trivial set theory by means of adaptive logic. Logic J. IGPL **21**, 108–125 (2012)
70. Verdée, P.: Non-monotonic set theory as a pragmatic foundation of mathematics. Found. Sci. **18**, 655–680 (2013)
71. Verdée, P.: A proof procedure for adaptive logics. Logic J. IGPL **21**, 743–766 (2013). doi:10.1093/jigpal/jzs046

Chapter 2
Round Squares Are No Contradictions (Tutorial on Negation Contradiction and Opposition)

Jean-Yves Beziau

Abstract We investigate the notion of contradiction taking as a central point the idea of a round square. After discussing the question of images of contradiction, related to the contest *Picturing Contradiction*, we explain why from the point of view of the theory of opposition, a round square is not a contradiction. We then draw a parallel between different kinds of oppositions and different kinds of negations. We explain why from this perspective, when we have a paraconsistent negation $\neg$, the formulas p and $\neg p$ cannot be considered as forming a contradiction. We finally introduce the notions of paranormal negation and opposition which may catch the concept of a round square.

Keywords Contradiction · Opposition · Negation · Paraconsistency · Round square

Mathematics Subject Classification (2000) Primary 03B53 · Secondary 03A05 · 03B45 · 03B50 · 03B60 · 03B05

Paraconsistent logic helps to clarify the concepts of negation and contradiction. On the one hand there are authors for whom contradictions play a quasi-mystical role, used to explain nearly everything in the universe, on the other hand excellent specialists think that contradiction is something unintelligible. Paraconsistent logic not only is useful to demystify contradiction but contributes to calm anyone who is afraid of it. Newton da Costa [33].

J.-Y. Beziau (✉)
UFRJ - Federal University of Rio de Janeiro, RJ, Brazil
e-mail: jyb@jyb-logic.org

J.-Y. Beziau
CNPq - Brazilian Research Council, South America, Brazil

J.-Y. Beziau
UCSD - University of California, San Diego, USA

J.-Y. Beziau et al. (eds.), *New Directions in Paraconsistent Logic*,
Springer Proceedings in Mathematics & Statistics 152,
DOI 10.1007/978-81-322-2719-9_2

Fig. 2.1 Round square (J)

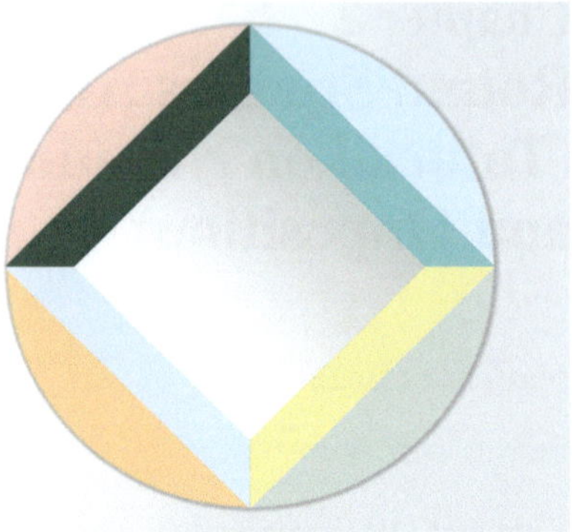

2.1 Picturing Contradiction

What is a contradiction? Contradiction is a famous notion. But do we have an idea, an image, or a definition of a contradiction? And what is the reality of contradiction, if any? In this paper we will investigate this notion considering the round square as a platform for developing a discussion about the trinity negation, contradiction, opposition.

What is a round square? A simple reply to this answer is the following Fig. 2.1

But this is not very satisfactory because this image is just a juxtaposition of a circle and a square. One may want to develop a logic of imagination considering $\bigcirc$ as a modal operator of imagination and taking as an axiom:

$$\bigcirc A \wedge \bigcirc B \rightarrow \bigcirc (A \wedge B)$$

But according to this logic of imagination, we can imagine lots of things.[1] It is not the same to imagine a man and a horse and a centaur, just compare classical mythology and modern mythology (Fig. 2.2).

Maybe the following image is a better representation of a round square, closer to the centaur construct, result of a blending (Fig. 2.3).

But according to standard plane geometry, this is indeed neither a square nor a circle. At the 5th World Congress on Paraconsistency in Kolkata, India, February 13–17, 2014, we organized the contest *Picturing Contradiction*. We asked people from all over the world to send us an image picturing contradiction. It was on the one hand a way to promote the participation of all the people, even those who were not able to come to Kolkata, and on the other hand a way to check if contradiction is not just a mere *flatus vocis*, if there is really something behind this word.

We received few interesting images. At the end the one which won the prize was entitled "Bridge to Nowhere," submitted by Daniel Strack, Associate Professor of

[1]The logic of imagination is still a quite new and open field. A starting point was a paper by Ilkka Niiniluoto in 1985 [44]; for a critical account of this paper see [30].

Fig. 2.2 Centaur versus man on a horse

Fig. 2.3 Round square (B)

the University of Kitakyushu, Japan (Fig. 2.4).[2] This is a juxtaposition of two objects representing two opposite ideas which are melting in some way, closer therefore to the second image of a round square above (but there the melting is purely material) rather than the first one.

One of the main themes of this 5th edition of the World Congress on Paraconsistency was quantum physics and we had chosen the Fig. 2.5 as a key image for the event (see, in particular, the web site http://www.paraconsistency.org/). This a poetic representation of the duality wave/particle. For the contest itself we chose the image Fig. 2.6, representing this duality in a still metaphoric but more conceptual way.

According to Fig. 2.7, the same object appears both as a circle and as a square. One could say that it is both a circle and as a square, from the point of view of 2-dimensional space. This figure corresponds to the spirit of the philosophy of David Bohm who has used the distinction between 2-dimensional and 3-dimensional space in various ways (see his book [26]), in particular to explain inseparability: a 3-dimensional fish is projected into two 2-dimensional fishes whose interaction seems difficult to understand at a the flat level.

---

[2]The president of the jury was Kuntal Ghosh, from the Indian Statistical Institute in Kolkata where the event was taking place.

Fig. 2.4 Bridge to nowhere

Fig. 2.5 Yemanjá playing with particles

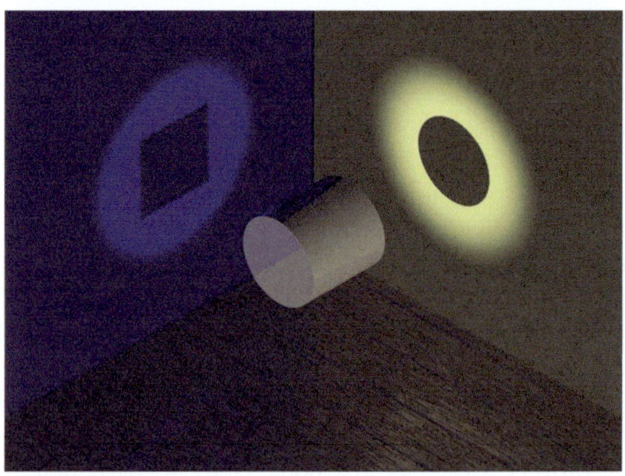

Fig. 2.6 A geometrical metaphor for the duality wave particle

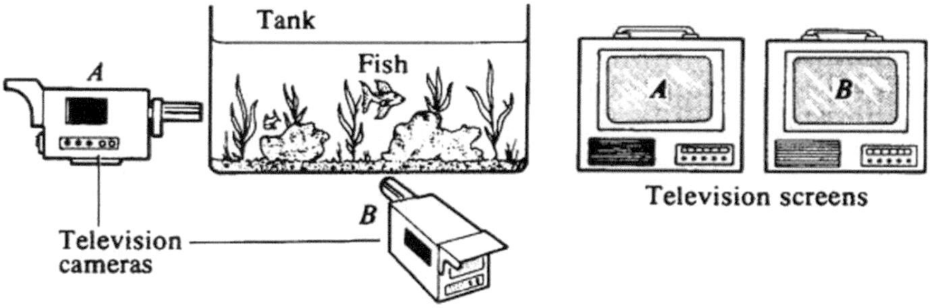

Fig. 2.7 David Bohm's metaphor for inseparability

Niels Bohr also had some ideas corresponding to Fig. 2.6. He wrote: "A complete elucidation of one and the same object may require diverse points of view which defy a unique description [27]." For both Bohm and Bohr the duality wave/particle can be interpreted as showing that reality is beyond wave and particle which are just appearances of it. This can be developed either in a Platonic perspective or in a Kantian perspective. The Kantian perspective has been emphasized by Bernard d'Espagnat (see, e.g., [34]), winner of the Templeton prize in 2009, under which I wrote a dissertation [3] at the Sorbonne in 1986 comparing Bohr, Heisenberg, and Bohm's views.[3]

In quantum physics we have a conceptual theory explaining reality but we do not have images of this reality. From this perspective one can argue that reality is beyond imagination, but that maybe our reason can catch it in some way. After developing the so-called Bohr's atom, inspired by the Rutherford's atom, a figure of microscopic reality establishing a parallel with macroscopic reality, Bohr rejected this approach and developed complementarity. He liked to wear on his jacket a picture of the Tao symbol. For him, this was not a picture of reality, but the symbol of his theory of complementarity (Fig. 2.8).

On the other hand Fig. 2.9 represents a more cosmic vision of the Tao, related with new age philosophy. It is not exactly clear what it means. The Tao can be interpreted as an intrinsic link between two contradictory notions, metaphorically represented by black and white. In Maoist philosophy, a blend of Marxism and Taosim, everything is inherently contradictory. Contradiction is understood as the unity and struggle of opposites and the law of contradiction is considered as the fundamental law governing nature and society. The unity and identity of all things is viewed as temporary and relative, while the struggle between opposites is considered as ceaseless and absolute (cf. Mao's 1937 essay *On contradiction* [43]). Such kind of theory, like the theory of evolution, can easily be used to justify war and conflict. First it is important to distinguish contradiction from conflict. Second we can consider that the world is always changing without seeing contradiction or/and conflict as a driving force. For

[3]I had the opportunity at this time to meet and discuss with David Bohm in London. After that I wrote a dissertation on Plato's cave [4] and later on I developed the paraconsistent logic Z inspired by Bohm's ideas. About this logic, see [9], and about how it was conceived, see [10].

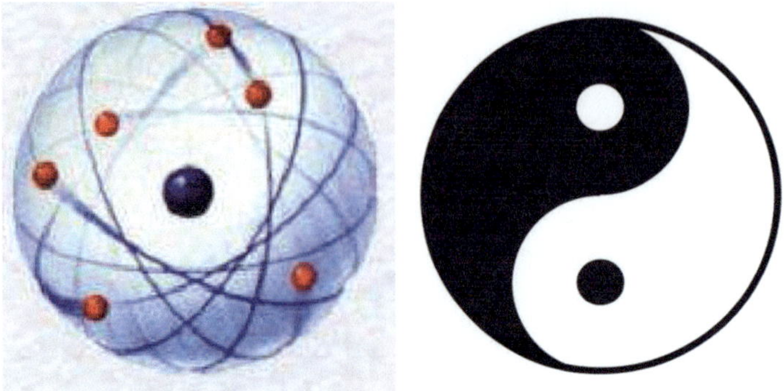

Fig. 2.8 Rutherford's atom versus the Tao of complementarity

Fig. 2.9 Taoist version of the universe

someone like Bergson contradiction is not the essence of reality but the result of the incapacity of our thought to catch the flux of reality, see e.g. [2].

2.2 Contradiction and the Square of Opposition

A standard and traditional definition of contradiction can be found in the square of opposition. Before entering into the details let us point out that we are using here the expression *square of opposition* as a name for the theory of opposition. This theory can be traced back to Aristotle, a no-square stage,[4] and is continuing to develop up to now, important stages in the development of this theory being the design of a square

[4]Larry Horn has, however, pointed out that even if we do not have a picture of the square of opposition by Aristotle, the Stagyrite suggested such a picture—see [39].

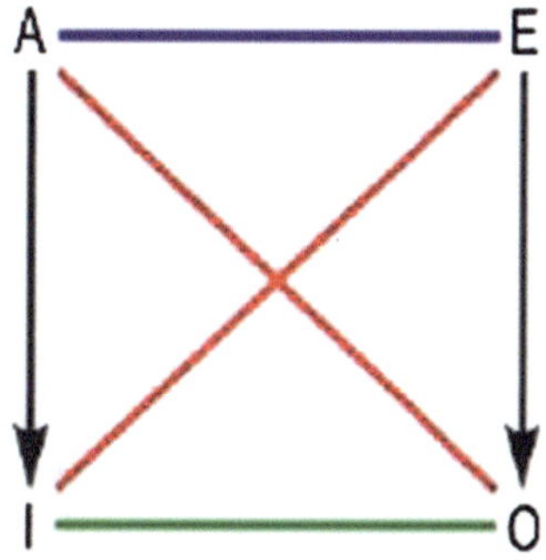

Fig. 2.10 Basic square of opposition

by Apuleius and Boethius, and the hexagon of Blanché.[5] This theory is not limited to a particular instantiation of the square figure, nor to the figure of the square itself. The backbone of this theory comprises three notions of opposition: contradiction, contrariety, and subcontrariety.[6]

These three notions can be defined as follows: two propositions are said to be *contradictories* iff they can neither be false, nor true together, *contraries*, iff they can be false together, but not true together, *subcontraries*, iff they can be true together, but not false together. These three notions of opposition can be applied directly or indirectly to concepts and properties in an intensional or extensional way. We can say that two concepts C and D are contradictories iff an object o cannot be at the same time C and D but has to be C or D. Putting this into propositions: "o is C" and "o is D" can neither be true nor false together. Extensionally speaking we can say that the sets of C-objects and D-objects are complement sets, a binary partition of the universe of objects. We can similarly adapt the two other notions of opposition, i.e. contrariety and subcontrariety, to concepts.

One of the basic figures presenting some relations between these three notions of opposition is the square represented by (Fig. 2.10). We have kept here the traditional names for the four corners, but these corners can be interpreted in many different ways: a variety of propositions and concepts, logical, metalogical, and of any field. Since 2003 [7] we have introduced colors for the three oppositions: red for contradiction, blue for contrariety, green for subcontrariety.[7] In black appears, besides the three notions of opposition, the notion of subalternation.

When we have a pair of contradictory concepts, we can talk about a contradiction. For example, in plane standard geometry a curved straight line is a contradiction. In other words: an object cannot be both a curve and a straight line.

---

[5]The work of Blanché has been published in [23–25], about the hexagon see [12].

[6]Since 2007 we are organizing a world congress on the square of opposition. The first edition happened in Montreux, the second in Corsica in 2010, the third in Beirut in 2013, the fourth in the Vatican in 2014, the next one is projected to happen in Easter Island in 2016—see http://www.square-of-opposition.org. Related publications are [11, 14, 18–22].

[7]These are the three primitive colors. The theory of opposition can also be applied to the theory of colors, see in particular the hexagon of colors of Dany Jaspers [40].

Figure 2.11 is not more a curved line than Fig. 2.1 is a round circle, but it is a juxtaposition of two images showing the contrast between the two concepts. The figure shows that something cannot be at the same time a curve and a straight line.

In the school of Pythagoras, there was the idea to explain everything by a series of pairs of concepts, considered as contradictory, listed in the Fig. 2.12.

What is interesting in this table is that the two sides of each pair are rather positive. One is not explicitly thought as the negation of the other, linguistically, and/or conceptually (excepted finite/infinite). One could argue that the idea of "classical" negation arose from that and not vice versa. Classical negation is perhaps an abstraction from a series of concrete contradictions. Plato, who had strongly been influenced by the Pythagoreans, developed the method of dichotomy, a way of thinking dividing everything in two. This method is strikingly presented in the dialogue *The Sophist*, where it is used to catch the animal of the same name. Pythagoreans were considering mathematics as the most important science. For Plato there was a further step. It

Fig. 2.11 A curved *straight line* is a contradiction

Fig. 2.12 Pythagoras' table of opposites

Odd	Even
Finite	Infinite
Straight	Crooked
Square	Oblonge
Right	Left
One	Many
In	Out
Happy	Sad
Close	Open
Rest	Motion
Good	Evil
Light	Darkness

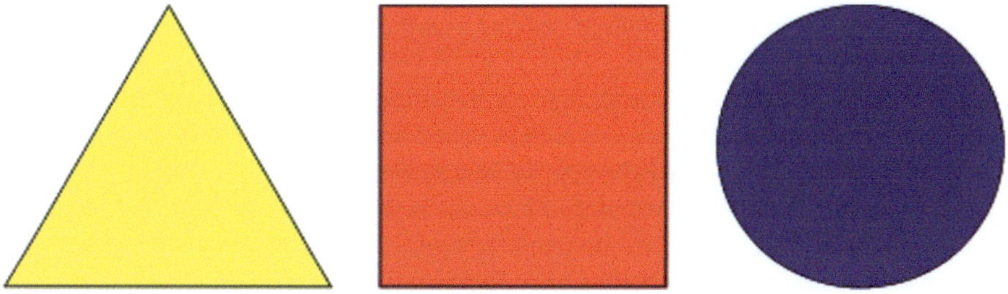

Fig. 2.13 A plane geometrical figure can be neither a *square* nor a *circle*

was *Dialectics*, a general methodology to think, reason and understand reality, the dichotomic procedure being a typical example of this methodology. Funny enough the word "dialectics" has been used later on by Hegel to denote something contrary (we use this word here in the technical sense defined above) to Plato's dialectics, the idea being that beyond the thesis and the antithesis, there is the synthesis. The table of opposites of Pythagoras is known in particular through Aristotle, but Aristotle went beyond the Pythagoro-Platonico dichotomy, long before Hegel and in a different way. He promoted the notion of contrariety (see, e.g., [1]). This is why he is considered as the father of the square of opposition. What is interesting in the square theory is that the dichotomy truth and falsity generates a trichotomy of oppositions.

Let us now come back to our mascot, the round square. Is its status the same as the curved line? No. From the point of view of standard plane geometry,[8] a figure can be neither a square nor a circle, for example, a triangle. Square and circle are not contradictory concepts, but contrary concepts: something cannot be at the same time a square and circle. A curved line is a true contradiction, a round square a fake contradiction (Fig. 2.13).

One may find two ways to explain the semantical sliding justifying naming a round square a contradiction, or qualifying it as such. The first justification is that contradiction and contrariety are both of the same family which can be labeled the *incompatibility* family: two propositions are incompatible if they cannot be true together, two concepts are incompatible if they have nothing in common. Maybe someone by saying that a round square is a contradiction has just in mind the notion of incompatibility. The second justification would be that a circle is considered as a typical representative of non-angular figures and a square as a typical representative of angular figures. But if non-angular is understood as with no angles at all, this would not work unless we define angular figures as figures having at least one angle. Angular and non-angular are in fact rather considered as a contrary pair of opposites of the same type as the famous pair which is at the the top of the square of quantification

[8]This context is important, not only to rule out other geometries—one may claim that a point is both a straight and a curved line, so that a curved line is not a contradiction, but in standard geometry a point is not a line—but also objects out of the scope of geometry, like an abstract concept such as beauty. It is possible to say that beauty is neither a square nor a circle, but this is not necessarily a convincing example to sustain that square and circle are not contradictory.

(all vs. none). In this case the "non" of "non-angular" is understood as a contrary negation (see next section).

These kinds of semantical slidings are quite common. One may consider that there are part of the semantical process which is based on variations of meanings leading sometimes to the situation where a word has at some stage a meaning opposite to a previous one. These semantical slidings can be explained in different ways, for example, by describing their mechanisms, a work which has been initiated by Bréal himself in his original 1897 book *Essai de sémantique—science des significations* [28], coining the word "sémantique" which has been later on increasingly popular. But a description of a phenomenon does mean that the phenomenon is right even if it is real. On the one hand one may want to justify some semantical variations with a theory of meaning explaining that they are coherent, this is for example the line of work developed by Larry Horn with the neo-Gricean notion of scalar implicature [38]. Some people may also argue that these slidings have a interesting creative aspect.[9] But such slidings can be consciously or unconsciously used in a dangerous way promoting confusion, this is common in advertisement and politics, part of the most monstrous creatures of the zoo of fallacies.

2.3 Negation and Contradiction

The notion of contradiction according to the square of opposition does not directly depend on the notion of negation, but only on the notions of truth and falsity. And we can define negation from the notion of contradiction, saying that two contradictory propositions or concepts are the negations of each other.

On the other hand it is also possible to define contradiction from negation, saying that the two propositions p and $\neg p$ form a contradiction. If we consider that $\neg$ is classical negation then this definition is equivalent to the square notion of contradiction. One of the most classical definition of classical negation is based on truth and falsity: p is true iff $\neg p$ is false. In this definition truth and falsity are considered as forming a dichotomy, the same dichotomy used to define the three notions of opposition of the square of opposition.

In the same way that this dichotomy can be used to define three types of oppositions, it can also be used to define three kinds of negations:

1. p is true iff $\neg p$ is false
2. if p is true then $\neg p$ is false, but not the converse
3. if p is false then $\neg p$ is true, but not the converse

[9] André Breton promoted as a key feature of surrealist writing the idea of "carambolage sémantique" [29]. But this is not the same as a "dérapage sémantique." The idea is to create a poetic effect by putting together opposed notions, leading to a sense of absurdity. Flaubert used systematically in his masterpiece *Bouvard et Pécuchet* [35] a process qualified as "antithetic juxtaposition" consisting of putting side by side two different opinions or theories. This was to show that human knowledge is not really coherent.

Note that these definitions are equivalent to the three following:

1. p and $\neg p$ cannot be true together, cannot be false together
2. p and $\neg p$ can be false together, but cannot be true together
3. p and $\neg p$ can be true together, but cannot be false together

And this second formulation clearly shows that there is a one-to-one correspondence between these three negations and the three oppositions of the square theory. To emphasize this connection and also to avoid words proliferation we can call these three negations:

1. contradictory negation
2. contrary negation
3. subcontrary negation

Let us apply these definitions to our mascot, the round square. If we have a contradictory negation, we cannot say that a square is a non-circle, we need a contrary negation and, yes, from this point of view a square can be considered as a non-circle and a circle as a non-square.

Someone may want to defend the idea that a "real" or "true" negation must be a contradictory negation. But what is the reality of negation, if any? One can claim that the word "negation" is, or, has been, used in correspondence with an operator behaving like a contradictory negation. This is ambiguous. Does this mean that the contradictory negation of classical logic is a good description of the way we use the word or that we should reason on the basis of such a negation? The ambiguity is also present the other way round. If someone rejects the classical position, does this mean that classical negation is not a good description of the way we are using negation in natural language and thought or does this mean that we shall use another negation?

Let us emphasize that it is a bit artificial to claim that classical logic is natural. Take the example of a classical non-cat. It is an abstract entity of which we do not have a positive idea or image, because the objects which are non-cats is a class of heterogeneous objects (ranging from dogs to cars through numbers). At the end we can produce an image only incorporating the abstract symbol of the cross (Fig. 2.14).[10]

On the other hand to say that classical negation is wrong, like Richard Routley, who liked to claim that every morning before breakfast, seems exaggerated.[11] Contradictory negation is the product of abstraction and abstraction is a fundamental power of human mind. The full strength of contradictory negation has to be recognized, this negation is not something which has to be rejected, but which has to be used with moderation. We do not support the idea that classical negation is the only negation and that we cannot use the word "negation" for other operators. This does not mean that we can use this word in an arbitrary way. We believe it is important to give the right name to the right thing, not based on a purely descriptive perspective, but by developing a theory which is, as any theory, relatively normative, keeping an

[10]For more discussion about the variety of symbolism, see [17].

[11]This was reported to me by Newton da Costa. He faced this phenomenon when visiting the Australopithecus in his own country in the 1970s.

Fig. 2.14 A non-cat is an abstract entity

equilibrium with description, the way the concept and the word are used. We defend the idea that the three above negations deserve to be qualified as negations. This is in particular coherent with the theory of the square of opposition. This is a also coherent with the development of modern logic where intuitionistic negation, which is a specific example of contrary negation, is called a "negation."

And to use the same symbol, "¬," for different negations corresponds to a natural procedure of "abus de language" common in mathematics where the same symbol, "0", is used for different numbers having different properties, to keep trace to their common properties. The idea of a perfect unambiguous language in science promoted by Frege (see [36]) and some neopositivists seems absurd to us nowadays.

If we want to put in the same bag contradictory and contrary negations, we can talk about *incompatible negations* or *negations of incompatibility*, the definition being that p and $\neg p$ cannot be true together. Someone may claim that a negation should be an incompatible negation that we have to exclude subcontrary negations. This is a kind of neo-Aristotelian position, because the Stagyrite rejected subcontrariety as an opposition. But there is a strong symmetry and duality between contrariety and subcontrariety that is clearly revealed by the picture of the square. In modern logic, if one admits a contrary negation, like intuitionistic negation, there is no good reasons to reject its dual, which is a subcontrary negation, part of the family of paraconsistent negations.

There are different ways of dualizing intuitionistic negation. I. Urbas presented a dualization based on sequent calculus considering restriction of one formula on the left instead as on the right [51]. I have myself worked on a dualization based on modal interpretation which can be extended to other contrary negations, defined

Fig. 2.15 Duality between contrary and subcontrary negations in modal logic

as "not possible," $\neg\Diamond$, where $\neg$ is classical negation, following the interpretation of intuitionistic negation in S4 by Gödel [37]. The dualization of $\neg\Diamond$ is $\neg\Box$, which is a subcontrary negation as illustrated by Fig. 2.15.[12]

2.4 Paraconsistent Logic and Contradiction

The starting point of paraconsistent logic is to reject the so-called law of explosion.[13] It means that we have a negation $\neg$ and propositions p and q such that:

$$p, \neg p \nvdash q$$

Considering a basic general Tarskian framework for consequence relation this is equivalent as to say that there is a proposition p, such that p and $\neg p$ can be true together—see [13, 41].

According to the theory of opposition, p and $\neg p$ do not therefore form a contradictory pair. They are at best a subcontrary pair, and paraconsistent negation at best a subcontrary negation. The place where there are contradictions is a logic with a classical negation. If there are contradictions in a paraconsistent logic it is because it is possible to define a classical negation within it, like in the paraconsistent logic $C1$ of Newton da Costa [31].

If someone says that given a paraconsistent negation $\neg$, p and $\neg p$ form a contradiction, she is changing the meaning of the word "contradiction," giving it a meaning opposite to the one it has in the theory of the square of opposition. The square is not a sacred cow and we do not necessarily need to be very strict with the use of the words, but bilateral exchange of meanings certainly leads to confusion: if someone calls a

[12]As explained in [11], not satisfied with this octagon, I split it in three stars that I put together in a three-dimensional polyhedron of opposition which also perfectly reflects the duality and symmetry between these two negations. The multidimensional theory of opposition has been further developed by Moretti [42], Smessaert [49] and Pélissier [45].

[13]For a detailed discussion about how to define a paraconsistent negation, see [5, 6].

square a circle and a circle a square, she will be able to claim that a circle has four corners and so on. Such claim may attract the attention, like many "tours de passe passe," but it is just a trick. G. Priest has gone somewhat in this direction, apparently not aware himself at first of the confusion, because he has even used the standard definitions of the square of opposition to claim that the negation of his system LP was a real negation, by contrast to the negation of da Costa system $C1$ (see [8, 46–48]). He has also introduced the word "dialetheia" to talk about a proposition p such that p and $\neg p$ can be true together. A dialetheia p is therefore not a contradiction considering that p and $\neg p$ do not form a contradictory pair.

To avoid any ambiguity it is better to call "paraconsistent" a formula such that p and $\neg p$ can be true together. A paraconsistent formula p and its paraconsistent negation $\neg p$ do not form a contradictory pair. And a paraconsistent formula is not a *trivial* formula, a formula from which everything follows. On the contrary it is a non trivial formula. From the point of view of a Tarskian consequence relation this definition of trivial formula is the same as the definition of a formula having no models, being always false.

Wittgenstein in the *Tractatus* [52] calls a trivial formula, a contradiction, by contrast to a tautology, a formula which is always true. In some sense it seems better to use the word "antilogy" to talk about a trivial formula, because the abstract idea of triviality does not depend on contradictory pair of formulas or/and on contradictory negation.[14] However, there is a relation and for Wittgenstein a typical example of a trivial formula is the formula of classical logic $p \wedge \neg p$, which can be seen as a pair of contradictory propositions. Tarski was at some point considering as an additional axiom of the consequence operator theory, the existence of at least a trivial formula (cf. Axiom 5 of [50]). Such kind of a formula is nowadays often singled out using the symbol $\perp$. What we know is that a trivial formula is related to negation. If we have a classical implication $\rightarrow$, the formula $p \rightarrow \perp$ has the bevahiour of a classical negation. And if we have an intuitionistic implication $\rightarrow$, the formula $p \rightarrow \perp$ has the bevahior of an intuitionistic negation. But we may have a logic with a negation and without a trivial formula, without contradiction, it is the case of the logic LP which has a subcontrary negation.

To finish let us explain why there is a good reason not to identify paraconsistent negation with subcontrary negation. This is because it is possible to have paraconsistent negations which are paranormal negations. A *paranormal* negation $\neg$ is a negation such that p and $\neg p$ can be true together and can be false together. Can we really still talk about negation for such an operator? A positive reply to this question is given by De Morgan logic, logic in which the four De Morgan laws hold as well as double negation, but where we do not have explosion, nor the validity of the law

[14] At the metalevel, tautology and antilogy form a contrary pair, see the metalogical hexagon presented in [16].

of excluded middle. A De Morgan negation seems to have enough properties to be called a negation.[15]

Now can we say that two propositions p and q are opposite if p and q can both be true and also can both be false? Yes if we put some additional properties corresponding to De Morgan laws and double negation. Adopting this "loose" perspective, we can defend the idea that a round square is a paranormal object. Because on the one hand, as we have pointed out, something can be neither a square, nor a circle, for example, a triangle and one the other hand something can appear as both a square and a circle, as illustrated by Fig. 2.6. At the end this figure is not a good metaphor for quantum physics, because a quanton may appear as a wave and as a particle, but may not be something else, so a quanton is rather a subcontrary object.

Acknowledgments The ideas presented in this paper have been developed over the years discussing with many people. A first version of this paper was presented as a tutorial I gave at the 5th World Congress on Paraconsistency, Kolkata, India, February 13–17, 2014. Later on I gave talks on this topic at the 3rd Latin American Analytic Philosophy Conference, May 27–30, 2014, Fortaleza, Brazil; University of California, Davis, February 6, 2015; University of Brasilia, March 17, 2015; Columbia University, New York, April 24, 2015; the University of Miami, Florida, April 29, 2015; the meeting Buddhism in Conversation at the State University of San Jose, California, May, 2, 2015; the University of California, San Diego, May 14, 18, 2015; the 43rd meeting of the SEP (Society for Exact Philosophy), MacMaster University, Hamilton, Ontario, Canada, May 22–24, 2015; the 5th Logic in Question, Sorbonne, Paris, June 8–9, 2015; the Logic Colloquium, Helsinki, August 3–8, 2015.
Thanks among others to Catherine Chantilly, Gila Sher, Larry Horn, Brad Dowden, Aldo Antonelli, Anand Vaidya, Otávio Bueno, Tamar Lando, Chris de Ronde, Alexandre Costa-Leite, Marcos Silva, Sandra Lapointe, Achille Varzi. My stay in the USA was supported by a CAPES grant.

References

1. Anton, J.P.: Aristotle's Theory of Contrariety. Routledge, London (1957)
2. Bergson, H.: La pensée et le mouvant. Alcan, Paris (1934). English Translation: the Creative Mind. Philosophical Library, New York (1946)
3. Beziau, J.-Y.: L'holomouvement selon David Bohm. Maîtrise of Philosophy, Université Paris 1 - Panthéon Sorbonne, Paris (1987)
4. Beziau, J.-Y.: D'une caverne à l'autre. Master thesis, Université Paris 1 - Panthéon Sorbonne, Paris (1988)
5. Beziau, J.-Y.: What is paraconsistent logic? in Frontiers of paraconsistent logic. In: Batens, D. et al. (eds.), pp. 95–111. Research Studies Press, Baldock (2000)
6. Beziau, J.-Y.: Are paraconsistent negations negations ? in Paraconsistency: the logical way to the inconsistent. In: Carnielli, W. et al. (eds.), pp. 465–486. Marcel Dekker, New-York (2002)
7. Beziau, J.-Y.: New light on the square of oppositions and its nameless corner. Log. Inv. **10**, 218–232 (2003)
8. Beziau, J.-Y.: Paraconsistent logic!. Sorites **17**, 17–26 (2006)
9. Beziau, J.-Y.: The paraconsistent logic Z-A possible solution to Jaśkowski's problem. Log. Log. Philos. **10**, 99–111 (2006)

[15]The expression "paranormal logic" was used in the paper [32] where a paranormal logic different from De Morgan logic was introduced. De Morgan logic is derived from De Morgan algebra, for details about this, see [15].

10. Beziau, J.-Y.: Adventures in the paraconsistent jungle. In: Handbook of Paraconsistency, pp. 63–80. College Publication, London (2007)
11. Beziau, J.-Y.: The new rising of the square. In: Beziau, J.-Y., Jacquette, D. (eds.) Around And Beyond the Square of Opposition, pp. 6–24. Birkhäuser, Basel (2012)
12. Beziau, J.-Y.: The power of the hexagon. Log. Univers. **6**, 1–43 (2012)
13. Beziau, J.-Y.: Paranormal logics and the theory of bivaluations. In: Universal Logic: an Anthology, pp. 361–372. Bikhäuser, Basel (2012)
14. Beziau, J.-Y. (ed.): Special issue on the hexagon of opposition. Logica Universalis **6**(1–2) (2012)
15. Beziau, J.-Y.: History of the concept of truth-value. In: Gabbay, D.M., Pelletier, F., Woods, J. (eds.) Handbook of the History of Logic, vol. 11. Elsevier, Amsterdam (2012)
16. Beziau, J.-Y.: The metalogical hexagon of opposition. Argumentos **10**, 111–122 (2013)
17. Beziau, J.-Y. (ed.): La pointure du symbole. Editions Pétra, Paris (2014)
18. Beziau, J.-Y., Jacquette, D. (eds.): Around and Beyond the Square of Opposition. Birkhäuser, Basel (2012)
19. Beziau, J.-Y., Payette, G. (eds.): Special issue on the square of opposition. Logica Universalis **2**(1) (2008)
20. Beziau, J.-Y., Payette, G. (eds.): The Square of Opposition-a General Framework for Cognition. Peter Lang, Bern (2012)
21. Beziau, J.-Y., Read, S. (eds.): Special issue of history and philosophy of logic, square of opposition: a diagram and a theory in historical perspective. **35**(4) (2014)
22. Beziau, J.-Y., Gan-Krzywoszynska, K. (eds.): New Dimensions of the Square of Opposition. Philosophia Verlag, Munich (2015)
23. Blanché, R.: Sur l'opposition des concepts. Theoria **19**, 89–130 (1953)
24. Blanché, R.: Opposition et négation. Revue Philosophique **167**, 187–216 (1957)
25. Blanché, R.: Structures intellectuelles. Essai sur lorganisation systématique des concepts, Vrin, Paris (1966)
26. Bohm, D.: Wholeness and the Implicate Order. Routledge, London (1980)
27. Bohr, N.,Wirkungsquantum und Naturbesreibung. Die Naturwissenschaften **17** 483–486 (1929). English translation: The quantum of action and the description of nature. Atomic Theory and the Description of Nature, pp. 92–101 University Press, Cambridge (1934)
28. Bréal, M.: Essai de sémantique : science des significations. Hachette, Paris (1897)
29. Breton, A.: Manifeste du surréalisme. Editions du Sagittaire, Paris (1924)
30. Costa-Leite, A.: Logical properties of imagination. Theoria **6**, 103–116 (2010)
31. da Costa, N.C.A.: Calculs propositionnels pour les systèmes formels inconsistants. Cr. R. Acad Sc. Paris **257**, 3790–3793 (1963)
32. da Costa, N.C.A.: Ensaio sobre os fundamentos da lógica. Hucitec, São Paulo, 1980. French translation: Logiques classiques et non classiques. Masson, Paris (1997)
33. da Costa, N.C.A., Beziau, J.Y.: Overclassical logic. Logique et Analyse, **157** 31–44 (1997)
34. d'Espagnat, B.: Le réel voilé, analyse des concepts quantiques. Fayard, Paris (1994): English translation: Veiled reality: an analysis of quantum mechanical concepts, 2003. Westview Press, Boulder, Colorado (2013)
35. Flaubert, G.: Bouvard et Pécuchet. Editions du Sagittaire, Paris (1881)
36. Frege, G.: Uber die wissenschaftlische Berechtigung einer Begriffsschrift. Zeitschrift für Philosophie und philosophische Kritik **81**, 48–56 (1892)
37. Gödel, K.: Eine Interpretation des intuitionistischen Aussagenkalküls. Ergebnisse eines mathematischen Kolloquiums **4**, 39–40 (1933)
38. Horn, L.R.: A Natural History of Negation. University Chicago Press, Chicago (1989)
39. Horn, L.R.: On the contrary: disjunctive syllogism and pragmatic strengthening. In: Koslow, A., Buchsbaum, A. (eds.) The Road to Universal Logic, Festschrift for 50th Birthday of Jean-Yves Béziau, vol. I, pp. 241–265. Birkhäuser, Basel (2015)
40. Jaspers, D.: Logic and colour. Log. Univers. **6**, 227–248 (2012)
41. Loparic, A., da Costa, N.C.A.: Paraconsistency, paracompleteness, and valuations. Logique et Analyse **106**, 119–131 (1984)

42. Moretti, A.: The geometry of logical opposition. PhD thesis, University of Neuchâtel, Neuchâtel (2009)
43. Tse-tung, M.: On contradiction (1937)
44. Niiniluoto, I.: Imagination and fiction. J. Semant. **4**, 209–222 (1985)
45. Pélissier, R.: "Setting" n-opposition. Log. Univers. **2**, 235–263 (2008)
46. Priest, G.: The logic of paradox. J. Philos. Logic **8**, 219–241 (1979)
47. Priest, G., Routley, R.: Systems of paraconsistent logic. In: Paraconsistent logic: essays on the Inconsistent, pp. 151–186. Philosophia, Munich (1989)
48. Salter, B.H.: Paraconsistent logic? J. Philos. Logic **24**, 451–454 (1995)
49. Smessaert, H.: On the 3D visualisation of logical relations. Log. Univers. **3**, 303–332 (2009)
50. Tarski, A.: Über einige fundamentale Begriffe der Metamathematik. Comptes Rendus des Séances de la Société des Sciences et des Lettres de Varsovie, Classe **III**(23), 22–29 (1930)
51. Urbas, I.: Dual intuitionistic logic. Notre Dame J. Formal Logic **37**, 440–451 (1996)
52. Wittgenstein, L.: "Logisch-philosophische Abhandlung", Annalen der Naturphilosophie **14** 185–262 (1921). Translated as Tractatus Logico-Philosophicus. Kegan Paul, London (1922)

Chapter 3
On the Philosophy and Mathematics of the Logics of Formal Inconsistency

Walter Carnielli and Abilio Rodrigues

Abstract The aim of this text is to present the philosophical motivations for the Logics of Formal Inconsistency (*LFI*s), along with some relevant technical results. The text is divided into two main parts (besides a short introduction). In Sect. 3.2, we present and discuss philosophical issues related to paraconsistency in general, and especially to logics of formal inconsistency. We argue that there are two basic and philosophically legitimate approaches to paraconsistency that depend on whether the contradictions are understood ontologically or epistemologically. *LFI*s are suitable to both options, but we emphasize the epistemological interpretation of contradictions. The main argument depends on the duality between paraconsistency and paracompleteness. In a few words, the idea is as follows: just as excluded middle may be rejected by intuitionistic logic due to epistemological reasons, explosion may also be rejected by paraconsistent logics due to epistemological reasons. In Sect. 3.3, some formal systems and a few basic technical results about them are presented.

Keywords Logics of Formal Inconsistency · Contradictions · Philosophy of paraconsistency

Mathematics Subject Classication (2000) Primary 03B53 · Secondary 03A05 · 03-01

The first author acknowledges support from FAPESP (São Paulo Research Council) and CNPq, Brazil (The National Council for Scientific and Technological Development). The second author acknowledges support from the Universidade Federal de Minas Gerais (project call 12/2011) and FAPEMIG (Fundação de Amparo à Pesquisa do Estado de Minas Gerais, research project 21308). We would like to thank Henrique Almeida, Marcos Silva and Peter Verdée for some valuable comments on a previous version of this text.

W. Carnielli (✉)
Centre for Logic, Epistemology and the History of Science and Department of Philosophy - State University of Campinas, Campinas, SP, Brazil
e-mail: walter.carnielli@cle.unicamp.br

A. Rodrigues
Department of Philosophy - Federal University of Minas Gerais, Belo Horizonte, MG, Brazil
e-mail: abilio@ufmg.br

J.-Y. Beziau et al. (eds.), *New Directions in Paraconsistent Logic*,
Springer Proceedings in Mathematics & Statistics 152,
DOI 10.1007/978-81-322-2719-9_3

3.1 Introduction

The aim of this text is to present the philosophical motivations for the Logics of Formal Inconsistency (*LFI*s), along with some relevant technical results. The target audience is mainly the philosopher and the logician interested in the philosophical aspects of paraconsistency.[1]

In Sect. 3.2, *On the philosophy of Logics of Formal Inconsistency*, we present and discuss philosophical issues related to paraconsistency in general, and especially to Logics of Formal Inconsistency. We argue that there are two basic and philosophically legitimate approaches to paraconsistency that depend on whether the contradictions are understood ontologically or epistemologically. *LFI*s are well suited to both options, but we emphasize the epistemological interpretation of contradictions. The main argument depends on the duality between paraconsistency and paracompleteness. In a few words, the idea is as follows: just as excluded middle may be rejected by intuitionistic logic due to epistemological reasons, explosion may also be rejected by paraconsistent logics due to epistemological reasons.

In Sect. 3.3, *On the mathematics of Logics of Formal Inconsistency*, some formal systems and a few of their basic technical results are presented. These systems are designed to fit the philosophical views presented in Sect. 3.2.

3.2 On the Philosophy of the Logics of Formal Inconsistency

It is a fact that contradictions appear in a number of real-life contexts of reasoning. Databases very often contain not only incomplete information but also conflicting (i.e. contradictory) information.[2] Since ancient Greece, paradoxes have intrigued logicians and philosophers, and, more recently, mathematicians as well. Scientific theories are another example of real situations in which contradictions seem to be unavoidable. There are several scientific theories, however, successful in their areas of knowledge that yield contradictions, either by themselves or when put together with other successful theories Contradictions are problematic when the principle of explosion holds:

---

[1]This paper corresponds, with some additions, to the tutorial on Logics of Formal Inconsistency presented in the 5th *World Congress on Paraconsistency* that took place in Kolkata, India, in February 2014. Parts of this material have already appeared in other texts by the authors, and other parts are already in print elsewhere [12, 13]. A much more detailed mathematical treatment can be found in Carnielli and Coniglio [10] and Carnielli et al. [15].

[2]We do not use the term 'information' here in a strictly technical sense. We might say, in an attempt not to define but rather to elucidate, that 'information' means any 'amount of data' that can be expressed by a sentence (or proposition) in natural language. Accordingly, there may be contradictory or conflicting information (in a sense to be clarified below), vague information, or lack of information.

$A \to (\sim A \to B)$.[3]

In this case, since anything follows from a contradiction, one may conclude anything whatsoever. In order to deal rationally with contradictions, explosion cannot be valid without restrictions, since triviality (that is, a circumstance such that everything holds) is obviously unacceptable. Given that in classical logic explosion is a valid principle of inference, the underlying logic of a contradictory context of reasoning cannot be classical.

In a few words, paraconsistency is the study of logical systems in which the presence of a contradiction does not imply triviality, that is, logical systems with a non-explosive negation $\neg$ such that a pair of propositions A and $\neg A$ does not (always) trivialize the system. However, it is not only the syntactic and semantic properties of these systems that are worth studying. Some questions arise that are perennial philosophical problems. The question about the nature of contradictions accepted by paraconsistent logics is where a good amount of the debate on the philosophical significance of paraconsistency has been concentrated.

In philosophical terminology, we say that something is ontological when it has to do with reality, the world in the widest sense, and that something is epistemological when it has to do with knowledge and the process of its acquisition. A central question for paraconsistency is the following: Are the contradictions that paraconsistent logic deals with ontological or epistemological? Do contradictions have to do with reality proper? That is, is reality intrinsically contradictory, in the sense that we really need some pairs of contradictory propositions in order to describe it correctly? Or do contradictions have to do with knowledge and thought? Contradictions of the latter kind would have their origin in our cognitive apparatus, in the failure of measuring instruments, in the interactions of these instruments with phenomena, in operations of thought, or even in simple mistakes that in principle could be corrected later on. Note that in all of these cases the contradiction does not belong to reality properly speaking.

The question about the nature of contradictions, in its turn, is related to another central issue in philosophy of logic, namely the nature of logic itself. As a theory of logical consequence, the task of logic is to formulate principles and methods for establishing when a proposition A follows from a set of premises Γ. But a question remains: What are the principles of logic about? Are they about language, thought, or reality? That logic is normative is not contentious, but its normative character may be combined both with an ontological and an epistemological approach.

The epistemological side of logic is present in the widespread (but not unanimous) characterization of logic as the study of laws of thought. This concept of logic, which acknowledges an inherent relationship between logic and human rationality, has been put aside since classical logic has acquired the status of the standard account of logical consequence—for example, in the work of Frege, Russell, Tarski, Quine, and many other influential logicians.

---

[3]The symbol $\sim$ will always denote the classical negation, while $\neg$ usually denotes a paraconsistent negation but sometimes a paracomplete (e.g. intuitionistic) negation. The context will make it clear in each case whether the negation is used in a paracomplete or paraconsistent sense.

Classical logic is a very good account of the notion of truth preservation, but it does not give a sustained account of rationality. This point shall not be developed in detail here, but it is well known that some classically valid inferences are not really applied in real-life contexts of reasoning, for example: from A, to conclude that anything implies A; from A, to conclude the disjunction of A and anything; from a contradiction, to conclude anything. The latter is the principle of explosion, and of course it is not rational to conclude that $2 + 2 = 5$ when we face some pair of contradictory propositions. Nevertheless, from the point of view of preservation of truth, given the classical meaning of sentential connectives, all the inferences above are irreproachable.

We assume here a concept of logic according to which logic is not restricted to the idea of truth preservation. Logical consequence is indeed the central notion of logic, but the task of logic is to tell us which conclusions can be drawn from a given set of premises, under certain conditions, in concrete situations of reasoning. We shall see that sometimes it may be the case that it is not only truth that is at stake.[4] Among the contexts of reasoning in which classical logic is not the most suitable tool, two are especially important: contexts with 'excess of information' and 'lack of information'. The logics suited to such contexts are, respectively, paraconsistent and paracomplete—in the former, explosion fails, in the latter excluded middle fails.

There are two basic approaches to paraconsistency. If some contradictions belong to reality, since it is not the case that everything holds, we do need an account of logical consequence that does not collapse in the face of a contradiction. On the other hand, if contradictions are epistemological, we argue that the rejection of explosion goes hand in hand with the rejection of excluded middle by intuitionistic logic. In the latter case, the formal system has an epistemological character and combines a descriptive with a normative approach.

This section is structured as follows. In Sect. 3.2.1 some basic concepts are presented in order to distinguish triviality from inconsistency. In addition, we make a first presentation of Logics of Formal Inconsistency, distinguishing paraconsistency and paracompleteness from the classical approach. In Sect. 3.2.2 we present a brief historical digression on the origins of paraconsistency and the forerunners of Logics of Formal Inconsistency. In Sect. 3.2.3 we examine the relationship between paraconsistency and the issue of the nature of logic. We argue that, like the rejection of excluded middle by intuitionistic logic, the rejection of explosion may be understood epistemologically. In Sect. 3.2.4 we discuss paraconsistency from the point of view of the issue of the nature of contradictions, and consider whether they should be understood ontologically or epistemologically. We argue that both positions are philosophically legitimate. Finally, in Sect. 3.2.5, we show how the simultaneous attribution of the value *0* to a pair of propositions A and $\neg A$ may be interpreted as conflicting evidence, not as truth and falsity of A.

---

[4]This idea has some consequences for Harman's arguments (see [31]) against non-classical logics, a point that we intend to develop elsewhere.

3.2.1 A First Look at Logics of Formal Inconsistency

We have seen that paraconsistent logics are able to deal with contradictory scenarios avoiding triviality by means of the rejection of the principle of explosion. Let us put these ideas more precisely. A theory is a set of propositions[5] closed under logical consequence. Given a set of propositions Γ in the language of a given logic L, let $T = \{A\colon \Gamma \vdash_L A\}$ be the theory whose non-logical axioms are the propositions of Γ and the underlying logic is L. Suppose the language of T has a negation $\sim$. We say that T is:

Contradictory: if and only if there is a proposition A in the language of T such that T proves A and T proves $\sim A$;

Trivial: if and only if for any proposition A in the language of T T proves A;

Explosive: if and only if T trivializes when exposed to a pair of contradictory formulas—i.e.: for all A and B, $T \cup \{A, \sim A\} \vdash B$.

In books of logic we find two different but classically equivalent notions of consistency with respect to a deductive system S with a negation $\sim$. S is consistent if and only if

i. There is a formula B such that $\nvdash_S B$;
ii. There is no formula A such that $\vdash_S A$ and $\vdash_S \sim A$.

What (i) says is that S is not trivial; and (ii) says that S is non-contradictory. In classical logic both are provably equivalent. A theory whose underlying logic is classical is contradictory if and only if it is trivial. But it is the case precisely because such a theory is explosive, since the principle of explosion holds in classical logic. It is clear, then, that *it is contradictoriness together with explosiveness that implies triviality.* The obvious move in order to deal with contradictions is, thus, to reject the unrestricted validity of the principle of explosion. This is a necessary condition if we want a contradictory but not-trivial theory.

The first formalization of paraconsistent logic to appear in the literature is to be found in [33]. In the beginning of the paper he presents three conditions that a contradictory but nontrivial logic must attend:

1. It must be non-explosive;
2. It should be “rich enough to enable practical inference”;
3. It should have “an intuitive justification”.

The condition (1), as we have seen, is a necessary condition for any paraconsistent system. We want to call attention to conditions (2) and (3). Indeed, the biggest challenge for a paraconsistentist is to devise a logical system compatible with what we intuitively think should follow (or not follow) from what. This is the idea expressed by the criteria (2) and (3) presented by Jáskowski. An intuitive and applicable notion

[5]Or sentences, if one prefers—here, we do not go into the distinction between sentences and propositions.

of logical consequence should be appropriate for describing and reconstructing real contexts of reasoning. An intuitive meaning for the logical connectives—more precisely, for paraconsistent negation—should be an integral part of such account of logical consequence. It follows that an intuitive interpretation of a paraconsistent notion of logical consequence depends essentially on an intuitive interpretation of negation.

For classical negation $\sim$ the following conditions hold:

1. $A \wedge \sim A \vDash$
2. $\vDash A \vee \sim A$

According to (1), there is no model M such that $A \wedge \sim A$ holds in M. (2) says that for every model M, $A \vee \sim A$ holds in M. Now, given the definition of classical consequence, $A \vee \sim A$ follows from anything, and anything follows from $A \wedge \sim A$.[6] We say that a negation is paracomplete if it disobeys (2), and that a negation is paraconsistent if it disobeys (1). From the point of view of rules of inference, the duality is not between non-contradiction and excluded middle, but rather between explosion and excluded middle. Notice that the notion of logical consequence has priority over the notion of logical truth: the latter must be defined in terms of the former, not the contrary. The principle of non-contradiction is usually taken as a claim that there can be no contradictions in reality. But we may well understand the principle of explosion as a stronger way of saying precisely the same thing: A and $\sim A$ cannot hold together, otherwise we get triviality. From the above considerations it is clear that in order to give a counterexample to the principle of explosion we need a weaker negation and a semantics in which there is a model M such that A and $\neg A$ holds in M ($\neg$ is a paraconsistent negation) but for some B, B does not hold in M. Dually, a paracomplete logic must have a model M such that both A and $\neg A$ do not hold in M (here, $\neg$ is a paracomplete negation).

A central feature of classical negation $\sim$ (but not of all negations, as we shall see) is that it is a contradictory forming operator. This is due to its semantic clause,

$$M(\sim A) = 1 \text{ iff } M(A) = 0,$$

that, in turn, holds because both (1) and (2) above hold. Applied to a proposition A, classical negation produces a proposition $\sim A$ such that A and $\sim A$ are contradictories in the sense that they cannot receive simultaneously the value *0*, nor simultaneously the value *1*. In classical logic the values *0* and *1* are understood, respectively, as false and true, but in non-classical logics this does not need to be the case. It is not necessary that a paracomplete logic takes a pair of formulas A and $\neg A$ as both false, nor that a paraconsistent logic takes them as both true.

Obviously, neither a paracomplete nor a paraconsistent negation is a contradictory forming operator, and neither is a truth-functional operator, since the value of $\neg A$ is not unequivocally determined by the value of A. Now a question arises: Can we say that such negations are really negations? Our answer is yes.

---

[6] For a more detailed explanation of the duality between paracompleteness and paraconsistency, see Marcos [36].

It should not be surprising that the meaning of a classical connective splits up into some alternative meanings when its use in natural language and real-life arguments is analyzed. Indeed, different meanings are sometimes attached to conditional, disjunction, and conjunction, and the connectives so obtained are still called conditional, disjunction, and conjunction, of course with some qualifications. What would be the reason by which the same cannot occur with negation? In fact, both paracomplete and paraconsistent negations do occur in real life. An example of the former is intuitionistic negation: it may be the case that we do have a classical proof of a proposition A but have no constructive proof of A. From the constructive point of view, we have neither A nor $\neg A$. On the other hand, sometimes it happens that we have to deal simultaneously with conflicting information about A. In these cases, we may have reasons to accept both A and $\neg A$, but we do not need to say that both are true. Finally, the above considerations show that a paraconsistent negation is a negation to the same extent that a paracomplete (including intuitionistic) negation is a negation. Nevertheless, what is of major importance is that the question of whether or not a paraconsistent negation may have an intuitive meaning has a positive answer.

Logics of Formal Inconsistency (from now on, *LFI*s) are a family of paraconsistent logics that encompass the majority of paraconsistent systems developed within the Brazilian tradition. In this section we present the basic ideas of *LFI*s without going into the technical details (this will be done in Sect. 3.3 of this text). *LFI*s have resources to express the notion of consistency inside the object language by means of a sentential unary connective: '$\circ A$' means that A *is consistent*. As in any other paraconsistent logic, explosion does not hold in *LFI*s. But it is handled in a way that allows distinguishing between contradictions that can be accepted from those that cannot. The point of this distinction is that no matter the nature of the contradictions a paraconsistentist is willing to accept, there are contradictions that cannot be accepted. In *LFI*s, negation is explosive only with respect to 'consistent' formulas:

$$A, \neg A \nvdash_{LFI} B, \text{ while } \circ A, A, \neg A \vdash_{LFI} B.$$

An *LFI* is thus a logic that separates the propositions for which explosion holds from those for which it does not hold. The former are marked with $\circ$. For this reason, they are called *gently explosive* (more on this point in Sect. 3.3.1.1). In the C_n hierarchy, introduced by da Costa [20], the so-called 'well-behavedness' of a formula A, in the sense that it is not the case that A and $\neg A$ hold, is also expressed inside the object language. However, in C_1, A° is an abbreviation of $\neg(A \wedge \neg A)$, which makes the 'well-behavedness' of a proposition A equivalent to saying that A is non-contradictory.[7]

We may say that a first step in paraconsistency is the distinction between triviality and contradictoriness. But there is a second step, namely the distinction between consistency and non-contradictoriness. In *LFI*s the consistency connective $\circ$ is not only primitive, but it is also not always logically equivalent to non-contradiction.

---

[7] Actually, da Costa has a hierarchy of systems, starting with the system C_1, where A° is an abbreviation of $\neg(A \wedge \neg A)$. A full hierarchy of calculi C_n, for n natural, is defined and studied in da Costa [20].

This is the most distinguishing feature of the Logics of Formal Inconsistency. Once we break up the equivalence between $\circ A$ and $\neg(A \wedge \neg A)$, some very interesting developments become available. Indeed, $\circ A$ may express notions different from consistency as freedom from contradiction.

3.2.2 A Very Brief Historical Digression: The Forerunners of Logics of Formal Inconsistency

The advent of paraconsistency occurred more than a century ago. In 1910 the Russian philosopher and psychologist Nicolai A. Vasiliev proposed the idea of a non-Aristotelian logic, free of the laws of excluded middle and non-contradiction. By analogy with the imaginary geometry of Lobachevsky, Vasiliev called his logic 'imaginary', meaning that it would hold in imaginary worlds. Despite publishing between 1912–13 some conceptual papers on the subject, Vasiliev was not concerned with formalizing his logic (cf. Gomes [28, pp. 307ff.]).

Jáskowski [33], trying to answer a question posed by Łukasiewicz, presented the first formal system for a paraconsistent logic, called 'discussive logic'. This system is connected to modalities, and later on came to be regarded as a particular member of the family of the Logics of Formal Inconsistency (cf. Carnielli [15, p. 22]).

Intending to study logical paradoxes from a formal perspective, Hállden [30] proposed a 'logic of nonsense' by means of three-valued logical matrices, closely related to the nonsense logic introduced in 1938 by the Russian logician A. Bochvar. Since a third truth-value is distinguished, Hállden's logic is paraconsistent, and it can also be considered as one of the first paraconsistent formal systems presented in the literature. In fact, like Jáskowski's logic, it is also a member of the family of the Logics of Formal Inconsistency.

Nelson [38] proposed an extension of positive intuitionistic logic with a new connective for 'constructible falsity' or 'strong negation', intended to overcome non-constructive features of intuitionistic negation. By eliminating the principle of explosion from this system, Nelson [39] obtained a first-order paraconsistent logic, although paraconsistency was not his primary concern (see Carnielli and Coniglio [10]).

Paraconsistency also has some early links to Karl Popper's falsificacionism. In 1954 (cf. Kapsner et al. [34]), Kalman Joseph Cohen, attending a suggestion of his supervisor Karl Popper, submitted to the University of Oxford a thesis entitled 'Alternative Systems of Logic' in which he intended to develop a logic dual to intuitionistic logic. In Cohen's logic, the law of explosion is no longer valid, while the law of excluded middle holds as a theorem. Cohen's thesis, according to Kapsner et al., escaped scholarly attention, having been only briefly mentioned in Popper's famous 'Conjectures and Refutations' [see [41], footnote 8, p. 321]. It did, however, in some sense anticipate more recent work on dual-intuitionist logics (which, as shown in Brunner and Carnielli [6], are paraconsistent).

In da Costa [19] we find a discussion of the status of contradiction in mathematics, introducing the *Principle of Non-Trivialization*, according to which nontriviality is more important than non-contradiction. The idea is that any mathematical theory is worth studying, provided it is not trivial. Although we do agree that mathematical (and logical) nontrivial systems are worth studying, on the other hand, an account of logical consequence needs a little bit more in order to be accepted as an account of reasoning. In 1963 da Costa presented his famous hierarchy of paraconsistent systems C_n (for $n \geq 1$), constituting the broadest formal study of paraconsistency proposed up to that time (cf. [20]). It is worth mentioning here what has been said by Newton da Costa, in private conversation. If we remember correctly, it goes more or less as follows: "As with the discovery of America, many people are said to have discovered paraconsistent logic before my work. I can only say that, as with Columbus, nobody has discovered paraconsistency after me, just as nobody discovered America after Columbus."

The Argentinian philosopher F. Asenjo introduced in 1966 a three-valued logic as a formal framework for studying antinomies. His logic is essentially defined by Kleene's three-valued truth-tables for negation and conjunction, where the third truth-value is distinguished. Asenjo's logic is structurally the same as the *Logic of Paradox*, presented in Priest [42], the essential difference being that in the latter there are two designated truth-values, intuitively understood as true and both true and false (see [4]).

From the 1970s on, after the Peruvian philosopher Francisco Miro Quesada coined the name 'paraconsistent logic' to encompass all these creations, several schools with different aims and methods have spread out around the world.[8]

3.2.3 *Paraconsistency and the Nature of Logic*

A central question in philosophy of logic asks about the nature of logical principles, and specifically whether these principles are about reality, thought, or language. We find this issue, brought forth, either implicitly or explicitly, in a number of places. In this section we shall discuss the relationship between paraconsistent logic and the problem of the nature of logic.

Aristotle formulates three versions of the principle of non-contradiction, each one corresponding to one of the aforementioned aspects of logic (more on this below). Tugendhat and Wolf [45, Chap. 1] present the problem mainly from a historical viewpoint, relating the three approaches (ontological, epistemological, and linguistic) to periods in the history of philosophy—respectively ancient and medieval, modern, and contemporary. Popper [41, pp. 206ff] presents the problem as follows. The central question is whether the principles of logic are:

[8]See 'Carta de Francisco Miro Quesada a Newton da Costa, 29.IX.1975' in Gomes [28, p. 609].

(I.a) laws of thought in the sense that they describe how we actually think;
(I.b) laws of thought in the sense that they are normative laws, i.e., laws that tell us how we should think;
(II) the most general laws of nature, i.e., laws that apply to any kind of object;
(III) laws of certain descriptive languages.

There are three basic options, which are not mutually exclusive: the laws of logic have (I) epistemological, (II) ontological, or (III) linguistic character. With respect to (I), they may be (I.a) descriptive or (I.b) normative. These aspects may be combined. Invariably, logic is taken as having a normative character, no matter whether it is conceived primarily as having to do with language, thought, or reality. The point of asking this question is not really to find a definitive answer. It is a perennial philosophical question, which, however, helps us to clarify and understand important aspects of paraconsistent logic.

We start with some remarks about the linguistic aspects of logic. According to widespread opinion, a linguistic conception of logic has prevailed during the twentieth century. From this perspective, logic has to do above all with the structure and functioning of certain languages. We do not agree with this view. For us, logic is primarily a theory about reality and thought.[9] The linguistic aspect appears only inasmuch as language is used in order to represent what is going on in reality and in thought. Although the linguistic aspect of logic is related to epistemology (since language and thought cannot be completely separated) and to ontology (by means of semantics), we do not think that a linguistic conception of logic is going to help much in clarifying a problem that is central for us here, that of whether contradictions have to do with reality or thought.

Aristotle, defending the principle of non-contradiction (*PNC*), makes it clear that it is a principle about reality, language, and thought, but there is a consensus among scholars that its main formulation is a claim about objects and properties: it cannot be the case that the same property belongs and does not belong to the same object. Put in this way, *PNC* is ontological in character. Like a general law of nature, space-time phenomena cannot disobey *PNC*, nor can mathematical objects.

The epistemological aspects of logic became clear in the modern period. A very illuminating passage can be found in the so-called *Logic of Port-Royal* (1662) [3, p. 23], where we read that logic has three purposes:

> The first is to assure us that we are using reason well.
> The second is to reveal and explain more easily the errors or defects that can occur in mental operations.
> The third purpose is to make us better acquainted with the nature of the mind by reflecting on its actions.

Notice how the passage above combines the normative character of logic with an analysis of mind. This view of logic does not fit very well with the account of logical consequence given by classical logic, but it has a lot to do with intuitionistic logic.

---

[9] A rejection of the linguistic conception of logic, and a defense of logic as a theory with ontological and epistemological aspects, can be found in Chateaubriand [17, Introduction].

Frege's *Begriffsschrift* [24] had an important role in establishing classical logic as the standard account of logical consequence. Although there is no semantics in Frege's work, it is well known that we find in the *Begriffsschrift* a complete and correct system of first-order classical logic. At first sight, Frege's approach is purely proof-theoretical, but one should not draw the conclusion that his system has no ontological commitments. We cannot lose sight of the fact that the idea of truth preservation developed by Frege, although worked out syntactically, is constrained by a realist notion of truth.

Frege had a realist concept of logic, according to which logic is independent of language and mind. In fact, since he was a full-blooded platonist with respect to mathematics, and his logicist project was to prove that arithmetic is a development of logic, he had to be a logical realist. For Frege, the laws of logic are as objective as mathematics, even though we may occasionally disobey them.[10] Frege's conception of logic is very well suited to the idea of truth-preservation. He indeed famously explains the task of logic as being 'to discern the laws of truth' [26], or more precisely, the laws of preservation of truth. Hence, it is not surprising that laws of logic cannot be obtained from concrete reasoning practices. In other words, logic cannot have a descriptive aspect, in the sense of (I.a) above.[11] It is worth noting that Frege proves the principle of explosion as a theorem of his system: it is proposition *36* of the *Begriffsschrift*.

It is important to emphasize the contrast between Frege's and Brouwer's conceptions of logic. This fact is especially relevant for our aims here because of the duality between paracompleteness and paraconsistency pointed out in Sect. 3.2.1 above. From the point of view of classical logic, the rejection of excluded middle by intuitionistic logic is like a mirror image of the rejection of explosion.

It is well known that for Brouwer mathematics is not a part of logic, as Frege wanted to prove. Quite the contrary, logic is abstracted from mathematical reasoning. Mathematics is a product of the human mind, and mathematical proofs are mental constructions that do not depend on language or logic. The role of logic in mathematics is only to describe methodically the constructions carried out by mathematicians.[12] We may say that intuitionistic logic has been obtained through an analysis of the functioning of mind in constructing mathematical proofs. To the extent that intuitionistic logic intends to avoid improper uses of excluded middle,

[10]Cf. Frege [25, p. 13]: 'they are boundary stones set in an eternal foundation, which our thought can overflow, but never displace.'

[11]There is a sense in which for Frege laws of logic are descriptive: they describe reality, as well as laws of physics and mathematics. But we say here that a logic is descriptive when it describes reasoning.

[12]Brouwer [5, pp. 51, 73–74]: "*Mathematics can deal with no other matter than that which it has itself constructed.* In the preceding pages it has been shown for the fundamental parts of mathematics how they can be built up from units of perception. (...) The words of your mathematical demonstration merely accompany a mathematical *construction* that is effected without words (...) While thus mathematics is independent of logic, logic does depend upon mathematics." A more acessible presentation of the motivations for intuitionistic logic is to be found in Heyting [32, Disputation].

it is normative, but it is descriptive precisely in the sense that, according to Frege, logic cannot be descriptive. Intuitionistic logic thus combines a descriptive with a normative character.

The view according to which intuitionistic logic has an epistemological character that contrasts with the ontological vein of classical logic is not new.[13] Note how the intuitionistic approach fits in well with the passage quoted above from logic of Port Royal. Furthermore, even if one wants to insist on an anti-realist notion of truth, the thesis that intuitionistic logic is not about truth properly speaking, but about mental constructions, is in line with the intuitionistic program as it was developed by Heyting and Brouwer.

Now we may ask: Does intuitionistic logic give an account of truth preservation? Our answer is in the negative: in our view, intuitionistic logic is not only about truth; it is about truth and something else. We may say that it is about constructive truth in the following sense: it is constrained by truth but it is not truth *simpliciter*; rather, it is about truth achieved in a constructive way. The notion of constructive provability is stronger than truth in the sense that if we have a constructive proof of A, we know that A is true, but the converse may not hold. Accordingly, not only the failure of excluded middle, but the whole enterprise of intuitionistic logic, may be seen from an epistemological perspective.[14]

An analogous interpretation can be made with respect to contradictions in paraconsistent logics. While in intuitionistic logic (and paracomplete logics in general) the failure of excluded middle may be seen as a kind of lack of information (no proof of A, no proof of $\neg A$), the failure of explosion may be interpreted epistemologically as excess of information (conflicting evidence both for A and for $\neg A$, but no evidence for B). The acceptance of contradictory propositions in some circumstances does not need to mean that reality is contradictory. It may be considered a step in the process of acquiring knowledge that, at least in principle, could be revised.

Suppose a context of reasoning such that there are some propositions well established as true (or as false) and some others that have not been conclusively established yet. Now, if among the latter there is a contradiction, one does not conclude that $2 + 2 = 5$, but, rather, one takes a more careful stance with respect to the specific contradictory proposition. On the other hand, the inferences allowed with respect to propositions already established as true are normally applied. In fact, what does happen is that the principle of explosion is not unrestrictedly applied. The contradictory propositions are still there, and it may happen that they are used in some inferences, but they are not taken as true propositions.

By means of a non-explosive negation and the consistency operator $\circ$, an *LFI* may formally represent this scenario. We will return to this point in more detail in

---

[13]See, for example, van Dalen [46, p. 225]: "two [logics] stand out as having a solid philosophical-mathematical justification. On the one hand, classical logic with its ontological basis and on the other hand intuitionistic logic with its epistemic motivation."

[14]It is worth noting that Brouwer's and Heyting's attempts to identify truth with a notion of proof have failed, as Raatikainen [44] shows, because the result is a concept of truth that goes against some basic intuitions about truth.

Sect. 3.2.5. For now, we want to emphasize that the sketch of a paraconsistent logic in which contradictions are epistemologically understood as conflicting evidence, and not as a pair of contradictory true propositions, is inspired by an analysis of real situations of reasoning in which contradictions occur. The notion of evidence is weaker than truth, in the sense that if we know that A is true, then there must be some evidence for A, but the fact that there is evidence for A does not imply that A is true. A paraconsistent logic may thus be obtained in a way analogous to the way that intuitionistic logic has been obtained.

3.2.4 Paraconsistency and the Nature of Contradictions

We now turn to a discussion of paraconsistency from the perspective of the problem of the nature of contradictions. The latter is a very old philosophical topic that can be traced back to the beginnings of philosophy in ancient Greece, and, as we have just seen, is closely related to the issue of the nature of logic. There is an extensive discussion and defense of the principle of non-contradiction (*PNC*) in Aristotle's *Metaphysics*, book Γ.[15] According to Aristotle, *PNC* is the most certain of all principles and has no other principle prior to it. Although *PNC* is, strictly speaking, indemonstrable, Aristotle presents arguments in defense of it. It is not in fact a problem, since these arguments may be considered as elucidations or informal explanations of *PNC*, rather than demonstrations in the strict sense. In *Metaphysics* Γ we find three versions of *PNC* that correspond to the three aspects of logic mentioned above, ontological, epistemological, and linguistic. We refer to them here respectively as *PNC-O*, *PNC-E*, and *PNC-L*.

> I. PNC-O (1005b19–20) [S]uch a principle is the most certain of all; which principle this is, we proceed to say. It is, that the same attribute cannot at the same time belong and not belong to the same subject in the same respect.
>
> II. PNC-E (1005b28–30) If it is impossible that contrary attributes should belong at the same time to the same subject (the usual qualifications must be presupposed in this proposition too), and if an opinion which contradicts another is contrary to it, obviously it is impossible for the same man at the same time to believe the same thing to be and not to be.
>
> III. PNC-L (1011b13–22) [T]he most indisputable of all beliefs is that contradictory statements are not at the same time true (...) If, then, it is impossible to affirm and deny truly at the same time, it is also impossible that contraries should belong to a subject at the same time.

The point is that *PCN-O* is talking about objects and their properties, *PCN-E* about beliefs, and *PCN-L* about propositions. These three versions are called by

[15] All passages from Aristotle referred to here are from [2].

Łukasiewicz [35] ontological, psychological, and semantic.[16] Łukasiewicz strongly attacks Aristotle's defense of *PNC*, and claims that the psychological (i.e. epistemological) version is simply false and that the ontological and the semantic (i.e. linguistic) versions have not been proven at all. He ends the paper by saying that Aristotle "might well have himself felt the weaknesses of his argument, and so he announced his principle a final axiom, an unassailable dogma" [35, p. 509]. We are not going to analyze Aristotle's arguments here, nor Łukasiewicz's criticisms in detail. Rather, we are interested in the following question: What should be the case in order to make true each one of the formulations of *PNC*? We will see that the weaknesses of Aristotle's arguments have a lot to reveal about contradictions.

The basic idea of *PNC-O* corresponds to a theorem of first-order logic: $\forall x \neg(Px \wedge \neg Px)$, i.e., the same property cannot both belong and not belong to the same object. An object may have different properties at different moments of time, or from two different perspectives, but obviously these cases do not qualify as counterexamples for *PNC* (cf. *Metaphysics*, 1009b1 and 1010b10). *PNC-O* depends on an ontological categorization of reality in terms of objects and properties. This categorization has been central in philosophy and is present in logic since its beginnings.[17] *PNC-O* has an ontological vein even if one is not sympathetic to the notion of property. It is enough to change 'the object *a* has the property *P*' to 'the object *a* satisfies the predicate *P*'. In any case, we are speaking in the broadest sense, which includes objects in space-time as well as mathematical objects.

The linguistic formulation here called *PNC-L*, although talking about language, also has an ontological vein because of the link between reality and the notion of truth. If there is a claim that is to a large extent uncontentious about truth, it is that if a proposition (or any other truth-bearer) is true, it is reality that makes it true; or, in other words, truth is grounded in reality. Understood in this way, *PNC-O* and *PNC-L* collapse, the only difference being that the former depends on the ontological categorization in terms of objects and properties, while the latter depends on language and an unqualified notion of truth. Note that Aristotle seems to conflate both, since in passage III quoted above *PNC-O* is the conclusion of an argument whose premise is *PNC-L*.

A violation of *PNC-O* would be an object *a* and a property *P* such that *a* has and does not have *P*. Hence, in order to show that *PNC-O* is true, one needs to show that there can be no such object. Now, this problem may be divided into two parts, one related to mathematics, the other related to empirical sciences. With respect to the former, a proof of *PNC-O* would be tantamount to showing that mathematics is consistent. But this cannot be proven, even with respect to arithmetic. With respect to the latter, there is an extensive literature on the occurrence of contradictions in empirical theories (see, for example, da Costa and French [23, Chap. 5] and Meheus [37]). However, up to the present day there is no indication that these contradictions

[16]This tripartite approach is also found in Gottlieb [29], where these three versions are called, respectively, ontological, doxastic, and semantic.

[17]For example, the issue of particulars/universals, the Fregean distinction between object and function, and even Quine's attacks to the notion of property.

are due to the nature of reality or belong to the theories, which are nothing but attempts to give a model of reality in order to predict its behavior. In other words, there is no clear indication, far less a conclusive argument, that these contradictions are ontological and not only epistemological.

The linguistic version of *PNC* is exactly the opposite of the dialetheist thesis as it is presented by Priest and Berto [43]:

> A dialetheia is a sentence, A, such that both it and its negation, $\neg A$, are true (...) Dialetheism is the view that there are dialetheias. (...) dialetheism amounts to the claim that there are true contradictions.

Thus, a proof of *PNC-L* would be tantamount to a disproof of dialetheism. Although dialetheism is far from being conclusively established as true, it has antecedents in the history of philosophy and is legitimate from the philosophical point of view. Further, if we accept that every proposition says something about something, a thesis that has not been rejected by logical analysis in terms of arguments and functions, what makes *PNC-O* true would also make *PNC-L* true, and vice versa. Our conclusion is that neither *PNC-O* nor *PNC-L* has been conclusively established as a true principle. And this is not because Aristotle's arguments, or any other philosophical arguments in defense of the two principles are not good. Rather, the point is that this issue outstrips what can be done a priori by philosophy itself. It seems to be useless for the philosopher to spend time trying to prove them.

Now we turn to *PNC-E*. As it stands, the principle says that the same person cannot believe in two contradictory propositions. Here, the point is not how it could be proved, because it really seems that there are sufficient reasons to suppose that it has already been disproved. It is a fact that in various circumstances people have contradictory beliefs. Even in the history of philosophy, as Łukasiewicz [35, p. 492] remarks, "contradictions have been asserted at the same time with full awareness." Indeed, since there are philosophers, like Hegel and the contemporary dialetheists, that defend the existence of contradictions in reality, this should be an adequate counterexample to *PNC-E*. Furthermore, if we take a look at some contexts of reasoning, we will find out that there are a number of situations in which one is justified in believing both A and $\neg A$. Sometimes we have simultaneous evidence for A and for $\neg A$, which does not mean that we have to take both as true, but we may have to deal simultaneously with both propositions. Nevertheless, the problem we have at hand may be put more precisely. *PNC-E* is somewhat naive and does not go to the core of the problem. The relevant question is whether the contradictions we find in real situations of reasoning—databases, paradoxes, scientific theories—belong to reality properly speaking, or have their origin in thought and/or in the process of acquiring knowledge.

Now, let us see what lessons may be taken from all of this. It is a fact that contradictions appear in several contexts of reasoning. Any philosophical attempt to give a conclusive answer to the question of whether there are contradictions that correctly describe reality is likely to be doomed to failure. However, the lack of such a conclusive answer does not imply that it is not legitimate to devise a formal system in which contradictions are interpreted as true. If there are some ontological contradictions, among the propositions that describe reality correctly we are going to find some true contradictions. But of course reality is not trivial, so we need a logic in which explosion does not hold. Therefore, if contradictions are ontological, a justification for paraconsistency is straightforward.

Regarding epistemological contradictions, even if some contradictions belong to reality, for sure it is not the case that every contradiction we face is not epistemological in the sense presented in Sect. 3.2. In general, conflicting information that is going to be corrected later, including contradictory results of scientific theories, may be taken as epistemological contradictions. It is perfectly legitimate, therefore, to devise formal systems in which contradictions are understood either epistemologically or ontologically. In the latter case, it may be that both A and $\neg A$ are true; in the former, we take A and $\neg A$ as meaning conflicting evidence about the truth-value of A. In both cases, explosion does not hold without restrictions; in both cases, the development of paraconsistent logics is line with the very nature of logic.

A philosophical justification for paraconsistent logics, and in particular for the Logics of Formal Inconsistency, depends essentially on showing that they are more than 'mathematical structures' with a language, a syntax, and a semantics, about which several technical properties can be proved. Working on the technical properties of formal systems helps us to understand various logical relations and properties of language and a number of concepts that are philosophically relevant. However, in order to justify a whole account of logical consequence it is necessary to show that such an account is committed with real situations of reasoning. From this perspective, given a formal system, the key question is whether or not it provides an intuitive account of what follows from what. Depending on the answer given, the logic at stake acquires a 'philosophical citizenship'.

In what follows, we show that Logics of Formal Inconsistency may be seen, on the one hand, as an account of contexts of reasoning in which contradictions occur because reality itself is contradictory, and, on the other hand, as an account of contexts in which contradictions are provisional states that (at least in principle) are going to be corrected later. *LFI*s are able to deal with contradictions, no matter whether they are understood epistemologically or ontologically. We may work out formal systems in which a contradiction means that there are propositions A and $\neg A$ such that both are true, as well as systems in which a contradiction is understood in a weaker sense as simultaneous evidence that A and $\neg A$ are true. In the latter case, faithful to the idea that contradictions are not ontological, the system does not tolerate a true contradiction—if it is the case that A and $\neg A$ are both true, triviality obtains.

3.2.5 Epistemological Contradictions

In this section we present the basic ideas of a paraconsistent formal system in which contradictions are understood epistemologically. From the viewpoint of a (semantical) intuitive interpretation, the duality between paraconsistency and paracompleteness may be understood as, respectively, 'excess of information' and 'lack of information'. Accordingly,

$v(A \wedge \neg A) = 1$ means that there is too much information about the truth value of A,

$v(A \vee \neg A) = 0$ means that there is too little information about the truth value of A.

Intuitionistic logic is paracomplete. Its 'lack of information' may be understood as the absence of a constructive proof, not as falsity in the strict sense. This approach allows a plausible and intuitive reading of Kripke models for intuitionistic logic. When a stage attributes the value *0* both to A and to $\neg A$, this may be taken, respectively, as 'A has not been proved yet' and '$\neg A$ has not been proved yet'. Notice that if there is a stage t_1 in a model M such that $M, t_1 \nvDash A \vee \neg A$, it means that $M, t_1 \nvDash A$ and $M, t_1 \nvDash \neg A$. But the latter is the case only if in M there is a stage $t_2 \geq t_1$ such that $M, t_2 \vDash A$ (remember that $\neg A$ holds in a stage t if and only if for all $t' \geq t, t' \nvDash A$). We suggest that the stage t_1 above may be intuitively interpreted as a scenario such that there is too little information about A and about $\neg A$ because both have not been proved yet, so both A and $\neg A$ receive the value *0*.

In Sect. 3.2.3 we mentioned that the identification of the notion of constructive proof with truth yields some difficult problems, and referred the reader to Raatikainen [44]. But it is easy to see some of these same problems with Kripke models if we identify *1* and *0* with, respectively, *true* and *false*. It is awkward if one has found out in stage t_2 that A is true, A being a mathematical theorem, that the truth of A is not transmitted to the past, to stage t_1. Likewise, since the whole story is designed for mathematical propositions, it is strange that in t_1 we regard a proposition A false, but that in a later stage t_2, conceived as a later moment in time, A becomes true.

Now we turn to paraconsistent logic, the dual situation in which both A and $\neg A$ receive the value *1*. This may be understood as the presence of simultaneous but non-conclusive evidence that A is true and $\neg A$ is true. 'Evidence for A' in the sense proposed here are 'reasons to believe in A'. One is justified in believing that A is true inasmuch one has evidence available that A is true. But of course it may be that there are also reasons for believing $\neg A$, and in this case the evidence is not conclusive. The following passage from da Costa [21, pp. 9–10] helps to clarify what we mean by evidence.

> Let us suppose that we want to define an operational concept of negation (...) $\neg A$, where A is atomic, is to be true if, and only if, the clauses of an appropriate criterion c are fulfilled, clauses that must be empirically testable (...) the same must be valid for the atomic proposition A, for the sake of coherence. Hence, there exists a criterion d for the truth of A. But clearly it may happen that the criteria c and d be such that they entail, under certain critical circumstances, the truth of both A and $\neg A$.

da Costa says that some 'certain critical circumstances', entail the truth and the falsity of A. However, it seems to us that it would be perfectly reasonable at this point not to draw the conclusion that A is both true and false. It is better to be more careful and to take the contradiction as a provisional state, a kind of excessive information that should, at least in principle, be eliminated by means of further investigation. Now, a counterexample for explosion is straightforward: we may have non-conclusive evidence for both A and $\neg A$, but no evidence for B at all.

It is remarkable that there is a kind of duality between the notions of evidence and constructive proof. They are, so to speak, respectively a 'weaker truth' and a 'stronger truth'. A constructive proof of A implies the truth of A, but the converse does not hold. Inversely, if one knows the truth of A, it implies the presence of evidence for A, but the converse does not hold.

We have thus given an intuitive interpretation for the paraconsistent negation that justifies the invalidity of explosion. So far so good. However, we cannot yet express that some proposition is true, because the notion of evidence is weaker than truth. With the help of the consistency operator this problem can be solved. We propose the following intuitive meaning for the consistency operator: $\circ A$ means informally that the truth-value of A has been conclusively established. Now we have resources to express not only that there is evidence that A is true but also that A has been established (by whatever means) as true: $\circ A \wedge A$. Notice that how the truth or falsity of a proposition is going to be established is not a problem of logic. Truth comes from outside the formal system. Let us see how we can express these ideas in a logic of formal inconsistency.

3.3 On the Mathematics of the Logics of Formal Inconsistency

3.3.1 mbC: A Minimal LFI

We start by presenting *mbC*, a basic *LFI*. This name stands for 'a minimal logic with the axiom bc1', and 'bc' stands for 'basic property of consistency'. *mbC* is an extension of classical positive propositional logic (from now on, *CPL+*) enriched with a non-explosive negation and a consistency operator, the unary operator $\circ$. *mbC* is interesting because it has a minimal apparatus and several technical properties that illustrate the main features of Logics of Formal Inconsistency. As we shall see, *mbC*:

i. permits us to define classical negation, and thus can be seen as an extension of classical logic;
ii. permits recovering classical consequence by means of a derivability adjustment theorem (*DAT*);
iii. distinguishes the consistency of a formula A from the non-contradiction of A, i.e., $\circ A$ and $\neg(A \wedge \neg A)$ are not equivalent;
iv. is gently explosive in the sense that it tolerates some pairs of formulas A and $\neg A$, while it is explosive with respect to others;
v. has a sound and complete bivalued semantics.[18]

[18] A more comprehensive and detailed presentation of *mbC* can be found in Carnielli et al. [15].

3.3.1.1 The Syntax of MbC

Let L_1 be a language with a denumerable set of sentential letters $\{p_1, p_2, p_3, \ldots\}$, the set of connectives $\{\circ, \neg, \wedge, \vee, \rightarrow\}$, and parentheses. The consistency operator $\circ$ is a primitive symbol and $\neg$ is a non-explosive negation. The set of formulas of L_1 is obtained recursively in the usual way; and Roman capitals stand for meta-variables for formulas of L_1. The logic *mbC* is defined over the language L_1 by the following Hilbert system:

Axiom-schemas:
Ax. 1. $A \rightarrow (B \rightarrow A)$
Ax. 2. $(A \rightarrow (B \rightarrow C)) \rightarrow ((A \rightarrow B) \rightarrow (A \rightarrow C))$
Ax. 3. $A \rightarrow (B \rightarrow (A \wedge B))$
Ax. 4. $(A \wedge B) \rightarrow A$
Ax. 5. $(A \wedge B) \rightarrow B$
Ax. 6. $A \rightarrow (A \vee B)$
Ax. 7. $B \rightarrow (A \vee B)$
Ax. 8. $(A \rightarrow C) \rightarrow ((B \rightarrow C) \rightarrow ((A \vee B) \rightarrow C))$
Ax. 9. $A \vee (A \rightarrow B)$
Ax. 10. $A \vee \neg A$
Ax. bc1. $\circ A \rightarrow (A \rightarrow (\neg A \rightarrow B))$
Inference rule: *modus ponens.*

Positive classical propositional logic, *CPL+*, is given by Axioms *1–9* plus modus ponens.[19] *mbC* is an extension of *CPL+*.[20] Due to the Axiom *bc1*, *mbC* is gently explosive[21]:

For some A and B:
$A, \neg A \nvdash B$,
$\circ A, A \nvdash B$,
$\circ A, \neg A \nvdash B$,
While for every A and B: $\circ A, A, \neg A \vdash B$

---

[19] An equivalent system is obtained by substituting Axiom *9* by Peirce's Law, $((A \rightarrow B) \rightarrow A) \rightarrow A$. Indeed, it is well known that Axioms *1* and *2* plus Peirce's Law define positive implicative classical logic.

[20] *LFI*s may also be obtained as extensions of positive intuitionistic propositional logic, *IPL+*, the system given by Axioms 1–8 plus modus ponens. Different formal systems may be obtained depending on the positive logic one starts with and the desired behavior of negation and the operator $\circ$. The inconsistency of a formula A may also be primitive, represented by $\bullet A$, that may or may not be equivalent to $\neg \circ A$ (see details in Carnielli and Coniglio [10]). We have chosen here to take *CPL+* as a basis and *mbC* as the first *LFI*, because this allows for a simpler and more didactic presentation, more suitable to the aims of this text.

[21] A precise characterization of the principle of gentle explosion is to be found in Carnielli et al. [15, pp. 19–20].

The idea, as we have seen, is that we should be able to separate the contradictions that do not lead to explosion from those that do. The Axiom *bc1* is also called the *gentle explosion law*, because it is explosive only with respect to formulas marked with $\circ$.

The definition of a derivation of A from a set of premises Γ ($\Gamma \vdash_{mbC} A$) is the usual one: a finite sequence of formulas $B_1 \ldots B_n$ such that A is B_n and each B_i, $1 \leq i \leq n$ (that is, each line of the proof) is an axiom, a formula that belongs to Γ, or a result of *modus ponens*. A theorem is a formula derived from the empty set of premises. The logic *mbC* satisfies the following properties:

P1. Reflexivity: if $A \in \Gamma$, then $\Gamma \vdash_{mbC} A$;
P2. Monotonicity: if $\Gamma \vdash_{mbC} B$, then $\Gamma, A \vdash_{mbC} B$, for any A;
P3. Cut: if $\Delta \vdash_{mbC} A$ and $\Gamma, A \vdash_{mbC} B$, then $\Delta, \Gamma \vdash_{mbC} B$;
P4. Deduction theorem: if $\Gamma, A \vdash_{mbC} B$, then $\Gamma \vdash_{mbC} A \to B$;
P5. Compactness: if $\Gamma \vdash_{mbC} A$, then there is $\Delta \subseteq \Gamma$, Δ finite, $\Delta \vdash_{mbC} A$.

Since the properties *P1*, *P2*, and *P3* hold, *mbC* is thus a standard logic (cf. Carnielli et al. [15] p. 6). The properties *P1*, *P2*, *P3*, and *P5* come directly from the definition of $\Gamma \vdash_{mbC} A$. The deduction theorem comes from Axioms 1 and 2 plus modus ponens. Since monotonicity holds, the converse of the deduction theorem also holds.

Classical logic may be recovered in *mbC* in two ways: by defining a negation that has the properties of the classical negation and by means of a derivability adjustment theorem (*DAT*).

Fact 3.1 *Classical negation is definable in* mbC

Proof We define $\bot := \circ A \wedge A \wedge \neg A$ and $\sim A := A \to \bot$. Now, we get explosion, $A \to (\sim A \to B)$, as a theorem, in a few steps from *bc1*. From the Axiom *9* we get excluded middle, $A \vee \sim A$. Classical propositional logic *CPL* is obtained by Axioms 1–8 plus explosion, excluded middle and modus ponens. □

The general purpose of derivability adjustment theorems is to establish a relationship between two logics **L1** and **L2**, in the sense of restoring inferences that are lacking in one of them.[22] The basic idea is that we have to 'add some information' to the premises in order to restore the inferences that are lacking. *DAT*s are especially interesting because they show what is needed in order to restore classical consequence in a paraconsistent scenario.

For the sake of stating precisely the *DAT* between *mbC* and *CPL*, we need to take into account the difference between the respective languages. The first step will be to translate one language into another. Let L_2 be a language with the set of

[22] As far as we know, *DAT*s were proposed for the first time by Diderik Batens, one of the main researchers in the field of paraconsistency. His inconsistency-adaptive logics are a kind of paraconsistent logic that restricts the validity of the principle of explosion according to the information available in some context. As we see the proposal of inconsistency-adaptive logics, they share with the Logics of Formal Inconsistency the possibility of interpreting contradictions epistemologically. However, an important difference is that in adaptive logics everything is supposed to be consistent unless proven otherwise. *LFI*s, in contrast, do not presuppose consistency.

connectives $\{\sim, \vee, \wedge, \rightarrow\}$. Instead of a paraconsistent negation $\neg$, L_2 has classical negation $\sim$. We present a simplified proof. A detailed proof may be found in Carnielli and Coniglio [10].

Fact 3.2 *Let* t *be a mapping which replaces* $\sim$ *by* $\neg$. *Then, the following holds: For all* Γ *and for all* B, $\Gamma \cup \{B\} \subseteq L_2$, *there is a* Δ, $\Delta \subseteq L_1$ *such that* $\Gamma \vdash_{CPL} B$ *iff* $t[\Gamma], \circ\Delta \vdash_{mbC} t[B]$, *where* $\circ\Delta = \{\circ A : A \in \Delta\}$.

Proof From left to right, suppose there is a derivation D of $\Gamma \vdash_{CPL} B$ (in the language L_2 of *CPL*). If we simply change the classical negations $\sim$ to $\neg$, such a derivation does not hold in *mbC*. We need to be concerned only with occurrences of explosion. The relevant point is that some information has to be available in order to reconstruct classical reasoning. An occurrence of a line

$i.\ A \rightarrow (\sim A \rightarrow B)$

in the derivation *D* has to be substituted by the following lines, obtaining a derivation *D'*:

$i_1.\ \circ A$
$i_2.\ \circ A \rightarrow (A \rightarrow (\neg A \rightarrow B))$
$i_3.\ A \rightarrow (\neg A \rightarrow B).$

From right to left, suppose there is a derivation *D'* of $t[\Gamma], \circ\Delta \vdash_{mbC} t[B]$. We get a derivation D of $\Gamma \vdash_{CPL} B$ just by deleting the occurrences of $\circ$ and changing $\neg$ to $\sim$. □

The reader should notice the difference between restoring classical consequence by means of a definition of a classical negation inside *mbC* and by means of a *DAT*. In the latter case, the central issue is the information that has to be available in order to restore classical reasoning. In each occurrence of classical explosion, $A \rightarrow (\sim A \rightarrow B)$, the information needed from the viewpoint of *mbC* is the consistency of *A*, represented by $\circ A$.

3.3.1.2 A Semantics for *mbC*

The sentential connectives of classical logic are truth-functional. That means that the truth-value of a molecular formula is functionally determined by its structure and by the truth-values of its components, which reduce to the truth-values of the atomic formulas. Truth-functionality as a property of the semantics of certain logics is a mathematical rendering of the *Principle of Compositionality*, which says that the meaning of a complex expression is functionally determined by the meanings of its constituent expressions and the rules used to combine them. This principle is also called Frege's Principle, since it can be traced back to Frege.

The truth-value of molecular formulas may be determined by using the familiar matrices (truth-tables) that any logic student is familiar with. These matrices have only two values (*true* and *false*, or *1* and *0*) in the case of classical propositional logic, but the idea can be generalized to any number of 'truth-values'.

A logic can be 'truth-functional' even if it is characterized semantically by a finite, or even by an infinite, number of 'truth-values'. Indeed, in most many-valued logics the 'truth-value' of a molecular formula is also functionally determined by the values of the atomic formulas.

However, instead of talking about truth-values, it would be better to talk about *semantic values*, since, as we have argued earlier, the values *0* and *1* attributed to formulas do not need to be *always* interpreted as false and true. The point is that we do not want to commit ourselves from the start to just the values true and false attributed to formulas. We have argued in Sect. 3.2.5 that in intuitionistic logic the value *1* attributed to A is better taken as an indication that a constructive proof of A is available, and our suggestion with respect to (some) paraconsistent logics is to interpret the value *1* as an indication that there is some evidence for the truth of A.

We will now try to put things in a more neutral and precise way. Let us say that a semantics for a logic **L** is called matrix-functional (instead of truth-functional) if the semantic value of a formula of **L** is functionally determined by means of a finite matrix. This is the case for classical logic, but not for intuitionistic logic, as Gödel [27] has proved, nor is the case for *mbC*.[23]

A non-matrix-functional semantics for paraconsistent logics was proposed by da Costa and Alves [22]. There, we find a bivalued semantics for da Costa's C_1. The semantic clause for a paraconsistent (but not paracomplete) negation has only 'half' of the clause for classical negation: if $v(\neg A) = 0$, then $v(A) = 1$. The idea is that it cannot be the case that A and $\neg A$ receive simultaneously the value *0*. But the possibility is open for both to receive the value *1*. This kind of semantics is described by the so-called *quasi-matrices*. The quasi-matrix for negation is as follows:

A	0	1	
$\neg A$	1	0	1

It is clear that the semantic value of $\neg A$ is not functionally determined by the semantic value of A: when $v(A) = 1$, $v(\neg A)$ may be *1* or *0*. For this reason, this semantics is clearly nonfunctional (i.e., the semantic value of $\neg A$ is not functionally determined by the semantic value of A). In Carnielli et al. [15, pp. 38ff], we find a bivalued semantics that is sound and complete for *mbC*, as described below:

Definition 3.3 An *mbC-valuation* is a function that assigns the values *0* and *1* to formulas of the language L_1, satisfying the following clauses:

(i) $v(A \wedge B) = 1$ if and only if $v(A) = 1$ and $v(B) = 1$;
(ii) $v(A \vee B) = 1$ if and only if $v(A) = 1$ or $v(B) = 1$;

---

[23] In fact, not only *mbC* but all logics of the da Costa hierarchy C_n, and most *LFI*s, are not characterizable by finite matrices (see Carnielli et al. [15, p. 74, Theorem 121]).

(iii) $v(A \to B) = 1$ if and only if $v(A) = 0$ or $v(B) = 1$;
(iv) $v(\neg A) = 0$ implies $v(A) = 1$;
(v) $v(\circ A) = 1$ implies $v(A) = 0$ or $v(\neg A) = 0$.

We say that a valuation v is a model of a set Γ if and only if every proposition of Γ receives the value *1* in v, and v is a model of a proposition A if and only if A receives the value *1* in v.

The notion of logical consequence is defined as usual: a formula A is an *mbC-consequence* of a set Γ ($\Gamma \vDash_{mbC} A$), if and only if for every valuation v, if v is a model of Γ, then v is a model of A (if $v \vDash \Gamma$, then $v \vDash A$; when there is no risk of ambiguity, we shall write simply $\vDash$ and $\vdash$).

Notice that the clauses for $\wedge$, $\vee$ and $\to$ are exactly the same as in classical logic. By clause (iv), the system is paraconsistent but not paracomplete, since excluded middle (for paraconsistent negation) holds. We suggest that the values *0* and *1* are not to be understood as (respectively) false and true, but rather as absence and presence of evidence. Thus:

$v(A) = 1$ means 'there is evidence that A is true';
$v(A) = 0$ means 'there is no evidence that A is true';
$v(\neg A) = 1$ means 'there is evidence that A is false';
$v(\neg A) = 0$ means 'there is no evidence that A is false'.

The same counterexample invalidates explosion and disjunctive syllogism: $v(A) = 1$, $v(B) = 0$ and $v(\neg A) = 1$. Non-contradiction is also invalid: $v(A) = v(\neg A) = 1$, hence $v(A \wedge \neg A) = 1$, but $v(\neg(A \wedge \neg A))$ may be *0*. Due to clause (v), it may be the case that $v(A) = 1$, $v(\neg A) = 0$ (or vice versa) but $v(\circ A) = 0$. In this valuation, $v(\neg(A \wedge \neg A)) = 1$, hence the non-equivalence between $\circ A$ and $\neg(A \wedge \neg A)$. Also, due to clause (v), it is clear that *mbC* does not admit a trivial model, i.e., a model such that $v(A) = 1$ for every formula A.

We would like to make some comments with respect to the validity of excluded middle in *mbC*, given the intended interpretation in terms of evidence. Indeed, we may have a situation such that there is no evidence at all, neither for the truth nor for the falsity of a proposition A, but this cannot be represented in *mbC*. In fact, *mbC* may be easily modified in order to be able to represent such a situation. On the other hand, the validity of excluded middle may be justified when we by default attribute evidence for $\neg A$ when there is no evidence at all. This happens, for instance, in a criminal investigation in which one starts considering everyone (in some group of people) not guilty until proof to the contrary.

Whether excluded middle should be valid from the start, or be recovered once some information has been added may be seen as a methodological decision that depends on the reasoning scenario we want to represent. However, the point we want to emphasize is that the fact that in *mbC* $\circ A$ and $\neg(A \wedge \neg A)$ are not logically equivalent makes *mbC* suitable to represent the basic features of the intuitive interpretation of contradiction as conflicting evidence.

3.3.1.3 Soundness and Completeness

The present section shows that the axiomatic system is sound and complete with respect to the bivalued semantics presented, that is, the relations $\vdash_{mbC}$ and $\vDash_{mbC}$ coincide.

Theorem 3.4 *Soundness: if* $\Gamma \vdash_{mbC} A$*, then* $\Gamma \vDash_{mbC} A$

Proof In order to show that the system is sound we have to check that every axiom is valid and that modus ponens preserves validity. This is an easy task, left as an exercise for the reader. We illustrate the procedure with Axiom *bc1* by means of the quasi-matrix below:

<table>
<tr><td>A</td><td colspan="4">0</td><td colspan="6">1</td></tr>
<tr><td>B</td><td colspan="2">0</td><td colspan="2">1</td><td colspan="3">0</td><td colspan="3">1</td></tr>
<tr><td>$\neg A$</td><td colspan="2">1</td><td colspan="2">1</td><td colspan="2">0</td><td>1</td><td colspan="2">0</td><td>1</td></tr>
<tr><td>$\circ A$</td><td>0</td><td>1</td><td>0</td><td>1</td><td>0</td><td>1</td><td>0</td><td>0</td><td>1</td><td>0</td></tr>
<tr><td>$\neg A \to B$</td><td>0</td><td>0</td><td>1</td><td>1</td><td>1</td><td>1</td><td>0</td><td>1</td><td>1</td><td>1</td></tr>
<tr><td>$A \to (\neg A \to B)$</td><td>1</td><td>1</td><td>1</td><td>1</td><td>1</td><td>1</td><td>0</td><td>1</td><td>1</td><td>1</td></tr>
<tr><td>$\circ A \to (A \to (\neg A \to B))$</td><td>1</td><td>1</td><td>1</td><td>1</td><td>1</td><td>1</td><td>1</td><td>1</td><td>1</td><td>1</td></tr>
</table>

We see that *bc1* receives the value *1* in every *mbC-valuation*. □

In order to obtain completeness we shall first prove some auxiliary lemmas.

Definition 3.5 A set of propositions Δ is *maximal with respect to A* iff:

(i) $\Delta \nvdash A$;
(ii) For any $B \notin \Delta$: $\Delta \cup \{B\} \vdash A$

The idea is that Δ does not imply A, but the set obtained by 'adding' to Δ anything outside it will entail A. We start by proving a Lindenbaum-style lemma.

Lemma 3.6 *Given a set* Γ *and a formula* A *such that* $\Gamma \nvdash A$*, then there is a set* Δ*,* $\Gamma \subseteq \Delta$*, such that* Δ *is maximal with respect to* A.

Proof The language L_1 has a denumerable number of formulas that can be put in a list: $B_0, B_1, B_2 \ldots$. We define a sequence of sets Γ_i whose union is maximal with respect to A.

$\Gamma_0 = \Gamma$
$\Gamma_{n+1} = \Gamma_n \cup \{B_n\}$, if $\Gamma_n \cup \{B_n\} \nvdash A$
Otherwise, $\Gamma_{n+1} = \Gamma_n$

Now, by taking the union of all Γ_n: $\Delta = \bigcup \{\Gamma_n : n \geq 0\}$, it remains to prove the following:

Δ is maximal with respect to A.

Indeed, we do this in three steps.
(i) for all n, $\Gamma_n \nvdash A$. The proof is by induction on n, given the method of construction of each Γ_n.
(ii) $\Delta \nvdash A$. Suppose $\Delta \vdash A$. Then there is a derivation D of A with premises in Δ. D is a finite sequence of formulas of L_1. The formulas that appear in the derivation D were in the list of formulas of L_1 and have been added to some Γ_n in the respective step of the construction of the sequence of sets Γ_0, Γ_1, Γ_2... Among these formulas, one has the greatest index, k. This formula A_k has been added to the set Γ_k, obtaining the set Γ_{k+1}. But Γ_{k+1} has not only A_k but also all formulas that appear in the derivation D with index less than k. Therefore, $\Gamma_{k+1} \vdash A$, which contradicts (i) above. Hence there is no derivation D, and $\Delta \nvdash A$. Notice that this proof depends essentially on the fact that compactness (property P5) holds for mbC.
(iii) For any $B \notin \Delta : \Delta \cup \{B\} \vdash A$. Suppose $B \notin \Delta$. B is in some position on the list of formulas of L_1, i.e., $B = A_n$ for some n. B has not been 'added' to the set Γ_n because it would make A derivable. Hence, $\Gamma_n \cup \{B\} \vdash A$. $\Gamma_n \subseteq \Delta$, so $\Delta \cup \{B\}$ $\vdash A$. □

Lemma 3.7 *If Δ is maximal with respect to* A, *then Δ is closed under derivability (i.e., $\Delta \vdash B$ iff $B \in \Delta$)*

Proof Suppose $B \notin \Delta$. From (iii) above, we get $\Delta \cup \{B\} \vdash A$. Now suppose $\Delta \vdash B$. So, by cut (property P3), $\Delta \vdash A$, which is impossible. Hence, $\Delta \nvdash B$. The converse is immediate, given reflexivity (property P1). □

Lemma 3.8 *Let Δ be a set maximal with respect to some proposition* A *in* mbC. *Then:*

(i) $(B \wedge C) \in \Delta$ iff $B \in \Delta$ and $C \in \Delta$;
(ii) $(B \vee C) \in \Delta$ iff $B \in \Delta$ or $C \in \Delta$;
(iii) $(B \rightarrow C) \in \Delta$ iff $B \notin \Delta$ or $C \in \Delta$;
(iv) $B \notin \Delta$ implies $\neg B \in \Delta$;
(v) $\circ B \in \Delta$ implies $B \notin \Delta$ or $\neg B \notin \Delta$.

Proof Item (i) depends on Axioms 3, 4, and 5; item (ii) depends on Axioms 6, 7, and 8; (iii) depends on Axioms 1, 9, and the deduction theorem; (iv) depends on Axiom 10; (v) depends on $bc1$. □

Corollary 3.9 *Let Δ be a set maximal with respect to some proposition* A. *The characteristic function of Δ defines an* mbC-*valuation.*

Proof Define a function $v : L_1 \longrightarrow \{0, 1\}$ such that for any formula B, $v(B) = 1$ iff $B \in \Delta$. It is easy to see that the valuation v satisfies clauses (i)–(v) above. □

This is the crucial step in the proof. Up to this point we have dealt only with syntax. Now, in obtaining a model for Δ, we have just related the syntax (i.e., the deductive apparatus) and the semantics in such a way that the proof of completeness will follow in a few steps.

Theorem 3.10 *Completeness: if* $\Gamma \vDash_{mbC} A$*, then* $\Gamma \vdash_{mbC} A$

Proof Suppose $\Gamma \nvdash A$. Then there is a set Δ maximal with respect to A such that $\Gamma \subseteq \Delta$. So, $\Delta \nvdash A$ and $A \notin \Delta$. The characteristic function v of Δ is such that $v(A) = 0$. Since $v \vDash \Delta$ and $\Gamma \subseteq \Delta$, it follows that $v \vDash \Gamma$. But $v \nvDash A$, hence $\Gamma \nvDash A$. By contraposition, completeness is readily established. □

3.3.2 Improving mbC: Towards a Logic of Evidence and Truth

In *mbC*, we have seen some important features of *LFI*s. However, the system may be improved in order to better express the intuitive interpretation of negation and the operator $\circ$ we have presented. This is achieved by the logic LET_K. This name stands for 'logic of evidence and truth based on *CPL+*'.

We have already remarked on the duality between paraconsistency and paracompleteness. Now, in a way analogous to that by which we recover explosion with respect to a formula *A*, in a paracomplete logic we may recover the validity of excluded middle with respect to *A*:

Ax. bd1. $\circ A \to (A \vee \neg A)$.

The semantic clause for the Axioms *bc1* and *bd1* is as follows:

(vi) if $v(\circ A) = 1$, then $[v(A) = 1 \text{ iff } v(\neg A) = 0]$.

If excluded middle holds for *A*, we say that *A* is determined. 'bd' stands for 'basic property of determinedness'. A system in which both *bd1* and *bc1* holds is thus paracomplete and paraconsistent. It is better to call $\circ$, in this context, not a consistent operator but rather a *classicality operator*, since $\circ A$ recovers classical truth conditions with respect to *A*. But $\circ A$ still may be informally understood as meaning that the truth-value of *A* has been conclusively established. We want to emphasize here that what makes possible these alternative readings of the operator $\circ$ is the fact that $\circ A$ may be independent of $\neg(A \wedge \neg A)$.

Now, with *bd1* and *bc1*, we have the resources to express the following situations: no evidence at all, non-conclusive evidence, conflicting evidence, and conclusive evidence (i.e., truth)—see the following table.

No evidence at all	$v(A) = 0, v(\neg A) = 0, v(\circ A) = 0$
Non-conclusive evidence for the truth of *A*	$v(A) = 1, v(\neg A) = 0, v(\circ A) = 0$
Non-conclusive evidence for the falsity of *A*	$v(A) = 0, v(\neg A) = 1, v(\circ A) = 0$
Conflicting evidence	$v(A) = 1, v(\neg A) = 1, v(\circ A) = 0$
Impossible valuation	$v(A) = 0, v(\neg A) = 0, v(\circ A) = 1$
A Has been conclusively established as true	$v(A) = 1, v(\neg A) = 0, v(\circ A) = 1$
A Has been conclusively established as false	$v(A) = 0, v(\neg A) = 1, v(\circ A) = 1$
Impossible valuation	$v(A) = 1, v(\neg A) = 1, v(\circ A) = 1$

We now improve negation by adding the axiom schemas below, which fit the idea of interpreting the values *0* and *1* as absence and presence of evidence. The logic LET_K is thus obtained by adding to the *CPL+* (Axioms 1–9) the Axioms 11–14 below, plus *bc1* and *bd1*.

Ax. 11. $A \leftrightarrow \neg\neg A$
Ax. 12. $\neg(A \wedge B) \leftrightarrow (\neg A \vee \neg B)$
Ax. 13. $\neg(A \vee B) \leftrightarrow (\neg A \wedge \neg B)$
Ax. 14. $\neg(A \rightarrow B) \leftrightarrow (A \wedge \neg B)$

One central question for paraconsistent logicians is that of specifying a non-explosive negation that still has enough properties to be called a negation. One way to approach this problem is to list the properties of classical negation and check whether or not each one of these properties fits the intuitive meaning we want to represent by a non-explosive negation. If we do that, we find out that Axioms *11–14* are well suited to express the ideas of absence and presence of evidence. Let us take a look at Axiom *12*. It is reasonable to conclude that if there is some evidence that a conjunction is false, that same evidence must be an evidence that one of the conjuncts is false. On the other hand, if there is some evidence that A is false, that same evidence must be evidence that $A \wedge B$ is false, for any B. Analogous reasoning applies for disjunction and implication.

3.3.2.1 A Semantics for LET_K

A bivalued semantics for LET_K is given by clauses (i)–(iii) of *mbC* (which give classical truth conditions for $\wedge$, $\vee$ and $\rightarrow$), plus the following:

(vi) if $v(\circ A) = 1$, then $[v(A) = 1$ if and only if $v(\neg A) = 0]$,
(vii) $v(A) = 1$ iff $v(\neg\neg A) = 1$,
(viii) $v(\neg(A \wedge B)) = 1$ iff $v(\neg A) = 1$ or $v(\neg B) = 1$,
(ix) $v(\neg(A \vee B)) = 1$ iff $v(\neg A) = 1$ and $v(\neg B) = 1$,
(x) $v(\neg(A \rightarrow B)) = 1$ iff $v(A) = 1$ and $v(\neg B) = 1$.

The logic LET_K can be proved without much trouble to be sound and complete with respect to the semantics above. The proof needs only to extend Lemma 3.8 of Sect. 3.3.1.3 to the new axioms, which can be done without difficulties. In LET_K, a *DAT* holds as in *mbC* and a classical negation is definable in the same way as in *mbC*, thus LET_K may be also seen as an extension of propositional classical logic. It is worth noting that according to the intuitive interpretation proposed, LET_K like *mbC* does not tolerate true contradictions: indeed, a true contradiction yields triviality, as in classical logic. If A is simultaneously true and false, this is expressed by $(\circ A \wedge A) \wedge (\circ A \wedge \neg A)$, that, in its turn, is equivalent to $\circ A \wedge A \wedge \neg A$, but the latter formula is nothing but a bottom particle.

3.3.3 The Logic Cila: Non-contradiction = Consistency

The logic *Cila* is equivalent (up to translations) to the system C_1 of da Costa's hierarchy. Aside from some differences in the choice of axioms, in C_1 A° is an abbreviation of $\neg(A \wedge \neg A)$, while in *Cila* the consistent operator $\circ$ is primitive. However, in *Cila*, differently from *mbC* and LET_K, $\circ A$ and $\neg(A \wedge \neg A)$ are logically equivalent.

Cila is well suited to interpret the values *0* and *1* as false and true. For this reason, *Cila* is *LFI* suitable for representing true contradictions. Since *Cila* is gently explosive, a distinction may be made between the contradictions that are explosive and those that are not.

Cila is obtained over the language L_1 as an extension of *mbC*. We add to the latter the axiom-schemas below:

Ax. cf. $\neg\neg A \rightarrow A$
Ax. ci. $\neg \circ A \rightarrow (A \wedge \neg A)$
Ax. cl. $\neg(A \wedge \neg A) \rightarrow \circ A$
Ax. ca1. $(\circ A \wedge \circ B) \rightarrow \circ(A \wedge B)$
Ax. ca2. $(\circ A \wedge \circ B) \rightarrow \circ(A \vee B)$
Ax. ca3. $(\circ A \wedge \circ B) \rightarrow \circ(A \rightarrow B)$

The Axioms *ci* and *cl*, together with *bc1*, give us the equivalence between $\circ A$ and $\neg(A \wedge \neg A)$. In *mbC* and LET_K we had already that $\circ A \vdash \neg(A \wedge \neg A)$ and $A \wedge \neg A \vdash \neg \circ A$. Now we get the converse of both.

Instead of having $\circ$ as primitive, we may adopt an inconsistency operator $\bullet$ ('$\bullet A$' means that A is inconsistent). Or we may define $\bullet A$ as $\neg \circ A$. In *Cila*, $\bullet A$ so defined and $A \wedge \neg A$ are logically equivalent. Indeed, in an *LFI* conceived to be able to express true contradictions, $\circ A$ should be logically equivalent to $\neg(A \wedge \neg A)$, as well as $\bullet A$ and $A \wedge \neg A$. The propagation of consistency holds in *Cila*, given by Axioms *ca1* to *ca3*. A classical negation is definable in Cila: $\sim A := \circ A \wedge \neg A$. Also, as well as in *mbC*, there is a bottom particle: $\bot := \circ A \wedge A \wedge \neg A$. We leave as an exercise for the reader the proof that $\vdash_{Cila} (A \rightarrow \bot) \leftrightarrow (\circ A \wedge \neg A)$.

3.3.3.1 A Semantics for Cila

A complete and correct semantics for *Cila* is given by the clauses (i)–(iv) of *mbC* plus the following:

(xiii) $v(\circ A) = 1$ iff $v(A) = 0$ or $v(\neg A) = 0$
(xiv) $v(\neg\neg A) = 1$ implies $v(A) = 1$

A completeness proof for *Cila* is like that for *mbC*. We need only to complement Lemma 3.8 of the completeness proof for *mbC*.

3.3.3.2 The Equivalence Between *Cila* and C_1

The logic C_1 is defined over the set of connectives $\{\wedge, \vee, \to, \neg\}$ in the following way. First, we drop from *Cila* the Axioms *9* (namely $A \vee (A \to B)$), *bc1*, *ci*, *cl*, *ca1*, *ca2* and *ca3*. Then we define

$$A^\circ := \neg(A \wedge \neg A)$$

and add the axioms below:

Ax. 12. $B^\circ \to ((A \to B) \to ((A \to \neg B) \to \neg A))$
Ax. 13. $(A^\circ \wedge B^\circ) \to ((A \wedge B)^\circ \wedge (A \vee B)^\circ \wedge (A \to B)^\circ)$

Since Axiom *9* is a theorem of C_1, the latter may be seen as an extension of *CPL+*.

A remarkable difference between C_1 and *Cila* is that the former recover classical reasoning by means of the principle of non-contradiction, while the latter recover it by means of principle of explosion, thus emphasizing the duality between explosion and excluded middle. We leave as an exercise for the reader to show the equivalence between C_1 and *Cila*, proving *bc1* in C_1 and the Axiom *12* in *Cila* (changing in each case the consistency operator).

3.4 Final Remarks

There is still a lot of work to be done in the field of Logics of Formal Inconsistency. They are a relatively new subject in paraconsistent research, having appeared in the literature for the first time in Carnielli and Marcos [11]. The unary operator $\circ$ (or its counterpart in da Costa's systems, $^\circ$) initially had the purpose of representing in the object language the metatheoretical notion of consistency. But the idea has been further developed in such a way that it may receive alternative meanings. Aside from freedom from contradiction, we have seen here two different ways of interpreting $\circ A$: (i) the truth value of A has been conclusively established; (ii) classical truth conditions for negation hold for A. According to (i), $\circ A$ says something about the justification of A; according to (ii), $\circ A$ recovers at once the validity of explosion and excluded middle with respect to A, and may be called, instead of a consistent operator, a classicality operator. Dually to logics in which explosion does not hold but may be recovered, which we call Logics of Formal Inconsistency, logics in which excluded middle does not hold but may be recovered are called Logics of Formal Undeterminedness (see Marcos [36]). 'Consistency' and 'determinedness' may be recovered together or one at a time, depending on the scenario we want to represent.

A first-order *LFI*, *QmbC*, has been investigated in full detail in Carnielli et al. [16]. *QmbC* is an extension of *mbC* where quantifiers are added with appropriate syntactical rules. A semantics is provided such that, as expected, there may be atomic formulas Fa and $\neg Fa$ such that both receive the value *1* without trivialization. A first-order extension of LET_K will be presented elsewhere.

It is a fact that paraconsistent logics have been gaining an increasingly important place in contemporary philosophical debate. On the other hand, there is still some resistance to recognizing their philosophical significance. As we have tried to show, although the view that maintains that contradictions belong to reality is legitimate and has antecedents in the history of philosophy, there is another way open, namely that of understanding contradictions from the epistemological viewpoint. We have argued that (and shown how) it can be done.

*LFI*s may be intuitively interpreted ontologically as well as epistemologically. We have presented formal systems that are amenable to both interpretations. *LFI*s are neutral with respect to the philosophical issues related to the nature of contradictions and to the nature of paraconsistency. This is not because *LFI*s are not concerned with an intuitive interpretation of formal systems, but rather because *LFI*s are appropriate for expressing the two basic philosophical views with respect to contradictions, the epistemological and the ontological.

There is another way to provide a semantics for *LFI*s, so-called possible-translation semantics (see Carnielli and Coniglio [8] and Carnielli [7]). There are also several applications of *LFI*s, such as in foundations of set theory (Carnielli and Coniglio [9]), databases (Carnielli et al. [14]), fuzzy logic (Coniglio et al. [18]), automatic theorem provers (Neto and Finger [40]), and quantum computation (Agudelo and Carnielli [1]), among others.

References

1. Agudelo, J.C., Carnielli, W.A.: Paraconsistent machines and their relation to quantum computing. J. Logic Comput. **20**, 573–595 (2010)
2. Aristotle.: The Complete Works of Aristotle. Oxford University Press (1996)
3. Arnauld, A., Nicole, P.: Logic or the Art of Thinking. Cambridge University Press (1996)
4. Asenjo, F.G.: A calculus for antinomies. Notre Dame J. Formal Logic **16**(1), 103–105 (1966)
5. Brouwer, L.E.J.: On the foundations of mathematics. In: Heyting, A. (ed.) Collected Works, vol. I. North-Holland Publishing Company (1975, 1907)
6. Brunner, A., Carnielli, W.A.: Anti-intuitionism and paraconsistency. J. Appl. Logic (2005)
7. Carnielli, W.A.: Possible-translations semantics for paraconsistent logics. In: Frontiers in Paraconsistent Logic: Proceedings of the I World Congress on Paraconsistency. College Publications, Ghent (2000)
8. Carnielli, W.A., Coniglio, M.E.: Splitting logics. In: We Will Show Them! Essays in Honour of Dov Gabbay, Volume One. College Publications (2005)
9. Carnielli, W.A., Coniglio, M.E.: Paraconsistent set theory by predicating on consistency. J. Logic Comput. (2013). doi:10.1093/logcom/ext020
10. Carnielli, W.A., Coniglio, M.E.: Paraconsistent Logic: Consistency, Contradiction and negation (2015)
11. Carnielli, W.A., Marcos, J.: A taxonomy of C-systems. In: Carnielli, W.A., Coniglio, M.E., D'Ottaviano, I. (eds.) Paraconsistency: The Logical Way to the Inconsistent. Proceedings of the Second World Congress on Paraconsistency. Marcel Dekker, New York (2001)
12. Carnielli, W.A., Rodrigues, A.: Towards a philosophical understanding of the logics of formal inconsistency. Submitted for publication (2015)
13. Carnielli, W.A., Rodrigues, A.: A logic for evidence and truth. Submitted for publication (2015)

14. Carnielli, W.A., Marcos, J., De Amo, S.: Formal inconsistency and evolutionary databases. Logic Logical Philos. **8**(2), 115–152 (2000)
15. Carnielli, W.A., Coniglio, M.E., Marcos, J.: Logics of formal inconsistency. In: Gabbay, D., Guenthner, F. (eds.) Handbook of Philosophical Logic, vol. 14. Springer (2007)
16. Carnielli, W.A., Coniglio, M.E., Podiacki, R., Rodrigues, T.: On the way to a wider model theory: Completeness theorems for first-order logics of formal inconsistency. Rev. Symbolic Logic (2014). doi:10.1017/S1755020314000148
17. Chateaubriand, O.: Logical Forms, vol. 1. UNICAMP-CLE, Campinas (2001)
18. Coniglio, M.E., Esteva, F., Godo, L.: Logics of formal inconsistency arising from systems of fuzzy logic. Logic J. IGPL (2014). doi:10.1093/jigpal/jzu016
19. da Costa, N.: Observações sobre o conceito de existência em matemática. Anuário da Sociedade Paranaense de Matemática (1959)
20. da Costa, N.: Sistemas Formais Inconsistentes, p. 1963. Editora da UFPR, Curitiba (1993)
21. da Costa, N.: The philosophical import of paraconsistent logic. J. Non-Class. Logics **1**, 1–19 (1982)
22. da Costa, N., Alves, E.H.: A semantical analysis of the calculi Cn. Notre Dame J. Formal Logic **18**, 621–630 (1977)
23. da Costa, N., French, S.: Science and Partial Truth: A Unitary Approach to Models and Scientific Reasoning. Oxford University Press, Oxford (2003)
24. Frege, G.: Begriffsschrift: a formula language, modeled upon that of arithmetic, for pure thought. In: From Frege to Gödel. Lincoln: toExcel Press (1979, 1879)
25. Frege, G.: The Basic Laws of Arithmetic. University of California Press (1893)
26. Frege, G.: The Thought. The Frege Reader. Blackwell Publishers, Oxford (1997, 1918)
27. Gödel, K.: Zum intuitionistischen aussagenkalkul. In: Anzeiger der Akademie der Wissenschaftischen in Wien, vol. 69, pp. 65-66 (1932). (Reprinted in Gödel, 1986, Collected Works vol. I, Oxford University Press)
28. Gomes, E.L.: Sobre a história da paraconsistência e a obra de da Costa: a instauração da lógica paraconsistente. PhD thesis, IFCH-UNICAMP, Universidade Estadual de Campinas, 2013
29. Gottlieb, P.: Aristotle on non-contradiction. The Stanford Encyclopedia of Philosophy (2013)
30. Hállden, S.: The Logic of Nonsense. Uppsala Universitets Arsskrift (1949)
31. Harman, G.: Change in View: Principles of Reasoning. The MIT Press, Massachussets (1986)
32. Heyting, A.: Intuitionism: An Introduction. North-Holland Publishing Company, London (1956)
33. Jáskowski, S.: Propositional calculus for contraditory deductive systems. Stud. Logica **24**(143–157), 1949 (1969)
34. Kapsner, A., Miller, D., Dyckhoff, R.: Report on Kalman Cohen's 1954 Oxford Thesis on Alternative Systems of Logic. In: Handbook of the 5th World Congress of Paraconsistency (2014). http://www.paraconsistency.org/book/Handbook-WCP5.pdf
35. Łukasiewicz, J.: On the principle of contradiction in Aristotle. Rev. Metaphys. (Mar., 1971) **24**(3), 485–509 (1910)
36. Marcos, J.: Nearly every normal modal logic is paranormal. Logique et Analyse **48**, 279–300 (2005)
37. Meheus, J. (ed.): Inconsistency in Science. Kluwer, Dordrecht (2002)
38. Nelson, D.: Constructible falsity. J. Symbolic Logic **14**(1), 16–26 (1949)
39. Nelson, D.: Negation and separation of concepts. In: Constructivity in Mathematics, pp. 208–225. North-Holland, Amsterdam (1959)
40. Neto, A.G.S.S., Finger, M.: Effective prover for minimal inconsistency logic. In:Artificial Intelligence in Theory and Practice: Conference and Symposium on Artificial Intelligence, Santiago, Chile, pp. 465–474. Springer, Boston (2006)
41. Popper, K.: Conjectures and Refutations. Harper, New York (1963)
42. Priest, G.: The logic of paradox. J. Philos. Logic **8**(1), 219–241 (1979)

43. Priest, G., Berto, F.: Dialetheism. Stanford Encyclopedia of Philosophy (2013). http://plato.stanford.edu/archives/sum2013/entries/dialetheism/
44. Raatikainen, P.: Conceptions of truth in intuitionism. History Philos. Logic **25**, 131–145 (2004)
45. Tugendhat, E., Wolf, U.: Logisch-semantische Propädeutik. Reclam, Ditzingen (1989)
46. van Dalen, D.: Intuitionistic logic. Handbook of Philosophical Logic: Volume III: Alternatives in Classical Logic. Synthese Library, vol. 166, pp. 225–339. Springer, Netherlands (1986)

Part II
Many-valued Systems of Paraconsistent Logic

Chapter 4
Three-Valued Paraconsistent Propositional Logics

Ofer Arieli and Arnon Avron

Abstract Three-valued matrices provide the simplest semantic framework for introducing paraconsistent logics. This paper is a comprehensive study of the main properties of propositional paraconsistent three-valued logics in general, and of the most important such logics in particular. For each logic in the latter group, we also provide a corresponding cut-free Gentzen-type system.

Keywords Paraconsistency · 3-valued matrices · Proof systems

Mathematics Subject Classification (2000) Primary 03B53, 03B50 · Secondary 03C90, 03F05

4.1 Introduction

It is well known that classical logic is not adequate for reasoning with inconsistent information. One of the oldest and the most common approaches to overcome this shortcoming of classical logic is to enrich the set of truth-values with a third element other than the two classical ones t and f. Indeed, since their introduction by Łukasiewicz [28] (see also [29]), three-valued logics have been extensively studied for uncertainty reasoning in general, and paraconsistent reasoning in particular (see, e.g., [4, 7, 16, 20], which in turn contain references to many other works). The goal of this work is to study this approach to paraconsistency in a systematic way, as well as to present what we believe to be the most important results concerning the better

This work is supported by The Israel Science Foundation under grant agreement No. 817-15.

O. Arieli (✉)
School of Computer Science, The Academic College of Tel-Aviv, Tel Aviv, Israel
e-mail: oarieli@mta.ac.il

A. Avron
School of Computer Science, Tel-Aviv University, Tel Aviv, Israel
e-mail: aa@cs.tau.ac.il

J.-Y. Beziau et al. (eds.), *New Directions in Paraconsistent Logic*,
Springer Proceedings in Mathematics & Statistics 152,
DOI 10.1007/978-81-322-2719-9_4

accepted logics that came out from this approach. However, it should be emphasized that the scope of the material we present is limited according to the following criteria:

1. The languages that are considered in the sequel are *propositional*, as this is the heart of every paraconsistent logic ever studied so far.
2. We confine ourselves to paraconsistent propositional *logics*, in which a propositional language is equipped with a structural and nontrivial Tarskian consequence relation. In particular, no form of nonmonotonic reasoning is considered in this paper.
3. We restrict ourselves here to logics which are based on *truth-functional* three-valued semantics.[1]

The rest of the paper is organized as follows: In the next section, we review some general definitions and basic concepts that are needed in the sequel. In Sect. 4.3, we define in precise terms what paraconsistent logics are, and what additional properties they expected to have. These properties are then investigated in the context of three-valued matrices in Sect. 4.4. The most important logics that are induced by these matrices are considered in Sect. 4.5, and corresponding proof systems are discussed in Sect. 4.6.

4.2 Preliminaries

4.2.1 Propositional Logics

In what follows a propositional language with a set $\mathsf{Atoms}(\mathcal{L}) = \{P_1, P_2, \ldots\}$ of atomic formulas is denoted by $\mathcal{L}$ and use p, q, r to vary over this set. The set of the well-formed formulas of $\mathcal{L}$ is denoted by $\mathcal{W}(\mathcal{L})$ and $\varphi, \psi, \phi, \sigma$ will vary over its elements. The set $\mathsf{Atoms}(\varphi)$ denotes the atomic formulas occurring in φ. Sets of formulas in $\mathcal{W}(\mathcal{L})$ are called *theories* and are denoted by $\mathcal{T}$ or $\mathcal{T}'$. Finite theories are denoted by Γ or Δ. Following the usual convention, we shall abbreviate $\mathcal{T} \cup \{\psi\}$ by $\mathcal{T}, \psi$. More generally, we shall write $\mathcal{T}, \mathcal{T}'$ instead of $\mathcal{T} \cup \mathcal{T}'$. A *rule* in a language $\mathcal{L}$ is a pair $\langle \Gamma, \psi \rangle$, where $\Gamma \cup \{\psi\}$ is a finite set of formulas in $\mathcal{L}$. We shall henceforth denote such a rule by Γ / ψ.

Definition 4.1 A (Tarskian) *consequence relation* for a language $\mathcal{L}$ (a tcr, for short) is a binary relation $\vdash$ between theories in $\mathcal{W}(\mathcal{L})$ and formulas in $\mathcal{W}(\mathcal{L})$, satisfying the following three conditions:

[1]When truth functionality is not required, further approaches based on nondeterministic semantics [10] are available. They give rise to another brand of useful three-valued logics, which includes many of the LFIs considered in [16]. We refer the reader to [12, 13] for further information on these logics and references to related papers.

Reflexivity: if $\psi \in \mathcal{T}$ then $\mathcal{T} \vdash \psi$.
Monotonicity: if $\mathcal{T} \vdash \psi$ and $\mathcal{T} \subseteq \mathcal{T}'$, then $\mathcal{T}' \vdash \psi$.
Transitivity (cut): if $\mathcal{T} \vdash \psi$ and $\mathcal{T}', \psi \vdash \phi$ then $\mathcal{T}, \mathcal{T}' \vdash \phi$.

Let $\vdash$ be a tcr for $\mathcal{L}$. We say that $\vdash$ is

- *structural*, if for every $\mathcal{L}$-substitution θ and every $\mathcal{T}$ and ψ, if $\mathcal{T} \vdash \psi$ then $\{\theta(\varphi) \mid \varphi \in \mathcal{T}\} \vdash \theta(\psi)$.
- *nontrivial*, if there exist some nonempty theory $\mathcal{T}$ and some formula ψ such that $\mathcal{T} \nvdash \psi$.
- *finitary*, if for every theory $\mathcal{T}$ and every formula ψ such that $\mathcal{T} \vdash \psi$ there is a *finite* theory $\Gamma \subseteq \mathcal{T}$ such that $\Gamma \vdash \psi$.

Definition 4.2 A (propositional) *logic* is a pair $\mathbf{L} = \langle \mathcal{L}, \vdash_{\mathbf{L}} \rangle$, such that $\mathcal{L}$ is a propositional language, and $\vdash$ is a structural and nontrivial[2] consequence relation for $\mathcal{L}$. A logic $\langle \mathcal{L}, \vdash_{\mathbf{L}} \rangle$ is *finitary* if so is $\vdash_{\mathbf{L}}$.

Definition 4.3 Let $\mathbf{L} = \langle \mathcal{L}, \vdash \rangle$ be a logic, and let S be a set of rules in $\mathcal{L}$. The *finitary* $\mathbf{L}$*-closure* $C_{\mathbf{L}}(S)$ of S is inductively defined as follows:

- $\langle \theta(\Gamma), \theta(\psi) \rangle \in C_{\mathbf{L}}(S)$, whenever θ is a uniform $\mathcal{L}$-substitution, Γ is a *finite* theory in $\mathcal{W}(\mathcal{L})$, and either $\Gamma \vdash \psi$ or $\Gamma/\psi \in S$.
- If the pairs $\langle \Gamma_1, \varphi \rangle$ and $\langle \Gamma_2 \cup \{\varphi\}, \psi \rangle$ are both in $C_{\mathbf{L}}(S)$, then so is the pair $\langle \Gamma_1 \cup \Gamma_2, \psi \rangle$.

Next we define what an extension of a logic means.

Definition 4.4 Let $\mathbf{L} = \langle \mathcal{L}, \vdash \rangle$ be a logic, and let S be a set of rules in $\mathcal{L}$.

- A logic $\mathbf{L}' = \langle \mathcal{L}, \vdash' \rangle$ is an *extension* of $\mathbf{L}$ (in the same language) if $\vdash \subseteq \vdash'$. We say that $\mathbf{L}'$ is a *proper extension* of $\mathbf{L}$, if $\vdash \subsetneq \vdash'$.
- The *extension of* $\mathbf{L}$ *by* S is the pair $\mathbf{L}^* = \langle \mathcal{L}, \vdash^* \rangle$, where $\vdash^*$ is the binary relation between theories in $\mathcal{W}(\mathcal{L})$ and formulas in $\mathcal{W}(\mathcal{L})$, defined by: $\mathcal{T} \vdash^* \psi$ if there is a finite $\Gamma \subseteq \mathcal{T}$ such that $\langle \Gamma, \psi \rangle \in C_{\mathbf{L}}(S)$.[3]
- Extending $\mathbf{L}$ by an axiom schema φ means extending it by the rule $\emptyset/\varphi$.

The usefulness of a logic strongly depends on the question that what kind of connectives are available in it. Three particularly important types of connectives are defined next.

[2]The condition of nontriviality is not always demanded in the literature, but we find it very convenient (and natural) to include it here.

[3]Note that $\mathbf{L}^*$ is a propositional logic unless $C_{\mathbf{L}}(S)$ contains *all* the pairs of finite theories in $\mathcal{W}(\mathcal{L})$ and formulas in $\mathcal{W}(\mathcal{L})$. Moreover, $\mathbf{L}^*$ is in that case the minimal extension of $\mathbf{L}$ such that $\Gamma \vdash^* \varphi$ whenever $\Gamma/\varphi \in S$.

Definition 4.5 Let $\mathbf{L} = \langle \mathcal{L}, \vdash_{\mathbf{L}} \rangle$ be a propositional logic.

- A binary connective $\supset$ of $\mathcal{L}$ is called an *implication for* $\mathbf{L}$ if the classical deduction theorem holds for $\supset$ and $\vdash_{\mathbf{L}}$:

$$\mathcal{T}, \varphi \vdash_{\mathbf{L}} \psi \text{ iff } \mathcal{T} \vdash_{\mathbf{L}} \varphi \supset \psi.$$

- A binary connective $\wedge$ of $\mathcal{L}$ is called a *conjunction for* $\mathbf{L}$ if it satisfies the following condition:

$$\mathcal{T} \vdash_{\mathbf{L}} \psi \wedge \varphi \text{ iff } \mathcal{T} \vdash_{\mathbf{L}} \psi \text{ and } \mathcal{T} \vdash_{\mathbf{L}} \varphi.$$

- A binary connective $\vee$ of $\mathcal{L}$ is called a *disjunction for* $\mathbf{L}$ if it satisfies the following condition:

$$\mathcal{T}, \psi \vee \varphi \vdash_{\mathbf{L}} \sigma \text{ iff } \mathcal{T}, \psi \vdash_{\mathbf{L}} \sigma \text{ and } \mathcal{T}, \varphi \vdash_{\mathbf{L}} \sigma.$$

- We say that $\mathbf{L}$ is *seminormal* if it has (at least) one of the three basic connectives defined above. We say that $\mathbf{L}$ is *normal* if it has *all* these three connectives.

The following lemma is easily verified:

Lemma 4.6 *Let* $\mathbf{L} = \langle \mathcal{L}, \vdash_{\mathbf{L}} \rangle$ *be a propositional logic.*

1. *If* $\supset$ *is an implication for* $\mathbf{L}$ *then the following three conditions hold for every* $\psi, \varphi \in \mathcal{W}(\mathcal{L})$*:*
   (a) $\varphi, \varphi \supset \psi \vdash_{\mathbf{L}} \psi$
   (b) $\vdash_{\mathbf{L}} \psi \supset \psi$
   (c) $\psi \vdash_{\mathbf{L}} \varphi \supset \psi$
2. $\wedge$ *is a conjunction for* $\mathbf{L}$ iff *the following three conditions hold for every* $\psi, \varphi \in \mathcal{W}(\mathcal{L})$*:*
   (a) $\psi \wedge \varphi \vdash_{\mathbf{L}} \psi$
   (b) $\psi \wedge \varphi \vdash_{\mathbf{L}} \varphi$
   (c) $\psi, \varphi \vdash_{\mathbf{L}} \psi \wedge \varphi$
3. *If* $\vee$ *is a disjunction for* $\mathbf{L}$ *then the following three conditions hold for every* $\psi, \varphi \in \mathcal{W}(\mathcal{L})$*:*
   (a) $\psi \vdash_{\mathbf{L}} \psi \vee \varphi$
   (b) $\varphi \vdash_{\mathbf{L}} \psi \vee \varphi$
   (c) $\varphi \vee \varphi \vdash_{\mathbf{L}} \varphi$

4.2.2 Many-Valued Matrices

The most standard semantic way of defining logics is by using the following type of structures (see, e.g., [26, 30, 39]).

Definition 4.7 A (multivalued) *matrix* for a language $\mathcal{L}$ is a triple $\mathcal{M} = \langle \mathcal{V}, \mathcal{D}, \mathcal{O} \rangle$, where

- $\mathcal{V}$ is a nonempty set of truth-values,
- $\mathcal{D}$ is a nonempty proper subset of $\mathcal{V}$, called the *designated* elements of $\mathcal{V}$, and
- $\mathcal{O}$ is a function that associates an n-ary function $\widetilde{\diamond}_{\mathcal{M}} : \mathcal{V}^n \to \mathcal{V}$ with every n-ary connective $\diamond$ of $\mathcal{L}$.

Definition 4.8 Let $\mathcal{M} = \langle \mathcal{V}, \mathcal{D}, \mathcal{O} \rangle$ be a matrix for a language $\mathcal{L}$, and let $\mathcal{L} \subseteq \mathcal{L}'$. A matrix $\mathcal{M}' = \langle \mathcal{V}', \mathcal{D}', \mathcal{O}' \rangle$ for $\mathcal{L}'$ is called an *expansion* of $\mathcal{M}$ to $\mathcal{L}'$ if $\mathcal{V} = \mathcal{V}'$, $\mathcal{D} = \mathcal{D}'$, and $\mathcal{O}'(\diamond) = \mathcal{O}(\diamond)$ for every connective $\diamond$ of $\mathcal{L}$.

In what follows, the elements in $\mathcal{V} \backslash \mathcal{D}$ are denoted by $\overline{\mathcal{D}}$. The set $\mathcal{D}$ is used for defining satisfiability and validity, as defined below:

Definition 4.9 Let $\mathcal{M} = \langle \mathcal{V}, \mathcal{D}, \mathcal{O} \rangle$ be a matrix for $\mathcal{L}$.

- An *$\mathcal{M}$-valuation* for $\mathcal{L}$ is a function $\nu : \mathcal{W}(\mathcal{L}) \to \mathcal{V}$ such that for every n-ary connective $\diamond$ of $\mathcal{L}$ and every $\psi_1, \ldots, \psi_n \in \mathcal{W}(\mathcal{L})$, $\nu(\diamond(\psi_1, \ldots, \psi_n)) = \widetilde{\diamond}_{\mathcal{M}} (\nu(\psi_1), \ldots, \nu(\psi_n))$. We denote the set of all the $\mathcal{M}$-valuations by $\Lambda_{\mathcal{M}}$.
- A valuation $\nu \in \Lambda_{\mathcal{M}}$ is an *$\mathcal{M}$-model* of a formula ψ (alternatively, ν *$\mathcal{M}$-satisfies* ψ), if it belongs to the set $mod_{\mathcal{M}}(\psi) = \{\nu \in \Lambda_{\mathcal{M}} \mid \nu(\psi) \in \mathcal{D}\}$. The $\mathcal{M}$-models of a theory $\mathcal{T}$ are the elements of the set $mod_{\mathcal{M}}(\mathcal{T}) = \bigcap_{\psi \in \mathcal{T}} mod_{\mathcal{M}}(\psi)$.
- A formula ψ is *$\mathcal{M}$-satisfiable* if $mod_{\mathcal{M}}(\psi) \neq \emptyset$. A theory $\mathcal{T}$ is $\mathcal{M}$-satisfiable if $mod_{\mathcal{M}}(\mathcal{T}) \neq \emptyset$.

In the sequel, we shall sometimes omit the prefix "$\mathcal{M}$" from the notions above. Also, when it is clear from the context, we shall omit the subscript "$\mathcal{M}$" in $\widetilde{\diamond}_{\mathcal{M}}$.

Definition 4.10 Given a matrix $\mathcal{M}$, the consequence relation $\vdash_{\mathcal{M}}$ that is *induced by* (or associated with) $\mathcal{M}$, is defined by $\mathcal{T} \vdash_{\mathcal{M}} \psi$ if $mod_{\mathcal{M}}(\mathcal{T}) \subseteq mod_{\mathcal{M}}(\psi)$. We denote by $\mathbf{L}_{\mathcal{M}}$ the pair $\langle \mathcal{L}, \vdash_{\mathcal{M}} \rangle$, where $\mathcal{M}$ is a matrix for $\mathcal{L}$ and $\vdash_{\mathcal{M}}$ is the consequence relation induced by $\mathcal{M}$.

Proposition 4.11 [36, 37] *For every propositional language $\mathcal{L}$ and a finite matrix $\mathcal{M}$ for $\mathcal{L}$, $\mathbf{L}_{\mathcal{M}} = \langle \mathcal{L}, \vdash_{\mathcal{M}} \rangle$ is a propositional logic. If $\mathcal{M}$ is finite, then $\vdash_{\mathcal{M}}$ is also finitary.*

We conclude this section with some simple, easily verified, results on the basic connectives (Definition 4.5) in the context of matrix-based logics.

Definition 4.12 Let $\mathcal{M} = \langle \mathcal{V}, \mathcal{D}, \mathcal{O} \rangle$ be a matrix for a language $\mathcal{L}$ and let $\mathcal{A} \subseteq \mathcal{V}$.

- An n-ary connective $\diamond$ of $\mathcal{L}$ is called *$\mathcal{A}$-closed*, if $\tilde{\diamond}(a_1, \ldots, a_n) \in \mathcal{A}$ for every $a_1, \ldots, a_n \in \mathcal{A}$.
- An n-ary connective $\diamond$ of $\mathcal{L}$ is called *$\mathcal{A}$-limited*, if for every $a_1, \ldots, a_n \in \mathcal{V}$, if $\tilde{\diamond}(a_1, \ldots, a_n) \in \mathcal{A}$ then $a_1, \ldots, a_n \in \mathcal{A}$.

Definition 4.13 Let $\mathcal{M} = \langle \mathcal{V}, \mathcal{D}, \mathcal{O} \rangle$ be a matrix for a language $\mathcal{L}$.

- A connective $\wedge$ in $\mathcal{L}$ is called an $\mathcal{M}$-*conjunction* if it is $\mathcal{D}$-closed and $\mathcal{D}$-limited, i.e., for every $a, b \in \mathcal{V}$, $a \tilde{\wedge} b \in \mathcal{D}$ iff $a \in \mathcal{D}$ and $b \in \mathcal{D}$.
- A connective $\vee$ in $\mathcal{L}$ is called an $\mathcal{M}$-*disjunction* if it is $\overline{\mathcal{D}}$-closed and $\overline{\mathcal{D}}$-limited, i.e., for every $a, b \in \mathcal{V}$, $a \tilde{\vee} b \in \mathcal{D}$ iff $a \in \mathcal{D}$ or $b \in \mathcal{D}$.
- A connective $\supset$ in $\mathcal{L}$ is called an $\mathcal{M}$-*implication* if for every $a, b \in \mathcal{V}$, $a \tilde{\supset} b \in \mathcal{D}$ iff either $a \notin \mathcal{D}$ or $b \in \mathcal{D}$.

Proposition 4.14 *Let $\mathcal{M} = \langle \mathcal{V}, \mathcal{D}, \mathcal{O} \rangle$ be a matrix for a language $\mathcal{L}$.*

1. *A connective of $\mathcal{L}$ is an $\mathcal{M}$-conjunction iff it is a conjunction for $\mathbf{L}_{\mathcal{M}}$.*
2. *A connective of $\mathcal{L}$ which is an $\mathcal{M}$-disjunction is also a disjunction for $\mathbf{L}_{\mathcal{M}}$.*
3. *A connective of $\mathcal{L}$ which is an $\mathcal{M}$-implication is also an implication for $\mathbf{L}_{\mathcal{M}}$.*

Corollary 4.15 *Let $\mathcal{M} = \langle \mathcal{V}, \mathcal{D}, \mathcal{O} \rangle$ be a matrix for a language $\mathcal{L}$, and let $\mathcal{M}'$ be an expansion of $\mathcal{M}$. Then*

1. *An $\mathcal{M}$-conjunction (respectively: $\mathcal{M}$-disjunction, $\mathcal{M}$-implication) is also a conjunction (respectively: disjunction, implication) of $\mathbf{L}_{\mathcal{M}'}$.*
2. *If $\mathcal{M}$ has either an $\mathcal{M}$-conjunction, or an $\mathcal{M}$-disjunction, or an $\mathcal{M}$-implication, then $\mathbf{L}_{\mathcal{M}'}$ is seminormal. If $\mathcal{M}$ has all of them then $\mathbf{L}_{\mathcal{M}'}$ is normal.*

4.3 Paraconsistent Logics

In this section, we define in precise terms the notion of *paraconsistency* which is used in this paper, as well some related desirable properties.

Definition 4.16 Let $\mathcal{L}$ be a language with a unary connective $\neg$, and let $\mathbf{L} = \langle \mathcal{L}, \vdash_{\mathbf{L}} \rangle$ be a logic for $\mathcal{L}$.

- $\mathbf{L}$ is called *pre-$\neg$-paraconsistent* if there are atoms p, q such that $p, \neg p \nvdash_{\mathbf{L}} q$.
- $\mathbf{L}$ is called *boldly pre-$\neg$-paraconsistent* if there are no formula σ and an atom $p \notin Atoms(\sigma)$ such that $p, \neg p \vdash_{\mathbf{L}} \sigma$ while $\nvdash_{\mathbf{L}} \sigma$.[4]

Since $\mathbf{L}$ is a *logic*, our definition of pre-$\neg$-paraconsistency can easily be seen to be equivalent to da-Costa's definition of paraconsistency [19], which requires that there would be a theory $\mathcal{T}$ and formulas ψ, φ in $\mathcal{W}(\mathcal{L})$ such that $\mathcal{T} \vdash_{\mathbf{L}} \psi$, $\mathcal{T} \vdash_{\mathbf{L}} \neg\psi$, but $\mathcal{T} \nvdash_{\mathbf{L}} \varphi$. Both of these definitions intend to capture the idea that a contradictory set of premises should not entail every formula. However, talking about "contradictory set" makes sense only if the underlying connective $\neg$ somehow represents a "negation" operation. This is assured by the condition of "coherence with classical logic," which is defined next. Intuitively, this condition states that a logic that has such a connective should not admit entailments that do not hold in classical logic.

[4] This is a variant of a notion from [16].

Definition 4.17 Let $\mathcal{L}$ be a language with a unary connective $\neg$. A *bivalent $\neg$-interpretation for* $\mathcal{L}$ is a function $\mathbf{F}$ that associates a two-valued truth-table with each connective of $\mathcal{L}$, such that $\mathbf{F}(\neg)$ is the classical truth-table for negation. We denote by $\mathcal{M}_{\mathbf{F}}$ the two-valued matrix for $\mathcal{L}$ induced by $\mathbf{F}$, that is, $\mathcal{M}_{\mathbf{F}} = \langle\{t, f\}, \{t\}, \mathbf{F}\rangle$ (see Definition 4.7).

Definition 4.18 Let $\mathbf{L} = \langle\mathcal{L}, \vdash\rangle$ be a propositional logic where $\mathcal{L}$ contains a unary connective $\neg$.

- Let $\mathbf{F}$ be a bivalent $\neg$-interpretation for $\mathcal{L}$. $\mathbf{L}$ is $\mathbf{F}$-*contained in classical logic* if the following holds for every $\varphi_1, \ldots, \varphi_n, \psi \in \mathcal{W}(\mathcal{L})$: if $\varphi_1, \ldots \varphi_n \vdash_{\mathbf{L}} \psi$ then $\varphi_1, \ldots, \varphi_n \vdash_{\mathcal{M}_{\mathbf{F}}} \psi$.
- [3] $\mathbf{L}$ is $\neg$ *-contained in classical logic*, if it is $\mathbf{F}$-contained in it for some bivalent $\neg$-interpretation $\mathbf{F}$.
- $\mathbf{L}$ is $\neg$*-coherent with classical logic*, if it has a seminormal fragment (Definition 4.5) which is $\neg$-contained in classical logic.

Definition 4.19 Let $\mathbf{L} = \langle\mathcal{L}, \vdash\rangle$ be a propositional logic where $\mathcal{L}$ contains a unary connective $\neg$. We say that $\neg$ is a *negation* of $\mathcal{L}$ if $\mathbf{L}$ is $\neg$-coherent with classical logic.

Note 4.20 If $\neg$ is a negation of $\mathbf{L} = \langle\mathcal{L}, \vdash_{\mathbf{L}}\rangle$, then for every atom p it holds that $p \nvdash_{\mathbf{L}} \neg p$ and $\neg p \nvdash_{\mathbf{L}} p$.

Definition 4.21 Let $\mathcal{L}$ be a language with a unary connective $\neg$, and let $\mathbf{L} = \langle\mathcal{L}, \vdash_{\mathbf{L}}\rangle$ be a logic for $\mathcal{L}$.

- $\mathbf{L}$ is called $\neg$*-paraconsistent* if it is pre-$\neg$-paraconsistent and $\neg$ is a negation of $\mathbf{L}$.
- $\mathbf{L}$ is called *boldly* $\neg$ *-paraconsistent* if it is boldly pre-$\neg$-paraconsistent, and $\neg$ is a negation of $\mathbf{L}$.

Henceforth, we shall frequently omit the $\neg$ sign (if it is clear from the context), and simply refer to (*boldly*) (*pre-*) *paraconsistent logics*.

Note 4.22 It should again be emphasized that our notion of paraconsistency has *two* components. In addition to the usual demand that a formula and its negation do not imply everything, we also demand that the "negation" connective under question can indeed be taken to be a sort of *negation*.

Paraconsistent logics reject the principle of explosion (known as *Ex Contradictione Sequitur Quodlibet*: $\mathcal{T}, \psi, \neg\psi \vdash \varphi$). Bold paraconsistency is a stronger version of this property. An even stronger demand is to reject explosion in *all* circumstances:

Definition 4.23 A logic $\langle\mathcal{L}, \vdash\rangle$ is *non-exploding* if for every theory $\mathcal{T}$ such that $\mathsf{Atoms}(\mathcal{T}) \neq \mathsf{Atoms}(\mathcal{L})$ there is a formula ψ such that $\mathcal{T} \nvdash \psi$.

Note 4.24 Obviously, every non-exploding logic which is $\neg$-coherent with classical logic is boldly paraconsistent.

There are many approaches to designing paraconsistent logics. One of the oldest and best known is Newton da-Costa's approach, which has led to the family of *Logics of Formal Inconsistency* (LFIs) [16]. Now, already in the early stages of investigating this topic, it has been acknowledged by da-Costa (and others) that pre-paraconsistency by itself is not sufficient. Further properties that an "ideal" paraconsistent logic is expected to have are defined in [3]. In the rest of this section, we briefly recall (with some improvements) these properties.

A. Reasonably Strong Language. Clearly, any logic (including paraconsistent ones) should have a sufficiently expressive language. The seminormality requirement (Definition 4.5) assures that in addition to negation, a useful paraconsistent logic should provide natural counterparts for all classical connectives:

Proposition 4.25 *Let* $\mathbf{L}$ *be a logic that is* $\mathbf{F}$*-contained in classical logic for some* $\mathbf{F}$*, and let* $\mathbf{F}(\diamond) = \diamond_{\mathbf{F}}$*. Then for every* $a, b \in \{t, f\}$ *we have*

1. *If* $\diamond$ *is an implication for* $\mathbf{L}$*, then* $a \diamond_{\mathbf{F}} b = f$ *if* $a = t$ *and* $b = f$*, otherwise* $a \diamond_{\mathbf{F}} b = t$*.*
2. *If* $\diamond$ *is a conjunction for* $\mathbf{L}$*, then* $a \diamond_{\mathbf{F}} b = t$ *if* $a = t$ *and* $b = t$*, otherwise* $a \diamond_{\mathbf{F}} b = f$*.*
3. *If* $\diamond$ *is a disjunction for* $\mathbf{L}$*, then* $a \diamond_{\mathbf{F}} b = t$ *if* $a = t$ *or* $b = t$*, otherwise* $a \diamond_{\mathbf{F}} b = f$*.*

Proof Let $\mathbf{F}$ be a bivalent interpretation for which $\mathbf{L}$ is $\mathbf{F}$-contained in classical logic.

1. Suppose that $\diamond$ is an implication for $\mathbf{L}$, and let $\mathbf{F}(\diamond) = \diamond_{\mathbf{F}}$. By Item (b) of Lemma 4.6–1, $\vdash_{\mathbf{L}} p \diamond p$. Hence, $\vdash_{\mathcal{M}_{\mathbf{F}}} p \diamond p$, and so necessarily $t \diamond_{\mathbf{F}} t = f \diamond_{\mathbf{F}} f = t$. Next, $p \vdash_{\mathbf{L}} q \diamond q$, and since $\diamond$ is an implication for $\mathbf{L}$, $\vdash_{\mathbf{L}} p \diamond (q \diamond q)$. Hence also $\vdash_{\mathcal{M}_{\mathbf{F}}} p \diamond (q \diamond q)$. Since $f \diamond_{\mathbf{F}} f = t$, this implies that $f \diamond_{\mathbf{F}} t = t$. Finally, by Item (a) of Lemma 4.6–1, $p \diamond q, p \vdash_{\mathbf{L}} q$. Hence, also $p \diamond q, p \vdash_{\mathcal{M}_{\mathbf{F}}} q$, and so $t \diamond_{\mathbf{F}} f = f$ (otherwise $\nu(p) = t, \nu(q) = f$ would be a counterexample).
2. Suppose that $\diamond$ is a conjunction for $\mathbf{L}$, and let $\mathbf{F}(\diamond) = \diamond_{\mathbf{F}}$. By Lemma 4.6–2, $p \diamond q \vdash_{\mathbf{L}} p$ and so also $p \diamond q \vdash_{\mathcal{M}_{\mathbf{F}}} p$. This implies that $f \diamond_{\mathbf{F}} t = f$ and $f \diamond_{\mathbf{F}} f = f$. Similarly, since $p \diamond q \vdash_{\mathbf{L}} q$, also $p \diamond q \vdash_{\mathcal{M}_{\mathbf{F}}} q$, and so $t \diamond_{\mathbf{F}} f = f$. Finally, by Lemma 4.6–2 again, $p, q \vdash_{\mathbf{L}} p \diamond q$ and so $p, q \vdash_{\mathcal{M}_{\mathbf{F}}} p \diamond q$, which implies that $t \diamond_{\mathbf{F}} t = t$ (otherwise, $\nu(p) = \nu(q) = t$ would be a counterexample).
3. Suppose that $\diamond$ is a disjunction for $\mathbf{L}$, and let $\mathbf{F}(\diamond) = \diamond_{\mathbf{F}}$. By Lemma 4.6–3, $p \vdash_{\mathbf{L}} p \diamond q$, and $q \vdash_{\mathbf{L}} p \diamond q$. Hence also $p \vdash_{\mathcal{M}_{\mathbf{F}}} p \diamond q$, and $q \vdash_{\mathcal{M}_{\mathbf{F}}} p \diamond q$, implying that $t \diamond_{\mathbf{F}} t = t \diamond_{\mathbf{F}} f = f \diamond_{\mathbf{F}} t = t$. Finally, by Item (c) of Lemma 4.6–3 the assumption that $\diamond$ is a disjunction for $\mathbf{L}$ implies that $p \diamond p \vdash_{\mathbf{L}} p$, and so $p \diamond p \vdash_{\mathcal{M}_{\mathbf{F}}} p$. It follows that $f \diamond_{\mathbf{F}} f = f$ (otherwise, $\nu(p) = f$ would be a counterexample). □

Corollary 4.26 *Let* $\mathcal{M} = \langle \mathcal{V}, \mathcal{D}, \mathcal{O} \rangle$ *be a matrix for* $\mathcal{L}$ *such that* $\mathbf{L}_{\mathcal{M}}$ *is* $\mathbf{F}$*-contained in classical logic.*

1. *If* $\wedge$ *is an* $\mathcal{M}$*-conjunction then* $\mathbf{F}(\wedge)$ *is the classical conjunction.*
2. *If* $\vee$ *is an* $\mathcal{M}$*-disjunction then* $\mathbf{F}(\vee)$ *is the classical disjunction.*
3. *If* $\supset$ *is an* $\mathcal{M}$*-implication then* $\mathbf{F}(\supset)$ *is the classical implication.*

Proof This follows from Propositions 4.25 and 4.14. □

Note 4.27 Let $\mathcal{M}$ be a martix for $\mathcal{L}$ such that $\mathbf{L}_{\mathcal{M}}$ is **F**-contained in classical logic. Suppose that $\mathcal{M}$ has a connective $\diamond$ of $\mathcal{L}$ which is either an $\mathcal{M}$-conjunction, an $\mathcal{M}$-disjunction, or an $\mathcal{M}$-implication. The last corollary implies that any two-valued function is then definable in terms of $\mathbf{F}(\diamond)$ and $\mathbf{F}(\neg)$. This shows the adequacy of the expressive power of such matrices.

B. Maximal Paraconsistency. A common requirement from a paraconsistent logic, which is already realized in da-Costa's seminal paper [19], is to "retain as much of classical logic as possible, while still allowing nontrivial inconsistent theories." As observed in [3, 4], this requirement has two different interpretations, corresponding to the two aspects of this demand:

B-1. Absolute Maximal Paraconsistency. Intuitively, this means that by trying to further extend the logic (without changing the language) we lose the property of paraconsistency.

Definition 4.28 Let $\mathbf{L} = \langle \mathcal{L}, \vdash \rangle$ be a $\neg$-paraconsistent logic

- We say that **L** is *maximally paraconsistent*, if every extension of **L** (in the sense of Definition 4.4) whose set of theorems *properly includes* that of **L**, is not pre-paraconsistent.
- We say that **L** is *strongly maximal*, if every proper extension of **L** (in the sense of Definition 4.4) is not pre-paraconsistent.

B-2. Maximality Relative to Classical Logic. The intuitive meaning of this property is that the logic is so close to classical logic, that any attempt to further extend it should necessarily end-up with classical logic.

Definition 4.29 Let **F** be a bivalent $\neg$-interpretation for a language $\mathcal{L}$ with a unary connective $\neg$.

- An $\mathcal{L}$-formula ψ is a *classical* **F** *-tautology*, if ψ is satisfied by every two-valued valuation which respects all the truth-tables (of the form $\mathbf{F}(\diamond)$) that **F** assigns to the connectives of $\mathcal{L}$.
- A logic $\mathbf{L} = \langle \mathcal{L}, \vdash \rangle$ is **F**-*complete*, if its set of theorems consists of all the classical **F**-tautologies.
- A logic **L** is **F**-*maximal relative to classical logic*, if the following hold:
  - **L** is **F**-contained in classical logic.
  - If ψ is a classical **F**-tautology not provable in **L**, then by adding ψ to **L** as a new axiom schema, an **F**-complete logic is obtained.
- A logic **L** is **F** *-maximally paraconsistent relative to classical logic*, if it is pre-paraconsistent and **F**-maximal relative to classical logic.

Definition 4.30 Let $\mathbf{L} = \langle \mathcal{L}, \vdash \rangle$ be a logic for a language with a unary connective $\neg$. We say that **L** is *maximally paraconsistent relative to classical logic* if there exists a bivalent $\neg$-interpretation **F** such that **L** is **F**-maximally paraconsistent relative to classical logic.

The two kinds of maximality are combined in the next definition.

Definition 4.31 We say that a seminormal finitary logic **L** is a *fully maximal* paraconsistent logic, if it is both maximally paraconsistent relative to classical logic and strongly maximal.

4.4 Three-Valued Paraconsistent Matrices

We now turn to the three-valued case, and investigate paraconsistent logics induced by three-valued matrices. We start with some general results.

Definition 4.32 Let $\mathcal{L}$ be a propositional language with a unary connective $\neg$. A matrix $\mathcal{M}$ for $\mathcal{L}$ is (*boldly, pre-*) $\neg$ *-paraconsistent* if so is $\mathbf{L}_{\mathcal{M}}$ (see Definitions 4.16 and 4.21).

Proposition 4.33 *Let $\mathcal{M} = \langle \mathcal{V}, \mathcal{D}, \mathcal{O} \rangle$ be a matrix for a language with $\neg$.*

1. *$\mathcal{M}$ is pre-paraconsistent iff there is an element $\top \in \mathcal{D}$, such that $\tilde{\neg}\top \in \mathcal{D}$.*
2. *If $\mathcal{M}$ is paraconsistent then there are three different elements t, f, and $\top$ in $\mathcal{V}$ such that $f = \tilde{\neg}t$, $f \notin \mathcal{D}$, and $\{t, \tilde{\neg}f, \top, \tilde{\neg}\top\} \subseteq \mathcal{D}$.*

Proof By its definition, $\mathcal{M}$ is pre-paraconsistent iff $p, \neg p \nvdash_{\mathcal{M}} q$. Obviously, this happens iff $\{p, \neg p\}$ has an $\mathcal{M}$-model. The latter, in turn, is possible iff there is some $\top \in \mathcal{D}$, such that $\tilde{\neg}\top \in \mathcal{D}$, as indicated in the first item of the proposition. For the second item, we may assume without loss in generality that $\mathcal{M}$ is $\neg$-contained in classical logic. We let $\mathbf{F}$ be a bivalent $\neg$-interpretation such that $\mathbf{L}_{\mathcal{M}}$ is $\mathbf{F}$-contained in classical logic. Since $p, \neg\neg p \nvdash_{\mathcal{M}_{\mathbf{F}}} \neg p$, also $p, \neg\neg p \nvdash_{\mathcal{M}} \neg p$, and so there is some $t \in \mathcal{D}$, such that $\tilde{\neg}t \notin \mathcal{D}$, while $\tilde{\neg}\tilde{\neg}t \in \mathcal{D}$. Let $f = \tilde{\neg}t$. Then t and f have the required properties, and together with the first item we are done. □

Corollary 4.34 *Any paraconsistent matrix is boldly paraconsistent.*

Proof Suppose that $\mathcal{M} = \langle \mathcal{V}, \mathcal{D}, \mathcal{O} \rangle$ is a paraconsistent matrix, σ is a formula in its language such that $\nvdash_{\mathcal{M}} \sigma$, and p is an atomic formula such that $p \notin \mathsf{Atoms}(\sigma)$. Then there is a valuation ν such that $\nu(\sigma) \notin \mathcal{D}$. Let $\top$ be an element of $\mathcal{V}$ like in the first item of Proposition 4.33. Define a valuation ν' by letting $\nu'(p) = \top$, and $\nu'(q) = \nu(q)$ for every atomic formula $q \neq p$. Then $\nu'(\sigma) = \nu(\sigma) \notin \mathcal{D}$. Hence ν' is an $\mathcal{M}$-model of $\{\neg p, p\}$ which is not an $\mathcal{M}$-model of σ, and so $\{\neg p, p\} \nvdash_{\mathcal{M}} \sigma$. It follows that $\mathcal{M}$ is boldly paraconsistent. □

By the second item of Proposition 4.33, we have

Corollary 4.35 *Every paraconsistent matrix has at least two designated elements, and so no two-valued matrix can be paraconsistent.*

The last corollary vindicates the general wisdom that truth-functional semantics of a reasonable paraconsistent logic should be based on at least three truth-values. The structure of paraconsistent matrices with exactly three values is characterized next.

Proposition 4.36 *Let $\mathcal{M}$ be a three-valued paraconsistent matrix. Then $\mathcal{M}$ isomorphic to a matrix $\mathcal{M}' = \langle \mathcal{V}, \mathcal{D}, \mathcal{O} \rangle$ in which $\mathcal{V} = \{t, f, \top\}$, $\mathcal{D} = \{t, \top\}$, $\neg t = f$, $\neg f = t$ and $\neg \top \in \mathcal{D}$.*

Proof By Proposition 4.33, we only need to show that $\tilde{\neg} f \neq \top$. Assume for contradiction that $\tilde{\neg} f = \top$. This implies that $\tilde{\neg}\tilde{\neg}\tilde{\neg}\top = \top$, no matter whether $\tilde{\neg}\top = \top$ or $\tilde{\neg}\top = t$. This and the facts that $\mathcal{D} = \{t, \top\}$ and $\tilde{\neg}\top \in \mathcal{D}$ imply that $p \vdash_{\mathcal{M}} \neg\neg\neg p$, which contradicts the $\neg$-coherence of $\mathcal{M}$ with classical logic. □

In the rest of the paper, we assume that any three-valued paraconsistent matrix has the form described in Proposition 4.36.

Next, we provide an effective necessary and sufficient criterion for checking which paraconsistent matrix is also non-exploding.

Proposition 4.37 *Let $\mathcal{M} = \langle \mathcal{V}, \mathcal{D}, \mathcal{O} \rangle$ be a paraconsistent 3-valued matrix. Then $\mathbf{L}_{\mathcal{M}}$ is non-exploding iff every connective $\diamond$ of $\mathcal{M}$ is $\{\top\}$-closed (i.e., $\tilde{\diamond}(\top, \ldots, \top) = \top$).*

Proof Suppose that every connective of $\mathcal{M}$ is $\{\top\}$-closed. Let $\mathcal{T}$ be a theory and q an atomic formula such that $q \notin \mathsf{Atoms}(\mathcal{T})$. Let ν be an assignment in $\mathcal{M}$ such that $\nu(p) = \top$ for every $p \in \mathsf{Atoms}(\mathcal{T})$, while $\nu(q) = f$. Since every connective of $\mathcal{M}$ is $\{\top\}$-closed, $\nu(\varphi) = \top$ for every $\varphi \in \mathcal{T}$. Hence ν is a model of $\mathcal{T}$ which is not a model of q. It follows that $\mathcal{T} \nvdash_{\mathcal{M}} q$.

For the converse, assume that there is an n-ary connective $\diamond$ of the language of $\mathcal{M}$ such that $\tilde{\diamond}$ is not $\{\top\}$-closed. Then $S = \{P_1, \neg P_1, \diamond(P_1, \ldots, P_1), \neg\diamond(P_1, \ldots, P_1)\}$ has no models in $\mathcal{M}$, and so $S \vdash_{\mathcal{M}} \varphi$ for every φ. Hence $\mathbf{L}_{\mathcal{M}}$ is not non-exploding. □

By Proposition 4.36, it follows that there are exactly two possible definitions for negation connectives in three-valued paraconsistent matrices:

- Kleene's negation [27], in which $\tilde{\neg} t = f$, $\tilde{\neg} f = t$, $\tilde{\neg}\top = \top$, and
- Sette's negation [35], in which $\tilde{\neg} t = f$, $\tilde{\neg} f = t$, $\tilde{\neg}\top = t$.

The other basic connectives are characterized by the following proposition.

Proposition 4.38 *Let $\mathcal{M} = \langle \mathcal{V}, \mathcal{D}, \mathcal{O} \rangle$ be a paraconsistent three-valued matrix.*

1. *A connective $\wedge$ is a conjunction for $\mathbf{L}_{\mathcal{M}}$ iff it is an $\mathcal{M}$-conjunction.*
2. *A connective $\vee$ is a disjunction for $\mathbf{L}_{\mathcal{M}}$ iff it is an $\mathcal{M}$-disjunction.*
3. *A connective $\supset$ is an implication for $\mathbf{L}_{\mathcal{M}}$ iff it is an $\mathcal{M}$-implication.*

Proof In all cases, the "if" direction is shown in Proposition 4.14. Below we prove the "only if" directions.

1. Immediate from Proposition 4.14.
2. Assume that $\vee$ is a disjunction for $\mathbf{L}_{\mathcal{M}}$. By Lemma 4.6–3, $\varphi \vdash_{\mathbf{L}_{\mathcal{M}}} \varphi \vee \psi$ and $\psi \vdash_{\mathbf{L}_{\mathcal{M}}} \varphi \vee \psi$. This implies that if either $a \in \mathcal{D}$ or $b \in \mathcal{D}$ then $a \tilde{\vee} b \in \mathcal{D}$. On the other hand, Item (c) of Lemma 4.6–3 entails that $\varphi \vee \varphi \vdash_{\mathbf{L}_{\mathcal{M}}} \varphi$, implying that $f \tilde{\vee} f = f$. It follows that $\vee$ is an $\mathcal{M}$-disjunction.
3. Assume that $\supset$ is an implication for $\mathbf{L}_{\mathcal{M}}$. By Item (a) in Lemma 4.6–1, $a \tilde{\supset} f = f$ for $a \in \mathcal{D}$, by Item (b) of the same lemma $f \tilde{\supset} f \in \mathcal{D}$, and by Item (c), $b \tilde{\supset} a \in \mathcal{D}$ for $a \in \mathcal{D}$. □

Corollary 4.39 *If $\mathcal{M}$ is a paraconsistent three-valued matrix, then $\mathbf{L}_{\mathcal{M}'}$ is paraconsistent for every expansion $\mathcal{M}'$ of $\mathcal{M}$. Moreover, if $\supset$ (respectively, if $\vee$, $\wedge$) is an implication (respectively, a disjunction, conjunction) for $\mathbf{L}_{\mathcal{M}}$, then it is also an implication (respectively, a disjunction, conjunction) for $\mathbf{L}_{\mathcal{M}'}$.*

Proof Immediate from Proposition 4.38 and Corollary 4.15. □

Corollary 4.40 *If $\mathcal{M}$ is a paraconsistent three-valued matrix, then $\mathbf{L}_{\mathcal{M}}$ is seminormal.*

Proof By definition of paraconsistency, if $\mathcal{M}$ is a paraconsistent then it has a paraconsistent seminormal fragment. Hence the claim follows from Corollary 4.39. □

We now give some general characterizations of logics which are induced by three-valued paraconsistent matrices, with particular emphasis on those which are actually $\neg$-contained in classical logic (and not just $\neg$-coherent with it). Our first result is the following:

Theorem 4.41 *Let $\mathcal{M} = \langle \mathcal{V}, \mathcal{D}, \mathcal{O} \rangle$ be a 3-valued $\neg$-paraconsistent matrix for a language $\mathcal{L}$. If $\mathcal{M}$ is $\neg$-contained in classical logic then $\mathcal{M}$ is classically closed (i.e., $\{t, f\}$-closed).*

Proof By Proposition 4.36, $\mathcal{V} = \{t, f, \top\}$, $\mathcal{D} = \{t, \top\}$, $f = \tilde{\neg} t$, $t = \tilde{\neg} f$, and $\tilde{\neg} \top \in \{t, \top\}$. Therefore, we have two cases to consider.

$\tilde{\neg}_{\mathcal{M}}$ is **Sette's negation**: Assume for contradiction that $\mathcal{M}$ is not classically closed, then $\tilde{\neg} \top = t$ and there is a connective $\diamond$ and $a_1, \ldots, a_n \in \{t, f\}$ such that $\tilde{\diamond}(a_1, \ldots, a_n) = \top$. For $i = 1, \ldots, n$ let $r_i = p_i$ if $a_i = t$ and $r_i = \neg p_i$ if $a_i = f$. Then, for every valuation $\nu \in \Lambda_{\mathcal{M}}$, if $\nu(p_i) = t$ for every $1 \leq i \leq n$ then $\nu(\diamond(r_1, \ldots, r_n)) = \top$. Let now $S = \{p_1, \neg\neg p_1, p_2, \neg\neg p_2, \ldots p_n, \neg\neg p_n\}$. Then $\upsilon \models_{\mathcal{M}} S$ iff $\nu(p_1) = \cdots = \nu(p_n) = t$. It follows that $S \vdash_{\mathcal{M}} \diamond(r_1, \ldots, r_n)$ and $S \vdash_{\mathcal{M}} \neg \diamond (r_1, \ldots, r_n)$. Since $\mathcal{M}$ is $\neg$-contained in classical logic, $S \vdash_{\mathcal{M}_{\mathbf{F}}} \diamond(r_1, \ldots, r_n)$ and $S \vdash_{\mathcal{M}_{\mathbf{F}}} \neg \diamond (r_1, \ldots, r_n)$ for some bivalent $\neg$-interpretation $\mathbf{F}$ for $\mathcal{L}$. This means that S is classically unsatisfiable, but this is false.

$\tilde{\neg}_{\mathcal{M}}$ is **Kleene's negation**: First, we show that the fact that $\mathcal{M}$ is seminormal (Corollary 4.40) entails in this case that it has an $\mathcal{M}$-disjunction. We do this by considering all the three possible cases.

- Suppose that $\mathbf{L}_{\mathcal{M}}$ has a disjunction connective $\vee$. Then $\vee$ is also an $\mathcal{M}$-disjunction by Proposition 4.38.
- Suppose that $\mathbf{L}_{\mathcal{M}}$ has a an $\mathcal{M}$-implication $\supset$. Then $\supset$ is an $\mathcal{M}$-implication by Proposition 4.38. This easily implies that the connective $\vee$ defined by $\varphi \vee \psi = (\varphi \supset \psi) \supset \psi$ is an $\mathcal{M}$-disjunction.
- Suppose that $\mathbf{L}_{\mathcal{M}}$ has a conjunction $\wedge$. Then $\wedge$ is an $\mathcal{M}$-conjunction by Proposition 4.38, and so we have $(*)$ $a\tilde{\wedge}b = f$ iff either $a = f$ or $b = f$.
First, we prove that $t\tilde{\wedge}t = t$. Assume otherwise. Then $t\tilde{\wedge}t = \top$ by $(*)$ above. Hence, if $\nu \in \Lambda_{\mathcal{M}}$ then $\nu(\neg(p \wedge p)) \in \mathcal{D}$ in case $\nu(p) = t$. By $(*)$ again, this implies that $\nu(\neg(p \wedge p)) \in \mathcal{D}$ in case $\nu(p) \in \{t, f\}$. On the other hand, if $\nu(p) = \top$ then $\nu(p) = \nu(\neg p)$, and so $\nu(\neg(p \wedge p)) = \nu(\neg(p \wedge \neg p))$. It follows that $\neg(p \wedge \neg p) \vdash_{\mathcal{M}} \neg(p \wedge p)$. Since $\mathcal{M}$ is $\neg$-contained in classical logic, $\neg(p \wedge \neg p) \vdash_{\mathcal{M}_{\mathrm{F}}} \neg(p \wedge p)$, which is false.
Next, we show that using $\neg$ and $\wedge$, it is possible to define in $\mathcal{L}$ an $\mathcal{M}$-disjunction $\vee$. We have two cases to consider:

  - $\top\tilde{\wedge}\top = t$:
  In this case we take $\varphi \vee \psi =_{Df} \neg(\neg(\varphi \wedge \varphi) \wedge \neg(\psi \wedge \psi))$. The fact that $t\tilde{\wedge}t = \top\tilde{\wedge}\top = t$ and (*) easily imply that this formula has the required property.
  - $\top\tilde{\wedge}\top = \top$:
  In this case we first let $\mathbf{t}_{\varphi,\psi}$ abbreviate $\neg(\varphi \wedge \neg\varphi \wedge \psi \wedge \neg\psi)$ (where association of conjunction is taken to the right,). Then $\nu(\mathbf{t}_{\varphi,\psi}) = \top$ in case that $\nu(\varphi) = \nu(\psi) = \top$, and $\nu(\mathbf{t}_{\varphi,\psi}) = t$ otherwise. Now, we take:

  $$\varphi \vee \psi =_{Df} \neg(\neg(\mathbf{t}_{\varphi,\psi} \wedge \varphi \wedge \mathbf{t}_{\varphi,\psi}) \wedge \neg(\mathbf{t}_{\varphi,\psi} \wedge \psi \wedge \mathbf{t}_{\varphi,\psi})).$$

  We show that this formula has in this case the required property:

    * Suppose first that $\nu(\varphi) = \nu(\psi) = f$. Since for every x, $x\tilde{\wedge}f = f\tilde{\wedge}x = f$, we have that $\nu(\mathbf{t}_{\varphi,\psi} \wedge \varphi \wedge \mathbf{t}_{\varphi,\psi}) = \nu(\mathbf{t}_{\varphi,\psi} \wedge \psi \wedge \mathbf{t}_{\varphi,\psi}) = f$. Since $\tilde{\neg}f = t$, $t\tilde{\wedge}t = t$, and $\tilde{\neg}t = f$, it follows that in this case $\nu(\varphi \vee \psi) = f$.
    * Suppose that $\nu(\varphi) = t$. Then $\nu(\mathbf{t}_{\varphi,\psi}) = t$. Since $t\tilde{\wedge}t = t$, $\nu(\mathbf{t}_{\varphi,\psi} \wedge \varphi \wedge \mathbf{t}_{\varphi,\psi}) = t$. Again, since $\tilde{\neg}t = f$, $f\tilde{\wedge}x = f$, and $\tilde{\neg}f = t$, we conclude that in this case $\nu(\varphi \vee \psi) = t$.
    * Suppose that $\nu(\psi) = t$. Then again $\nu(\mathbf{t}_{\varphi,\psi}) = t$. Like in the previous case, this implies that $\nu(\varphi \vee \psi) = t$.
    * Suppose that $\nu(\varphi) = \nu(\psi) = \top$. Then $\nu(\mathbf{t}_{\varphi,\psi}) = \top$. Since $\tilde{\neg}\top = \top$ and $\top\tilde{\wedge}\top = \top$, $\nu(\sigma) = \top$ for every sub-formula σ of $\varphi \vee \psi$. Hence $\nu(\varphi \vee \psi) = \top$ as well.
    * Suppose that $\nu(\varphi) = f$, $\nu(\psi) = \top$. Then $\nu(\mathbf{t}_{\varphi,\psi}) = t$, and so we haver that $\nu(\varphi \vee \psi) = \tilde{\neg}(t \mathbin{\tilde{\wedge}} \tilde{\neg}((t \mathbin{\tilde{\wedge}} \top) \mathbin{\tilde{\wedge}} t))$. If $(t \mathbin{\tilde{\wedge}} \top) \mathbin{\tilde{\wedge}} t) = t$ (which is the case if either $t \mathbin{\tilde{\wedge}} \top = t$ or $\top \mathbin{\tilde{\wedge}} t = t$) then $\nu((\varphi \vee \psi) = t$, and if $(t \mathbin{\tilde{\wedge}} \top) \mathbin{\tilde{\wedge}} t) = \top$ (which is the case if $t \mathbin{\tilde{\wedge}} \top = \top \mathbin{\tilde{\wedge}} t = \top$) then $\nu(\varphi \vee \psi) = \top$. In both cases we are done.
    * The case where $\nu(\varphi) = \top$ and $\nu(\psi) = f$ is similar to the previous case.

We therefore have shown that $\mathcal{M}$ has an $\mathcal{M}$-disjunction. We show that this implies that $\mathcal{M}$ is classically closed. Assume for contradiction that it is not, then there is a connective $\diamond$ and elements $a_1, \dots, a_n \in \{t, f\}$, such that $\tilde{\diamond}(a_1, \dots, a_n) = \top$. For $i = 1, \dots, n$ let $r_i = p_i$ if $a_i = t$, $r_i = \neg p_i$ if $a_i = f$. Then for every $\nu \in \Lambda_{\mathcal{M}}$, if $\nu(p_i) = t$ for every $1 \le i \le n$ then $\nu(\diamond(r_1, \dots, r_n)) = \tilde{\diamond}(a_1, \dots, a_n) = \top$. Hence $\tilde{\neg}\tilde{\diamond}(a_1, \dots, a_n)$ is in $\{t, \top\}$. These two facts imply:

$$p_1, \dots, p_n \vdash_{\mathcal{M}} \neg p_1 \vee \dots \vee \neg p_n \vee \diamond(r_1, \dots, r_n)$$
$$p_1, \dots, p_n \vdash_{\mathcal{M}} \neg p_1 \vee \dots \vee \neg p_n \vee \neg \diamond (r_1, \dots, r_n)$$

Indeed, let ν be a model of $\{p_1, \dots, p_n\}$. If $\nu(p_i) \neq t$ for some i then $\nu(\neg p_i) \in \mathcal{D}$, and so ν is a model of the disjunctions on the right-hand sides. If $\nu(p_i) = t$ for all i then ν is a model of both $\tilde{\diamond}(r_1, \dots, r_n)$ and $\tilde{\neg}\tilde{\diamond}(r_1, \dots, r_n)$, and so again ν is a model of both right-hand sides. Now, since $\mathcal{M}$ is $\neg$-contained in classical logic, Corollary 4.26 entails that the above two facts remain true if we replace $\vdash_{\mathcal{M}}$ by $\vdash_{\mathcal{M}_\mathsf{F}}$ and interpret $\vee$ and $\neg$ as the classical disjunction and negation (respectively). However, this is impossible for any two-valued interpretation of $\diamond$. □

The next theorems characterize all the three-valued matrices which induce paraconsistent logics that are $\neg$-contained in classical logic and show how to construct all such matrices which induce (semi)normal logics in a language that contains an implication $\supset$, a conjunction $\wedge$, and a disjunction $\vee$.

Theorem 4.42 *There are exactly* 2^{13} (8192) *distinct normal paraconsistent logics in the language* $\mathcal{L}_{CL} = \{\neg, \wedge, \vee, \supset\}$ *which are* $\neg$*-contained in classical logic, induced by three-valued matrices, and in which* $\supset$ *is an implication,* $\wedge$*—a conjunction, and* $\vee$*—a disjunction. The corresponding matrices are those that belong to the following family* 8Kb *of matrices from* [16][5] *(where the notation "*$x \wr y$*" means that* x *and* y *are two optional values):*

$\tilde{\wedge}$	t	f	$\top$
t	t	f	$t \wr \top$
f	f	f	f
$\top$	$t \wr \top$	f	$t \wr \top$

$\tilde{\vee}$	t	f	$\top$
t	t	t	$t \wr \top$
f	t	f	$t \wr \top$
$\top$	$t \wr \top$	$t \wr \top$	$t \wr \top$

$\tilde{\supset}$	t	f	$\top$
t	t	f	$t \wr \top$
f	t	t	$t \wr \top$
$\top$	$t \wr \top$	f	$t \wr \top$

	$\tilde{\neg}$
t	f
f	t
$\top$	$t \wr \top$

Proof That the matrices above indeed exhaust all the possible cases follows from Propositions 4.36, 4.38, and Theorem 4.41. That all of them induce paraconsistent logics which are $\neg$-contained in classical logic easily follows from Proposition 4.33

[5] In [16] the language is extended with a consistency operator $\circ$, defined by $\tilde{\circ} t = t$, $\tilde{\circ} f = t$, and $\tilde{\circ} \top = f$.

(second part) and the fact that the $\{t, f\}$-reductions of the connectives yield bivalent $\neg$-interpretations. That they are all normal follows from Proposition 4.14.

It is also not difficult to show that all of the logics in **8Kb** are indeed different. For instance, suppose that $\vdash_1$ and $\vdash_2$ are two consequence relations induced by matrices with different interpretations for a disjunction. Below we check the possible cases for such different interpretations and show that in each case the logics that are obtained are indeed different. First, note that by Theorem 4.41 and Corollary 4.26, the two matrices coincide on $a\tilde{\vee}b$ whenever $a \in \{t, f\}$ and $b \in \{t, f\}$. Now,

1. Suppose that $\top\tilde{\vee}_1\top = \top$ while $\top\tilde{\vee}_2\top = t$.
   In this case $p, \neg p, q, \neg q \vdash_1 \neg(p \vee q)$, but this is not true for $\vdash_2$ (since in both cases all the models of the right-hand side assign $\top$ to p and q).
2. Suppose that $\top\tilde{\vee}_1\top = \top\tilde{\vee}_2\top \in \{t, \top\}$ and that $f\tilde{\vee}_1\top = \top$ while $f\tilde{\vee}_2\top = t$. Then $q, \neg q, \neg(p \vee q) \vdash_2 p$ (a model of the right-hand side must assign $\top$ to q, and since $f\tilde{\vee}_2\top = t$ it cannot assign f to p), while this is not true for $\vdash_1$ (a counter-model in this case assigns f to p and $\top$ to q).
3. The remaining cases are dual to the ones in the previous cases. □

Theorem 4.43 *Let $\mathcal{M}$ be a three-valued matrix for a language with a unary connective $\neg$.*

1. $\mathcal{M}$ induces a $\neg$-paraconsistent logic which is $\neg$-contained in classical logic iff it is isomorphic to a matrix of the form $\langle\{t, f, \top\}, \{t, \top\}, \mathcal{O}\rangle$ which satisfies the following conditions:

   (a) It has as its interpretation of $\neg$ one of the two tables for $\neg$ given in Theorem 4.42;
   (b) It has a (possibly definable) connective whose interpretation is either one of the 2^3 possible interpretations for conjunction ($\wedge$) given in Theorem 4.42, or one of the 2^5 interpretations for disjunction ($\vee$) given there, or one of the 2^4 interpretations for implication ($\supset$) given there;
   (c) All its connectives are classically closed: $\tilde{\diamond}(a_1, \ldots, a_n) \in \{t, f\}$ for all $a_1, \ldots, a_n \in \{t, f\}$.

2. $\mathcal{M}$ induces a $\neg$-paraconsistent logic iff it is isomorphic to a matrix of the form $\langle\{t, f, \top\}, \{t, \top\}, \mathcal{O}\rangle$ which satisfies Conditions (a) and (b) above.

3. $\mathcal{M}$ induces a normal *$\neg$-paraconsistent logic which is $\neg$-contained in classical logic iff it is isomorphic to a matrix of the form $\langle\{t, f, \top\}, \{t, \top\}, \mathcal{O}\rangle$ which satisfies the following conditions:*

   (a) It has its interpretation of $\neg$ one of the two tables for $\neg$ given in Theorem 4.42;
   (b) It has a (possibly definable) connective whose interpretation is one of the 2^3 possible interpretations for $\wedge$ given in Theorem 4.42, and a connective whose interpretation is one of the 2^5 interpretations for $\vee$ given there, and a connective whose interpretation is one of the 2^4 interpretations for $\supset$ given there;
   (c) All its connectives are classically closed (i.e., $\{t, f\}$-closed).

Proof That the conditions in Parts 1 and 3 are necessary again follows from Propositions 4.36, 4.38, and Theorem 4.41. That they are sufficient again follows from Proposition 4.33 (second part), the fact that the $\{t, f\}$-reductions of the connectives yield bivalent $\neg$-interpretations and Proposition 4.14. Part 2 follows from Part 1 and Corollary 4.39. □

Note 4.44 Although all the logics which are induced by the matrices in the family $\mathsf{8Kb}$ are different from each other, some of them have the same expressive power. For instance, consider any paraconsistent matrix for the language of $\{\neg, \supset\}$ in which $\tilde{\neg}\top = \top$ and $\tilde{\supset}$ is D'Ottaviano and da-Costa's implication [19, 21], defined as follows:

$$a \supset b = \begin{cases} b & \text{if } a \neq f, \\ t & \text{if } a = f. \end{cases}$$

In this language, the formulas $\neg(\varphi \supset \neg\psi)$ and $\neg(\psi \supset \neg\varphi)$ define two different conjunctions. Hence the corresponding matrices in the family $\mathsf{8Kb}$ are equivalent in their expressive power.

We conclude this section with a theorem about the desirable maximal paraconsistency properties (Sect. 4.3) that three-valued $\neg$-paraconsistent logics enjoy:

Theorem 4.45 *Let $\mathcal{M}$ be a three-valued paraconsistent matrix. Then*

1. $\mathbf{L}_{\mathcal{M}}$ *is strongly maximal.*
2. *If $\mathcal{M}$ is $\neg$-contained in classical logic then it is also maximally paraconsistent relative to classical logic (and so it is fully maximal).*

Note 4.46 The first part of Theorem 4.45 is a generalization of [3, Theorem 2] (see also [4, Theorem 3.2]) and the second part of the theorem is a generalization of [3, Theorem 1]. In both cases, the proofs given below are similar to the ones given in [3]. To keep this paper complete, we repeat those proofs and adjust them to the more general case considered here.

Proof Let $\mathcal{M}$ be a three-valued paraconsistent matrix for a language $\mathcal{L}$. To see the first item of the theorem, note first that Theorem 4.43 implies that $\mathcal{M}$ has a classically closed binary connective $\diamond$ (from those listed in Theorem 4.42), which is either an $\mathcal{M}$-disjunction, or an $\mathcal{M}$-conjunction, or an $\mathcal{M}$-implication. Let $\Psi(p)$ be $\neg p \diamond p$ in the first case, $\neg(\neg p \diamond p)$ in the second one, and $p \diamond p$ in the third case. Then for all $\nu \in \Lambda_{\mathcal{M}}$, $\nu(\Psi) = t$ if $\nu(p) \neq \top$.

Now let $\langle \mathcal{L}, \vdash \rangle$ be a proper extension of $\mathbf{L}_{\mathcal{M}}$ by some set of rules. We show that $\langle \mathcal{L}, \vdash \rangle$ is not pre-paraconsistent. Let Γ be a finite theory and ψ a formula in $\mathcal{L}$ such that $\Gamma \vdash \psi$ but $\Gamma \nvdash_{\mathcal{M}} \psi$. In particular, there is a valuation $\nu \in mod_{\mathcal{M}}(\Gamma)$ such that $\nu(\psi) = f$. Consider the substitution θ, defined for every $p \in \mathsf{Atoms}(\Gamma \cup \{\psi\})$ by

$$\theta(p) = \begin{cases} q_0 & \text{if } \nu(p) = t, \\ \neg q_0 & \text{if } \nu(p) = f, \\ p_0 & \text{if } \nu(p) = \top, \end{cases}$$

where p_0 and q_0 are two different atoms in $\mathcal{L}$. Note that $\theta(\Gamma)$ and $\theta(\psi)$ contain (at most) the variables p_0, q_0, and that for every valuation $\mu \in \Lambda_{\mathcal{M}}$ where $\mu(p_0) = \top$ and $\mu(q_0) = t$ it holds that $\mu(\theta(\phi)) = \nu(\phi)$ for every formula ϕ such that $\mathsf{Atoms}(\{\phi\}) \subseteq \mathsf{Atoms}(\Gamma \cup \{\psi\})$. Thus,

($\star$) *any* $\mu \in \Lambda_{\mathcal{M}}$ such that $\mu(p_0) = \top$ and $\mu(q_0) = t$ is an $\mathcal{M}$-model of $\theta(\Gamma)$ but not of $\theta(\psi)$.

Now, consider the following two cases:

Case I There is a formula $\phi(p, q)$ (i.e., $\mathsf{Atoms}(\phi) = \{p, q\}$, where $p \neq q$) such that for every $\mu \in \Lambda_{\mathcal{M}}$, $\mu(\phi) \neq \top$ if $\mu(p) = \mu(q) = \top$.
In this case, let $\mathsf{tt} = \Psi(\phi(p_0, p_0))$. Note that $\mu(\mathsf{tt}) = t$ for every $\mu \in \Lambda_{\mathcal{M}}$ such that $\mu(p_0) = \top$. Now, as $\vdash$ is structural, $\Gamma \vdash \psi$ implies that

$$\theta(\Gamma)\,[\mathsf{tt}/q_0] \vdash \theta(\psi)\,[\mathsf{tt}/q_0]. \tag{4.1}$$

Also, by the above property of tt and by ($\star$), any $\mu \in \Lambda_{\mathcal{M}}$ for which $\mu(p_0) = \top$ is a model of $\theta(\Gamma)\,[\mathsf{tt}/q_0]$ but does not $\mathcal{M}$-satisfy $\theta(\psi)\,[\mathsf{tt}/q_0]$. Thus,

- $p_0, \neg p_0 \vdash_{\mathcal{M}} \theta(\gamma)\,[\mathsf{tt}/q_0]$ for every $\gamma \in \Gamma$. As $\langle \mathcal{L}, \vdash \rangle$ is stronger than $\langle \mathcal{L}, \vdash_{\mathcal{M}} \rangle$, this implies that
$$p_0, \neg p_0 \vdash \theta(\gamma)\,[\mathsf{tt}/q_0] \text{ for every } \gamma \in \Gamma. \tag{4.2}$$

- The set $\{p_0, \neg p_0, \theta(\psi)[\mathsf{tt}/q_0]\}$ is not $\mathcal{M}$-satisfiable, thus $p_0, \neg p_0, \theta(\psi)\,[\mathsf{tt}/q_0] \vdash_{\mathcal{M}} q_0$. Again, as $\langle \mathcal{L}, \vdash \rangle$ is stronger than $\langle \mathcal{L}, \vdash_{\mathcal{M}} \rangle$, we have that
$$p_0, \neg p_0, \theta(\psi)\,[\mathsf{tt}/q_0] \vdash q_0. \tag{4.3}$$

By (4.1)–(4.3) $p_0, \neg p_0 \vdash q_0$, thus $\langle \mathcal{L}, \vdash \rangle$ is not pre-paraconsistent.

Case II For every formula $\phi(p, q)$ and for every $\mu \in \Lambda_{\mathcal{M}}$, if $\mu(p) = \mu(q) = \top$ then $\mu(\phi) = \top$.
Again, as $\vdash$ is structural, and since $\Gamma \vdash \psi$,

$$\theta(\Gamma)\,[\Psi(q_0)/q_0] \vdash \theta(\psi)\,[\Psi(q_0)/q_0]. \tag{4.4}$$

In addition, ($\star$) above entails that any valuation $\mu \in \Lambda_{\mathcal{M}}$ such that $\mu(p_0) = \top$ and $\mu(q_0) \in \{t, f\}$ is a model of $\theta(\Gamma)\,[\Psi(q_0)/q_0]$ which is not a model of $\theta(\psi)\,[\Psi(q_0)/q_0]$. Thus, the only $\mathcal{M}$-model of $\{p_0, \neg p_0, \theta(\psi)\,[\Psi(q_0)/q_0]\}$ is the one in which both of p_0 and q_0 are assigned the value $\top$. It follows that $p_0, \neg p_0, \theta(\psi)\,[\Psi(q_0)/q_0] \vdash_{\mathcal{M}} q_0$. Thus,

$$p_0, \neg p_0, \theta(\psi)\,[\Psi(q_0)/q_0] \vdash q_0. \tag{4.5}$$

By using ($\star$) again (for $\mu(q_0) \in \{t, f\}$) and the condition of Case II (for $\mu(q_0) = \top$), we have

$$p_0, \neg p_0 \vdash \theta(\gamma)\,[\Psi(q_0)/q_0] \text{ for every } \gamma \in \Gamma. \tag{4.6}$$

Again, by (4.4)–(4.6) above we have that $p_0, \neg p_0 \vdash q_0$, and so $\langle \mathcal{L}, \vdash \rangle$ is not pre-paraconsistent in this case either.

For the second part of the theorem we need the following lemma.

Lemma 4.47 *Let $\mathcal{M}$ be a paraconsistent three-valued matrix, and suppose that there is some bivalent $\neg$-interpretation* **F** *such that* $\mathbf{L}_{\mathcal{M}}$ *is* **F***-contained in classical logic, but* $\mathbf{L}_{\mathcal{M}}$ *is not* **F***-maximal relative to classical logic. Then $\mathcal{M}$ is classically closed.*

Proof The assumption about **F** implies that there is some classical **F**-tautology (Definition 4.29) ψ_0 which is not provable in $\mathbf{L}_{\mathcal{M}}$, and by adding it as an axiom to $\mathbf{L}_{\mathcal{M}}$ we get a logic $\mathbf{L}^*$ that is *not* **F**-complete. Since $\mathbf{L}_{\mathcal{M}}$ is strongly maximal by the first part of this theorem (and $\mathbf{L}^*$ is an extension by a rule of $\mathbf{L}_{\mathcal{M}}$), $\varphi, \neg\varphi \vdash_{\mathbf{L}^*} \phi$ for every φ, ϕ. It follows that

$$\mathcal{S}^*, \varphi, \neg\varphi \vdash_{\mathcal{M}} \phi \text{ for every } \varphi, \phi \tag{4.7}$$

where $\mathcal{S}^*$ is the set of all substitution instances of ψ_0. Now, let σ be some classical **F**-tautology not provable in $\mathbf{L}^*$. So $\nvdash_{\mathbf{L}^*} \sigma$, and so $\mathcal{S}^* \nvdash_{\mathcal{M}} \sigma$. Hence there is a valuation $\nu \in \Lambda_{\mathcal{M}}$ which is a model of $\mathcal{S}^*$, but $\nu(\sigma) = f$. We show that there is no formula ψ for which $\nu(\psi) = \top$. Assume for contradiction that this is not the case for some ψ. Since ν is a model of $\mathcal{S}^*$, it is also a model of $\mathcal{S}^* \cup \{\psi, \neg\psi\}$, and so it is a model of σ by (4.7) above. This contradicts the fact that $\nu(\sigma) = f$. It follows that $\nu(\psi) \in \{t, f\}$ for all ψ. We show that this implies that all the operations of $\mathcal{M}$ are classically closed. Let $\diamond$ be some n-ary connective of $\mathcal{L}$ and let $a_1, \ldots, a_n \in \{t, f\}$. For $i = 1, \ldots, n$, define $\varphi_i = P_i$ if $\nu(p_i) = a_i$, and $\varphi_i = \neg P_i$ otherwise. Thus $\nu(\varphi_i) = a_i$, and $\tilde{\diamond}(a_1, \ldots, a_n) = \tilde{\diamond}(\nu(\varphi_1), \ldots, \nu(\varphi_n)) = \nu(\diamond(\varphi_1, \ldots, \varphi_n)) \in \{t, f\}$. □

Now we can show the second part of Theorem 4.45. The assumption that $\mathcal{M}$ is $\neg$-contained in classical logic entails that it is **F**-contained in classical logic for some **F**. If $\mathbf{L}_{\mathcal{M}}$ is **F**-maximal relative to classical logic, then we are done. Otherwise, $\mathcal{M}$ is classically closed by Lemma 4.47, and so we can consider the bivalent $\neg$-interpretation induced by $\mathcal{M}$, defined by $\mathbf{F}_{\mathcal{M}}(\diamond) = \tilde{\diamond}_{\mathcal{M}}/\{t, f\}^n$ (where n is the arity of $\diamond$, and $\tilde{\diamond}_{\mathcal{M}}/\{t, f\}^n$ is the reduction of $\tilde{\diamond}_{\mathcal{M}}$ to $\{t, f\}^n$). As the next lemma shows, **F** must be identical to this interpretation.

Lemma 4.48 $\mathbf{F} = \mathbf{F}_{\mathcal{M}}$

Proof Suppose otherwise.Then there is some n-ary connective $\diamond$ of $\mathcal{L}$ such that $\tilde{\diamond}/\{t, f\} = \mathbf{F}_{\mathcal{M}}(\diamond) \neq \mathbf{F}(\diamond)$. Hence there are some elements $a_1, \ldots, a_n \in \{t, f\}$ such that $\tilde{\diamond}(a_1, \ldots, a_n) \neq \mathbf{F}(\diamond)(a_1, \ldots, a_n)$. Because **F** and $\mathbf{F}_{\mathcal{M}}$ are both bivalent $\neg$-interpretations, we may assume without loss of generality that $\mathbf{F}(\diamond)(a_1, \ldots, a_n) = t$ and $\tilde{\diamond}(a_1, \ldots, a_n) = f$ (otherwise we consider $\neg\diamond$ instead of $\diamond$). Next, for $i = 1, \ldots, n$ we define $\varphi_i = p$ if $a_i = t$ and $\varphi_i = \neg p$ otherwise, thus $p, \diamond(\varphi_1, \ldots, \varphi_n) \vdash_{\mathcal{M}} \neg p$, while $p, \diamond(\varphi_1, \ldots, \varphi_n) \nvdash_{\mathcal{M}_\mathbf{F}} \neg p$ (because $\nu(p) = t$ provides a counterexample). This contradicts the **F**-containment of $\mathbf{L}_{\mathcal{M}}$ in classical logic. □

Now, by the lemma above, $\mathbf{L}_{\mathcal{M}}$ is $\mathbf{F}_{\mathcal{M}}$-contained in classical logic. We end by showing that $\mathbf{L}_{\mathcal{M}}$ is $\mathbf{F}_{\mathcal{M}}$-maximal relative to classical logic. The proof of this is very similar to the proof of Lemma 4.47: Let ψ' be a classical $\mathbf{F}_{\mathcal{M}}$-tautology not provable in $\mathbf{L}_{\mathcal{M}}$, and let $\mathcal{S}'^*$ be the set of all of its substitution instances. Let $\mathbf{L}'^*$ be the logic obtained by adding ψ' as a new axiom to $\mathbf{L}_{\mathcal{M}}$. Then for every theory $\mathcal{T}$ we have that $\mathcal{T} \vdash_{\mathbf{L}'^*} \phi$ iff $\mathcal{T}, \mathcal{S}'^* \vdash_{\mathcal{M}} \phi$. In particular, since $\mathbf{L}_{\mathcal{M}}$ is strongly maximal, Condition (4.7) from the proof of Lemma 4.47 holds for $\mathcal{S}'^*$. Suppose for contradiction that there is some classical $\mathbf{F}_{\mathcal{M}}$-tautology σ not provable in $\mathbf{L}'^*$. Since $\nvdash_{\mathbf{L}'^*} \sigma$, also $\mathcal{S}'^* \nvdash_{\mathcal{M}} \sigma$. Hence, there is a valuation $\nu \in \Lambda_{\mathcal{M}}$ which is a model of $\mathcal{S}'^*$, but $\nu(\sigma) = f$. If there is some ψ such that $\nu(\psi) = \top$, then since ν is a model of $\mathcal{S}'^*$, it is also a model of $\mathcal{S}'^* \cup \{\psi, \neg\psi\}$, and so by (4.7) it is a model of σ, in contradiction to the fact that $\nu(\sigma) = f$. Otherwise, $\nu(\psi) \in \{t, f\}$ for all ψ, and so ν is an $\mathcal{M}_{\mathbf{F}_{\mathcal{M}}}$-valuation, which assigns f to σ. This contradicts the fact that $\vdash_{\mathcal{M}_{\mathbf{F}_{\mathcal{M}}}} \sigma$. Hence, all classical $\mathbf{F}_{\mathcal{M}}$-tautologies are provable in $\mathbf{L}'^*$, and so $\mathbf{L}_{\mathcal{M}}$ is $\mathbf{F}_{\mathcal{M}}$-maximal relative to classical logic. □

Note 4.49 Suppose that $\mathcal{M}$ is a three-valued paraconsistent matrix which is $\neg$-contained in classical logic. Then any three-valued expansion of it which is obtained by enriching the language of $\mathcal{M}$ with extra classically closed connectives necessarily has the same properties (see Theorem 4.43). It follows that not only is $\mathbf{L}_{\mathcal{M}}$ fully maximal, but so must be also all the logics induced by its expansions that are so obtained.

4.5 The Most Important Paraconsistent Three-Valued Logics

As shown in the previous section, there are exactly eight ways of defining conjunctions in three-valued paraconsistent matrices. Of these eight operations, only four are symmetric. Of these four, only two are $\{\top\}$-closed, and to the best of our knowledge, only three (including these two) have been seriously investigated in the literature. In this section we examine in greater detail the properties of the most important (and famous) three-valued paraconsistent logics that are based on these three symmetric conjunctions and the two possible negations. Then in the next section, we shall show that each of these logics has a corresponding cut-free Gentzen-type system, which is very close to the classical one.

Our main criterion here for "importance" of three-valued paraconsistent matrices is having a *natural* set of connectives that can be characterized by a combination of potentially desirable properties. The most important such property is of course $\{t, f\}$-closure, which by Theorem 4.43 is equivalent to $\neg$-containment in classical logic. Another important property is $\{\top\}$-closure, which by Proposition 4.37 is equivalent to being non-exploding. Other properties are introduced and used in the sequel.

4.5.1 *The Logic* $\mathbf{P_1}$

Sette's logic $\mathbf{P_1} = \langle \mathcal{L}_{\mathsf{P}_1}, \vdash_{\mathsf{P}_1} \rangle$ [35] is induced by the matrix $\mathsf{P}_1 = \langle \{t, f, \top\}, \{t, \top\}, \{\tilde{\wedge}, \tilde{\neg}\} \rangle$,[6] where the operations are defined as follows:

$\tilde{\wedge}$	t	f	$\top$
t	t	f	t
f	f	f	f
$\top$	t	f	t

$\tilde{\neg}$	
t	f
f	t
$\top$	t

Proposition 4.50 $\mathbf{P_1}$ *is boldly paraconsistent, normal,* $\neg$*-contained in classical logic, and fully maximal.*

Proof Define $\psi \vee \phi = \neg(\neg(\psi \wedge \psi) \wedge \neg(\phi \wedge \phi))$ and $\psi \supset \phi = \neg((\psi \wedge \psi) \wedge \neg(\phi \wedge \phi))$. The corresponding interpretations are the following:

$\tilde{\vee}$	t	f	$\top$
t	t	t	t
f	t	f	t
$\top$	t	t	t

$\tilde{\supset}$	t	f	$\top$
t	t	f	t
f	t	t	t
$\top$	t	f	t

Therefore, Item 3 of Theorem 4.43 implies that $\mathbf{P_1}$ is a normal paraconsistent logic, which is $\neg$-contained in classical logic. The other properties follow from Corollary 4.34 and Theorem 4.45. □

Note 4.51 As far as we know, $\mathbf{P_1}$ was the first paraconsistent logic for which a maximality property has been stated and proved (in [35]). Therefore, it is frequently referred to as "Sette maximal paraconsistent logic." However, the results in Sect. 4.4 show that there is nothing special about $\mathbf{P_1}$ in this respect. Its maximality is just one (out of thousands) instances of Theorem 4.45.

The next theorem characterizes the expressive power of the language of $\mathbf{P_1}$.

Theorem 4.52 *A function* $g : \{t, f, \top\}^n \to \{t, f, \top\}$ *is representable in* $\mathcal{L}_{\mathsf{P}_1}$ *iff its range is* $\{t, f\}$.

Proof Obviously, the condition is necessary. To show that it is also sufficient, define:

$$\psi_a(p) = \begin{cases} p \wedge \neg\neg p & \text{if } a = t \\ \neg(p \wedge p) & \text{if } a = f \\ p \wedge \neg p & \text{if } a = \top \end{cases}$$

It is easy to check that if ν is a valuation in $\mathbf{P_1}$, then $\nu \models_{\mathbf{P_1}} \psi_a(p)$ iff $\nu(p) = a$. Now, given a function $g : \{t, f, \top\}^n \to \{t, f\}$, it is not difficult to see that g is represented in

[6] Note that in our notations $\mathbf{P_1}$ is also denoted $\mathbf{L}_{\mathsf{P}_1}$.

$\mathcal{L}_{\mathsf{P}_1}$ by the disjunction (as defined in the proof of Proposition 4.50) of all the formulas of the form $\psi_{a_1}(P_1) \wedge \psi_{a_2}(P_2) \wedge \cdots \wedge \psi_{a_n}(P_n)$ such that $g(a_1, a_2, \ldots, a_n) = t$ (and by the formula $\neg\neg P_1 \wedge \neg P_1$ if no such $a_1, a_2, \ldots, a_n$ exist). □

Corollary 4.53 *The connectives defined in the proof of Proposition 4.50 are the* only *disjunction and implication definable in* $\mathbf{P_1}$.

Proof This easily follows from Theorems 4.42 and 4.52. □

Note 4.54 As noted previously, the Logic $\mathbf{P_1}$ has all the desirable properties mentioned in the previous section. Nevertheless, $\mathbf{P_1}$ also has the following two severe drawbacks:

- It is paraconsistent only with respect to atomic formulas (that is, for a nonatomic formula ψ we have that $\psi, \neg\psi \vdash_{\mathsf{P}_1} \varphi$, since nonatomic formulas get only values in $\{t, f\}$).
- The conjunction–negation combination does not always behave as expected, e.g., $\neg p \nvdash_{\mathsf{P}_1} \neg(p \wedge q)$.

The main source of these problematic features is the fact that Sette's negation (which is the negation used in $\mathbf{P_1}$) has the following drawbacks in comparison to Kleene's negation:

- It is explosive with respect to negated data: $\neg\varphi, \neg\neg\varphi \vdash_{\mathsf{P}_1} \psi$ for every φ, ψ.
- It is not right involutive: $p \nvdash_{\mathsf{P}_1} \neg\neg p$.

These drawbacks should be the reason why $\mathbf{P_1}$ is (to the best of our knowledge) the only three-valued paraconsistent logic considered in the literature whose negation is Sette's negation. Accordingly, all the other logics described in this section use Kleene's negation.

4.5.2 *The Logic* $\mathbf{SRM}_{\tilde{\rightarrow}}$

Another conjunction of the eight possible conjunctions listed in Sect. 4.4 has (implicitly) been used by Sobociński in his three-valued matrix [38]. This is the matrix $\mathcal{A}_1 = \langle\{t, f, \top\}, \{t, \top\}, \{\tilde{\otimes}, \tilde{\neg}\}\rangle$ for the language $\mathcal{IL} = \{\neg, \otimes\}$ in which $\tilde{\neg}$ is Kleene's negation, and $\tilde{\otimes}$ is Sobociński's conjunction, defined below:

$\tilde{\otimes}$	t	f	$\top$
t	t	f	t
f	f	f	f
$\top$	t	f	$\top$

$\tilde{\neg}$	
t	f
f	t
$\top$	$\top$

We denote by $\mathbf{SRM}_{\tilde{\rightarrow}}$ (or $\mathbf{SRMI}^{\mathbf{1}}_{\tilde{\rightarrow}}$) the logic that is induced by $\mathcal{A}_1$.

Note 4.55 The official language that was used in [38] (as well as in the literature on relevance logic) is $\{\neg, \rightarrow\}$, and the interpretation of $\rightarrow$ there was the following *Sobociński's implication*:

$$a \rightarrow_S b \;=\; \begin{cases} \top & \text{if } a = b = \top, \\ f & \text{if } a = t \text{ and } b \neq t, \text{ or } b = f \text{ and } a \neq f, \\ t & \text{otherwise.} \end{cases}$$

It is easy to see that $a \rightarrow_S b = \tilde{\neg}(a \,\tilde{\otimes}\, \tilde{\neg} b)$, while $a \,\tilde{\otimes}\, b = \tilde{\neg}(a \rightarrow_S \tilde{\neg} b)$. Hence, $\mathcal{IL}$ and $\mathcal{A}_1$ are equivalent to Sobociński's original language and matrix (respectively).

Note 4.56 It should be emphasized that $\mathbf{SRM}_{\tilde{\rightarrow}}$ is *not* identical to the logic introduced by Sobociński in [38]. That logic has only been *motivated* by the matrix $\mathcal{A}_1$. What Sobociński actually did in [38] is to axiomatize *the set of valid formulas of* $\mathcal{A}_1$ using a Hilbert-type system with Modus Ponens for $\rightarrow$ as the single rule of inference. In other words: his system is only *weakly* complete for $\mathcal{A}_1$. Thus, one cannot derived in it φ from $\varphi \otimes \psi$, even though $\varphi \otimes \psi \vdash_{\mathbf{SRM}_{\tilde{\rightarrow}}} \varphi$.[7]

The connective $\rightarrow$ of $\mathbf{SRM}_{\tilde{\rightarrow}}$ is not an implication for that logic (since $\varphi \rightarrow (\psi \rightarrow \varphi)$ is not valid in $\mathcal{A}_1$). Despite this we have

Proposition 4.57 $\mathbf{SRM}_{\tilde{\rightarrow}}$ *is non-exploding, normal, $\neg$-contained in classical logic, and fully maximal.*

Proof Define $\varphi \supset \psi = \varphi \rightarrow (\varphi \otimes \psi)$, where (as above) $\varphi \rightarrow \psi = \neg(\varphi \otimes \neg\psi)$. Then $\supset$ has in $\mathcal{A}_1$ the following interpretation:

$\tilde{\supset}$	t	f	$\top$
t	t	f	t
f	t	t	t
$\top$	t	f	$\top$

It follows that $\supset$ is an $\mathcal{A}_1$-implication. This implies that the connective $\vee$, defined by $\psi \vee \varphi = (\psi \supset \varphi) \supset \varphi$, is an $\mathcal{A}_1$-disjunction. Finally, $\otimes$ is an $\mathcal{A}_1$-conjunction. Therefore, Item 3 of Theorem 4.43 implies that $\mathbf{SRM}_{\tilde{\rightarrow}}$ is a normal paraconsistent logic which is $\neg$-contained in classical logic. The other properties follow from Theorem 4.45 and Proposition 4.37. □

The following theorem characterizes the expressive power of the language of $\mathbf{SRM}_{\tilde{\rightarrow}}$:

Theorem 4.58 [9] *The connectives that are definable in the language of* $\mathbf{SRM}_{\tilde{\rightarrow}}$ *are those that are both $\{\top\}$-closed and $\{\top\}$-limited (Definition 4.12).*

Note that by the last theorem, it follows that Kleene's conjunction (see next section) is *not* definable in the language of $\mathbf{SRM}_{\tilde{\rightarrow}}$ (since Kleene's conjunction is not $\{\top\}$-limited).

---

[7] Meyer has shown (see [1]) that Sobociński's system induces the $\{\neg, \rightarrow, \otimes\}$-fragment of the semi-relevant logic **RM**.

4.5.3 The Logic **LP** *and Its Main Monotonic Expansions*

The most popular conjunction used in three-valued paraconsistent logics and three-valued logics in general is Kleene's (strong) conjunction (see the truth-table below), and the most basic paraconsistent logic which is based on it is Asenjo–Priest's three-valued logic **LP** [5, 31–33]. This is the logic induced by the three-valued matrix $\mathsf{LP} = \langle\{t, f, \top\}, \{t, \top\}, \{\tilde{\wedge}, \tilde{\neg}\}\rangle$, where the truth-tables for $\neg$ and $\wedge$ are the following:

$\tilde{\wedge}$	t	f	$\top$
t	t	f	$\top$
f	f	f	f
$\top$	$\top$	f	$\top$

$\tilde{\neg}$	
t	f
f	t
$\top$	$\top$

The matrix LP also has a disjunction, defined by $\psi \vee \varphi = \neg(\neg\psi \wedge \neg\varphi)$. What is obtained is one of the possible interpretations of disjunction given in Theorem 4.43: the strong Kleene's disjunction, whose truth-table is the following:

$\tilde{\vee}$	t	f	$\top$
t	t	t	t
f	t	f	$\top$
$\top$	t	$\top$	$\top$

Note 4.59 A common way of defining and understanding the disjunction, conjunction, and negation of LP is with respect to total order $\leq_{\mathsf{t}}$ on $\{t, f, \top\}$, in which t is the maximal element and f is the minimal one. This order may be intuitively understood as reflecting differences in the amount of *truth* that each element exhibits. Here, $\tilde{\wedge}$ and $\tilde{\vee}$ are the meet and the join (respectively) of $\leq_{\mathsf{t}}$, and $\tilde{\neg}$ is order reversing with respect to $\leq_{\mathsf{t}}$.

Next, we introduce the simplest expansions of **LP**: those that are obtained by adding to its language the propositional constants which correspond to the truth-values that are used. We shall denote by f the one for which $\forall\nu \in \Lambda\ \nu(\mathsf{f}) = f$ and by $\top$ the constant for which $\forall\nu \in \Lambda\ \nu(\top) = \top$. (There is no need to consider also a constant for t, because such a constant and f are definable in terms of each other and $\neg$.)

Definition 4.60

- $\mathbf{LP}^{\mathsf{f}}$ is the logic induced by the expansion of the matrix LP to the language $\{\neg, \wedge, \vee, \mathsf{f}\}$ (or just $\{\neg, \wedge, \mathsf{f}\}$).
- $\mathbf{LP}^{\top}$ is the logic induced by the expansion of the matrix LP to the language $\{\neg, \wedge, \vee, \top\}$.
- $\mathbf{LP}^{\mathsf{f},\top}$ is the logic induced by the expansion of the matrix LP to the language $\{\neg, \wedge, \vee, \mathsf{f}, \top\}$

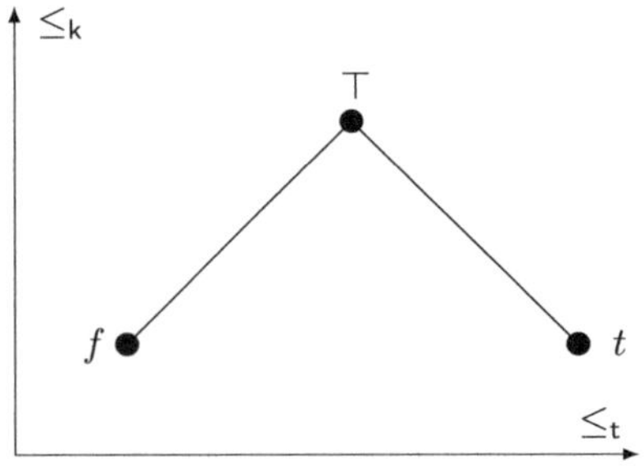

Fig. 4.1 $\mathcal{THREE}$

For characterizing the expressive power of the languages of **LP** and its above expansions, it is convenient to order the truth-values in a partial order $\leq_k$ that intuitively reflects differences in the amount of *knowledge* (or *information*) that the truth values convey. According to this relation $\top$ is the maximal element, while neither of the remaining truth-values is greater than the other. Therefore, $\langle \mathcal{V}, \leq_k \rangle$ is an upper semilattice. A Double-Hasse diagram representing the structure $\mathcal{THREE}$ which is induced by $\leq_k$ and $\leq_t$ (Note 4.59) is given in Fig. 4.1. In this diagram b is an immediate $\leq_t$-successor of a iff b is on the right-hand side of a, and there is an edge between them; Similarly, b is an immediate $\leq_k$-successor of a iff b is above a, and there is an edge between them.[8]

Definition 4.61 A function $g : \{t, f, \top\}^n \to \{t, f, \top\}$ is $\leq_k$*-monotonic* if $g(a_1, \ldots, a_n) \leq_k g(b_1, \ldots, b_n)$ in case $a_i \leq_k b_i$ for every $1 \leq i \leq n$.

Now we are able to characterize the expressive power of **LP** and its expansions:

Theorem 4.62 [8] *Let* $g : \{t, f, \top\}^n \to \{t, f, \top\}$.

1. *g is representable in the language of* $\mathbf{LP}^{\mathsf{f},\top}$ *iff it is* $\leq_k$*-monotonic.*
2. *g is representable in the language of* $\mathbf{LP}^{\top}$ *iff it is* $\leq_k$*-monotonic and* $\{\top\}$*-closed.*
3. *g is representable in the language of* $\mathbf{LP}^{\mathsf{f}}$ *iff it is* $\leq_k$*-monotonic and classically (i.e.,* $\{t, f\}$*-) closed.*
4. *g is representable in the language of* **LP** *iff it is* $\leq_k$*-monotonic,* $\{\top\}$*-closed, and classically closed.*

Next we turn to the main properties of the four logics considered in Theorem 4.62.

Proposition 4.63

1. **LP**, $\mathbf{LP}^{\mathsf{f}}$, $\mathbf{LP}^{\top}$, *and* $\mathbf{LP}^{\top,\mathsf{f}}$ *are all boldly paraconsistent and strongly maximal paraconsistent logics.*
2. **LP** *is* $\neg$*-contained in classical logic and fully maximal. The same is true for* $\mathbf{LP}^{\mathsf{f}}$ *(but not for* $\mathbf{LP}^{\top}$ *or* $\mathbf{LP}^{\top,\mathsf{f}}$*).*
3. **LP** *is non-exploding. The same is true for* $\mathbf{LP}^{\top}$*, but not for* $\mathbf{LP}^{\mathsf{f}}$ *or* $\mathbf{LP}^{\top,\mathsf{f}}$*.*

[8]We refer to [2, 15, 23, 25] for further motivation and discussions on algebraic structures that combine order relations about truth and knowledge.

Proof Immediate from Corollary 4.34, Theorems 4.43, 4.45, 4.62, and Proposition 4.37. □

Perhaps the most remarkable property of **LP** (and $\mathbf{LP^f}$) is given in the next proposition.

Proposition 4.64 [31] *The tautologies of* **LP** *and* $\mathbf{LP^f}$ *are the same as those of classical logic in their languages: if* ψ *is a formula in the language of* $\{\neg, \wedge, \vee\}$ $(\{\neg, \wedge, \vee, \mathsf{f}\})$ *then* $\vdash_{\mathsf{LP}} \psi$ $(\vdash_{\mathbf{LP^f}} \psi)$ *iff* $\vdash_{\mathcal{M}_{CL}} \psi$*, where* $\mathcal{M}_{CL}$ *is the two-valued matrix for classical logic.*

Proof One direction is trivial. For the converse, suppose, e.g., that ν is an LP-valuation (the proof in the case of $\mathbf{LP^f}$ is similar). Let μ be the $\mathcal{M}_{CL}$-valuation such that for every $p \in \mathsf{Atoms}$, $\mu(p) = t$ iff $\nu(p) \in \{t, \top\}$. It is easy to prove by induction on the complexity of ψ that if $\mu(\psi) = t$ then $\nu(\psi) \in \{t, \top\}$, and if $\mu(\psi) = f$ then $\nu(\psi) \in \{f, \top\}$. It follows that if for every $\mathcal{M}_{CL}$-valuation μ it holds that $\mu(\psi) = t$, then for every LP-valuation ν, $\nu(\psi)$ is designated. □

Note 4.65 Despite of having the same set of valid formulas, **LP** is paraconsistent, while classical logic (in the language of $\{\neg, \wedge, \vee\}$) is not. The difference between the two is due to their *consequence relations*.

The main drawback of **LP** and the other logics studied in this section is given in the next proposition.

Proposition 4.66 [3] *Suppose that* $\mathcal{M}$ *is a three-valued paraconsistent matrix which has only* $\leq_k$*-monotonic connectives. Then* $\mathbf{L}_{\mathcal{M}}$ *does not have an implication connective.*

Proof Suppose for contradiction that $\supset$ is a definable implication for $\mathbf{L}_{\mathcal{M}}$. By Lemma 4.6 this implies that (i) $\vdash_{\mathcal{M}} p \supset p$, and (ii) $p, p \supset q \vdash_{\mathcal{M}} q$. Now, (i) entails that $\tilde{\supset}(f, f) \in \{t, \top\}$. Therefore it follows from the $\leq_k$-monotonicity of $\supset$ that $\tilde{\supset}(\top, f) \in \{t, \top\}$. This contradicts (ii), since it is refuted by any assignment ν such that $\nu(p) = \top$ and $\nu(q) = f$. □

Corollary 4.67 *The logics* **LP**, $\mathbf{LP}^{\top}$, $\mathbf{LP^f}$, *and* $\mathbf{LP}^{\mathsf{f},\top}$ *are not normal, but only semi-normal.*

Proof Immediate from Corollary 4.40, Theorem 4.62, and Proposition 4.66. □

4.5.4 The Logics PAC ($\mathbf{RM_3}$) *and Its Main Expansions*

The most straightforward way to turn **LP** into a normal logic is to extend LP by an implication connective. A natural candidate for this is D'Ottaviano and da-Costa's implication [19, 21], considered in Note 4.44. Because of its nice properties (to be presented below), this is the main implication connective (in the sense of Definition 4.5) which has been used in three-valued paraconsistent logics. The logic that is obtained by extending LP with $\supset$ is called **PAC** (also known as **RM$_3$**) [6, 7, 14, 20, 22, 34]. Thus, **PAC** is the logic which is induced by the three-valued matrix $\mathsf{PAC} = \langle\{t, f, \top\}, \{t, \top\}, \{\tilde{\wedge}, \tilde{\vee}, \tilde{\supset}, \tilde{\neg}\}\rangle$, where $\tilde{\wedge}$, $\tilde{\vee}$, and $\tilde{\neg}$ are like in LP, while $\tilde{\supset}$ is given by the following truth-table:

$\tilde{\supset}$	t	f	$\top$
t	t	f	$\top$
f	t	t	t
$\top$	t	f	$\top$

Note 4.68 Since $a \tilde{\to}_S b = (a \tilde{\supset} b) \tilde{\wedge} (\tilde{\neg} b \tilde{\supset} \tilde{\neg} a)$, while $a \tilde{\supset} b = b \tilde{\vee} (a \tilde{\to}_S b)$, another way that leads to **PAC** is to extend $\mathcal{A}_1$, and with it $\mathbf{SRM}_{\tilde{\to}}$, with Kleene's conjunction (which, as indicated at the end of Sect. 4.5.2, is not definable in their language).

Again, the simplest expansions of **PAC** are those that are obtained by adding to its language the propositional constants $\top$ and f.

Definition 4.69

- **J$_3$** [20, 22] is the logic induced by the expansion of the matrix PAC to the language $\{\neg, \wedge, \vee, \supset, \mathsf{f}\}$.
- **PAC$^\top$** is the logic induced by the expansion of the matrix PAC to the language $\{\neg, \wedge, \vee, \supset, \top\}$.
- $\mathbf{J}_3^\top$ is the logic induced by the expansion of the matrix PAC to the language $\{\neg, \wedge, \vee, \supset, \mathsf{f}, \top\}$

Note 4.70 Instead of the propositional constant f it is common in the literature on **J$_3$** to use as the extra connective the *consistency* operator $\circ$, whose interpretation $\tilde{\circ}$ is given by: $\tilde{\circ}(t) = \tilde{\circ}(f) = t$, and $\tilde{\circ}(\top) = f$. This does not make much difference, since $\tilde{\circ}(a) = (a \tilde{\wedge} \tilde{\neg} a) \tilde{\supset} \mathsf{f}$, while $\mathsf{f} = \tilde{\circ}(a) \tilde{\wedge} \tilde{\neg}\tilde{\circ}(a)$. As a logic in the language of $\{\neg, \wedge, \vee, \supset, \circ\}$, **J$_3$** is *the strongest logic* in the family of LFIs (Logics of Formal Inconsistency, [16]) in this language. Recently, **J$_3$** and its weaker versions have also been considered in the context of epistemic logics, where in [17, 18] it is shown that these logics can be encoded in a simple fragment of the modal logic **KD**, containing only modal formulas without nesting.

The following theorem characterizes the expressive power of the languages of PAC and its expansions:

Theorem 4.71 [8] *Let $g : \{t, f, \top\}^n \to \{t, f, \top\}$.*

1. *g is representable in the language of $\mathbf{J}_3^\top$.*
2. *g is representable in the language of **J$_3$** iff it is $\{t, f\}$-closed (i.e., iff it is classically closed).*

3. *g is representable in the language of* $\mathbf{PAC}^{\top}$ *iff it is* $\{\top\}$*-closed.*
4. *g is representable in the language of* **PAC** *iff it is both* $\{t, f\}$*-closed and* $\{\top\}$*-closed.*

Note 4.72 In [8] it is also shown that by adding to PAC *any* classically closed connective not available in it, we get a matrix in which exactly the classically closed connectives are available. Similarly, by adding to PAC *any* $\{\top\}$-closed connective not available in it, we get a matrix in which exactly the $\{\top\}$-closed connectives are available. It follows that there is no intermediate expansion of **PAC** between **PAC** and $\mathbf{J_3}$, or between **PAC** and $\mathbf{PAC}^{\top}$. From the results of [8], it also follows that there is no intermediate expansion of $\mathbf{J_3}$ or $\mathbf{PAC}^{\top}$ between these logics and $\mathbf{J_3^{\top}}$.

The main properties of the four logics discussed above are considered next.

Proposition 4.73

1. **PAC**, $\mathbf{J_3}$, $\mathbf{PAC}^{\top}$, *and* $\mathbf{J_3^{\top}}$ *are all normal, boldly paraconsistent, and strongly maximal paraconsistent logics.*
2. **PAC** *and* $\mathbf{J_3}$ *are* $\neg$*-contained in classical logic and fully maximal. This is false for* $\mathbf{PAC}^{\top}$ *and* $\mathbf{J_3^{\top}}$.
3. **PAC** *and* $\mathbf{PAC}^{\top}$ *are non-exploding. This is false for* $\mathbf{J_3}$ *and* $\mathbf{J_3^{\top}}$.

Proof Follows from Corollary 4.34, Theorems 4.71, 4.43, 4.45, and Proposition 4.37. □

Corollary 4.74

1. *Every three-valued paraconsistent logic can be embedded in* $\mathbf{J_3^{\top}}$.
2. $\mathbf{J_3}$ *is the strongest three-valued paraconsistent logic which is* $\neg$*-contained in classical logic (i.e., every other logic with these properties, like* $\mathbf{P_1}$*, can be embedded in it).*
3. $\mathbf{PAC}^{\top}$ *is the strongest three-valued paraconsistent logic which is non-exploding.*
4. **PAC** *is the strongest three-valued paraconsistent logic which is both* $\neg$*-contained in classical logic and non-exploding.*

Proof Immediate from Theorems 4.71 and 4.43, and from Propositions 4.73 and 4.37. □

4.5.5 *The Logic* $\mathbf{PAC}_{\tilde{\supset}}^{\tilde{\neg}}$

One more interesting paraconsistent three-valued logic is given by the $\{\neg, \supset\}$-fragment of **PAC**. We call this fragment $\mathbf{PAC}_{\tilde{\supset}}^{\tilde{\neg}}$, and it is the logic induced by the matrix $\mathsf{PAC}_{\tilde{\supset}}^{\tilde{\neg}} = \langle\{t, f, \top\}, \{t, \top\}, \{\tilde{\supset}, \tilde{\neg}\}\rangle$, where $\tilde{\supset}$ and $\tilde{\neg}$ are like in PAC.

Proposition 4.75 [6] *The matrix* $\mathsf{PAC}_{\tilde{\supset}}$ *is (equivalent to) a proper expansion of* $\mathcal{A}_1$.

Proof The matrix $\mathsf{PAC}_{\tilde{\supset}}$ is (equivalent to) an expansion of $\mathcal{A}_1$, since:

$$a \tilde{\otimes} b = \tilde{\neg}(\tilde{\neg}(a \tilde{\supset} \tilde{\neg} b) \tilde{\supset} (\tilde{\neg}(\tilde{\neg} b \tilde{\supset} a))).$$

The expansion is proper by Proposition 4.58 and by the fact that $\tilde{\supset}$ is not $\{\top\}$-limited. □

Corollary 4.76 $\mathbf{PAC}_{\tilde{\supset}}$ *is non-exploding, normal,* $\neg$*-contained in classical logic, and fully maximal.*

Proof The normality of $\mathbf{PAC}_{\tilde{\supset}}$ follows from Propositions 4.75, 4.57, and Corollary 4.39. The other properties follow, as usual, from Theorems 4.43, 4.45, and Proposition 4.37. □

Next, we characterize the expressive power of the language of $\mathbf{PAC}_{\tilde{\supset}}$.

Theorem 4.77 *A function* $g : \{t, f, \top\}^n \to \{t, f, \top\}$ *is representable in the language of* $\mathbf{PAC}_{\tilde{\supset}}$ *iff it is* $\{\top\}$*-closed, and there is* $1 \leq i \leq n$ *such that* $g(a_1, \ldots, a_n) = \top$ *only if* $a_i = \top$.

Proof For a formula φ in the language of $\{\neg, \supset\}$, we define $\varphi_\top$ recursively as follows: $p_\top = p$ if p is atomic, $(\neg\psi)_\top = \psi_\top$, and $(\varphi \supset \psi)_\top = \psi_\top$. It is easy to verify that for every φ, $\varphi_\top$ is an atom such that $\nu(\varphi_\top) = \top$ whenever ν is a valuation in $\mathsf{PAC}_{\tilde{\supset}}$ such that $\nu(\varphi) = \top$. This easily implies that if g representable in the language of $\mathbf{PAC}_{\tilde{\supset}}$ then it satisfies the condition given above. Obviously, such g is also $\{\top\}$-closed. This prove the "only if" part of the proposition.

For the converse, let $\mathsf{f}_n = \neg P_1 \otimes P_1 \otimes \neg P_2 \otimes P_2 \otimes \cdots \otimes \neg P_n \otimes P_n$. For $a \in \{t, f, \top\}$ we define:

$$\psi_a(p) = \begin{cases} \neg p \supset \mathsf{f}_n & \text{if } a = t, \\ p \supset \mathsf{f}_n & \text{if } a = f, \\ p \otimes \neg p & \text{if } a = \top. \end{cases}$$

It is easy to check that for every valuation ν such that $\nu(P_j) \neq \top$ for some $1 \leq j \leq n$, it holds that $\nu(\psi_a(p)) \neq f$ iff $\nu(p) = a$. Next, for $\mathbf{a} = (a_1, \ldots, a_n) \in \{t, f, \top\}^n$ we let $\psi_{\mathbf{a}} = \psi_{a_1}(P_1) \otimes \cdots \otimes \psi_{a_n}(P_n)$. Then for every valuation ν, $\nu(\psi_{\mathbf{a}}) \neq f$ iff $\nu(P_i) = a_i$ for every $1 \leq i \leq n$, or $\nu(P_i) = \top$ for every $1 \leq i \leq n$.

Now, suppose that $g : \{t, f, \top\}^n \to \{t, f, \top\}$ has the above two properties, and let $1 \leq i \leq n$ have the property that $g(a_1, \ldots, a_n) = \top$ only if $a_i = \top$. It is not difficult to check that g is represented by the $\otimes$-conjunction of all the formulas which either has the form $\psi_{\mathbf{a}} \supset \mathsf{f}_n$ where $g(\mathbf{a}) = f$, or the form $\psi_{\mathbf{a}} \supset (P_i \supset P_i)$ where $g(\mathbf{a}) = \top$. (Note that since g is $\{\top\}$-closed, there is at least one formula of the latter form). □

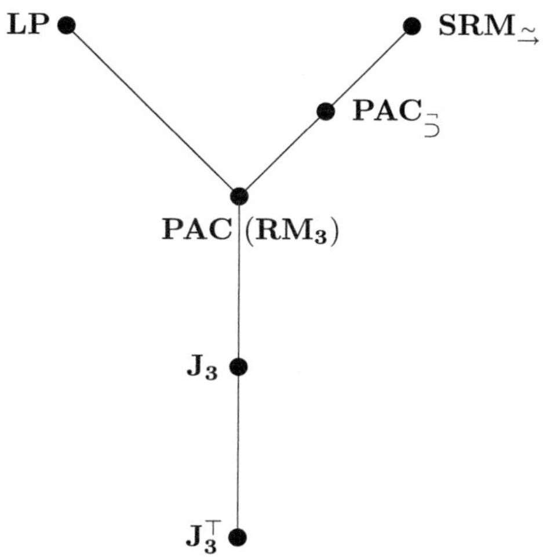

Fig. 4.2 Relative strength of some logics with Kleene's negation

Corollary 4.78 **PAC** *is a proper expansion of* **PAC**$_{\underset{\supset}{\neg}}$.

Proof This follows from the previous proposition and the fact that Kleene's conjunction does not satisfy the second condition given there. □

Figure 4.2 shows the relative expressive power of six of the three-valued logics with Kleene's negation which are considered in this section (in the figure, if two logics are connected, the lower one is the stronger).

4.6 Proof Systems

4.6.1 Gentzen-Type Systems

In this section, we provide an explicit and concise presentation of Gentzen-type systems which correspond to the logics discussed in Sect. 4.5, as well as direct proofs of their completeness and the admissibility of the cut rule in them.[9] We start by recalling the notions of derivation and provability in a Gentzen-type sequent calculi. Below, we denote a sequent in a language $\mathcal{L}$ by s, or more explicitly by $\Gamma \Rightarrow \Delta$, where Γ and Δ are finite sets of formulas in $\mathcal{L}$ and $\Rightarrow$ is a new symbol, not used in $\mathcal{L}$.

Definition 4.79 Let G be a Gentzen-type sequent calculus.

- A *proof* (or *derivation*) in G of a sequent s from a set $\mathcal{S}$ of sequents is a finite sequence of sequents which ends with s, and every element in it either belongs to

[9]In [11] a general algorithm has been given for deriving sound and complete, cut-free Gentzen-type systems for finite-valued logics which have sufficiently expressive languages. That algorithm in fact works for *all* three-valued paraconsistent logics, but we shall not describe it here.

Axioms: $\psi \Rightarrow \psi$

Structural Rules:

Weakening: $$\frac{\Gamma \Rightarrow \Delta}{\Gamma, \Gamma' \Rightarrow \Delta, \Delta'}$$

Cut: $$\frac{\Gamma_1 \Rightarrow \Delta_1, \psi \quad \Gamma_2, \psi \Rightarrow \Delta_2}{\Gamma_1, \Gamma_2 \Rightarrow \Delta_1, \Delta_2}$$

Logical Rules:

$$[\wedge\Rightarrow]\ \frac{\Gamma, \psi, \varphi \Rightarrow \Delta}{\Gamma, \psi \wedge \varphi \Rightarrow \Delta} \qquad [\Rightarrow\wedge]\ \frac{\Gamma \Rightarrow \Delta, \psi \quad \Gamma \Rightarrow \Delta, \varphi}{\Gamma \Rightarrow \Delta, \psi \wedge \varphi}$$

$$[\vee\Rightarrow]\ \frac{\Gamma, \psi \Rightarrow \Delta \quad \Gamma, \varphi \Rightarrow \Delta}{\Gamma, \psi \vee \varphi \Rightarrow \Delta} \qquad [\Rightarrow\vee]\ \frac{\Gamma \Rightarrow \Delta, \psi, \varphi}{\Gamma \Rightarrow \Delta, \psi \vee \varphi}$$

$$[\supset\Rightarrow]\ \frac{\Gamma \Rightarrow \psi, \Delta \quad \Gamma, \varphi \Rightarrow \Delta}{\Gamma, \psi \supset \varphi \Rightarrow \Delta} \qquad [\Rightarrow\supset]\ \frac{\Gamma, \psi \Rightarrow \varphi, \Delta}{\Gamma \Rightarrow \psi \supset \varphi, \Delta}$$

Fig. 4.3 The proof system LK^+

Axioms: $\varphi \Rightarrow \varphi$

Rules: All the rules of LK^+, and the following rules for negation:

$$\frac{\Gamma, \varphi \Rightarrow \Delta}{\Gamma \Rightarrow \Delta, \neg\varphi}$$

$$\frac{\Gamma, \varphi \Rightarrow \Delta, \psi}{\Gamma, \neg(\varphi \supset \psi) \Rightarrow \Delta} \qquad \frac{\Gamma \Rightarrow \Delta, \varphi, \psi}{\Gamma, \neg(\varphi \vee \psi) \Rightarrow \Delta}$$

$$\frac{\Gamma \Rightarrow \Delta, \varphi \quad \Gamma \Rightarrow \Delta, \psi}{\Gamma, \neg(\varphi \wedge \psi) \Rightarrow \Delta} \qquad \frac{\Gamma \Rightarrow \Delta, \neg\varphi}{\Gamma, \neg\neg\varphi \Rightarrow \Delta}$$

Fig. 4.4 The proof system $G_{\mathbf{P_1}}$

$\mathcal{S}$, or is an axiom of G, or is obtained from previous elements of the sequence by one of the rules of G.

- We say that s *follows* from $\mathcal{S}$ in G (notation: $\mathcal{S} \vdash_{\mathsf{G}} s$), if there is a proof in G of s from $\mathcal{S}$.
- A sequent s is *provable* in G (notation: $\vdash_{\mathsf{G}} s$), if it follows in G from the empty set of sequents.
- The tcr $\vdash_{\mathsf{G}}$ induced by G is defined by $\mathcal{T} \vdash_{\mathsf{G}} \varphi$, if there exists a finite Γ such that $\vdash_{\mathsf{G}} \Gamma \Rightarrow \varphi$, and Γ consists only of elements of $\mathcal{T}$.[10]

In what follows, Fig. 4.3 presents a well-known version of Gentzen's proof system LK^+ for positive classical logic [24], on which all the Gentzen-type calculi presented here are based. Figure 4.4 describes a Gentzen-type system $G_{\mathbf{P_1}}$ for Sette's logic $\mathbf{P_1}$,

[10] Although the notation $\vdash_{\mathsf{G}}$ is overloaded in this definition, this should not cause any confusion in what follows.

Axioms: $\varphi \Rightarrow \varphi \qquad \Rightarrow \neg\varphi, \varphi$

Rules: All the rules of LK^+, and the following rules for $\neg$, f, and $\top$:

$$[\neg\neg\Rightarrow]\ \frac{\Gamma, \varphi \Rightarrow \Delta}{\Gamma, \neg\neg\varphi \Rightarrow \Delta} \qquad [\Rightarrow\neg\neg]\ \frac{\Gamma \Rightarrow \Delta, \varphi}{\Gamma \Rightarrow \Delta, \neg\neg\varphi}$$

$$[\neg\wedge\Rightarrow]\ \frac{\Gamma, \neg\varphi \Rightarrow \Delta \quad \Gamma, \neg\psi \Rightarrow \Delta}{\Gamma, \neg(\varphi \wedge \psi) \Rightarrow \Delta} \qquad [\Rightarrow\neg\wedge]\ \frac{\Gamma \Rightarrow \Delta, \neg\varphi, \neg\psi}{\Gamma \Rightarrow \Delta, \neg(\varphi \wedge \psi)}$$

$$[\neg\vee\Rightarrow]\ \frac{\Gamma, \neg\varphi, \neg\psi \Rightarrow \Delta}{\Gamma, \neg(\varphi \vee \psi) \Rightarrow \Delta} \qquad [\Rightarrow\neg\vee]\ \frac{\Gamma \Rightarrow \Delta, \neg\varphi \quad \Gamma \Rightarrow \Delta, \neg\psi}{\Gamma \Rightarrow \Delta, \neg(\varphi \vee \psi)}$$

$$[\neg\supset\Rightarrow]\ \frac{\Gamma, \varphi, \neg\psi \Rightarrow \Delta}{\Gamma, \neg(\varphi \supset \psi) \Rightarrow \Delta} \qquad [\Rightarrow\neg\supset]\ \frac{\Gamma \Rightarrow \varphi, \Delta \quad \Gamma \Rightarrow \neg\psi, \Delta}{\Gamma \Rightarrow \neg(\varphi \supset \psi), \Delta}$$

$$[\mathsf{f}\Rightarrow]\ \Gamma, \mathsf{f} \Rightarrow \Delta \qquad [\Rightarrow\neg\mathsf{f}]\ \Gamma \Rightarrow \Delta, \neg\mathsf{f}$$

$$[\Rightarrow\top]\ \Gamma \Rightarrow \Delta, \top$$

$$[\Rightarrow\neg\top]\ \Gamma \Rightarrow \Delta, \neg\top$$

Fig. 4.5 The proof system $G_{\mathbf{J}_3^\top}$

and Fig. 4.5 describes a Gentzen-type system $G_{\mathbf{J}_3^\top}$ for $\mathbf{J}_3^\top$. A Gentzen-type system $G_{\mathbf{L}}$ for every $\mathbf{L} \in \{\mathbf{LP}, \mathbf{LP}^{\mathsf{f}}, \mathbf{LP}^{\top}, \mathbf{LP}^{\mathsf{f},\top}, \mathbf{PAC}, \mathbf{J_3}, \mathbf{PAC}^{\top}, \mathbf{PAC}_{\tilde{\supset}}\}$ is obtained from $G_{\mathbf{J}_3^\top}$ by deleting from it the irrelevant rules (e.g., the rules for $\supset$ and f in the case of $\mathbf{LP}^\top$). Finally, Fig. 4.6 describes a Gentzen-type system $G_{\mathbf{SRM}_{\tilde{\rightarrow}}}$ for $\mathbf{SRM}_{\tilde{\rightarrow}}$ in the primitive language of this logic.

Note 4.80 Here are some important remarks about the Gentzen-type systems presented in this section:

- The last four rules in Fig. 4.4 can be combined into one rule: Infer $\neg\varphi, \Gamma \Rightarrow \Delta$ from $\Gamma \Rightarrow \Delta, \varphi$ (which is the rule $[\neg\Rightarrow]$, introducing negation on the left-hand side, of Gentzen's system LK for classical logic) with the constraint that the active formula (φ) should not be atomic.
- It is possible to take as axioms of $G_{\mathbf{J}_3^\top}$ only $p \Rightarrow p$, $\neg p \Rightarrow \neg p$, and $\Rightarrow p, \neg p$, where p is atomic (and the rules for f and $\top$, which are really axioms). All other instances of the axioms are then derivable using the logical rules of the system. The same is true for $G_{\mathbf{SRM}_{\tilde{\rightarrow}}}$ and for the various fragments of $G_{\mathbf{J}_3^\top}$.
- The first rule for $G_{\mathbf{P_1}}$ shown in Fig. 4.4 (which is also the rule $[\Rightarrow\neg]$ of LK, introducing negation on the right-hand sides of sequents) is valid for every logic which is induced by a three-valued paraconsistent matrix, and the extra axioms of $G_{\mathbf{J}_3^\top}$ are derivable by it from the standard identity axioms. Therefore, we could have included this rule in the definition of $G_{\mathbf{J}_3^\top}$ instead of its new axioms (note that this rule is derivable from these axioms using a cut). We prefer our official formulation in Fig. 4.5, because all of its logical rules are *invertible* (see Lemma 4.88). This

Axioms: $\Gamma, \varphi \Rightarrow \Delta, \varphi \qquad \Gamma \Rightarrow \Delta, \varphi, \neg\varphi$

Rules: Exchange, Contraction, and the following logical rules:

$$\frac{\Gamma, \varphi \Rightarrow \Delta}{\Gamma, \neg\neg\varphi \Rightarrow \Delta} \qquad \frac{\Gamma \Rightarrow \Delta, \varphi}{\Gamma \Rightarrow \Delta, \neg\neg\varphi}$$

$$\frac{\Gamma, \varphi, \psi \Rightarrow \Delta}{\Gamma, \varphi \otimes \psi \Rightarrow \Delta} \qquad \frac{\Gamma \Rightarrow \Delta, \varphi \quad \Gamma \Rightarrow \Delta, \psi}{\Gamma \Rightarrow \Delta, \varphi \otimes \psi}$$

$$\frac{\Gamma \Rightarrow \Delta, \varphi \quad \Gamma \Rightarrow \Delta, \psi \quad \Gamma, \neg\varphi, \neg\psi \Rightarrow \Delta}{\Gamma, \neg(\varphi \otimes \psi) \Rightarrow \Delta} \qquad \frac{\Gamma, \varphi \Rightarrow \Delta, \neg\psi \quad \Gamma, \psi \Rightarrow \Delta, \neg\varphi}{\Gamma \Rightarrow \Delta, \neg(\varphi \otimes \psi)}$$

$$\frac{\Gamma \Rightarrow \Delta, \varphi \quad \Gamma \Rightarrow \Delta, \neg\psi \quad \Gamma, \neg\varphi, \psi \Rightarrow \Delta}{\Gamma, \varphi \to \psi \Rightarrow \Delta} \qquad \frac{\Gamma, \varphi \Rightarrow \Delta, \psi \quad \Gamma, \neg\psi \Rightarrow \Delta, \neg\varphi}{\Gamma \Rightarrow \Delta, \varphi \to \psi}$$

$$\frac{\Gamma, \varphi, \neg\psi \Rightarrow \Delta}{\Gamma, \neg(\varphi \to \psi) \Rightarrow \Delta} \qquad \frac{\Gamma \Rightarrow \Delta, \varphi \quad \Gamma \Rightarrow \Delta, \neg\psi}{\Gamma \Rightarrow \Delta, \neg(\varphi \to \psi)}$$

Fig. 4.6 The proof system $G_{\mathbf{SRM}_{\tilde{\to}}}$

is a very useful property in proof search and for other goals (as the proofs given below show).

- Actually, we could have formulated $G_{\mathbf{P_1}}$ too by using only invertible rules. This can be done by adding to it the new axioms of $G_{\mathbf{J}_3^{\top}}$, and limiting the applications of $[\Rightarrow\neg]$ to the case where the active formula is not atomic. Again, we can have only $p \Rightarrow p, \neg p \Rightarrow \neg p$, and $\Rightarrow p, \neg p$ as axioms in these versions of the system, where p is atomic.

Our next goal is to show the strong soundness and completeness of all these Gentzen-type systems. Our first step toward this goal is to define the semantics of sequents in the context of matrices.

Definition 4.81 Let $\mathcal{M}$ be a matrix for $\mathcal{L}$ and let $\nu \in \Lambda_{\mathcal{M}}$.

- We say that ν is an $\mathcal{M}$-*model* of a sequent $\Gamma \Rightarrow \Delta$, or that ν $\mathcal{M}$-*satisfies* $\Gamma \Rightarrow \Delta$ (notation: $\nu \models_{\mathcal{M}} \Gamma \Rightarrow \Delta$) if $\nu \not\models_{\mathcal{M}} \varphi$ for some φ in Γ, or $\nu \models_{\mathcal{M}} \psi$ for some ψ in Δ.
- We say that a sequent s $\mathcal{M}$-*follows* from a set $\mathcal{S}$ of sequents (notation: $\mathcal{S} \vdash_{\mathcal{M}} s$) if every $\mathcal{M}$-model of $\mathcal{S}$ is also an $\mathcal{M}$-model of s.
- A sequent s is $\mathcal{M}$-valid (notation: $\vdash_{\mathcal{M}} s$) if $\nu \models_{\mathcal{M}} s$ for every $\nu \in \Lambda_{\mathcal{M}}$ (i.e., if $\emptyset \vdash_{\mathcal{M}} s$).

By Definition 4.81 and Proposition 4.36, we have

Proposition 4.82 *Let $\mathcal{M}$ be a three-valued paraconsistent matrix, and let ν be an assignment in $\mathcal{M}$. Then $\nu \models_{\mathcal{M}} \Gamma \Rightarrow \Delta$ (where $\Gamma \Rightarrow \Delta$ is a sequent in the language* *of $\mathcal{M}$) iff either $\nu(\varphi) = f$ for some $\varphi \in \Gamma$, or $\nu(\psi) \neq f$ (i.e., $\nu(\psi) \in \{t, \top\}$) for some $\psi \in \Delta$.*

Note 4.83 It is easy to see that $\vdash_{\mathcal{M}} \Gamma \Rightarrow \psi$ iff $\Gamma \vdash_{\mathcal{M}} \psi$.

Definition 4.84 Let $\mathbf{L} = \mathbf{L}_{\mathcal{M}}$ be one of the logics discussed in Sect. 4.5, and let $G_{\mathbf{L}}$ be the corresponding Gentzen-type calculus. We say that $G_{\mathbf{L}}$ is (strongly) sound and complete for $\mathbf{L}$ if for every $\mathcal{T}$ and ψ it holds that $\mathcal{T} \vdash_{G_{\mathbf{L}}} \psi$ (Definition 4.79) iff $\mathcal{T} \vdash_{\mathbf{L}} \psi$.

To show soundness and completeness of our various systems, we first need some lemmas.

Lemma 4.85 *Let $\mathcal{M}$ be a three-valued paraconsistent matrix, and let $\Gamma \Rightarrow \Delta$ be a sequent which consists of literals (i.e., atomic formulas or negations of atomic formulas).*

1. *$\vdash_{\mathcal{M}} \Gamma \Rightarrow \Delta$ iff either $\Gamma \cap \Delta \neq \emptyset$, or there is an atomic formula p such that $\{p, \neg p\} \subseteq \Delta$, or $\mathsf{f} \in \Gamma$, or $\neg\mathsf{f} \in \Delta$, or $\top \in \Delta$, or $\neg\top \in \Delta$.*
2. *If $\mathbf{L}$ is one of the logics discussed in Sect. 4.5 and $\vdash_{\mathbf{L}} \Gamma \Rightarrow \Delta$, then $\vdash_{G_{\mathbf{L}}} \Gamma \Rightarrow \Delta$.*

Proof Suppose Γ and Δ consist only of literals.

1. From Proposition 4.36, it follows that if Γ and Δ satisfies one of the six conditions, then $\vdash_{\mathcal{M}} \Gamma \Rightarrow \Delta$. Suppose now that $\Gamma \Rightarrow \Delta$ does not satisfy any of them. Define

$$\nu(p) = \begin{cases} f & \text{if } p \in \Delta, \\ t & \text{if } \neg p \in \Delta, \\ \top & \text{otherwise.} \end{cases}$$

   Then ν is well-defined, and $\nu \not\models_{\mathcal{M}} \Gamma \Rightarrow \Delta$. Hence $\not\vdash_{\mathcal{M}} \Gamma \Rightarrow \Delta$ in this case.
2. This follows from the first part and the fact that every sequent which satisfies the condition given in that part is obviously provable in $G_{\mathbf{L}}$ (except in the case that $\mathbf{L} = \mathbf{P_1}$ such a sequent is simply an axiom of $G_{\mathbf{L}}$. In the case of $\mathbf{P_1}$ we use the rule for introducing negation on the right). □

Lemma 4.86 *Let $\mathbf{L}$ be one of the logics discussed in Sect. 4.5, and let $\mathcal{M}$ be the three-valued paraconsistent matrix which induces $\mathbf{L}$. Then every logical rule of $G_{\mathbf{L}}$ is strongly sound for $\vdash_{\mathcal{M}}$: if $\mathcal{S}$ is the set of premises of (an application of) such a rule, and s is its conclusion, then $\mathcal{S} \vdash_{\mathcal{M}} s$.*

Proof Easy. As an example, we show the case of the rule of $G_{\mathbf{SRM}_{\vec{\sim}}}$ for introducing $\neg(\varphi \otimes \psi)$ on the right. So assume that $\nu \models_{\mathcal{A}_1} \Gamma, \varphi \Rightarrow \Delta, \neg\psi$ and $\nu \models_{\mathcal{A}_1} \Gamma, \psi \Rightarrow \Delta, \neg\varphi$. We show that $\nu \models_{\mathcal{A}_1} \Gamma \Rightarrow \Delta, \neg(\varphi \otimes \psi)$. If $\nu \models_{\mathcal{A}_1} \Gamma \Rightarrow \Delta$ we are done. Otherwise, we have that either $\nu(\varphi) = f$ or $\nu(\psi) \neq t$, and either $\nu(\psi) = f$ or $\nu(\varphi) \neq t$. This gives us four possibilities, and it is easy to check that in all of them $\nu(\varphi \otimes \psi) \neq t$, i.e., $\nu(\neg(\varphi \otimes \psi)) \neq f$. □

Lemma 4.87 *Let* **L** *be one of the logics discussed in Sect. 4.5. Then* $G_{\mathbf{L}}$ *is strongly sound for* **L**.

Proof By Lemma 4.86 we need only to check that the axioms of $G_{\mathbf{L}}$ are valid in **L**. This is obvious. □

Lemma 4.88 *Let* **L** *and* $\mathcal{M}$ *be like in Lemma 4.86, and let r be a logical rule of* $G_{\mathbf{L}}$. *If r is not the rule of* $G_{\mathbf{P_1}}$ *for introducing* $\neg$ *on the right, then r is strongly invertible in* $\vdash_{\mathcal{M}}$*: If* s_c *is the conclusion of r and* s_p *is any of its premises, then* $s_c \vdash_{\mathcal{M}} s_p$. *This is true also for every application of the exceptional rule in which the active formula is not atomic (I.e., if* φ *is not atomic then* $\Gamma \Rightarrow \Delta, \neg\varphi \vdash_{\mathsf{P}_1} \varphi, \Gamma \Rightarrow \Delta$*).*

Proof Again we do as an example the case in which r is the rule of $G_{\mathbf{SRM}_{\vec{\sim}}}$ for introducing $\neg(\varphi \otimes \psi)$ on the right. So assume that $\nu \models_{\mathcal{A}_1} \Gamma \Rightarrow \Delta, \neg(\varphi \otimes \psi)$. We show, e.g., that $\nu \models_{\mathcal{A}_1} \Gamma, \psi \Rightarrow \Delta, \neg\varphi$. If $\nu \models_{\mathcal{A}_1} \Gamma \Rightarrow \Delta$ we are done. Otherwise $\nu(\neg(\varphi \otimes \psi)) \neq f$, and so $\nu(\varphi \otimes \psi) \neq t$. This implies that either $\nu(\varphi) = f$, or $\nu(\psi) = f$, or $\nu(\varphi) = \nu(\psi) = \top$, and so either $\nu(\psi) = f$ or $\nu(\neg\varphi) \neq f$. In both cases, we have that $\nu \models_{\mathcal{A}_1} \Gamma, \psi \Rightarrow \Delta, \neg\varphi$.

As for the exceptional rule, suppose that φ is not atomic, and $\nu \models_{\mathsf{P}_1} \Gamma \Rightarrow \Delta, \neg\varphi$. We show that $\nu \models_{\mathsf{P}_1} \varphi, \Gamma \Rightarrow \Delta$. If $\nu \models_{\mathsf{P}_1} \Gamma \Rightarrow \Delta$ we are done. Otherwise $\nu(\neg\varphi) \neq f$, and so $\nu(\varphi) \in \{f, \top\}$. Since φ is not atomic, $\nu(\varphi) \neq \top$. It follows that $\nu(\varphi) = f$, and so $\nu \models_{\mathsf{P}_1} \varphi, \Gamma \Rightarrow \Delta$. □

Lemma 4.89 *Let* **L** *and* $\mathcal{M}$ *be like in Lemma 4.86, and let s be a sequent in the language of* **L**. *If* $\vdash_{\mathcal{M}} s$ *then s has a cut-free proof in* $G_{\mathbf{L}}$.

Proof It is easy to check that by applying the logical rules of $G_{\mathbf{L}}$ backward, and using Lemma 4.88, we can construct for every sequent s a finite set $\mathcal{S}(s)$ with the following properties:

1. Each element of $\mathcal{S}(s)$ is a sequent which consists only of literals.
2. $s \vdash_{\mathcal{M}} s'$ for every element s' of $\mathcal{S}(s)$.
3. There is a cut-free proof of s from $\mathcal{S}(s)$.

Suppose now that $\vdash_{\mathcal{M}} s$. By Lemma 4.88 and the second property of $\mathcal{S}(s)$ this implies that $\vdash_{\mathcal{M}} s'$ for every element s' of $\mathcal{S}(s)$. By Lemma 4.85 and the first and third properties of $\mathcal{S}(s)$ it follows that s has a cut-free proof in $G_{\mathbf{L}}$. □

Now we are ready to prove the two main results of this section.

Theorem 4.90 *Let* **L** *be one of the logics discussed in Sect. 4.5.*

1. $G_{\mathbf{L}}$ *is sound and complete for* **L**.
2. $\mathcal{T} \vdash_{G_{\mathbf{L}}} \psi$ *iff* $\mathcal{T} \vdash_{\mathbf{L}} \psi$.

Proof The first part is immediate from Lemmas 4.87 and 4.89; The second part follows from the first part and the fact that by Proposition 4.11, **L** is finitary. □

Inference Rule: [MP] $\dfrac{\psi \quad \psi \supset \varphi}{\varphi}$

Axioms:

[⊃1]	$\psi \supset \varphi \supset \psi$
[⊃2]	$(\psi \supset \varphi \supset \tau) \supset (\psi \supset \varphi) \supset (\psi \supset \tau)$
[⊃3]	$((\psi \supset \varphi) \supset \psi) \supset \psi$
[∧⊃]	$\psi \wedge \varphi \supset \psi, \quad \psi \wedge \varphi \supset \varphi$
[⊃∧]	$\psi \supset \varphi \supset \psi \wedge \varphi$
[⊃∨]	$\psi \supset \psi \vee \varphi, \quad \varphi \supset \psi \vee \varphi$
[∨⊃]	$(\psi \supset \tau) \supset (\varphi \supset \tau) \supset (\psi \vee \varphi \supset \tau)$

Fig. 4.7 The proof system HCL^+

Theorem 4.91 *Let* **L** *be one of the logics discussed in Sect. 4.5. Then* $G_{\mathbf{L}}$ *admits cut-elimination (i.e., every sequent that is provable in* $G_{\mathbf{L}}$ *has a proof in which the cut rule is not used).*

Proof Suppose that $\vdash_{G_{\mathbf{L}}} s$. By Lemma 4.87, $\vdash_{\mathbf{L}} s$. By Lemma 4.89 this implies that s has a cut-free proof in $G_{\mathbf{L}}$. □

4.6.2 Hilbert-Type Systems

To complete the picture, in this final subsection we present Hilbert-type proof systems with MP for ⊃ as the sole rule of inference for all the logics studied in Sect. 4.5 in which ⊃ is a primitive connective.[11] Again, these systems are based on some sound and complete proof system of the same type for positive classical logic ($\mathbf{CL}^+$). Such a system, denoted HCL^+, is presented in Fig. 4.7.[12]

Definition 4.92 Figures 4.8 and 4.9 contain Hilbert-type proof systems for the logic $\mathbf{P_1}$ and the logics **PAC** and $\mathbf{J_3}$, respectively. Hilbert-type proof systems $H_{\mathbf{PAC}^{\top}}$ and $H_{\mathbf{J}_3^{\top}}$ for the logics $\mathbf{PAC}^{\top}$ and $\mathbf{J}_3^{\top}$ (respectively) are obtained by adding to $H_{\mathbf{PAC}}$ and $H_{\mathbf{J_3}}$ (respectively) the axioms $\top$ and $\neg\top$. A Hilbert-type proof system $H_{\mathbf{PAC}_{\overline{\supset}}}$ for $\mathbf{PAC}_{\overline{\supset}}$ is obtained from $H_{\mathbf{PAC}}$ by replacing [t] with either $(\neg\varphi \supset \varphi) \supset \varphi$ or $(\psi \supset \varphi) \supset (\neg\psi \supset \varphi) \supset \varphi$, changing [⇒¬⊃] to $\varphi \supset (\neg\psi \supset \neg(\varphi \supset \psi))$, and deleting all axioms that mention ∧ or ∨.

[11] Note that by Proposition 4.66, the four $\leq_k$-monotonic expansions of **LP** (including **LP** itself) have no implication, and so they cannot have a corresponding Hilbert-type system of the above type. In contrast, by Proposition 4.57 $\mathbf{SRM}_{\tilde{\to}}$ can be defined using such a system, but the resulting system does not look very natural. A natural Hilbert-type system for $\mathbf{SRM}_{\tilde{\to}}$ in its primitive language (but with two inference rules) can be found in [9].

[12] As usual, in the formulation of the axioms of the systems the association of nested implications is taken to the right.

Inference Rule: [MP] $\dfrac{\psi \quad \psi \supset \varphi}{\varphi}$

Axioms: The axioms of HCL^+ and:

[t]	$\neg\psi \vee \psi$
$[\neg{\supset}{\Rightarrow}]$	$(\varphi \supset \psi) \supset \neg(\varphi \supset \psi) \supset \tau$
$[\neg{\vee}{\Rightarrow}]$	$(\varphi \vee \psi) \supset \neg(\varphi \vee \psi) \supset \tau$
$[\neg{\wedge}{\Rightarrow}]$	$(\varphi \wedge \psi) \supset \neg(\varphi \wedge \psi) \supset \tau$
$[\neg{\Rightarrow}]$	$\neg\varphi \supset \neg\neg\varphi \supset \tau$

Fig. 4.8 The proof system $H_{\mathbf{P_1}}$

Inference Rule: [MP] $\dfrac{\psi \quad \psi \supset \varphi}{\varphi}$

Axioms of $H_{\mathbf{PAC}}$: The axioms of HCL^+ and:

[t]	$\neg\psi \vee \psi$
$[\neg\neg{\Rightarrow}]$	$\neg\neg\varphi \supset \varphi$
$[{\Rightarrow}\neg\neg]$	$\varphi \supset \neg\neg\varphi$
$[\neg{\supset}{\Rightarrow}1]$	$\neg(\varphi \supset \psi) \supset \varphi$
$[\neg{\supset}{\Rightarrow}2]$	$\neg(\varphi \supset \psi) \supset \neg\psi$
$[{\Rightarrow}\neg{\supset}]$	$(\varphi \wedge \neg\psi) \supset \neg(\varphi \supset \psi)$
$[\neg{\vee}{\Rightarrow}1]$	$\neg(\varphi \vee \psi) \supset \neg\varphi$
$[\neg{\vee}{\Rightarrow}2]$	$\neg(\varphi \vee \psi) \supset \neg\psi$
$[{\Rightarrow}\neg{\vee}]$	$(\neg\varphi \wedge \neg\psi) \supset \neg(\varphi \vee \psi)$
$[\neg{\wedge}{\Rightarrow}]$	$\neg(\varphi \wedge \psi) \supset (\neg\varphi \vee \neg\psi)$
$[{\Rightarrow}\neg\wedge 1]$	$\neg\varphi \supset \neg(\varphi \wedge \psi)$
$[{\Rightarrow}\neg\wedge 2]$	$\neg\psi \supset \neg(\varphi \wedge \psi)$

Axioms of $H_{\mathbf{J_3}}$: The axioms of $H_{\mathbf{PAC}}$ and:

$[\mathsf{f}{\supset}]$	$\mathsf{f} \supset \psi$
$[{\supset}\mathsf{f}]$	$\psi \supset \neg\mathsf{f}$

Fig. 4.9 The proof systems $H_{\mathbf{PAC}}$ and $H_{\mathbf{J_3}}$

Theorem 4.93 *Let* $\mathbf{L} \in \{\mathbf{P_1}, \mathbf{PAC}_{\overset{\neg}{\supset}}, \mathbf{PAC}, \mathbf{PAC}^{\top}, \mathbf{J_3}, \mathbf{J}_3^{\top}\}$. *Then* $\vdash_{H_{\mathbf{L}}} = \vdash_{G_{\mathbf{L}}}$.

Proof Using cuts and the fact that $\vdash_{LK^+} \psi, \psi \supset \varphi \Rightarrow \varphi$, it is easy to show by induction on length of proofs in $H_{\mathbf{L}}$ that if $\Gamma \vdash_{H_{\mathbf{L}}} \varphi$ (where Γ is finite) then $\Gamma \vdash_{G_{\mathbf{L}}} \varphi$. All one needs to do is to show that $\vdash_{G_{\mathbf{L}}} \varphi$ for every axiom φ of $H_{\mathbf{L}}$, and this is a straightforward exercise. It immediately follows that $\vdash_{H_{\mathbf{L}}} \subseteq \vdash_{G_{\mathbf{L}}}$.

For the converse, it would be more convenient to use the versions of the Gentzen-type systems which employ lists of formulas rather than finite sets,[13] and to treat each of the six logics separately.

L = PAC. In this case, it is easy to prove (either syntactically, using the cut-elimination theorem for $G_{\mathbf{PAC}}$, or semantically, using the soundness theorem for it)

[13] In such a case we need also the structural rules of Permutation, Contraction, and Expansion that assure that the underlying consequence relation remains the same.

that a sequent $s = \varphi_1, \ldots, \varphi_n \Rightarrow \psi_1, \ldots, \psi_m$ is provable in $G_{\mathbf{PAC}}$ only if $m > 0$. For each such sequent s we define a translation $Tr_{\mathbf{L}}(s)$ by $Tr_{\mathbf{L}}(s) = \varphi_1 \wedge \cdots \wedge \varphi_n \supset \psi_1 \vee \cdots \vee \psi_m$ (in particular: $Tr_{\mathbf{L}}(\Rightarrow \psi_1, \ldots, \psi_m) = \psi_1 \vee \cdots \vee \psi_m$). Obviously, to show that $\vdash_{G_{\mathbf{L}}} \subseteq \vdash_{H_{\mathbf{L}}}$ it suffices to prove that if $\vdash_{G_{\mathbf{L}}} s$ then $\vdash_{H_{\mathbf{L}}} Tr_{\mathbf{L}}(s)$. We prove this claim by induction on length of proofs in $G_{\mathbf{L}}$. This is a routine (though tedious) induction, and here we shall do as examples three of the various possible cases that should be considered.

- Suppose s is an axiom of the form $\Rightarrow \neg\varphi, \varphi$. Then $Tr_{\mathbf{L}}(s)$ is an instance of the axiom **[t]** of **L** (= **PAC**).
- Suppose s is inferred from s_1 and s_2 using $[\supset\Rightarrow]$. Then there are formulas φ, ψ, τ_2, and (perhaps) τ_1 such that $Tr_{\mathbf{L}}(s) = \tau_1 \wedge (\varphi \supset \psi) \supset \tau_2$, $Tr_{\mathbf{L}}(s_1) = \tau_1 \supset \tau_2 \vee \varphi$, and $Tr_{\mathbf{L}}(s_2) = \tau_1 \wedge \psi \supset \tau_2$ (the case where $Tr_{\mathbf{L}}(s) = (\varphi \supset \psi) \supset \tau_2$, $Tr_{\mathbf{L}}(s_1) = \tau_2 \vee \varphi$, and $Tr_{\mathbf{L}}(s_2) = \psi \supset \tau_2$ is similar, but easier). By induction hypothesis, $\vdash_{H_{\mathbf{L}}} Tr_{\mathbf{L}}(s_1)$ and $\vdash_{H_{\mathbf{L}}} Tr_{\mathbf{L}}(s_2)$. Now
$$P_1 \supset P_2 \vee P_3,\ P_1 \wedge P_4 \supset P_2 \vdash_{\mathbf{CL}^+} P_1 \wedge (P_3 \supset P_4) \supset P_2.$$
Since HCL^+ is complete for $\mathbf{CL}^+$ and $H_{\mathbf{L}}$ is an extension of HCL^+, it follows (by substituting τ_1 for P_1, τ_2 for P_2, φ for P_3, and ψ for P_4) that $Tr_{\mathbf{L}}(s_1), Tr_{\mathbf{L}}(s_2) \vdash_{H_{\mathbf{L}}} Tr_{\mathbf{L}}(s)$. Hence $\vdash_{H_{\mathbf{L}}} Tr_{\mathbf{L}}(s)$.
- Suppose s is inferred from s_1 using $[\neg\supset\Rightarrow]$. Then there are formulas φ, ψ, τ_2, and (perhaps) τ_1 such that $Tr_{\mathbf{L}}(s) = \tau_1 \wedge \neg(\varphi \supset \psi) \supset \tau_2$, while $Tr_{\mathbf{L}}(s_1) = \tau_1 \wedge \varphi \wedge \neg\psi \supset \tau_2$ (again the case where there is no τ_1 is easier). By induction hypothesis, $\vdash_{H_{\mathbf{L}}} Tr_{\mathbf{L}}(s_1)$. Now
$$P_5 \supset P_3,\ P_5 \supset P_4,\ P_1 \wedge P_3 \wedge P_4 \supset P_2 \vdash_{\mathbf{CL}^+} P_1 \wedge P_5 \supset P_2.$$
Since HCL^+ is complete for $\mathbf{CL}^+$ and $H_{\mathbf{L}}$ is an extension of HCL^+, it follows (by substituting τ_1 for P_1, τ_2 for P_2, φ for P_3, $\neg\psi$ for P_4, and $\neg(\varphi \supset \psi)$ for P_5) that
$$\neg(\varphi \supset \psi) \supset \varphi,\ \neg(\varphi \supset \psi) \supset \neg\psi,\ Tr_{\mathbf{L}}(s_1) \vdash_{H_{\mathbf{L}}} Tr_{\mathbf{L}}(s).$$
Using the axioms $[\neg\supset\Rightarrow 1]$ and $[\neg\supset\Rightarrow 2]$ of $H_{\mathbf{L}}$, it follows from the induction hypothesis for s_1 that $\vdash_{H_{\mathbf{L}}} Tr_{\mathbf{L}}(s)$.

The proofs in the other cases are similar. One should only note that in some of the cases (e.g., when s is inferred from s_1 using weakening on the right) there are four subcases to consider (rather than just two as in the cases handled above): that we have both τ_1 and τ_2; that we have τ_1 but not τ_2; that we have τ_2 but not τ_1; and that we have neither τ_1 nor τ_2.

L = PAC$^{\top}$. The proof in this case is very similar to that in the previous one, and is left to the reader.

L = J$_3$. The proof in this case is again similar to that in case of **PAC**. The main difference is that now also sequents of the form $\Gamma \Rightarrow$ may be proved in $G_{\mathbf{L}}$, and so the translation of sequents into formulas should be extended to these type of sequents. This is done by letting $Tr_{\mathbf{L}}(\varphi_1, \ldots, \varphi_n \Rightarrow)$ be $\varphi_1 \wedge \cdots \wedge \varphi_n \supset \mathsf{f}$. Details are left to the reader.

$\mathbf{L} = \mathbf{J}_3^{\top}$. The proof in this case is very similar to that in the case of $\mathbf{J}_3$, and is left to the reader.

$\mathbf{L} = \mathbf{P}_1$. The proof in this case is similar to the case $\mathbf{L} = \mathbf{J}_3$, but instead of f we use $\neg P_1 \wedge \neg\neg P_1$ (say).

$\mathbf{L} = \mathbf{PAC}_{\supset}^{\neg}$. This time there is another problem: $\wedge$ and $\vee$ are not included in the language of $\mathbf{PAC}_{\supset}^{\neg}$, and so we cannot employ the translation function that was used in the case of $\mathbf{PAC}$. However, we can use the facts that $\varphi \vee \psi$ is equivalent in $\mathbf{CL}^+$ to $(\varphi \supset \psi) \supset \psi$ and $\varphi \wedge \psi \supset \tau$ is equivalent in $\mathbf{CL}^+$ to $\varphi \supset \psi \supset \tau$. With the help of this fact we can transform the definition of $Tr_{\mathbf{PAC}}$ into an equivalent (in $\mathbf{CL}^+$) definition in which $\wedge$ and $\vee$ are not used:

$$Tr_{\mathbf{PAC}_{\supset}^{\neg}}(\varphi_1, \ldots, \varphi_n \Rightarrow \psi_1, \ldots, \psi_m)$$
$$= \varphi_1 \supset \ldots \varphi_n \supset (\ldots((\psi_1 \supset \psi_2) \supset \psi_2) \supset \cdots \supset \psi_m) \supset \psi_m$$

With this definition, and using instead of HCL^+ the Hilbert-type system consisting only of [MP], [$\supset$1], [$\supset$2], and [$\supset$3] (this proof system is sound and complete with respect to the $\{\supset\}$-fragment of classical logic), one can proceed in a way which is very similar to that used in the case $\mathbf{L} = \mathbf{PAC}$. □

Theorem 4.94 *For every logic* $\mathbf{L} \in \{\mathbf{P}_1, \mathbf{PAC}_{\supset}^{\neg}, \mathbf{PAC}, \mathbf{PAC}^{\top}, \mathbf{J}_3, \mathbf{J}_3^{\top}\}$, *the proof system* $H_{\mathbf{L}}$ *is strongly sound and complete for* $\mathbf{L}$, *i.e.,* $\mathcal{T} \vdash_{H_{\mathbf{L}}} \psi$ *iff* $\mathcal{T} \vdash_{\mathbf{L}} \psi$ *for each such* $\mathbf{L}$.

Proof This is a direct corollary of Theorems 4.93 and 4.90. □

References

1. Anderson, A., Belnap, N.: Entailment, vol. 1. Princeton University Press (1975)
2. Arieli, O., Avron, A.: The value of the four values. Artif. Intell. **102**(1), 97–141 (1998)
3. Arieli, O., Avron, A., Zamansky, A.: Ideal paraconsistent logics. Stud. Logica. **99**(1–3), 31–60 (2011)
4. Arieli, O., Avron, A., Zamansky, A.: Maximal and premaximal paraconsistency in the framework of three-valued semantics. Stud. Logica. **97**(1), 31–60 (2011)
5. Asenjo, F.G.: A calculus of antinomies. Notre Dame J. Formal Logic **7**, 103–106 (1966)
6. Avron, A.: On an implication connective of RM. Notre Dame J. Formal Logic **27**, 201–209 (1986)
7. Avron, A.: Natural 3-valued logics: characterization and proof theory. J. Symbolic Logic **56**(1), 276–294 (1991)
8. Avron, A.: On the expressive power of three-valued and four-valued languages. J. Logic Comput. **9**(6), 977–994 (1999)
9. Avron, A.: Combining classical logic, paraconsistency and relevance. J. Appl. Logic **3**, 133–160 (2005)

10. Avron, A., Lev, I.: Non-deterministic multi-valued structures. J. Logic Comput. **15**, 241–261 (2005)
11. Avron, A., Ben-Naim, J., Konikowska, B.: Cut-free ordinary sequent calculi for logics having generalized finite-valued semantics. Log. Univers. **1**, 41–69 (2006)
12. Avron, A., Konikowska, B., Zamansky, A.: Cut-free sequent calculi for C-systems with generalized finite-valued semantics. J. Logic Comput. **23**, 517–540 (2013)
13. Avron, A., Zamansky, A.: Non-deterministic semantics for logical systems—a survey. In: Gabbay, D., Guenther, F. (eds) Handbook of Philosophical Logic, vol. 16, pp. 227–304. Springer (2011)
14. Batens, D.: Paraconsistent extensional propositional logics. Logique et Analyse **90–91**, 195–234 (1980)
15. Belnap, N.: How a computer should think. In: Ryle, G. (ed.) Contemporary Aspects of Philosophy, pp. 30–56. Oriel Press (1977)
16. Carnielli, W., Coniglio, M., Marcos, J.: Logics of formal inconsistency. In: Gabbay, D., Guenthner, F. (eds.) Handbook of Philosophical Logic, 2nd edn., pp, vol. 14, pp. 1–93. Springer (2007)
17. Ciucci, D., Dubois, D.: A modal theorem-preserving translation of a class of three-valued logics of incomplete information. J. Appl. Non-Class. Logics **23**(4), 321–352 (2013)
18. Ciucci, D., Dubois, D.: From possibility theory to paraconsistency (2015). A chapter in this book
19. da Costa, N.: On the theory of inconsistent formal systems. Notre Dame J. Formal Logic **15**, 497–510 (1974)
20. D'Ottaviano, I.: The completeness and compactness of a three-valued first-order logic. Revista Colombiana de Matematicas **XIX**(1–2), 31–42 (1985)
21. D'Ottaviano, I., da Costa, N.C.: Sur un problèm de Jakowśki. C. R. Acad Sc. Paris **270**(Sèrie A), 1349–1353 (1970)
22. Epstein, R.L.: The Semantic Foundation of Logic, vol. I, Propositional Logics. Kluwer (1990)
23. Fitting, M.: Kleene's three valued logics and their children. Fundamenta Informaticae **20**(1–3), 113–131 (1994)
24. Gentzen, G.: Investigations into Logical Deduction (1934). In: Szabo, M.E. (ed.) In German. An English translation appears in 'The Collected Works of Gerhard Gentzen'. North-Holland (1969)
25. Ginsberg, M.: Multi-valued logics: a uniform approach to reasoning in AI. Comput. Intell. **4**, 256–316 (1988)
26. Gottwald, S.: A treatise on many-valued logics. In: Studies in Logic and Computation, vol. 9. Research Studies Press, Baldock (2001)
27. Kleene, S.C.: Introduction to Metamathematics. Van Nostrand (1950)
28. Łukasiewicz, J.: On 3-valued logic. Ruch Filosoficzny **5**, 169–171 (1920). English translation: McCall, S. (ed.) Polish Logic 1920–1939, pp. 15–18. Oxford University Press, Oxford (1967)
29. Łukasiewicz, J.: Selected Works. North Holland (1970). L. Borkowski (ed)
30. Malinowski, G.: Many-Valued Logics. Clarendon Press (1993)
31. Priest, G.: Logic of paradox. J. Philos. Logic **8**, 219–241 (1979)
32. Priest, G.: Reasoning about truth. Artif. Intell. **39**, 231–244 (1989)
33. Priest, G.: Minimally inconsistent LP. Stud. Logica. **50**, 321–331 (1991)
34. Rozoner, L.I.: On interpretation of inconsistent theories. Inf. Sci. **47**, 243–266 (1989)
35. Sette, A.M.: On propositional calculus P_1. Mathematica Japonica **16**, 173–180 (1973)
36. Shoesmith, D.J., Smiley, T.J.: Deducibility and many-valuedness. J. Symbolic Logic **36**, 610–622 (1971)
37. Shoesmith, D.J., Smiley, T.J.: Multiple Conclusion Logic. Cambridge University Press (1978)
38. Sobociński, B.: Axiomatization of a partial system of three-value calculus of propositions. J. Comput. Syst. **1**, 23–55 (1952)
39. Urquhart, A.: Many-valued logic. In: Gabbay, D., Guenthner, F. (eds.) Handbook of Philosophical Logic, 2nd edn., vol. II, pp. 249–295. Kluwer (2001)

Chapter 5
Strong Three-Valued Paraconsistent Logics

Jean-Yves Beziau and Anna Franceschetto

Abstract After describing the two formulations of the principle of non contradiction in modern logic $T \vdash \neg(p \wedge \neg p)$ (NC) and $T, p, \neg p \vdash q$ (EC) and explaining that three-valued matrices can be used to easily prove their independence, we investigate the possibilities to construct strong paraconsistent negations, i.e., for which neither (NC) nor (EC) holds, using three-valued logical matrices.

Keywords Principle of non contradiction · Negation · Paraconsistent logic · Many-valued logic

Mathematics Subject Classification (2000) Primary 03B53 · Secondary 03B50

5.1 Introduction

A paraconsistent negation can roughly speaking be defined as a negation not obeying the principle of non contradiction.[1]

In modern logic there are two main formulations of this principle:

(NC) $T \vdash \neg(p \wedge \neg p)$
(EC) $T, p, \neg p \vdash q$

It is possible to have several readings or interpretations of these expressions, depending on which framework we are working with. Here we consider that $\vdash$ is a

[1]For a general discussion about how a paraconisstent negation can be defined, see [7, 8].

J.-Y. Beziau (✉)
UFRJ - Federal University of Rio de Janeiro,
CNPq - Brazilian Research Council,Rio de Janeiro, Brazil
e-mail: jyb@jyb-logic.org

A. Franceschetto
Department of Mathematics, University of Padua, Padua, Italy
e-mail: anna.franceschetto@studenti.unipd.it

J.-Y. Beziau et al. (eds.), *New Directions in Paraconsistent Logic*,
Springer Proceedings in Mathematics & Statistics 152,
DOI 10.1007/978-81-322-2719-9_5

Tarksian structural consequence relation,[2] T is any theory (set of formulas), p and q any formulas. We work here with a language with only negation $\neg$, disjunction $\vee$, and conjunction Û. We require conjunction and disjunction to behave classically, so for us $T, p, \neg p \vdash q$ is equivalent to $T, p \wedge \neg p \vdash q$.

There are some paraconsistent negations for which neither (NC) nor (EC) is valid like the paraconsistent logic C1 of Newton da Costa [11], but generally the emphasis is put on the rejection of (EC), so that it is common to consider as paraconsistent a negation not obeying (EC) but obeying (NC). It is not clear if it really makes sense. We define here a *strong* paraconsistent negation as a paraconsistent negation obeying neither (NC) nor (EC) and we systematically examine what kind of strong paraconsistent negations can be constructed using three-valued matrices.

5.2 Independency of EC and NC and Basic Framework

We first present a simple but interesting result concerning three-valued matrices and the two forms of the principle of noncontradiction. We consider an additional value $\frac{1}{2}$ besides truth 1 and falsity 0 and we consider the following tables for negation and conjunction:

	$\neg$
0	1
$\frac{1}{2}$	$\frac{1}{2}$
1	0

3-valued table M for Negation

$\wedge$	0	$\frac{1}{2}$	1
0	0	0	1
$\frac{1}{2}$	0	$\frac{1}{2}$	$\frac{1}{2}$
1	0	$\frac{1}{2}$	1

3-valued table M for Conjunction

Then we have the following table:

p	$\neg p$	$p \wedge \neg p$	$\neg(p \wedge \neg p)$
0	1	0	1
$\frac{1}{2}$	$\frac{1}{2}$	$\frac{1}{2}$	$\frac{1}{2}$
1	0	0	1

[2] This means that reflexivity, monotonicity, transitivity hold as well as substitution, see [17].

Using the standard framework to define logical truth and consequence, if we consider that $\frac{1}{2}$ is undesignated, then (NC) is not valid, since $\neg(p \wedge \neg p)$ can be undesignated, but (EC) is valid, since $p \wedge \neg p$ is always undesignated. Now if we consider that $\frac{1}{2}$ is designated we have exactly the reverse situation: (NC) is valid and (EC) is not valid.

The two above tables therefore do not allow us to construct a strong paraconsistent negation.

Remark Asenjo's logic of antinomy [1] (probably the first three-valued paraconsistent logic), da Costa/D'Ottaviano's logic $J3$ [13] and Priest's logic LP [20] have the above tables for negation and conjunction and they are therefore not strong paraconsistent logics. $J3$ was originally presented with disjunction, not conjunction. The presentation with conjunction can be found in Chap. 9 of Epstein's book [14].

These tables are the same as the one for Łukasiewicz's logic L3 [18] and Kleene's logic K3 [16]. The difference is that in these logics $\frac{1}{2}$ is considered as undesigned; (NC) is therefore not valid but (EC) is valid. L3 and K3 are not strong paraconsistent logics and they are in fact generally not considered as paraconsistent logic because in modern logic paraconstistency has been defined on the basis of the rejection of (EC).

Let us point out that the notation $\frac{1}{2}$ suggests that the third value, whether it is designed or undesigned, is in between 0 and 1 and that the table of $\wedge$ is built according to the idea that the value of the conjunction is the smallest value of the two terms of the conjunction based on the linear order 0, $\frac{1}{2}$, 1. Asenjo chose 0 for 1, 1 for 0 and 2 for $\frac{1}{2}$. This is rather confusing. In our paper on the history of truth-values [9], we are discussing the question of ordering of truth-values.

Let us note that due to a general result we have presented in [6], Asenjo's logic of antinomy, $J3$ and LP are not algebraizable. This result states that a logic where we do not have (EC) but where we have (NC) and the double negation law, is not algebraizable.

In the present paper we show that it is possible to proceed in a different way. First we need to set up a general framework we will work with.

It is not difficult to prove that if we have a third value any truth-table for conjunction obeying the condition that the value of the conjunction is designated iff the two terms of the conjunction are designated defines a classical conjunction, i.e., a conjunction obeying the following three laws[3]:

($\wedge$1) $T, p \wedge q \vdash p$ ($\wedge$2) $T, p \wedge q \vdash q$ ($\wedge$3) $T, p, q \vdash p \wedge q$

Also if we have a disjunction defined by a truth-table according to which the value of the disjunction is undesignated iff the two terms of the disjunction are undesignated, then this disjunction is classical, i.e., it is a disjunction obeying the following three laws[4]:

($\vee$1) If $T, p \vdash r$ and $T, q \vdash r$ then $T, p \vee q \vdash r$

($\vee$2) $T, p \vdash p \vee q$ ($\vee$3) $T, q \vdash p \vee q$

---

[3] We are working in abstract logic, not in proof theory, so we are not considering that these are rules.

[4] Same remark as in the previous footnote.

We will call *neo-classical* tables for conjunction and disjunction obeying the above conditions (note that these definitions work for all many-valued logics: three or any number of values). These conditions can be represented by the following two tables where $+$ holds for designated and $-$ holds for undesignated:

$\wedge$	$-$	$+$
$-$	$-$	$-$
$+$	$-$	$+$

$\vee$	$-$	$+$
$-$	$-$	$+$
$+$	$+$	$+$

Neo-classical tables for
Conjunction and Disjunction

The result then goes as follows: *Neo-classical tables for conjunction and disjunction define classical conjunction and disjunction.*

In this work we will moreover consider only truth-tables which are *conservative extensions* of the classical bivalent ones, that is to say the classical parts of those truth-tables remain unchanged. We also require *symmetry*, i.e., $\wedge(x, y) = \wedge(y, x)$ and $\vee(x, y) = \vee(y, x)$, for any truth-values x and y. Note however that the above result does not depend on these additional conditions.

Let us now called a neo-classical truth-table for negation, a table as follows:

	$\neg$
$-$	$+$
$+$	$-$

Neo-classical table for Negation

It is easy to show that a negation defined by a neo-classical table is necessarily classical. This means that:

(RA) If $T, \neg p \vdash q$ and $T, \neg p \vdash \neg q$ then $T \vdash p$

In this case (EC) holds—see [5] for the different laws for pure negation and their interrelations. And if we have a classical conjunction, (NC) also holds.

Moreover if we have a (scheme of) formula s (for example, the conjunction of two formulas: $a \wedge b$) such that $v(s)$ is designated iff $v(\neg s)$ is undesignated, then we have:

(RA-s) If $T, \neg s \vdash q$ and $T, \neg s \vdash \neg q$ then $T \vdash s$

and therefore (NC-s) $T \vdash \neg(s \wedge \neg s)$ and (EC-s) $T, s, \neg s \vdash p$

In this case we say that s *behaves classically*.

To construct a 3-valued paraconsistent negation which obeys neither (NC) nor (EC), based on the logic of classical conjunction and disjunction, we have to change the above M-tables for negation and/or for conjunction. Furthermore we need a table of negation which is not neo-classical (but we will consider only conservative extensions of the truth-table of classical negation). And finally it is sufficient to have tables for conjunction and disjunction which are neo-classical.

The three-valued paraconsistent logics we are presenting here are part of the family 8 K of logics described by João Marcos [19] (but we are not explicitly dealing with implication). A recent systematic approach to paraconsistent three-valued logics is the paper by Arielli and Avron [2] included in this volume. However, in both cases the authors are not focusing on strong paraconsistent negations and are not dealing explicitly with the two systems of three-valued paraconsistent logic we are distinguishing here.

5.3 Changing the Standard 3-Valued Table for Negation

Let us first see what we can do by changing the table for negation. We will now use k for the third value to stay neutral about its position in relation with 0 and 1. If we choose the table:

	$\neg$
0	1
k	0
1	0

whether k is designated or undesignated, $p \wedge \neg p$ is undesignated when the value of p is k since 0 is undesignated (taking into consideration that the conjunction should be defined by a neo-classical table). This can be represented by the following table (where undesignated is represented by the sign $-$):

p	$\neg p$	$p \wedge \neg p$
0	1	0
k	0	$-$
1	0	0

Therefore (EC) holds, so $\neg$ is not a strong paraconsistent negation.

The other option is the following:

	$\neg$
0	1
k	1
1	0

If k is undesignated, then the value of $p \wedge \neg p$ has to be k when the value of p is k. The value of $p \wedge \neg p$ is always undesigned, therefore (EC) holds and the value of $\neg(p \wedge \neg p)$ is always 1, therefore (NC) holds, so $\neg$ is not a strong paraconsistent negation. This can be represented by the following table where we use the notation $k-$ to specify that k is undesignated:

p	$\neg p$	$p \wedge \neg p$	$\neg(p \wedge \neg p)$
0	1	0	1
$k-$	1	$k-$	1
1	0	0	1

We have therefore to consider that k is designated. We use the notation $k+$ and we have the following table:

	$\neg$
0	1
$k+$	1
1	0

We have then two options to define conjunction. The first option is described by the following table where $+$ denotes a designated value (k or 1) and where we use the notation $k+$ to emphasize that k is designated (we apply here neo-classical and symmetry conditions):

$\wedge$	0	$k+$	1
0	0	0	0
$k+$	0	$+$	$k+$
1	0	$k+$	1

Note that if the designated value $+$ at the middle of this table is $k+$ we have a similar table as the table M for conjunction presented in the first section. Anyway in both cases ($+$ being $k+$ or 1) we have:

p	$\neg p$	$p \wedge \neg p$	$\neg(p \wedge \neg p)$
0	1	0	1
$k+$	1	$k+$	1
1	0	0	1

This shows that $\neg(p \wedge \neg p)$ is always 1 and that (NC) holds, so $\neg p$ is not a strong paraconsistent negation.

The second option is described by the following table (we also apply here neo-classical and symmetry conditions):

$\wedge$	0	$k+$	1
0	0	0	0
$k+$	0	$+$	1
1	0	1	1

We have then:

p	$\neg p$	$p \wedge \neg p$	$\neg(p \wedge \neg p)$
0	1	0	1
$k+$	1	1	0
1	0	0	1

This shows that neither (EC) nor (NC) holds, hence we have a strong paraconsistent negation.

According to this table $p \wedge \neg p$ behaves classically. Moreover the formula $\neg p \wedge \neg(p \wedge \neg p)$ behaves as a classical negation as shown by the following table:

p	$\neg p$	$p \wedge \neg p$	$\neg(p \wedge \neg p)$	$\neg p \wedge \neg(p \wedge \neg p)$
0	1	0	1	1
$k+$	1	1	0	0
1	0	0	1	0

This is a similar phenomenon as in the paraconsistent logic $C1$ of Newton da Costa [12].

In the above table for conjunction the designated value + can be interpreted as $k+$ or 1, corresponding to the two following tables:

$\wedge$	0	$k+$	1
0	0	0	0
$k+$	0	1	1
1	0	1	1

$\wedge$	0	$k+$	1
0	0	0	0
$k+$	0	$k+$	1
1	0	1	1

If we choose the left table, this means that the conjunction $a \wedge b$ of any formulas a and b behaves classically.

To minimize molecularization (molecular propositions behaving classically) it is better to choose the right table. We will call $L3A$ the 3-valued logic constructed using this right table for conjunction and the modified table for negation.

Remark Sette's logic $P1$ has the left table for conjunction (cf [21], p.12) and the same truth-table for negation as discussed here (cf [21], p.7). Sette uses T_0 for 1, T_1 for $k+$, F for 0.

5.4 Not Changing the Standard 3-Valued Table for Negation

Let us present again the M-table for negation, but using k instead of $\frac{1}{2}$ to avoid any speculation:

	$\neg$
0	1
k	k
1	0

If we consider that k is undesignated (we use the notation $k-$), then (EC) is valid as shown by the following table:

p	$\neg p$	$p \wedge \neg p$
0	1	0
$k-$	$k-$	$k-$
1	0	0

We have therefore to choose k as designated (we use the notation $k+$):

	$\neg$
0	1
$k+$	$k+$
1	0

This generally what has been chosen by the people working in three-valued paraconsistent logic, but without changing the table for conjunction, which, as we have seen in Sect. 5.2, necessarily leads to paraconsistent negations which are not strong.

First let us note that the table:

$\wedge$	0	$k+$	1
0	0	0	0
$k+$	0	$k+$	$+$
1	0	$+$	1

together with the above table for negation lead to the following table:

p	$\neg p$	$p \wedge \neg p$	$\neg(p \wedge \neg p)$
0	1	0	1
$k+$	$k+$	$k+$	$k+$
1	0	0	1

showing that (NC) is valid. So if we consider that the conjunction of two terms whose both values are $k+$ is $k+$, we have no strong paraconsistent negation.

We have therefore to choose the following situation:

$\wedge$	0	$k+$	1
0	0	0	0
$k+$	0	1	+
1	0	+	1

Independently of the interpretation of +, we have:

p	$\neg p$	$p \wedge \neg p$	$\neg(p \wedge \neg p)$
0	1	0	1
$k+$	$k+$	1	0
1	0	0	1

showing that neither (NC) nor (EC) holds, a good situation for strong paraconsistent negations. This table shows that, as in the previous section, $p \wedge \neg p$ behaves classically. And the following table shows that $\neg p \wedge \neg(p \wedge \neg p)$ behaves also here like a classical negation.

p	$\neg p$	$p \wedge \neg p$	$\neg(p \wedge \neg p)$	$\neg p \wedge \neg(p \wedge \neg p)$
0	1	0	1	1
$k+$	$k+$	1	0	0
1	0	0	1	0

We have then two options (respecting symmetry):

$\wedge$	0	$k+$	1
0	0	0	0
$k+$	0	1	1
1	0	1	1

$\wedge$	0	$k+$	1
0	0	0	0
$k+$	0	1	$k+$
1	0	$k+$	1

The option on the left leads to molecularization. We will call $L3B$ the 3-valued logic constructed using the right table for conjunction and the M-table for negation with the third value as designated.

5.5 Two Interesting Strong 3-Valued Paraconsistent Logics

Let us present here the two 3-valued logics with a strong paraconsistent negation we have selected, minimizing molecularization, one changing the standard M-table for negation, the logic $L3A$, one not changing the standard M-table for negation, the logic $L3B$, but in both cases changing the M-table for conjunction:

$L3A$

	$\neg$
0	1
$k+$	1
1	0

$\wedge$	0	$k+$	1
0	0	0	0
$k+$	0	$k+$	1
1	0	1	1

$L3B$

	$\neg$
0	1
$k+$	$k+$
1	0

$\wedge$	0	$k+$	1
0	0	0	0
$k+$	0	1	$k+$
1	0	$k+$	1

In $L3A$ and $L3B$ it is possible to define a classical negation as $\neg p \wedge \neg(p \wedge \neg p)$, like in da Costa's logic $C1$, therefore classical logic is translatable into these logics, giving more examples of the translation's paradox [15]. Also, as a consequence, in these logics not every formula can be considered as a so-called "dialetheia" by contrast to the three-valued logic LP (cf. [10]).

Let us note that the option $L3B$ is better to avoid molecularization, since in $L3A$ we lost the third value when applying negation, which means that a negated formula behaves classically. According to the truth-table for conjunction of $L3A$ we can interpret $k+$ as "super-true" that could be denoted by 2. If we consider the linear order between 0, 1, 2, then this truth-table for conjunction respects the idea to take the smallest value. The strategy of the smallest doesn't work for the truth-table for conjunction of $L3B$ whether we interpret $k+$ as in-between or above. On the other hand its truth-table for negation is quite normal. The case of the one of $L3A$ is not really clear, the negation transforming super-true into true.

Let us also point out that the law of double negation holds for $L3B$ but not for $L3A$, as shown by the following tables:

p	$\neg p$	$\neg\neg p$
0	1	0
$k+$	1	0
1	0	0

$L3A$

p	$\neg p$	$\neg\neg p$
0	1	0
$k+$	$k+$	$k+$
1	0	1

$L3B$

In $L3A$ we have $T, \neg\neg p \vdash p$ but $T, p \nvdash \neg\neg p$. For example a being an atomic formula, we have $a \nvdash \neg\neg a$.

But this does not mean that $L3B$ is stronger than $L3A$. Because, due to molecularization, in $L3A$ we have $\neg a, \neg\neg a \vdash q$, which is not the case in $L3B$. So $L3A$ and $L3B$ are incomparable.

We will now develop a further comparative study of these two strong paraconsistent logics introducing disjunction and studying De Morgan laws.

5.6 Disjunction and De Morgan Laws

5.6.1 *Law of Excluded Middle*

Considering any neo-classical disjunction, we can see that the law of excluded middle holds for both $L3A$ and $L3B$:

p	$\neg p$	$p \vee \neg p$
0	1	1
$k+$	1	$+$
1	0	1

$L3A$

p	$\neg p$	$p \vee \neg p$
0	1	1
$k+$	$k+$	$+$
1	0	1

$L3B$

We have used here neo-classical truth-tables for disjunction without specifying the value of $+$. In the next subsections we will examine the best solutions to build truth-tables of disjunction for these logics relatively to De Morgan laws

5.6.2 *De Morgan Laws*

We remember and label the four basic De Morgan laws in the following table:

$D1a$	$\neg(p \wedge q) \vdash \neg p \vee \neg q$	$D1b$	$\neg p \vee \neg q \vdash \neg(p \wedge q)$
$D2a$	$\neg(p \vee q) \vdash \neg p \wedge \neg q$	$D2b$	$\neg p \wedge \neg q \vdash \neg(p \vee q)$

De Morgan Laws

5.6.3 *De Morgan Laws and Disjunction for $L3A$*

Using the tables of conjunction and negation of $L3A$ as well as the neo-classical conditions for disjunction, we have the following table[5]:

p	q	$\neg p$	$\neg q$	$p \wedge q$	$\neg(p \wedge q)$	$\neg p \vee \neg q$	$\neg p \wedge \neg q$	$\neg(p \vee q)$	$p \vee q$
0	$k+$	1	1	0	1	1	1	?	$+$
$k+$	0	1	1	0	1	1	1	?	$+$
$k+$	$k+$	1	1	$k+$	1	1	1	?	$+$
$k+$	1	1	0	1	0	1	0	?	$+$
1	$k+$	0	1	1	0	1	0	?	$+$

[5]Since we are working with truth-tables which are conservative extensions of the classical ones, we omit the classical parts in all tables built to check De Morgan laws hereafter.

In this table + is any designated value and ? is a value that depends on how we construct the table for disjunction. We see that De Morgan Law $D1b$ is not valid as shown by the 4th and 5th lines of values of the table. On the other hand whatever our choice for ? is, $D1a$ will hold.

If we want the De Morgan Laws $D2$ to hold, we have to fill the penultimate column as follows (remembering that the only undesignated value is 0):

p	q	$\neg p$	$\neg q$	$p \wedge q$	$\neg(p \wedge q)$	$\neg p \vee \neg q$	$\neg p \wedge \neg q$	$\neg(p \vee q)$	$p \vee q$
0	$k+$	1	1	0	1	1	1	+	+
$k+$	0	1	1	0	1	1	1	+	+
$k+$	$k+$	1	1	$k+$	1	1	1	+	+
$k+$	1	1	0	1	0	1	0	0	+
1	$k+$	0	1	1	0	1	0	0	+

Taking into consideration the table for negation, the two last columns should be as follows:

p	q	$\neg p$	$\neg q$	$p \wedge q$	$\neg(p \wedge q)$	$\neg p \vee \neg q$	$\neg p \wedge \neg q$	$\neg(p \vee q)$	$p \vee q$
0	$k+$	1	1	0	1	1	1	1	$k+$
$k+$	0	1	1	0	1	1	1	1	$k+$
$k+$	$k+$	1	1	$k+$	1	1	1	1	$k+$
$k+$	1	1	0	1	0	1	0	0	1
1	$k+$	0	1	1	0	1	0	0	1

The table for disjunction must therefore be as follows:

$\vee$	0	$k+$	1
0	0	$k+$	1
$k+$	$k+$	$k+$	1
1	1	1	1

This is the standard table, the same as in L3, K3, Asenjo's logic of antinomy, J3, LP.

5.6.4 De Morgan Laws and Disjunction for L3B

Using the tables of conjunction and negation of $L3B$ as well as the neo-classical conditions for disjunction, we have the following table:

p	q	$\neg p$	$\neg q$	$p \wedge q$	$\neg(p \wedge q)$	$\neg p \vee \neg q$	$\neg p \wedge \neg q$	$\neg(p \vee q)$	$p \vee q$
0	$k+$	1	$k+$	0	1	$+$	$k+$	?	$+$
$k+$	0	$k+$	1	0	1	$+$	$k+$	?	$+$
$k+$	$k+$	$k+$	$k+$	1	0	$+$	1	?	$+$
$k+$	1	$k+$	0	$k+$	$k+$	$+$	0	?	$+$
1	$k+$	0	$k+$	$k+$	$k+$	$+$	0	?	$+$

In this table $+$ is any designated value and ? is a value that depends on how we construct the table for disjunction. We see that De Morgan law $D1b$ is not valid as shown by the 3rd line of values of the table. On the other hand whatever our choice for ? is, $D1a$ will hold.

If we want the De Morgan laws $D2$ to hold, we have to fill the penultimate column as follows (remembering that the only undesignated value is 0):

p	q	$\neg p$	$\neg q$	$p \wedge q$	$\neg(p \wedge q)$	$\neg p \vee \neg q$	$\neg p \wedge \neg q$	$\neg(p \vee q)$	$p \vee q$
0	$k+$	1	$k+$	0	1	$+$	$k+$	$+$	$+$
$k+$	0	$k+$	1	0	1	$+$	$k+$	$+$	$+$
$k+$	$k+$	$k+$	$k+$	1	0	$+$	1	$+$	$+$
$k+$	1	$k+$	0	$k+$	$k+$	$+$	0	0	$+$
1	$k+$	0	$k+$	$k+$	$k+$	$+$	0	0	$+$

Taking into consideration the table for negation, the two last columns should be as follows:

p	q	$\neg p$	$\neg q$	$p \wedge q$	$\neg(p \wedge q)$	$\neg p \vee \neg q$	$\neg p \wedge \neg q$	$\neg(p \vee q)$	$p \vee q$
0	$k+$	1	$k+$	0	1	$+$	$k+$	$k+$	$k+$
$k+$	0	$k+$	1	0	1	$+$	$k+$	$k+$	$k+$
$k+$	$k+$	$k+$	$k+$	1	0	$+$	1	$k+$	$k+$
$k+$	1	$k+$	0	$k+$	$k+$	$+$	0	0	1
1	$k+$	0	$k+$	$k+$	$k+$	$+$	0	0	1

The case is similar to $L3A$ in the sense that we must have a standard truth-table for disjunction:

$\vee$	0	$k+$	1
0	0	$k+$	1
$k+$	$k+$	$k+$	1
1	1	1	1

So at the end the table for $L3B$ is as follows:

p	q	$\neg p$	$\neg q$	$p \wedge q$	$\neg(p \wedge q)$	$\neg p \vee \neg q$	$\neg p \wedge \neg q$	$\neg(p \vee q)$	$p \vee q$
0	$k+$	1	$k+$	0	1	1	$k+$	$k+$	$k+$
$k+$	0	$k+$	1	0	1	1	$k+$	$k+$	$k+$
$k+$	$k+$	$k+$	$k+$	1	0	$k+$	1	$k+$	$k+$
$k+$	1	$k+$	0	$k+$	$k+$	$k+$	0	0	1
1	$k+$	0	$k+$	$k+$	$k+$	$k+$	0	0	1

5.6.5 De Morgan Laws and the Replacement Theorem

So at the end the situation of $L3A$ and $L3B$ is the same relatively to De Morgan laws. The fact that the law $D1b$ $\neg p \vee \neg q \vdash \neg(p \wedge q)$ is not valid is not necessarily surprising if we consider that in both $L3A$ and $L3B$, we have $\vdash p \vee \neg p$ but $\nvdash \neg(p \wedge \neg p)$ and that through double negation and replacement we have from $D1b$, $p \vee \neg p \vdash \neg(\neg p \wedge p)$.

However is the replacement theorem valid in $L3A$ and $L3B$? We have no counterexamples to provide here, but it is not difficult to prove that both $L3A$ and $L3B$ are extensions of the paraconsistent logic $C1$ (even of the extension $C1+$ of $C1$ introduced in [3], see also [4]) and I. Urbas has proven that there are no self-extensional extensions of $C1$ (cf. [22]). We will develop this topic in a further paper.

Acknowledgments This work starts during a seminar at the Federal University of Rio de Janeiro (August 2013–December 2013) conducted by J.Y. Beziau during the visit of Anne Franceschetto who was visiting Brazil to know more about paraconsistent logic. Other students, in particular Rodrigo de Almeida and Edson Vinicius Bezerra, had an active participation to this seminar

References

1. Asenjo, F.G.: A calculus of antinomies. Notre Dame J. Form. Log. **7**, 103–105
2. Arieli, O., Avron, A.: Three-valued paraconsistent propositional logics. In: Beziau, J.-Y., Chakraborty, M., Dutta, S. (eds.) New Directions in Paraconsistent Logic. Springer India (2015), pp. 91–129
3. Beziau, J.-Y.: Logiques construites suivant les méthodes de da Costa. Log. Anal. **131–132**, 259–272 (1990)
4. Beziau, J.-Y.: Nouveaux résultats et nouveau regard sur la logique paraconsistante C1. Log. Anal. **141–142**, 45–48 (1993)
5. Beziau, J.-Y.: Théorie législative de la négation pure. Log. Anal. **147–148**, 209–225 (1994)
6. Beziau, J.-Y.: Idempotent full paraconsistent negations are not algebraizable. Notre Dame J. Form. Log. **39**, 135–139 (1998)
7. Beziau, J.-Y.: What is paraconsistent logic?. In: Batens, D. et al. (eds.) Frontiers of Paraconsistent Logic, pp. 95–111. Research Studies Press, Baldock (2000)
8. Beziau, J.-Y.: Are paraconsistent negations negations? In: Carnielli, W. et al. (eds.) Paraconsistency: the Logical Way to the Inconsistent, pp. 465–486. Marcel Dekker, New-York (2002)
9. Beziau, J.-Y.: History of the concept of truth-value. In: Gabbay, D.M., Pelletier, J., Woods, J. (eds.) Handbook of the History of Logic, vol. 11. Elsevier, Amsteradm (2012)
10. Beziau, J.-Y.: Trivial dialetheism and the logic of paradox. Log. Logic. Philos. **25** (2016)
11. da Costa, N.C.A.: Calculs propositionnels pour les systémes formels inconsistants. Cr. R. Acad. Sc. Paris **257**, 3790–3793 (1963)

12. da Costa, N.C.A., Guillaume, M.: Négations composées et Loi de Peirce dans les systémes Cn. Portugalia Mathematica **24**, 201–210 (1965)
13. D'Ottaviano, I.M.L., da Costa, N.C.A.: Sur un probléme de Jaśkowski. Cr. R. Acad. Sc. Paris **270**, 1349–1353 (1970)
14. Epstein, R.L.: The Semantic Foundations of Logic. Volume 1: Propositional Logics. Kluwer, Dordrecht (1990)
15. Humberstone, L.: Beziau's translation paradox. Theoria **71**, 138–181 (2005)
16. Kleene, S.: On a notation for ordinal numbers. J. Symbolic Log. **3**, 150–155 (1938)
17. Łoś, J., Suszko, R.: Remarks on sentential logics. Indigationes Math. **10**, 177–183 (1958)
18. Łukasiewicz, J.: O logice trójwartościowej. Ruch Filozoficny **5**, 170–171 (1920)
19. Marcos, J.: 8K solutions and semi-solutions to a problem of da Costa
20. Priest, G.: The logic of paradox. J. Philos. Log. **8**, 219–241 (1979)
21. Sette, A.M.: On the propositional calculus P1. Notas e comunicacões de matemática, **17** (1971)
22. Urbas, I.: Paraconsistency and the C-ysytems of da Costa. Notre Dame J. Form. Log. **30**, 583–597 (1989)

Chapter 6
An Approach to Paraconsistent Multivalued Logic: Evaluation by Complex Truth Values

J. Nescolarde-Selva, J.L. Usó-Doménech and K. Alonso-Stenberg

Abstract The main purpose of the paper is to connect some kind of dialetheism to the use of complex truth values, with new definitions of basic truth-functional connectives that allow for p, \not p to both be true. 'True' is interpreted as $|p| = 1$, 'False' as $|p| = 0$; other values are dispensed with. New definitions of basic truth-functional connectives then allow for "p and not p" to be true. A propositional logic is discussed with the set of connectives including negation, conjunction, disjunction, implication, concordance, discordance, complementary, and equivalence. The authors introduce truth values of propositions, which belong to a subset E, of an uncountable semi-ring F and valuations of propositions, which can be obtained from truth values with the help of a function $V : E \to [0, 1]$ satisfying simple properties. Finally, a paraconsistent Boolean logic is introduced.

Keywords Circle of truth · Contradiction · Complex number · Denier · Logic coordinations · Paraconsistency · Propositions · Truth values

Mathematics Subject Classification (2000) Primary 03B50 · Secondary 03B45

6.1 Introduction

The main objective of the authors is to establish a theory of truth value evaluation for paraconsistent logics, unlike others who are in the literature (Asenjo [2]; Avron [3]; Belnap [4]; Bueno [5]; Carnielli et al. [7]; Dunn [10]; Tanaka et al. [20]), with the goal of using that paraconsistent logic in analyzing ideological, mythical,

J. Nescolarde-Selva (✉) · J.L. Usó-Doménech · K. Alonso-Stenberg
Department of Applied Mathematics, University of Alicante, Alicante, Spain
e-mail: josue.selva@ua.es

J.L. Usó-Doménech
e-mail: kishukaze@hotmail.com

K. Alonso-Stenberg
e-mail: kas1@alu.ua.es

J.-Y. Beziau et al. (eds.), *New Directions in Paraconsistent Logic*,
Springer Proceedings in Mathematics & Statistics 152,
DOI 10.1007/978-81-322-2719-9_6

religious, and mystic belief systems (Nescolarde-Selva and Usó-Doménech [11–14]; Usó-Doménech and Nescolarde-Selva [21]). The doctrine of *coincidentia oppositorum*, the interpenetration, interdependence, and unification of opposites has long been one of the defining characteristics of *mystical* (as opposed to philosophical) thought. Mystics of various persuasions generally held that such paradoxes are the best means of expressing within language, truths about a whole that is sundered by the very operation of language itself. Any effort, it is said, to analyze these paradoxes and provide them with logical sense is doomed from the start because logic itself rests upon assumptions, such as the principles PNC and PEM, that are violated by the mystical ideas. The *coincidentia oppositorum* is a common trope in many religious traditions, particularly those with a mystical or initiatory aspect.

In the current paradigm of consciousness, duality is perceived be a binary state of mutual exclusion. One sees this notion reflected in human thought and language where something must be "*either X or Y*", but not "*both X and Y*". A new paradigm of consciousness is required that no longer operates in a "*dualistic*" notion of "*either/or*", but one that conveys a "*holistic*" notion of "*both/and*". We currently view duality as a disjunctive rather than a conjunctive aspect of being. The difference between this dualistic and holistic paradigm of consciousness can be symbolically expressed in the language of logic. Current dualistic paradigm of consciousness $X \neq Y$; $X \cup Y$ (something is either X or Y); a disjunctive exclusion. Emerging holistic paradigm of consciousness $X = Y$; $X \cap Y$ (something is both X and Y); a conjunctive inclusion. This conceptual paradigm of viewing the world in such an exclusionary and disjunctive dualistic state has been programmed into us by an outdated Cartesian philosophical worldview and a Newtonian scientific view of the universe. This modern paradigm of dualistic thought has been prevalent ever since René Descartes proclaimed. That tradition clings to the principle of noncontradiction, but understood as *rejection of contradiction* hereinafter abbreviated as **RC**. Well, is any approach paraconsistent that rejects this same RC, i.e., admit that certain contradictions can be true (not necessarily all, of course). In particular, today is paraconsistent a treatment of problems such as a philosophy of religion that accommodates certain antinomian assertions and in doing so, offered as underlying logic to build theory, not classical logic is a logic of Aristotelian stamp, but one of the denominated precisely paraconsistent logics.

It seems to us that the philosophy of paraconsistency can propose a concept of modern rationality which will enable us to restore and gradually elaborate in never ending self-criticism "*the vision of the whole*" as a coevolutionary unity of mankind and Nature. To the basics of this modern rationality would belong of the nonexclusive relation between analytical and dialectical thinking, their developmental unity. The desirable unifying can be conceived of in various ways. It follows from this paper that we are skeptical about the proposal to unify analytical and dialectical thinking through a kind of reduction of the latter to the first by applying the idea of paraconsistency. It would mean to reduce the whole to a part. What we propose is to conceive analytical thinking as a part of and a derivative from something more complex and more fundamental.

Numerous paraconsistent logical calculi have been constructed which allow the formula $P \wedge \neg P$ to be true (derivable) under some special conditions and thus tolerate $P \wedge \neg P$ without becoming trivial. To provide some grounds for this theory let us take a look at Aristotle's theory of contrariety from the point of view of modern dialectic. This is detailed account of what Aristotle calls *Antifasis* is to be found in *Metaphysics Book 4*. Aristotle's examples of the four kinds of opposites are: *double* and *half*, *bad* and *good*, *blindness* and *sight*, and *he sits* and *he does not sit*. Aristotle was deeply interested in investigating the modes of opposition and their ontological relevance in the early, middle, and late periods of his philosophizing. He ascribed to the opposites an important role in almost all fields of reality, in Nature, in society as well as in thought, but disagreed with that ontological overestimation of the role of opposites, which he found in many preceding Greek thinkers. The second is his misinterpretation of Heraclitus in the sense of Protagoras' relativism hereby not only the sophistic relativism, but also the Heraclitian anticipations of dialectical ontology.

Aristotle is right in insisting that the denial of this principle would lead to a kind of total trivialization of human thinking and people would become prisoners of a helpless tenet "*which prevents a thing from being made definite by thought*." Now let us compare three following allegedly synonymous formulations. Aristotle took all three as stating the same principle and in different places mutually argues the truth of each of them from the presupposed evidence of each of them.

1. "*Contradictory propositions are not true simultaneously*". This statement is, as already mentioned, acceptable and respected on the new ontology.
2. "*Contradictories cannot be predicated at the same time*". This statement would be unacceptable if interpreted in the following way: (in the European tradition translated as "*contradictio*") is for Aristotle sometimes the conjunction of two sentences (or statements, propositions) of which one affirms what the other denies; sometimes either part of this conjunction; sometimes the negation of any given subject, property, relation, action, etc. (e.g., *man–not-man*, *changing–unchanging*).
3. "*Contraries cannot at the same time belong to the same subject*" if taken, as Aristotle did, as a general principle valid for all entities.

These opposites are, for Heraclitus, to be taken in unity, as constituting in their opposition and unity something identical. If sometimes in the dialectical tradition Heraclitus' position was characterized as claiming not only the unity, but even the identity of opposites, never was the Leibnizian identity meant, allowing us to replace one of the identical expressions and/or concepts by the other mutually and thus to remove completely the opposition.

Some of the adherents of this trend in contemporary logic investigate explicitly also its philosophical presuppositions and implications (Bueno [6]; Carnielli and Marcos [8]). Among other problems, the question of the relationship between the idea of paraconsistency and the traditional and/or contemporary forms of dialectical thinking is being examined. It seems to us that in the philosophy of paraconsistency a differentiation can be observed today. One of the tendencies, represented by Arruda

(Arruda et al. [1]), da Costa (da Costa and Wolf [9]), Quesada [19] while assessing highly important philosophical implications of the logic of paraconsistency, insists upon the view that paraconsistency is closely linked with the theory of logical calculi. The philosophizing logicians of this tendency give, as a rule, only modest hypothetical accounts of the relationship between paraconsistency and dialectic. The other tendency, represented by Priest [15–17]; Priest et al. [18], dares to defend vehemently more radical and ambitious assumptions about the philosophical and scientific implications of paraconsistent logic, concerning not only the relation to dialectic, but also the conception of rationality in general. Let us have a closer critical look at some main claims of the philosophy of paraconsistency from a special point of view, namely, from the point of view of secular (ontopraxeological) dialectic which aims at elaborating a theory of modern rationality taking inspiration from Hegel's critique of Kant and Marx's critique of Hegel. Needless to say, no simple reception of any philosophy of the past is able to cope with our contemporary problems of rationality. References to Kant and Hegel remain mostly mere decoration. Priest's use of the calculus-oriented notion of inconsistency in his interpretation of the so-called Kant/Hegel thesis about the inherently inconsistent Nature of human reason seems to us to be misleading. Supposing we accept Kant's argumentation in his "*Transcendental Dialectics*" as a justification of the statement that our thinking is in its very Nature (apparently, but necessarily) inconsistent, then Hegel's critique of Kant's antinomies should be taken as an attempt at a new consistency which corrects the antinomic dialectic of (apparent, but necessary) inconsistency of human reason in a section of its usage. The dictum of a unitary "*Kant/Hegel thesis*" hides this difference.

Following Priest, we will say that a logical system is *paraconsistent*, if and only if its relation of logical consequence is not "*explosive*", i.e., iff it is not the case that for every formula P and Q, P and not-P entail Q; and we will say a system is *dialectical*, iff it is paraconsistent and yields (or "endorses") *true contradictions*, named "*dialetheias*". A paraconsistent system enables to model theories which in spite of being (classically) inconsistent are not trivial, while a dialectical system goes further, since it permits *dialetheias*, namely contradictions as true propositions. Still following Priest, semantics of dialectical systems provide truth value *gluts* (its worlds or setups are overdetermined); however, truth value *gaps* (opened by worlds or setups which are underdetermined) are considered by Priest to be irrelevant or even improper for dialectical systems. Besides that, sometimes the distinction is drawn between weak and strong paraconsistency, the latter considered as equivalent with dialectics. A reader of recent literature in this field may have an impression that dialectics as strong paraconsistency is more a question of ontology than of logic itself, namely that it states the existence of "*inconsistent facts*" (in our actual world) which should verify dialetheias. One more introductory remark has to be put here: in recent literature of paraconsistency there are no quite unanimous, among paraconsistent logicians generally accepted distinction between paraconsistent and dialectical logical systems. But it remains an open question whether semantically paradoxes express any "*inconsistent facts*".

6.2 Contradiction and Deniers

Let F be an uncountable set whose algebraic structure is at least that of a semi-ring. F is a semi-ring, a ring, or a group.

We lay the following definitions and fundamental axioms:

Axiom 1 Any proposition P has a truth value p, element of a set E which is a part, not countable, and stable for multiplication of the set F.

Axiom 2 Any proposition P is endowed with a valuation $v \in [0, 1]$ such that $v = V(p)$, V application of E on $[0, 1]$ subject to the following conditions:

(1) $V^{-1}(0) = 0$
(2) $V(p_1 p_2) = V(p_1) V(p_2)$

being p_1 and p_2 two truth values.

Axiom 3 Truth value p^* denotes the negation or contradiction of P denoted as $\neg P$ and $V(p + p^*) = 1$.

Let $\boldsymbol{P}_i$ be n propositions, $i = 1, 2, \ldots, n$ of p_i and p_i^* be the truth values of their contradictories. Then

Definition 1 A compound proposition (or logical coordination or logical expression) of order n is a proposition whose truth value c is a function f_n of p_i and p_i^*.

$$c = f_n\left(p_1, p_1^*, p_2, p_2^*, \ldots, p_n, p_n^*\right)$$

f_n values in F; it determines a truth value if $c \in E$. The condition of existence of a compound proposition defined by f_n is $c \in E$, or what is equivalent, $V(c) \in [0, 1]$.

Axiom 4 f_n is a polynomial in which each index $1, 2, \ldots, n$ must be at least once and that all coefficients are equal to unity.

Definition 2 Let $p + p^* = u$ be. u is a denier of the proposition P if the following three conditions are fulfilled:

(a) $u \in E$
(b) $V(u) = 1$; u unitary truth value (from Axiom 3)
(c) $u - p = p^* \in E$ (from Axiom 1)

Paraconsistent logic admits that the contradiction can be true. Then

$$v(P \wedge \neg P) = p(1 - p) = 1 \Rightarrow p - p^2 - 1 = 0 \Rightarrow p^2 - p + 1 = 0$$

This equation has no real roots but admits complex roots $p = e^{\pm i \frac{\pi}{3}}$. This is the result which leads to develop a multivalued logic to complex truth values. The sum of truth values being isomorphic to the vector of the plane, it is natural to relate the function V

to the metric of the vector space R^2. We will adopt as valuations the norms of vectors. E is the set of complex numbers of modulus less or equal to 1 and the function V is such that $V(p) = |p|^2$ and it has satisfied Axiom 2.

Let P be a proposition of fixed truth value $p = |p|\, e^{i\alpha}$. If $p = 0$, α is indeterminate, we agree to take $\alpha = 0$. According to Definition 2, here a denier is a unitary complex number $u = e^{i\theta}$ such that $|u - p| \leq 1$. Putting $\phi = \theta - \alpha$ (Fig. 6.1) this inequality entails $\cos\phi \geq \frac{|p|}{2}$ to be $|\phi| \leq \theta,\ \ \theta = \left|\arccos\frac{|p|}{2}\right|$.
In summary

$$\phi \in [-\theta, \theta];\ \cos\theta = \frac{|p|}{2};\ \theta \in \left[\frac{\pi}{3}, \frac{\pi}{2}\right] \tag{6.1}$$

Deniers u of P form a continuous set: the sector of the *circle of truth* (trigonometric circle) of angle 2θ whose vector p is collinear to the bisector (Fig. 6.2).

A denier is determined by the angle θ. $u(\phi)$ is a bijective function. Contradictory proposition $\neg P$ provides, fixed p, a continuous set of truth values $p^*(\phi)$, $p^* = |p^*|\, e^{i\alpha^*}$, and then

$$\left|p^*\right| \in [1 - |p|, 1];\ \ \phi = 0 \Leftrightarrow \left|p^*\right| = 1 - |p|;\ \ \phi = \pm\theta \Leftrightarrow \left|p^*\right| = 1 \tag{6.2}$$

Putting $\omega - \alpha^* = \phi^*$, $\phi = -\theta \Rightarrow \phi^* = \pi - 2\theta$; $\phi = 0 \Rightarrow \phi^* = 0 \Rightarrow \phi^* = 2\theta - \pi$; in summary:

$$\phi^* \in [2\theta - \pi, \pi - 2\theta] \Rightarrow \phi - \phi^* \in [\theta - \pi, \pi - \theta] \tag{6.3}$$

On the other hand, we have $\alpha^* - \alpha = \phi - \phi^*$.

The truth contradiction $v(P \wedge \neg P) = V\,|pp^*| = 1$ requires $|p| = 1$ where $\theta = \pm\frac{\pi}{3}$ and also $|p^*| = 1$ where $\phi = \pm\frac{\pi}{3}$ and $\phi^* = \pm\frac{\pi}{3}$:

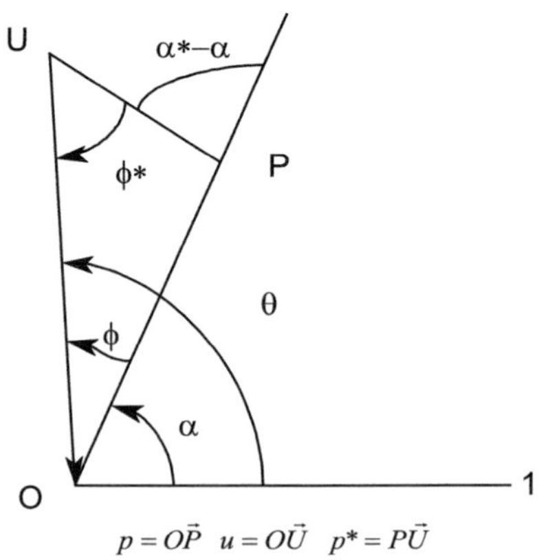

Fig. 6.1 Circle of truth

Fig. 6.2 Location of deniers U of P

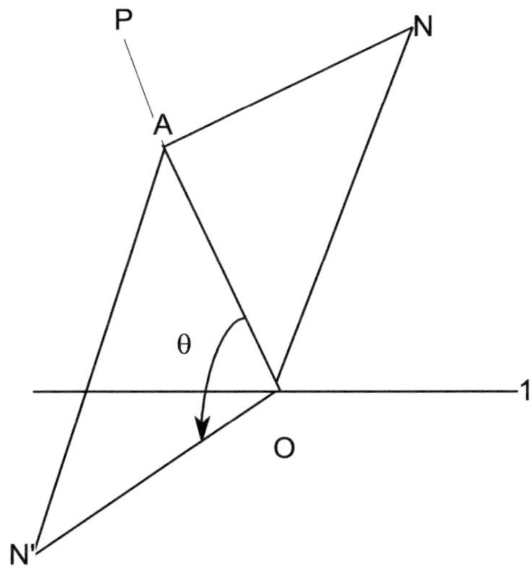

Location of U: arc NPN'; NN' mediatrix of OA

Fig. 6.3 Solutions of P and no P

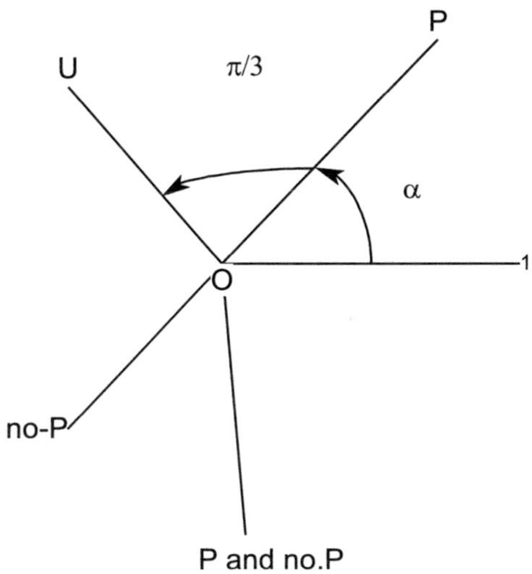

$$\phi - \phi^* = \pm\frac{2\pi}{3} = \alpha^* - \alpha \tag{6.4}$$

Solutions of $\upsilon\,(P \wedge \neg P) = 1$ are finally (Fig. 6.3): $p = e^{i\alpha}$, α any one; $p^* = e^{i\left(\alpha \pm \frac{2\pi}{3}\right)}$.

Multivalued logic with complex truth values is paraconsistent.

Another characteristic of paraconsistent logic is that the negation of the negation does not necessarily leads back to the original proposition as Hegel said. If $u' \neq u, \ \ v(\neg\neg P) = V\left(u' - p^*\right) \neq V(p); \ \ v(\neg\neg P) \neq v(P)$.

6.3 Conditions

6.3.1 Condition 1

It is written $|u_1 u_2 - p_1 p_2| \leq 1$ with the above notation:

$$\left|e^{i(\phi_1+\phi_2)} - |p_1||p_2|\right| \leq 1$$

Condition 1 back to:

$$\cos(\phi_1 + \phi_2) \geq \frac{|p_1||p_2|}{2}$$

There is a continuous set of deniers $u_1(\phi_1)$ and other deniers $u_2(\phi_2)$ that satisfy (e.g., $\phi_1 + \phi_2 = 0$). As well the $\neg P_1 \vee \neg P_2$ incompatibility exists, provides p_1 and p_2 fixed, on a continuous set of truth values $u_1 u_2 - p_1 p_2$ and we have:

$$v(\neg P_1 \vee \neg P_2) = v(\neg(P_1 \wedge P_2))$$

Similarly, $P_1 \vee P_2$ does exist, provided that

$$\cos\left(\phi_1^* + \phi_2^*\right) \geq \frac{|p_1^*||p_2^*|}{2}$$

satisfied, for example, if $\phi_1 + \phi_2 = 0$.

Similarly, the implication $P_1 \Rightarrow P_2$ on condition that

$$\cos\left(\phi_1^* + \phi_2^*\right) \geq \frac{|p_1||p_2|}{2}$$

6.3.2 Condition 2

Posing $\alpha_1 - \alpha_2 = \alpha$, that $|p_1 + p_2| = \left||p_1| e^{i\alpha} + |p_2|^2\right|$ it results that $\forall p_1 \neq 0, \forall p_2 \neq 0, \ \left|\alpha \geq \frac{2\pi}{3}\right| \Rightarrow |p_1 + p_2| \leq 1$; then $|p_1 + p_2| > 1 \Rightarrow |\alpha| < \frac{2\pi}{3}$.

If $|p_1 + p_2| > 1$, which requires nonzero p_1 and p_2, complementarity $P_1 \Im P_2$ does not exist and $\neg P_1 \Im \neg P_2$ must exist.

Theorem 1 *Can be found deniers u_1 and u_2, such that* $\left|p_1^* + p_2^*\right| \le 1$

Proof Just for this inequation is satisfied that $\left|\alpha_1^* - \alpha_2^*\right| \ge \frac{2\pi}{3}$. After (6.4) $\alpha_1^* - \alpha_2^* = (\phi_1 - \phi_1^*) - (\phi_2 - \phi_2^*) + \alpha$.

Let $0 \le \alpha = \frac{2\pi}{3} - \beta$ be. After (6.3) the maximum value of $\phi_1 - \phi_1^*$ is $\pi - \theta_1 > 0$ and the one of $-(\phi_2 - \phi_2^*)$ is $\pi - \theta_2 > 0$ and therefore:

$$\sup\left|\alpha_1^* - \alpha_2^*\right| = 2\pi - (\theta_1 + \theta_2) + \frac{2\pi}{3} - \beta > \frac{5\pi}{3} - \beta$$

because $\theta_1 + \theta_2 < \pi$.

The result is $\sup\left|\alpha_1^* - \alpha_2^*\right| \ge \pi$ since $0 \le \beta \le \frac{2\pi}{3}$; or sufficient condition is $\left|\alpha_1^* - \alpha_2^*\right| \ge \frac{2\pi}{3}$.

There is a continuous set of values of $\left|\alpha_1^* - \alpha_2^*\right|$ and therefore of deniers u_1 (ϕ_1) and u_2 (ϕ_2) which satisfy this condition. □

6.3.3 Condition 3

It is written $\left|p_1 p_2 + p_1^* p_2^*\right| \le 1$. As Condition 2, it is sufficient for it is fulfilled that the angle of nonzero vectors $p_1 p_2$ and $p_1^* p_2^*$ is $\ge \frac{2\pi}{3}$ therefore $\left|\alpha_1 + \alpha_2 - (\alpha_1^* + \alpha_2^*)\right| \ge \frac{2\pi}{3}$ or else after (6.4) that:

$$\left|\phi_1^* - \phi_1 + \phi_2^* - \phi_2\right| \ge \frac{2\pi}{3} \tag{6.5}$$

or in according (6.3):

$$\sup\left|\phi_1^* - \phi_1 + \phi_2^* - \phi_2\right| = 2\pi - (\theta_1 + \theta_2) > \pi \tag{6.6}$$

6.3.4 Condition 4

It is written $\left|p_1 p_2^* + p_1^* p_2\right| \le 1$. Studied by the same method it proves to be satisfied if (sufficient condition):

$$\left|\phi_1 - \phi_1^* + \phi_2 - \phi_2^*\right| \ge \frac{2\pi}{3} \tag{6.7}$$

It is an inequation whose solution is the same of (6.5).

The result is that concordance $P_1 \Leftrightarrow P_2$ and discordance $P_1 \Updownarrow P_2$ can exist together; then one is the negation of the other by denier $u_1 u_2$.

6.4 Propositional Strong Paraconsistent Algebra

Propositional algebra can be built on the set of complex truth values. The main normal binary propositions are the following:

1. *Conjunction*:

$$v\left(P_1 \wedge P_2\right) = \left|p_1 p_2\right|^2 = \left|p_1\right|^2 \left|p_2\right|^2 \tag{6.8}$$

2. *Incompatibility*:

$$\begin{cases} v\left(\neg P_1 \wedge \neg P_2\right) = \left|u_1 u_2 - p_1 p_2\right|^2 = \left|e^{i(\phi_1+\phi_2)} - \left|p_1\right|\left|p_2\right|\right|^2 \\ v\left(\neg P_1 \vee \neg P_2\right) = v\left(\neg\left(P_1 \wedge P_2\right)\right), \;\; denier\ u_1 u_2 \end{cases} \tag{6.9}$$

3. *Disjunction*:

$$\begin{cases} v\left(P_1 \vee P_2\right) = \left|u_2 p_1 + u_1 p_2 - p_1 p_2\right|^2 = \left|\left|p_1\right| e^{i\phi_2} + \left|p_2\right| e^{\phi_1} - \left|p_1\right|\left|p_2\right|\right|^2 \\ v\left(P_1 \vee P_2\right) = v\left(\neg\left(\neg P_1 \wedge \neg P_2\right)\right), \;\; denier\ u_1 u_2 \end{cases} \tag{6.10}$$

4. *Implication*:

$$\begin{cases} v\left(P_1 \Rightarrow P_2\right) = \left|u_1 u_2 - p_1\left(u_2 - p_2\right)\right|^2 = \left|e^{i(\phi_1+\phi_2)} - \left|p_1\right|\left(e^{i\phi_2} - \left|p_2\right|\right)\right|^2 \\ v\left(P_1 \vee P_2\right) = v\left(\neg\left(P_1 \wedge \neg P_2\right)\right), \;\; denier\ u_1 u_2 \end{cases} \tag{6.11}$$

5. *Concordance*:

$$\begin{cases} v\left(P_1 \Leftrightarrow P_2\right) = \left|u_1 u_2 - u_2 p_1 - u_1 p_2 + 2 p_1 p_2\right|^2 \\ \qquad = \left|e^{i(\phi_1+\phi_2)} - \left|p_1\right| e^{i\phi_2} - \left|p_2\right| e^{i\phi_1} + 2\left|p_1\right|\left|p_2\right|\right|^2 \\ v\left(P_1 \Leftrightarrow P_2\right) = v\left(\neg\left(P_1 \Updownarrow P_2\right)\right), \;\; denier\ u_1 u_2 \end{cases} \tag{6.12}$$

6. *Discordance*:

$$\begin{cases} v\left(P_1 \Updownarrow P_2\right) = \left|u_2 u_1 + u_1 p_2 - 2 p_1 p_2\right|^2 \\ \qquad = \left|\left|p_1\right| e^{i\phi_2} + \left|p_2\right| e^{i\phi_1} - 2\left|p_1\right|\left|p_2\right|\right|^2 \\ v\left(P_1 \Updownarrow P_2\right) = v\left(\neg\left(P_1 \Leftrightarrow P_2\right)\right), \;\; denier\ u_1 u_2 \end{cases} \tag{6.13}$$

7. *Complementarity*:

$$v\left(P_1 \Im P_2\right) = \left|p_1 + p_2\right|^2 = \left|\left|p_1\right| e^{i(\alpha_1-\alpha_2)} + \left|p_2\right|\right|^2 \tag{6.14}$$

8. *Inverse complementarity*:

$$\begin{aligned} v\left(\neg P_1 \Im \neg P_2\right) &= \left|u_1 + u_2 - p_1 - p_2\right|^2 \\ &= \left|\left(e^{i\phi_1} - \left|p_1\right|\right) e^{i(\alpha_1-\alpha_2)} + e^{i\phi_2} - \left|p_2\right|\right|^2 \end{aligned} \tag{6.15}$$

Fig. 6.4 The geometric representation

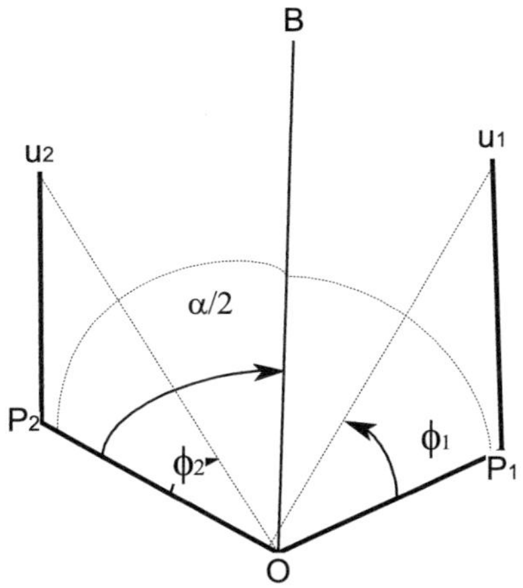

9. *Equivalence*:

$$v\left(P_2 \wp P_1\right) = |p_1 + u_2 - p_2|^2 = \left||p_1|\, e^{i(\alpha_1-\alpha_2)} + e^{i\phi_2} - |p_2|\right|^2 \tag{6.16}$$

Here intervenes the angle $\alpha_1 - \alpha_2$ of vectors p_1, p_2.

Seek what deniers should be chosen so that if $v\left(P_1\right) = v\left(P_2\right)$, that is to say, if $|p_1| = |p_2| = |p|\,, \alpha_1 \neq \alpha_2$, we have: $v\left(P_2 \wp P_1\right) = 1 = v\left(P_1 \wp P_2\right)$.

We then: $v\left(P_2 \wp P_1\right) = \left||p|\left(e^{i\alpha} - 1\right) + e^{i\phi_2}\right|^2$ where $\alpha = \alpha_1 - \alpha_2$. So that $v\left(P_2 \wp P_1\right) = 1$, the necessary and sufficient condition is:

$$\sin\left(\frac{\alpha}{2} - \phi_2\right) = |p| \sin\frac{\alpha}{2} \tag{6.17}$$

Similarly, for $v\left(P_1 \wp P_2\right) = 1$ the necessary and sufficient condition is:

$$\sin\left(\frac{\alpha}{2} + \phi_1\right) = |p| \sin\frac{\alpha}{2} \tag{6.18}$$

of which $\phi_1 = \phi_2$, ϕ_2 solution of (6.17).

Figure. 6.4 shows the geometric representation.

6.4.1 Normal Propositions of Order N

1. *Conjunction*:

$$v\left(P_1 \wedge P_2 \wedge \cdots \wedge P_n\right) = |p_1 p_2 \ldots p_n|^2 \tag{6.19}$$

2. *Incompatibility*:

$$\begin{aligned} &\upsilon\left(\neg P_1 \vee \neg P_2 \vee \cdots \vee \neg P_n\right) = \left|u_1 u_2 \ldots u_n - p_1 p_2 \ldots p_n\right|^2 \\ &\upsilon\left(\neg P_1 \vee \neg P_2 \vee \cdots \vee \neg P_n\right) = \upsilon(P_1 \wedge P_2 \wedge \cdots \wedge P_n) \end{aligned} \tag{6.20}$$

3. *Disjunction*:

$$\begin{aligned} &\upsilon\left(P_1 \vee P_2 \vee \cdots \vee P_n\right) = \left|u_1 u_2 \ldots u_n - p_1^* p_2^* \ldots p_n^*\right|^2 \\ &\upsilon\left(P_1 \vee P_2 \vee \cdots \vee P_n\right) = \upsilon\left(\neg P_1 \wedge \neg P_2 \wedge \cdots \wedge \neg P_n\right) \end{aligned} \tag{6.21}$$

4. *Complementarity*:

$$\upsilon\left(P_1 \Im P_2 \Im \ldots \Im P_n\right) = \left|p_1 + p_2 + \cdots + p_n\right|^2 \tag{6.22}$$

5. *Inverse complementarity*:

$$\upsilon\left(\neg P_1 \Im \neg P_2 \Im \ldots \Im \neg P_n\right) = \left|u_1 + u_2 + \cdots + u_n - (p_1 + p_2 + \cdots + p_n)\right|^2 \tag{6.23}$$

6.5 Paraconsistent Boolean Logic

It is the Boolean reduction of strong paraconsistent logic; modules of complex truth values there can be only 0 or 1. The circle of truth is there reduced to its center and its circumference. Although Boolean, this logic differs radically from the classical logic: it remains paraconsistent. The contradiction can be true there. We may have verified all the normal binary propositions that the propositional algebra of the paraconsistent Boolean logic contains well beyond the classical logic as a special case.

Since $\upsilon\,(P) = 0 \Rightarrow \upsilon\,(\neg P) = 1$, p^* indeterminate, and $\upsilon\,(P \wedge \neg P) = 0$.

Since $\upsilon\,(P) = 1 \Rightarrow \theta = \pm\frac{\pi}{3}$. Since $|p^*|$ must be Boolean ϕ can only take two values:

$$\begin{aligned} &\phi = 0 \Leftrightarrow \upsilon\,(\neg P) = 0 \Rightarrow \upsilon\,(P \wedge \neg P) = 0 \\ &\phi = \pm\frac{\pi}{3} \Leftrightarrow \upsilon\,(\neg P) = 1 \Rightarrow \upsilon\,(P \wedge \neg P) = 1 \end{aligned} \tag{6.24}$$

It has always: $\phi^* = -\phi \Rightarrow \alpha^* - \alpha = 2\phi$.

1. *Conjunction*: The truth table of the *conjunction* is identical to that of classical logic.

Table 6.1 Truth table of disjunction $P_1 \vee P_2$

	P_1		
P_2		1	0
	1	1/0*	1
	0	1	0

2. *Disjunction*: From (6.10), it is false if $v(P_1) = v(P_2) = 0$ and true if $v(P_1) = 0$ and $v(P_2) = 1$ or if $v(P_1) = 1$ and $v(P_2) = 0$.
Condition 1 is written $\cos\left(\phi_1^* + \phi_2^*\right) \geq \frac{1}{2}$; $\phi^* = -\phi$ thus the disjunction exists only if $\phi_1 + \phi_2 = -\frac{\pi}{3}, 0, \frac{\pi}{3}$; we have: $v(P_1 \vee P_2) = \left|e^{i\phi_2} + e^{i\phi_1} - 1\right|$ then:

$$v(P_1) = v(P_2) = 1 \Rightarrow \begin{cases} v(P_1 \vee P_2) = 0 \;\; if \;\; \phi_1 = \pm\dfrac{\pi}{3}, \;\; \phi_2 = \pm\dfrac{\pi}{3} \\ v(P_1 \vee P_2) = 1 \;\; if \;\; \phi_1 + \phi_2 = \pm\dfrac{\pi}{3}, \;\; or \;\; \phi_1 = \phi_2 = 0 \end{cases} \tag{6.25}$$

Hence the truth table of disjunction (Table 6.1):

Not conform evaluation to the classical logic is indicated by *.

It coincides with that of classical logic, in case $v(P_1) = v(P_2) = 1$, it is chosen ϕ_1 and ϕ_2 such that $\phi_1\phi_2 = 0$.

3. *Implication*: From (6.11) it is true if $\forall p_2$, $v(P_1) = 0$. If $v(P_1) = v(P_2) = 1$ the condition of existence is $\cos(\phi_1 - \phi_2) \geq \frac{1}{2}$; such as $v(P_1 \Rightarrow P_2) = \left|e^{i(\phi_1+\phi_2)} - e^{i\phi_2} + 1\right|^2$, were $v(P_1 \Rightarrow P_2) = 1$ in all cases permitted by the condition of existence except where $\phi_1 = \phi_2 = \pm\frac{\pi}{3}$ for which $v(P_1 \Rightarrow P_2) = 0$. If $v(P_1) = 1$ and $v(P_2) = 0$ were $v(P_1 \Rightarrow P_2) = \left|e^{i\phi_1} - 1\right|$; $v(P_1 \Rightarrow P_2) = 0$ if $\phi_1 = 0$, $v(P_1 \Rightarrow P_2) = 1$ if $\phi_1 = \pm\frac{\pi}{3}$.
Hence the truth table of implication (Table 6.2):

It coincides with that of classical logic if one rejects the case $v(P_1) = v(P_2) = 1$ the choice $\phi_1 = \phi_2 = \pm\frac{\pi}{3}$ and the case $v(P_1) = 1$ and $v(P_2) = 0$ the choice $\phi_1 = \pm\frac{\pi}{3}$.

These rejections are required to conduct a rigorous deduction in paraconsistent Boolean logic: the fundamental articulation of the deduction is indeed true implication denoted $\Rightarrow$, that if P_1 is true requires true P_2.

Table 6.2 Truth table of implication $P_1 \Rightarrow P_2$

	P_1		
P_2		1	0
	1	1/0*	1
	0	0/1*	1

Table 6.3 Truth table of concordance $P_1 \Leftrightarrow P_2$

	P_1		
P_2		1	0
	1	1	0/1*
	0	0/1*	1

4. *Concordance*: From (6.12) it is true if $v(P_1) = v(P_2) = 0$. If $v(P_1) = 1$ and $v(P_2) = 0$ then $v(P_1 \Leftrightarrow P_2) = \left|e^{i\phi_1} - 1\right|^2$ therefore $v(P_1 \Leftrightarrow P_2) = 0$ if $v(P_1 \Leftrightarrow P_2) = 1\ \phi_1 = 0$ and $v(P_1 \Leftrightarrow P_2) = 0$ if $\phi_1 = \pm\frac{\pi}{3}$; same if $v(P_1) = 0$ and $v(P_2) = 1$ we have $v(P_1 \Leftrightarrow P_2) = 0$ if $\phi_2 = 0$ and $v(P_1 \Leftrightarrow P_2) = 1$ if $\phi_2 = \pm\frac{\pi}{3}$. If $v(P_1) = v(P_2) = 1$ the concordance does not exist when $\phi_1 = \pm\frac{\pi}{3}, \phi_2 = \pm\frac{\pi}{3}$, but it is true in all other cases.

Hence the truth table of concordance (Table 6.3):

It coincides with that of the equivalence of classical logic if when $v(P_1) = 1$ and $v(P_2) = 0$ is chosen, $\phi_1 = 0$ and when $v(P_1) = 0$ and $v(P_2) = 1$ is chosen, $\phi_2 = 0$.

Importantly, to conduct a rigorous reasoning with these choices, the concordance becomes identical to the deductive equivalence.

6.6 Reflections

The intention of the authors in carrying out this study is not to conclude this approximation to a paraconsistent logic. As we have already explained in the first part, the objective of this research is to provide a coherent response to the contradictions observed in mystical and mythical belief systems, specifically to the problem of *coincidentia oppositorum*, which can be observed in the works of Nicholas de Cusa, in the Hebrew Kabbalah, in the works of Jung or to a somewhat lesser extent in J. Derrida. The authors are developing a multivalued paraconsistent logic called NR logic (Numinous-Religious logic), where for purposes of the bipolar proposition $\boldsymbol{P}$ **and no**$-\boldsymbol{P}$ are attributed a truth value called K and that $p = |\beta|\, e^{\pm\alpha\frac{\pi}{3}}; \left|\beta \in \left[0, \frac{1}{e}\right]\right|$ and $p \in [0, 1]$ being 0 the absolute falsehood and 1 the absolute truth. Modulus $|\beta| \in \left[0, \frac{1}{e}\right]$ defines the $\boldsymbol{P}$ truth value and complex part $e^{\pm i\alpha\frac{\pi}{3}}$ defines the $\neg\boldsymbol{P}$ pole, being α the $\neg\boldsymbol{P}$ truth value. $\neg\boldsymbol{P}$ truth value α only can take values $\alpha \in \left[0, \frac{3i}{\pi}\right]$ or $\alpha \in \left[-\frac{3i}{\pi}, 0\right]$. Therefore, $\neg\boldsymbol{P}$ truth values only can take values between $\left[-\frac{3i}{\pi}, 0\right]$ and $\left[0, \frac{3i}{\pi}\right]$. Values between $\left[-\frac{3i}{\pi}, 0\right]$ will denominate *right*$-\neg\boldsymbol{P}$ *truth values* and values between $\left[0, \frac{3i}{\pi}\right]$ will denominate *left*$-\neg\boldsymbol{P}$ *truth values*.

The argument concerning belief systems may be circumvented if one claims that ordinary belief is not deductively closed. That is, at least, controversial, but an ideal reasoner should aspire to closure. Considering the case of a paraconsistent system being used as a metalanguage to analyze a belief system, it is also the task of paraconsistent logic to define paraconsistent contradictions, that is, contradictions that are so threatening to this belief system that they really compromise rational inference making within the belief system. This "*bad*" kind of inconsistency can be quantitative (too many classic contradictions may be a sign that even paraconsistency cannot save the belief system) or qualitative—that is, the classic contradiction in question is so strong (for example, a proof that all statements of the belief system can be proved both true and false) that it is also a paraconsistent contradiction, a contradiction that even a paraconsistent logician cannot accept. This argument implies the idea that the set of paraconsistent contradictions is a subset of the set of classic contradictions and that is indeed a rather intuitive idea. But we cannot think of any conclusive argument against the existence of a paraconsistent contradiction that is not a classic contradiction, so this idea is only a plausible conjecture.

Paraconsistent logic was in some sense born of the realization that consistency, in its classical sense, was not a good enough criterions to discriminate between good and bad belief system, exactly because our actual reasoning is, it seems, much more able to cope with inconsistent premises than classical logic. Indeed, it has become a motto in many circles of nonclassical logic that classical logic simply is not an accurate model of human rationality.

So as a minimum, paraconsistent approaches have shown that there are other avenues to explore, there are other alternatives, other viable options, which cannot be ruled out in principle as it used to from the Aristotelian perspective, which has continued to dominate for so long, even among those who did not wish to see themselves as Aristotelian.

Annex A. Truth Table of Principal Normal Binary Propositions

We will represent in the following table a comparison between three logics: classical (CL), quasi-paraconsistent (QPL), and strong paraconsistent (SPL) (Table 6.4).

Table 6.4 Truth table of principal normal binary propositions

Notation	Name	CL truth values $p_1, p_2 \in \{0, 1\}$	QPL truth values $p_1, p_2 \in [0, 1]$
$P_1 \wedge P_2$	Conjunction	$p_1 p_2$	$p_1 p_2$
$\neg P_1 \vee \neg P_2$	Incompatibility	$1 - p_1 p_2$	$u_1 u_2 - p_1 p_2$
$P_1 \vee P_2$	Disjunction	$p_1 + p_2 - p_1 p_2$	$u_2 p_1 + u_1 p_2 - p_1 p_2$
$P_1 \Rightarrow P_2$	Implication	$1 - p_1 + p_1 p_2$	$u_1 u_2 - p_1 (u_2 - p_2)$
$P_1 \Leftrightarrow P_2$	Concordance	$1 - p_1 - p_2 + 2 p_1 p_2$	$u_1 u_2 - (u_2 p_1 + u_1 p_2) + 2 p_1 p_2$
$P_1 \Updownarrow P_2$	Discordance	$p_1 + p_2 - 2 p_1 p_2$	$u_2 p_1 + u_1 p_2 - 2 p_1 p_2$
$P_1 \Im P_2$	Complementarity	$p_1 + p_2$	$p_1 + p_2$
$\neg P_1 \Im \neg P_2$	Inverse complementarity	$2 - p_1 - p_2$	$u_1 + u_2 - p_1 - p_2$
$P_2 \wp P_1$	Equivalence	$1 + p_1 - p_2$	$p_1 + u_2 - p_2$
$P_1 \wp P_2$	Inverse equivalence	$1 - p_1 + p_2$	$u_1 - p_1 + p_2$
Notation	**Name**	**SPL truth values $p_1, p_2 \in [0, 1]$**	
$P_1 \wedge P_2$	Conjunction	$\lvert p_1 \rvert^2 \lvert p_2 \rvert^2$	
$\neg P_1 \vee \neg P_2$	Incompatibility	$\left\lvert e^{i(\phi_1+\phi_2)} - \lvert p_1 \rvert \lvert p_2 \rvert \right\rvert$	
$P_1 \vee P_2$	Disjunction	$\left\lvert \lvert p_1 \rvert e^{i\phi_2} + \lvert p_2 \rvert e^{\phi_1} - \lvert p_1 \rvert \lvert p_2 \rvert \right\rvert^2$	
$P_1 \Rightarrow P_2$	Implication	$\left\lvert e^{i(\phi_1+\phi_2)} - \lvert p_1 \rvert \left(e^{i\phi_2} - \lvert p_2 \rvert\right) \right\rvert^2$	
$P_1 \Leftrightarrow P_2$	Concordance	$\left\lvert e^{i(\phi_1+\phi_2)} - \lvert p_1 \rvert e^{i\phi_2} - \lvert p_2 \rvert e^{i\phi_1} + 2 \lvert p_1 \rvert \lvert p_2 \rvert \right\rvert^2$	
$P_1 \Updownarrow P_2$	Discordance	$\left\lvert \lvert p_1 \rvert e^{i\phi_2} + \lvert p_2 \rvert e^{i\phi_1} - 2 \lvert p_1 \rvert \lvert p_2 \rvert \right\rvert^2$	
$P_1 \Im P_2$	Complementarity	$\left\lvert \lvert p_1 \rvert e^{i(\alpha_1-\alpha_2)} + \lvert p_2 \rvert \right\rvert^2$	
$\neg P_1 \Im \neg P_2$	Inverse complementarity	$\left\lvert \left(e^{i\phi_1} - \lvert p_1 \rvert\right) e^{i(\alpha_1-\alpha_2)} + e^{i\phi_2} - \lvert p_2 \rvert \right\rvert^2$	
$P_2 \wp P_1$	Equivalence	$\left\lvert \lvert p_1 \rvert e^{i(\alpha_1-\alpha_2)} + e^{i\phi_2} - \lvert p_2 \rvert \right\rvert^2$	
$P_1 \wp P_2$	Inverse equivalence	$\left\lvert \lvert p_2 \rvert e^{i(\alpha_1-\alpha_2)} + e^{i\phi_1} - \lvert p_1 \rvert \right\rvert^2$	

References

1. Arruda, A.I., Chuaqui, R., da Costa, N.C.A. (eds.): Mathematical Logic in Latin America. North-Holland Publishing Company, Amsterdam, New York, Oxford (1980)
2. Asenjo, F.G.: A calculus of antinomies. Notre Dame J. Form. Log. **7**, 103–105 (1966)
3. Avron, A.: Combining classical logic, paraconsistency and relevance. J. Appl. Log. **3**(1), 133–160 (2005)
4. Belnap, N.D.: How a computer should think. In: Ryle, G. (ed.) Contemporary Aspects of Philosophy, pp. 30–55. Oriel Press, Boston (1977)
5. Bueno, O.: True, Quasi-true and paraconsistency. Contemp. Math. **39**, 275–293 (1999)
6. Bueno, O.: Philosophy of logic. In: Allhoff, F. (ed.) Philosophies of the Sciences: a Guide, pp. 55. Wiley (2010)
7. Carnielli, W.A., Coniglio, M., Lof D'ottaviano, I.M.: Paraconsistency: the Logical Way to the Inconsistent. Marcel Dekker, Inc., New York (2002)
8. Carnielli, W., Marcos, J.: Ex contradictione non sequitur quodlibet. Proceeding of 2nd Conference on Reasoning and Logic. July 2000, Bucharest (2001)

9. Da Costa, N.C.A., Wolf, R.G.: Studies in Paraconsistent Logic I: the Dialectical Principle of the Unity of Opposites. Philosophia Phil. Q. Israel **9**, 189–217 (1980)
10. Dunn, J.M.: Intuitive semantics for first-degree entailments and coupled trees. Phil. Stud. **29**, 149–168 (1976)
11. Nescolarde-Selva, J., Usó-Doménech, J.L.: Semiotic vision of ideologies. Found. Sci. **19**(3), 263–282 (2014). doi:10.1007/s10699-013-9329-8
12. Nescolarde-Selva, J., Usó-Doménech, J.L.: Reality, Systems and Impure Systems. Found. Sci. **19**(3), 289–306. (2014). doi:10.1007/s10699-013-9337-8
13. Nescolarde-Selva, J., Usó-Doménech, J.: Topological Structures of Complex Belief Systems. Complexity. **19**(1), 46–62 (2013). doi:10.1002/cplx.21455
14. Nescolarde-Selva, J., Usó-Doménech, J.: Topological structures of complex belief systems (ii): textual materialization. **19**(2), 50–62 (2013). doi:10.1002/cplx.21476
15. Priest, G.: In Contradiction. A Study of the Transconsistent, Martinus Nijhoff Publishers, Dordrecht, Boston, Lancaster (1987)
16. Priest, G.: Beyond the Limits of Thought. Cambridge University Press (1995)
17. Priest, G.: Dialetheism. Stanford Encyclopedia of Philosophy (1998)
18. Priest, G., Routley, R., Norman, J. (eds.): Paraconsistent logic. Essays on the Inconsistent. Philosophia Verlag, Munchen, Hamden, Wien (1989)
19. Quesada F.M.: Paraconsistent logic: some philosophical issues, in: Priest et al., 627–651 (1989)
20. Tanaka, K., Berto, F., Mares, E., Paoli, F. (eds.): Paraconsistency: Logic and Applications. Logic, Epistemology and Unity of Science, vol. 26. Springer Dordrecht, Heidelberg, London, New York (2013)
21. Usó-Doménech, J.L., Nescolarde-Selva, J.A.: Mathematic and Semiotic Theory of Ideological Systems. A systemic vision of the Beliefs. LAP LAMBERT Academic Publishing, Saarbrücken, Germany (2012). ISBN 978-3-8443-9815-1

Chapter 7
A Paraconsistent Logic Obtained from an Algebra-Valued Model of Set Theory

Sourav Tarafder and Mihir Kr. Chakraborty

Abstract This paper presents a three-valued paraconsistent logic obtained from some algebra-valued model of set theory. *Soundness* and *completeness* theorems are established. The logic has been compared with other three-valued paraconsistent logics.

Keywords Paraconsistent logic · Three-valued matrices · Models of set theory

Mathematics Subject Classification (2000) Primary 03B53 · Secondary 02K10, 03C90

7.1 Introduction and Overview

The idea of *Boolean-valued models* $\mathbf{V}^{(\mathbb{B})}$ where $\mathbf{V}$ is the standard class model of classical set theory and $\mathbb{B}$ is any arbitrary but fixed complete Boolean algebra was introduced in 1960s (cf. [2]). The motivation behind this construction was to understand in a different way Cohen's method of *forcing* [5] that is used to prove the

We are thankful to Prof. Benedikt Löwe for his support in the set theoretic investigations indicated in this paper. We also acknowledge the anonymous referee for making valuable comments. The first author's research is funded by CSIR Ph.D. fellowship, Govt. of India, grant number: 09/028(0754)/2009 -EMR-1.

S. Tarafder (✉)
St. Xavier's College, 30 Mother Teresa Sarani, Kolkata 700016, India
e-mail: souravt09@gmail.com

S. Tarafder
Department of Pure Mathematics, University of Calcutta,
35 Ballygunge Circular Road, Kolkata 700019, India

M.K. Chakraborty
School of Cognitive Science, Jadavpur University,
188 Raja S. C. Mullik Road, Kolkata 700032, India
e-mail: mihirc4@gmail.com

J.-Y. Beziau et al. (eds.), *New Directions in Paraconsistent Logic*,
Springer Proceedings in Mathematics & Statistics 152,
DOI 10.1007/978-81-322-2719-9_7

consistency and independence results. Afterward, this kind of construction has come into practice as an autonomous mathematical technique from the angle of generalization of $\mathbf{V}^{(\mathbb{B})}$ and for some other reasons. As for example, we get *Heyting-valued model* of *Intuitionistic set theory* [7] where the complete Boolean algebra is replaced by a complete *Heyting algebra*, *Orthomodular-valued model* of *Quantum set theory* [12] where the same is replaced by a complete *orthomodular lattice*. Generally speaking, an $\mathbb{A}$-valued model $\mathbf{V}^{(\mathbb{A})}$ are models of all (or some) of the ZF-axioms where $\mathbb{A}$ is the algebraic model for the underlying logic on which ZF-axioms are situated. For instance, Boolean algebras are algebraic models of classical logic, Heyting algebras are algebraic models of *intuitionistic logic*, and orthomodular lattices are algebraic models for *quantum logic*. Accordingly, $\mathbf{V}^{(\mathbb{A})}$ is called a model for classical set theory, intuitionistic set theory, and quantum set theory when $\mathbb{A}$ is a Boolean algebra, a Heyting algebra and an orthomodular lattice, respectively.

This paper may be considered as a part of a larger project of constructing a paraconsistent set theory. As a step an algebra $\mathbb{A}$ is defined so that $\mathbf{V}^{(\mathbb{A})}$ becomes a model of some version of ZF-axioms (cf. Sect. 7.2.1) [9]. This algebra $\mathbb{A}$ is sufficiently general to give rise to some non-classical logics other than paraconsistent logics. In this paper, we have chosen a three-valued matrix PS_3 as an instance of the above-mentioned algebra $\mathbb{A}$. PS_3 is shown to be a semantics of a paraconsistent logic. We would like to mention some of the results obtained in [9] to justify the selection of the particular algebra PS_3 here. In fact, in this paper we shall discuss the reason for the choice of the particular algebra PS_3 and investigate the logical properties that it represents. As another part of the mentioned bigger project, Tarafder investigated some properties of ordinals in the model $\mathbf{V}^{(PS_3)}$ of some paraconsistent set theory [13].

It should be focused that claiming an algebra to be the model of some logic means that the logic should be *sound* and *complete* with respect to the semantics given in the algebra. Thus in the algebra some elements need to be designated, that is, some 0-ary operations have to be present. In the three cases mentioned above the designated element is only the top element of the bounded lattice. There can be more than one designated elements as will be in our case. An algebra with designated elements is usually called a matrix in logic–literature.

This paper has been designed as follows. In the next section (viz. Sect. 7.2) the requirements to the algebra shall be specified and a three-element matrix will be identified for the purpose. Section 7.3 will deal with the paraconsistent logic (in Hilbert style axiomatic presentation) which is sound and complete with respect to the matrix. In Sect. 7.4 some comparisons of the present logic with other paraconsistent logics will be made. Section 7.5 contains the concluding remarks. All the proofs are given as an appendix.

7.2 The Algebra for the Proposed Algebra-Valued Models

7.2.1 Algebraic Properties Required

Let us consider an algebra $\mathbb{A} = (A, \wedge, \vee, \Rightarrow, ^*)$ where $(A, \wedge, \vee)$ is a complete distributive lattice such that the following conditions are satisfied:

P1: $x \wedge y \leq z$ implies $x \leq y \Rightarrow z$.
P2: $y \leq z$ implies $x \Rightarrow y \leq x \Rightarrow z$.
P3: $y \leq z$ implies $z \Rightarrow x \leq y \Rightarrow x$.
P4: $(x \wedge y) \Rightarrow z = x \Rightarrow (y \Rightarrow z)$.

Let us construct $\mathbf{V}^{(\mathbb{A})}$ in the same way as the Boolean-valued model construction of classical set theory. Then extend the language of set theory by adding a constant corresponding to every element of $\mathbf{V}^{(\mathbb{A})}$. Following the standard way (cf. [2]), every formula of the *extended language* of set theory is associated with a value of A by the mapping $[\![.]\!]$ as follows:
for any u, v in $\mathbf{V}^{(\mathbb{A})}$,

$$[\![u \in v]\!] = \bigvee_{x \in dom(v)} (v(x) \wedge [\![x = u]\!])$$

$$[\![u = v]\!] = \bigwedge_{x \in dom(u)} (u(x) \Rightarrow [\![x \in v]\!]) \wedge \bigwedge_{y \in dom(v)} (v(y) \Rightarrow [\![y \in u]\!]).$$

Now for any well-formed formulas σ and τ we define

$$[\![\sigma \wedge \tau]\!] = [\![\sigma]\!] \wedge [\![\tau]\!]$$
$$[\![\sigma \vee \tau]\!] = [\![\sigma]\!] \vee [\![\tau]\!]$$
$$[\![\sigma \rightarrow \tau]\!] = [\![\sigma]\!] \Rightarrow [\![\tau]\!]$$
$$[\![\neg\sigma]\!] = [\![\sigma]\!]^*$$

for any formula $\varphi(x)$ with one free variable x,

$$[\![\forall x \varphi(x)]\!] = \bigwedge_{u \in \mathbf{V}^{(\mathbb{A})}} [\![\varphi(u)]\!]$$
$$[\![\exists x \varphi(x)]\!] = \bigvee_{u \in \mathbf{V}^{(\mathbb{A})}} [\![\varphi(u)]\!]$$

A formula φ will be called valid in $\mathbf{V}^{(\mathbb{A})}$ iff $[\![\varphi]\!] \in D$, where D is the designated set of $\mathbb{A}$. It is well known for a Boolean-valued model or Heyting-valued model that for any formula $\varphi(x)$ with one free variable and any constant u of the extended language,

$$[\![\forall x(x \in u) \rightarrow \varphi(x)]\!] = \bigwedge_{x \in dom(u)} (u(x) \Rightarrow [\![\varphi(x)]\!]). \qquad (\mathrm{BQ}_\varphi)$$

But BQ_φ does not hold generally in $\mathbf{V}^{(\mathbb{A})}$. A formula φ will be called a *negation-free formula* (NFF) if no subformula of it contains the unary connective $\neg$. In [9] it is proved that if BQ_φ holds in $\mathbf{V}^{(\mathbb{A})}$ for each negation-free formula φ then the ZF-axioms *Extensionality, Pairing, Infinity, Union*, and *Powerset* are valid in $\mathbf{V}^{(\mathbb{A})}$. Moreover, under the same conditions *NFF-Separation* and *NFF-Replacement* are also valid in $\mathbf{V}^{(\mathbb{A})}$ where NFF-Separation and NFF-Replacement means, respectively, the *Separation* and *Replacement* axiom schemas only for negation-free formulas.

It should be clear that although a negation operator is present in the algebra it did not play any role so far. Now, in order to incorporate the paraconsistency factor, negation comes into the picture. It is known that the key feature of paraconsistency lies in the non-approval of the *law of explosion* ($\{\varphi, \neg\varphi\} \vdash \psi$ for all well-formed formulas φ, ψ) which is present in the classical as well as intuitionistic logics. So paraconsistency is characterized by the fact that there are well-formed formulas φ and ψ such that

$$\{\varphi, \neg\varphi\} \nvdash \psi. \qquad (\mathrm{Par})$$

The reflection of (Par) in the algebra should be as follows: while both φ and $\neg\varphi$ receive some designated values, ψ does not. The algebra $(A, \wedge, \vee, \Rightarrow)$ is to be enhanced with a * so as to fulfill the above condition. Can we find such a matrix? The two-element lattice cannot be a good choice since for satisfying properties **P1**, **P2**, **P3**, and **P4** the truth table of $\Rightarrow$ has either to be the same as the two-valued Boolean implication table or it becomes degenerated. So we look into a three-element lattice.

7.2.2 The Three-Valued Matrix $\mathbf{PS_3}$

In this section, we introduce a three-valued matrix $\mathrm{PS}_3 = \langle\{1, 1/2, 0\}, \wedge, \vee, \Rightarrow, ^*\rangle$ having the following truth tables:

$\wedge$	1	1/2	0
1	1	1/2	0
1/2	1/2	1/2	0
0	0	0	0

$\vee$	1	1/2	0
1	1	1	1
1/2	1	1/2	1/2
0	1	1/2	0

$\Rightarrow$	1	1/2	0
1	1	1	0
1/2	1	1	0
0	1	1	1

$*$	
1	0
1/2	1/2
0	1

and $\{1, 1/2\}$ as the designated set.

PS_3 is a *complete distributive lattice* relative to $\wedge, \vee$. It is easy to check that PS_3 satisfies **P1**, **P2**, **P3**, and **P4**. By asserting values from 1/2 to φ and 0 to ψ one can check that (Par) is satisfied. Moreover, in [9] it is proved that BQ_φ holds in $\mathbf{V}^{(\mathrm{PS}_3)}$ for every negation-free formula φ.

This matrix is included in the collection of 2^{13} three-valued matrices of the *Logic of Formal Inconsistencies* (cf. [4]) after exclusion of the *inconsistency operator* "$\bullet$".

For our purpose we do not need the operator for inconsistency which acts for internalizing *inconsistency* within the *object language*. Now it is important to explain why we have chosen PS_3. First of all $(\{1, 1/2, 0\}, \wedge, \vee)$ has to be a complete distributive lattice for which $\wedge$ and $\vee$ have to be the operators *minimum* and *maximum*, respectively. Second, for satisfying properties **P1**, **P2**, **P3**, and **P4** the only possibilities for the implication are given below:

$\Rightarrow_1$	1	1/2	0
1	1	1	1
1/2	1	1	1
0	1	1	1

$\Rightarrow_2$	1	1/2	0
1	0	0	0
1/2	0	0	0
0	0	0	0

$\Rightarrow_3$	1	1/2	0
1	1	1/2	0
1/2	1	1	0
0	1	1	1

$\Rightarrow_4$	1	1/2	0
1	1	1	0
1/2	1	1	0
0	1	1	1

The implications $\Rightarrow_1$ and $\Rightarrow_2$ cannot produce a reasonable logic as these two are degenerated. The implication $\Rightarrow_3$ produces converse of **P1** also which we do not require. Besides, $\Rightarrow_3$ together with the above-mentioned operators $\wedge$ and $\vee$ produce the three-valued Heyting algebra. As a consequence we are interested in $\Rightarrow_4$ which is the implication of PS_3. Before fixing the truth table of * it should be noticed, since $1 \Rightarrow 1/2 = 1$ in the chosen truth table for $\Rightarrow$, for getting *Modus Ponens* as a valid rule in our system the designated set has to be fixed as $\{1, 1/2\}$. For the truth table of * we can go for any of the following tables which can produce a paraconsistent logic provided the designated set is fixed as $\{1, 1/2\}$.

$*_1$	
1	0
1/2	1/2
0	1

$*_2$	
1	0
1/2	1
0	1

$*_3$	
1	1/2
1/2	0
0	1

$*_4$	
1	0
1/2	1/2
0	1/2

$*_5$	
1	0
1/2	1
0	1/2

$*_6$	
1	1/2
1/2	0
0	1/2

We are interested in taking $1^* = 0$ and $0^* = 1$ so that it does not violate the third criterion of Jaśkowski for being a paraconsistent logic (cf. Sect. 7.4). So we could choose any one of $^*{}_1$ and $^*{}_2$. Since we want to have the rule of *double negation*, as in many of the other well-known paraconsistent logics (shown in Sect. 7.4) the only choice for * is $^*{}_1$. However, it may be mentioned that in [11] a three-valued paraconsistent logic G'_3 having connectives $\wedge, \vee, \Rightarrow$ and * has been intensively investigated in which $\wedge$ and $\vee$ are same as PS_3 but $\Rightarrow$ and * are taken as $\Rightarrow_3$ and $^*{}_2$, respectively. It is to be noted that PS_3 is a fixed, particular algebra of type (2, 2, 2, 1, 0, 0) or a matrix that satisfies the conditions **P1**, **P2**, **P3**, **P4**, and (Par). $\mathbf{V}^{(PS_3)}$ will have all the desired properties provided it can be made an algebraic model of some paraconsistent logic.

7.3 The Logic $\mathbb{L}PS_3$

In this section we introduce an axiom system for the propositional logic $\mathbb{L}PS_3$ having the matrix PS_3 as the three-valued semantics. The alphabet of $\mathbb{L}PS_3$ consists of propositional letters $p_1, p_2, \ldots$; logical connectives $\neg, \wedge, \vee, \rightarrow$. The well-formed formulas are constructed in the usual way.

7.3.1 *The Axiom System for* $\mathbb{L}\mathrm{PS}_3$

The following formulas are taken as the axioms for $\mathbb{L}\mathrm{PS}_3$:

(Ax1) $\varphi \to (\psi \to \varphi)$
(Ax2) $(\varphi \to (\psi \to \gamma)) \to ((\varphi \to \psi) \to (\varphi \to \gamma))$
(Ax3) $\varphi \wedge \psi \to \varphi$
(Ax4) $\varphi \wedge \psi \to \psi$
(Ax5) $\varphi \to \varphi \vee \psi$
(Ax6) $(\varphi \to \gamma) \wedge (\psi \to \gamma) \to (\varphi \vee \psi \to \gamma)$
(Ax7) $(\varphi \to \psi) \wedge (\varphi \to \gamma) \to (\varphi \to \psi \wedge \gamma)$
(Ax8) $\varphi \leftrightarrow \neg\neg\varphi$
(Ax9) $\neg(\varphi \wedge \psi) \leftrightarrow (\neg\varphi \vee \neg\psi)$
(Ax10) $(\varphi \wedge \neg\varphi) \to (\neg(\psi \to \varphi) \to \gamma)$
(Ax11) $(\varphi \to \psi) \to (\neg(\varphi \to \gamma) \to \psi)$
(Ax12) $(\neg\varphi \to \psi) \to (\neg(\gamma \to \varphi) \to \psi)$
(Ax13) $\bot \to \varphi$
(Ax14) $(\varphi \wedge (\psi \to \bot)) \to \neg(\varphi \to \psi)$
(Ax15) $(\varphi \wedge (\neg\varphi \to \bot)) \vee (\varphi \wedge \neg\varphi) \vee (\neg\varphi \wedge (\varphi \to \bot))$

where φ, ψ, γ are any well-formed formulas and $\bot$ is the abbreviation for $\neg(\theta \to \theta)$ for any arbitrary formula θ.

The rules for $\mathbb{L}\mathrm{PS}_3$ are the following:

1. $\frac{\varphi,\ \psi}{\varphi \wedge \psi}$

2. $\frac{\varphi,\ \varphi \to \psi}{\psi}$

Let $\vdash$ and $\models$ be the syntactic and semantic consequence relations, respectively, defined in the usual way with respect to the above-mentioned axiom system and the matrix PS_3. It will be proved that the above propositional axiom system is *sound* and *(weak)complete* with respect to PS_3.

7.3.2 *Soundness*

Theorem 7.3.1 *For any formula φ and a set of formulas Γ, if $\Gamma \vdash \varphi$ then $\Gamma \models \varphi$.*

Proof It is immediate that under any valuation the values of the axioms are either 1 or $1/2$ and all the rules preserve designatedness of well-formed formulas. So the theorem follows. □

7.3.3 *Completeness*

For the proof of *completeness* we need a few lemmas.

Lemma 7.3.2 *For any formula φ, $\vdash \varphi \to \varphi$ holds.*

Lemma 7.3.3 (Deduction Theorem) *If* $\Gamma \cup \{\varphi\} \vdash \psi$ *then* $\Gamma \vdash \varphi \to \psi$.

Proof The proof is in the usual procedure by induction on the length of derivation of ψ from $\Gamma \cup \{\varphi\}$. □

Using the Deduction theorem one can prove the following lemma.

Lemma 7.3.4 *For any formulas* φ, ψ *and* γ *the following formulas are theorems.*

(i) $(\varphi \to \psi) \to ((\varphi \wedge \gamma) \to \psi)$
(ii) $(\varphi \to \psi) \to ((\psi \to \gamma) \to (\varphi \to \gamma))$
(iii) $(\varphi \to \psi) \wedge (\psi \to \gamma) \to (\varphi \to \gamma)$

Lemma 7.3.5 is the most important step to prove the completeness theorem.

Lemma 7.3.5 *For any formula* φ *and a given valuation* v *with respect to* PS_3 *let* φ' *be defined by*

$$\varphi' = \begin{cases} \varphi \wedge (\neg\varphi \to \bot) & if \quad v(\varphi) = 1; \\ \varphi \wedge \neg\varphi & if \quad v(\varphi) = 1/2; \\ \neg\varphi \wedge (\varphi \to \bot) & if \quad v(\varphi) = 0. \end{cases}$$

If $p_{i_1}, p_{i_2}, \ldots, p_{i_k}$ *are the propositional letters in* φ *then* $\{p'_{i_1}, p'_{i_2}, \ldots, p'_{i_k}\} \vdash \varphi'$.

Proof The proof is obtained by Kalmer's method by induction on the complexity of φ. For the detail proof we refer Appendix A. □

Now we are in the position to prove the weak completeness theorem.

Theorem 7.3.6 (Completeness) *For any formula* φ*, if* $\models \varphi$ *then* $\vdash \varphi$.

Proof Let φ be a formula such that $\models \varphi$. Moreover, let $p_{i_1}, p_{i_2}, \ldots, p_{i_n}$ be n propositional letters in φ. By the Lemma 7.3.5, for any arbitrarily fixed valuation we have $\{p'_{i_1}, p'_{i_2}, \ldots, p'_{i_n}\} \vdash \varphi'$. Since, $\models \varphi$, by the definition, φ' is either $\varphi \wedge (\neg\varphi \to \bot)$ or $\varphi \wedge \neg\varphi$. So in any case $\{p'_{i_1}, p'_{i_2}, \ldots, p'_{i_n}\} \vdash \varphi$ can be derived by *Axiom 3* and using *M.P.* Hence, deduction theorem gives

$$\{p'_{i_1}, p'_{i_2}, \ldots, p'_{i_{n-1}}\} \vdash p'_{i_n} \to \varphi.$$

Now, since the valuation was arbitrary, we get

$$\begin{aligned} &\{p'_{i_1}, p'_{i_2}, \ldots, p'_{i_{n-1}}\} \vdash [p_{i_n} \wedge (\neg p_{i_n} \to \bot)] \to \varphi, \\ &\{p'_{i_1}, p'_{i_2}, \ldots, p'_{i_{n-1}}\} \vdash (p_{i_n} \wedge \neg p_{i_n}) \to \varphi \\ \text{and} \quad &\{p'_{i_1}, p'_{i_2}, \ldots, p'_{i_{n-1}}\} \vdash [\neg p_{i_n} \wedge (p_{i_n} \to \bot)] \to \varphi. \end{aligned}$$

Hence, the following derivation can be established:

$\{p'_{i_1}, p'_{i_2}, \ldots, p'_{i_{n-1}}\} \vdash$

1. $[(p_{i_n} \wedge (\neg p_{i_n} \rightarrow \bot)) \rightarrow \varphi] \wedge [(p_{i_n} \wedge \neg p_{i_n}) \rightarrow \varphi] \wedge$
   $[(\neg p_{i_n} \wedge (p_{i_n} \rightarrow \bot)) \rightarrow \varphi]$ Rule 1
2. $[(p_{i_n} \wedge (\neg p_{i_n} \rightarrow \bot)) \vee (p_{i_n} \wedge \neg p_{i_n}) \vee$
   $(\neg p_{i_n} \wedge (p_{i_n} \rightarrow \bot))] \rightarrow \varphi$ Ax6
3. $(p_{i_n} \wedge (\neg p_{i_n} \rightarrow \bot)) \vee (p_{i_n} \wedge \neg p_{i_n}) \vee (\neg p_{i_n} \wedge (p_{i_n} \rightarrow \bot))$ Ax15
4. φ M.P. 2, 3

Repeating this process for each of the remaining p'_{i_j} $\{j = n-1, n-2, \ldots, 1\}$ we get $\vdash \varphi$.

Hence, the (weak) completeness theorem is proved. □

It is worthwhile to mention that another axiomatization of the matrix PS_3 together with an extra operator • can be found in [6]. But as described earlier we do not need the logical operator • for our purpose. It can be proved that the •-less axioms cannot form a complete axiom system for PS_3. For the •-free fragment of this logic our axioms give a •-free set of axioms.

7.4 $\mathbb{L}PS_3$ and Other Three-Valued Paraconsistent Logics

In this section, some important properties of $\mathbb{L}PS_3$ will be discussed and comparisons between $\mathbb{L}PS_3$ and some other well-known three-valued paraconsistent logics will be pointed out with respect to some logical properties.

7.4.1 Maximality Relative to Classical Propositional Logic

Maximality is an important issue in the study of paraconsistent logics (cf. [1, 4]).

Definition 7.4.1 A logic $\mathbf{L_1} = \langle \mathscr{L}, \vdash_1 \rangle$ is said to be *maximal relative to* a logic $\mathbf{L_2} = \langle \mathscr{L}, \vdash_2 \rangle$ iff

(i) $\vdash_1 \varphi$ implies $\vdash_2 \varphi$ for any φ, and
(ii) if $\nvdash_1 \varphi$, $\vdash_2 \varphi$ and $\vdash_1$ is enhanced to $\vdash'_1$ by adding φ to the theorem set of $\mathbf{L_1}$ then $\langle \mathscr{L}, \vdash'_1 \rangle = \langle \mathscr{L}, \vdash_2 \rangle$.

Definition 7.4.1 is more demanding than what is there in [3]. The change is in the condition (ii) of the Definition 7.4.1. In [3] the condition (ii) was taken as if $\nvdash_1 \varphi$, $\vdash_2 \varphi$ and $\vdash_1$ is enhanced to $\vdash'_1$ by adding φ to the theorem set of $\mathbf{L_1}$ then the set of theorems in $\langle \mathscr{L}, \vdash'_1 \rangle$ is same as the set of theorems in $\langle \mathscr{L}, \vdash_2 \rangle$.

The following theorem shows the relationship between $\mathbb{L}PS_3$ and the classical propositional logic.

Theorem 7.4.2 $\mathbb{L}PS_3$ *is maximal relative to the classical propositional logic* (CPL).

Proof The proof is given in Appendix B.

7.4.2 Comparison with Other Logics

The three-valued paraconsistent logics chosen for comparisons are G. PRIEST's *Logic of Paradox* (LP), *Logic of Formal Inconsistency 1* (LFI1) and *Logic of Formal Inconsistency 2* (LFI2) by W.A. CARNIELLI, J. MARCOS, and S. DE AMO, I.M.L. D'OTTAVIANO's logic (J_3), The Logic RM_3 by the ENTAILMENT school, A. M. SETTE's three-valued paraconsistent logic P^1 and C. MORTENSEN's paraconsistent logic $C_{0,2}$. Moreover, we will also make a comparison table with respect to S. JAŚKOWSKI's and N. DA COSTA's criteria for paraconsistent logics.

In 1948 S. Jaśkowski proposed three conditions for a paraconsistent propositional logic, and the simplified versions of which are as follows (cf. [8, 10]):

Jas1 the logic does not satisfy the implicational law of overfilling:

$$\varphi \to (\neg\varphi \to \psi);$$

Jas2 the logic should be rich enough to enable practical inferences: it satisfies modus ponens and the following formulas:

$$\varphi \to \varphi,$$
$$(\varphi \to \psi) \to ((\gamma \to \varphi) \to (\gamma \to \psi)),$$
$$(\varphi \to (\psi \to \gamma)) \to (\psi \to (\varphi \to \gamma));$$

Jas3 it should have an intuitive justification: restriction to $\{0, 1\}$ gives the classical valuation.

Driven by some different motivations in 1963 Newton da Costa wanted to characterize paraconsistency by proposing a whole hierarchy of paraconsistent propositional calculi, known as C_n, for $0 < n < \omega$. The following four conditions are the basic requirements for this calculi (cf. [10]):

NdaC1 the *law of non-contradiction*, $\neg(\varphi \wedge \neg\varphi)$, should not be a valid schema;
NdaC2 from the set of formulae, $\{\varphi, \neg\varphi\}$, not all formulas should be derived in general;
NdaC3 extensions to the predicate calculi (with or without equality) of these propositional calculi are simple;
NdaC4 without violating *NdaC1*, the calculi should contain the most part of the schemata and rules of the classical propositional calculus.

We will first produce a comparison table below where the following abbreviations are used:

- DN for $\vdash \neg\neg\varphi \leftrightarrow \varphi$.
- DM1 for $\vdash \neg(\varphi \wedge \psi) \leftrightarrow (\neg\varphi \vee \neg\psi)$.
- DM2 for $\vdash \neg(\varphi \vee \psi) \leftrightarrow (\neg\varphi \wedge \neg\psi)$.
- LEM for $\vdash \varphi \vee \neg\varphi$.
- C for $\vdash (\neg\varphi \rightarrow \neg\psi) \rightarrow (\psi \rightarrow \varphi)$.
- C1 for $\vdash (\neg\varphi \rightarrow \psi) \rightarrow (\neg\psi \rightarrow \varphi)$.
- C2 for $\vdash (\varphi \rightarrow \neg\psi) \rightarrow (\psi \rightarrow \neg\varphi)$.
- C3 for $\vdash (\varphi \rightarrow \psi) \rightarrow (\neg\psi \rightarrow \neg\varphi)$.
- HS for $\vdash (\varphi \rightarrow \psi) \wedge (\psi \rightarrow \gamma) \rightarrow (\varphi \rightarrow \gamma)$.
- MP for the rule Modus Ponens.
- DT for the Deduction Theorem.

(✓)-mark indicates that the property holds and (x)-mark indicates that the property does not hold in the corresponding logical system.

	DN	DM1	DM2	LEM	C	C1	C2	C3	HS	MP	DT
$\mathbb{L}PS_3$	✓	✓	✓	✓	x	x	x	x	✓	✓	✓
LP	✓	✓	✓	✓	✓	✓	✓	✓	x	x	✓
LFI1	✓	✓	✓	✓	x	x	x	x	✓	✓	✓
LFI2	✓	x	x	✓	x	x	x	x	✓	✓	✓
J_3	✓	✓	✓	✓	x	x	x	x	✓	✓	✓
RM_3	✓	✓	✓	✓	✓	✓	✓	✓	✓	✓	x
P^1	x	x	x	✓	x	x	x	x	✓	✓	✓
$C_{0,2}$	✓	x	x	✓	x	x	x	x	✓	✓	✓

We also present the following comparison table with respect to Jas1, Jas2, Jas3, NdaC1, and NdaC2.

	Jas1	Jas2	Jas3	NdaC1	NdaC2
$\mathbb{L}PS_3$	✓	✓	✓	x	✓
LP	x	x	✓	x	✓
LFI1	✓	✓	✓	x	✓
LFI2	✓	✓	✓	x	✓
J_3	✓	✓	✓	x	✓
RM_3	✓	✓	✓	x	✓
P^1	✓	✓	✓	✓	✓
$C_{0,2}$	✓	✓	✓	✓	✓

Note. All the observations for LP and $C_{0,2}$ are semantical.

7.4.3 *Some Other Differences*

This section shows some more differences between some of the aforesaid three-valued logics and $\mathbb{L}PS_3$ which are not included in the comparison table.

Comparison with LFI1 **and** LFI2: From the tables we see that the entries for $\mathbb{L}PS_3$ and LFI1 are the same. In fact, though the structure for LFI1 contains an extra unary operator $\bullet$ named as inconsistency, this structure and PS_3 are interdefinable. But as described earlier we do not need the logical operator $\bullet$ for our purpose. Second, with respect to the $\Rightarrow$ defined in PS_3, conditions **P1**, **P2**, **P3**, and **P4** are satisfied but not with respect to the implication operator of the matrix for LFI1.

Some differences are marked in the table with LFI2. Another major difference lies in the fact that the matrix for LFI2 does not form a lattice.

Comparison between $\mathbb{L}PS_3$ **and** RM_3: The comparison table shows that there are formulas which are theorems of RM_3 but not of $\mathbb{L}PS_3$, whereas the comparison tables do not explain anything about the converse. But a simple observation shows that the formula $\varphi \to (\psi \to \varphi)$ which is one of the axioms of $\mathbb{L}PS_3$ is not a theorem of RM_3.

$\mathbb{L}PS_3$ and RM_3 agree on the important fact that neither $(\varphi \to \psi) \leftrightarrow (\neg\varphi \vee \psi)$ nor $(\varphi \to \psi) \leftrightarrow \neg(\varphi \wedge \neg\psi)$ is a theorem.

Comparison between $\mathbb{L}PS_3$ **and** P^1: Like the three-valued semantics of LFI2 the three-valued semantics of P^1 also does not form a lattice which makes a difference with PS_3. The following interesting similarities and dissimilarities of these two logics may also be noted.

- $(\varphi \to \psi) \to (\neg\varphi \vee \psi)$ is a theorem in both.
- The reverse, $(\neg\varphi \vee \psi) \to (\varphi \to \psi)$ is not a theorem in any of them.
- $(\varphi \to \psi) \to \neg(\varphi \wedge \neg\psi)$ is a theorem of $\mathbb{L}PS_3$ but not a theorem in P^1.
- But the reverse, $\neg(\varphi \wedge \neg\psi) \to (\varphi \to \psi)$ is a theorem of P^1 but not a theorem of $\mathbb{L}PS_3$.

7.5 Conclusion

In the literature one can get several paraconsistent logics having algebraic semantics. Some of them are mentioned in this paper. These logics and their algebraic semantics were developed from various motivations. Our motivation is to construct models of some paraconsistent set theories.

With respect to the algebraic properties discussed in Sect. 7.2 which are needed for making an algebra-valued model of a paraconsistent set theory, the following comparison with other paraconsistent logics is made. The algebras of the three-valued semantics for LFI2, P^1 and $C_{0,2}$ are not lattices and for the others we make the following comparison table.

	P1	P2	P3	P4
$\mathbb{L}PS_3$	✓	✓	✓	✓
LP	x	✓	✓	✓
LFI1	x	✓	✓	✓
J_3	x	✓	✓	✓
RM_3	x	✓	✓	x

Note. The comparisons are made with respect to the corresponding three-valued semantics of the respective logics

Thus PS_3 differs from the other three-valued matrices mentioned here in forming an algebraic structure suitable to construct some model of set theory within **V** with an underlying paraconsistent logic. It is yet unknown whether other logics are suitable for this purpose. The full development of the corresponding predicate logic by extending $\mathbb{L}PS_3$ is still pending.

Appendix A

Proof of the Lemma 7.3.5:
As it was indicated that the proof will be by induction on the complexity of φ. Let, $\Gamma = \{p'_{i_1}, p'_{i_2}, \ldots, p'_{i_k}\}$.

Basis step: It is obvious when the complexity is 0.

Induction hypothesis: Assume the lemma holds well for formulas with complexity less than n.

Induction step: Let the complexity of φ be n.

Case 1: Let $\varphi = \neg\psi$.
Clearly, the complexity of ψ is less than n and the propositional letters in ψ are exactly same as the propositional letters in φ.

Subcase 1.1: If $v(\psi) = 1$ then $v(\varphi) = 0$. Hence, by our construction,

$$\psi' = \psi \wedge (\neg\psi \to \bot) \text{ and } \varphi' = \neg\varphi \wedge (\varphi \to \bot).$$

Here we get

$\Gamma \vdash$

1. ψ'	induction hypothesis
2. ψ	Ax3 and M.P.
3. $\neg\psi \to \bot$	Ax4 and M.P.
4. $\psi \to \neg\neg\psi$	Ax8
5. $\neg\neg\psi$	M.P. 2, 4
6. $\neg\varphi \wedge (\varphi \to \bot)$	Rule 1 on 3 and 5

Hence, in Subcase 1.1, $\Gamma \vdash \varphi'$.

Subcase 1.2: If $v(\psi) = 1/2$ then $v(\varphi) = 1/2$. So by the construction,

$$\psi' = \psi \wedge \neg\psi \text{ and } \varphi' = \varphi \wedge \neg\varphi.$$

So we have

$\Gamma \vdash$

1. ψ'	induction hypothesis
2. ψ	Ax3 and M.P.
3. $\neg\psi$	Ax4 and M.P.
4. $\psi \to \neg\neg\psi$	Ax8
5. $\neg\neg\psi$	M.P. 2, 4
6. $\varphi \wedge \neg\varphi$	Rule 1 on 3 and 5

Hence, $\Gamma \vdash \varphi'$ holds here.

Subcase 1.3: If $v(\psi) = 0$ then $v(\varphi) = 1$. Hence,

$$\psi' = \neg\psi \wedge (\psi \to \bot) \text{ and } \varphi' = \varphi \wedge (\neg\varphi \to \bot).$$

The following derivation can be made

$\Gamma \vdash$

1. ψ'	induction hypothesis
2. $\neg\psi$	Ax3 and M.P.
3. $\psi \to \bot$	Ax4 and M.P.
4. $(\neg\neg\psi \to \psi) \to [(\psi \to \bot) \to (\neg\neg\psi \to \bot)]$	Theorem 7.3.4(ii)
5. $\neg\neg\psi \to \psi$	Ax8
6. $(\psi \to \bot) \to (\neg\neg\psi \to \bot)$	M.P. 4, 5
7. $\neg\neg\psi \to \bot$	M.P. 3, 6
8. φ'	Rule 1 on 2, 7

Hence, in *Case 1* we always get $\Gamma \vdash \varphi'$.

Case 2: Let $\varphi = \psi \wedge \gamma$.

Obviously, both the complexities of ψ and γ are less than n and the sets of propositional letters in φ and ψ are proper subsets of $\{p'_{i_1}, p'_{i_2}, \ldots, p'_{i_k}\}$, the set of propositional letters in φ. Hence, clearly by the induction hypothesis and monotonicity property we get

$$\Gamma \vdash \psi' \quad \text{and} \quad \Gamma \vdash \gamma'.$$

Subcase 2.1: If any one of $v(\psi)$ and $v(\gamma)$ is 0 then it can be proved $\Gamma \vdash \varphi'$. Without loss of generality, let $v(\psi) = 0$, then $v(\varphi) = 0$. Hence, we get the following:

$$\psi' = \neg\psi \wedge (\psi \to \bot) \quad \text{and} \quad \varphi' = \neg\varphi \wedge (\varphi \to \bot).$$

Since $\Gamma \vdash \psi'$,

$\Gamma \vdash$

1. $\neg\psi$	Ax3 and M.P.
2. $\psi \to \bot$	Ax4 and M.P.
3. $\neg\psi \to (\neg\psi \vee \neg\gamma)$	Ax5
4. $\neg\psi \vee \neg\gamma$	M.P. 1, 3
5. $(\neg\psi \vee \neg\gamma) \to \neg(\psi \wedge \gamma)$	Ax9
6. $\neg(\psi \wedge \gamma)$	M.P. 4, 5
7. $(\psi \to \bot) \to (\psi \wedge \gamma \to \bot)$	Theorem 7.3.4(i)
8. $\psi \wedge \gamma \to \bot$	M.P. 2, 7
9. φ'	Rule 1 on 6, 8

Subcase 2.2: If $v(\psi) = 1/2$ and $v(\gamma) = 1/2$ then $v(\varphi) = 1/2$ also. So by the definition,

$$\psi' = \psi \wedge \neg\psi, \quad \gamma' = \gamma \wedge \neg\gamma \quad \text{and} \quad \varphi' = \varphi \wedge \neg\varphi.$$

Now for proving $\Gamma \vdash \varphi'$, i.e., $\Gamma \vdash (\psi \wedge \gamma) \wedge \neg(\psi \wedge \gamma)$ we go through the following derivation, using $\Gamma \vdash \psi'$ and $\Gamma \vdash \gamma'$.

$\Gamma \vdash$

1. ψ	Ax3 and M.P.
2. $\neg\psi$	Ax4 and M.P.
3. γ	Ax3 and M.P.
4. $\psi \wedge \gamma$	Rule 1 on 1, 3
5. $\neg\psi \to (\neg\psi \vee \neg\gamma)$	Ax5
6. $(\neg\psi \vee \neg\gamma)$	M.P. 2, 5
7. $(\neg\psi \vee \neg\gamma) \to \neg(\psi \wedge \gamma)$	Ax9
8. $\neg(\psi \wedge \gamma)$	M.P. 6, 7
9. φ'	Rule 1 on 4, 8

Subcase 2.3: If $v(\psi) = 1/2$ and $v(\gamma) = 1$ then $v(\varphi) = 1/2$ also. Hence,

$$\psi' = \psi \wedge \neg\psi, \quad \gamma' = \gamma \wedge (\neg\gamma \to \bot) \quad \text{and} \quad \varphi' = \varphi \wedge \neg\varphi.$$

Since $\Gamma \vdash \psi'$ and $\Gamma \vdash \gamma'$, using Axiom 1, 2 and rule M.P. we get

$$\Gamma \vdash \psi, \quad \Gamma \vdash \neg\psi \quad \text{and} \quad \Gamma \vdash \gamma.$$

Now following the same derivation as above we can prove $\Gamma \vdash \psi'$.

Subcase 2.4: If $v(\psi) = 1$ and $v(\gamma) = 1$ then $v(\varphi) = 1$. Therefore, by our construction,

$$\psi' = \psi \wedge (\neg\psi \to \bot), \quad \gamma' = \gamma \wedge (\neg\gamma \to \bot) \quad \text{and} \quad \varphi' = \varphi \wedge (\neg\varphi \to \bot).$$

So we have to prove $\Gamma \vdash \varphi'$, i.e., $\Gamma \vdash (\psi \wedge \gamma) \wedge [\neg(\psi \wedge \gamma) \to \bot]$. The derivation is as follows:

$\Gamma \vdash$

1. ψ Ax3 and M.P.
2. γ Ax3 and M.P.
3. $\neg\psi \to \bot$ Ax4 and M.P.
4. $\neg\gamma \to \bot$ Ax4 and M.P.
5. $\psi \wedge \gamma$ Rule 1 on 1 and 2
6. $(\neg\psi \to \bot) \wedge (\neg\gamma \to \bot)$ Rule 1 on 3 and 4
7. $(\neg\psi \to \bot) \wedge (\neg\gamma \to \bot) \to (\neg\psi \vee \neg\gamma \to \bot)$ Ax6
8. $\neg\psi \vee \neg\gamma \to \bot$ M.P. 6, 7
9. $\neg(\psi \wedge \gamma) \to (\neg\psi \vee \neg\gamma)$ Ax9
10. $[\neg(\psi \wedge \gamma) \to (\neg\psi \vee \neg\gamma)] \to$
    $[((\neg\psi \vee \neg\gamma) \to \bot) \to (\neg(\psi \wedge \gamma) \to \bot)]$ Theorem 7.3.4(ii)
11. $\neg(\psi \wedge \gamma) \to \bot$ M.P. repeatedly on 10, 9, 8
12. φ' Rule 1 on 5, 11

Subcase 2.5: If $v(\psi) = 1$ and $v(\gamma) = 1/2$ then by the same derivation in Subcase 2.3 it can be proved $\Gamma \vdash \varphi'$.

Hence, in *Case 2* we can always prove $\Gamma \vdash \varphi'$.

Case 3: Let $\varphi = \psi \vee \gamma$.
Since $\varphi \vee \psi$ can be abbreviated as $\neg(\neg\varphi \wedge \neg\psi)$ therefore using Case 1 and Case 2, $\Gamma \vdash \varphi'$ can be proved in this case also.

Case 4: Let $\varphi = \psi \to \gamma$.
Obviously, both the complexities of ψ and γ are less than n and the sets of propositional letters in φ and ψ are subsets of $\{p'_{i_1}, p'_{i_2}, \ldots, p'_{i_k}\}$, the set of propositional letters in φ. Hence, clearly by the induction hypothesis and monotonicity property we get

$$\Gamma \vdash \psi' \quad \text{and} \quad \Gamma \vdash \gamma'.$$

Subcase 4.1: If $v(\gamma) = 1$ then no matter what $v(\psi)$ is, $v(\varphi) = 1$ always. So we get $\gamma' = \gamma \wedge (\neg\gamma \to \bot)$ and $\varphi' = \varphi \wedge (\neg\varphi \to \bot) = (\psi \to \gamma) \wedge [\neg(\psi \to \gamma) \to \bot]$.

Now for proving $\Gamma \vdash \varphi'$ we go through the following derivation:

$\Gamma \vdash$

1. γ Ax3 and M.P.
2. $\neg\gamma \to \bot$ Ax4 and M.P.
3. $\gamma \to (\psi \to \gamma)$ Ax1
4. $\psi \to \gamma$ M.P. 1, 3
5. $(\neg\gamma \to \bot) \to [\neg(\psi \to \gamma) \to \bot]$ Ax12
6. $\neg(\psi \to \gamma) \to \bot$ M.P. 2, 5
7. φ' Rule 1 on 4, 6

Subcase 4.2: If $v(\gamma) = 1/2$ then always $v(\varphi) = 1$. Hence, by the definition,

$$\gamma' = \gamma \wedge \neg\gamma \quad \text{and} \quad \varphi' = \varphi \wedge (\neg\varphi \to \bot) = (\psi \to \gamma) \wedge [\neg(\psi \to \gamma) \to \bot].$$

Hence, we get the following:

$\Gamma \vdash$

1. $\gamma \wedge \neg\gamma$	induction hypothesis
2. γ	Ax3 and M.P.
3. $\gamma \rightarrow (\psi \rightarrow \gamma)$	Ax1
4. $\psi \rightarrow \gamma$	M.P. 1, 2
5. $(\gamma \wedge \neg\gamma) \rightarrow [\neg(\psi \rightarrow \gamma) \rightarrow \bot]$	Ax10
6. $\neg(\psi \rightarrow \gamma) \rightarrow \bot$	M.P. 1, 5
7. φ'	Rule 1 on 3, 6

Subcase 4.3: If $v(\gamma) = 1/2$ and $v(\psi) = 0$ then $v(\varphi) = 1$. So by the construction,

$$\gamma' = \neg\gamma \wedge (\gamma \rightarrow \bot),\ \psi' = \neg\psi \wedge (\psi \rightarrow \bot) \text{ and}$$

$$\begin{aligned}\varphi' &= \varphi \wedge (\neg\varphi \rightarrow \bot)\\ &= (\psi \rightarrow \gamma) \wedge [\neg(\psi \rightarrow \gamma) \rightarrow \bot].\end{aligned}$$

Now the following derivation shows that $\Gamma \vdash \varphi'$ holds in this subcase also.

$\Gamma \vdash$

1. $\psi \rightarrow \bot$	Ax4 and M.P.
2. $\bot \rightarrow \gamma$	Ax13
3. $(\psi \rightarrow \bot) \rightarrow [(\bot \rightarrow \gamma) \rightarrow (\psi \rightarrow \gamma)]$	Theorem 7.3.4(ii)
4. $(\bot \rightarrow \gamma) \rightarrow (\psi \rightarrow \gamma)$	M.P. 1, 3
5. $\psi \rightarrow \gamma$	M.P. 2, 4
6. $(\psi \rightarrow \bot) \rightarrow [\neg(\psi \rightarrow \gamma) \rightarrow \bot]$	Ax11
7. $\neg(\psi \rightarrow \gamma) \rightarrow \bot$	M.P. 1, 6
8. φ'	Rule 1 on 5, 7

Subcase 4.4: If $v(\gamma) = 0$ and $v(\psi) = 1$ then $v(\varphi) = 0$. Therefore,

$$\gamma' = \neg\gamma \wedge (\gamma \rightarrow \bot),\ \psi' = \psi \wedge (\neg\psi \rightarrow \bot) \text{ and}$$

$$\begin{aligned}\varphi' &= \neg\varphi \wedge (\varphi \rightarrow \bot)\\ &= \neg(\psi \rightarrow \gamma) \wedge [(\psi \rightarrow \gamma) \rightarrow \bot].\end{aligned}$$

Deduction theorem will be used here for proving $\Gamma \vdash \varphi'$. Since we know $\Gamma \vdash \psi'$ and $\Gamma \vdash \gamma'$

$\Gamma \cup \{\psi \rightarrow \gamma\} \vdash$

1. ψ'	monotonicity
2. ψ	Ax3 and M.P. with 1
3. γ'	monotonicity
4. $\gamma \rightarrow \bot$	Ax4 and M.P. with 3
5. $\psi \rightarrow \gamma$	assumption
6. γ	M.P. 2, 5
7. $\bot$	M.P. 4, 6

Now applying Deduction theorem we get $\Gamma \vdash (\psi \rightarrow \gamma) \rightarrow \bot$.

Again for proving $\Gamma \vdash \neg(\psi \rightarrow \gamma)$ we do the following derivation:

$\Gamma \vdash$

1. ψ'	induction hypothesis
2. ψ	Ax3 and M.P. with 1
3. γ'	induction hypothesis
4. $\gamma \rightarrow \bot$	Ax4 and M.P. with 3
5. $\psi \wedge (\gamma \rightarrow \bot)$	Rule 1 on 2 and 4
6. $\psi \wedge (\gamma \rightarrow \bot) \rightarrow \neg(\psi \rightarrow \gamma)$	Axiom 14
7. $\neg(\psi \rightarrow \gamma)$	M.P. 3, 4

Hence, again by *Rule 1* it is derived $\Gamma \vdash \neg(\psi \rightarrow \gamma) \wedge [(\psi \rightarrow \gamma) \rightarrow \bot]$ i.e., $\Gamma \vdash \varphi'$.

Subcase 4.5: If $v(\gamma) = 0$ and $v(\psi) = 1/2$ then $v(\varphi) = 0$. Therefore by definition,

$$\gamma' = \neg\gamma \wedge (\gamma \rightarrow \bot),\ \psi' = \psi \wedge \neg\psi \text{ and}$$
$$\varphi' = \neg\varphi \wedge (\varphi \rightarrow \bot) = \neg(\psi \rightarrow \gamma) \wedge [(\psi \rightarrow \gamma) \rightarrow \bot].$$

Since $\Gamma \vdash \psi'$ and $\Gamma \vdash \gamma'$, by *Axiom 1, 2* and using *M.P.* we get

$$\Gamma \vdash \psi \text{ and } \Gamma \vdash \gamma \rightarrow \bot.$$

Therefore, in this subcase $\Gamma \vdash \varphi'$ can be proved by following the same steps used in *Subcase 4.4*.

Hence, combining all the cases the Lemma 7.3.5 is proved. □

Appendix B

Proof of the theorem 7.4.2:

LEMMA. First, we show that by adding any theorem of CPL (Classical Propositional Logic) which is not a theorem of $\mathbb{L}PS_3$, as an axiom schema in $\mathbb{L}PS_3$, all the theorems of CPL can be proved.

Let $\varphi(p_{i_1}, p_{i_2}, \ldots, p_{i_n})$ be a theorem of CPL but not a theorem of $\mathbb{L}PS_3$, where $p_{i_1}, p_{i_2}, \ldots, p_{i_n}$ are the propositional variables. Hence, for any valuation v from the set of all formulas of $\mathbb{L}PS_3$ to PS_3 for which

$$v(\varphi(p_{i_1}, p_{i_2}, \ldots, p_{i_n})) = 0$$

there must exist some p_{i_l}, $1 \leq l \leq n$ such that $v(p_{i_l}) = 1/2$. Now using this fact without loss of generality we may assume that for any given valuation v we have $v(\varphi(p_{i_1}, p_{i_2}, \ldots, p_{i_n})) = 0$ iff $v(p_{i_l}) = 1/2$ for all $l \in \{1, \ldots, n\}$. It is guaranteed by the following fact: Suppose a formula $\psi(p_{r_1}, p_{r_2}, \ldots p_{r_{t+1}})$ is such that $v(p_{r_l}) = 1/2$

for all $l \in \{1, \ldots, t\}$ but $v(p_{r_{t+1}}) \neq 1/2$. We then replace the propositional variable $p_{r_{t+1}}$

- by $\neg(p_{r_1} \to p_{r_1})$ if $v(p_{r_{t+1}}) = 0$
- by $(p_{r_1} \to p_{r_1})$ if $v(P_{r_{t+1}}) = 1$

in the formula $\psi(p_{r_1}, p_{r_2}, \ldots p_{r_{t+1}})$ and therefore after replacing, the formula will get the value 0 iff all its propositional variables take the value 1/2. In this way we always get such a formula $\varphi(p_{i_1}, p_{i_2}, \ldots, p_{i_n})$.

Let us now assume $\sigma(p_{k_1}, p_{k_2}, \ldots, p_{k_m})$ be another arbitrarily chosen theorem of CPL which is not a theorem of $\mathbb{L}PS_3$, where $p_{k_1}, p_{k_2}, \ldots, p_{k_m}$ are the propositional variables. It will be proved that $\sigma(p_{k_1}, p_{k_2}, \ldots, p_{k_m})$ can be derived from the axiom system of $\mathbb{L}PS_3$ if it is extended by the new axiom schema $\varphi(p_{i_1}, p_{i_2}, \ldots, p_{i_n})$. Let $\varphi(p_{k_j})$ be the formula replacing each propositional variable of $\varphi(p_{i_1}, p_{i_2}, \ldots, p_{i_n})$ by p_{k_j}, for all $j \in \{1, \ldots, m\}$.

Claim: $\varphi(p_{k_1}) \wedge \varphi(p_{k_2}) \wedge \cdots \wedge \varphi(p_{k_m}) \to \sigma(p_{k_1}, p_{k_2}, \ldots, p_{k_m})$ is a theorem of $\mathbb{L}PS_3$.

Proof of the claim. Let v be any valuation. Two cases could happen: either $v(\sigma(p_{k_1}, p_{k_2}, \ldots, p_{k_m})) = 0$ or $v(\sigma(p_{k_1}, p_{k_2}, \ldots, p_{k_m})) \neq 0$.

If $v(\sigma(p_{k_1}, p_{k_2}, \ldots, p_{k_m})) = 0$ and since $\sigma(p_{k_1}, p_{k_2}, \ldots, p_{k_m})$ is a theorem of CPL there must exist p_{k_j} such that $v(p_{k_j}) = 1/2$ for some $j \in \{1, \ldots, m\}$. Hence, $v(\varphi(p_{k_j})) = 0$ and therefore

$$v(\varphi(p_{k_1}) \wedge \varphi(p_{k_2}) \wedge \cdots \wedge \varphi(p_{k_m}) \to \sigma(p_{k_1}, p_{k_2}, \ldots, p_{k_m})) = 1.$$

Again if $v(\sigma(p_{k_1}, p_{k_2}, \ldots, p_{k_m})) \neq 0$ then by the truth tables of PS_3

$$v(\varphi(p_{k_1}) \wedge \varphi(p_{k_2}) \wedge \cdots \wedge \varphi(p_{k_m}) \to \sigma(p_{k_1}, p_{k_2}, \ldots, p_{k_m})) = 1.$$

Hence, for any valuation the formula

$$\varphi(p_{k_1}) \wedge \varphi(p_{k_2}) \wedge \cdots \wedge \varphi(p_{k_m}) \to \sigma(p_{k_1}, p_{k_2}, \ldots, p_{k_m})$$

always get the value 1. So by the completeness theorem of $\mathbb{L}PS_3$ the claim is proved.

Now let us extend the axiom system of $\mathbb{L}PS_3$ by including $\varphi(p_{i_1}, \ldots, p_{i_n})$ as an axiom schema. Let this system be denoted by $\mathbb{L}'PS_3$. Then by the Rule 1, $\varphi(p_{k_1}) \wedge \varphi(p_{k_2}) \wedge \cdots \wedge \varphi(p_{k_m})$ is a theorem of $\mathbb{L}'PS_3$. Now using M.P. we have $\sigma(p_{k_1}, p_{k_2}, \ldots, p_{k_m})$ as a theorem of the new system. Hence, the lemma is proved.

From this lemma it follows that for the enhanced system $\mathbb{L}'PS_3$,

$$\Gamma \vdash_{\mathbb{L}'PS_3} \varphi \text{ iff } \Gamma \vdash_{CPL} \varphi.$$

□

References

1. Arieli, O., Avron, A., Zamansky, A.: Maximal and premaximal paraconsistency in the framework of three-valued semantics. Stud. Logica **97**, 31–60 (2011)
2. Bell, J.L.: Boolean-Valued Models and Independence Proofs in Set Theory. Oxford University Press (1985)
3. Carnielli, W.A., Marcos, J., Amo, S.: Formal inconsistency and evolutionary databases. Logic Logical Philos **8**, 115–152 (2000)
4. Carnielli, W.A., Coniglio, M., Marcos, J.: Logics of formal inconsistency. In: Gabbay, D., Guenthner, F. (eds.) Handbook of Philosophycal Logic, 2nd edn., vol. 14, pp. 1–93. Springer (2007)
5. Cohen, P.J.: The independence of the continuum hypothesis. I., Proc. Natl. Acad. Sci. USA **50**, 1143–1148 (1963)
6. Coniglio, M.E., da Cruz Silvestrini, L.H.: An alternative approach for quasi-truth. Logic J. IGPL **22**(2), 387–410 (2014)
7. Grayson, R.J.: Heyting-valued models for intuitionistic set theory. Lect. Notes Math. **753**, 402–414 (1979)
8. Karpenko, A.S.: Jaśkowsi's criterion and three-valued paraconsistent logics. Logic Logical Philos. **7**, 81–86 (1999)
9. Löwe, B., Tarafder, S.: Generalized algebra-valued models of set theory. Rev. Symbolic Logic **8**(01), 192–205 (2015)
10. Marcos, J.: On a Problem of da Costa, pp. 53–69. Polimetrica International Scientific Publisher Monza/Italy, Essays on the Foundations of Mathematics and Logic (2005)
11. Osorio, M., Carballido, J.: Brief Study of G'_3 logic. J. Appl. Non-Class. Logics **18**(4), 475–499 (2008)
12. Ozawa, M.: Orthomodular-Valued Models for Quantum Set Theory. arXiv:0908.0367 (2009)
13. Tarafder, S.: Ordinals in an algebra-valued model of a paraconsistent set theory, in Banerjee, M., Krishna, S. (eds.), Logic and Its Applications, 6th International Conference, ICLA 2015, Mumbai, India, 8–10 Jan. 2015, Lecture Notes in Computer Science, vol. 8923, pp. 195–206. Springer, Berlin

Chapter 8
Two Consistent Many-Valued Logics for Paraconsistent Phenomena

Esko Turunen and J. Tinguaro Rodríguez

Abstract In this reviewing paper, we recall the main results of our papers [24, 31] where we introduced two paraconsistent semantics for Pavelka style fuzzy logic. Each logic formula α is associated with a 2×2 matrix called evidence matrix. The two semantics are consistent if they are seen from 'outside'; the structure of the set of the evidence matrices M is an MV-algebra and there is nothing paraconsistent there. However, seen from 'inside,' that is, in the construction of a single evidence matrix paraconsistency comes in, truth and falsehood are not each others complements and there is also contradiction and lack of information (unknown) involved. Moreover, we discuss the possible applications of the two logics in real-world phenomena.

Keywords Mathematical fuzzy logic · Paraconsistent logic · MV-algebra

Mathematics Subject Classification (2000) 03-02 · 03620 · 06D35

8.1 Introduction

Quoting from Stanford Encyclopedia of Philosophy [20] '*The contemporary logical orthodoxy has it that, from contradictory premises, anything can be inferred. To be more precise, let $\models$ be a relation of logical consequence, defined either semantically*

The first author acknowledges the support by the Czech Technical University in Prague under project SGS12/ 187/ OHK3/ 3T/ 13. The second author has been partially supported by grant TIN2012-32482 from the Government of Spain and excellence network S2013/ICE–2845 of the Region of Madrid.

E. Turunen (✉)
University of Technology, Vienna, Wiedner Hauptstrasse 8–10, A-1040 Wien, Austria
e-mail: esko.turunen@tut.fi

J.T. Rodríguez
Faculty of Mathematics, Complutense University, Plaza de Ciencias 3,
28040 Madrid, Spain
e-mail: jtrodrig@mat.ucm.es

J.-Y. Beziau et al. (eds.), *New Directions in Paraconsistent Logic*,
Springer Proceedings in Mathematics & Statistics 152,
DOI 10.1007/978-81-322-2719-9_8

or proof theoretically. Call $\models$ explosive *if it validates* $\{A, \neg A\} \models B$ *for every A and B* (ex contradictione quodlibet). *The contemporary orthodoxy, i.e., classical logic, is explosive, but also some non-classical logics such as intuitionist logic and most other standard logics are explosive. The major motivation behind paraconsistent logic is to challenge this orthodoxy. A logical consequence relation,* $\models$, *is said to be* paraconsistent *if it is not explosive. Thus, if* $\models$ *is paraconsistent, then even if we are in certain circumstances where the available information is inconsistent, the inference relation does not explode into triviality. Thus, paraconsistent logic accommodates inconsistency in a sensible manner that treats inconsistent information as informative.*'

During the last decades, the application potential of paraconsistent logic has been noted in many areas of science and technology; besides our own works, we refer here to the following papers, and this list is not exhaustive [2, 7, 12, 14].

It should be noted that the concept of paraconsistent logic is not unique, rather there is a number of different approaches. In Belnap's paraconsistent logic [3], four possible values associated with atomic formulas α are interpreted as `told only True`, `told only False`, `both told True and told False`, and `neither told True nor told False`, respectively. However, we call them for simplicity `true`, `false`, `contradictory`, and `unknown`: if there is evidence for α and no evidence against α, then α obtains the value `true` and if there is no evidence for α and evidence against α, then α obtains the value `false`. A value `contradictory` corresponds to a situation where there is simultaneous evidence for α and against α and, finally, α is labeled by value `unknown` if there is no evidence for α nor evidence against α. More formally, the values are associated with ordered couples $T = \langle 1, 0\rangle$, $F = \langle 0, 1\rangle$, $K = \langle 1, 1\rangle$, and $U = \langle 0, 0\rangle$, respectively.

In [17, 19], a continuous valued extension of Belnap's logic was studied. The authors imposed reasonable conditions that this continuous valued extension should obey and, after a careful analysis, they came to the conclusion that the graded values are to be computed via

$$t(\alpha) = \min\{a, 1 - b\}, \tag{8.1}$$

$$k(\alpha) = \max\{a + b - 1, 0\}, \tag{8.2}$$

$$u(\alpha) = \max\{1 - a - b, 0\}, \tag{8.3}$$

$$f(\alpha) = \min\{1 - a, b\}, \tag{8.4}$$

where an ordered couple $\langle a, b\rangle \in [0, 1]^2$, called *evidence couple*, is given. The intuitive meaning of a and b is the degree of evidence for a statement α and against α, respectively. Moreover, the set of 2×2 *evidence matrices* of the form

$$\begin{bmatrix} f(\alpha) & k(\alpha) \\ u(\alpha) & t(\alpha) \end{bmatrix}$$

is denoted by M. The values $f(\alpha)$, $k(\alpha)$, $u(\alpha)$, and $t(\alpha)$ are values on the real unit interval [0, 1] such that $f(\alpha) + k(\alpha) + u(\alpha) + t(\alpha) = 1$. Their intuitive meaning is *f*(α) = *falsehood, k*(α) = *contradictory, u*(α) = *unknown*, and *t*(α) = *truth* of the statement α. In [19] it is shown how such a fuzzy version of Belnap's logic can be applied in preference modeling. In [30] we show how paraconsistency is related to data mining.

We showed in [31] that, instead of a Boolean structure (suggested originally by Öztürk and Tsoukiás), the valuation domain M should be equipped with a more general algebraic structure called injective MV-algebra. The standard Łukasiewicz algebra on the real unit interval is an example of an injective MV-algebra. This associates fuzzy extensions of four valued paraconsistent logics with Pavelka style fuzzy sentential logic [18], giving rise to concepts such as *fuzzy set of axioms, provability degree, degree of theoremhood*, and *evaluated proof*. As a consequence a complete truth calculus is obtained. Our basic observation in [31] was that the algebraic operations in (8.1)–(8.4) are expressible only by the Łukasiewicz t-norm and the corresponding residuum, i.e., in the standard MV-algebra. This fact was implicitly shown in the analysis done in [17, 19]. Thus, if we would start with some other t-norm conjunction and involutive negation then the reasonable conditions a continuous valued extension of paraconsistent logic should obey would cease to hold.

The present first author has generalized in [28] Pavelka's ideas by introducing a Pavelka style fuzzy sentential logic with truth values in an injective MV-algebra, thus generalizing [0, 1]-valued logic. Injective MV-algebras **L** are complete when seen as lattices; all suprema and infima exist in **L**, and **L** satisfies a certain divisibility condition. Indeed, in [28] it is proved that Pavelka style fuzzy sentential logic is a complete logic in the sense that if the truth value set forms an injective MV-algebra **L**, then the set of a-tautologies and the set of a-provable formulas coincide for all $a \in L$. For a complete description, see the textbook [29] Chap. 3. Recently, the present first author proved (cf. [32]) the most general semantic completeness theorem: Pavelka style semantic completeness holds if and only if the set of truth values is a complete (as a lattice) MV-algebra; thus, the divisibility condition of the truth value structure is redundant.

As a consequence of the above results we now have that, given a set (of evidence values) which is a complete MV-algebra, it is possible to transfer a complete MV-structure to the set M, too. The corresponding paraconsistent sentential logic is Pavelka style fuzzy logic with new semantics. Thus, a rich semantics and syntax is available. For example, Łukasiewicz tautologies as well as intuitionistic tautologies can be expressed in the framework of this logic. This follows by the fact that we have two sorts of logical connectives conjunction, disjunction, implication, and negation interpreted either by the monoidal operations $\odot, \oplus, \longrightarrow, ^*$ or by the lattice operations $\wedge, \vee, \Rightarrow, ^\sim$, respectively (however, neither $^\sim$ nor * is a lattice complementation). Besides, there are many other logical connectives available.

The association of many-valued Belnap's logic and Pavelka sentential logic crucially depends on the fact that the product of two complete MV-algebras (and thus the set of evidence pairs) can also be naturally equipped with a complete MV-algebra structure. In [24], we have shown how to develop a continuous paraconsistent logic on truth scales that cannot be viewed as the product of MV-algebras, but still having a complete MV-algebra structure. In this sense, we focused on the study of possible MV-algebra structures over the unit triangle

$$\triangledown = \{(a, b) \in [0, 1]^2; a + b \leq 1\},$$

thus introducing an approach to the study of the logical properties of this scale different from but complementary to that of triangle algebras considered in [33]. The intuitive idea here is that evidence for a formula α (the value a) and evidence against α (the value b) are mutually restricted by the condition $a + b \leq 1$, based on such an idea a decision support system focused on a context in which incomplete as well as conflicting information often appears is studied in [22].

Recall that $\triangledown$ is usually equipped with a lattice structure through the partial order $\leq_t$ defined by $\mathbf{x} \leq_t \mathbf{y}$ iff $x_1 \leq y_1$ and $y_2 \leq x_2$, where $\mathbf{x} = \langle x_1, x_2 \rangle, \mathbf{y} = \langle y_1, y_2 \rangle$ are element of the unit triangle $\triangledown$. Furthermore, in [6] it is proven that the lattice $(\triangledown, \leq_t)$ cannot be endowed with an MV-algebra structure if we use t-norms (and thus also t-conorms) as logical operators. Nevertheless, it does not mean that the set $\triangledown$ cannot be endowed with an MV-algebra structure at all. On the contrary, we proved in [22] that different MV-algebra structures can be achieved for $\triangledown$, though then the canonical order of such an MV-algebra cannot coincide with $\leq_t$. The reason why it is interesting to develop a paraconsistent logic on the unit triangle $\triangledown$ is because it is the natural valuation set that arises when evidence for several categories is computed from the same data. Particularly, when two of these categories are regarded as opposite, the presence of information for both categories can manifest a certain type of inconsistency. The proposed paraconsistent logic on $\triangledown$ is applied in this sense in [24] in the context of disaster management for the development of a classification methodology that allows considering some of the classes as opposite, a relevant modeling choice in some application contexts.

In this review, we recall the main results of our papers [24, 31]. We want to emphasize that the two logics are consistent if they are seen from 'outside'; the structure of the set of the evidence matrices M is an MV-algebra and there is nothing paraconsistent there. However, seen from 'inside', the construction of a single evidence matrix paraconsistency comes in. This justifies the name of the paper. Moreover, we discuss possible applications of the two logics in real-world phenomena.

Finally, we wish to stress the following aspect. Dubois [8] published a critical study on Belnap's approach, to which our work is linked to. According to Dubois, the main difficulty lies in the confusion between truth values and information states. We emphasize that we study paraconsistent logic from a purely formal point of view without any deeper philosophical interpretation.

This paper is organized as follows. In Sect. 8.2, we recall all the algebraic definitions and properties of MV-algebras that are necessary to understand our approach. In Sect. 8.3 we introduce evidence couples and evidence matrices, first on the unit square and then on the unit triangle. In Sect. 8.4 we present a brief outline of the main features of Pavelka logic. Then, we show how paraconsistent logic is associated with it; some illustrative examples are also given. Section 8.5 concludes the paper.

8.2 Algebraic Preliminaries

We recall some basic definitions and properties of MV-algebras that are needed to understand and justify our results; all detail can be found in [5, 29, 31]. An *MV-algebra* $\mathbf{L} = \langle L, \oplus, ^*, \mathbf{0}\rangle$ is a structure such that $\langle L, \oplus, \mathbf{0}\rangle$ is a commutative monoid, i.e.,

$$x \oplus y = y \oplus x, \tag{8.5}$$
$$x \oplus (y \oplus z) = (x \oplus y) \oplus z, \tag{8.6}$$
$$x \oplus \mathbf{0} = x \tag{8.7}$$

holds for all elements $x, y, z \in L$ and, moreover,

$$x^{**} = x, \tag{8.8}$$
$$x \oplus \mathbf{0}^* = \mathbf{0}^*, \tag{8.9}$$
$$(x^* \oplus y)^* \oplus y = (y^* \oplus x)^* \oplus x. \tag{8.10}$$

Denote $x \odot y = (x^* \oplus y^*)^*$ and $\mathbf{1} = \mathbf{0}^*$. Then $\langle L, \odot, \mathbf{1}\rangle$ is another commutative monoid and hence

$$x \odot y = y \odot x, \tag{8.11}$$
$$x \odot (y \odot z) = (x \odot y) \odot z, \tag{8.12}$$
$$x \odot \mathbf{1} = x \tag{8.13}$$

holds for all elements $x, y, z \in L$. It is obvious that $x \oplus y = (x^* \odot y^*)^*$, thus the triple $\langle \oplus, ^*, \odot\rangle$ satisfies De Morgan laws. A partial order on the set L is introduced by

$$x \leq y \text{ iff } x^* \oplus y = \mathbf{1} \text{ iff } x \odot y^* = \mathbf{0}. \tag{8.14}$$

By setting

$$x \vee y = (x^* \oplus y)^* \oplus y, \tag{8.15}$$
$$x \wedge y = (x^* \vee y^*)^*[= (x^* \odot y)^* \odot y] \tag{8.16}$$

for all $x, y, z \in L$ the structure $\langle L, \wedge, \vee\rangle$ is a lattice. Moreover, $x \vee y = (x^* \wedge y^*)^*$ holds and therefore the triple $\langle \wedge, ^*, \vee\rangle$, too, satisfies the De Morgan laws. However, the unary operation * called *complementation* is not a lattice complementation. By stipulating

$$x \rightarrow y = x^* \oplus y \tag{8.17}$$

the structure $\langle L, \leq \wedge, \vee, \odot, \rightarrow, \mathbf{0}, \mathbf{1}\rangle$ is a residuated lattice with the bottom and top elements $\mathbf{0}$, $\mathbf{1}$, respectively. In particular, a residuation

$$x \odot y \leq z \text{ iff } x \leq y \rightarrow z \tag{8.18}$$

holds for all $x, y, z \in L$. The couple $\langle \odot, \rightarrow\rangle$ is an *adjoint couple*. Notice that the lattice operations on L can be expressed also via

$$x \vee y = (x \rightarrow y) \rightarrow y, \tag{8.19}$$

$$x \wedge y = x \odot (x \rightarrow y). \tag{8.20}$$

The standard example of an MV-algebra is the *Łukasiewicz structure L*, also called *standard MV-algebra*; the underlying set is the real unit interval [0, 1] equipped with the usual order and, for each $x, y \in [0, 1]$,

$$x \oplus y = \min\{x + y, 1\}, \tag{8.21}$$

$$x^* = 1 - x. \tag{8.22}$$

Moreover,

$$x \odot y = \max\{0, x + y - 1\}, \tag{8.23}$$

$$x \vee y = \max\{x, y\}, \tag{8.24}$$

$$x \wedge y = \min\{x, y\}, \tag{8.25}$$

$$x \rightarrow y = \min\{1, 1 - x + y\}, \tag{8.26}$$

$$x \odot y^* = \max\{x - y, 0\}. \tag{8.27}$$

For any natural number $m \geq 2$, a finite chain $0 < \frac{1}{m} < \cdots < \frac{m-1}{m} < 1$ can be equipped with MV-algebra operations by defining $\frac{n}{m} \oplus \frac{k}{m} = \min\{\frac{n+k}{m}, 1\}$ and $(\frac{n}{m})^* = \frac{m-n}{m}$. Finally, a structure $L \cap \mathbb{Q}$ with the Łukasiewicz operations is an example of a countable MV-algebra called *rational Łukasiewicz structure*. All these examples are linear MV-algebras, i.e. the corresponding order is a total order. Moreover, they are MV-subalgebras of the structure L. A Boolean algebra is an MV-algebra such that the monoidal operations $\oplus, \odot$ and the lattice operations $\vee, \wedge$ coincide, respectively.

An MV-algebra $\mathbf{L}$ is called *complete* if $\bigvee\{a_i \mid i \in \Gamma\}, \bigwedge\{a_i \mid i \in \Gamma\} \in L$ for any subset $\{a_i : i \in \Gamma\} \subseteq L$. Complete MV-algebras are infinitely distributive, that is, they satisfy

$$x \wedge \bigvee_{i \in \Gamma} y_i = \bigvee_{i \in \Gamma} (x \wedge y_i), \quad x \vee \bigwedge_{i \in \Gamma} y_i = \bigwedge_{i \in \Gamma} (x \vee y_i), \tag{8.28}$$

for any $x \in L$, $\{y_i \mid i \in \Gamma\} \subseteq L$. Thus, in a complete MV-algebra we can define another adjoint couple $\langle \wedge, \Rightarrow\rangle$, where the operation $\Rightarrow$ is defined via

$$x \Rightarrow y = \bigvee \{z|\ x \wedge z \leq y\}. \tag{8.29}$$

In particular, $x^{\sim} = x \Rightarrow \mathbf{0}$ defines another complementation (called *weak complementation*) in complete MV-algebras. However, weak complementation needs not to be lattice complementation. A *Heyting algebra* H is a bounded lattice such that for all $a, b \in H$ there is a greatest element x in H such that $a \wedge x \leq b$. Thus, to any complete MV-algebra $\langle L, \oplus, ^*, \mathbf{0}\rangle$ there is an associated Heyting algebra $\langle L, \wedge, \Rightarrow, ^{\sim}, \mathbf{0}, \mathbf{1}\rangle$ with an adjoint couple $\langle \wedge, \Rightarrow\rangle$. In fact, even more is true; the structure is a Gödel algebra, a prelinear Heyting algebra. The Łukasiewicz structure and all finite MV-algebras are complete as well as the direct product of complete MV-algebras is a complete MV-algebra. However, the rational Łukasiewicz structure is not complete.

In [31] following three propositions were proved.

Proposition 8.1 *In an MV-algebra* $\mathbf{L}$ *the following holds for all* $x, y \in L$*:*

$$(x \odot y) \wedge (x^* \odot y^*) = \mathbf{0}, \tag{8.30}$$
$$(x^* \wedge y) \oplus (x \odot y) \oplus (x^* \odot y^*) \oplus (x \wedge y^*) = \mathbf{1}. \tag{8.31}$$

Proposition 8.2 *Assume* x, y, a, b *are elements of an MV-algebra* $\mathbf{L}$ *such that the following system of equations holds:*

$$(A) \begin{cases} x^* \wedge y &= a^* \wedge b, \\ x \odot y &= a \odot b, \\ x^* \odot y^* &= a^* \odot b^*, \\ x \wedge y^* &= a \wedge b^*. \end{cases}$$

Then $x = a$ *and* $y = b$.

Proposition 8.3 *Assume* x, y *are elements of an MV-algebra* $\mathbf{L}$ *such that*

$$(B) \begin{cases} x^* \wedge y &= f, \\ x \odot y &= k, \\ x^* \odot y^* &= u, \\ x \wedge y^* &= t. \end{cases}$$

Then (C) $x = t \oplus k,\ y = f \oplus k$ *and (D)* $x = (f \oplus u)^*,\ y = (t \oplus u)^*$.

Propositions 8.2 and 8.3 put ordered couples $\langle x, y\rangle$ and values f, k, u, t defined by (B) into a one-to-one correspondence.

8.3 Evidence Couples and Evidence Matrices

8.3.1 Evidence Couples on the Unit Square

Let $\mathbf{L} = \langle L, \oplus, ^*, \mathbf{0}\rangle$ be an MV-algebra. The product set $L \times L$ can be equipped with an MV-structure by setting

$$\langle a, b\rangle \otimes \langle c, d\rangle = \langle a \oplus c, b \odot d\rangle, \tag{8.32}$$
$$\langle a, b\rangle^{\perp} = \langle a^*, b^*\rangle, \tag{8.33}$$
$$\overline{\mathbf{0}} = \langle \mathbf{0}, \mathbf{1}\rangle \tag{8.34}$$

for each ordered couple $\langle a, b\rangle, \langle c, d\rangle \in L \times L$. Indeed, the axioms (8.5)–(8.9) hold trivially and, to prove that the axiom (8.10) holds, it is enough to realize that

$$\begin{aligned}(\langle a, b\rangle^{\perp} \otimes \langle c, d\rangle)^{\perp} \otimes \langle c, d\rangle &= \langle a \vee c, b \wedge d\rangle = \langle c \vee a, d \wedge b\rangle \\ &= (\langle c, d\rangle^{\perp} \otimes \langle a, b\rangle)^{\perp} \otimes \langle a, b\rangle.\end{aligned}$$

It is routine to verify that the order on $L \times L$ is defined via

$$\langle a, b\rangle \leq \langle c, d\rangle \text{ if and only if } a \leq c, d \leq b, \tag{8.35}$$

the lattice operations by

$$\langle a, b\rangle \vee \langle c, d\rangle = \langle a \vee c, b \wedge d\rangle, \tag{8.36}$$
$$\langle a, b\rangle \wedge \langle c, d\rangle = \langle a \wedge c, b \vee d\rangle, \tag{8.37}$$

and an adjoint couple $\langle \star, \mapsto\rangle$ by

$$\langle a, b\rangle \star \langle c, d\rangle = \langle a \odot c, b \oplus d\rangle, \tag{8.38}$$
$$\langle a, b\rangle \mapsto \langle c, d\rangle = \langle a \to c, (d \to b)^*\rangle. \tag{8.39}$$

Notice that $a \to c = a^* \oplus c$ and $(d \to b)^* = (d^* \oplus b)^* = d \odot b^* = b^* \odot d$.

Definition 8.4 Given an MV-algebra $\mathbf{L}$, denote the structure described via (8.32)–(8.39) by $\mathbf{L}_{EC}$ and call it the MV-algebra of evidence couples induced by $\mathbf{L}$.

Definition 8.5 Given an MV-algebra $\mathbf{L}$, denote

$$M = \left\{ \begin{bmatrix} a^* \wedge b & a \odot b \\ a^* \odot b^* & a \wedge b^* \end{bmatrix} | \langle a, b\rangle \in L \times L \right\}$$

and call it the set of evidence matrices induced by evidence couples.

By Propositions 8.2 and 8.3 we have the following theorem:

Theorem 8.6 *There is a one-to-one correspondence between $L \times L$ and M: if $A, B \in M$ are two evidence matrices induced by evidence couples $\langle a, b\rangle$ and $\langle x, y\rangle$, respectively, then $A = B$ if and only if $a = x$ and $b = y$.*

The MV-structure descends from $\mathbf{L}_{EC}$ to M in a natural way: if $A, B \in M$ are two evidence matrices induced by evidence couples $\langle a, b\rangle$ and $\langle x, y\rangle$, respectively, then the evidence couple $\langle a \oplus x, b \odot y\rangle$ induces an evidence matrix

$$C = \begin{bmatrix} (a \oplus x)^* \wedge (b \odot y) & (a \oplus x) \odot (b \odot y) \\ (a \oplus x)^* \odot (b \odot y)^* & (a \oplus x) \wedge (b \odot y)^* \end{bmatrix}.$$

Thus, we may define a binary operation $\bigoplus$ on M by

$$\begin{bmatrix} a^* \wedge b & a \odot b \\ a^* \odot b^* & a \wedge b^* \end{bmatrix} \bigoplus \begin{bmatrix} x^* \wedge y & x \odot y \\ x^* \odot y^* & x \wedge y^* \end{bmatrix} = C.$$

Similarly, if $A \in M$ is an evidence matrix induced by an evidence couple $\langle a, b \rangle$, then the evidence couple $\langle a^*, b^* \rangle$ induces an evidence matrix

$$A^{\perp} = \begin{bmatrix} a \wedge b^* & a^* \odot b^* \\ a \odot b & a^* \wedge b \end{bmatrix}.$$

In particular, the evidence couple $\langle \mathbf{0}, \mathbf{1} \rangle$ induces the following evidence matrix:

$$\mathbb{F} = \begin{bmatrix} \mathbf{0}^* \wedge \mathbf{1} & \mathbf{0} \odot \mathbf{1} \\ \mathbf{0}^* \odot \mathbf{1}^* & \mathbf{0} \wedge \mathbf{1}^* \end{bmatrix} = \begin{bmatrix} \mathbf{1} & \mathbf{0} \\ \mathbf{0} & \mathbf{0} \end{bmatrix}.$$

Moreover, it is easy to verify that the evidence couples $\langle \mathbf{1}, \mathbf{0} \rangle$, $\langle \mathbf{1}, \mathbf{1} \rangle$, and $\langle \mathbf{0}, \mathbf{0} \rangle$ induce the following evidence matrices:

$$\mathbb{T} = \begin{bmatrix} \mathbf{0} & \mathbf{0} \\ \mathbf{0} & \mathbf{1} \end{bmatrix}, \mathbb{K} = \begin{bmatrix} \mathbf{0} & \mathbf{1} \\ \mathbf{0} & \mathbf{0} \end{bmatrix}, \mathbb{U} = \begin{bmatrix} \mathbf{0} & \mathbf{0} \\ \mathbf{1} & \mathbf{0} \end{bmatrix},$$

respectively. In [31] we proved the following.

Theorem 8.7 *Let* $\mathbf{L}$ *be an MV-algebra. The structure* $M = \langle M, \bigoplus, ^{\perp}, F \rangle$ *as defined above is an MV-algebra (called* the MV-algebra of evidence matrices*).*

$$\text{Assume } A = \begin{bmatrix} a^* \wedge b & a \odot b \\ a^* \odot b^* & a \wedge b^* \end{bmatrix}, B = \begin{bmatrix} x^* \wedge y & x \odot y \\ x^* \odot y^* & x \wedge y^* \end{bmatrix} \in M$$

Then it is obvious that the lattice operations $\wedge$, $\vee$, the monoidal operation $\bigodot$, and the residual operation $\longrightarrow$ are defined via

$$A \wedge B = \begin{bmatrix} (a \wedge x)^* \wedge (b \vee y) & (a \wedge x) \odot (b \vee y) \\ (a \wedge x)^* \odot (b \vee y)^* & (a \wedge x) \wedge (b \vee y)^* \end{bmatrix},$$

$$A \vee B = \begin{bmatrix} (a \vee x)^* \wedge (b \wedge y) & (a \vee x) \odot (b \wedge y) \\ (a \vee x)^* \odot (b \wedge y)^* & (a \vee x) \wedge (b \wedge y)^* \end{bmatrix},$$

$$A \bigodot B = \begin{bmatrix} (a \odot x)^* \wedge (b \oplus y) & (a \odot x) \odot (b \oplus y) \\ (a \odot x)^* \odot (b \oplus y)^* & (a \odot x) \wedge (b \oplus y)^* \end{bmatrix},$$

$$A \longrightarrow B = \begin{bmatrix} (a \to x)^* \wedge (y \to b)^* & (a \to x) \odot (y \to b)^* \\ (a \to x)^* \odot (y \to b) & (a \to x) \wedge (y \to b) \end{bmatrix}.$$

If the original MV-algebra $\mathbf{L}$ is complete, then the structure M is a complete MV-algebra, too, and suprema and infima are defined by evidence couples

$$\bigvee_{i\in\Gamma}\{\langle a_i, b_i\rangle\} = \langle \bigvee_{i\in\Gamma} a_i, \bigwedge_{i\in\Gamma} b_i\rangle, \quad \bigwedge_{i\in\Gamma}\{\langle a_i, b_i\rangle\} = \langle \bigwedge_{i\in\Gamma} a_i, \bigvee_{i\in\Gamma} b_i\rangle.$$

Thus, we may define another residual operation $\Rightarrow$ on M via

$$A \Rightarrow B = \begin{bmatrix} (a \Rightarrow x)^* \wedge (b^* \Rightarrow y^*)^* & (a \Rightarrow x) \odot (b^* \Rightarrow y^*)^* \\ (a \Rightarrow x)^* \odot (b^* \Rightarrow y^*) & (a \Rightarrow x) \wedge (b^* \Rightarrow y^*) \end{bmatrix}.$$

To verify this last claim, assume $\langle a, b\rangle \wedge \langle x, y\rangle \leq \langle c, d\rangle$ in $\mathbf{L}_{EC}$, which is equivalent to

$$a \wedge x \leq c \text{ and } d \leq b \vee y, \text{ that is,}$$

$$a \leq x \Rightarrow c \text{ and } (b \vee y)^* = b^* \wedge y^* \leq d^*, \text{ i.e.,}$$

$$a \leq x \Rightarrow c \text{ and } b^* \leq y^* \Rightarrow d^*, \text{ or equivalently,}$$

$$a \leq x \Rightarrow c \text{ and } (y^* \Rightarrow d^*)^* \leq b, \text{ i.e.,}$$

$\langle a, b\rangle \leq \langle x \Rightarrow c, (y^* \Rightarrow d^*)^*\rangle$ in $\mathbf{L}_{EC}$. Therefore, if A is induced by $\langle a, b\rangle$ and B is induced by $\langle x, y\rangle$ then the evidence matrix $A \Rightarrow B$ is induced by the evidence couple $\langle a \Rightarrow x, (b^* \Rightarrow y^*)^*\rangle$. In particular, the weak complementation * on M is defined via $A^\star = A \Rightarrow \mathbb{F}$ and induced by

$$\begin{array}{l} \langle \mathbf{1}, \mathbf{0}\rangle \text{ if } a = \mathbf{0}, b = \mathbf{1}, \text{ then } A^\star = \mathbb{T}, \\ \langle \mathbf{0}, \mathbf{0}\rangle \text{ if } a > \mathbf{0}, b = \mathbf{1}, \text{ then } A^\star = \mathbb{U}, \\ \langle \mathbf{1}, \mathbf{1}\rangle \text{ if } a = \mathbf{0}, b < \mathbf{1}, \text{ then } A^\star = \mathbb{K}, \\ \langle \mathbf{0}, \mathbf{1}\rangle \text{ if } a > \mathbf{0}, b < \mathbf{1}, \text{ then } A^\star = \mathbb{F} \end{array}$$

The matrices $\mathbb{F}, \mathbb{T}, \mathbb{K}, \mathbb{U}$ correspond to Belnap's original values *false, true, contradictory, unknown*, respectively.

If in particular, the original MV-algebra $\mathbf{L}$ is the standard MV-algebra, then the evidence couples $\langle a, b\rangle$ are points on the unit square $[0, 1]^2$. In the next section they will be points on the unit triangle $\triangledown = \{(a, b) \in [0, 1]^2; a + b \leq 1\}$, see the following indicative figure.

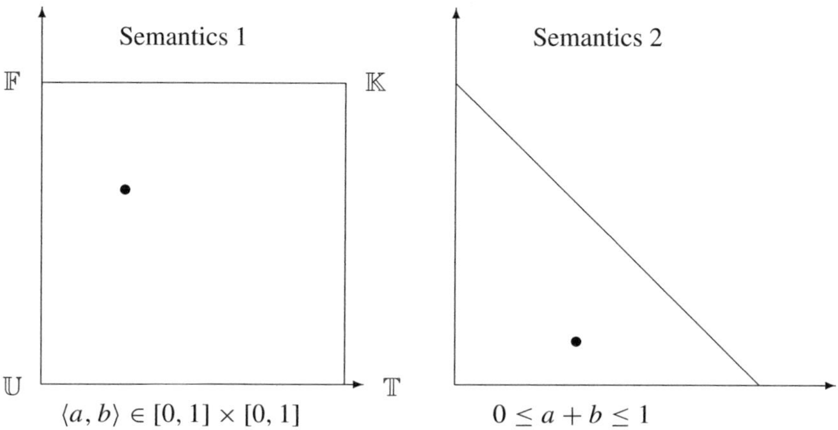

8.3.2 Evidence Couples on the Unit Triangle

It is important to notice that the reasoning above, leading to conclude that the set of evidence matrices *M* can be endowed with a complete MV-algebra structure, critically depends on the fact that the set of evidence couples $\mathbf{L}_{EC}$ naturally presents a complete MV-algebra structure when it is obtained as the product $L \times L$ of a complete MV-algebra structure $\mathbf{L} = \langle L, \oplus, ^*, \mathbf{0}\rangle$. Once $\mathbf{L}_{EC}$ is equipped with the desired structure, this is naturally transferred to *M* by means of the bijective mapping that unequivocally assigns evidence couples to evidence matrices and *vice versa*. Therefore, if the set of evidence couples L_{EC} (understood here as a general set of evidence couples $\langle \texttt{pro}, \texttt{con}\rangle$, not necessarily the unit square) cannot be considered as a product MV-algebra, *M* does not necessarily present a complete MV-algebra structure, even when it is obtained from L_{EC} through a one-to-one correspondence. However, as shown in [24], we can overcome this obstacle by transferring the structure to L_{EC} from an adequate complete MV-algebra. For instance, let us now consider evidence couples on the unit triangle

$$\triangledown = \{(a, b) \in [0, 1]^2; a + b \leq 1\},$$

that is, let us take $L_{EC} = \triangledown$. This approach is based on the idea that evidence for (value a) and evidence against (value b) of a given statement α are not completely independent of each other, but they are bound by the condition $a + b \leq 1$. In [24] it is shown that, similar to expressions (8.1)–(8.4), we can define another continuous extension of Belnap's logic through the formulas

$$T(\alpha) = \max\{a - b, 0\}, \tag{8.40}$$
$$F(\alpha) = \max\{b - a, 0\}, \tag{8.41}$$
$$K(\alpha) = 2\min\{a, b\}, \tag{8.42}$$
$$U(\alpha) = 1 - (a + b), \tag{8.43}$$

where now an evidence couple $\langle a, b\rangle \in \triangledown$ is given. It is easy to see that again $T + F + K + U = 1$ holds for any $\langle a, b\rangle \in \triangledown$, as well as that now the value `contradictory` is associated with the pair $K = \langle \frac{1}{2}, \frac{1}{2}\rangle$.

Let us denote by $M_{\triangledown}$ the set of evidence matrices of the form

$$\begin{bmatrix} F(\alpha) & K(\alpha) \\ U(\alpha) & T(\alpha) \end{bmatrix}$$

with $\langle a, b\rangle \in \triangledown$. Note that the extension in (8.40)–(8.43) is also expressible in terms of Łukasiewicz algebraic operations, since it holds that

$$\begin{bmatrix} F(\alpha) & K(\alpha) \\ U(\alpha) & T(\alpha) \end{bmatrix} = \begin{bmatrix} a^* \odot b & (a \wedge b) \oplus (a \wedge b) \\ a^* \odot b^* & a \odot b^* \end{bmatrix}.$$

Moreover, the following two results presented in [24] show that there is a one-to-one correspondence between evidence couples in $\triangledown$ and evidence matrices in $M_{\triangledown}$.

Proposition 8.8 *Let* $\mathbf{v} = \langle a, b\rangle$ *and* $\mathbf{s} = \langle c, d\rangle$ *be two evidence couples in* $\triangledown$ *such that*

$$\begin{bmatrix} c^* \odot d & (c \wedge d) \oplus (c \wedge d) \\ c^* \odot d^* & c \odot d^* \end{bmatrix} = \begin{bmatrix} a^* \odot b & (a \wedge b) \oplus (a \wedge b) \\ a^* \odot b^* & a \odot b^* \end{bmatrix}.$$

Then $\mathbf{v} = \mathbf{s}$, *that is* $a = c$ *and* $b = d$.

Proposition 8.9 *If an evidence matrix*

$$\begin{bmatrix} F(\alpha) & K(\alpha) \\ U(\alpha) & T(\alpha) \end{bmatrix} = \begin{bmatrix} a^* \odot b & (a \wedge b) \oplus (a \wedge b) \\ a^* \odot b^* & a \odot b^* \end{bmatrix}.$$

is given, then $\langle a, b\rangle = \langle T \oplus \frac{1}{2}K, F \oplus \frac{1}{2}K\rangle \in \triangledown$ *is the evidence couple associated to such a matrix.*

Thus, at this point, in order to obtain a complete MV-algebra structure for $M_{\triangledown}$, we just need to endow the set of evidence couples, in this case the unit triangle $\triangledown$, with a complete MV-algebra structure. However, it is obvious that we cannot obtain $\triangledown$ as the product of two MV-algebras. Therefore, we need to equip $\triangledown$ with the desired structure by means of a different argument. To this aim, the following general result was proved in [24]:

Theorem 8.10 *Let $\mathbf{L} = \langle L, \oplus, ^*, \mathbf{0}\rangle$ be a complete MV-algebra, and let $H : L \curvearrowright L_{EC}$ be a bijective mapping, where L_{EC} is a general set of evidence couples $\langle \texttt{pro}, \texttt{con}\rangle$, not necessarily the unit square. Then $\overline{\mathbf{L}}_{EC} = \langle L_{EC}, \oplus_{EC}, ^{*_{EC}}, \mathbf{0}_{EC}\rangle$ is a complete MV-algebra with the operations $\oplus_{EC}, ^{*_{EC}}$ and the neutral element $\mathbf{0}_{EC}$ defined for all $\mathbf{v}, \mathbf{w} \in L_{EC}$ as follows:*

$$\mathbf{v} \oplus_{EC} \mathbf{w} = H(H^{-1}(\mathbf{v}) \oplus H^{-1}(\mathbf{w})), \tag{8.44}$$

$$\mathbf{v}^{*_{EC}} = H(H^{-1}(\mathbf{v})^*), \tag{8.45}$$

$$\mathbf{0}_{EC} = H(\mathbf{0}). \tag{8.46}$$

As a consequence of Theorem 8.10, any bijective mapping H between a complete MV-algebra $\mathbf{L} = \langle L, \oplus, ^*, \mathbf{0}\rangle$ and the unit triangle $\triangledown$ defines a complete MV-algebra structure on $M_\triangledown$. Particularly, in [24] we studied the MV-algebra of evidence matrices $M_\triangledown$ that results from considering $L = [0, 1]^2$ (equipped with the complete MV-algebra structure obtained through (8.32)–(8.39) as a product of the standard MV-algebra) and the bijective mapping $H : [0, 1]^2 \curvearrowright \triangledown$ given by the expression

$$H(\mathbf{x}) = \langle x_1 - \frac{1}{2}(x_1 \wedge x_2), x_2 - \frac{1}{2}(x_1 \wedge x_2)\rangle \tag{8.47}$$

with inverse $H^{-1} : \triangledown \curvearrowright [0, 1]^2$ given by

$$H^{-1}(\mathbf{v}) = \langle v_1 + (v_1 \wedge v_2), v_2 + (v_1 \wedge v_2)\rangle \tag{8.48}$$

for all $\mathbf{x} = \langle x_1, x_2\rangle \in [0, 1]^2$ and $\mathbf{v} = \langle v_1, v_2\rangle \in \triangledown$. Remarkably, the canonical partial orders of $\triangledown$ and $M_\triangledown$ as MV-algebras (obtained through expression (8.14)) show a nice connection with the extension (8.40)–(8.43), as it was proven in [24] that for any two evidence couples $\mathbf{v}_1, \mathbf{v}_2 \in \triangledown$ and their associated evidence matrices

$$M_1 = \begin{bmatrix} F_1 & K_1 \\ U_1 & T_1 \end{bmatrix}, M_2 = \begin{bmatrix} F_2 & K_2 \\ U_2 & T_2 \end{bmatrix} \in M_\triangledown$$

it holds that

$$\mathbf{v}_1 \leq_\triangledown \mathbf{v}_2 \text{ iff } M_1 \leq_{M_\triangledown} M_2 \text{ iff } \begin{cases} K_1 \leq K_2 \text{ and } U_1 \leq U_2 & \text{if } T_1 = T_2 = 0, \\ K_1 \geq K_2 \text{ and } U_1 \geq U_2 & \text{if } F_1 = F_2 = 0, \\ 1 - U_1 \geq K_2 \text{ and } 1 - K_1 \leq U_2 & \text{if } T_1 = F_2 = 0. \end{cases}$$

Let us emphasize that, as proven in [6], the partial order $\leq_\triangledown$ cannot coincide with the partial order $\leq_t$ defined by $\mathbf{x} \leq_t \mathbf{y}$ iff $x_1 \leq y_1$ and $y_2 \leq x_2$, where $\mathbf{x} = \langle x_1, x_2\rangle, \mathbf{y} = \langle y_1, y_2\rangle$ are elements of the unit triangle $\triangledown$. That is, it is not possible to obtain the order $\leq_t$ through formula (8.14) for any MV-algebra defined on the unit triangle $\triangledown$, in particular for any MV-algebra structure on $\triangledown$ obtained by means of Theorem 8.10 with any base MV-algebra **L** and the bijective mapping $H : [0, 1]^2 \curvearrowright \triangledown$.

At this respect, some authors have defined and studied the notion of *triangle algebra* (cf. [33]), which discards the MV-algebra structure and analyzes the properties of the structure resulting from using t-norms as logical operations on the lattice $(\triangledown, \leq_\triangledown)$. In this sense, in [33] it is shown that triangle algebras retain some of the main properties of MV-algebras, therefore constituting interesting structures for the development of a formal logic on $(\triangledown, \leq_t)$. However, instead of studying possible structures on the lattice $(\triangledown, \leq_t)$, the approach proposed in [24] enable studying those lattices $(\triangledown, \leq_\triangledown)$ compatible with an MV-algebra structure, therefore presenting a complementary approach to the study of the unit triangle $\triangledown$ to that of triangle algebras.

8.4 Paraconsistent Pavelka Style Fuzzy Logic

Now we recall Pavelka style fuzzy logic to be associated separately with the above discussed two different paraconsistent semantics. For a comprehensive presentation of complete MV-algebra valued Pavelka style sentential logic, see [29, 32].

8.4.1 Pavelka Style Fuzzy Sentential Logic

A standard approach in mathematical sentential logic is the following. After introducing atomic formulas, logical connectives and the set of well-formed formulas, these formulas are semantically interpreted in suitable algebraic structures. In classical logic these structures are Boolean algebras, in Hájek's Basic fuzzy logic [11], for example, the suitable structures are standard BL-algebras. *Tautologies* of a logic are those formulas that obtain the top value **1** in all interpretations in all suitable algebraic structures; for this reason, tautologies are sometimes called **1**-tautologies. For example, tautologies in Basic fuzzy logic are exactly the formulas that obtain value **1** in all interpretations in all standard BL-algebras. The next step in usual mathematical logic is to fix the axiom schemata and the rules of inference: a well-formed formula is a *theorem* if it is either an axiom or obtained recursively from axioms using finitely many times rules of inference. *Completeness* of the logic with respect to some semantics means that tautologies and theorems coincide; Classical sentential logic and Basic fuzzy sentential logic, for example, are complete logics with respect to their algebraic semantics. For now on, the symbols F and T stand for language and fuzzy theory, respectively.

In Pavelka style fuzzy sentential logic [18] the situation is somewhat different. We start by fixing a set of truth values, in fact an algebraic structure. In Pavelka's own approach this structure is the standard MV-algebra L on the real unit interval, while in [32] the structure is a more general (but fixed!) complete MV-algebra $\mathbf{L} = \langle L, \oplus, ^*, \mathbf{0} \rangle$.

Consider a zero-order language F with a set of infinitely many propositional variables `p, q, r, ...`, and a set of *truth constants* $\{\mathbf{a} \mid a \in L\}$ corresponding to the elements in the set L. As proved in [11], if the set of truth values is the whole real interval [0, 1] then it is enough to include truth constants corresponding to rationals belonging to [0, 1]. In two-valued logic, truth constants correspond to the truth constants $\perp$ and $\top$. Propositional variables and truth constants constitute the set F_a of *atomic formulas*. The elementary logical connectives are *implication* `imp` and *conjunction* `and`. The set of all well- formed formulas (wffs) is obtained in the natural way: atomic formulas are wffs and if α, β are wffs, then α `imp` β, α `and` β are wffs.

As shown in [29], we can introduce many other logical connectives by abbreviations, e.g., *negation* `not` α stands for (α `imp` $\mathbf{0}$), *disjunction* α `or` β stands for `not (not` α `and not` β). Also, *equivalence* `equiv` and *exclusive or* `xor` are possible. Moreover, the connectives *weak implication* $\overline{\texttt{imp}}$, *weak conjunction* $\overline{\texttt{and}}$, *weak disjunction* $\overline{\texttt{or}}$, *weak negation* $\overline{\texttt{non}}$, *weak equivalence* $\overline{\texttt{equiv}}$, and *weak exclusive or* $\overline{\texttt{xor}}$ are available in this logic. We call the logical connectives without bar *Łukasiewicz* connectives, and those with bar are *Intuitionistic* connectives.

Semantics in Pavelka style fuzzy sentential logic is introduced in the following way: any mapping $v : F_a \curvearrowright L$ such that $v(\mathbf{a}) = a$ for all truth constants $\mathbf{a}$ can be extended recursively into the whole F by setting $v(\alpha$ `imp` $\beta) = v(\alpha) \rightarrow v(\beta)$ and $v(\alpha$ `and` $\beta) = v(\alpha) \odot v(\beta)$. Such mappings v are called *valuations*. The *degree of tautology* of a wff α is the infimum of all values $v(\alpha)$, that is,

$$C^{sem}(\alpha) = \bigwedge \{v(\alpha) \mid v \text{ is a valuation}\}.$$

We may fix some set $T \subseteq F$ of wffs and make it fuzzy by associating each formula $\alpha \in T$ with a value $T(\alpha)$ in L, and $T(\beta) = \mathbf{0}$ for wffs β not in T, and call the result *a fuzzy theory T*. Then we consider valuations v such that $T(\alpha) \leq v(\alpha)$ for all wffs $\alpha \in F$. If such a valuation exists, then the fuzzy theory T is *satisfiable*. We say that formulas $\alpha \in F$ such that $T(\alpha) \neq 0$ are the *special axioms* of the fuzzy theory T. Then we consider values

$$C^{sem}(T)(\alpha) = \bigwedge \{v(\alpha) \mid v \text{ is a valuation}, T \text{ satisfies } v\}.$$

The set of logical axioms, denoted by `A`, is composed by the following thirteen forms of formulas (see p. 88 in [29] and also [32]).

(Ax.1) α imp α,
(Ax.2) $(\alpha$ imp $\beta)$ imp $[(\beta$ imp $\gamma)$ imp $(\alpha$ imp $\gamma)]$,
(Ax.3) $(\alpha_1$ imp $\beta_1)$ imp $\{(\beta_2$ imp $\alpha_2)$ imp $[(\beta_1$ imp $\beta_2)$ imp $(\alpha_1$ imp $\alpha_2)]\}$,
(Ax.4) α imp $\mathbf{1}$,
(Ax.5) $\mathbf{0}$ imp α,
(Ax.6) $(\alpha$ and not $\alpha)$ imp $\mathbf{0}$,
(Ax.7) $\mathbf{a}$,
(Ax.8) α imp $(\beta$ imp $\alpha)$,
(Ax.9) $(\mathbf{1}$ imp $\alpha)$ imp α,
(Ax.10) $[(\alpha$ imp $\beta)$ imp $\beta]$ imp $[(\beta$ imp $\alpha)$ imp $\alpha]$,
(Ax.11) (not α imp not β) imp $(\beta$ imp $\alpha)$,
(Ax.12) $[\alpha$ or (not α and $\beta)]$ imp$[(\alpha$ imp $\beta)$ imp $\beta]$,
(Ax.13) $\mathbf{a}$ imp $\mathbf{b}$.

All the axiom formulas δ in (Ax.1)–(Ax.6) and (Ax.8)–(Ax.12) are 1−tautologies, that is $C^{sem}(\delta) = 1$, for axioms (Ax.7), $C^{sem}(\mathbf{a}) = a \in L$ and for axioms (Ax.13), $C^{sem}(\mathbf{a}$ imp $\mathbf{b}) = a \to b \in L$. Notice that, corresponding to different (non-isomorphic) complete MV-algebra valued fuzzy logics, the set of logical axioms is not different except, of course, axioms (Ax.7) and (Ax.13).

A *fuzzy rule of inference* is a scheme

$$\frac{\alpha_1, \ldots, \alpha_n}{r^{\mathrm{syn}}(\alpha_1, \ldots, \alpha_n)}, \frac{a_1, \ldots, a_n}{r^{\mathrm{sem}}(a_1, \ldots, a_n)},$$

where the wffs $\alpha_1, \ldots, \alpha_n$ are *premises* and the wff $r^{\mathrm{syn}}(a_1, \ldots, a_n)$ is the *conclusion*. The values $a_1, \ldots, a_n$ and $r^{\mathrm{sem}}(a_1, \ldots, a_n) \in L$ are the corresponding truth values. The mappings $r^{\mathrm{sem}} : L^n \curvearrowright L$ are semi-continuous, i.e.,

$$r^{\mathrm{sem}}(a_1, \ldots, \bigvee_{i\in\Gamma} a_{k_i}, \ldots, a_n) = \bigvee_{i\in\Gamma} r^{\mathrm{sem}}(a_1, \ldots, a_{k_i}, \ldots, a_n)$$

holds for all $1 \leq k \leq n$. Moreover, the fuzzy rules are required to be *sound* in a sense that

$$r^{\mathrm{sem}}(v(\alpha_1), \ldots, v(\alpha_n)) \leq v(r^{\mathrm{syn}}(\alpha_1, \ldots, \alpha_n))$$

holds for all valuations v. The following are examples of fuzzy rules of inference, denoted by a set R:

Generalized Modus Ponens (GMP):

$$\frac{\alpha, \alpha \ \text{imp} \ \beta}{\beta}, \frac{a, b}{a \odot b}$$

a–Consistency testing rules (**a**–CTR):

$$\frac{\mathbf{a}}{\mathbf{0}}, \frac{b}{c}$$

where $\mathbf{a}$ is a truth constant and $c = \mathbf{0}$ if $b \leq a$ and $c = \mathbf{1}$ elsewhere.

a–Lifting rules (**a**–LR):

$$\frac{\alpha}{\mathbf{a}\ \mathtt{imp}\,\alpha}\ ,\ \frac{b}{a \to b}$$

where **a** is an truth constant.

Rule of Bold Conjunction (RBC):

$$\frac{\alpha, \beta}{\alpha\ \mathtt{and}\ \beta}\ ,\ \frac{b}{a \odot b}$$

Rule of Bold Disjunction (RBD):

$$\frac{\alpha, \beta}{\alpha\ \mathtt{or}\ \beta}\ ,\ \frac{b}{a \oplus b}$$

An R-*proof* w of a wff α in a fuzzy theory T is a finite sequence

$$\begin{array}{cc} \alpha_1 & , \; a_1 \\ \vdots & \vdots \\ \alpha_m & , \; a_m \end{array}$$

where

(i) $\alpha_m = \alpha$,
(ii) for each i, $1 \le i \le m$, α_i is a logical axiom, or is a special axiom, or there is a fuzzy rule of inference in R and wff formulas $\alpha_{i_1}, \ldots, \alpha_{i_n}$ with $i_1, \ldots, i_n < i$ such that $\alpha_i = r^{\mathtt{syn}}(\alpha_{i_1}, \ldots, \alpha_{i_n})$,
(iii) for each i, $1 \le i \le m$, the value $a_i \in L$ is given by

$$a_i = \begin{cases} a & \text{if } \alpha_i \text{ is the axiom Ax.7} \\ a \to b & \text{if } \alpha_i \text{ is the axiom Ax.13} \\ \mathbf{1} & \text{if } \alpha_i \text{ is some other logical axiom in the set A} \\ T(\alpha_i) & \text{if } \alpha_i \text{ is a special axiom} \\ r^{\mathtt{sem}}(a_{i_1}, \ldots, a_{i_n}) & \text{if } \alpha_i = r^{\mathtt{syn}}(\alpha_{i_1}, \ldots, \alpha_{i_n}) \end{cases}$$

The value a_m is called the *degree* of the R–proof w. Since a wff α may have various R–proofs with different degrees, we define the *degree of deduction* or *provability degree* of a formula α to be the supremum of all such values, i.e.,

$$C^{syn}(T)(\alpha) = \bigvee \{a_m |\ w \text{ is a R– proof for } \alpha \text{ in the fuzzy theory } T\}.$$

A fuzzy theory T is *consistent* if $C^{sem}(T)(\mathbf{a}) = a$ for all truth constants $\mathbf{a}$. By Proposition 94 in [29], any satisfiable fuzzy theory is consistent. Theorem 25 in [29] (see also [32]) now states the completeness of Pavelka style fuzzy sentential logic:

If a fuzzy theory T is consistent, then $C^{sem}(T)(\alpha) = C^{syn}(T)(\alpha)$ for any wff α.

Thus, in Pavelka style fuzzy sentential logic we may talk about tautologies of a degree a and theorems of a degree a for all truth values $a \in L$, and these concepts coincide. Notice that the axiom schemata (Ax.1)–(Ax.13) and the set R of fuzzy rules of inference are sufficient to guarantee the completeness theorem irrespective of the choice of the complete MV-algebra.

This completeness result remains valid if we extend the language to contain intuitionistic connectives $\overline{\texttt{and}}$ or $\overline{\texttt{or}}$. However, it does not hold if the language is extended by the intuitionistic connectives $\overline{\texttt{imp}}$ or $\overline{\texttt{non}}$.

8.4.2 Semantics of Paraconsistent Pavelka Logic

The above Pavelka style construction of fuzzy logic can be carried out in any complete MV-algebra thus, in particular, in the complete MV-algebras M and $M_\triangledown$ of evidence matrices as introduced in Sect. 8.3 from a complete MV-algebra $\mathbf{L}$. Indeed, semantics is introduced by associating to each atomic formula p an evidence couple $\langle \texttt{pro}, \ \texttt{con} \rangle$ or simply $\langle a, b \rangle \in L_{EC}$. The evidence couple $\langle a, b \rangle$ induces a unique evidence matrix $A \in M$ (or $A \in M_\triangledown$) and therefore *valuations* are mappings v such that $v(\mathrm{p}) = A$ for propositional variables and $v(\mathbf{I}) = I$ for truth constants ($\in M$ or $\in M_\triangledown$). A valuation v is then extended recursively to whole F via

$$v(\alpha \ \ \texttt{imp} \ \ \beta) = v(\alpha) \longrightarrow v(\beta), \qquad v(\alpha \ \texttt{and} \ \beta) = v(\alpha) \bigodot v(\beta). \tag{8.49}$$

Similar to the procedure in [29], Sect. 3.1, we can show that

$$v(\alpha \ \texttt{or} \ \beta) = v(\alpha) \bigoplus v(\beta), \qquad v(\texttt{not}\alpha) = [v(\alpha)]^{\perp}, \tag{8.50}$$

$$v(\alpha \ \texttt{equiv} \ \beta) = [v(\alpha) \longrightarrow v(\beta)] \wedge [v(\beta) \longrightarrow v(\alpha)], \tag{8.51}$$

$$v(\alpha \ \texttt{xor} \ \beta) = [v(\alpha) \bigoplus v(\beta)] \wedge [v(\beta) \longrightarrow v(\alpha)^{\perp}] \wedge [v(\alpha) \longrightarrow v(\beta)^{\perp}], \tag{8.52}$$

$$v(\alpha \ \overline{\texttt{and}} \ \beta) = v(\alpha) \wedge v(\beta), \quad v(\alpha \ \overline{\texttt{or}} \ \beta) = v(\alpha) \vee v(\beta), \tag{8.53}$$

$$v(\alpha \ \overline{\texttt{imp}} \ \beta) = v(\alpha) \Rightarrow v(\beta), \qquad v(\overline{\texttt{not}}\alpha) = v(\alpha)^{\sim}, \tag{8.54}$$

$$v(\alpha \ \overline{\texttt{equiv}} \ \beta) = [v(\alpha) \Rightarrow v(\beta)] \wedge [v(\beta) \Rightarrow v(\alpha)]. \tag{8.55}$$

Example 1. Evidence Couples and Evidence Matrices on the Unit Real Square. Let **L** be the Łukasiewicz structure L. Assume α and β are associated with evidence couples $\langle 0.8, 0.4\rangle$, $\langle 0.7, 0.2\rangle \in [0, 1]^2$, respectively. They induce the following evidence matrices in M, where the corresponding values T, F, K, U related to formulas α and β can be read.

$$v(\alpha) = \begin{bmatrix} 0.2 & 0.2 \\ 0 & 0.6 \end{bmatrix}, \quad v(\texttt{not}\,\alpha) = \begin{bmatrix} 0.6 & 0 \\ 0.2 & 0.2 \end{bmatrix}$$

$$v(\beta) = \begin{bmatrix} 0.2 & 0 \\ 0.1 & 0.7 \end{bmatrix}, \quad v(\texttt{not}\,\beta) = \begin{bmatrix} 0.7 & 0.1 \\ 0 & 0.2 \end{bmatrix}$$

$$v(\alpha \;\; \texttt{and} \;\; \beta) = \begin{bmatrix} 0.5 & 0.1 \\ 0 & 0.4 \end{bmatrix}, \quad v(\alpha \;\; \texttt{or} \;\; \beta) = \begin{bmatrix} 0 & 0 \\ 0 & 1 \end{bmatrix} (= \mathbb{T})$$

$$v(\alpha \;\; \texttt{imp} \;\; \beta) = \begin{bmatrix} 0 & 0 \\ 0.1 & 0.9 \end{bmatrix}, \quad v(\alpha \;\; \texttt{imp} \;\; \texttt{not} \;\; \beta) = \begin{bmatrix} 0.4 & 0 \\ 0.1 & 0.5 \end{bmatrix}$$

$$v(\beta \;\; \texttt{imp} \;\; \alpha) = \begin{bmatrix} 0 & 0.2 \\ 0 & 0.8 \end{bmatrix}, \quad v(\beta \;\; \texttt{imp} \;\; \texttt{not} \;\; \alpha) = \begin{bmatrix} 0.4 & 0 \\ 0.1 & 0.5 \end{bmatrix}$$

$$v(\alpha \;\; \texttt{equiv} \;\; \beta) = \begin{bmatrix} 0.1 & 0.1 \\ 0 & 0.8 \end{bmatrix}, \quad v(\alpha \; \texttt{xor} \; \beta) = \begin{bmatrix} 0.4 & 0 \\ 0.1 & 0.5 \end{bmatrix}$$

$$v(\overline{\texttt{not}}\alpha) = \begin{bmatrix} 1 & 0 \\ 0 & 0 \end{bmatrix}, \quad v(\overline{\texttt{not}}\beta) = \begin{bmatrix} 1 & 0 \\ 0 & 0 \end{bmatrix}$$

$$v(\alpha \; \overline{\texttt{and}} \; \beta) = \begin{bmatrix} 0.3 & 0.1 \\ 0 & 0.6 \end{bmatrix}, \quad v(\alpha \; \overline{\texttt{or}} \; \beta) = \begin{bmatrix} 0.2 & 0 \\ 0 & 0.8 \end{bmatrix}$$

$$v(\alpha \; \overline{\texttt{imp}} \; \beta) = \begin{bmatrix} 0 & 0 \\ 0.3 & 0.7 \end{bmatrix}, \quad v(\beta \; \overline{\texttt{imp}} \; \alpha) = \begin{bmatrix} 0 & 0.4 \\ 0 & 0.6 \end{bmatrix}$$

$$v(\alpha \; \overline{\texttt{equiv}} \; \beta) = \begin{bmatrix} 0.3 & 0.1 \\ 0 & 0.6 \end{bmatrix}.$$

Example 2. Evidence Couples and Evidence Matrices on the Unit Real Triangle. Assume now α and β are associated with evidence couples $\langle 0.6, 0.2\rangle$, $\langle 0.6, 0.1\rangle \in \triangledown$, respectively. Then $H^{-1}(\langle 0.6, 0.2\rangle) = \langle 0.8, 0.4\rangle$ and $H^{-1}(\langle 0.6, 0.1\rangle) = \langle 0.7, 0.2\rangle$; through the adequate translation of expressions (8.49)–(8.55) by means of the method provided in Theorem 8.10, the following evidence matrices on $M_{\triangledown}$ are obtained:

$$v(\alpha) = \begin{bmatrix} 0 & 0.4 \\ 0.2 & 0.4 \end{bmatrix}, \quad v(\texttt{not}\,\alpha) = \begin{bmatrix} 0.4 & 0.2 \\ 0.4 & 0 \end{bmatrix}$$

$$v(\beta) = \begin{bmatrix} 0 & 0.2 \\ 0.3 & 0.5 \end{bmatrix}, \quad v(\texttt{not}\,\beta) = \begin{bmatrix} 0.5 & 0.3 \\ 0.2 & 0 \end{bmatrix}$$

$$v(\alpha \quad \texttt{and} \quad \beta) = \begin{bmatrix} 0.5 & 0.3 \\ 0.2 & 0 \end{bmatrix}, v(\alpha \quad \texttt{or} \quad \beta) = \begin{bmatrix} 0 & 0 \\ 0 & 1 \end{bmatrix} (= \mathbb{T})$$

$$v(\alpha \quad \texttt{imp} \quad \beta) = \begin{bmatrix} 0 & 0 \\ 0 & 1 \end{bmatrix}, \quad v(\alpha \quad \texttt{imp} \quad \texttt{not}\beta) = \begin{bmatrix} 0 & 0.4 \\ 0.5 & 0.1 \end{bmatrix}$$

$$v(\beta \quad \texttt{imp} \quad \alpha) = \begin{bmatrix} 0 & 0.2 \\ 0 & 0.8 \end{bmatrix}, \quad v(\beta \quad \texttt{imp} \quad \texttt{not} \ \alpha) = \begin{bmatrix} 0 & 0.4 \\ 0.5 & 0.1 \end{bmatrix}$$

$$v(\alpha \quad \texttt{equiv} \quad \beta) = \begin{bmatrix} 0 & 0.2 \\ 0.1 & 0.7 \end{bmatrix}, v(\alpha \ \texttt{xor} \ \beta) = \begin{bmatrix} 0 & 0.4 \\ 0.5 & 0.1 \end{bmatrix}$$

$$v(\overline{\texttt{not}}\alpha) = \begin{bmatrix} 1 & 0 \\ 0 & 0 \end{bmatrix}, \quad v(\overline{\texttt{not}}\beta) = \begin{bmatrix} 1 & 0 \\ 0 & 0 \end{bmatrix}$$

$$v(\alpha \ \overline{\texttt{and}} \ \beta) = \begin{bmatrix} 0 & 0.4 \\ 0.3 & 0.3 \end{bmatrix}, \quad v(\alpha \ \overline{\texttt{or}} \ \beta) = \begin{bmatrix} 0 & 0.2 \\ 0.2 & 0.6 \end{bmatrix}$$

$$v(\alpha \ \overline{\texttt{imp}} \ \beta) = \begin{bmatrix} 0 & 0 \\ 0.3 & 0.7 \end{bmatrix}, \quad v(\beta \ \overline{\texttt{imp}} \ \alpha) = \begin{bmatrix} 0 & 0.4 \\ 0 & 0.6 \end{bmatrix}$$

$$v(\alpha \ \overline{\texttt{equiv}} \ \beta) = \begin{bmatrix} 0 & 0.4 \\ 0.3 & 0.3 \end{bmatrix}.$$

Example 2 ends here.

The obtained continuous valued paraconsistent logic is a complete logic in the Pavelka sense. The logical axioms are the axiom schemata (Ax.1)–(Ax.13) and rules of inference are the fuzzy rules of inference in the set R. Thus, we have a solid syntax available and for example all the many-valued extensions of classical rules of inference are available; 25 such rules are listed in [29]. For example, the following are sound rules of inference.

Generalized Modus Tollendo Tollens (GMTT);

$$\frac{\texttt{not}\,\beta, \alpha \quad \texttt{imp} \quad \beta}{\texttt{not}\,\alpha}, \ \frac{A, B}{A \odot B}$$

Generalized Simplification Law 1 (GSL1);

$$\frac{\alpha \quad \texttt{and} \quad \beta}{\alpha}, \ \frac{A}{A}$$

Generalized Simplification Law 2 (GSL2);

$$\frac{\alpha \quad \mathtt{and} \quad \beta}{\beta}\, , \frac{\underline{A}}{A}$$

Generalized De Morgan Law 1 (GDeML1);

$$\frac{(\mathtt{not}\ \alpha) \quad \mathtt{and} \quad (\mathtt{not}\ \beta)}{\mathtt{not}\ (\alpha \quad \mathtt{or} \quad \beta)}\, , \frac{\underline{A}}{A}$$

Generalized De Morgan Law 2 (GDML2);

$$\frac{\mathtt{not}\ (\alpha \quad \mathtt{or} \quad \beta)}{(\mathtt{non}\ \alpha) \quad \mathtt{and} \quad (\mathtt{not}\ \beta)}\, , \frac{\underline{A}}{A}$$

To illustrate the use of these two paraconsistent logics and the differences between them, we present two examples.

Example 3 First, consider a fuzzy theory T with the following four special axioms, where the evidence couples are on the unit real square $[0, 1]^2$.

Statement	Formally	Evidence
(1) If wages rise or prices rise there will be inflation	$(p \ \mathtt{or} \ q) \ \mathtt{imp} \ r$	$\langle 1, 0 \rangle$
(2) If there will be inflation, the Government will stop it or people will suffer	$r \ \mathtt{imp} \ (s \ \mathtt{or} \ t)$	$\langle 0.9, 0.1 \rangle$
(3) If people will suffer the Government will lose popularity	$t \ \mathtt{imp} \ w$	$\langle 0.8, 0.1 \rangle$
(4) The Government will not stop inflation and will not lose popularity	$\mathtt{not}\, s \ \mathtt{and} \ \mathtt{not}\, w$	$\langle 1, 0 \rangle$

We have interpreted the logical connectives or and and to be the Lukasiewicz ones; however, they could be intuitionistic $\overline{\mathtt{or}}$ and $\overline{\mathtt{and}}$ too. Moreover, the inclusive or connective could be the exclusive disjunction xor as well.

1° We show that T is satisfiable and therefore consistent. By Theorem 8.6 it is enough to consider evidence couples; focus on the following:

Statement	Atomic formula	Evidence couple
Wages rise	p	$\langle 0.3, 0.8 \rangle$
Prices rise	q	$\langle 0, 1 \rangle$
There will be inflation	r	$\langle 0.3, 0.8 \rangle$
Government will stop inflation	s	$\langle 0, 1 \rangle$
People will suffer	t	$\langle 0.2, 0.9 \rangle$
Government will lose popularity	w	$\langle 0, 1 \rangle$

By direct computation we realize that they lead to the same evidence couples as in the fuzzy theory T. Indeed, for example the evidence for the first special axiom $[(p \;\texttt{ or }\; q) \;\texttt{ imp }\; r]$ is $(0.3 \oplus 0) \rightarrow 0.3 = 1$ and evidence against the axiom $[(p \;\texttt{ or }\; q) \;\texttt{ imp }\; r]$ is $(0.8 \odot 1)^* \odot 0.8 = 0$, and similarly for the other axioms. Thus, T is satisfiable and consistent.

2° What can be said on logical grounds about the claim *wages will not rise*, formally expressed by $\texttt{not}\; p$? The above consideration on evidence couples associates with $(\texttt{not}\; p)$ an evidence couple $\langle 0.3, 0.8\rangle^{\perp} = \langle 0.7, 0.2\rangle$ and the corresponding valuation v is given by the evidence matrix

$$v(\texttt{not}\; p) = \begin{bmatrix} 0.7^* \wedge 0.2 & 0.7 \odot 0.2 \\ 0.7^* \odot 0.2^* & 0.7 \wedge 0.2^* \end{bmatrix} = \begin{bmatrix} 0.2 & 0 \\ 0.1 & 0.7 \end{bmatrix},$$

and the degree of tautology of $(\texttt{not}\; p)$ is less than or equal to $v(\texttt{not}\; p)$.

3° We prove that the degree of tautology of the wff $(\texttt{not}\; p)$ cannot be less that $v(\texttt{not}\; p)$, thus it is equal to $v(\texttt{not}\; p)$. To this end consider the following $\texttt{R}$-proof:

(1)	$(p \;\texttt{ or }\; q) \;\texttt{ imp }\; r$	$\langle 1, 0\rangle$	special axiom
(2)	$r \;\texttt{ imp }\; (s \;\texttt{ or }\; t)$	$\langle 0.9, 0.1\rangle$	special axiom
(3)	$t \;\texttt{ imp }\; w$	$\langle 0.8, 0.1\rangle$	special axiom
(4)	$\texttt{not}\, s \;\texttt{and}\; \texttt{not}\, w$	$\langle 1, 0\rangle$	special axiom
(5)	$\texttt{not}\, w$	$\langle 1, 0\rangle$	(4), GS2
(6)	$\texttt{not}\, s$	$\langle 1, 0\rangle$	(4), GS1
(7)	$\texttt{not}\, t$	$\langle 0.8, 0.1\rangle$	(5), (3), GMTT
(8)	$\texttt{not}\, s \;\texttt{and}\; \texttt{not}\, t$	$\langle 0.8, 0.1\rangle$	(6), (7), RBC
(9)	$\texttt{not}(s \;\texttt{ or }\; t)$	$\langle 0.8, 0.1\rangle$	(8), GDeML1
(10)	$\texttt{not}\, r$	$\langle 0.7, 0.2\rangle$	(9), (2), GMTT
(11)	$\texttt{not}(p \;\texttt{ or }\; q)$	$\langle 0.7, 0.2\rangle$	(10), (1) GMTT
(12)	$\texttt{not}\, p \;\texttt{and}\; \texttt{not}\, q$	$\langle 0.7, 0.2\rangle$	(11), GDeML2
(13)	$\texttt{not}\, p$	$\langle 0.7, 0.2\rangle$	(12), GS1

4° By completeness of T we conclude

$$C^{sem}(T)(\texttt{not}\; p) = C^{syn}(T)(\texttt{not}\; p) = \begin{bmatrix} 0.2 & 0 \\ 0.1 & 0.7 \end{bmatrix}.$$

We interpret this result by saying that, from a logical point of view, the claim *wages will not rise* is much more true $(t(\texttt{not}\; p) = 0.7)$ than false $(f(\texttt{not}\; p) = 0.2)$, and is not contradictory $(k(\texttt{not}\; p) = 0)$ but lacks some information $(u(\texttt{not}\; p) = 0.1)$.

Example 4 Consider the fuzzy theory T with the same four special axioms introduced in Example 3, but with the difference that the evidence couples are on the unit real triangle $\triangledown$. Assume also that the new evidence couples are given as follows.

Statement	Formally	Evidence
(1) If wages rise or prices rise there will be inflation	$(p$ `or` $q)$ `imp` r	$\langle 1, 0\rangle$
(2) If there will be inflation, the Government will stop it or people will suffer	r `imp` $(s$ `or` $t)$	$\langle 0.85, 0.05\rangle$
(3) If people will suffer the Government will lose popularity	t `imp` w	$\langle 0.75, 0.05\rangle$
(4) The Government will not stop inflation and will not lose popularity	`not` s `and` `not` w	$\langle 1, 0\rangle$

1° Again, it is easy to verify by direct computation that the fuzzy theory T is satisfiable (and therefore consistent) when the following evidence couples are assigned to the atomic formulas.

Statement	Atomic formula	Evidence couple
Wages rise	p	$\langle 0.15, 0.65\rangle$
Prices rise	q	$\langle 0, 1\rangle$
There will be inflation	r	$\langle 0.15, 0.65\rangle$
Government will stop inflation	s	$\langle 0, 1\rangle$
People will suffer	t	$\langle 0.1, 0.8\rangle$
Government will lose popularity	w	$\langle 0, 1\rangle$

For instance, the evidence couple for the first special axiom can be obtained as

$$\begin{aligned} H([H^{-1}(\langle 0.15, 0.65\rangle) \oplus H^{-1}(\langle 0, 1\rangle)] \odot H^{-1}(\langle 0.15, 0.65\rangle)) &= H([\langle 0.3, 0.8\rangle \oplus \langle 0, 1\rangle] \odot \langle 0.3, 0.8\rangle) \\ &= H(\langle 0.3, 0.8\rangle) \\ &= \langle 0.15, 0.65\rangle. \end{aligned}$$

and similarly for the other three axioms.

2° Note that the statement `not` p is now associated with the evidence couple

$$\langle 0.15, 0.65\rangle^{*\triangledown} = H([H^{-1}(\langle 0.15, 0.65\rangle^{\perp}) = H(\langle 0.3, 0.8\rangle^{\perp}) = H(\langle 0.7, 0.2\rangle) = \langle 0.6, 0.1\rangle$$

and the corresponding valuation v is given by the evidence matrix

$$v(\texttt{not}\ p) = \begin{bmatrix} 0.6^* \odot 0.1 & 2(0.6 \wedge 0.1) \\ 0.6^* \odot 0.1^* & 0.1^* \odot 0.6 \end{bmatrix} = \begin{bmatrix} 0 & 0.2 \\ 0.3 & 0.5 \end{bmatrix} \in M_{\triangledown},$$

thus, the degree of tautology of (`not` p) has to be less than or equal to $v(\texttt{not}\ p)$.

2° In order to prove that the degree of tautology of the wff (`not` p) cannot be less that $v(\texttt{not}\ p)$, we conduct a similar `R`-proof as before::

(1)	$(p$ `or` $q)$ `imp` r	$\langle 1, 0\rangle$	special axiom
(2)	r `imp` $(s$ `or` $t)$	$\langle 0.85, 0.05\rangle$	special axiom
(3)	t `imp` w	$\langle 0.75, 0.05\rangle$	special axiom
(4)	`not` s `and` `not` w	$\langle 1, 0\rangle$	special axiom
(5)	`not` w	$\langle 1, 0\rangle$	(4), GS2
(6)	`not` s	$\langle 1, 0\rangle$	(4), GS1
(7)	`not` t	$\langle 0.75, 0.05\rangle$	(5), (3), GMTT
(8)	`not` s `and` `not` t	$\langle 0.75, 0.05\rangle$	(6), (7), RBC
(9)	`not`$(s$ `or` $t)$	$\langle 0.75, 0.05\rangle$	(8), GDeML1
(10)	`not` r	$\langle 0.6, 0.1\rangle$	(9), (2), GMTT
(11)	`not`$(p$ `or` $q)$	$\langle 0.6, 0.1\rangle$	(10), (1) GMTT
(12)	`not` p `and` `not` q	$\langle 0.6, 0.1\rangle$	(11), GDeML2
(13)	`not` p	$\langle 0.6, 0.1\rangle$	(12), GS1

4° Then, by completeness of T if follows that

$$C^{sem}(T)(\texttt{not}\ p) = C^{syn}(T)(\texttt{not}\ p) = \begin{bmatrix} 0 & 0.2 \\ 0.3 & 0.5 \end{bmatrix}.$$

Now, we interpret this result in similar terms as before, i.e., the claim *wages will not rise* is half-true ($T(\texttt{not p}) = 0.5$) and not false ($F(\texttt{not p}) = 0$), though some of the available information is conflicting ($K(\texttt{not p}) = 0.2$) at the same time that we lack some further pieces of information ($U(\texttt{not p}) = 0.3$).

8.5 Conclusion and Future Work

We have reviewed two Pavelka style fuzzy logical systems that make it possible to handle and analyze information that is not consistent but still relevant to make meaningful conclusions. In general, paraconsistent logic may constitute a valuable tool for data mining and decision support, as it provides a formal language able to address information lacks and inconsistencies, which are often found in many real-life applications in which several sources of information have to be combined and exploited. In this sense, the paraconsistent logic in the unit square, introduced through formulae (8.1)–(8.4), has been found to be a particularly attractive tool to expand the power and expressiveness of the GUHA data mining methodology (see [30]). Similarly, also the proposed paraconsistent logic in the unit triangle, introduced through formulae (8.40)–(8.43), has shown a quite promising performance in the development of a decision support system focused on the initial assessment of the consequences of natural disasters, a context in which incomplete as well as conflicting information often appears (see [22]). The introduction of a paraconsistent logic in this context allows not only handling the low-quality information usually available after disaster strikes, but also to introduce on the mathematical models some of the requirements of the decision makers, as for instance worst-case scenario analysis.

Moreover, following [23] some efforts are being placed on developing a general classification methodology based on this last logic. Future works in this direction will address the potential interest of paraconsistent logic for the development of more sophisticated machine learning techniques, therefore pointing to close the gap between computers and humans learning and reasoning abilities in the presence of inconsistent information.

References

1. Arieli, O., Avron, A.: Reasoning with logical bilattices. J. Logic Lang. Inform. **5**, 25–63 (1996)
2. Barbon, J.S.; Guido, R.C.; Vieira, L.S.: A neural-network approach for speech features classication based on paraconsistent logic. In: Proceedings of the 11th IEEE conference International Symposium Multimedia, pp. 567–570. San Dego, CA, 14–16 Dec. 2009
3. Belnap, N.D.: A useful four-valued logic. In: Epstein, G., Dumme, J. (eds.) Modern Uses of Multiple Valued Logics, pp. 8–37. D. Reidel, Dordrecht (1977)
4. Chang, C.C.: Algebraic analysis of many-valued logics. Trans. Am. Math. Soc. **88**, 476–490 (1958)
5. Cignoli, R., D'Ottaviano, I.M.L., Mundici, D.: Algebraic Foundations of Many-valued Reasoning, Trends in Logic, vol. 7. Kluwer Academic Publishers, Dordrecht (2000)
6. Cornelis, C., Deschrijver, G., Kerre, E.E.: Advances and challenges in interval-valued fuzzy logic. Fuzzy Sets Syst. **157**(5), 622–627 (2006)
7. da Silva Lopes, H.F., Abe, J.M., Anghinah, R.: Application of Paraconsistent artificial neural networks as a method of aid in the diagnosis of alzheimer disease. J. Med. Syst. **34**, 1073–81 (2010)
8. Dubois, D.: On ignorance and contradiction considered as truth-values. Logic J. IGPL **16**(2), 195–216 (2008)
9. Ginsberg, M.K.: Multivalued logics: a uniform approach to inference in artificial intelligence. Comput. Intell. **4**, 265–316 (1988)
10. Gluschankof, D.: Prime deductive systems and injective objects in the algebras of Łukasiewicz infinite-valued calculi. Algebra Univers. **29**, 354–377 (1992)
11. Hájek, P.: Metamathematics of Fuzzy Logic. Kluwer (1998)
12. Jun, M., Zhang, G., Lu, J.: A state-based knowledge representation approach for information logical inconsistency detection in warning systems. J. Knowl.-Based Syst. **23**, 125–131 (2010)
13. Kukkurainen, P., Turunen, E.: Many-valued similarity reasoning. An axiomatic approach. Int. J. Multiple Valued Logic **8**, 751–760 (2002)
14. Nakamatsu, K., Suito, H., Abe, J.M, Suzuki, A.: Paraconsistent logic program based safety verification for air traffic control. IEEE Int. Conf. Syst. Man Cybern. **6**, 5 (2002)
15. Novák, V., Perfilieva, I., Mockor, J.: Mathematical Principles of Fuzzy Logic. Kluwer, Boston (1999)
16. Odintsov, S.P.: On axiomatizing Shramko-Wansing's logic. Stud. Logica **93**, 407–428 (2009)
17. Öztürk, M., Tsoukiás, A.: Modeling uncertain positive and negative reasons in decision aiding. Decis. Support Syst. **43**, 1512–1526 (2007)
18. Pavelka, J.: On fuzzy logic I, II, III. Zeitsch. f. Math. Logik **25**, 45–52, 119–134, 447–464 (1979)
19. Perny, P., Tsoukiás, A.: On the continuous extensions of four valued logic for preference modeling. In: Proceedings of the IPMU Conference, pp. 302–309 (1998)
20. Priest, G., Tanaka, K., Weber, Z.: 'Paraconsistent Logic', The Stanford Encyclopedia of Philosophy (Fall 2013 Edition), Edward N. Zalta (ed.), URL = ⟨
21. Rivieccio, U.: Neutrosophic logics: prospects and problems. Fuzzy Sets Syst. **159**, 1860–1868 (2008)

22. Rodríguez, J.T., Vitoriano, B., Montero, J.: Classification of disasters and emergencies under bipolar knowledge representation. In: Vitoriano, B., Montero, J., Ruan, D. (eds.) Decision Aid Models for Disaster Management and Emergencies. Atlantis Computational Intelligence Systems, vol. 7, pp. 209–232. Atlantis Press, Paris (2013)
23. Rodríguez, J.T., Vitoriano, B., Montero, J.: Fuzzy Dissimilarity–based classification for disaster initial assessment. In: Proceedings of the 8th Conference of the European Society for Fuzzy Logic and Technologies, pp. 448–455 (2013)
24. Rodríguez, J.T., Turunen, E., Ruan, D., Montero, J.: Another paraconsistent algebraic semantics for Łukasiewicz-Pavelka logic. Another paraconsistent algebraic semantics for Lukasiewicz-Pavelka logic. Fuzzy Sets Syst. **242**, 132–147 (2014)
25. Shramko, Y., Wansing, H.: Hypercontradictions, generalized truth values, and logics of truth and falsehood. J. Logic Lang. Inform. **15**, 403–424 (2006)
26. Shramko, Y., Wansing, H.: Some useful 16-valued logics: how a computer network should think. J. Philos. Logic **34**, 121–153 (2005)
27. Tsoukiás, A.: A first order, four valued, weakly Paraconsistent logic and its relation to rough sets semantics. Foundations Comput. Dec. Sci. **12**, 85–108 (2002)
28. Turunen, E.: Well-defined fuzzy sentential logic. Math. Logic Q. **41**, 236–248 (1995)
29. Turunen, E.: Mathematics behind Fuzzy Logic. Springer (1999)
30. Turunen, E.: Interpreting GUHA data mining logic in paraconsistent fuzzy logic framework. In: Tsoukiás, A., Rossi, F. (eds.) Algorithmic Decision Theory, LNCS, vol. 5783. Springer, Berlin/Heidelberg. pp. 284–293 (2009)
31. Turunen, E., Öztürk, M., Tsoukiás, A.: Paraconsistent semantics for Pavelka style fuzzy sentential logic. Fuzzy Sets Syst. **161**, 1926–1940 (2010)
32. Turunen, E.: Complete MV–algebra Valued Pavelka Logic. (submitted)
33. Van Gasse, B., Cornelis, C., Deschrijver, G., Kerre, E.E.: Triangle algebras: a formal logic approach to interval-valued residuated lattices. Fuzzy Sets Syst. **159**(9), 1042–1060 (2008)

Part III
Paraconsitency and Modality

Chapter 9
On Modal Logics Defining Jaśkowski-Like Discussive Logics

Marek Nasieniewski and Andrzej Pietruszczak

Abstract The present paper concerns Jaśkowski-like discussive logics which arise by modification of Jaśkowski's original translation of discussive conjunction. In each case, we indicate the smallest modal logic defining a given Jaśkowski-like discussive logic.

Keywords Modal logic · Jaśkowski logic $\mathbf{D_2}$ · Jaśkowski-like discussive logics · Minimal modal logics defining $\mathbf{D_2}$ · Minimal modal logics defining Jaśkowski-like discussive logics · Jaśkowski's problem

Mathematics Subject Classification (2000) Primary 03B45 · Secondary 03B53

9.1 Introduction

Jaśkowski's discussive logic $\mathbf{D_2}$, [5, 6], has been formulated with the help of the modal logic **S5** as follows (see Sects. 9.2 and 9.3): $A \in \mathbf{D_2}$ iff $\ulcorner \Diamond A^\bullet \urcorner \in \mathbf{S5}$, where $(-)^\bullet$ is a translation of discussive formulas into the modal language such that: the function $(-)^\bullet$ from the set of all discussive formulas $\mathrm{For}^{\mathrm{d}}$ to the set of all modal formulas $\mathrm{For}_{\mathrm{m}}$ such that:

1. $(a)^\bullet = a$, for any propositional variable a,
2. for all discussive formulas A, and B:

    (a) $(\neg A)^\bullet = \ulcorner \neg A^\bullet \urcorner$,
    (b) $(A \vee B)^\bullet = \ulcorner A^\bullet \vee B^\bullet \urcorner$,
    (c) $(A \wedge^{\mathrm{d}} B)^\bullet = \ulcorner A^\bullet \wedge \Diamond B^\bullet \urcorner$,

M. Nasieniewski (✉) · A. Pietruszczak
Department of Logic, Nicolaus Copernicus University in Toruń,
ul. Stanisława Moniuszki 16, 87–100 Toruń, Poland
e-mail: mnasien@umk.pl

A. Pietruszczak
e-mail: pietrusz@umk.pl

J.-Y. Beziau et al. (eds.), *New Directions in Paraconsistent Logic*,
Springer Proceedings in Mathematics & Statistics 152,
DOI 10.1007/978-81-322-2719-9_9

(d) $(A \to^{\mathrm{d}} B)^{\bullet} = \ulcorner \Diamond A^{\bullet} \to B^{\bullet} \urcorner$,
(e) $(A \leftrightarrow^{\mathrm{d}} B)^{\bullet} = \ulcorner (\Diamond A^{\bullet} \to B^{\bullet}) \wedge \Diamond(\Diamond B^{\bullet} \to A^{\bullet}) \urcorner$.

Thus, the key role in the definition of the logic $\mathbf{D_2}$ is played by the logic **S5**. There are considered other modal logics that are also defining the same logic $\mathbf{D_2}$. In particular, the smallest normal, regular, monotonic, congruential, cm-, rte-, and generally, the smallest modal logic defining $\mathbf{D_2}$ were given (cf. [8, 9]).

In the literature, there are also considered translations that are determining other Jaśkowski-like logics. In [4, 7] for example, instead of the original, right discussive conjunction, the left discussive conjunction is treated as Jaśkowki's one (the other connectives are defined by the same conditions as in the case of the transformation $(-)^{\bullet}$):

$$(A \wedge^{\mathrm{d}} B)^{*} = \ulcorner \Diamond A^{*} \wedge B^{*} \urcorner.$$

In [3], it has been shown that the transformation $(-)^{*}$ yields a logic different from $\mathbf{D_2}$. Ciuciura denotes the obtained logic by '$\mathbf{D}_2^{*}$'. There are two other possibilities as regards the internal translation of conjunction:

$$(A \wedge^{\mathrm{d}} B)^{\frown} = \ulcorner A^{\frown} \wedge B^{\frown} \urcorner.$$
$$(A \wedge^{\mathrm{d}} B)^{\times} = \ulcorner \Diamond A^{\times} \wedge \Diamond B^{\times} \urcorner.$$

The question arises (which has been stated by João Marcos), what does it change if we consider the weakest in the mentioned classes, modal logics that determine the obtained logics, and in this paper we will give an answer to this question.

9.1.1 Meaning of Discussive Conjunctions

We can try to intuitively explain these new understandings of conjunction. For example, while the original Jaśkowski's conjunction can be understood as saying that two assertions $\mathfrak{P}$ and $\mathfrak{Q}$ have been articulated by two participants during a discussion, where the conjunctive statement has been made from the point of view of the first of these participants (see for example [10]), the meaning stipulated by the function $(-)^{*}$ could be seen as an expression of the point of view of the second participant. On the other hand, the function $(-)^{\frown}$ can be justified to be a report that assertions $\mathfrak{P}$ and $\mathfrak{Q}$ have been made by a single participant—it is an intuitive understanding of classical conjunction considered in the context of the model of discussion (see [10]). And finally, the last translation could be treated as a narration of participant who has made none of two assertions.

9.2 Some Facts—Modal Logic

Modal language. As in [2], modal formulas are formed in the standard way from propositional variables: 'p', 'q', 'p_0', 'p_1', 'p_2', ...; truth-value operators: '$\neg$',

'$\vee$', '$\wedge$', '$\rightarrow$', and '$\leftrightarrow$' (connectives of negation, disjunction, conjunction, material implication, and material equivalence, respectively); modal connectives: '$\Box$' and '$\Diamond$' (necessity and possibility operators); and brackets. By $\mathrm{For_m}$ we denote the set of modal formulas. Of course, the set $\mathrm{For_m}$ includes the set of all classical formulas (without '$\Box$' and '$\Diamond$'); let **Taut** be the set of all classical tautologies. Besides, for any $\varphi, \psi, \chi \in \mathrm{For_m}$, let $\chi[^{\varphi}/_{\psi}]$ be any formula that results from χ by replacing none, one or more than one occurrence of φ in χ, by ψ. Finally, let $\psi[^{\Diamond}/_{\neg\Box\neg}]$ denote a result of replacing in ψ of none, one or more than one occurrence of any formula of the form $\Diamond\varphi$, by $\neg\Box\neg\varphi$.

For every $\varphi \in \mathrm{For_m}$ let $\mathrm{Sub}(\varphi)$ be the set of all modal formulas being substitution instances of φ. For any $\Phi \subseteq \mathrm{For_m}$ let $\mathrm{Sub}(\Phi) := \bigcup_{\varphi\in\Phi} \mathrm{Sub}(\varphi)$. We have $\psi \in \mathrm{Sub}(\psi)$ and $\Phi \subseteq \mathrm{Sub}(\Phi)$. Moreover, we put $\Diamond\Phi := \{\psi : \exists_{\varphi\in\Phi}\ \psi = \ulcorner\Diamond\varphi\urcorner\} = \{\ulcorner\Diamond\varphi\urcorner : \varphi \in \Phi\}$.

Modal logics. Modal logics are certain sets of formulas. As in [1], we define a *modal logic* as any set $\boldsymbol{L}$ of modal formulas satisfying the following conditions:

- $\mathbf{Taut} \subseteq \boldsymbol{L}$,
- $\boldsymbol{L}$ includes the following set of formulas

$$\left\{\ulcorner\chi[^{\neg\Box\neg\varphi}/_{\Diamond\varphi}] \leftrightarrow \chi\urcorner \ :\ \varphi, \chi \in \mathrm{For_m}\right\}. \qquad (\mathtt{rep}^{\Box})$$

- $\boldsymbol{L}$ is closed under the following two rules: *modus ponens* for '$\rightarrow$':

$$\frac{\varphi \qquad \varphi \rightarrow \psi}{\psi} \qquad (\mathrm{mp})$$

  and *uniform substitution*:

$$\frac{\varphi}{\mathrm{s}\,\varphi} \qquad (\mathrm{sb})$$

  where $\mathrm{s}\,\varphi$ is the result of uniform substitution of formulas for propositional variables in φ.

By (sb), every modal logic includes the set **PL** of modal formulas which are instances of classical tautologies (i.e., instances of elements of **Taut**).

An element of a given logic is called its *thesis*. By ($\mathtt{rep}^{\Box}$), every modal logic has the following thesis:

$$\Diamond p \leftrightarrow \neg\Box\neg p \qquad (\mathtt{df}\,\Diamond)$$

Chosen classes of logics. We say that a modal logic $\boldsymbol{L}$ is an *rte-logic* iff $\boldsymbol{L}$ is closed under replacement of tautological equivalents, i.e., for any $\varphi, \psi, \chi \in \mathrm{For_m}$

$$\text{if } \ulcorner\varphi \leftrightarrow \psi\urcorner \in \mathbf{PL} \text{ and } \chi \in \boldsymbol{L}, \text{ then } \chi[^{\varphi}/_{\psi}] \in \boldsymbol{L}. \qquad (\mathrm{rte})$$

Equivalently, a modal logic is an rte-logic iff it contains the following set

$$\{\ulcorner \chi[{}^{\varphi}/_{\psi}] \leftrightarrow \chi \urcorner : \varphi, \psi, \chi \in \mathrm{For_m} \text{ and } \ulcorner \varphi \leftrightarrow \psi \urcorner \in \mathbf{PL}\}. \qquad (\mathtt{rep}_{\mathbf{PL}})$$

Similarly as in the case of $(\mathtt{rep}^{\Box})$ also $(\mathtt{rep}_{\mathbf{PL}})$ is closed on (sb).

Remark 9.2.1 Apart from a possibility of replacing of '$\neg\Box\neg$' by '$\Diamond$' valid for every modal logic, in any thesis of any rte-logic we can replace one or more occurrences of '$\Box\neg$' (resp. '$\neg\Box$', '$\neg\Diamond\neg$', '$\neg\Diamond$', '$\Diamond\neg$') by '$\neg\Diamond$' (resp. '$\Diamond\neg$','$\Box$', '$\Box\neg$', '$\neg\Box$') and vice versa. Thus, every rte-logic has the following thesis

$$\Box p \leftrightarrow \neg\Diamond\neg p \qquad (\mathtt{df}\,\Box)$$

Lemma 9.2.2 *An rte-logic contains the following formula*:

$$\Box(p \wedge q) \leftrightarrow (\Box p \wedge \Box q) \qquad (\mathtt{R})$$

iff it contains its dual form:

$$\Diamond(p \vee q) \leftrightarrow (\Diamond p \vee \Diamond q) \qquad (\mathtt{R}^{\Diamond})$$

In [1] a modal logic is called *classical modal* (*cm-logic* for short) iff it is an rte-logic which contains

$$\Box(p \to q) \to (\Box p \to \Box q) \qquad (\mathtt{K})$$

and

$$\Box(p \to p) \qquad (\mathtt{N})$$

Thus, every cm-logic contains the set $\Box\mathbf{PL} := \{\Box\tau : \tau \in \mathbf{PL}\}$.

We say that a modal logic is *congruential* iff it is closed under the congruence rule

$$\frac{\varphi \leftrightarrow \psi}{\Box\varphi \leftrightarrow \Box\psi} \qquad (\mathrm{cgr})$$

Equivalently, a modal logic is congruential iff it is closed under replacement

$$\frac{\varphi \leftrightarrow \psi}{\chi[{}^{\varphi}/_{\psi}] \leftrightarrow \chi} \qquad (\mathrm{rep})$$

Every congruential logic is an rte-logic.

We say that a modal logic is *monotonic* iff it is closed under the monotonicity rule:

$$\frac{\varphi \to \psi}{\Box\varphi \to \Box\psi} \qquad (\mathrm{mon})$$

Every monotonic logic is congruential.

We say that a modal logic is *regular* iff it is closed under the regularity rule:

$$\frac{\varphi \wedge \psi \to \chi}{\Box\varphi \wedge \Box\psi \to \Box\chi} \qquad \text{(reg)}$$

Equivalently, a modal logic is regular iff it contains the formula (K) and is closed under (mon) iff it contains the formula (R) and is closed under (cgr).

A modal logic is *normal* iff it contains (K) and is closed under the necessitation rule

$$\frac{\varphi}{\Box\varphi} \qquad \text{(nec)}$$

Similarly, a modal logic is *normal* iff it contains (N) and (K), and it is closed under the rule (cgr). Every normal logic is a cm-logic.

9.3 The Logics $\mathbf{D_2}$, $\mathbf{D_2^*}$, $\mathbf{D_2^-}$, and $\mathbf{D_2^{**}}$

Discussive language. The logics $\mathbf{D_2}$, $\mathbf{D_2^*}$, $\mathbf{D_2^-}$, and $\mathbf{D_2^{**}}$ are defined as certain sets of discussive formulas. These formulas are formed in the standard way from propositional variables: 'p', 'q', 'p_0', 'p_1', 'p_2', ...; truth-value operators: '$\neg$' and '$\vee$' (negation and disjunction); discussive connectives: '$\wedge^{\mathrm{d}}$', '$\to^{\mathrm{d}}$', '$\leftrightarrow^{\mathrm{d}}$' (conjunction, implication, and equivalence); and brackets. Let $\mathrm{For}^{\mathrm{d}}$ be the set of all such expressions.

Definition of discussive logic $\mathbf{D_2}$. As we mentioned in Sect. 9.1, the logic $\mathbf{D_2}$ is formulated with the help of the modal logic **S5** as follows:

$$\mathbf{D_2} := \{ A \in \mathrm{For}^{\mathrm{d}} : \ulcorner \Diamond A^{\bullet} \urcorner \in \mathbf{S5} \},$$

where $(-)^{\bullet}$ is a translation given on p. 1, of discussive formulas into the modal language. Of course, $\mathbf{D_2}$ is closed under uniform substitution of discussive formulas. Moreover, $\mathbf{D_2}$ is closed under modus ponens for '$\to^{\mathrm{d}}$':

$$\frac{A, \quad A \to^{\mathrm{d}} B}{B} \qquad (\mathrm{mp_d})$$

because **S5** is closed under the following rule:

$$\frac{\Diamond\varphi, \quad \Diamond(\Diamond\varphi \to \psi)}{\Diamond\psi} \qquad (\mathrm{RC})$$

Definitions of Jaśkowski-like logics. We already mentioned the logic $\mathbf{D_2^*}$. Let us now define it together with other two Jaśkowski-like discussive logics $\mathbf{D_2^-}$ and $\mathbf{D_2^{**}}$

as follows:

$$\mathbf{D}_2^* := \{ A \in \mathrm{For}^{\mathrm{d}} : \ulcorner \Diamond A^* \urcorner \in \mathbf{S5} \},$$
$$\mathbf{D}_2^- := \{ A \in \mathrm{For}^{\mathrm{d}} : \ulcorner \Diamond A^\frown \urcorner \in \mathbf{S5} \},$$
$$\mathbf{D}_2^{**} := \{ A \in \mathrm{For}^{\mathrm{d}} : \ulcorner \Diamond A^\times \urcorner \in \mathbf{S5} \},$$

where $(-)^*$, $(-)^\frown$, $(-)^\times : \mathrm{For}^{\mathrm{d}} \longrightarrow \mathrm{For}_{\mathrm{m}}$ and for any $A, B \in \mathrm{For}^{\mathrm{d}}$:

(c*) $(A \wedge^{\mathrm{d}} B)^* = \ulcorner \Diamond A^* \wedge B^* \urcorner$,
(e)* $(A \leftrightarrow^{\mathrm{d}} B)^* = \ulcorner \Diamond(\Diamond A^* \to B^*) \wedge (\Diamond B^* \to A^*) \urcorner$,
(c⌢) $(A \wedge^{\mathrm{d}} B)^\frown = \ulcorner A^\frown \wedge B^\frown \urcorner$,
(e)⌢ $(A \leftrightarrow^{\mathrm{d}} B)^\frown = \ulcorner (\Diamond A^\frown \to B^\frown) \wedge (\Diamond B^\frown \to A^\frown) \urcorner$.
(c×) $(A \wedge^{\mathrm{d}} B)^\times = \ulcorner \Diamond A^\times \wedge \Diamond B^\times \urcorner$,
(e)× $(A \leftrightarrow^{\mathrm{d}} B)^\times = \ulcorner \Diamond(\Diamond A^\times \to B^\times) \wedge \Diamond(\Diamond B^\times \to A^\times) \urcorner$,

and in the case of every considered above translation, conditions for other connectives stay unchanged with respect to the function $(-)^\bullet$. As a matter of fact all defined logics differ only with respect to the condition for conjunction, because for any translation $\$ \in \{\bullet, *, \frown, \times\}$, for all $A, B \in \mathrm{For}^{\mathrm{d}}$ we have: $(A \leftrightarrow^{\mathrm{d}} B)^\$ = ((A \to^{\mathrm{d}} B) \wedge^{\mathrm{d}} (B \to^{\mathrm{d}} A))^\$$.

9.4 Modal Logics for $\mathbf{D}_2$, $\mathbf{D}_2^*$, $\mathbf{D}_2^-$, and $\mathbf{D}_2^{**}$

It is known that aside from **S5** there are also other modal logics that define $\mathbf{D}_2$. The same holds for the other three discussive logics. In particular, there is a procedure (see [8]) that for a given class of logics fulfilling rather natural conditions, gives as an outcome a logic $\boldsymbol{L}$ which is minimal in the considered class among logics which have the same as **S5** theses beginning with $\Diamond$. The same procedure can be applied for $\mathbf{D}_2^*$, $\mathbf{D}_2^-$ and $\mathbf{D}_2^{**}$.

We say that a modal logic $\boldsymbol{L}$ *defines* $\mathbf{D}_2$ (resp. $\mathbf{D}_2^*$, $\mathbf{D}_2^-$, $\mathbf{D}_2^{**}$) iff $\mathbf{D}_2 = \{A \in \mathrm{For}^{\mathrm{d}} : \ulcorner \Diamond A^\bullet \urcorner \in \boldsymbol{L}\}$ (resp. $\mathbf{D}_2^* = \{A \in \mathrm{For}^{\mathrm{d}} : \ulcorner \Diamond A^* \urcorner \in \boldsymbol{L}\}$, $\mathbf{D}_2^- = \{A \in \mathrm{For}^{\mathrm{d}} : \ulcorner \Diamond A^\frown \urcorner \in \boldsymbol{L}\}$, $\mathbf{D}_2^{**} = \{A \in \mathrm{For}^{\mathrm{d}} : \ulcorner \Diamond A^\times \urcorner \in \boldsymbol{L}\}$).

We see that while expressing logic $\mathbf{D}_2$ we refer to modal logics which

have the same theses beginning with '$\Diamond$' as **S5**. (†)

Let $\mathsf{S5}_\diamond$ be the set of all these logics, i.e.,

$$\boldsymbol{L} \in \mathsf{S5}_\diamond \quad \text{iff} \quad \forall_{A \in \mathrm{For}_{\mathrm{m}}} (\ulcorner \Diamond \varphi \urcorner \in \boldsymbol{L} \iff \ulcorner \Diamond \varphi \urcorner \in \mathbf{S5}).$$

By the definition we see:

Fact 9.4.1 *For any $\boldsymbol{L} \in \mathsf{S5}_\diamond$:*

1. $\{\ulcorner \Diamond \varphi \urcorner : \ulcorner \Diamond \varphi \urcorner \in \mathbf{S5}\} \subseteq \boldsymbol{L}$,
2. *If $\boldsymbol{L} \in \mathsf{S5}_\diamond$, then $\boldsymbol{L}$ defines $\mathbf{D}_2$, $\mathbf{D}_2^*$, $\mathbf{D}_2^-$ and $\mathbf{D}_2^{**}$.*

Let us recall (see [8]) that rte$\mathbf{S5^M}$, cm$\mathbf{S5^M}$, e$\mathbf{S5^M}$, m$\mathbf{S5^M}$, r$\mathbf{S5^M}$ and $\mathbf{S5^M}$ are respectively, the smallest rte-, cm-, congruential, monotonic, regular, and normal logic in $\mathsf{S5}_\diamond$. Thus, by Fact 9.4.1 each of them defines also logics $\mathbf{D}_2^*$, $\mathbf{D}_2^-$ and $\mathbf{D}_2^{**}$.

Let $(-)^{\mathrm{any}}$ be any translation of discussive formulas into the modal language, that is, the function $(-)^{\mathrm{any}}$ from $\mathrm{For^d}$ into $\mathrm{For_m}$. And let

$$\mathbf{D}_2^{\mathrm{any}} := \{\, A \in \mathrm{For^d} : \ulcorner \Diamond A^{\mathrm{any}} \urcorner \in \mathbf{S5} \,\},$$

Corollary 9.4.2 *The logics* rte$\mathbf{S5^M}$*,* cm$\mathbf{S5^M}$*,* e$\mathbf{S5^M}$*,* m$\mathbf{S5^M}$*,* r$\mathbf{S5^M}$*, and* $\mathbf{S5^M}$ *are the smallest rte-, cm-, congruential, monotonic, regular, and normal logic in* $\mathsf{S5}_\diamond$ *defining* $\mathbf{D}_2^{\mathrm{any}}$*, respectively.*

One could ask whether there are modal logics defining discussive logics, which do not belong to $\mathsf{S5}_\diamond$. For the case of the largest among classes considered in Corollary 9.4.2—the class of rte-logics—the answer is 'no':

Fact 9.4.3 ([8]) *For any rte-logic* $\boldsymbol{L}$*:* $\boldsymbol{L}$ *defines* $\mathbf{D_2}$ *iff* $\boldsymbol{L} \in \mathsf{S5}_\diamond$*.*

The similar result can be obtained for the other considered here Jaśkowski-like logics:

Fact 9.4.4 *For any rte-logic* $\boldsymbol{L}$*:* $\boldsymbol{L}$ *defines* $\mathbf{D}_2^*$ *iff* $\boldsymbol{L} \in \mathsf{S5}_\diamond$*.*

Proof "$\Rightarrow$" Let $\boldsymbol{L}$ be any rte-logic. We define a function $(-)^\circ : \mathrm{For_m} \longrightarrow \mathrm{For^d}$ which $\ll$ un-modalizes $\gg$ every modal formula:

1. $(a)^\circ = a$, for any propositional variable a,
2. for any $\varphi, \psi \in \mathrm{For_m}$:
   (a) $(\neg\varphi)^\circ = \ulcorner \neg\varphi^\circ \urcorner$,
   (b) $(\varphi \vee \psi)^\circ = \ulcorner \varphi^\circ \vee \psi^\circ \urcorner$,
   (c) $(\varphi \wedge \psi)^\circ = \ulcorner \neg(\neg\varphi^\circ \vee \neg\psi^\circ) \urcorner$,
   (d) $(\varphi \to \psi)^\circ = \ulcorner \neg\varphi^\circ \vee \psi^\circ \urcorner$,
   (e) $(\varphi \leftrightarrow \psi)^\circ = \ulcorner \neg(\neg(\neg\varphi^\circ \vee \psi^\circ) \vee \neg(\neg\psi^\circ \vee \varphi^\circ)) \urcorner$,
   (f) $(\Diamond\varphi)^\circ = \ulcorner \varphi^\circ \wedge^{\mathrm{d}} (p \vee \neg p) \urcorner$,
   (g) $(\Box\varphi)^\circ = \ulcorner \neg\varphi^\circ \to^{\mathrm{d}} \neg(p \vee \neg p) \urcorner$.

Notice that for any $\varphi, \psi \in \mathrm{For_m}$ we have the following equalities:

$$\begin{aligned}
(\neg\varphi)^{\circ *} &= \ulcorner \neg\varphi^{\circ *} \urcorner, \\
(\varphi \vee \psi)^{\circ *} &= \ulcorner \varphi^{\circ *} \vee \psi^{\circ *} \urcorner, \\
(\varphi \wedge \psi)^{\circ *} &= \ulcorner \neg(\neg\varphi^{\circ *} \vee \neg\psi^{\circ *}) \urcorner, \\
(\varphi \to \psi)^{\circ *} &= \ulcorner \neg\varphi^{\circ *} \vee \psi^{\circ *} \urcorner, \\
(\varphi \leftrightarrow \psi)^{\circ *} &= \ulcorner \neg(\neg(\neg\varphi^{\circ *} \vee \psi^{\circ *}) \vee \neg(\neg\psi^{\circ *} \vee \varphi^{\circ *})) \urcorner, \\
(\Diamond\varphi)^{\circ *} &= \ulcorner \Diamond\varphi^{\circ *} \wedge (p \vee \neg p) \urcorner, \\
(\Box\varphi)^{\circ *} &= \ulcorner \Diamond\neg\varphi^{\circ *} \to \neg(p \vee \neg p) \urcorner.
\end{aligned}$$

Thus, for any φ, $\psi \in \mathrm{For_m}$, $\S \in \{\neg, \Diamond\}$ and $\underline{*} \in \{\wedge, \vee, \rightarrow, \leftrightarrow\}$ the following formulas belong to **PL**:

$$\begin{aligned} (\S\,\varphi)^{\circ *} &\leftrightarrow \S\,\varphi^{\circ *} \\ (\varphi \,\underline{*}\, \psi)^{\circ *} &\leftrightarrow (\varphi^{\circ *} \,\underline{*}\, \psi^{\circ *}) \\ (\Box\varphi)^{\circ *} &\leftrightarrow \neg \Diamond \neg \varphi^{\circ *} \end{aligned} \qquad (\star)$$

Therefore for any φ, ψ, $\chi \in \mathrm{For_m}$:

$$\begin{gathered} \ulcorner \chi \leftrightarrow \chi[^{(\S\,\varphi)^{\circ *}}/_{\S\,\varphi^{\circ *}}]\urcorner \in \boldsymbol{L}\,, \\ \ulcorner \chi \leftrightarrow \chi[^{(\varphi \underline{*} \psi)^{\circ *}}/_{(\varphi^{\circ *} \underline{*} \psi^{\circ *})}]\urcorner \in \boldsymbol{L}\,, \\ \ulcorner \chi \leftrightarrow \chi[^{(\Box\varphi)^{\circ *}}/_{\neg \Diamond \neg \varphi^{\circ *}}]\urcorner \in \boldsymbol{L}\,. \end{gathered}$$

Thus, for any modal formulas $\varphi_1, \ldots, \varphi_n, \chi$ we obtain:

$$\chi^{\circ *} \in \boldsymbol{L} \ \text{ iff } \ \chi[^{\Box\varphi_1}/_{\neg \Diamond \neg \varphi_1}, \ldots, {}^{\Box\varphi_n}/_{\neg \Diamond \neg \varphi_n}] \in \boldsymbol{L}\,,$$

since for every propositional variable a we have $a^{\circ *} = a$. Hence we get:

$$\chi^{\circ *} \in \boldsymbol{L} \ \text{ iff } \ \chi \in \boldsymbol{L}\,, \qquad (\star\star)$$

since we can replace every occurrence of '$\neg \Diamond \neg$' by '$\Box$', thanks to Remark 9.2.1.

If additionally $\boldsymbol{L}$ defines $\mathbf{D}_2^*$, we obtain that for any $\varphi \in \mathrm{For_m}$: $\ulcorner \Diamond\varphi \urcorner \in \boldsymbol{L}$ iff (by $(\star\star)$) $\ulcorner (\Diamond\varphi)^{\circ *} \urcorner \in \boldsymbol{L}$ iff (by $(\star)$) $\ulcorner \Diamond\varphi^{\circ *} \urcorner \in \boldsymbol{L}$ iff (since $\boldsymbol{L}$ defines $\mathbf{D}_2^*$) $\varphi^\circ \in \mathbf{D}_2^*$ iff $\ulcorner \Diamond\varphi^{\circ *} \urcorner \in \mathbf{S5}$ iff (by $(\star)$ and $(\star\star)$, and the fact that **S5** is an rte-logic) $\ulcorner \Diamond\varphi \urcorner \in \mathbf{S5}$. So $\boldsymbol{L} \in \mathsf{S5}_\Diamond$. □

Fact 9.4.5 *For any rte-logic* $\boldsymbol{L}$: $\boldsymbol{L}$ *defines* $\mathbf{D}_2^-$ *iff* $\boldsymbol{L} \in \mathsf{S5}_\Diamond$.

Proof "$\Rightarrow$" Let $\boldsymbol{L}$ be any rte-logic. We redefine the function $(-)^\circ$ used in the proof of Fact 9.4.4 by stipulating that for any φ, $\psi \in \mathrm{For_m}$:

(f) $(\Diamond\varphi)^\circ = \ulcorner \neg(\varphi^\circ \rightarrow^{\mathrm{d}} \neg(p \vee \neg p))$,

while other cases stay unchanged with respect to the definition of the function $(-)^\circ$.

Notice that for any φ, $\psi \in \mathrm{For_m}$ we have the following equalities:

$$\begin{aligned} (\neg\varphi)^{\circ\frown} &= \ulcorner \neg \varphi^{\circ\frown} \urcorner, \\ (\varphi \vee \psi)^{\circ\frown} &= \ulcorner \varphi^{\circ\frown} \vee \psi^{\circ\frown} \urcorner, \\ (\varphi \wedge \psi)^{\circ\frown} &= \ulcorner \neg(\neg \varphi^{\circ\frown} \vee \neg \psi^{\circ\frown}) \urcorner, \\ (\varphi \rightarrow \psi)^{\circ\frown} &= \ulcorner \neg \varphi^{\circ\frown} \vee \psi^{\circ\frown} \urcorner, \\ (\varphi \leftrightarrow \psi)^{\circ\frown} &= \ulcorner \neg(\neg(\neg \varphi^{\circ\frown} \vee \psi^{\circ\frown}) \vee \neg(\neg \psi^{\circ\frown} \vee \varphi^{\circ\frown})) \urcorner, \\ (\Diamond\varphi)^{\circ\frown} &= \ulcorner \neg(\Diamond\varphi^\circ \rightarrow \neg(p \vee \neg p)), \\ (\Box\varphi)^{\circ\frown} &= \ulcorner \Diamond \neg \varphi^{\circ\frown} \rightarrow \neg(p \vee \neg p) \urcorner. \end{aligned}$$

The rest of the proof goes as previously. □

Fact 9.4.6 *For any rte-logic* $\mathbf{L}$: $\mathbf{L}$ *defines* $\mathbf{D}_2^{**}$ *iff* $\mathbf{L} \in \mathsf{S5}_\Diamond$.

Proof "⇒" Let $\mathbf{L}$ be any rte-logic. We use the function $(-)^\circ : \mathrm{For_m} \longrightarrow \mathrm{For^d}$, where for any $\varphi, \psi \in \mathrm{For_m}$:

(f) $(\Diamond\varphi)^\circ = \ulcorner \varphi^\circ \wedge^{\mathrm{d}} \varphi^\circ \urcorner$,

and again we leave unchanged the other conditions given in the proof of Fact 9.4.4.

Notice that for any $\varphi, \psi \in \mathrm{For_m}$ we have the following equalities:

$$
\begin{aligned}
(\neg\varphi)^{\circ\times} &= \ulcorner \neg\varphi^{\circ\times} \urcorner,\\
(\varphi \vee \psi)^{\circ\times} &= \ulcorner \varphi^{\circ\times} \vee \psi^{\circ\times} \urcorner,\\
(\varphi \wedge \psi)^{\circ\times} &= \ulcorner \neg(\neg\varphi^{\circ\times} \vee \neg\psi^{\circ\times}) \urcorner,\\
(\varphi \rightarrow \psi)^{\circ\times} &= \ulcorner \neg\varphi^{\circ\times} \vee \psi^{\circ\times} \urcorner,\\
(\varphi \leftrightarrow \psi)^{\circ\times} &= \ulcorner \neg(\neg(\neg\varphi^{\circ\times} \vee \psi^{\circ\times}) \vee \neg(\neg\psi^{\circ\times} \vee \varphi^{\circ\times})) \urcorner,\\
(\Diamond\varphi)^{\circ\times} &= \ulcorner \Diamond\varphi^{\circ\times} \wedge \Diamond\varphi^{\circ\times} \urcorner,\\
(\Box\varphi)^{\circ\times} &= \ulcorner \Diamond\neg\varphi^{\circ\times} \rightarrow \neg(p \vee \neg p) \urcorner.
\end{aligned}
$$

Thus, for any $\varphi, \psi \in \mathrm{For_m}$, $\S \in \{\neg, \Diamond\}$ and $\underline{\times} \in \{\wedge, \vee, \rightarrow, \leftrightarrow\}$ the following formulas belong to **PL**:

$$
\begin{aligned}
(\S\varphi)^{\circ\times} &\leftrightarrow \S\varphi^{\circ\times}\\
(\varphi \,\underline{\times}\, \psi)^{\circ\times} &\leftrightarrow (\varphi^{\circ\times} \,\underline{\times}\, \psi^{\circ\times})\\
(\Box\varphi)^{\circ\times} &\leftrightarrow \neg\Diamond\neg\varphi^{\circ\times}
\end{aligned}
\qquad (\star)
$$

Analogously, from that the thesis of the fact follows. □

Corollary 9.4.7 *The logic* rte$\mathbf{S5^M}$ (*resp.* cm$\mathbf{S5^M}$, $\mathbf{eS5^M}$, $\mathbf{mS5^M}$, $\mathbf{rS5^M}$, $\mathbf{S5^M}$) *is the smallest rte- (resp. cm-, congruential, monotonic, regular, normal modal) logic defining the logics* $\mathbf{D}_2$, $\mathbf{D}_2^*$, $\mathbf{D}_2^-$, *and* $\mathbf{D}_2^{**}$.

Thus, the difference as regards modal logics defining respective Jaśkowski-like discussive logics can appear only in the case of weaker than rte-logics. So, as in the case of $\mathbf{D}_2$ we will indicate the weakest modal logics defining logics $\mathbf{D}_2^*$, $\mathbf{D}_2^-$, and $\mathbf{D}_2^{**}$. For each of these modal logics, there are formulas of the form $\ulcorner\Diamond\varphi\urcorner$ that belong to **S5**, but do not belong to those logics.

9.5 Relations Between $\mathbf{D}_2$, $\mathbf{D}_2^*$, $\mathbf{D}_2^-$, and $\mathbf{D}_2^{**}$

In [3] Ciuciura observed that $\mathbf{D}_2^* \not\subseteq \mathbf{D}_2$. It was shown that one of the axioms of the logic $\mathbf{D}_2^*$ that had three variables, is not a thesis of the logic $\mathbf{D}_2$.

One can indicate some other formulas which show that this inclusion does not hold, that is, $\mathbf{D}_2^* \not\subseteq \mathbf{D}_2$. To be able to find out what are relations between four discussive logics, we consider the following formulas:

$$\neg(p \wedge^{\mathrm{d}} q) \rightarrow^{\mathrm{d}} (p \rightarrow^{\mathrm{d}} \neg q) \tag{ncon}$$

We see that

$$\Diamond(\neg(p \wedge^{\mathrm{d}} q) \rightarrow^{\mathrm{d}} (p \rightarrow^{\mathrm{d}} \neg q))^{\bullet} = \Diamond(\Diamond \neg(p \wedge \Diamond q) \rightarrow (\Diamond p \rightarrow \neg q))$$

On the basis of **S5** it is equivalent to:

$$\Diamond \neg p \vee \neg \Diamond q \rightarrow (\Diamond p \rightarrow \Diamond \neg q)$$

which is not a thesis of **S5**. So (ncon) $\notin \mathbf{D_2}$.

We have

$$\Diamond(\neg(p \wedge^{\mathrm{d}} q) \rightarrow^{\mathrm{d}} (p \rightarrow^{\mathrm{d}} \neg q))^{*} = \Diamond(\Diamond \neg(\Diamond p \wedge q) \rightarrow (\Diamond p \rightarrow \neg q))$$

On the basis of **S5** it is equivalent to:

$$\neg \Diamond p \vee \Diamond \neg q \rightarrow (\Diamond p \rightarrow \Diamond \neg q)$$

which is a thesis of **S5**. So (ncon) $\in \mathbf{D_2^*}$. Besides,

$$\Diamond(\neg(p \wedge^{\mathrm{d}} q) \rightarrow^{\mathrm{d}} (p \rightarrow^{\mathrm{d}} \neg q))^{\frown} = \Diamond(\Diamond \neg(p \wedge q) \rightarrow (\Diamond p \rightarrow \neg q))$$

On the basis of **S5** the formula it is equivalent to:

$$\Diamond \neg p \vee \Diamond \neg q \rightarrow (\Diamond p \rightarrow \Diamond \neg q)$$

Again, this formula is not a thesis of **S5**. So (ncon) $\notin \mathbf{D_2^-}$. While

$$\Diamond(\neg(p \wedge^{\mathrm{d}} q) \rightarrow^{\mathrm{d}} (p \rightarrow^{\mathrm{d}} \neg q))^{\times} = \Diamond(\Diamond \neg(\Diamond p \wedge \Diamond q) \rightarrow (\Diamond p \rightarrow \neg q))$$

This formula is equivalent on the basis of **S5** to:

$$\neg \Diamond p \vee \neg \Diamond q \rightarrow (\Diamond p \rightarrow \Diamond \neg q)$$

This formula is a thesis of **S5**. So (ncon) $\in \mathbf{D_2^{**}}$. Thus, $\mathbf{D_2^*} \not\subseteq \mathbf{D_2}$, $\mathbf{D_2^*} \not\subseteq \mathbf{D_2^-}$. $\mathbf{D_2^{**}} \not\subseteq \mathbf{D_2}$, $\mathbf{D_2^{**}} \not\subseteq \mathbf{D_2^-}$.

Additionally, we see that $\mathbf{D_2} \not\subseteq \mathbf{D_2^*}$. Indeed, consider the formula:

$$\neg p \rightarrow^{\mathrm{d}} \neg(p \wedge^{\mathrm{d}} q) \tag{scontr}$$

We have:

$$\Diamond(\neg p \rightarrow^{\mathrm{d}} \neg(p \wedge^{\mathrm{d}} q))^{\bullet} = \Diamond(\Diamond \neg p \rightarrow \neg(p \wedge \Diamond q))$$

This formula is equivalent on the basis of **S5** to:

$$\Diamond\neg p \to \Diamond\neg p \vee \neg\Diamond q$$

which is a thesis of **PL**. So (scontr) $\in \mathbf{D_2}$, since **PL** $\subseteq$ **S5**. We also have:

$$\Diamond(\neg p \to^{\mathrm{d}} \neg(p \wedge^{\mathrm{d}} q))^{*} = \Diamond(\Diamond\neg p \to \neg(\Diamond p \wedge q))$$

This formula is equivalent to:

$$\Diamond\neg p \to \neg\Diamond p \vee \Diamond\neg q$$

which is not a thesis of **S5**. So (scontr) $\notin \mathbf{D}_2^{*}$.

We also have:

$$\Diamond(\neg p \to^{\mathrm{d}} \neg(p \wedge^{\mathrm{d}} q))^{\frown} = \Diamond(\Diamond\neg p \to \neg(p \wedge q))$$

This formula is equivalent to:

$$\Diamond\neg p \to \Diamond\neg p \vee \Diamond\neg q$$

which is a thesis of **S5**. So (scontr) $\in \mathbf{D}_2^{-}$.

We have:

$$\Diamond(\neg p \to^{\mathrm{d}} \neg(p \wedge^{\mathrm{d}} q))^{\times} = \Diamond(\Diamond\neg p \to \neg(\Diamond p \wedge \Diamond q))$$

This formula is equivalent to:

$$\Diamond\neg p \to \neg\Diamond p \vee \neg\Diamond q$$

which is not a thesis of **S5**. So (scontr) $\notin \mathbf{D}_2^{**}$.

Thus, as it has been announced $\mathbf{D_2} \not\subseteq \mathbf{D}_2^{*}$, but also $\mathbf{D_2} \not\subseteq \mathbf{D}_2^{**}$, $\mathbf{D}_2^{-} \not\subseteq \mathbf{D}_2^{*}$, and $\mathbf{D}_2^{-} \not\subseteq \mathbf{D}_2^{**}$.

Now consider

$$\neg p \to^{\mathrm{d}} \neg((p \vee \neg p) \wedge^{\mathrm{d}} p) \qquad \text{(contr-)}$$

We have:

$$\Diamond(\neg p \to^{\mathrm{d}} \neg((p \vee \neg p) \wedge^{\mathrm{d}} p))^{\bullet} = \Diamond(\Diamond\neg p \to \neg((p \vee \neg p) \wedge \Diamond p))$$

On the basis of **S5**, it is equivalent to:

$$\Diamond\neg p \to (\Diamond\neg(p \vee \neg p) \vee \neg\Diamond p),$$

which is not a thesis of **S5**. So (contr-) $\notin \mathbf{D_2}$. We have:

$$\Diamond(\neg p \to^{\mathrm{d}} \neg((p \vee \neg p) \wedge^{\mathrm{d}} p))^{*} = \Diamond(\Diamond \neg p \to \neg(\Diamond(p \vee \neg p) \wedge p))$$

This formula on the basis of **S5** is equivalent to:

$$\Diamond \neg p \to (\neg \Diamond(p \vee \neg p) \vee \Diamond \neg p)$$

which is a thesis of **S5**. So (contr-) $\in \mathbf{D_2^*}$. We also have:

$$\Diamond(\neg p \to^{\mathrm{d}} \neg((p \vee \neg p) \wedge^{\mathrm{d}} p))^{\frown} = \Diamond(\Diamond \neg p \to \neg((p \vee \neg p) \wedge p))$$

This formula on the basis of **S5** is equivalent to:

$$\Diamond \neg p \to (\Diamond \neg(p \vee \neg p) \vee \Diamond \neg p)$$

which is a thesis of **S5**. So (contr-) $\in \mathbf{D_2^-}$.

We have:

$$\Diamond(\neg p \to^{\mathrm{d}} \neg((p \vee \neg p) \wedge^{\mathrm{d}} p))^{\times} = \Diamond(\Diamond \neg p \to \neg(\Diamond(p \vee \neg p) \wedge \Diamond p))$$

This formula on the basis of **S5** is equivalent to:

$$\Diamond \neg p \to (\Box \neg(p \vee \neg p) \vee \Box \neg p)$$

which is not a thesis of **S5**. So (contr-) $\notin \mathbf{D_2^{**}}$.

Thus, $\mathbf{D_2^-} \nsubseteq \mathbf{D_2}$ and $\mathbf{D_2^*} \nsubseteq \mathbf{D_2^{**}}$, Consider again the following formula of two variables:

$$\neg(q \wedge^{\mathrm{d}} p) \to^{\mathrm{d}} (p \to^{\mathrm{d}} \neg q) \tag{ncon$'$}$$

We see that

$$\Diamond(\neg(q \wedge^{\mathrm{d}} p) \to^{\mathrm{d}} (p \to^{\mathrm{d}} \neg q))^{\bullet} = \Diamond(\Diamond \neg(q \wedge \Diamond p) \to (\Diamond p \to \neg q))$$

On the basis of **S5** it is equivalent to:

$$\Diamond \neg q \vee \neg \Diamond p \to (\Diamond p \to \Diamond \neg q)$$

which is a thesis of **S5**. So (ncon$'$) $\in \mathbf{D_2}$. Besides,

$$\Diamond(\neg(q \wedge^{\mathrm{d}} p) \to^{\mathrm{d}} (p \to^{\mathrm{d}} \neg q))^{\frown} = \Diamond(\Diamond \neg(q \wedge p) \to (\Diamond p \to \neg q))$$

On the basis of **S5** this formula is equivalent to:

$$\Diamond \neg q \vee \Diamond \neg p \to (\Diamond p \to \Diamond \neg q)$$

This formula is again not a thesis of **S5**. So (ncon′) $\notin \mathbf{D}_2^-$. While

$$\Diamond(\neg(q \wedge^{\mathrm{d}} p) \rightarrow^{\mathrm{d}} (p \rightarrow^{\mathrm{d}} \neg q))^{\times} = \Diamond(\Diamond \neg(\Diamond q \wedge \Diamond p) \rightarrow (\Diamond p \rightarrow \neg q))$$

and this formula is equivalent on the basis of **S5** to:

$$\neg \Diamond q \vee \neg \Diamond p \rightarrow (\Diamond p \rightarrow \Diamond \neg q)$$

This formula is a thesis of **S5**. So (ncon′) $\in \mathbf{D}_2^{**}$.

$$\Diamond(\neg(q \wedge^{\mathrm{d}} p) \rightarrow^{\mathrm{d}} (p \rightarrow^{\mathrm{d}} \neg q))^{*} = \Diamond(\Diamond \neg(\Diamond q \wedge p) \rightarrow (\Diamond p \rightarrow \neg q))$$

This formula on the basis of **S5** it is equivalent to:

$$\neg \Diamond q \vee \Diamond \neg p \rightarrow (\Diamond p \rightarrow \Diamond \neg q)$$

and this is not a thesis of **S5**. So (ncon′) $\notin \mathbf{D}_2^*$.

Summarizing, $\mathbf{D}_2 \not\subseteq \mathbf{D}_2^-$ and $\mathbf{D}_2^{**} \not\subseteq \mathbf{D}_2^*$.

All four discussive logics have some thesis in common, take as an example $p \rightarrow^{\mathrm{d}} p$.

Corollary 9.5.1 *In the set* $\{\mathbf{D}_2, \mathbf{D}_2^*, \mathbf{D}_2^-, \mathbf{D}_2^{**}\}$ *every two logics cross each other.*

Of course, none of $\mathbf{D}_2, \mathbf{D}_2^*, \mathbf{D}_2^-, \mathbf{D}_2^{**}$ has Duns Scotus law $p \rightarrow^{\mathrm{d}} (\neg p \rightarrow^{\mathrm{d}} q)$ as its thesis. Thus, each of these logics can be treated as a solution to Jaśkowski's problem (see [5]).

As regards some other properties of considered discussive logics let us mention that for example, Jaśkowski's methodological Theorem 1 given in [5, 6] stays valid for $\mathbf{D}_2^*$ and $\mathbf{D}_2^{**}$. The proof goes as the original Jaśkowski's proof. While this theorem does not hold for $\mathbf{D}_2^-$ (even in the weaker version from [6]) due to our, 'discussive' formulation of equivalence given on p. 6. Consider a formula $(p \rightarrow^{\mathrm{d}} q) \rightarrow^{\mathrm{d}} ((q \rightarrow^{\mathrm{d}} p) \rightarrow^{\mathrm{d}} (p \leftrightarrow^{\mathrm{d}} q))$. We see that $\Diamond((p \rightarrow^{\mathrm{d}} q) \rightarrow^{\mathrm{d}} ((q \rightarrow^{\mathrm{d}} p) \rightarrow^{\mathrm{d}} (p \leftrightarrow^{\mathrm{d}} q)))^{\frown} = \Diamond((\Diamond p \rightarrow q) \rightarrow ((\Diamond q \rightarrow p) \rightarrow ((\Diamond p \rightarrow q) \wedge (\Diamond q \rightarrow p))))$, and the last formula is not a thesis of **S5**, so $p \rightarrow^{\mathrm{d}} (q \rightarrow^{\mathrm{d}} (p \leftrightarrow^{\mathrm{d}} q)) \notin \mathbf{D}_2^-$.

9.6 The Smallest Logics Defining $\mathbf{D}_2^*$, $\mathbf{D}_2^-$, $\mathbf{D}_2^{**}$

We consider the following sets of modal formulas:

$$\begin{aligned}
\mathrm{Gen} &:= \{\varphi \in \mathrm{For_m} : \exists_{A \in \mathbf{D}_2}\ \varphi = \ulcorner \Diamond A^{\bullet} \urcorner\} = \{\ulcorner \Diamond A^{\bullet} \urcorner \in \mathrm{For_m} : A \in \mathbf{D}_2\},\\
\mathrm{Gen}^* &:= \{\varphi \in \mathrm{For_m} : \exists_{A \in \mathbf{D}_2^*}\ \varphi = \ulcorner \Diamond A^{*} \urcorner\} = \{\ulcorner \Diamond A^{*} \urcorner \in \mathrm{For_m} : A \in \mathbf{D}_2^*\},\\
\mathrm{Gen}^{\frown} &:= \{\varphi \in \mathrm{For_m} : \exists_{A \in \mathbf{D}_2^-}\ \varphi = \ulcorner \Diamond A^{\frown} \urcorner\} = \{\ulcorner \Diamond A^{\frown} \urcorner \in \mathrm{For_m} : A \in \mathbf{D}_2^-\},\\
\mathrm{Gen}^{\times} &:= \{\varphi \in \mathrm{For_m} : \exists_{A \in \mathbf{D}_2^{**}}\ \varphi = \ulcorner \Diamond A^{\times} \urcorner\} = \{\ulcorner \Diamond A^{\times} \urcorner \in \mathrm{For_m} : A \in \mathbf{D}_2^*\}.
\end{aligned}$$

Lemma 9.6.1 *Every modal logic defining* $\mathbf{D_2}$ *(resp.* $\mathbf{D}_2^*$, $\mathbf{D}_2^-$, $\mathbf{D}_2^{**}$*) includes the set* Sub(Gen) *(resp.* Sub(Gen*), Sub(Gen$^\frown$), Sub(Gen$^\times$)).

We say that a set $\boldsymbol{L} \subseteq \mathrm{For_m}$ is *axiomatized by a set* R *of rules and a set* S *of formulas* iff $\boldsymbol{L}$ is the smallest set containing S which is closed on all rules from R.

Let $\mathrm{Ax}_{\mathbf{PL}}$ be a set of modal formulas such that the pair $\langle \mathrm{Ax}_{\mathbf{PL}}, \{(\mathrm{mp})\}\rangle$ is an axiomatization of the modal logic **PL**.

Let us recall (see [9]), that **A** is *the smallest modal logic defining* $\mathbf{D_2}$. Since similarly as in the case of the logic $\mathbf{D_2}$, the family of modal logics defining $\mathbf{D}_2^*$ (resp. $\mathbf{D}_2^-$, $\mathbf{D}_2^{**}$) is closed under arbitrary intersections, there is the smallest modal logic defining $\mathbf{D}_2^*$ (resp. $\mathbf{D}_2^-$, $\mathbf{D}_2^{**}$). Let $\mathbf{A}^*$ (resp. $\mathbf{A}^-$, $\mathbf{A}^\times$) be *the smallest modal logic defining* $\mathbf{D}_2^*$ (resp. $\mathbf{D}_2^-$, $\mathbf{D}_2^{**}$).

Since sets ($\mathtt{rep}^\Box$), Sub(Gen), Sub(Gen*), Sub(Gen$^\frown$), and Sub(Gen$^\times$) are closed under substitution and the considered axiomatization of $\mathrm{Ax}_{\mathbf{PL}}$ relies only on modus ponens rule, we easily see that:

Fact 9.6.2 **A** *(respectively* $\mathbf{A}^*$, $\mathbf{A}^-$, $\mathbf{A}^\times$*) is the smallest modal logic including the set* Gen *(resp.* Gen$^\frown$, Gen$^\frown$, Gen$^\times$*). Consequently,* **A** *(resp.* $\mathbf{A}^*$, $\mathbf{A}^-$, $\mathbf{A}^\times$*) is axiomatized by the rule* (mp) *and the sum of the sets* $\mathrm{Ax}_{\mathbf{PL}}$, ($\mathtt{rep}^\Box$)*, and* Sub(Gen) *(respectively* Sub(Gen*), Sub(Gen$^\frown$), Sub(Gen$^\times$)).

One can observe (see [9]) that:

Lemma 9.6.3 $\mathbf{A} \cap \Diamond\mathrm{For_m} \subseteq \mathrm{Sub}(\Diamond\{\varphi : \exists_{\psi \in (\mathrm{For^d})^\bullet}\ \varphi = \psi[{}^\Diamond/{}_{\neg\Box\neg}]\})$.

A similar observation holds for other considered discussive logics.

Lemma 9.6.4 1. $\mathbf{A}^* \cap \Diamond\mathrm{For_m} \subseteq \mathrm{Sub}(\Diamond\{\varphi : \exists_{\psi \in (\mathrm{For^d})^*}\ \varphi = \psi[{}^\Diamond/{}_{\neg\Box\neg}]\})$.
2. $\mathbf{A}^- \cap \Diamond\mathrm{For_m} \subseteq \mathrm{Sub}(\Diamond\{\varphi : \exists_{\psi \in (\mathrm{For^d})^\frown}\ \varphi = \psi[{}^\Diamond/{}_{\neg\Box\neg}]\})$.
3. $\mathbf{A}^\times \cap \Diamond\mathrm{For_m} \subseteq \mathrm{Sub}(\Diamond\{\varphi : \exists_{\psi \in (\mathrm{For^d})^\times}\ \varphi = \psi[{}^\Diamond/{}_{\neg\Box\neg}]\})$.

Proof Let v be any valuation from $\mathrm{For_m}$ into $\{0, 1\}$ such that it preserves classical truth conditions for classical constants and for any $\varphi \in \mathrm{For_m}$:

- $v(\Diamond\varphi) = 1$ iff $\exists_{\psi \in (\mathrm{For^d})^*}\ \varphi \in \mathrm{Sub}(\psi[{}^\Diamond/{}_{\neg\Box\neg}])$,
- $v(\Box\varphi) = 0$ iff $\exists_{\psi \in (\mathrm{For^d})^*}\ \varphi \in \mathrm{Sub}(\ulcorner\neg\psi[{}^\Diamond/{}_{\neg\Box\neg}]\urcorner)$.

We show that for any $\varphi \in \mathbf{A}^*$: $v(\varphi) = 1$.

For any φ from $\mathbf{PL} \cup (\mathtt{rep}^\Box) \cup \mathrm{Sub(Gen)}$ we have that $v(\varphi) = 1$. Thus, by the induction on the length of the proof, relative to the axiomatization considered in Fact 9.6.2, we obtain: if $\varphi \in \mathbf{A}^*$, then $v(\varphi) = 1$.

Proofs of the other cases proceed analogously. □

Using modal formulas given in Sect. 9.5, by the above lemma we obtain:

Lemma 9.6.5 *In the family of sets* $\{\mathbf{A} \cap \Diamond\mathrm{For_m}, \mathbf{A}^* \cap \Diamond\mathrm{For_m}, \mathbf{A}^- \cap \Diamond\mathrm{For_m}, \mathbf{A}^\times \cap \Diamond\mathrm{For_m}\}$ *each pair of sets cross each other.*

Corollary 9.6.6 *Every two logics among* **A**, $\mathbf{A}^*$, $\mathbf{A}^-$, $\mathbf{A}^\times$ *cross each other.*

We know that:

Fact 9.6.7 ([9]) *The logic* $\mathbf{A}$ *is not an rte-logic, in particular* $\mathbf{A} \subsetneq \mathrm{rte}\mathbf{S5}^{\mathbf{M}}$.

From Facts 9.4.4, 9.4.5, 9.4.6, and Corollary 9.6.6 and the above fact we have:

Corollary 9.6.8 *None of logics* $\mathbf{A}^{*}$, $\mathbf{A}^{-}$, $\mathbf{A}^{\times}$ *is an rte-logic.*

9.7 Conclusion

In this paper, we considered discussive logics denoted as $\mathbf{D}_2^{*}$, $\mathbf{D}_2^{-}$ and $\mathbf{D}_2^{**}$ which are obtained by a slight modification of Jaśkowski's original understanding of discussive conjunction given in [6]. These logics are defined with the help of the logic $\mathbf{S5}$, however it can be also done by the use of other modal logics. In particular, we have given the smallest modal logics defining respective Jaśkowski-like discussive logics. We have shown that although all four discussive logics can be defined by a joint rte-logic $\mathrm{rte}\mathbf{S5}^{\mathbf{M}}$ being the smallest rte-logic that can define $\mathbf{D}_2$ (and $\mathbf{D}_2^{*}$, $\mathbf{D}_2^{-}$, and $\mathbf{D}_2^{**}$), the smallest logics in the family of all modal logics defining respectively $\mathbf{D}_2$, $\mathbf{D}_2^{*}$, $\mathbf{D}_2^{-}$, and $\mathbf{D}_2^{**}$ differ among each other. And finally, it is worth to mention that all new discussive logics are paraconsistent (at least is some of possible ways of understanding of this notion), seems to have some philosophical motivations and can be viewed as solutions to Jaśkowski's problem.

Acknowledgments We would like to thank anonymous referee for her/his useful suggestions on a previous version of this paper.

References

1. Bull, R.A., Segerberg, K.: Basic modal logic. In: Gabbay, D.M., Guenthner, F. (eds.) Handbook of Philosophical Logic, vol. II, pp. 1–88. D. Reidel Publishing Company, Dordrecht (1984)
2. Chellas, B.F.: Modal Logic. An Introduction. Cambridge University Press, Cambridge (1980)
3. Ciuciura, J.: On the da Costa, Dubikajtis and Kotas' system of the discursive logic, $\mathbf{D}_2^{*}$. Logic Log. Philos. **14**, 235–252 (2005)
4. da Costa, N.C.A., Dubikajtis, L.: On Jaśkowski's discussive logic. In: Arruda, A.I., da Costa, N.C.A., Chuaqui, R. (eds.) Non-Classical Logics, Model Theory and Computability, pp. 57–73. North-Holland Publishing, Amsterdam (1977)
5. Jaśkowski, S.: Rachunek zdań dla systemów dedukcyjnych sprzecznych. Studia Societatis Scientiarum Torunensis, Sect. A, I(5), pp. 57–77 (1948). The first English version: Propositional calculus for contradictory deductive systems. Studia Logica, vol. 24, pp. 143–157 (1969). The second English version: Logic and Logical Philosophy, vol. 7, pp. 35–56 (1999)
6. Jaśkowski, S.: O koniunkcji dyskusyjnej w rachunku zdań dla systemów dedukcyjnych sprzecznych. Studia Societatis Scientiarum Torunensis, Sect. A, vol. I, no. 8, pp. 171–172 (1949). The English version: On the discussive conjunction in the propositional calculus for inconsistent deductive systems. Logic and Logical Philosophy, vol. 7, pp. 57–59 (1999)

7. Kotas, J., da Costa, N.C.A.: On some modal logical systems defined in connexion with Jaśkowski's problem. In: Arruda, A.I., da Costa, N.C.A., Chuaqui, R. (eds.) Non Classical Logics, Model Theory and Computability, pp. 57–73. North-Holland Publishing, Amsterdam (1977)
8. Nasieniewski, M., Pietruszczak, A.: A method of generating modal logics defining Jaśkowski's discussive logic D_2. Stud. Logica **97**(1), 161–182 (2011)
9. Nasieniewski, M., Pietruszczak, A.: On the weakest modal logics defining Jaśkowski's logic D_2 and the D_2-consequence. Bull. Sect. Logic **41**, 215–232 (2012)
10. Nasieniewski, M., Pietruszczak, A.: On modal logics defining Jaśkowski's D2-consequence. In: Tanaka, K., Berto, F., Mares, E., Paoli, F. (eds.) Paraconsistency: Logic and Applications, chapter IX, Logic, Epistemology and the Unity of Science Series, vol. 26, pp. 141–160. Springer (2013)

Chapter 10
From Possibility Theory to Paraconsistency

Davide Ciucci and Didier Dubois

Abstract The significance of three-valued logics partly depends on the interpretation of the third truth-value. When it refers to the idea of *unknown*, we have shown that a number of three-valued logics, especially Kleene, Łukasiewicz, and Nelson, can be encoded in a simple fragment of the modal logic KD, called MEL, containing only modal formulas without nesting. This is the logic of possibility theory, the semantics of which can be expressed in terms of all-or-nothing possibility distributions representing an agent's epistemic state. Here we show that this formalism can also encode some three-valued paraconsistent logics, like Priest, Jaśkowski, and Sobociński's, where the third truth-value represents the idea of contradiction. The idea is just to change the designated truth-values used for their translations. We show that all these translations into modal logic are very close in spirit to Avron's early work expressing natural three-valued logics using hypersequents. Our work unifies a number of existing formalisms and the translation also highlights the perfect symmetry between three-valued logics of contradiction and three-valued logics of incomplete information, which corresponds to a swapping of modalities in MEL.

Keywords Three-valued logics · Paraconsistent logic · Possibility theory

Mathematics Subject Classification (2000) Primary 03B53 · Secondary 03B50, 03B45, 03B42

D. Ciucci (✉)
DISCo, Università degli Studi di Milano – Bicocca,
Viale Sarca 336 – U14, 20126 Milano, Italy
e-mail: ciucci@disco.unimib.it

D. Dubois
IRIT, CNRS and Université Paul Sabatier,
118 route de Narbonne, 31062 Toulouse cedex 9, France
e-mail: dubois@irit.fr

J.-Y. Beziau et al. (eds.), *New Directions in Paraconsistent Logic*,
Springer Proceedings in Mathematics & Statistics 152,
DOI 10.1007/978-81-322-2719-9_10

10.1 Introduction

The program of paraconsistent logics after Jaśkowski [17] is to find a logic which can manage contradictions and satisfies three requirements:

1. when applied to the contradictory, systems would not always entail their over-completeness[1];
2. would be rich enough to enable practical inference;
3. would have an intuitive justification.

Nowadays, a consensual definition of what a paraconsistent logic is does not seem to exist, but a necessary condition is that the logical consequence relation is not explosive, that is, there can exist contradictions in the logic without implying that everything is true (point 1 above)—see the introduction of [23]. Several proposals to define such a logic have been studied in literature, following different lines: discursive logic, preservationism, adaptive logics, relevant logics, and many-valued logics [23]. Jaśkowski used a third value to express paraconsistency, but he seems not to have a clear definition of what a contradiction stands for. He refers to theories with conflicting hypotheses, each one able to only partially explain the result of some experiment, or to facts that are not predictable a-priori. D'Ottaviano and da Costa [14] cite, as a justification for their three-valued logic J_3, the existence of contradictory theories in empirical disciplines for which we are (at the moment) not able to say which theory is the correct one. In [2], the third value represents antinomies, that is, propositions that are *at the same time* true and false. Priest logic of paradox [20] uses the same connectives as in Kleene logic but it has two designated truth-values one of them standing for *both true and false*, as in Belnap setups [8]. Although its meaning is not always expressed clearly, it seems that, in all these logics, the third truth-value refers to this view.

In the following, we try to provide a unified view of three-valued logics of paraconsistency, using the latter interpretation of the third value, and adopting a society semantics in the style of Carnielli and Lima-Marques [10], where several sources of information assign a different (Boolean) truth-value to a proposition. We first recall the translation of three-valued logics of incomplete information into a simplified epistemic logic called MEL [6], which we recently carried out [12]. This is a fragment of KD, containing only modal formulas of depth 1 (no nested modalities), and a simplified semantics in terms of nonempty subsets of propositional interpretations. We show that our approach to the translation of Kleene, Nelson, and Łukasiewicz three-valued logics bears similarities with their reconstruction by means of hypersequents after Avron [5] who calls them *natural* three-valued logics. Next, we show that we can translate several three-valued paraconsistent logics into MEL, including Priest logic of paradox (PLP), J3, and RM3, in a way that parallels the translations of the three-valued logics of incomplete information. We also do it for Sette logic. The well-known symmetry between unknown and contradictory stemming from the Belnap bilattice structure is here expressed by the fact that the MEL logic can capture

[1] A system is over-complete if any formula is a theorem.

both three-valued logics of incomplete information and paraconsistent three-valued logics, just changing the meaning of MEL models, viewing them either as epistemic states of a single ill-informed agent or as complete epistemic states of sets of totally informed agents. This translation highlights the epistemic meaning of sentences in three-valued paraconsistent logics and facilitates a comparison among them.

10.2 From Three-Valued Logic to Modal Logic: Incomplete Information

Let us consider a set of propositional variables $\mathcal{V} = \{a, b, c, \dots\}$ and a standard propositional language $\mathcal{L}$ built on these symbols with the Boolean connectives of conjunction ($\wedge$), disjunction ($\vee$), negation ($'$), and implication ($\rightarrow$), plus tautology symbol ($\top$). Let Ω be the set of interpretations of $\mathcal{L}$: $\{w : \mathcal{V} \rightarrow \{0, 1\}\}$. The set of models of $p \in \mathcal{L}$ is denoted by $[p] \subseteq \Omega$.

10.2.1 The Possibilistic Modal Logic of Incomplete Information

Consider a higher level propositional language $\mathcal{L}_\Box$ defined by

$$\phi \in \mathcal{L}_\Box \iff \phi = \Box p | \phi' | \phi \wedge \psi$$

where $\Box$ denotes the necessity modality. Note that atoms of $\mathcal{L}_\Box$ are of the form $\Box p$, $p \in \mathcal{L}$. As usual, the possibility modality $\Diamond p$ is short for $(\Box(p'))'$. $\mathcal{L}_\Box$ is a very elementary fragment of a KD modal language proposed by Banerjee and Dubois under the name MEL [6, 7]. It is viewed as a minimalist epistemic logic containing only modal formulas of depth 1 (without nested modalities). In particular, it contains no non-modal formulas from $\mathcal{L}$. $\Box p$ expresses the idea that based on its epistemic state, an agent is bound to believe p. This statement can be expressed in various uncertainty theories, $\Box p$, respectively, meaning $P([p]) = 1$ in probability theory, and $N([p]) = 1$ in possibility theory. In the latter equality, N is a set function with values in $\{0, 1\}$, expressing the idea of certainty. Its basic axiom is *minitivity*: $N([p \wedge q]) = \min(N([p]), N([q]))$ which is at the root of possibility theory [16], that is, being certain of p and of q amounts to being certain of their conjunction, along with $N([\bot]) = N(\emptyset) = 0$ and $N([\top]) = N(\Omega) = 1$. Likewise, the intended meaning of the sentence $\Diamond p$ is that the agent has no evidence suggesting that p is false. The corresponding set function $\Pi([p]) = 1 - N([p'])$ is a possibility measure such that $\Pi([p \vee q]) = \max(\Pi([p]), \Pi([q]))$.

In the finite setting, the set function N is equivalent to the existence of a *non-empty* subset $E \subseteq S$ of propositional models, understood as an epistemic state: the information possessed by an agent that only knows that the real world lies in E. Its non-emptiness means that the agent has a consistent epistemic state. MEL formulas can be evaluated on the set $\mathcal{E}$ of nonempty subsets of interpretations as follows:

- $E \models \Box p$, if and only if $E \subseteq [p]$;
- $E \models \phi'$, if and only if $E \not\models \phi$;
- $E \models \phi \wedge \psi$, if and only if $E \models \phi$; and $E \models \psi$, where ϕ, ψ are any $\mathcal{L}_\Box$-formulae.
- So, $E \models \Diamond p$ if and only if $E \cap [p] \neq \emptyset$.

The following KD axioms and inference rule are valid for these semantics:

- All axioms of propositional logic for $\mathcal{L}_\Box$-formulas.
- $(K) : \Box(p \rightarrow q) \rightarrow (\Box p \rightarrow \Box q)$.
- $(N) : \Box\top$.
- $(D) : \Box p \rightarrow \Diamond p$.
- **Modus Ponens**: If ϕ, $\phi \rightarrow \psi$ then ψ.

This axiom system implies axiom (C): $\Box(p \wedge q) \equiv \Box p \wedge \Box q$, which is the minitivity axiom. Sets E that represent MEL models cannot be empty due to axiom D, which means the agent has a consistent epistemic state. The satisfaction of MEL-formulae is then defined recursively: For any set $\Gamma \cup \{\phi\}$ of $\mathcal{L}_\Box$-formulae, ϕ is a semantic consequence of Γ, written $\Gamma \models \phi$, provided for every epistemic state E, $E \models \Gamma$ implies $E \models \phi$.

This Boolean possibilistic logic, equipped with modus ponens, (the $\mathcal{L}_\Box$-fragment of KD) is sound and complete w.r.t. this semantics [7]. In particular, it does not require the use of accessibility relations. In fact, MEL is a (higher-order) propositional logic and the deduction theorem is valid in it.

In the sequel, a useful fragment of MEL is when modalities are only applied to literals. The reduced language is then $\mathcal{L}^\ell_\Box$ defined by

$$\phi \in \mathcal{L}^\ell_\Box \iff \phi = (\Box a|\Box a', a \in \mathcal{V})|\phi'|\phi \wedge \psi.$$

The MEL logic restricted to this language is sound and complete with respect to the above epistemic semantics where the epistemic states E are restricted to partial Boolean models [12]. A partial model is a conjunction of literals of the form $\wedge_{a\in A} a \bigwedge \wedge_{b\in B} b'$, for some pair (A, B) of disjoint subsets of variables. Let $\mathcal{L}^\ell_\Diamond$ be the language obtained from $\mathcal{L}^\ell_\Box$ by replacing modality $\Box$ by $\Diamond$. Clearly, the two languages have the same expressive power, since $\Box$ and $\Diamond$ are mutually definable.

10.2.2 Translating Three-Valued Logics of Incomplete Information

Now, let us consider a three-valued logic. A three-valued interpretation is a mapping $v : \mathcal{V} \rightarrow L_3 = \{\mathbf{0}, \frac{1}{2}, \mathbf{1}\}$. We assume $\mathbf{0} < \frac{1}{2} < \mathbf{1}$. If we interpret the three truth-values

$\mathbf{0}, \frac{1}{2}, \mathbf{1}$ as *certainly false*, *unknown*, and *certainly true*, respectively, then $v(a)$ is called the *epistemic truth-value* of the variable a, as held by one agent. More generally, the assertion "$v(a) \in T$," where $T \subseteq L_3$ informs about the knowledge state of the agent regarding a Boolean variable, which we also denote as a. For instance, $v(a) \geq \frac{1}{2}$ expresses that the agent believes a is true or that (s)he ignores whether a is true or false. Strictly speaking, we should not use the same notations for three-valued propositional variables and Boolean ones. However, we will do it for the sake of simplicity.

Let us encode in MEL the assignment of L_3 truth-values to a propositional variable a. We denote by $\mathcal{T}(v(a) \in T)$ the translation into MEL of the statement $v(a) \in T$. This translation of the assignment of a subset of ternary truth-values to an atom is a function $\mathcal{T} : 2^{(\{\mathbf{0},\mathbf{1},\frac{1}{2}\}^{\mathcal{V}})} \to \mathcal{L}_\Box$ from subsets of ternary valuations to the modal language $\mathcal{L}_\Box$. The following translation is in agreement with our understanding of the third truth-value [12] as *unknown for the agent*.

$$\mathcal{T}(v(a) \geq \tfrac{1}{2}) = \Diamond a \qquad \mathcal{T}(v(a) \leq \tfrac{1}{2}) = \Diamond a' \tag{10.2.1}$$
$$\mathcal{T}(v(a) = \mathbf{1}) = \Box a \qquad \mathcal{T}(v(a) = \mathbf{0}) = \Box a' \tag{10.2.2}$$
$$\mathcal{T}(v(a) = \tfrac{1}{2}) = \Diamond a \wedge \Diamond a' \qquad \mathcal{T}(v(a) \in \emptyset) = \Box\bot \tag{10.2.3}$$
$$\mathcal{T}(v(a) \in \{\mathbf{0}, \mathbf{1}\}) = \Box a \vee \Box a' \qquad \mathcal{T}(v(a) \geq \mathbf{0}) = \Diamond\top \tag{10.2.4}$$

Note that this is NOT a syntactic translation from one logic into another: it provides a tool for expressing the semantics of one logic into the syntax of another one. We remark that these definitions clarify the debate on the acceptability or not of the excluded middle law and the contradiction principle in the presence of *unknown* value: a is always true or false, but $\Box a \vee \Box a'$ is not a tautology nor is $\Diamond a \wedge \Diamond a'$ a contradiction.

On this basis, we can translate into MEL the assignment of epistemic truth-values to complex propositions in a given three-valued logic using the truth-tables of the available connectives, as extensively described in [12]. It is clear that the result of these translations will be formulas in the restricted language $\mathcal{L}^\ell_\Box$. Moreover, three-valued valuations v are in one-to-one correspondence with partial Boolean models $\wedge_{a\in\mathcal{V}:v(a)=\mathbf{1}} a \bigwedge \wedge_{b\in\mathcal{V}:v(b)=\mathbf{0}} b'$ whose sets of propositional models correspond to special cases of epistemic states that serve as interpretations of the language $\mathcal{L}_\Box$ of MEL.

10.2.3 The Case of Kleene Logic

The best known and often used logic to represent uncertainty due to incomplete information is Kleene logic. The connectives are simply the min $\sqcap$, the max $\sqcup$, and the involutive negation $\neg$. A material implication $a \to_K b := \neg a \sqcup b$ is then derived. The involutive negation preserves the De Morgan laws between $\sqcap$ and $\sqcup$. The syntax of Kleene logic is the same as the one of the propositional logics (replacing $\wedge, \vee, '$ by $\sqcap, \sqcup, \neg$). Besides, it is known that Kleene logic does not have any tautology (there is no formula p such that $\forall v$, $v(p) = \mathbf{1}$).

Asserting the truth of a formula p in Kleene logic comes down to writing $v(p) = \mathbf{1}$. Then, it is clear that [12]

$$\begin{aligned} \mathcal{T}(v(p \sqcap q) = \mathbf{1}) &= \mathcal{T}(v(p) = \mathbf{1}) \wedge \mathcal{T}(v(q) = \mathbf{1}) \\ \mathcal{T}(v(p \sqcup q) = \mathbf{1}) &= \mathcal{T}(v(p) = \mathbf{1}) \vee \mathcal{T}(v(q) = \mathbf{1}) \\ \mathcal{T}(v(\neg a) = \mathbf{1}) &= \mathcal{T}(v(a) = \mathbf{0}) = \Box a'. \end{aligned}$$

The translation of Kleene implication $\neg a \sqcup b$ is then $\Box a' \vee \Box b$. The translation into MEL lays bare the meaning of Kleene implication: $a \to_K b$ is "true" means that b is certain if a is possible, which suggests a very strong implication.

We can always put a Kleene logic formula in conjunctive normal form (CNF), that is, a conjunction of disjunction of literals (without simplifying terms of the form $a \sqcup \neg a$). Its translation into MEL consists of the same set of clauses, where we put the modality $\Box$ in front of each literal. Finally, we see that the fragment of MEL that exactly captures the language of Kleene logic contains only the set (conjunctions) of disjunctions of elementary formulae of the form $\Box a$ or $\Box a'$:

$$\mathcal{L}_{\Box}^{K} = \Box a | \Box a' | \phi \vee \psi | \phi \wedge \psi \subset \mathcal{L}_{\Box}^{\ell}.$$

It does not include expressions containing the formula $(\Box \ell)'$ where ℓ is a literal. We remark that the modal axioms of MEL cannot be expressed in this fragment. It is also easy to see that the translation of any propositional tautology (if we replace each literal ℓ by $\Box \ell$ in its CNF) will no longer be a tautology in MEL.

A knowledge base B in Kleene logic is a conjunction of formulae (clauses for simplicity) supposed to have designated truth-value **1**. Reasoning in Kleene logic corresponds to the following semantic inference:

$$B \models p \text{ if and only if whenever } v(p_i) = \mathbf{1} \text{ for all } p_i \in B \text{ then } v(p) = \mathbf{1} \qquad \text{(I1)}$$

Informally, $B \models p$ means that if the agent is certain that all p_i's in B are true, then this agent is also certain that p is true. A proof system for this logic is described by Avron [5], who notices that $\neg$ is not an internal negation,[2] nor $\to_K$ an internal implication (in particular we do not have that $B \cup \{a\} \models b$ implies $B \models a \to_K b$.) Nevertheless, Kleene logic is a basic natural three-valued logic because it is based on the only truth-preserving inference in the structure defined by the involutive negation $\neg$ and combining conjunction $\sqcap$ ($B \models p \sqcap q$ if and only if $B \models p$ and $B \models q$).

We can use MEL to reason in Kleene logic. Denote by $\mathcal{T}(B)$ the translation into MEL of $v(B) = \mathbf{1}$, where B is a Kleene base. We note that a form of modus ponens applies to literals (since from $\Box a$ and $\Box a' \vee \Box b$, we can derive $\Box b$ in MEL).

---

[2] A negation $\neg$ is *internal* if "$\Gamma, A \vdash \Delta$ iff $\Gamma \vdash \Delta, \neg A$" or equivalently if "$\Gamma \vdash \Delta, A$ iff $\Gamma, \neg A \vdash \Delta$." An implication $\to$ is *internal* if "$\Gamma, A \vdash \Delta, B$ iff $\Gamma \vdash \Delta, A \to B$."

The same counterpart of the resolution principle is also valid. Kleene logic appears like a propositional logic without tautologies but with standard rules of inference. Here, as we have seen, a three-valued valuation v corresponds to a Boolean partial model $\wedge_{a \in \mathcal{V}: v(a)=\mathbf{1}} a \bigwedge \wedge_{b \in \mathcal{V}: v(b)=\mathbf{0}} b'$. Let E_v be the corresponding special kind of epistemic state. Conversely, to each epistemic state E can be assigned a unique three-valued valuation v_E such that $v_E(a) = \mathbf{1}$ if $E \subseteq [a]$, and $\mathbf{0}$ if $E \subseteq [a']$. We have that $E \subseteq E_{v_E}$. At the semantic level we can prove the following result [12].

Proposition 10.2.1 *Let p be a formula in Kleene logic. For each three-valued valuation v such that $v(p) = \mathbf{1}$, the epistemic state E_v is a model (in the sense of MEL) of $\mathcal{T}(v(p) = \mathbf{1})$. Conversely, for each model in the sense of MEL (epistemic state) E of $\mathcal{T}(v(p) = \mathbf{1})$, the three-valued interpretation v_E is a model of p in the sense that $v_E(p) = \mathbf{1}$.*

We can easily verify that the inference $B \models p$ in Kleene logic can be expressed in the Kleene fragment of MEL (using models of the form E_v at the semantic level) in the sense that $\mathcal{T}(B) \vdash \mathcal{T}(v(p) = \mathbf{1})$ in MEL if and only if $B \models_K p$.

10.2.4 Two Other Three-Valued Logics of Incomplete Information

Avron [5] constructs two additional well-known three-valued logics, first by introducing an internal implication inside Kleene logic; the resulting framework captures Nelson's logic. Then by forming the conjunction of this implication and its contraposition obtained by the Kleene negation (thus getting Łukasiewicz logic). In [12] we have shown that both logics can be captured in MEL and correspond exactly to the fragment $\mathcal{L}^{\ell}_{\Box}$ of the MEL language where the only restriction is to have $\Box$ in front of literals only. Here we explain this situation using the same approach as Avron, albeit in MEL.

Indeed, we can easily augment the fragment $\mathcal{L}^{K}_{\Box}$ of the MEL language capturing Kleene logic, by introducing an implication that obeys the deduction theorem, which is valid in MEL. By applying this valid property to reasoning with the Kleene fragment of MEL, we generate the translation of a three-valued implication that is not part of Kleene logic.

In MEL, the deduction theorem is valid, namely $K \cup \{\Box a\} \vdash \Box b$ if and only if $K \vdash (\Box a)' \vee \Box b$. The expression $(\Box a)' \vee \Box b \in \mathcal{L}^{\ell}_{\Box}$ means that "a is not certain or b is certain." But $(\Box a)' \vee \Box b \notin \mathcal{L}^{K}_{\Box}$, it is not the translation of some Kleene logic formula declared true. The idea then is that $(\Box a)' \vee \Box b$ can stand for the modal translation of an internal implication (one that obeys the deduction theorem).

Expressed in terms of epistemic truth-values, the truth of $(\Box a)' \vee \Box b$ corresponds to a three-valued implication $a \Rightarrow b$ that takes value $\mathbf{1}$ when $v(a) < \mathbf{1}$ or $v(b) = \mathbf{1}$, which describes condition "a is not certain or b is certain". Besides, if $\Box a$ is true, then $(\Box a)' \vee \Box b$ is equivalent to $\Box b$. It means that if $v(a) = \mathbf{1}$ then $v(a \Rightarrow b) = v(b)$. So the MEL setting generates the implication

$$v(a \to_N b) = \begin{cases} \mathbf{1} \text{ if } v(a) < \mathbf{1}; \\ v(b) \text{ otherwise.} \end{cases} \quad (10.2.5)$$

This is Nelson [19] implication $\to_N$ also introduced by Monteiro [18]. Three-valued Nelson logic is based on Kleene connectives plus Nelson implication $\to_N$, inducing an additional negation given by $-a = a \to_N \mathbf{0}$ (an internal negation in the sense of Avron). The logic based on Kleene connectives ($\sqcap, \sqcup$) and the internal connectives ($\to_N, -$) satisfies the axioms of classical logic, to which must be added specific axioms for $\neg$. A syntactic inference $B \vdash_N p$ uses these axioms and modus ponens. The semantic inference in Nelson's logic is the same as in Kleene logic, based on the preservation of the designated truth-value $\mathbf{1}$.

It is shown in [12] that Nelson logic can be captured in MEL, in the sense that all axioms in Nelson logic translate into MEL tautologies, where by translation, we mean $\mathcal{T}(v(p) = \mathbf{1})$ for a Nelson logic formula. The same results hold as for Kleene logic except that the attained fragment of the MEL language is $\mathcal{L}^{\ell}_{\Box}$ with modalities prefixing literals only (thus including negations $(\Box\ell)'$, contrary to the Kleene fragment). Moreover, we again have the completeness of the Nelson fragment in MEL with respect to the models of the form E_v. The translation of Nelson logic into MEL is such that if B is a set of propositions in Nelson logic, $B \vdash_N p$ if and only if $\mathcal{T}(B) \vdash \mathcal{T}(v(p) = \mathbf{1})$ in MEL.

Łukasiewicz three-valued implication can be captured as well inside MEL restricted to $\mathcal{L}^{\ell}_{\Box}$. Notice that Nelson implication is not contrapositive in the sense that $a \to_N b$ is not equivalent to $\neg b \to_N \neg a$. The MEL translation of the latter is $\Box b' \to \Box a'$, which clearly differs from the translation $\Box a \to \Box b$ of Nelson implication. However, considering the implication $(\Box a \to \Box b) \wedge (\Box b' \to \Box a')$ restores a form of contraposition.[3] In the three-valued logic, it corresponds to an implication $\to_L$ defined by $(a \to_N b) \sqcap (\neg b \to_N \neg a)$ that is provably equal to Łukasiewicz three-valued implication. This symmetrization process is shown by Avron [5] to be very general, starting from a general consequence relation with an internal negation, and a so-called combining conjunction (recalled in the previous subsection). Using MEL, the translation $(\Box a \to \Box b) \wedge (\Box b' \to \Box a')$ of Łukasiewicz implication immediately lays bare its symmetrized form [12].

Łukasiewicz three-valued logic $Ł_3$ is based on the connectives $\to_L$ and $\neg$ ($\neg a$ is identified with $a \to_L 0$), from which all connectives in Kleene and Nelson logic can be recovered. In fact, both Łukasiewicz and Nelson logics have the same expressive power. The proof-theoretic machinery of Łukasiewicz logic relies on Wajsberg axioms and modus ponens (it is not recalled here). The syntactic consequence $B \vdash_L p$ in $Ł_3$ can be captured in MEL, inside the sublanguage $\mathcal{L}^{\ell}_{\Box}$ again in the sense that, if B is a set of propositions in $Ł_3$, then $B \vdash_L p$ if and only if $\mathcal{T}(B) \vdash \mathcal{T}(v(p) = \mathbf{1})$ in MEL [12].

[3]Not to be confused with the MEL contrapositive form of $\Box a \to \Box b$, i.e., $(\Box b)' \to (\Box a)'$, of course equivalent to the former.

10.3 From Three-Valued Paraconsistent Logics to MEL

In this section, we proceed to the same translation as above for some three-valued paraconsistent logics that appear as mirror images of Kleene, Nelson, and Łukasiewicz logics (as already pointed out by Avron [5] in a different context). The only difference will be the choice of designated truth-values ($\frac{1}{2}$ and $\mathbf{1}$ instead of just $\mathbf{1}$), and the understanding of epistemic models of MEL formulas. First, we provide our version of the multisource information semantics that we shall use to interpret these three-valued logics.

10.3.1 Modal and Three-Valued Logics for Several Totally Informed Agents

Instead of one agent having incomplete information, consider the case of several logically sophisticated agents forming a set $\mathcal{A}$ sharing the same propositional language $\mathcal{L}$, each capable to decide whether any proposition $p \in \mathcal{L}$ is true or false, but possibly disagreeing. In other words each agent possesses complete Boolean knowledge, albeit not in full agreement with other agents. It comes down to assuming that agent i believes that the real world is $w_i \in \Omega$ (and can be assimilated to it). For each formula $p \in \mathcal{L}$, agent i can say whether it is true ($w_i(p) = T$) or false ($w_i(p) = F$). Agent i's knowledge is described by $E_i = \{w_i\}$. Let A be the set of interpretations $w_i, i \in \mathcal{A}$.

Let us show that the MEL logic can account for this situation as well. We are interested to express the following statement in the formal language of MEL: "at least one agent asserts that p is true." It is clear that this statement means $A \cap [p] \neq \emptyset$ that reads $\Diamond p$ in the MEL language.

Now, let us define the meaning of the modal symbols in agreement with the above remarks[4]

- $\Diamond p$ stands for "at least one agent asserts p."
- As usual, we define $\Box p := (\Diamond(p'))'$, which here means that "all agents assert p" (since $A \subseteq [p]$).

This kind of multisource setting was first introduced by Belnap [8] with agents having incomplete knowledge about atomic propositions. It leads to a four-valued truth-functional logic where the four values include *unknown* and *contradictory*. It also corresponds to the so-called *society semantics* [10].

Now, let us consider a three-valued logic. We denote again by $v(a)$ the (epistemic) truth-value of the variable a, $v(a) \in \{\mathbf{0}, \mathbf{1}, \frac{1}{2}\}$. However, here, these three values will be interpreted in the light of the joint assertions of n fully informed agents:

---

[4]The convention differs from the ones in the preliminary version of this paper [11] where we used $\Box p$ to stand for "at least one source asserts p." The latter convention leads to a logic where $\Box$ has the same properties as $\Diamond$ in a KD system, which may be misleading.

- $v(a) = \mathbf{1}$ means that all agents say that a is true;
- $v(a) = \mathbf{0}$ means that all agents say that a is false;
- $v(a) = \frac{1}{2}$ means that some agents say a is true, the other ones say a is false;
- $v(a) \geq \frac{1}{2}$ means that at least some agents say a is true.

Remember that those agents that do not say that a is true say it is false. None can claim ignorance in our setting. We can then encode the assignment of a subset of truth-values to a (Boolean) propositional variable a by means of the modalities $\Box$ and $\Diamond$. It is easy to see that the same translation rules (10.2.1)–(10.2.4), as described in the previous section, are valid. What changes is the interpretation of the epistemic truth-values, e.g., $\frac{1}{2}$ means that some sources assert a and the other ones a'. So this truth-value corresponds to some paraconsistent assertion of truth. Under this interpretive view, it is natural to consider $\frac{1}{2}$ and $\mathbf{1}$ as designated truth-values, contrary to the incomplete information setting. This is the usual assumption in three-valued paraconsistent logics. As a consequence, we see that asserting a in a paraconsistent logic corresponds to writing $\Diamond a$ in MEL.

10.3.2 Translation of Priest Logic of Paradox PLP

The logic PLP [20] attaches to the third truth-value the meaning of a paradox; it refers to sentences that are "both true and false". Priest's intention is to "isolate paradoxes and prevent them from contaminating everything else". A similar intuition of the third value is given by Asenjo [2] to deal with *antinomies*. However, in this paper, we just consider this logic in the light of the multisource semantics outlined above.

Formally, PLP is the same as Kleene logic, but for the designated truth-values and the ensuing semantic inference. It uses for connectives the minimum $\sqcap$, the maximum $\sqcup$, the involutive negation $\neg$, and the material implication $\neg a \sqcup b$ of a Kleene lattice. However, in his system both $\mathbf{1}$ and $\frac{1}{2}$ are designated truth-values. In fact, the notion of paraconsistent semantic inference is defined as follows:

Definition 10.3.1 If B is a set of propositions in a three-valued logic, then $B \models_{para} p$ if and only if whenever $v(q) \geq \frac{1}{2}, \forall q \in B$ then $v(p) \geq \frac{1}{2}$.

In other words, there does not exist an interpretation v such that $v(p) = \mathbf{0}$ and for all $q \in B$, $v(q) \in \{\mathbf{1}, \frac{1}{2}\}$. All Boolean tautologies are valid in Priest logic, that is, $v(p) \geq \frac{1}{2}$, for all LPL formulas having the form of a propositional tautology, but modus ponens does not apply. Of course, contradictions are accommodated in the sense that $p \sqcap \neg p \not\models_{para} q$.

If a knowledge base B in PLP contains an atom a, it means that $v(a) \geq \frac{1}{2}$, that is, we should write $\Diamond a$ in MEL. For a negated atom, $v(\neg a) \geq \frac{1}{2}$ if and only if $v(a) \leq \frac{1}{2}$ hence we should write $\Diamond a'$ in MEL. If B contains the conjunction of two atoms $a \sqcap b$, this is translated as

$$\mathcal{T}(v(a \sqcap b) \geq \tfrac{1}{2}) = \Diamond a \wedge \Diamond b$$

Table 10.1 Jaśkowski implication

$\to_J$	0	$\frac{1}{2}$	1
0	1	1	1
$\frac{1}{2}$	0	$\frac{1}{2}$	1
1	0	$\frac{1}{2}$	1

and similarly for the disjunction. A material implication statement $\neg a \sqcup b \in B$ thus translates into $\Diamond a' \vee \Diamond b$.

Let p be a PLP formula in conjunctive normal form (without simplifying the terms of the form $a \sqcap \neg a$). Now, since designated truth-values are $\mathbf{1}$ and $\frac{1}{2}$, its translation into MEL consists in the same classical conjunction of disjunctions where we put a modality $\Diamond$ in front of all literals [12]. Clearly, the fragment of MEL we can capture is just given by the conjunction and disjunction of literals preceded by a $\Diamond$, that is, $\mathcal{L}^P_\Diamond = \Diamond a | \Diamond a' | \psi \vee \phi | \psi \wedge \phi$.

Note that if we replace all literals l by $\Diamond l$ in a propositional tautology, it remains a tautology in MEL. As a result, we can explain via the translation why all Boolean tautologies are still valid in PLP. Moreover, it is clear that from $\Diamond a' \vee \Diamond b$ and $\Diamond a$ one cannot infer $\Diamond b$ in MEL. It explains why modus ponens does not hold any longer in PLP. Similarly, $a \sqcap \neg a \vDash_{para} b$ does not hold, and this is expressed by the fact that $\Diamond a \wedge \Diamond a'$ is not a contradiction in MEL.

The symmetry between Priest and Kleene logics is clear, since going from a Kleene logic base to a PLP base in MEL comes down to turning $\Box$ into $\Diamond$. We can then easily verify that the inference $B \vDash_{para} p$ can be expressed in the Priest fragment $\mathcal{L}^P_\Diamond$ of MEL (using models of the form E_v at the semantic level) in the sense that $\{\mathcal{T}(v(q) \geq \frac{1}{2}) : q \in B\} \vdash \mathcal{T}(v(p) \geq \frac{1}{2})$ in MEL if and only if $B \vDash_{para} p$ [12].

10.3.3 Translation of **A** *and* J_3 *Logics*

The logic **A** proposed by Asenjo and Tamburino in [3] has for connectives the ones of Kleene–Priest logics plus Jaśkowski implication[5] defined on Table 10.1.

In this logic, the designated values are $\mathbf{1}$ and $\frac{1}{2}$. So, the only difference with respect to PLP is the introduction of the new implication $\to_J$. The negation $\sim a := a \to_J \mathbf{0}$, such that $v(\sim a) = \mathbf{1}$ if and only if $v(a) = \mathbf{0}$, only takes values $\mathbf{0}, \mathbf{1}$ and is an "intuitionistic" one (in the sense that it violates the excluded middle law and obeys the contradiction law). We have the equality $p \to_J q = \sim p \sqcup q$. Clearly, in this logic modus ponens does hold. Note the analogy with the situation in Nelson's logic, in which one implication is added to Kleene connectives and all tautologies of classical logic hold in the fragment without $\neg$ (and where $a \to_N b = -a \sqcup b$).

[5]Note that Jaśkowski traces back this implication to Słupecki.

In [14] the authors introduce another logic J_3 with a view to address issues raised by Jaśkowski and recalled in the introduction. The primitive connectives are Kleene's conjunction, disjunction, and negation, plus the unary *strengthening operator* ∇ such that $\nabla \mathbf{0} = \mathbf{0}$, $\nabla \frac{1}{2} = \nabla \mathbf{1} = \mathbf{1}$. Among the derived operations, it is obvious that Jaśkowski implication is recovered as $a \rightarrow_J b := \neg \nabla a \vee b$, and we have $\sim a = \neg \nabla a$. Further, it is obvious that ∇ is also definable in $\mathbf{A}$ as $\nabla a = \neg(a \rightarrow_J \mathbf{0}) = \neg \sim a$. So $\mathbf{A}$ and J_3 have the same connectives and all the results proved for $\mathbf{A}$ also hold for J_3. Not all classical tautologies are valid in these logics. But if we consider the fragment of $\mathbf{A}$ without $\neg$ we have that all classical tautologies are still valid in $\mathbf{A}$ (this has been stated in [14] for the equivalent logic J_3).

Now we can consider the translation of any of these logics in MEL. We are basically interested in computing $\mathcal{T}(v(p) \geq \frac{1}{2})$ since inference from a knowledge base in any of these logics comes down to preserving at least the truth-value $\frac{1}{2}$. To the translation principles of PLP we must add the following definition (clear from the truth-table of $\rightarrow_J$):

$$\mathcal{T}(v(p \rightarrow_J q) \geq \tfrac{1}{2}) = \mathcal{T}(v(p) \geq \tfrac{1}{2}) \rightarrow \mathcal{T}(v(q) \geq \tfrac{1}{2}) = \mathcal{T}(v(p) \geq \tfrac{1}{2})' \vee \mathcal{T}(v(q) \geq \tfrac{1}{2})$$

If $p = a$, $q = b$ are atoms, we obtain $\Diamond a \rightarrow \Diamond b$ as the translation of Jaśkowski's implication. Unsurprizingly, we get modus ponens back, an inference rule that is not valid in Priest logic. The translation into MEL of the negation $\sim a$ is

$$\mathcal{T}(v(\sim a) \geq \tfrac{1}{2}) = \mathcal{T}(v(\sim a) = \mathbf{1}) = \mathcal{T}(v(a) = 0) = \Box a'.$$

Likewise $\mathcal{T}(v(\nabla a) \geq \frac{1}{2})$ translates into $\Diamond a$. So the fragment in MEL corresponding to the $\mathbf{A}$ logic is $\mathcal{L}^{\ell}_{\Diamond}$ that is equivalent to $\mathcal{L}^{\ell}_{\Box}$.

In fact, adding connective $\rightarrow_J$ corresponds to adding an internal implication to Priest logic, following the lines of Avron [5], that is an implication $\Rightarrow$ such that $v(p \Rightarrow q) \geq \frac{1}{2}$ if and only if $p \models_{para} q$. He shows that adding symmetry conditions to this implication enforces the truth-table of $\rightarrow_J$.

The internal implication is obtained for free in MEL where, by the deduction theorem, $\vdash \mathcal{T}(v(p) \geq \frac{1}{2}) \rightarrow \mathcal{T}(v(q) \geq \frac{1}{2})$, where $\rightarrow$ denotes a material implication, which is equivalent to $\mathcal{T}(v(p) \geq \frac{1}{2}) \vdash \mathcal{T}(v(q) \geq \frac{1}{2})$. The only price we pay is that we move out of the target language $\mathcal{L}^{P}_{\Diamond}$ of PLP. Namely, for atoms, the formula $\Diamond a \rightarrow \Diamond b \notin \mathcal{L}^{P}_{\Diamond}$ since it is $(\Diamond a)' \vee \Diamond b$.

So our approach is to consider that the formula $\Diamond a \rightarrow \Diamond b$ is the translation into MEL of $v(a \Rightarrow b) \geq \frac{1}{2}$ for an implication $\Rightarrow$ added to PLP. Besides, remember that $v(\neg p) \geq \frac{1}{2}$ if and only if $v(p) \leq \frac{1}{2}$, and in particular the translation of $v(\neg a) \geq \frac{1}{2}$ is $\Diamond a'$. So we must also assume that the translation of $v(a \Rightarrow b) \leq \frac{1}{2}$ is $\Diamond a \wedge \Diamond b'$, which expresses that a is paraconsistently true while b is paraconsistently false. On this basis, we can retrieve the truth-table of Jaśkowski's implication from the properties of classical logic inside MEL, as we did with Nelson.

Proposition 10.3.2 *Suppose that the three-valued implication $\Rightarrow$ added to PLP translates into MEL as follows:*

$\mathcal{T}(v(a \Rightarrow b) \geq \frac{1}{2}) = \Diamond a \to \Diamond b; \mathcal{T}(v(\neg(a \Rightarrow b)) \geq \frac{1}{2}) = \mathcal{T}(v(a \Rightarrow b) \leq \frac{1}{2}) = \Diamond a \wedge \Diamond b'$

Then implication $\Rightarrow$ is Jaśkowski's implication.

Proof The deduction theorem in MEL comes down to $v(a \Rightarrow b) \geq \frac{1}{2}$ if and only if $v(a) \geq \frac{1}{2}$ implies $v(b) \geq \frac{1}{2}$. The latter condition equivalently reads $v(a) = \mathbf{0}$ or $v(b) \geq \frac{1}{2}$. Negating it reads $v(b) = \mathbf{0}$ and $v(a) \geq \frac{1}{2}$, for which one must have $v(a \Rightarrow b) = \mathbf{0}$, and $v(a \Rightarrow b) \geq \frac{1}{2}$ otherwise. It yields part of the truth-table for $\Rightarrow$.

The other condition reads $v(a \Rightarrow b) \leq \frac{1}{2}$ if and only if $v(a) \geq \frac{1}{2}$ and $v(b) \leq \frac{1}{2}$. From this equivalence it clearly follows that $v(a \Rightarrow b) = \mathbf{1}$ if and only if $v(b) = \mathbf{1}$ or $v(a) = \mathbf{0}$. Moreover, if $v(b) = \frac{1}{2}$, and $v(a) \geq \frac{1}{2}$, the only possibility is that $v(a \Rightarrow b) = \frac{1}{2}$, as the deduction theorem in MEL enforces $v(a \Rightarrow b) \geq \frac{1}{2}$ in that case. □

Jaśkowski's implication also writes

$$v(a \to_J b) = \begin{cases} 1 \text{ if } v(a) = \mathbf{0}; \\ v(b) \text{ otherwise.} \end{cases} \quad (10.3.1)$$

which highlights its similarity with Nelson's implication (10.2.5).

Remark 10.3.3 In the logic **A** [3], atoms are divided in two groups both at the semantic and at the syntactic level: antinomic and non-antinomic. In the semantics, non-antinomic atoms can have value **0** or **1**, whereas antinomic atoms are those whose truth-value is $\frac{1}{2}$ and compound statements involving antinomic atoms can have any value. However, we cannot have such a thing as antinomic atoms in MEL as it relies on classical logic, where no provision is made for an intrinsic notion of antinomy. In this paper, the antinomy arises from sources claiming a and other ones its opposite ($a \sqcap \neg a$ is expressed by $\Diamond a \wedge \Diamond \neg a$); hence, it is an epistemic form of antinomy. The purpose of our translation process is limited to an underlying society semantics. It has nothing to do with mathematical paradoxes, nor statements that would be contradictory in all possible worlds. In particular, we cannot translate the logical constant $\frac{1}{2}$ into MEL, while precisely, antinomic atoms would be such that they ever take this truth-value. Namely, there is no Boolean atom a for which the formula $\Diamond a \wedge \Diamond a'$ would be a MEL tautology.

10.3.4 Translation of RM3 (Sobociński) Logic

The same connectives that define **A** also define the relevance logic $RM_3^{\to_J}$ [4] ($\to_J$ is denoted by $\supset$ in the original paper), which is also equivalent to RM3 [1, 9], that is, to Sobociński [4, 22] logic, through the following mutual definitions:

$$p \to_S q := (p \to_J q) \wedge (\neg q \to_J \neg p)$$
$$p \to_J q := q \vee (p \to_S q)$$

Thus, even if Sobociński's and RM3 logics were not originally meant to deal with paraconsistency, they can play a role in this framework: in [4], it is claimed that $RM_3^{\to_J}$ "might be considered an optimal paraconsistent logic". Interestingly, implication $\to_S$ is obtained as the symmetrization of Jaśkowski's implication, just like Łukasiewicz implication is the symmetrized form of Nelson's implication for logics of incomplete information. Here, we must again assume that **1** and $\frac{1}{2}$ are designated truth-values in order to get the paraconsistent behavior. Then RM3 has the same expressive power as the logic **A**.

As a consequence of all these equivalences, all above translation results also apply to RM3 and Sobociński logic. In particular, Sobociński implication whose truth-table is in Table 10.2 can be translated as

$$\mathcal{T}(v(p \to_S q) \geq \tfrac{1}{2}) = \mathcal{T}(v(p \to_J q) \geq \tfrac{1}{2}) \wedge \mathcal{T}(v(\neg q \to_J \neg p) \geq \tfrac{1}{2})$$

For atoms, it yields $(\Diamond a \to \Diamond b) \wedge (\Box a \to \Box b)$ in MEL, the same translation as Łukasiewicz implication with designated truth-value **1**. Note that $a \to_S \mathbf{0} = \sim a$. Sobociński conjunction and disjunction can be translated into MEL from Table 10.2 or by means of the definition $p \wedge_S q = \neg(p \to_S \neg q)$ and de Morgan properties. On atoms they read as

$$\mathcal{T}(v(a \wedge_S b) \geq \tfrac{1}{2}) = \Diamond a \wedge \Diamond b$$
$$\mathcal{T}(v(a \vee_S b) \geq \tfrac{1}{2}) = \Box a \vee \Box b \vee (\Diamond a \wedge \Diamond a' \wedge \Diamond b \wedge \Diamond b')$$

As far as the axiom system is concerned and due to the equivalence between RM3 and Sobociński logics, we just consider the axiom system of RM3 [9].

(R1) $p \to_S p$;
(R2) $(p \sqcap (p \to_S q)) \to_S q$;
(R3) $p \sqcap q \to_S p$;
(R4) $p \sqcap q \to_S q$;
(R5) $((p \to_S q) \sqcap (p \to_S r)) \to_S (p \to_S (q \sqcap r))$;
(R6) $(p \to_S \neg p) \to_S \neg p$;
(R7) $(p \to_S \neg q) \to_S (q \to_S \neg p)$;
(R8) $(\neg\neg p \to_S p)$;
(R9) $(\neg p \sqcap q) \to_S (p \to_S q)$;
(R10) $\neg p \to_S (p \sqcup (p \to_S q))$;

Inference rules are

Table 10.2 Sobociński implication, conjunction, and disjunction

$\to_S$	0	$\frac{1}{2}$	1
0	1	1	1
$\frac{1}{2}$	0	$\frac{1}{2}$	1
1	0	0	1

$\wedge_S$	0	$\frac{1}{2}$	1
0	0	0	0
$\frac{1}{2}$	0	$\frac{1}{2}$	1
1	0	1	1

$\vee_S$	0	$\frac{1}{2}$	1
0	0	0	1
$\frac{1}{2}$	0	$\frac{1}{2}$	1
1	1	1	1

1. p and $p \to_S q$ implies q;
2. p and q implies $p \sqcap q$; and
3. $p \to_S q$ and $r \to_S t$ implies $(q \to_S r) \to_S (p \to_S t)$.

Lemma 10.3.4 *If p is a formula in the RM3 logic, then $\mathcal{T}(v(p) \geq \frac{1}{2}) \vee \mathcal{T}(v(p) \leq \frac{1}{2})$ is a tautology in MEL.*

Proof The proof is by induction on the structure of p. □

- $p = a$. We have $\mathcal{T}(v(a) \geq \frac{1}{2}) \vee \mathcal{T}(v(a) \leq \frac{1}{2}) = \Diamond a \vee \Diamond a'$ that is an axiom of the logic.
- $\mathcal{T}(v(\neg p) \geq \frac{1}{2}) \vee \mathcal{T}(v(\neg p) \leq \frac{1}{2}) = \mathcal{T}(v(p) \leq \frac{1}{2}) \vee \mathcal{T}(v(p) \geq \frac{1}{2})$ and so, it is sufficient to use the induction.
- $p = p_1 \sqcap p_2$. Thus, $\mathcal{T}(v(p) \geq \frac{1}{2}) \vee \mathcal{T}(v(p) \leq \frac{1}{2}) =$ $[\mathcal{T}(v(p_1) \geq \frac{1}{2}) \wedge \mathcal{T}(v(p_2) \geq \frac{1}{2})] \vee [\mathcal{T}(v(p_1) \leq \frac{1}{2}) \wedge \mathcal{T}(v(p_2) \leq \frac{1}{2})]$.
  Now, it is sufficient to apply the distributivity and the induction. For the disjunction $\sqcup$ the proof is similar.
- $p = p_1 \to_S p_2$. So, $\mathcal{T}(v(p_1 \to_S p_2) \geq \frac{1}{2}) \vee \mathcal{T}(v(p_1 \to_S p_2) \leq \frac{1}{2})$ is translated as $\{[\mathcal{T}(v(p_1) \geq \frac{1}{2}) \to \mathcal{T}(v(p_2) \geq \frac{1}{2})] \wedge [\mathcal{T}(v(p_1) = \mathbf{1}) \to \mathcal{T}(v(p_2) = \mathbf{1}]\} \vee$ $[\mathcal{T}(v(p) \geq \frac{1}{2}) \to \mathcal{T}(v(q) = \mathbf{1})]'$. Then, by distributivity, we get
  $\psi \vee [\mathcal{T}(v(p) \geq \frac{1}{2}) \to \mathcal{T}(v(q) = \mathbf{1})] \vee [\mathcal{T}(v(p) \geq \frac{1}{2}) \to \mathcal{T}(v(q) = \mathbf{1})]'$
  (for the sake of simplicity we do not write the complete development of ψ) which is a Boolean tautology.

Proposition 10.3.5 *If A is an axiom in the RM3 logic, then $\mathcal{T}(v(A) \geq \frac{1}{2})$ is a tautology in MEL.*

Proof Axioms (R1), (R3), (R4), and (R8) are easily proved.
Axiom (R2). It is the conjunction of two tautologies. The first one follows from the fact that (R2) holds in Boolean logic and the second one by Lemma 10.3.4.
Axioms (R5) and (R7) are the conjunctions of two tautologies that easily follows by the fact that (R5) is a Boolean theorem.
Axiom (R6) is the conjunction of two tautologies. The first one is just Lemma 10.3.4 and the second one the Boolean version of axiom (R6).
Axiom (R9) is the conjunction of two tautologies. The second one is just axiom R9 in Boolean logic. The first half is $(\mathcal{T}(v(\neg p) \geq \frac{1}{2}) \wedge \mathcal{T}(v(q) \geq \frac{1}{2})) \to [(\mathcal{T}(v(p) \geq \frac{1}{2}) \to \mathcal{T}(v(q) \geq \frac{1}{2})) \wedge (\mathcal{T}(v(p) = \mathbf{1}) \to \mathcal{T}(v(q) = \mathbf{1}))]$,
that is, $\mathcal{T}(v(p) = \mathbf{1}) \vee (\mathcal{T}(v(q) \geq \frac{1}{2}))' \vee [(\mathcal{T}(v(p) \geq \frac{1}{2})' \vee \mathcal{T}(v(q) \geq \frac{1}{2})) \wedge (\mathcal{T}(v(p) = \mathbf{1})' \vee \mathcal{T}(v(q) = \mathbf{1})]$.

Then by distributivity, we can simplify as
$\mathcal{T}(v(p) = \mathbf{1}) \vee (\mathcal{T}(v(q) \geq \frac{1}{2}))' \vee (\mathcal{T}(v(p) = \mathbf{1})' \vee \mathcal{T}(v(q) = \mathbf{1}))$,
which is a Boolean tautology.

Table 10.3 Sette logic connectives

$\to_{Se}$	0	$\frac{1}{2}$	1
0	1	1	1
$\frac{1}{2}$	0	1	1
1	0	1	1

x	− x
0	1
$\frac{1}{2}$	1
1	0

$\wedge_{Se}$	0	$\frac{1}{2}$	1
0	0	0	0
$\frac{1}{2}$	0	1	1
1	0	1	1

$\vee_{Se}$	0	$\frac{1}{2}$	1
0	0	1	1
$\frac{1}{2}$	1	1	1
1	1	1	1

Axiom (10) is translated into the following two tautologies:
$\mathcal{T}(v(p) = \mathbf{1}) \vee \mathcal{T}(v(p) \geq \frac{1}{2}) \vee [((\mathcal{T}(v(p) \geq \frac{1}{2})' \vee \mathcal{T}(v(q \geq \frac{1}{2})) \wedge (\mathcal{T}(v(p) = 1)'$
$\vee \mathcal{T}(v(q) = \mathbf{1}))]$
and $\mathcal{T}(v(p) \geq \frac{1}{2}) \vee \mathcal{T}(v(p) = \mathbf{1}) \vee (\mathcal{T}(v(p) \geq \frac{1}{2})' \vee \mathcal{T}(v(q) = \mathbf{1})$. □

Inference rules of RM3 can be checked to hold in their translation into MEL. So, again we can simulate RM3 in MEL.

10.4 On Sette Logic

Sette [21] independently introduced a paraconsistent logic with the aim to deal with "inconsistent (but not absolutely inconsistent) formal systems." Its basic connectives are one implication $\to_{Se}$ and the already met "paraconsistent"[6] negation − in Table 10.3. In this logic, the intuitionistic negation is definable as $\sim a := a \to_{Se} \mathbf{0}$.

The peculiarity of this logic is that all these connectives yield Boolean results. Special Boolean conjunction and disjunction can be expressed from implication and negation as follows (they are not min and max):

$$x \wedge_{Se} y := (((x \to_{Se} x) \to_{Se} x) \to_{Se} -((y \to_{Se} y) \to_{Se} y)) \to_{Se} -(x \to_{Se} -y)$$
$$x \vee_{Se} y := (x \to_{Se} --x) \to_{Se} (-x \to_{Se} y)$$

Note that the table of $\to_{Se}$ results from Jaśkowski's implication when turning $\frac{1}{2}$ into **1**. Sette conjunction and disjunction tables are obtained likewise from Sobociński conjunction and disjunction.

An axiom system together with modus ponens is given:

(S1) $p \to_{Se} (q \to_{Se} p)$
(S2) $(p \to_{Se} (q \to_{Se} \gamma)) \to_{Se} ((p \to_{Se} q) \to_{Se} (p \to_{Se} \gamma))$
(S3) $(-p \to_{Se} -q) \to_{Se} ((-p \to_{Se} --q) \to_{Se} p)$
(S4) $-(p \to_{Se} --p) \to_{Se} p$
(S5) $(p \to_{Se} q) \to_{Se} --(p \to_{Se} q)$

This logic (named P_1) is complete and (to quote [21]) "cannot be strengthened (i.e., there is no propositional calculus between P_1 and P_0, where P_0 is the classical

[6]So-called, as it violates the contradiction law and satisfies the excluded middle law.

propositional calculus)". In other words, if we add to P_1 any tautology which holds in P_0 but not in P_1 we get P_0.

The meaning of the third value in Sette logic is not discussed, but its connectives (see Table 10.3), except for negation, suggest that no difference is made between **1** and $\frac{1}{2}$. The translation of Sette connectives into MEL is thus the same in the two cases $\geq \frac{1}{2}$ and $= \mathbf{1}$:

$$\mathcal{T}(v(p \to_{Se} q) = \mathbf{1}) = \mathcal{T}(v(p) \geq \tfrac{1}{2}) \to \mathcal{T}(v(q) \geq \tfrac{1}{2}) = \mathcal{T}(v(p \to_J q) \geq \tfrac{1}{2})$$
$$\mathcal{T}(v(p \wedge_{Se} q) = \mathbf{1}) = \mathcal{T}(v(p) \geq \tfrac{1}{2}) \wedge \mathcal{T}(v(q) \geq \tfrac{1}{2}) = \mathcal{T}(v(p \sqcap q) \geq \tfrac{1}{2})$$
$$\mathcal{T}(v(p \vee_{Se} q) = \mathbf{1}) = \mathcal{T}(v(p) \geq \tfrac{1}{2}) \vee \mathcal{T}(v(q) \geq \tfrac{1}{2}) = \mathcal{T}(v(p \sqcup q) \geq \tfrac{1}{2})$$

For atoms we have $\mathcal{T}(v(-a) = \mathbf{1}) = \Diamond a'$ and $\mathcal{T}(v(a \to_{Se} b) = \mathbf{1}) = \Diamond a \to \Diamond b$; $\mathcal{T}(v(a \wedge_{Se} b) = \mathbf{1}) = \Diamond a \wedge \Diamond b$; $\mathcal{T}(v(a \vee_{Se} b) = \mathbf{1}) = \Diamond a \vee \Diamond b$. Note that this is the same translation as conjunctions and disjunctions of Priest logic (the min and the max) and of Jaśkowski implication.

Modus ponens also holds in the translation. Indeed, since we consider both **1** and $\frac{1}{2}$ as designated values, it corresponds to "from $\mathcal{T}(v(p) \geq \frac{1}{2})$ and $\mathcal{T}(v(p) \geq \frac{1}{2}) \to \mathcal{T}(v(q) \geq \frac{1}{2})$ it follows $\mathcal{T}(v(q) \geq \frac{1}{2})$," which clearly holds in MEL.

Proposition 10.4.1 *If A is an axiom in Sette logic, then* $\mathcal{T}(v(A) \geq \frac{1}{2}) = \mathcal{T}(v(A) = \mathbf{1})$ *is a tautology in MEL.*

Proof Lemma 10.3.4 applies to Sette Logic. Axioms (S1) and (S2) are Boolean axioms, thus they easily follow. For axiom (S3), we use recursive translation rules and come down to a formula of the form $\mathcal{T}(v(p) \leq \frac{1}{2}) \vee \mathcal{T}(v(p) \geq \frac{1}{2})$, which is a tautology by Lemma 10.3.4. Axioms (S4, S5) are proved in a similar manner. □

The fragment of the MEL language capturing Sette logic is thus again $\mathcal{L}^{\ell}_{\Diamond}$, the same as for J_3 and Sobociński's logics.

Sette logic connectives are definable in Łukasiewicz three-valued logic as follows [13]: $p \wedge_{Se} q := \neg(p \wedge q) \to_L (p \wedge q)$ and $p \to_{Se} q := J_0(q) \to_L J_0(p)$, where $J_0(p) := \neg p \wedge_{Se} \neg(p \wedge_{Se} \neg p)$ and $\wedge$ denotes the minimum; J_0 returns 1 if and only if $v(p) = \mathbf{0}$. The converse does not hold since with Sette connectives is not possible to obtain the value $\frac{1}{2}$(nor can the latter be translated into MEL anyway).

Likewise, we can define Sette logic in Sobocinski's as the former implication is of the form

$$x \to_{Se} y = \sim y \to_S \sim x \qquad -x = \sim\sim x \vee_S x.$$

The other way round is again not possible for the same reason as above. So why would we obtain the same MEL fragment from both Sette and Sobocinski's logic? One answer is that we do not translate into MEL the full truth-tables of these connectives. For instance, Sette and Jaskowski's implications translate the same since $v(a \to_J b) \geq \frac{1}{2}$ if and only if $v(a \to_{Se} b) = \mathbf{1}$.

So, even if the three three-valued logics J_3, $RM3$, and Sette are not equivalent, they share the same paraconsistent behavior: they can be expressed in the same

MEL fragment where the only language restriction is the presence of modalities only in front of literals, and conjunction, disjunction, and implication have the same translation.

10.5 Conclusion

This paper suggests that the modal logic MEL, originally the logic of incomplete information in the sense of possibility theory, is also a natural framework for reasoning under contradictory information. Not only many three-valued logics of incomplete information extending Kleene logic but also several existing three-valued paraconsistent logics, when the third truth-value means both true and false, can be translated into MEL. The restriction to the fragment of MEL putting modalities only in front of literals is the price paid by the truth-functionality of all these logics.

In contrast with our approach based on several completely informed agents, Carnielli and Lima-Marques (see also Dubois [15]) indicate that only two incompletely informed agents are needed to render Belnap's four truth-values without any loss of generality. In the multiple-agent semantics of paraconsistent logics discussed here, on the basis of the translation of three-valued calculi into MEL, we put no restriction on the number of completely informed agents. So one open question is whether we can also assume there are only two completely informed agents. It comes down to restricting MEL epistemic models of formulas in $\mathcal{L}_{\Box}^{\ell}$ to subsets containing at most two interpretations. Whether it is an alternative semantics for MEL restricted to this sublanguage is an interesting question to be investigated.

Note that Belnap [8] epistemic truth-values *unknown* and *contradictory* play symmetric roles in the bilattice structure. Likewise here, the two streams of three-valued logics studied here (Kleene-Nelson-Łukasiewicz vs. PLP-J3-RM3) form pairs of logics (one for paraconsistency, one for incomplete information: PLP-Kleene, J3-Nelson, RM3-Łukasiewicz), whose translations into MEL are in one-to-one correspondence, exchanging $\Diamond$ and $\Box$. Clearly, we can deal with both flaws in information, namely, incomplete and contradictory knowledge, together. In this case, we need four values, in the style of Dunn–Belnap logic, and another interesting prospect is to formally show that the MEL-like fragment of a classical modal logic (EMN) introduced in [15] can play for Belnap logic the same role as MEL plays for Kleene and for Priest logics.

Acknowledgments The authors wish to thank Philippe Besnard for remarks on the first draft that led us to improve the presentation.

References

1. Anderson, A.R., Belnap, N.D.: Entailment. The Logic of Relevance and Necessity. Princeton University Press, Princeton (1975)
2. Asenjo, F.G.: A calculus of antinomies. Notre Dame J. Form. Log. **7**, 103–105 (1966)
3. Asenjo, F.G., Tamburino, J.: Logic of antinomies. Notre Dame J. Form. Log. **16**, 17–44 (1975)
4. Avron, A.: On an implication connective of RM. Notre Dame J. Form. Log. **27**, 201–209 (1986)
5. Avron, A.: Natural 3-valued logics—characterization and proof theory. J. Symb. Log. **56**(1), 276–294 (1991)
6. Banerjee, M., Dubois, D.: A simple modal logic for reasoning about revealed beliefs. In: Sossai, C., Chemello, G. (eds.) Proceedings ECSQARU 2009, Verona, Italy, LNAI 5590. Springer, pp. 805–816 (2009)
7. Banerjee, M., Dubois, D.: A simple logic for reasoning about incomplete knowledge. Int. J. Approx. Reason. **55**, 639–653 (2014)
8. Belnap, N.D.: A useful four-valued logic. In: Dunn, J.M., Epstein, G. (eds.) Modern Uses of Multiple-Valued Logic. D. Reidel, pp. 8–37 (1977)
9. Brady, R.: Completeness proofs for the systems RM3 and BN4. Logique et Analyse **25**, 51–61 (1982)
10. Carnielli, W., Lima-Marques, M.: Society semantics for multiple-valued logics. In: Advances in Contemporary Logic and Computer Science, Contemporary Mathematics, vol. 235, pp. 33–52. American Mathematical Society (1999)
11. Ciucci, D., Dubois, D.: From paraconsistent three-valued logics to multiple-source epistemic logic. In: Proceedings 8th Conference of the European Society for Fuzzy Logic and Technology, EUSFLAT-13. Atlantis Press (2013)
12. Ciucci, D., Dubois, D.: A modal theorem-preserving translation of a class of three-valued logics of incomplete information. J. Appl. Non-Class. Log. **23**(4), 321–352 (2013)
13. da Costa, N.C.A., Alves, E.H.: Relations between paraconsistent logics and many-valued logic. Bull. Sec. Logic Polish Acad. Sci. **10**, 185–191 (1981)
14. D'Ottaviano, I.M.L., da Costa, N.C.A.: Sur un problème de Jaśkowski. Comptes Rendus de l'Académie des Science **270**, 1349–1353 (1970)
15. Dubois, D.: Reasoning about ignorance and contradiction: many-valued logics versus epistemic logic. Soft Comput. **16**(11), 1817–1831 (2012)
16. Dubois, D., Prade, H.: Possibility theory: qualitative and quantitative aspects. In: Gabbay,D.M., Smets, Ph. (eds.) Quantified Representation of Uncertainty and Imprecision. Handbook of Defeasible Reasoning and Uncertainty Management Systems, vol. 1, pp. 169–226. Kluwer Academic Publications (1998)
17. Jaśkowski, S.: Propositional calculus for contradictory deductive systems. Stud. Logica **24**, 143–160 (1969)
18. Monteiro, D.: Construction des algèbres de łukasiewicz trivalentes dans les algèbres de Boole monadiques. Math. Japon. **12**, 1–23 (1967)
19. Nelson, D.: Constructible falsity. J. Symb. Logic **14**, 16–26 (1949)
20. Priest, G.: The Logic of Paradox. J. Philos. Logic **8**, 219–241 (1979)
21. Sette, A.M.: On propositional calculus P_1. Math. Japon. **16**, 173–180 (1973)
22. Sobociński, B.: Axiomatization of a partial system of three-value calculus of propositions. J. Comput. Syst. **1**, 23–55 (1952)
23. Tanaka, K., Berto, F., Mares, E., Paoli, F. (eds.): Paraconsistency: Logic and Applications. Springer (2013)

Chapter 11
Modality, Potentiality, and Contradiction in Quantum Mechanics

Christian de Ronde

Abstract In da Costa and de Ronde (Found Phys 43:845–858, 2013), Newton da Costa together with the author of this paper argued in favor of the possibility to consider quantum superpositions in terms of a paraconsistent approach. We claimed that, even though most interpretations of Quantum Mechanics (QM) attempt to escape contradictions, there are many hints that indicate it could be worth while to engage in a research of this kind. Recently, Arenhart and Krause (New dimensions of the square of opposition, Philosophia Verlag, Munich, 2014; Logique et Analyse, 2014; The Road to Universal Logic (volume II), Springer, 2014) have raised several arguments against this approach and claimed that—taking into account the square of opposition—quantum superpositions are better understood in terms of *contrariety propositions* rather than *contradictory propositions*. In de Ronde (Los Alamos 2014) we defended the Paraconsistent Approach to Quantum Superpositions (PAQS) and provided arguments in favor of its development. In the present paper we attempt to analyze the meaning of *modality*, *potentiality*, and *contradiction* in QM, and provide further arguments of why the PAQS is better suited, than the Contrariety Approach to Quantum Superpositions (CAQS) proposed by Arenhart and Krause, to face the interpretational questions that quantum technology is forcing us to consider.

Keywords Modality · Potentiality · Contradiction · Quantum mechanics

Mathematics Subject Classification (2000) Primary 03G12 · Secondary 03A10

This work was partially supported by the following grants: FWO project G.0405.08 and FWO-research community W0.030.06. CONICET RES. 4541-12 (2013–2014).

C. de Ronde (✉)
Philosophy Institute "Dr. A. Korn",
Buenos Aires University, CONICET, Buenos Aires, Argentina
e-mail: cderonde@vub.ac.be

C. de Ronde
Center Leo Apostel and Foundations of the Exact Sciences,
Brussels Free University, Brussels, Belgium

J.-Y. Beziau et al. (eds.), *New Directions in Paraconsistent Logic*,
Springer Proceedings in Mathematics & Statistics 152,
DOI 10.1007/978-81-322-2719-9_11

11.1 Introduction

In [11], Newton da Costa together with the author of this paper argued in favor of the possibility to consider quantum superpositions in terms of a paraconsistent approach. We claimed that, even though most interpretations of QM attempt to escape contradictions, there are many hints that indicate it could be worth while to engage in a research of this kind. Recently, Arenhart and Krause [1, 2] have raised several arguments against this approach. More specifically, taking into account the square of opposition, they have argued that quantum superpositions are better understood in terms of *contrariety propositions* rather than in terms of *contradictory propositions*. In [17] we defended the PAQS and provided arguments in favor of its development. We showed that: *(i) Arenhart and Krause placed their obstacles from a specific metaphysical stance, which we characterized in terms of what we call the Orthodox Line of Research (OLR).* And also, *(ii) This is not necessarily the only possible line, and that a different one, namely, a Constructive Metaphysical Line of Research (CMLR) provides a different perspective in which PAQS can be regarded as a valuable prospect.* Furthermore, we explained how, within the CMLR, the problems and obstacles raised by Arenhart and Krause disappear. More specifically, we argued that the OLR is based on two main principles:

1. **Quantum to Classical Limit**: The principle that one must find a continuous bridge between classical mechanics and QM, i.e., that the main notions of classical physics must be used in order to explain quantum theory.
2. **Classical Physical Representation**: The principle that one needs to presuppose the classical physical representation—structured through the metaphysics of entities together with the mode of being of actuality—in any interpretation of QM.

In this context, regarding quantum superpositions, the Measurement Problem (MP) is one of the main questions imposed by the OLR. Given the fact that QM describes mathematically the state in terms of a superposition, the question is why do we observe a single result instead of a superposition of them? Although the MP accepts the fact that there is something very weird about quantum superpositions, leaving aside their problematic meaning, it focuses on the justification of the actualization process.

Taking distance from the OLR, the CMLR is based on three main presuppositions already put forward in [14, pp. 56–57].

1. **Closed Representational Stance**: Each physical theory is closed under its own formal and conceptual structure and provides access to a specific set of phenomena. The theory also provides the constraints to consider, explain, and understand physical phenomenon. The understanding of a phenomena is always *local* for it refers to the closed structure given by the physical theory from which observations are determined.
2. **Formalism and Empirical Adequacy**: The formalism of QM is able to provide (outstanding) empirically adequate results. Empirical adequacy determines the success of a theory and not its commitment to a certain presupposed conception

of the world. Thus, it seems to us that the problem is not to find a new formalism. On the contrary, the 'road signs' point in the direction that *we must stay close to the orthodox quantum formalism.*

3. **Constructive Stance**: To learn about what the formalism of QM is telling us about reality we might be in need of *creating new (non-classical) physical concepts.*

Changing the metaphysical standpoint, the CMLR presents a different questioning which assumes right from the start the need of bringing into stage a different metaphysical scheme to the one assumed by the OLR—based on the representation provided by classical physics. What is needed, according to this stance, is a radical inversion of orthodoxy and its problems. Regarding quantum superpositions, instead of considering the MP we should invert the questioning—changing the perspective—and concentrate in the analysis of what we have called: 'the superposition problem (SP)'.

In a more recent paper [3], Arenhart and Krause have continued their analysis arguing against the notion of potentiality and power presented in [16] and discussed in [17]. In the present paper we attempt to analyze the notions of *modality*, *potentiality*, *power*, and *contradiction* in QM, and provide further arguments of why the PAQS is better suited, than the Contrariety Approach to Quantum Superpositions CAQS proposed by Arenhart and Krause, to face the interpretational questions that quantum technology is forcing us to consider. The paper is organized as follows. In Sect. 11.2, we discuss the physical representation of quantum superpositions. Section 11.3 analyzes the meaning of modality in QM and puts forward our interpretation in terms of 'ontological potentiality'. In Sect. 11.4, we discuss the meaning of the notion of 'power' as a real physical existent. In Sect. 11.5, we analyze two different approaches to quantum superpositions, the PAQS and the CAQS. In Sect. 11.6, we provide the conclusions of the paper.

11.2 The Physical Representation of Quantum Superpositions

In [17] we made it clear why we are interested—through the CMLR—in attacking the SP, which attempts to develop a physical representation of quantum superpositions, instead of discussing the famous MP which—following the OLR—attempts to justify the actual non-contradictory realm of existence. The idea that quantum superpositions cannot be physically represented was stated already in 1930 by Paul Dirac in the first edition of his famous book: *The Principles of Quantum Mechanics.*

> The nature of the relationships which the superposition principle requires to exist between the states of any system is of a kind that cannot be explained in terms of familiar physical concepts. One cannot in the classical sense picture a system being partly in each of two states and see the equivalence of this to the system being completely in some other state. There is an entirely new idea involved, to which one must get accustomed and in terms of which *one must proceed to build up an exact mathematical theory, without having any detailed classical picture.* [24, p. 12] (emphasis added)

Also Niels Bohr was eager to defend the classical physical representation of our world and set the limits of such representation in classical physics itself [8]. Bohr would set the problems of the present OLR by claiming explicitly that: [41, p. 7] "[...] the unambiguous interpretation of any measurement must be essentially framed in terms of classical physical theories, and we may say that in this sense the language of Newton and Maxwell will remain the language of physicists for all time." At the same time he closed any further conceptual development by arguing that "it would be a misconception to believe that the difficulties of the atomic theory may be evaded by eventually replacing the concepts of classical physics by new conceptual forms." Even Erwin Schrödinger, who was one of the first to see the implications of **the superposition principle** exposed through his famous 'cat experiment', did not dare to think beyond the representation of classical physics [37].

Unfortunately, these ideas have sedimented in the present foundational research regarding QM. Indeed, the strategy of the OLR has been to presuppose the classical representation provided by classical Newtonian mechanics in terms of an *Actual State of Affairs (ASA)*. There are two main problems which block such a classical type representation. The first problem is the so-called 'basis problem' which attempts to explain how nature 'chooses' a single basis—between the many possible ones—when an experimental arrangement is determined in the laboratory—this also relates to the problem of contextuality which we have analyzed in detail in [18]. The second problem is the already mentioned MP: given the fact that QM describes mathematically the state in terms of a superposition of states, the question is why do we observe a single result instead of a superposition of them? It should be clear that the MP presupposes the determination of a basis and is not related to contextuality nor the BP. Although the MP accepts the fact that there is something very weird about quantum superpositions, leaving aside their problematic physical meaning, it focuses on the justification of the actualization process. Taking the single outcome as a standpoint it asks: how do we get to a single measurement result from the quantum superposition?[1] The MP attempts to justify why, regardless of QM, we only observe actuality. The problem places the result in the origin, and what needs to be justified is the already known answer.

Our interest, contrary to the OLR, focuses instead on what we have called the SP. According to it, in case one attempts to provide a realist account of QM, one should concentrate in finding a set of physical concepts which provide a physical representation of quantum superpositions. But in order to do so we need to go beyond the question regarding measurement outcomes. Before we can understand actualization we first need to explain what a quantum superposition *is* or *represents*. As we have argued elsewhere [15], there is no self evident path between the superposition and its outcome for there are multiple ways of understanding the *projection postulate* (see for a discussion [20]).

Our research is focused on the idea that quantum superpositions relate to something physically real that exists in nature, and that in order to understand QM we

---

[1] The questioning is completely analogous to the one posed by the quantum to classical limit problem: how do we get from contextual weird QM into our safe classical physical description of the world?

need to provide a physical representation of such existence (see for a more detailed discussion [18]). But why do we think we have good reasons to believe that quantum superpositions exist? Mainly because quantum superpositions are one of the main sources used by present experimental physicists to develop the most outstanding technical developments and experiments of the last decades. Indeed, there are many characteristics and behaviors we have learnt about superpositions: we know about *their existence regardless of the effectuation of one of its terms*, as shown, for example, by the interference of different possibilities in *welcher-weg* type experiments [10, 33], *their reference to contradictory properties*, as in Schrödinger cat states [34], we also know about *their nonstandard route to actuality*, as explicitly shown by the MKS theorem [20, 25], and we even know about *their nonclassical interference with themselves and with other superpositions*, used today within the latest technical developments in quantum information processing [5]. In spite of the fact we still cannot say what a quantum superposition *is* or *represents*, we must admit that they seem ontologically robust. If the terms within a quantum superposition are considered as quantum possibilities (of being actualized) then we must also admit that such quantum possibilities *interact*—according to the Schrödinger equation. It is also well known that one can produce interactions between multiple superpositions (entanglement) and then calculate the behavior of all terms as well as the ratio of all possible outcomes. It then becomes difficult not to believe that these terms that 'interact', 'evolve', and 'can be predicted' according to the theory, are not real (in some way).

Disregarding these known facts, most interpretations of QM do not consider quantum superpositions as related to physical reality. For example, the so-called Copenhagen interpretation remains agnostic with respect to the mode of existence of properties *prior* to measurement. The same interpretation is endorsed by van Fraassen in his Copenhagen modal variant.[2] Much more extreme is the instrumentalist perspective put forward by Fuchs and Peres [30, p. 1] who claim that: "[…] quantum theory does not describe physical reality. What it does is provide an algorithm for computing probabilities for the macroscopic events ('detector clicks') that are the consequences of experimental interventions." In Dieks' realistic modal version quantum superpositions are not considered as physical existents, only one of them is real (actual), namely, the one written as a single term, while all other superpositions of more than one term are considered as possible (in the classical sense). It seems then difficult to explain, through this interpretation, what is happening to the rest of non-actual terms which can be also predicted—even though not with certainty. In a similar vein, the consistent histories interpretation developed by Griffiths and Omnès also argues

[2]According to Van Fraassen [39, p. 280]: "The interpretational question facing us is exactly: in general, which value attributions are true? The response to this question can be very conservative or very liberal. Both court later puzzles. I take it that the Copenhagen interpretation—really, a roughly correlated set of attitudes expressed by members of the Copenhagen school, and not a precise interpretation—introduced great conservatism in this respect. Copenhagen scientists appeared to doubt or deny that observables even have values, unless their state forces to say so. I shall accordingly refer to the following very cautious answer as the *Copenhagen variant* of the modal interpretation. It is the variant I prefer."

against quantum superpositions [31].[3] Bohmian mechanics proposes the change the formalism and talk instead of a quantum field that governs the evolution of particles.

One might also argue that some interpretations, although not explicitly, leave space to consider superpositions as existent in a potential, propensity, dispositional, or latent realm. The Jauch and Piron School, Popper or Margenau's interpretations, are good examples of such proposals (see for discussion [14] and references therein). However, within such interpretations the collapse is accepted and potentialities, propensities, or dispositions are only defined in terms of 'their becoming actual'—mainly because, forced by the OLR, they have been only focused in providing an answer to the MP. In any case, such realms are not articulated nor physically represented beyond their meaning in terms of the actual realm. Only the many worlds' interpretation goes as far as claiming that all terms in the superposition are real in actuality. However, the quite expensive metaphysical price to pay is to argue that there is a multiplicity of unobservable worlds (branches) in which each one of the terms is actual. Thus, the superposition expresses the multiplicity of such classical actual worlds.

Instead of taking one of these two paths which force us either into the abandonment of representation and physical reality or to the exclusive account of physical representation in terms of an ASA, we have proposed through the CMLR to develop a new path which concentrates in developing radically new (non-classical) concepts.

11.3 Modality and Ontological Potentiality in Quantum Mechanics

QM has been related to modality since its origin, when Max Born interpreted Schrödinger's quantum wave function, Ψ, as a 'probability wave'. However, it was very clear from the very begining that the meaning of modality and probability in the context of QM was something completely new. As remarked by Heisenberg himself:

> [The] concept of the probability wave [in quantum mechanics] was something entirely new in theoretical physics since Newton. Probability in mathematics or in statistical mechanics means a statement about our degree of knowledge of the actual situation. In throwing dice we do not know the fine details of the motion of our hands which determine the fall of the dice and therefore we say that the probability for throwing a special number is just one in six. The probability wave function, however, meant more than that; it meant a tendency for something. [32, p. 42]

Today, it is well known that quantum probability does not allow an interpretation in terms of ignorance [36]—even though many papers in the literature still use probability uncritically in this way. Instead, as we mentioned above, the quantum formalism seems to imply some kind of weird *interaction of possibilities* governed by the Schrödinger equation.

[3]For a detailed analysis of the arguments provided by Dieks and Griffiths see: [18].

So they say, we do not understand QM and trying to do so almost makes no sense since it is too difficult problem to be solved. If Einstein, Bohr, Heisenberg, Schrödinger, and many of the most brilliant minds in the last century could not find an answer to this problem, maybe it is better to leave it aside. In line with these ideas, the problems put forward by the OLR have left behind the development of a new physical representation of QM and have instead concentrated their efforts in justifying our classical world of entities in the actual mode of existence. Only when leaving behind the OLR, one might be able to consider the possibility to provide a new non-classical physical representation of QM. Of course this implies reconsidering the meaning of existence itself and the abandonment of another presupposed dogma: existence and reality are represented by actuality either as an observation *hic et nunc* (empiricism) or as a mode of existence (classical realism).

Following the CMLR, we believe that a reasonable strategy would be to begin with what we know works perfectly well, namely, the orthodox formalism of QM and advance in the metaphysical principles which constitute our understanding of the theory. Escaping the metaphysics of actuality and starting from the formalism, a good candidate to develop a mode of existence is of course *quantum possibility*. In several papers, together with Domenech and Freytes, we have analyzed how to understand possibility in the context of the orthodox formalism of QM [25–28]. From this investigation there are several conclusions which can be drawn. We started our analysis with a question regarding the contextual aspect of possibility. As it is well known, the Kochen–Specker (KS) theorem does not talk about probabilities, but rather about the constraints of the formalism to actual definite-valued properties considered from multiple contexts (see for an extensive discussion regarding the meaning of contextuality [18]). What we found via the analysis of possible families of valuations is that a theorem which we called—for obvious reasons—the Modal KS (MKS) theorem can be derived which proves that quantum possibility, contrary to classical possibility, is also contextually constrained [25]. This means that regardless of its use in the literature, *quantum possibility* is not *classical possibility*. In a recent paper, [20] we have concentrated in the analysis of *actualization* within the orthodox frame and interpreted, following the structure, the logical realm of possibility in terms of potentiality.

Once we accept we have two distinct realms of existence: potentiality and actuality, we must be careful about the way in which we define *contradictions*. Certainly, contradictions cannot be defined in terms of truth valuations in the actual realm, simply because the physical notion that must be related to quantum superpositions must be, according to our research, an existent in the potential realm—not in the actual one. The MKS theorem shows explicitly that a quantum wave function implies multiple incompatible valuations which can be interpreted as *potential contradictions*. Our analysis has always kept in mind the idea that contradictions—by definition—are never found in the actual realm. Our attempt is to turn things upside down: we do not need to explain the actual via the potential but rather, we need to use the actual in order to develop the potential [14, p. 148]. Leaving aside the paranoia against contradictions, the PAQS does the job of allowing a further formal development of a realm in which all terms of a superposition exist, regardless of actuality. In the sense just discussed the PAQS opens possibilities of development which have not yet been fully investigated. It should be also clear that we are not claiming that all terms in the superposition are actual—as in the many worlds interpretations—overpopulating existence with unobservable actualities. What we claim is that PAQS opens the door to consider all terms as existent in potentiality—independently of actuality. We claim that just like we need all properties to characterize a physical object, all terms in the superposition are needed for a proper characterization of what exists according to QM. Contrary to Arenhart and Krause we do not agree that our proposal is subject of Priest's razor, the metaphysical principle according to which we should not populate the world with contradictions beyond necessity [35]. The PAQS does not overpopulate metaphysically the world with contradictions, rather it attempts to take into account what the quantum formalism and present experiments seem to be telling us about physical reality.[4]

Modal interpretations are difficult to define within the literature.[5] We understand that modal interpretations have two main desiderata that must be fulfilled by any interpretation which deserves being part of the club. The first is to stay close to the standard formalism of QM, the second is to investigate the meaning of modality and existence within the orthodox formalism of the theory. The modal interpretation that we have proposed [14] attempts to develop—following these two desiderata and the CMLR—a physical representation of the formalism based on two main notions: the notion of 'ontological potentiality' and notion of 'power'. The notion of ontological potentiality has been explicitly developed taking into account what we have learnt from the orthodox formalism about quantum possibility, taking potentiality to its limit, and escaping the dogmatic ruling of actuality. Contrary to the teleological notion of potentiality used within many interpretations of QM our notion of ontological potentiality is not defined in terms of actuality [38]. Such perspective has determined not only the need to consider what we call a *Potential State of Affairs (PSA)*—in analogous fashion to the *ASA* considered within physical theories—but also the distinction between *actual effectuations*, which is the effectuation of potentiality in the actual realm, and *potential effectuations* which happens in the potential realm regardless of actuality [15, 16, 20]. Actualization only discusses the actual effectuation of the potential, while potential effectuations remain in the potential realm evolving according to QM. The question we would like to discuss in the following section is: what is that which exists in the potential realm?

---

[4]Regarding observation it is important to remark that such contradictory potentialities are observable just in the same way as actual properties can be observed in an object. Potentialities can be observed through actual effectuations in analogous fashion to physical objects—we never observe all perspectives of an object simultaneously, instead, we observe at most a single set of actual properties.

[5]As we have discussed in [13] modal interpretations range from empiricist positions such as that of Van Fraassen [39] to realist ones such as the one endorsed in different ways by Dieks [23], Bub [9], and Bacciagaluppi [4]. There are even different strategies and ideas regarding what should be considered to be a coherent interpretation within this group.

11.4 Powers as Real Quantum Physical Existents

Entities are composed by properties which exist in the actual mode of being. But what is that which exists in the ontological potential realm? We have argued that an interesting candidate to consider is the notion of *power*. Elsewhere [14, 16], we have put forward such an ontological interpretation of powers. In the following we summarize such ideas and provide an axiomatic characterization of QM in line with these concepts.

The mode of being of a power is potentiality, not thought in terms of classical possibility (which relies on actuality) but rather as a mode of existence—i.e., in terms of ontological potentiality. To possess the power of *raising my hand*, does not mean that in the future 'I *will* raise my hand' *or* that in the future 'I *will not* raise my hand'; what it means is that, here and now, I possess a power which exists in the mode of being of potentiality, *independently of what will happen in actuality*. Powers do not exist in the mode of being of actuality, they are not actual existents, they are undetermined potential existents. Powers, like classical properties, preexist to observation, unlike them, preexistence is not defined in the actual mode of being as an ASA, instead we have a *potential preexistence* of powers which determines a PSA. *Powers are indetermined.* Powers are a conceptual machinery which can allow us to compress experience into a picture of the world, just like entities such as particles, waves, and fields, allow us to do so in classical physics. We cannot 'see' powers in the same way we see objects.[6] Powers are experienced in actuality through *elementary processes*. A power is sustained by a logic of actions which do not necessarily take place, it *is* and *is not*, *hic et nunc*.

A basic question which we have posed to ourselves regards the ontological meaning of a *quantum superposition* [14]. What does it mean to have a mathematical expression such as: $\alpha \mid \uparrow\rangle + \beta \mid \downarrow\rangle$, which allows us to predict precisely, according to the Born rule, experimental outcomes? Our theory of powers has been explicitly developed in order to try to answer this particular question. Given a superposition in a particular basis, $\Sigma\, c_i|\alpha_i\rangle$, the powers are represented by the elements of the basis, $|\alpha_i\rangle$, while the coordinates in square modulus, $|c_i|^2$, are interpreted as the potentia of each respective power. *Powers can be superposed to different—even contradictory—powers.* We understand a quantum superposition as encoding a set of powers, each of which possesses a definite *potentia*. This is what we call a *Quantum Situation* (QS). For example, the QS represented by the superposition $\alpha \mid \uparrow\rangle + \beta \mid \downarrow\rangle$, combines the contradictory powers, $\mid \uparrow\rangle$ and $\mid \downarrow\rangle$, with their potentia, $|\alpha|^2$ and $|\beta|^2$, respectively. Contrary to the orthodox interpretation of the quantum state, we do not assume the metaphysical identity of the multiple mathematical representations given by different basis [22]. Each superposition is basis dependent and must be considered as a distinct quantum situation. For example, the superpositions $c_{x1}| \uparrow_x\rangle + c_{x2}| \downarrow_x\rangle$ and $c_{y1}| \uparrow_y\rangle + c_{y2}| \downarrow_y\rangle$, which are both representations of the same Ψ and can be derived

[6]It is important to notice that there is no difference in this point with the case of entities: we cannot 'see' entities—not in the sense of having a complete access to them. We only see perspectives which are unified through the notion of object.

from one another via a change in basis, are considered as two different and distinct quantum situations, QS_{Ψ,B_x} and QS_{Ψ,B_y}.

The logical structure of a superposition is such that a power and its opposite can exist at one and the same time, violating the principle of non-contradiction [11]. Within the faculty of raising my hand, both powers (i.e., the power 'I am *able to* raise my hand' and the power 'I am *able not to* raise my hand') coexist. A QS is *compressed activity*, something which *is* and *is not* the case, *hic et nunc*. It cannot be thought in terms of identity but is expressed as a difference, as a *quantum*.

Our interpretation can be condensed in the following eight postulates.

I. **Hilbert Space**: QM is represented in a vector Hilbert space.

II. **Potential State of Affairs (PSA)**: A specific vector Ψ with no given mathematical representation (basis) in Hilbert space represents a PSA, i.e., the definite existence of a multiplicity of powers, each one of them with a specific potentia.

III. **Actual State of Affairs (ASA)**: Given a PSA and the choice of a definite basis $B, B', B'', \ldots$ (or equivalently a C.S.C.O.) a context is defined in which a set of powers, each one of them with a definite potentia, is univocally determined as related to a specific experimental arrangement (which in turn corresponds to a definite ASA). The context builds a bridge between the potential and the actual realms, between quantum powers and classical objects. The experimental arrangement (in the ASA) allows the powers (in the PSA) to express themselves in actuality through elementary processes which produce actual effectuations.

IV. **Quantum Situations, Powers, and Potentia**: Given a PSA, Ψ, and the context we call a QS to any superposition of one or more than one power. In general given the basis $B = \{|\alpha_i\rangle\}$ the quantum situation $QS_{\Psi,B}$ is represented by the following superposition of powers:

$$c_1|\alpha_1\rangle + c_2|\alpha_2\rangle + \cdots + c_n|\alpha_n\rangle \tag{11.4.1}$$

We write the QS of the PSA, Ψ, in the context B in terms of the order pair given by the elements of the basis and the coordinates in square modulus of the PSA in that basis:

$$QS_{\Psi,B} = (|\alpha_i\rangle, |c_i|^2) \tag{11.4.2}$$

The elements of the basis, $|\alpha_i\rangle$, are interpreted in terms of *powers*. The coordinates of the elements of the basis in square modulus, $|c_i|^2$, are interpreted as the *potentia* of the power $|\alpha_i\rangle$, respectively. Given the PSA and the context, the quantum situation, $QS_{\Psi,B}$, is univocally determined in terms of a set of powers and their respective potentia. (Notice that in contradistinction with the notion of quantum state the definition of a QS is basis dependent.)

V. **Elementary Process**: In QM one can observe discrete shifts of energy (quantum postulate). These discrete shifts are interpreted in terms of *elementary processes* which produce actual effectuations. An elementary process is the path which undertakes a power from the potential realm to its actual effectuation. This path is governed by the *immanent cause* which allows the power to remain preexistent in the potential realm independently of its actual effectuation. Each power $|\alpha_i\rangle$ is univocally related to an elementary process represented by the projection operator $P_{\alpha_i} = |\alpha_i\rangle\langle\alpha_i|$.

VI. **Actual Effectuation of Powers (Measurement)**: Powers exist in the mode of being of ontological potentiality. An *actual effectuation* is the expression of a specific power in actuality. Different actual effectuations expose the different powers of a given QS. In order to learn about a specific PSA (constituted by a set of powers and their potentia) we must measure repeatedly the actual effectuations of each power exposed in the laboratory. (Notice that we consider a laboratory as constituted by the set of all possible experimental arrangements that can be related to the same Ψ.)

VII. **Potentia (Born Rule)**: A *potentia* is the strength of a power to exist in the potential realm and to express itself in the actual realm. Given a PSA, the potentia is represented via the Born rule. The potentia p_i of the power $|\alpha_i\rangle$ in the specific PSA, Ψ, is given by:

$$Potentia\ (|\alpha_i\rangle, \Psi) = \langle\Psi|P_{\alpha_i}|\Psi\rangle = Tr[P_\Psi P_{\alpha_i}] \quad (11.4.3)$$

In order to learn about a QS we must observe not only its powers (which are exposed in actuality through actual effectuations) but we must also measure the potentia of each respective power. In order to measure the potentia of each power we need to expose the QS statistically through a repeated series of measurements . The potentia, given by the Born rule, coincides with the probability frequency of repeated measurements when the number of observations goes to infinity.

VIII. **Potential Effectuation of Powers (Schrödinger Evolution)**: Given a PSA, Ψ, powers and potentia evolve deterministically, independently of actual effectuations, producing *potential effectuations* according to the following unitary transformation:

$$i\hbar\frac{d}{dt}|\Psi(t)\rangle = H|\Psi(t)\rangle \quad (11.4.4)$$

While *potential effectuations* evolve according to the Schrödinger equation, *actual effectuations* are particular expressions of each power (that constitutes the PSA, Ψ) in the actual realm. The ratio of such expressions in actuality is determined by the potentia of each power.

According to our interpretation, just like classical physics talks about entities composed by properties that preexist in the actual realm, QM talks about powers that preexist in the (ontological) potential realm, independently of the specific actual context of inquiry. This interpretational move allows us to define powers independently of the context regaining an objective picture of physical reality independent of measurements and subjective choices. The price we are willing to pay is the abandonment of a metaphysical equation that has been presupposed in the analysis of the interpretation of QM: 'actuality = reality'. In the following section, taking into account a typical quantum experience, we discuss in what sense powers are to be considered in terms of 'contradiction' or 'contrariety'.

11.5 Contradiction and Contrariety in Quantum Superpositions

Arenhart and Krause have called the attention to the understanding of contradiction via the Square of Opposition.

> States in QM such as the one describing the famous Schrödinger cat—which is in a superposition between the states 'the cat is dead' and 'the cat is alive'—present a challenge for our understanding which may be approached via the conceptual tools provided by the square. According to some interpretations, such states represent contradictory properties of a system (for one such interpretation see, for instance, da Costa and de Ronde [6]). On the other hand, we have advanced the thesis that states such as 'the cat is dead' and 'the cat is alive' are contrary rather than contradictory (see Arenhart and Krause [1, 2]). [3, p. 2]

Within their CAQS, Arenhart and Krause have argued in [3] against the concept of potentiality and its relation to contradiction concluding "that contrariety is still a more adequate way to understand superpositions." Elsewhere, together with Domenech and Freytes, we have analyzed via the Square of Opposition the meaning of quantum possibility. We argued that the notion of possibility needs to be discussed in terms of the formal structure of the theory itself and that, in such case, one should not study the Classical Square of Opposition but rather an Orthomodular Square of Opposition. In [29] we developed such a structure and in [21] we provided an interpretation of the Orthomodular Square of Opposition in terms of the notion of potentiality. Furthermore, according to the author of this paper, the development should also consider the analysis of the hexagon, paraconsistent negation, and modalities provided by Béziau in [6, 7]. In this section we argue that Arenhart and Krause have misinterpreted our notion of 'potentiality' and 'power' and explain why the PAQS is better suited to account for quantum superpositions than the CAQS.

Let us begin our analysis recalling the traditional definitions of the famous square of opposition and the meaning of contradiction and contrariety.

Contradiction Propositions: α and β are *contradictory* when both cannot be true and both cannot be false.
Contrariety Propositions: α and β are *contrary* when both cannot be true, but both can be false.
Subcontrariety Propositions: α and β are *subcontraries* when both can be true, but both cannot be false.
Subaltern Propositions: α is *subaltern* to proposition β if the truth of β implies the truth of α.

Discussing the inadequacy of the notion of power, Arenhart and Krause provide the following analysis:

> First of all, a property, taken by itself as a power (a real entity not actual), is not affirmed nor denied of anything. To take properties such as 'to have spin up in the x direction' and 'to have spin down in the x direction' by themselves does not affirm nor deny anything. To say 'to have spin up in the x direction' is not even a statement, it is analogous to speak 'green' or 'red hair'. To speak of a contradiction, it seems, one must have complete statements, where properties or relations are attributed to something. That is, one must have something like 'spin up is measured in a given direction', or 'Mary is red haired', otherwise there will be no occasion for truth and falsehood, and consequently, no occasion for a contradiction. So, the realm of the potential must be also a realm of attribution of properties to something if contradiction is to enter in it. However, this idea of attribution of properties seems to run counter the idea of a merely potential realm. On the other hand, the idea of a contradiction seems to require that we speak about truth and falsehood. [3]

The idea that potentiality determines a contradictory realm goes back to Aristotle himself who claimed that contradictions find themselves in potentiality. Of course, as remarked by Arenhart and Krause, the square of opposition is discussing about actual truth and falsehood. Thus, potentiality is not considered in terms of a mode of existence but rather as mere logical possibility. The interesting question is if our representation of quantum superpositions in terms of powers is compatible with the square. We believe it easy to see that such is the case provided special attention is given to the realms involved in the discussion. Furthermore, it is also easy to see that the CAQS is incompatible with QM due to its empirical inadequacy. Some remarks go in order.

First, we must stress the fact that a power is not—as claimed by Arenhart and Krause [*Op. cit.*, p. 7]—an entity. A physical entity exists in the mode of being of actuality and is represented by three main logical and ontological principles: **the principle of existence**, **the principle of noncontradiction**, and **the principle of identity** (see for discussion [40]). As discussed in the previous section, quite independently of such principles we have defined the notion of power in terms of **the principle of indetermination**, **the principle of superposition**, and **the principle of difference**. The adequacy or not of powers to interpret QM needs to be analyzed taking into account this specific scheme [16]. Instead of doing so, Arenhart and Krause have criticized a notion of power which they have not specified in rigorous terms.

Second, truth and falsehood are related to actuality, since in the orthodox view this is the only exclusive realm considered as real. However, our notion of ontological potentiality is completely independent of actuality, and since powers are real objective existents it makes perfect sense to extend 'truth' and 'falsity' to this mode of being. It is the PSA which determines the specific set of powers which potentially preexist in a given quantum situation. Thus, in analogous fashion to the way an ASA determines the set of properties which are 'true' and 'false', a PSA determines a set of powers which are 'true' and 'false', namely, those powers which potentially preexist and can be exposed through the choice of different quantum situations (i.e., the multiplicity of possible contexts). Our redefinition of truth and falsehood with respect to potentiality escapes any subjective choice and regains an objective description of physical reality.

In a given situation all the powers which determine possible actual effectuations compose a PSA. For example, a Stern–Gerlach apparatus in a laboratory which can be placed in the x, y, or z direction determines the existence of the powers: $|\uparrow_x>$, $|\downarrow_x>$, $|\uparrow_y>$, $|\downarrow_y>$, $|\uparrow_z>$, and $|\downarrow_z>$ irrespectively of the actual choice of the particular context (i.e., the particular actual direction in which the Stern–Gerlach is placed). We can say that even though the PSA is defined independently of the context of inquiry, QS are indeed *contextual existents*.

Third, let us investigate, provided we grant, for the sake of the argument, that powers do exist. Which is then the most suitable notion to account for two powers that can be actualized in a typical quantum experiment? Consider we have a Stern–Gerlach apparatus placed in the x direction, if we have the following quantum superposition: $\alpha|\uparrow_x> + \beta|\downarrow_x>$, this means we have the power of having spin up in the x-direction, $|\uparrow_x>$, with potentia $|\alpha|^2$ and the power of having spin down in the x-direction, $|\downarrow_x>$, with potentia $|\beta|^2$. Both powers can become actual. Is it contradiction or contrariety the best notion suited to account for powers in this quantum experiment? Given this QS, it is clear that both actualizations of the powers (elementary processes) $|\uparrow_x>$ and $|\downarrow_x>$ *cannot be simultaneously 'true' in actuality*, since only one of them will become actual; it is also the case that both actualizations of the powers (elementary processes) $|\uparrow_x>$ and $|\downarrow_x>$ *cannot be simultaneously 'false' in actuality*, since when we measure this QS we know we will obtain either the elementary process 'spin up in the x-direction', $|\uparrow_x><\uparrow_x|$, or the elementary process 'spin down in the x-direction', $|\downarrow_x><\downarrow_x|$. Now, if we consider the CAQS, contrary propositions are determined when *both cannot be true*, but *both can be false*. But this is not the case in QM, in particular, it is not the case for the example we have just considered. Given a measurement on the quantum superposition, $\alpha|\uparrow_x> + \beta|\downarrow_x>$, one of the two terms will become actual (true) while the other term will not be actual (false), which implies that *both cannot be false*. Thus, while the PAQS is able to describe what we know about what happens in a typical quantum measurement, the CAQS of Arenhart and Krause is not capable of fulfilling empirical adequacy.

In the conclusion of their paper, Arenhart and Krasue discuss what happens when the state is in a superposition. They argue that one possibility is to claim that "when not in an eigenstate the system does not have any of the properties associated with the superposition." According to them: "This option is compatible with the claim that states in a superposition are contraries: both fail to be the case." But as we have seen in the last section, given a superposition state such as $\alpha|\uparrow_x> + \beta|\downarrow_x>$, we know with certainty that one of the terms will become actual if measured. Thus, it makes no sense to claim that both will 'fail to be the case'. The CAQS, fails to provide the empirical adequacy needed to account for basic quantum experiments. A second possibility is to "assume another interpretation [...] and hold that even in a superposition one of the associated properties hold, even if not in an eigenstate." According to Arenhart and Krasue: "Following this second option, notice, the understanding of superpositions as contraries still hold: even when one of the properties in a superposition hold, the other must not be the case." However, if only one of the properties is *true* then it seems difficult to explain how a property that *is the case* can interact with a property that *is not the case*. As we know, the interaction of superpositions happens between all terms in the superposition, the possibilities contained in the superposition may interfere between each other. The question then raises: how can something that exists interact with something else which does not exist?[7]

11.6 Final Remarks

Although we agree with Arenhart and Krause regarding the fact that the formal approach that we provided in [11] was not completely adequate to the idea discussed here, we must also remark that we never claimed that this was the final formal description of quantum superpositions but rather a very first step in such paraconsistent development. In this respect, we believe that this approach is still in need of further development.[8] However, we must also remark that the approach provides a suitable answer to the existence of the multiple terms in a quantum superposition, something that is needed in order to make sense about present and future quantum experiments and technical developments. We believe that the possibilities it might open deserve not only attention but also criticism. We thank Arenhart and Krause for their careful and incisive analysis.

References

1. Arenhart, J.R., Krause, D.: Oppositions in quantum mechanics. In: Béziau, J.-Y., Katarzyna, G.-K. (eds.) New Dimensions of the Square of Opposition, pp. 337–356. Philosophia Verlag, Munich (2014)
2. Arenhart, J.R., Krause, D.: Contradiction, quantum mechanics, and the square of opposition. Logique et Analyse (2014)
3. Arenhart, J.R., Krause, D.: Potentiality and contradiction in quantum mechanics. In: Koslow, A., Buchsbaum, A. (eds.) The Road to Universal Logic (volume II). Springer (2014)
4. Bacciagaluppi, G.: Topics in the modal interpretation of quantum mechanics. Doctoral Dissertation. University of Cambridge, Cambridge (1996)
5. Bernien, H., Hensen, B., Pfaff, W., Koolstra, G., Blok, M.S., Robledo, L., Taminiau, T.H., Markham, M., Twitchen, D.J., Childress, L., Hanson, R.: Heralded entanglement between solid-state qubits separated by three metres. Nature **497**, 86–90 (2013)
6. Béziau, J.-Y.: The power of the hexagon. Log. Univers. **6**, 1–43 (2012)
7. Béziau, J.-Y.: Paraconsistent logic and contradictory viewpoint. To appear in Rev. Bras. de Filosofia **241** (2014)
8. Bokulich, P., Bokulich, A.: Niels Bohr's generalization of classical mechanics. Found. Phys. **35**, 347–371 (2005)
9. Bub, J.: Interpreting the Quantum World. Cambridge University Press, Cambridge (1997)
10. Clausen, C., Usmani, I., Bussières, F., Sangouard, N., Afzelius, M., de Riedmatten, H., Gisin, N.: Quantum storage of photonic entanglement in a crystal. Nature **469**, 508–511 (2011)

[7]For a detailed analysis of the relation between quantum superpositions and physical reality see: [19].

[8]A possible development in line with the interpretation presented in this paper will be analyzed in [12].

11. da Costa, N., de Ronde, C.: The paraconsistent logic of quantum superpositions. Found. Phys. **43**, 845–858 (2013)
12. da Costa, N., de Ronde, C.: The Paraconsistent Approach to Quantum Superpositions Reloaded: Formalizing Contradictory Powers in the Potential Realm, in preparation (2015)
13. de Ronde, C.: For and against metaphysics in the modal interpretation of quantum mechanics. Philosophica **83**, 85–117 (2010)
14. de Ronde, C.: The contextual and modal character of quantum mechanics: a formal and philosophical analysis in the foundations of physics. PhD dissertation, Utrecht University (2011)
15. de Ronde, C.: Quantum Superpositions and Causality: on the Multiple Paths to the Measurement Result. Los Alamos Archive (2013). arXiv:1310.4534
16. de Ronde, C.: Representing Quantum Superpositions: Powers, Potentia and Potential Effectuations. Los Alamos Archive (2013). arXiv:1312.7322
17. de Ronde, C.: A defense of the paraconsistent approach to quantum superpositions (Answer to Arenhart and Krause). Los Alamos Archive (2014). arXiv:1404.5186
18. de Ronde, C.: Hilbert space quantum mechanics is contextual. (Reply to R.B. Griffiths). Los Alomas Archive (2015). arxiv:1502.05396
19. de Ronde, C.: Quantum superpositions do exist! but 'quantum physical reality $\neq$ actuality'. Los Alamos Archive (2015). arxiv:1502.05311
20. de Ronde, C., Freytes, H., Domenech, G.: Interpreting the modal Kochen-Specker theorem: possibility and many worlds in quantum mechanics. Stud. Hist. Philos. Mod. Phys. **45**, 11–18 (2014)
21. de Ronde, C., Freytes, H., Domenech, G.: Quantum mechanics and the interpretation of the orthomodular square of opposition. In: Béziau, Jean-Yves, Gan-Krzywoszynska, Katarzyna (eds.) New Dimensions of the Square of Opposition, pp. 223–242. Philosophia Verlag, Munich (2014)
22. de Ronde, C., Massri, C.: Revisiting the Orthodox Interpretation of 'Physical States' in Quantum Mechanics. Los Alamos Archive (2014). arXiv:1412.2701
23. Dieks, D.: Quantum mechanics without the projection postulate and its realistic interpretation. Found. Phys. **19**, 1397–1423 (1989)
24. Dirac, P.A.M.: The Principles of Quantum Mechanics, 4th edn. Oxford University Press, London (1974)
25. Domenech, G., Freytes, H., de Ronde, C.: Scopes and limits of modality in quantum mechanics. Ann. Phys. **15**, 853–860 (2006)
26. Domenech, G., Freytes, H., de Ronde, C.: A topological study of contextuality and modality in quantum mechanics. Int. J. Theor. Phys. **47**, 168–174 (2008)
27. Domenech, G., Freytes, H., de Ronde, C.: Modal-type orthomodular logic. Math. Log. Q. **3**, 307–319 (2009)
28. Domenech, G., Freytes, H., de Ronde, C.: Many worlds and modality in the interpretation of quantum mechanics: an algebraic approach. J. Math. Phys. **50**, 072108 (2009)
29. Freytes, H., de Ronde, C., Domenech, G.: The square of opposition in orthodmodular logic. In: Béziau, Jean-Yves, Jacquette, Dale (eds.) Around and Beyond the Square of Opposition: Studies in Universal Logic, pp. 193–201. Springer, Basel (2012)
30. Fuchs, C., Peres, A.: Quantum theory needs no 'interpretation'. Phys. Today **53**, 70 (2000)
31. Griffiths, R.B.: Hilbert space quantum mechanics is non contextual. Stud. Hist. Philos. Mod. Phys. **44**, 174–181 (2013)
32. Heisenberg, W.: Physics and Philosophy, World Perspectives. George Allen and Unwin Ltd., London (1958)
33. Ma, X., Zotter, S., Kofler, J., Ursin, R., Jennewein, T., Brukner, C., Zeilinger, A.: Experimental delayed-choice entanglement swapping. Nat. Phys. **8**, 480–485 (2012)
34. Ourjoumtsev, A., Jeong, H., Tualle-Brouri, R., Grangier, P.: Generation of optical 'Schrödinger cats' from photon number states. Nature **448**, 784–786 (2007)
35. Priest, G.: In Contradiction. Nijhoff, Dordrecht (1987)
36. Rédei, M.: Some historical and philosophical aspects of quantum probability theory and its interpretation. In: Dieks, D. et al. (eds.) Probabilities, Laws, and Structures, pp. 497–506. Springer (2012)

37. Schrödinger, E.: The present situation in quantum mechanics. Naturwiss **23**, 807 (1935). Translated to english. In: Wheeler, J.A., Zurek, W.H. (eds.) Quantum Theory and Measurement. Princeton University Press, Princeton (1983)
38. Smets, S.: The modes of physical properties in the logical foundations of physics. Log. Log. Philos. **14**, 37–53 (2005)
39. Van Fraassen, B.C.: Quantum Mechanics: an Empiricist View. Clarendon, Oxford (1991)
40. Verelst, K., Coecke, B.: Early Greek thought and perspectives for the interpretation of quantum mechanics: preliminaries to an ontological approach. In: Aerts, D. (ed.) The Blue Book of Einstein Meets Magritte, pp. 163–196. Kluwer Academic Publishers, Dordrecht (1999)
41. Wheeler, J.A., Zurek, W.H. (eds.): Theory and Measurement. Princeton University Press, Princeton (1983)

Part IV
Tools and Framework

Chapter 12
Consequence–Inconsistency Interrelation: In the Framework of Paraconsistent Logics

Soma Dutta and Mihir K. Chakraborty

Abstract This paper deals with a relativized notion of inconsistency, which turns out to be equivalent to a non-explosive consequence under certain sets of axiomatization in a propositional language. The paper also shows that several existing paraconsistent systems fall under this characterization.

Keywords Non-explosive consequence · Inconsistency · Paraconsistent logics

Mathematics Subject Classification (2000) 03B53

12.1 Introduction

The notions of consequence and consistency, and hence inconsistency too, are interwoven in the context of classical logic. In [16] Surma presented two sets of axioms characterizing the notions of consequence and consistency, and showed that taking any one of them as primitive the other can be obtained. This equivalence greatly depends on the idea that inconsistency, also called negation inconsistency (i.e. a set of formulae yielding a formula and its negation together), and triviality, also called absolute inconsistency (i.e. every formula derivable from a set of formulae)

The first author of this paper acknowledges the support obtained from the Institute of Mathematical Sciences, Chennai, India, during the initial phase of preparation of this paper; the final preparation of this paper has been carried out during the tenure of an ERCIM Alain Bensoussan fellowship.

S. Dutta (✉)
Faculty of Mathematics, Informatics, and Mechanics,
University of Warsaw, Warsaw, Poland
e-mail: somadutta9@gmail.com

M.K. Chakraborty
School of Cognitive Science, Jadavpur University, Kolkata, India
e-mail: mihirc4@gmail.com

J.-Y. Beziau et al. (eds.), *New Directions in Paraconsistent Logic*,
Springer Proceedings in Mathematics & Statistics 152,
DOI 10.1007/978-81-322-2719-9_12

are equivalent. This is not the case in the context of paraconsistent logics. As a result consequence–inconsistency equivalence does not work here.

From the notion of absolute inconsistency it is clear that classical notion of inconsistency is global in nature. That is, all inconsistent sets of formulae are identical in the sense that their sets of consequences are the same viz. the set of all formulae. In paraconsistent logics inconsistency means negation inconsistency, which requires a formula and its negation to be followed together. That is, it depends on a formula. So, we introduce a relativized notion of inconsistency. More specifically inconsistency in the context of paraconsistent logics, let us call para-inconsistency or in short PI, is a binary relation between a set of formulae and a single formula. We say $(X, \alpha) \in$ PI to represent that X is inconsistent with respect to α. A similar notion one can find in [6] where authors talked about α-contradictory set, though it was not the focus of the discussion of the paper [6]. In our earlier paper [11] this relativized notion of inconsistency has been studied first time as an alternative notion of classical inconsistency, which could fit well with a non-explosive notion of consequence. In this paper [11] we have concentrated only on the negation fragment of a language. This paper is a continuation of the earlier one, and here we shall consider the study of consequence–inconsistency duo over the entire propositional fragment of a language.

12.2 Non-explosive Consequence and Corresponding Inconsistency

The notion of inconsistency, in a syntactic framework of a logic, is defined in terms of the notion of consequence and the object language negation ($\neg$). In classical logic, logical connectives are so related that its explosive nature does not solely depend on the connective negation ($\neg$). In this section we shall explore different fragments of a propositional language, and exclude possible ways of explosion. In other words, we shall look for those conditions which are necessitated by the law of non-explosion, i.e. there exists some wff α such that $\{\alpha, \neg\alpha\} \nvdash \beta$ for some β.

Let us first concentrate on the $\neg$-fragment of a language.

12.2.1 Consequence and Inconsistency in the ¬-fragment

In our presentation we need to refer to the standard sequent calculus presentation of classical logic [12] at several points of time. We consider a sequent as an expression of the form $X \vdash Y$, where X, Y are sets of formulae. With the progress of the presentation of this article, according to our requirement, gradually we shall impose constraints on the notion of sequent. For the sake of reference we stick to the following sequent calculus presentation for the structural and non-structural rules.

axiom	$\alpha \vdash \alpha$	
cut	$\dfrac{\Gamma \vdash \Pi, \alpha \quad \alpha, \Sigma \vdash \Delta}{\Gamma, \Sigma \vdash \Pi, \Delta}$	
	right rule	left rule
dilution	$\dfrac{\Gamma \vdash \Delta}{\Gamma \vdash \Delta, \alpha}$	$\dfrac{\Gamma \vdash \Delta}{\alpha, \Gamma \vdash \Delta}$

	right rule	left rule
&	$\dfrac{\Gamma \vdash \Delta, \alpha \quad \Gamma \vdash \Delta, \beta}{\Gamma \vdash \Delta, \alpha \& \beta}$	$\dfrac{\alpha, \Gamma \vdash \Delta}{\alpha \& \beta, \Gamma \vdash \Delta} \quad \dfrac{\beta, \Gamma \vdash \Delta}{\alpha \& \beta, \Gamma \vdash \Delta}$
$\vee$	$\dfrac{\Gamma \vdash \Delta, \alpha}{\Gamma \vdash \Delta, \alpha \vee \beta} \quad \dfrac{\Gamma \vdash \Delta, \beta}{\Gamma \vdash \Delta, \alpha \vee \beta}$	$\dfrac{\alpha, \Gamma \vdash \Delta \quad \beta, \Gamma \vdash \Delta}{\alpha \vee \beta, \Gamma \vdash \Delta}$
$\supset$	$\dfrac{\alpha, \Gamma \vdash \Delta, \beta}{\Gamma \vdash \Delta, \alpha \supset \beta}$	$\dfrac{\Gamma \vdash \Delta, \alpha \quad \beta, \Pi \vdash \Sigma}{\alpha \supset \beta, \Gamma, \Pi \vdash \Delta, \Sigma}$
$\neg$	$\dfrac{\alpha, \Gamma \vdash \Delta}{\Gamma \vdash \Delta, \neg\alpha}$	$\dfrac{\Gamma \vdash \Delta, \alpha}{\neg\alpha, \Gamma \vdash \Delta}$

We shall also refer to the following rules of $\neg$, which are derivable from consecutive use of $\neg$-left and $\neg$-right.

($\neg\neg$-I) $\alpha \vdash \neg\neg\alpha$ ($\neg\neg$-E) $\neg\neg\alpha \vdash \alpha$.

Let us present some derivations generating explosion in the $\neg$-fragment of a language.

(a) $\neg$-left + dilution-right imply explosion.

(D1)
$$\dfrac{\dfrac{\alpha \vdash \alpha}{\alpha, \neg\alpha \vdash}\ (\neg\text{-left})}{\alpha, \neg\alpha \vdash \beta}\ (\text{dil-right})$$

(D2)
$$\dfrac{\dfrac{\alpha \vdash \alpha}{\alpha \vdash \alpha, \beta}\ (\text{dil-right})}{\alpha, \neg\alpha \vdash \beta}\ (\neg\text{-left})$$

(b) $\neg$-right + $\neg$-left + dilution-left + cut imply explosion.

(D3)
$$\dfrac{\dfrac{\dfrac{\alpha \vdash \neg\neg\alpha \ (\neg\neg\text{-I}) \quad \dfrac{\dfrac{\dfrac{\dfrac{\alpha \vdash \alpha}{\vdash \alpha, \neg\alpha}\ (\neg\text{-right})}{\neg\beta \vdash \alpha, \neg\alpha}\ (\text{dilution-left})}{\neg\beta, \neg\alpha \vdash \neg\alpha}\ (\neg\text{-left})}{\neg\alpha, \neg\neg\alpha \vdash \neg\neg\beta}\ (\neg\text{-left, right})}{\alpha, \neg\alpha \vdash \neg\neg\beta}\ (\text{cut}) \quad \neg\neg\beta \vdash \beta \ (\neg\neg\text{-E})}{\alpha, \neg\alpha \vdash \beta}\ (\text{cut})$$

So, from the above derivations we can conclude that for non-explosion it is necessary to (a) not to take $\neg$-left and dilution-right together, and (b) not to take $\neg$-right, $\neg$-left, dilution-left and cut together.

From (a) and (b) it is clear that to avoid explosion we have to drop some of the rules from the above-mentioned set of rules. We prefer to keep the structural properties of a consequence relation intact. We know that a sequent of the form $X \vdash Y$ never represents Y as a set of consequences of X; rather it represents that at least one formula of Y is a consequence of X. We hence represent consequence, which is presented by a sequent, as $X \vdash \alpha$, and set of all consequences of X as $C(X)$, where $C(X) = \{\alpha : X \vdash \alpha\}$. This consideration automatically helps us to eliminate $\neg$-left, $\neg$-right from our system of concern. Instead of $\neg$-left, $\neg$-right we may consider $\neg\neg$-I and $\neg\neg$-E, the weaker versions of $\neg$-left, $\neg$-right, from now onwards.

In classical logic $\neg$-right and $\neg$-left are the conditions, through which the behaviour of classical negation is captured. Let us explore in the absence of these two properties what else can be added to $\neg\neg$-I and $\neg\neg$-E to keep the properties of negation as close as possible to the classical one, at the same time disrespecting explosion. Following Michael Dunn's kite diagram for negation [9, 10] let us list some properties of negation. As $\neg\neg$-I and $\neg\neg$-E are already mentioned above, we list the other properties.

$$\frac{X, \alpha \vdash \beta}{X, \neg\beta \vdash \neg\alpha} \quad \text{(Subminimal)}$$

$$\frac{X \vdash \alpha \quad X \vdash \neg\alpha}{X \vdash \beta} \quad \text{(Ex Contradictionae Quodlibet/ECQ)}$$

$$\frac{X, \alpha \vdash \beta \quad X, \neg\alpha \vdash \beta}{X \vdash \beta} \quad (\mathrm{R}\neg)$$

It can be shown that (i) subminimal + $\neg\neg$-I imply $\alpha, \neg\alpha \vdash \neg\beta$,
(ii) subminimal + $\neg\neg$-E imply explosion and
(iii) subminimal + $\neg\neg$-E + R$\neg$ imply explosion.

It is visible that the subminimal property of negation is the common factor causing explosion. So, the idea is to drop the subminimal property of $\neg$, and consider the other properties excluding explosion (i.e. ECQ). Following this line of thought in [11] we have proposed sets of axioms characterizing a non-explosive notion of consequence and its corresponding notion of inconsistency. The sets of axioms in the $\neg$-fragment of a language [11] are given below.

$C : P(F) \mapsto P(F)$, denoting a consequence operator, satisfies the following conditions.

(C1): $X \subseteq C(X)$. (reflexivity)
(C2): $X \subseteq Y$ implies $C(X) \subseteq C(Y)$. (monotonicity/dilution)
(C3): $C(C(X)) = C(X)$. (idempotence/cut)
(C4): For some α, $C(\{\alpha, \neg\alpha\}) \neq F$. (non-explosion)
(C5): $C(X \cup \{\alpha\}) \cap C(X \cup \{\neg\alpha\}) = C(X)$. (R $\neg$)
(C6): $C(\{\alpha\}) \subseteq C(\{\neg\neg\alpha\})$. ($\neg\neg$-E)
(C7): $C(\{\neg\neg\alpha\}) \subseteq C(\{\alpha\})$. ($\neg\neg$-I)

PI $\subseteq P(F) \times F$, denoting the notion of inconsistency, is a binary relation such that it satisfies

(PI1) $\{(X \cup \{\neg\alpha\}, \alpha) : X \subseteq F, \alpha \in X\} \subseteq$ PI,
(PI2) If $X \subseteq Y$ then $(X, \alpha) \in$ PI implies $(Y, \alpha) \in$ PI,
(PI3) If for all $\alpha \in Y$, $(X \cup \{\neg\alpha\}, \alpha) \in$ PI, then $(X \cup Y, \beta) \in$ PI implies $(X, \beta) \in$ PI,
(PI4) For some α, there is some β such that $(\{\alpha, \neg\alpha\}, \beta) \notin$ PI,

(PI5) $(X \cup \{\alpha\}, \beta) \in \text{PI}$ and $(X \cup \{\neg\alpha\}, \beta) \in \text{PI}$ imply $(X, \beta) \in \text{PI}$,
(PI6) $(X, \neg\alpha) \in \text{PI}$ implies $(X, \alpha) \in \text{PI}$, and
(PI7) $(X \cup \{\neg\neg\alpha\}, \beta) \in \text{PI}$ implies $(X \cup \{\alpha\}, \beta) \in \text{PI}$.

Following theorems [11] establish that C, the notion of non-explosive consequence, and PI, the relativized notion of inconsistency, are equivalent in the sense that considering one of the notions as primitive the other can be derived.

(i) Theorem: Let PI be given, and C be defined as follows:
$\alpha \in C(X)$ if $(X \cup \{\neg\alpha\}, \alpha) \in \text{PI}$. Then C satisfies (C1) to (C7) axioms.
(ii) Theorem: Let C be given, and PI be defined as follows:
$(X, \alpha) \in \text{PI}$ if $\{\alpha, \neg\alpha\} \subseteq C(X)$. Then PI satisfies (PI1) to (PI7) axioms.

Some of the logics [1–4, 6, 13, 17, 19] satisfying the consequence axioms, and hence inconsistency axioms too, are the following.

(1) D_2 (discussive logic, Jaśkowski, 1977)
(2) J_n, $1 \leq n \leq 5$ (Arruda, da Costa, 1968)
(3) $J3$ (da Costa, D'Ottavino, 1970)
(4) Calculus of antinomies (Asenjo, 1966)
(5) LP (Logic of paradox, Priest, 1979)
(6) Pac (Avron, 1991)
(7) C_{ie} systems (Carnielli, Coniglio, Marcos, 2003).

12.2.2 Consequence and Inconsistency in the $\{\neg, \vee\}$-fragment

From now onwards, for each connective $*$, we shall consider two kinds of properties of $*$; one is related to its basic nature, i.e. right/left rules, and the other is related to its interaction with $\neg$. In case of the connective $\vee$, usual interpretation of which is linguistic 'or', disjunctive syllogism (DS) is a property which presents the interaction of $\vee$ with respect to the $\neg$. DS is given by $\alpha, \neg\alpha \vee \beta \vdash \beta$.

Below, some derivations generating explosion in the $\{\neg, \vee\}$-fragment are presented. Though while characterizing the notion of consequence in the $\neg$-fragment we have decided to consider sequent of the form $X \vdash \alpha$, and hence eliminated $\neg$-left, $\neg$-right from our concern of presentation, we still continue to refer to the forms of the table of Sect. 2.1 in the further sections also. The reason behind this is first to show various ways leading to explosion in the standard presentation of sequent calculus, and then to format the notion of consequence in our framework based on that observations.

(a) $\vee$-right + cut + DS imply explosion

(D4)
$$\dfrac{\dfrac{\neg\alpha \vdash \neg\alpha}{\neg\alpha \vdash \neg\alpha \vee \beta}\,(\vee\text{-right}) \qquad \alpha, \neg\alpha \vee \beta \vdash \beta\,(\text{DS})}{\alpha, \neg\alpha \vdash \beta}\,(\text{cut})$$

(b) $\vee$-right + cut imply (R$\vee$) $\neg$-left + $\vee$-left imply DS

(D5) $$\dfrac{\dfrac{X, \alpha \vee \beta \vdash \gamma \ \text{(assume)}}{\alpha \vdash \alpha \vee \beta \ (\vee\text{-right})}}{X, \alpha \vdash \gamma \ \text{(cut)}}$$

(D6) $$\dfrac{\dfrac{\dfrac{\alpha \vdash \alpha}{\alpha, \neg\alpha, \vdash \ (\neg\text{-left})}}{\beta \vdash \beta}}{\alpha, \neg\alpha \vee \beta \vdash \beta \ (\vee\text{-left})}$$

(R$\vee$) + DS imply explosion

(D7) $$\dfrac{\alpha, \neg\alpha \vee \beta \vdash \beta \ \text{(DS)}}{\alpha, \neg\alpha \vdash \beta \ (\text{R}\vee)}$$

From (a) it is clear that non-explosion necessitates that $\vee$-right and DS cannot be taken together, and from (b) that $\vee$-right and $\vee$-left cannot be taken together in the presence of $\neg$-left. As we have already abandoned $\neg$-left from our consideration, we need to only ensure non-occurence of $\vee$-right and DS together.

So, there are two possibilities to extend the axiomatization obtained at the $\neg$-fragment. Let us denote C characterized by (C1) to (C7) by $C_{\{\neg\}}$.

$C_{\{\neg\}}$
↙ ↘
($\vee$-right) + ($\vee$-left) ($\vee$-left) + (DS)

So, the extensions are
(**b**$_1$) $C_{\{\neg\}}$ +

(C8) $\alpha \in C(X)$ implies $\alpha \vee \beta \in C(X)$ ($\vee$-right)
(C9) $C(X \cup \{\alpha\}) \cap C(X \cup \{\beta\}) \subseteq C(X \cup \{\alpha \vee \beta\})$ ($\vee$-left)

(**b**$_2$) $C_{\{\neg\}}$ +

(C9) $C(X \cup \{\alpha\}) \cap C(X \cup \{\beta\}) \subseteq C(X \cup \{\alpha \vee \beta\})$ ($\vee$-left)
(C10) $\beta \in C(\{\alpha, \neg\alpha \vee \beta\})$ (DS)

Correspondingly, the axioms for PI also are extended in two wings.
(**b**$_1$) PI$_{\{\neg\}}$ +

(PI8) $(X \cup \{\neg\alpha\}, \alpha) \in PI$ implies $(X \cup \{\neg(\alpha \vee \beta)\}, \alpha \vee \beta) \in PI$
(PI9) $(X \cup \{\alpha\}, \gamma) \in PI$ and $(X \cup \{\beta\}, \gamma) \in PI$ imply $(X \cup \{\alpha \vee \beta\}, \gamma) \in PI$

(**b**$_2$) PI$_{\{\neg\}}$ +

(PI9) $(X \cup \{\alpha\}, \gamma) \in PI$ and $(X \cup \{\beta\}, \gamma) \in PI$ imply $(X \cup \{\alpha \vee \beta\}, \gamma) \in PI$
(PI10) $(X \cup \{\beta\}, \gamma) \in PI$ implies $(X \cup \{\alpha, \neg\alpha \vee \beta\}, \gamma) \in PI$

In the same way as given in theorems (i) and (ii) of Sect. 2.1, the following theorem can be established.

Theorem: $C_{\{\neg,\vee\}}$, the notion of non-explosive consequence, and $PI_{\{\neg,\vee\}}$, the relativized notion of inconsistency, are equivalent for both the extensions (b_1) and (b_2).

The bifurcation of $C_{\{\neg\}}$ into two branches b_1 and b_2 is evident from the following theorem.

Theorem: In the presence of (C1) to (C4)

(i) if (C8) holds, then (C10) does not hold, and
(ii) if (C10) holds, then (C8) does not hold.

Proof Below we give a proof outline of (i).
(C4) ensures the presence of an α, β such that $\beta \notin C(\{\alpha, \neg\alpha\})$.
By (C1) we have $\neg\alpha \in C(\{\neg\alpha\})$. Hence by (C8) $\neg\alpha \vee \beta \in C(\{\neg\alpha\})$.
Therefore, using (C2) and (C3) we have $C(\{\alpha, \neg\alpha \vee \beta\}) \subseteq C(\{\alpha, \neg\alpha\})$.
Hence as $\beta \notin C(\{\alpha, \neg\alpha\})$, $\beta \notin C(\{\alpha, \neg\alpha \vee \beta\})$. □

12.2.3 Consequence and Inconsistency in {¬, ∨, &}-fragment

There are two versions for non-explosion in a $\{\neg, \&\}$-fragment of a language. One is the usual one, i.e. $\{\alpha, \neg\alpha\} \nvdash \beta$ for some α, β; and the other is $\alpha \& \neg\alpha \nvdash \beta$ for some α, β. Some logical systems accept true contradiction, i.e. $\alpha \& \neg\alpha$ may not be false there. Those systems, as a result, may validate $\alpha \& \neg\alpha \nvdash \beta$ for some α, β. Believing in true contradiction is considered as dialetheism. There are some logical systems which endorse both $\alpha \& \neg\alpha \nvdash \beta$ for some α, β, as well as $\{\alpha, \neg\alpha\} \nvdash \beta$ for some α, β. These are known as strong paraconsistent logics. The systems, where only $\{\alpha, \neg\alpha\} \vdash \beta$ for any β is violated, are known as weak paraconsistent systems.

In this section we shall consider these two kinds of explosion. $\alpha \& \neg\alpha \vdash \beta$ for any β will be referred to as &-explosion.

(a) (&-explosion) + &-right + cut imply explosion

(D8) $\alpha \& \neg\alpha \vdash \beta$

$$\dfrac{\alpha\&\neg\alpha \vdash \beta \qquad \dfrac{\alpha \vdash \alpha \quad \neg\alpha \vdash \neg\alpha}{\alpha, \neg\alpha \vdash \alpha\&\neg\alpha}\ (\&\text{-right})}{\alpha, \neg\alpha \vdash \beta}\ (\text{cut})$$

(b) ¬-left + &-left + dilution-right + cut imply &-explosion

(D9)
$$\dfrac{\dfrac{\dfrac{\alpha \vdash \alpha \quad \alpha \vdash \alpha \quad \neg\alpha \vdash \neg\alpha}{\alpha, \neg\alpha \vdash \quad \alpha\&\neg\alpha \vdash \alpha \quad \alpha\&\neg\alpha \vdash \neg\alpha}\ (\neg\text{-left}, \&\text{-left})}{\alpha, \neg\alpha \vdash \beta}\ (\text{dil-right})}{\alpha\&\neg\alpha \vdash \beta}\ (\text{cut})$$

(c) $\alpha \& \beta \equiv \neg(\neg\alpha \vee \neg\beta)$ + subminimal + ¬¬-E + LEM + dilution + cut imply &-explosion

$$
\begin{array}{ll}
\text{(D10)} \ \underline{\vdash \neg\alpha \vee \neg\neg\alpha} & \text{(LEM)} \\
\underline{\neg\beta \vdash \neg\alpha \vee \neg\neg\alpha} & \text{(dilution)} \\
\underline{\neg(\neg\alpha \vee \neg\neg\alpha) \vdash \neg\beta} & \text{(subminimal)} \\
\underline{\alpha\&\neg\alpha \vdash \neg\neg\beta} & \text{(definition of \&)} \\
\underline{\neg\neg\beta \vdash \beta} & (\neg\neg\text{-E}) \\
\alpha\&\neg\alpha \vdash \beta & \text{(cut)}
\end{array}
$$

Derivation of (a) suggests that non-explosion does not endorse &-right and &-explosion to occur together. There are well-known paraconsistent systems, known as non-adjunctive logics [1, 19], where &-explosion is allowed but non-explosion is achieved by rejecting &-right. In derivations of (b) and (c), &-explosion is achieved via some logical rules which include $\neg$-left (in b), and subminimal property of $\neg$ (in c). Both of these properties are excluded from our axiomatization. So, one can safely go with two possible extensions of $C_{\{\neg,\vee\}}$; one is to include &-right, &-left, and the other is to include &-left, &-explosion. LEM is equivalent to (R$\neg$), i.e. (C5) in the presence of $\vee$-right and $\vee$-left. So, in the branch (b_1) of $\{\neg, \vee\}$-fragment LEM can be obtained. One can also define &in terms of $\neg$ and $\vee$, as given in (c). But this interdefinability does not ensure the derivability of &-right from the rules for $\vee$-right/left and axioms of $C_{\{\neg\}}$.

So, we propose a pair of simple extensions of $C_{\{\neg,\vee\}}$, each for both the branches (b_1) and (b_2).

$$
\begin{array}{ccc}
 & C_{\{\neg,\vee\}} & \\
\swarrow & & \searrow \\
(\&\text{-right}) + (\&\text{-left}) & & (\&\text{-left}) + (\&\text{-explosion})
\end{array}
$$

The extensions are given below for $i = 1, 2$.

($\mathbf{b}_{i1}$) $C_{\{\neg,\vee\}}$ +

(C11) $\alpha, \beta \in C(X)$ implies $\alpha\&\beta \in C(X)$ (&-right)

(C12) $\gamma \in C(\{\alpha\})$ implies $\gamma \in C(\{\alpha\&\beta\})$ (&-left)

($\mathbf{b}_{i2}$) $C_{\{\neg,\vee\}}$ +

(C12) $\gamma \in C(\{\alpha\})$ implies $\gamma \in C(\{\alpha\&\beta\})$ (&-left)

(C13) $\beta \in C(\{\alpha\&\neg\alpha\})$ (&-explosion)

Accordingly, the extensions for $\mathrm{PI}_{\{\neg,\vee\}}$ are given below.

($\mathbf{b}_{i1}$) $\mathrm{PI}_{\{\neg,\vee\}}$ +

(PI11) $(X \cup \{\neg\alpha\}, \alpha) \in PI$ and $(X \cup \{\neg\beta\}, \beta) \in PI$ imply $(X \cup \{\neg(\alpha\&\beta)\}, \alpha\&\beta) \in PI$

(PI12) $(X \cup \{\neg(\alpha\&\beta)\}, \alpha\&\beta) \in PI$ implies $(X \cup \{\neg\alpha\}, \alpha) \in PI$

($\mathbf{b}_{i2}$) $\mathrm{PI}_{\{\neg,\vee\}}$ +

(PI12) $(X \cup \{\neg(\alpha\&\beta)\}, \alpha\&\beta) \in PI$ implies $(X \cup \{\neg\alpha\}, \alpha) \in PI$

(PI13) For any α, β, $(\{\alpha\&\neg\alpha\}, \beta) \in PI$.

As given in Sects. 2.1 and 2.2, here also we have proved the following theorem.

Theorem: For each of the branches b_{ij}, $i, j = 1, 2$, the respective notions of $C_{\{\neg,\vee,\&\}}$ and $\mathrm{PI}_{\{\neg,\vee,\&\}}$ are equivalent.

The bifarcation of $C_{\{\neg,\vee\}}$ for each of its branch is endorsed by the following theorem.

Theorem: In the presence of (C1) to (C4)

(i) if (C11) holds, then (C13) does not hold, and
(ii) if (C13) holds, then (C11) does not hold.

12.2.4 *Consequence and Inconsistency in* {¬, ∨, &, ⊃}*-fragment*

Modus ponens (MP) is a rule which is almost everywhere acceptable in the context of reasoning. In many logical systems the connective implication (⊃) is defined in terms of ∨ and ¬; or vice-versa, i.e. ∨ is defined in terms of ¬ and ⊃. In such cases, MP and DS are equivalent to each other. Specifically, MP + $\alpha \supset \beta \equiv \neg\alpha \vee \beta$ imply DS, and DS + $\alpha \supset \beta \equiv \neg\alpha \vee \beta$ imply MP. Let us see some derivations generating explosion in the {¬, ∨, &, ⊃}-fragment of a language.
(a) ∨-right + cut + MP + $\alpha \supset \beta \equiv \neg\alpha \vee \beta$ imply explosion.

$$\text{(D11)}\quad \begin{array}{ll} \underline{X, \alpha \vee \beta \vdash \gamma} & \\ \underline{\alpha \vdash \alpha} & \\ \underline{\alpha \vdash \alpha \vee \beta} & (\vee\text{-right}) \\ X, \alpha \vdash \gamma & (\text{cut}) \end{array}$$

Call the property obtained above $(\vee_N)$: $\dfrac{X, \alpha \vee \beta \vdash \gamma}{X, \alpha \vdash \gamma}$

$$\text{(D12)}\quad \begin{array}{ll} \underline{\alpha, \alpha \supset \beta \vdash \beta} & (\text{MP}) \\ \underline{\alpha, \neg\alpha \vee \beta \vdash \beta} & (\text{definition of } \vee) \\ \alpha, \neg\alpha \vdash \beta & (\vee_N) \end{array}$$

(b) DT + $\alpha \vee \beta \equiv \neg\alpha \supset \beta$ imply ∨-right:

$$\text{(D13)}\quad \begin{array}{ll} \underline{\beta \vdash \beta} & \\ \underline{\neg\alpha, \beta \vdash \beta} & (\text{dilution}) \\ \underline{\beta \vdash \neg\alpha \supset \beta} & (\text{DT}) \\ \beta \vdash \alpha \vee \beta & \end{array}$$

(¬¬-I) + MP + $\alpha \vee \beta \equiv \neg\alpha \supset \beta$ imply DS.

$$\text{(D14)}\quad \begin{array}{ll} \underline{\neg\neg\alpha, \neg\neg\alpha \supset \beta \vdash \beta} & (\text{MP}) \\ \underline{\alpha \vdash \neg\neg\alpha} & (\neg\neg\text{-I}) \\ \underline{\alpha, \neg\neg\alpha \supset \beta \vdash \beta} & (\text{cut}) \\ \alpha, \neg\alpha \vee \beta \vdash \beta & \end{array}$$

∨-right + DS imply explosion.

So, from (a) we can conclude that in the presence of $\vee$-right non-explosion implies either $\alpha \supset \beta \equiv \neg\alpha \vee \beta$ has to be dropped or MP has to be dropped. On the other hand, (b) shows that MP and DT cannot be taken together if $\vee$ is defined by $\alpha \vee \beta \equiv \neg\alpha \supset \beta$.

In the last two sections, for each connective apart from the right/left rule one more rule is considered; the rule relates the concerned connective with $\neg$. In the context of interaction of $\supset$ with $\neg$, the first natural property is contraposition, i.e. $\alpha \supset \beta \vdash \neg\beta \supset \neg\alpha$ or its inverse i.e. $\neg\beta \supset \neg\alpha \vdash \alpha \supset \beta$. Contraposition seems to be a variant form of subminimal property, which has been abandoned at the very beginning of $C_{\{\neg\}}$ axiomatization. In this context it is relevant to consider the Logic of Paradox (LP) of Priest [13]. As mentioned in Sect. 2.1, LP satisfies all axioms of $C_{\{\neg\}}$. Furthermore, LP satisfies (C8)-(C9), and (C11)-(C12), that is, LP can be considered in the branch (b_{11}). It is to be noted that LP satisfies DT, and contraposition but does not satisfy MP, that is, in LP subminimal property does not hold but contraposition holds. This claim can also be established from the truth tables of $\neg$ and $\supset$ defined in LP [13].

So, we here consider three properties of $\supset$; One is DT ($\supset$-right), the second is MP ($\supset$-left), and the third is contraposition along with its inverse, which reflects interrelation between $\supset$ and $\neg$, that is, additional axioms for $C_{\{\neg,\vee,\&,\supset\}}$ are as follows.

(C14) $\beta \in C(X \cup \{\alpha\})$ implies $\alpha \supset \beta \in C(X)$ ($\supset$-right/DT)
(C15) $\alpha \supset \beta \in C(X)$ implies $\beta \in C(X \cup \{\alpha\})$ ($\supset$-left/MP)
(C16) $C(X \cup \{\alpha \supset \beta\}) = C(X \cup \{\neg\beta \supset \neg\alpha\})$ (contraposition and its inverse)

The following theorem shows that $C_{\{\neg,\vee,\&\}}$ can be extended in three possible directions.

Theorem: In the presence of (C1) to (C4)

(i) if (C14) and (C15) hold, then (C16) does not hold,
(ii) if (C15) and (C16) hold, then (C14) does not hold, and
(iii) if (C14) and (C16) hold, then (C15) does not hold.

Proof We give the proof of (iii) only.
By (C4) we have for some α, β, $\beta \notin C(\{\alpha, \neg\alpha\})$.
By (C1), $\neg\alpha \in C(\{\alpha, \neg\alpha, \neg\beta\})$. Hence $\neg\beta \supset \neg\alpha \in C(\{\alpha, \neg\alpha\})$ (by (C14)).
By (C16) we have, $\alpha \supset \beta \in C(\{\neg\beta \supset \neg\alpha\})$. Therefore using (C3) we have $\alpha \supset \beta \in C(\{\alpha, \neg\alpha\})$. Hence by (C2), (C3), $C(\{\alpha, \alpha \supset \beta\}) \subseteq C(\{\alpha, \neg\alpha\})$.
Now as, $\beta \notin C(\{\alpha, \neg\alpha\})$, $\beta \notin C(\{\alpha, \alpha \supset \beta\})$, that is, MP does not hold. □

So we obtain the following directions.

$$C_{\{\neg,\vee,\&\}}$$
$$\swarrow \quad \downarrow \quad \searrow$$
$$\swarrow \quad \downarrow \quad \searrow$$
$$(C14) + (C15) \quad (C15) + (C16) \quad (C14) + (C16)$$

Inconsistency axioms are also accordingly divided into three branches (b_{ij1}), (b_{ij2}),

and (b_{ij3}), for $i, j = 1, 2$. The extensions are as follows.
($\mathbf{b}_{ij1}$) $\mathrm{PI}_{\{\neg,\vee,\&\}}$ +

(PI14) $(X \cup \{\alpha, \neg\beta\}, \beta) \in PI$ implies $(X \cup \{\neg(\alpha \supset \beta\}, \alpha \supset \beta) \in PI$
(PI15) $(X \cup \{\neg(\alpha \supset \beta\}, \alpha \supset \beta) \in PI$ implies $(X \cup \{\alpha, \neg\beta\}, \beta) \in PI$

($\mathbf{b}_{ij2}$) $\mathrm{PI}_{\{\neg,\vee,\&\}}$ +

(PI15) $(X \cup \{\neg(\alpha \supset \beta\}, \alpha \supset \beta) \in PI$ implies $(X \cup \{\alpha, \neg\beta\}, \beta) \in PI$
(PI16) $(X \cup \{\alpha \supset \beta\}, \gamma) \in PI$ iff $C(X \cup \{\neg\beta \supset \neg\alpha\}, \gamma) \in PI$

($\mathbf{b}_{ij3}$) $\mathrm{PI}_{\{\neg,\vee,\&\}}$ +

(PI14) $(X \cup \{\alpha, \neg\beta\}, \beta) \in PI$ implies $(X \cup \{\neg(\alpha \supset \beta\}, \alpha \supset \beta) \in PI$
(PI16) $(X \cup \{\alpha \supset \beta\}, \gamma) \in PI$ iff $C(X \cup \{\neg\beta \supset \neg\alpha\}, \gamma) \in PI$

In the same manner, as proved for all the other fragments of the propositional language, for $\{\neg, \vee, \&, \supset\}$-fragment also the following theorem can be proved.

Theorem: $C_{\{\neg,\vee,\&,\supset\}}$ and $\mathrm{PI}_{\{\neg,\vee,\&,\supset\}}$ are equivalent for each of the respective extensions b_{ijk}, $i, j = 1, 2$, and $k = 1, 2, 3$.

This study leads us towards the following diagram.

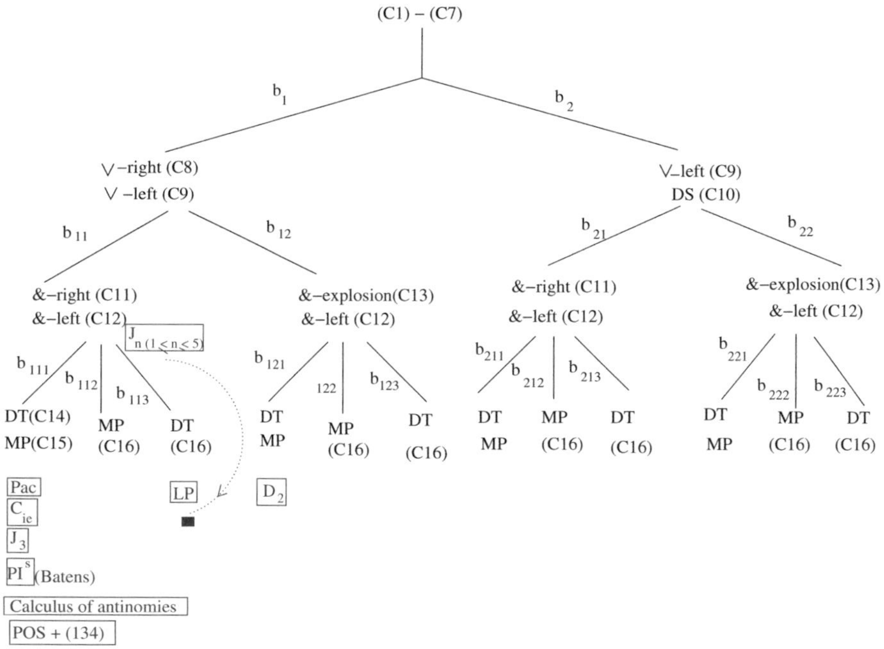

The above diagram shows different possible branches of paraconsistent logics, for which the notions of consequence and inconsistency are interwoven. Moreover, it exhibits which paraconsistent system [1–6, 13–15, 18, 19] is lying where. What else this study can yield that we shall discuss in the concluding section.

Before ending this section we here present a concrete example to visualize the use of this consequence–inconsistency relationship in the context of paraconsistent logics. We consider a particular known logic, viz. Logic of Paradox [13], whose notion of consequence is well defined. Through this example our attempt is to show that the logic also has a well-defined theory for inconsistency, which can determine, given $(X_i, \alpha_i) \in$ PI for finitely many i's, whether $(X, \alpha) \in$ PI or not, where X and α are related to X_i, α_i in some sense.

Example: The following truth table is based on the truth functions for $\supset$ and $\neg$ of the Logic of Paradox [13]. The value set and the designated set are $\{t, p, f\}$ and $\{t, p\}$, respectively.

α	β	$\neg\alpha$	$\neg\beta$	$\alpha \supset \beta$	$\beta \supset \neg\alpha$
t	t	f	f	t	f
t	p	f	p	p	p
t	f	f	t	f	t
p	t	p	f	t	p
p	p	p	p	p	p
p	f	p	t	p	t
f	t	t	f	t	t
f	p	t	p	t	t
f	f	t	t	t	t

We consider the following sets; $X = \{\alpha, \neg\alpha, \beta\}$, $Y = \{\alpha, \beta \supset \neg\alpha, \beta\}$, and $Z = \{\alpha, \beta, \neg(\alpha \supset \beta)\}$. From the above truth table the following can be verified.

(i) $X \vdash \alpha$, $X \vdash \neg\alpha$, $X \nvdash \neg\beta$. That is, $(X, \alpha) \in$ PI, $(X, \beta) \notin$ PI.
(ii) $Y \vdash \alpha$, $Y \nvdash \neg\alpha$, $Y \nvdash \neg\beta$. That is, $(Y, \alpha) \notin$ PI, $(Y, \beta) \notin$ PI.
(iii) $Z \vdash \alpha$, $Z \vdash \neg\alpha$, $Z \vdash \beta$, $Z \vdash \neg\beta$, $Z \vdash \alpha \supset \beta$, and $Z \vdash \neg(\alpha \supset \beta)$.

That is, $(Z, \alpha) \in$ PI, $(Z, \beta) \in$ PI, $(Z, \alpha \supset \beta) \in$ PI.

Now using the theory for PI from (i) we obtain the following.
We have $(\{\alpha\} \cup \{\beta, \neg\alpha\}, \alpha) \in$ PI. Hence by (PI14) we can conclude $(\{\alpha, \neg(\beta \supset \alpha)\}, \beta \supset \alpha) \in$ PI. Also by (PI8), for $\{\neg, \vee\}$-fragment, we conclude $(\{\alpha, \beta, \neg(\alpha \vee \gamma)\}, \alpha \vee \gamma) \in$ PI, but cannot claim $(\{\alpha, \beta, \neg(\beta \vee \gamma)\}, \beta \vee \gamma) \in$ PI.

On the other hand, in case of (iii) we have $(\{\alpha, \beta, \neg(\alpha \supset \beta)\}, \alpha \supset \beta) \in$ PI, but as in LP (PI15) is absent we cannot claim $(\{\alpha, \beta\} \cup \{\neg\beta\}, \beta) \in$ PI using (PI15); but the same can be claimed directly using (PI1).

12.3 Concluding Remarks and Future Directions

One of the pioneering works where a systematic attempt is made to develop a class of paraconsistent logics satisfying some properties of the notion of consequence, as close as possible to the classical one, is due to Newton da costa [7, 8]. Newton da Costa proposed a hierarchy of C-systems imposing some constraints on the notion of consequence such that the logical inference of those systems remains as close as possible to that of the classical system, and at the same time rejects the law of contradiction. Later, da Costa along with other researchers [1] proposed clauses for non-truth functional semantics for negation, some of which give the semantic justification for da Costa's C-systems. In the C-systems da Costa kept a room open for expressing 'a formula behaves consistently' in the syntax of the object language. The same attempt is also found in the literature of logics of formal inconsistency [6]. Logics of formal inconsistency (LFI) are also developed in a systematic manner where one by one properties have been added to the meta-logical notion of consequence to obtain a general scheme for generating paraconsistent logics. The basic system of this class of LFI's is known as $C_{\min}$; da Costa's C-systems turn out to be LFI's proposed in [6]. By imposing one by one conditions to the consequence operator of $C_{\min}$, a hierarchy of LFI's is obtained. Some of these LFI's are proposed to have an operator $\circ$ in the object language so that 'α is consistent' can be expressed by $\circ\alpha$ in the object language. In both of these approaches, the idea was to designate some formulas which *behave consistently*, and allow them to explode in the presence of their negated formulas. Thus the target was to restrict explosion for a class of formulas, and make the system generally non-explosive. In order to achieve this objective, the meta-logical notion of 'consistency' was brought in the object language of the logical system. We, in contrast, concentrate on a study of consequence and inconsistency simultaneously, and explore how the properties of one meta-logical notion influence the property of the other meta-logical notion in order to have a one-to-one correspondence between consequence and inconsistency for paraconsistent logics. This study leads us towards a general scheme for generating paraconsistent logics where the notions of consequence and inconsistency are interwoven.

The tree diagram, given in the last section, works like a scheme for generating a class of paraconsistent logics following a common pattern. Each system lying at the end of a branch should satisfies the following. For each binary connective, say $*$, either the system satisfies left and right rule for $*$, or it satisfies any one of them along with a rule specifying interaction of $*$ with $\neg$. From the diagram, it is visible that most of the known existing systems are lying in the branches of b_1. We yet have not found systems under the other branches. The search may lead to rediscovering existing systems under this consequence–inconsistency characterization, or generating new paraconsistent systems following the scheme.

We could have a different diagram if we consider different rules specifying interaction of a connective with $\neg$. Let us take the instance of $\supset$, for which we have considered contraposition and its inverse. The systems J_n, $1 \leq n \leq 5$, satisfy a variant of (C16), and we are not sure whether these five systems lie in the branch b_{123} or not. So, a little variation to (C16) may give another diagram as a model for some other paraconsistent logics. These all are new challenges to explore.

Finding an alternative theoretical perspective for paraconsistent logics where consequence and inconsistency are interderivable is one of the future aims for this study. The importance of this future research direction, perhaps, will be clear from the discussion below. The usual way of proving completeness theorem of the first order logic exploits the interrelation between the notions of consequence and inconsistency, and the bivalent nature of the connective negation. Both of these are missing in the context of paraconsistent logics. The result of this missing link is also evident from the fact that Diderik Batens had to take a different route in order to prove the completeness theorem for a class of paraconsistent logics. The following lines are referred to [5] in this regard.

The traditional proof that the classical propositional calculus (PC) is strongly complete (i.e. if $\alpha \models A$ then $\alpha \vdash A$) is based on the notion of a maximal consistent set of formulas, and hence on certain properties of strong (i.e. PC) negation. I present a completeness-proof method which does not refer to maximal consistent sets, but only to sets which are (i) non-trivial (not all formulas are members), (ii) deductively closed (all syntactical consequences are members) and (iii) implication saturated (for all B, $A \supset B$ is a member if A is not a member).

In the completeness-proof, mentioned above, sets with the properties (i)–(iii) play the similar role which maximal consistent sets play in ordinary context. Completeness-proof is not the only case; interrelation between consequence and consistency/inconsistency makes it smooth to prove the meta-theory of a logic. In paraconsistent literature different approaches are taken to design a non-explosive consequence; but the notion of inconsistency has not been modified accordingly to fit suitably with such inconsistency tolerant notion of consequence. We expect that this relativized notion of inconsistency and its interrelation with non-explosive consequence may help us to get an alternative route for proving meta-theorems of a paraconsistent system.

References

1. Arruda, A.I.: Aspects of the historical development of paraconsistent logic. In: Priest, G., Routley, R., Norman, J. (eds.) Paraconsistent Logic: Essays on the Inconsistent, pp. 99–129. Philosophia Verlag, Nunchen, Handen, Wien (1989)
2. Asenjo, F.G.: A calculus of antinomies. Notre Dame J. Form. Logic **VII**(1), 103–106 (1966)
3. Avron, A.: Natural 3-valued logics: characterization and proof theory. J. Symb. Logic **56**, 276–294 (1991)
4. Avron, A.: Simple consequence relations. Inf. Comput. **92**, 105–139 (1991)
5. Batens, D.: A completeness-proof method for extensions of the implicational fragment of the propositional calculus. Notre Dame J. Form. Logic **21**(3), 509–517 (1981)
6. Carnielli, A.W., Coniglio, M.E., Marcos, J.: Logics of formal inconsistency. In: Gabbay, D., Guenthner, F. (eds.) Handbook of Philosophical Logic, vol. 14, pp. 1–93 (2003)
7. da Costa, N.: Calculus propositionnels pour less systèmas formels inconsistants. Comptes Rendus Helodomadaires des Séances de I' Academie des Sciences Paris, Series A **257**, 3790–3792 (1963)
8. da Costa, N.: On the theory of inconsistent formal systems. Notre Dame J. Form. Logic **XV**, 497–510 (1974)
9. Dunn, M.J.: Star and perp: two treatments of negation. Philos. Perspect. Lang. Log. **7**, 331–357 (1993)
10. Dunn, M.J.: A comparative study of various model theoretic treatments of negation: a history of formal negation. In: Gabbay, D.M., Wansing, H. (eds.) What is Negation?, pp. 23–51 (1999)
11. Dutta, S., Chakraborty, M.K.: Negation and paraconsistent logics. Log. Univers. **5**(1), 165–176 (2011)

12. Gentzen, G.: Investigations into logical deductions. In: The Collected Papers of Gentzen, G., Szabo, M.E. (eds.), pp. 68–131. North Holland Publications, Amsterdam (1969)
13. Priest, G.: The logic of paradox. J. Philos. Logic **8**, 219–241 (1979)
14. Priest, G.: Paraconsistent logic. In: The Handbook of Philosophical Logic, vol. 6, pp. 287–393 (2002)
15. Restall, G.: Laws of non-contradiction, laws of the excluded middle, and logics. In: Priest, G., Beall, J.C., Garb-Armour, J.C. (eds.) The Law of Non-contradiction, pp. 73–84 (2004)
16. Surma, S.J.: The growth of logic out of the foundational research in mathematics. In: Agazzi, E. (ed.) Modern Logic-A Survey, pp. 15–33. D. Reidel Publishing co., Dordrecht (1981)
17. Tennant, N.: An anti-realist critique to dialetheism. In: Priest, Beall, and Garb-Armour (eds.) The Law of Non-contradiction, pp. 355–384 (2004)
18. Tuziak, R.: Paraconsistent extensions of positive logic. Bull. Sect. Logic **25**(1), 15–20 (1996)
19. Vasyukov, V.L.: A new axiomatization of Jaśkowski's discussive logic. Logic Log. Philos. **9**, 35–46 (2001)

Chapter 13
Univalent Foundations of Mathematics and Paraconsistency

Vladimir L. Vasyukov

Abstract Vladimir Voevodsky in his Univalent Foundations Project writes that univalent foundations can be used both for constructive and for non-constructive mathematics. The last is of extreme interest since this project would be understood in a sense that this means an opportunity to extend univalent approach on non-classical mathematics. In general, Univalent Foundations Project allows the exploitation of the structures on homotopy types instead of structures on sets or structures on categories as in case of set-level mathematics or category-level mathematics. Non-classical mathematics should be respectively considered either as non-classical set-level mathematics or as non-classical category-level (toposes-level) mathematics. Since it is possible to directly formalize the world of homotopy types using in particular Martin-Lof type systems then the task is to pass to non-classical type systems e.g. da Costa paraconsistent type systems in order to formalize the world of non-classical homotopy types. Taking into account that the univalent model takes values in the homotopy category associated with a given set theory and to construct this model one usually first chooses a locally cartesian closed model category (in the sense of homotopy theory) then trying to extend this scheme for a case of non-classical set theories (e.g. paraconsistent ones) we need to evaluate respective non-classical homotopy types not in cartesian closed categories but in more suitable ones. In any case it seems that such Non-Classical Univalent Foundations Project should be directly developed according to Logical Pluralism paradigma and and it seems that it is difficult to find counter-argument of logical or mathematical character against such an opportunity except the globality and complexity of a such enterprise.

Keywords Homotopy types · Univalent foundations · Logical pluralism · Non-classical mathematics · Paraconsistent sets · Paraconsistent categories

The work was done under financial support of Russian Foundation of Humanities in the framework of the project No 13-02-00384 “Axiomatic Thought as the Tool of Scientific Knowledge”.

V.L. Vasyukov (✉)
Institute of Philosophy, Russian Academy of Science,
Volkhonka 14, 119991 Moscow, Russia
e-mail: vasyukov4@gmail.com

J.-Y. Beziau et al. (eds.), *New Directions in Paraconsistent Logic*,
Springer Proceedings in Mathematics & Statistics 152,
DOI 10.1007/978-81-322-2719-9_13

Mathematics Subject Classification 03A05 · 03B62 · 97E30

13.1 Introduction

A few years ago Vladimir Voevodsky have come up with an idea for a new semantics for dependent type theories—"univalent semantics"—which unlike of the usual semantics interpretation of types as sets interprets types as homotopy types. The key property of the univalent interpretation was that it satisfies the univalence axiom which makes it possible to automatically transport constructions and proofs between types which are connected by appropriately defined weak equivalences. According to Voevodsky [6] the key features of these "univalent foundations" are as follows:

1. Univalent foundations naturally include "axiomatization" of the categorical and higher categorical thinking.
2. Univalent foundations can be conveniently formalized using the class of languages called dependent type systems.
3. Univalent foundations are based on direct axiomatization of the "world" of homotopy types instead of the world of sets.
4. Univalent foundations can be used both for constructive and for non-constructive mathematics.

The central concept of the univalent foundations is a *homotopy*. A homotopy between continuous maps $f, g : X \to Y$ is a continuous map $\vartheta : X \times [0; 1] \to Y$ satisfying $\vartheta(x; 0) = f(x)$ and $\vartheta(x; 1) = g(x)$. Such a homotopy ϑ can be thought of as a "continuous deformation" of f into g. Two spaces X and Y are said to be homotopy-equivalent if there are continuous maps going back and forth $X \underset{g}{\overset{f}{\rightleftarrows}} Y$, the compositions of which are homotopical to the respective identity mappings (which is tantamount to saying that there exist homotopies $fg \times [0; 1] \to 1_X$ and $gf \times [0; 1] \to 1_Y$). When this latter condition holds spaces X and Y are called *homotopy equivalent*, or interchangeably, *belonging to the same homotopy type*. It is natural to also consider homotopies between homotopies, referred to as *higher homotopies*. When we consider a space X, a distinguished point $p \in X$, and the paths in X beginning and ending at p, and identify such paths up to homotopy, the result is the fundamental group $\pi(X; p)$ of the space at the point. If we remove the dependence on the base-point p by considering the fundamental groupoid $\pi(X)$,[1] consisting of all points and all paths up to homotopy. Next, rather than identifying homotopic paths, we can consider the homotopies between paths as distinct, new objects of a higher dimension (just as the paths themselves are homotopies between points). Continuing in this way, we obtain a structure consisting of the points of X, the paths in X, the

[1] A groupoid is like a group, but with a partially-defined composition operation Precisely, a groupoid can be defined as a category in which every arrow has an inverse. A group is thus a groupoid with only one object.

homotopies between paths, the higher homotopies between homotopies, and so on for even higher homotopies.

There is a groupoid model of Martin-Löf's type theory, where a given basic type A (a judgements of the form $\vdash A : type$) is groupoid A, term x of type A (judgement $\vdash x : A$) is object x of groupoid A, and dependent type $B(x)$ (judgement $x : A \vdash B(x) : type$) is fibration[2] of the form $B \to A$. Identity type $Id_A(x; y)$ in this model is the *arrow groupoid* of groupoid A, which is a functor category of the form $[I; A]$ where I is the connected groupoid having exactly two non-identical objects and a single non-identity isomorphism between these objects. The crucial idea here was to replace families of sets indexed by sets by families of groupoids indexed by groupoids.

Members of Voevodsky's hierarchy at low levels are as follows (A is a space of h-level $n+1$ if for all its points $x; y$ path spaces $paths_A(x; y)$ are of h-level n):

- *Level 0*: up to homotopy equivalence there is just one contractible space[3] that we call "point" and denote pt;
- *Level 1*: up to homotopy equivalence there are two spaces at this level: the empty space $\varnothing$ and the point pt. We call $\varnothing; pt$ truth values; we also refer to types of this level as properties and propositions. Notice that h-level n corresponds to the logical level $n-1$: the propositional logic (i.e., the propositional segment of our type theory) lives at h-level 1.
- *Level 2*: Types of this level are characterized by the following property: their path spaces are either empty or contractible. So such types are disjoint unions of contractible components (points), or in other words sets of points. This will be our working notion of set available in this framework.
- *Level 3*: Types of this level are characterized by the following property: their path spaces are sets (up to homotopy equivalence). These are obviously (ordinary at) groupoids (with path spaces hom-sets).
- *Level 4*: Here we get 2-groupoids.
- *Level $n+2$*: n-groupoids.

It is interesting to notice that like Euclid Voevodsky begins constructing his hierarchical universe of homotopy types with a point and then applies a simple inductive procedure for generating from this point the rest of this universe (cf. [3]).

The last feature of univalent foundations above looks very attractive for those who are interested in so-called non-classical mathematics paradigma which considers mathematics based on various non-classical logics. The last thesis becomes more comprehensive indeed if we take into account that such "mathematics" should obviously be non-constructive by their nature because within this paradigma they all differ drastically from mathematics based on intuitionistic logic.

---

[2]Fibrations here are functors $p : E \to B$ between grupoids E, B such that for each object e from E and any isomorphism $i : p(e) \leftrightarrow b$ from B there exists an isomorphism $j : e \leftrightarrow e'$ such that $p(j) = i$.

[3]A space A is called *contractible* when there is point $x : A$ connected by a path with each point $y : A$ in such a way that all these paths are homotopic.

At first glance it seems that considerations of such a kind are too abstract, too global and obscure to provide arguments for extending Voevodsky's program in such a manner. But they are indispensable for studies of the foundations of mathematics being the part of the quest for answering the question of proven uniqueness of mathematic existed which would not be taken for granted due to the lack of the worthwhile rivals. And here is one more tendency in univalent foundations program and like consisting not just in incorporating logic to mathematical structures but in deriiving logic from mathematical considerations, in an "internalization" of logic and making it the secondary thing. But why one believe that the result always will be the same? We will try to show that there are some other possibilities which deserve to be taken into account in future investigations.

13.2 Logical Pluralism and Non-classical Mathematics

The motto of the first conference on non-classical mathematics (Hejnice, Czech republic, 2009) was: "The 20th century has witnessed several attempts to build (parts of) mathematics on different grounds than those provided by classical logic. The original intuitionist and constructivist renderings of set theory, arithmetic, analysis, etc. were later accompanied by those based on relevant, paraconsistent, non-contractive, modal, and other non-classical logical frameworks. The subject studying such theories can be called non-classical mathematics and formally understood as a study of (any part of) mathematics that is, or can in principle be, formalized in some logic other than classical".

The featured topics included in program of this conference, but were not limited to, were, in particular, the following:

- *Intuitionistic mathematics*: Heyting arithmetic, intuitionistic set theory, topos-theoretical foundations of mathematics, etc.
- *Constructive mathematics*: constructive set or type theories, pointless topology, etc.
- *Substructural mathematics*: relevant arithmetic, non-contractive naive set theories, axiomatic fuzzy set theories, etc.
- *Inconsistent mathematics*: calculi of infinitesimals, inconsistent set theories, etc.
- *Modal mathematics*: arithmetic or set theory with epistemic, alethic, or other modalities, modal comprehension principles, modal treatment of vague objects, modal structuralism, etc.

An issue arising in connection with these topics would be formulated as follows: it is evident that there are not one but many true mathematics (such point of view could be called a mathematical pluralism) but what is their mutual relationship—they are rivals, amicable with each other, complementary or mutually exclusive? It reminds us the situation with non-euclidean geometry when after Lobachevsky and Riemann discoveries it turns out to be that there are many equivalent systems of geometry and the matter is just their relationship. If to draw an analogy then taking into account modern state of affairs in field of geometry one can in case of non-classical mathematics suggests an opposition of classical and non-classical mathematics: is our mathematics globally classical and locally non-classical (that is, have nonclassical parts) or, vice versa, it is globally non-classical being at the same time locally classical?

Conception of mathematical pluralism is mostly inspired by the situation in logics. In modern philosophy of logic very popular is the point of view of correctness not one but a great number of true logics. Namely this standpoint is being known as logical pluralism. Contemporary debate has led to a re-examination of some older views, especially the pluralism resulting from Carnap's famous tolerance for different linguistic frameworks and, for example, the work of Scottish/French logician Hugh McColl (1837–1909), who some have claimed was an early logical pluralist.

But what is the impact of foundational logic (or rather the change of foundational logic) in non-classical mathematics on the mathematical constructions themselves? Does it really matter?

13.3 Paraconsistent Sets and Homotopies

On the one hand, we can say that foundational logic serves not so much "to prop up the house of mathematics as to clarify the principles and methods by which the house was built in the first place. 'Foundations' as a discipline that can be seen as a branch of mathematics standing apart from the rest of the subject in order to describe the world in which the working mathematician lives" [2, p. 14]. But, on the other hand, set theory being the *lingua universalis* for mathematical foundations grows on the base of foundational logic and as mathematical practice shows the change of logical basis is not unnoticed for set theory. As a consequence we have now intuitionistic set theory, paraconsistent set theory, fuzzy set theory, quantum set theory etc. nucleating the foundational frameworks of the respective non-classical mathematics.

The last thesis would be reinforced by the non-classical attitude in homotopy theory consideration. For this aim let us consider the definitions which take place in the usual homotopy type theory. But firstly we recall the inductive definition (see [7]) which is used for describing homotopy type hierarchy:

- (i) Given space A is called *contractible* when there is point $x : A$ connected by a path with each point $y : A$ in such a way that all these paths are homotopic.
- (ii) We say that A is a space of h-level $n + 1$ if for all its points $x; y$ path spaces $paths_A(x; y)$ are of h-level n.

This completes the defnition.

And now return to the main definitions [6, p. 1]:

- A (homotopy) type T is said to be of h-level 0 if it is contractible,

- A (homotopy) type T is said to be of h-level 1 if for any two points of T the space of paths between these two points is contractible,
- A (homotopy) type T is said to be of h-level $n+1$ if for any two points of T the space of paths between these two points is of h-level n.

Then we have:

- There is only one (up to a homotopy equivalence) type of h-level 0—the one point type pt.
- There are exactly two types of h-level 1, pt and $\emptyset$; i.e. types of h-level 1 are the truth values.
- Types of h-level 2 are types such that the space of paths between any two points is either empty or contractible. Such a type is a disjoint union of contractible components i.e. (up to an equivalence) types of h-level 2 are sets.
- Types of h-level 3 are (homotopy types of nerves of) groupoids.
- More generally, types of h-level $n+2$ can be seen as equivalence classes of n-groupoids.

For $\emptyset$ condition (ii) is satisfied vacuously; for pt (ii) is satisfied because in pt there exists only one path, which consists of this very point. Usually $\emptyset$, pt are called *truth values* and also refer to types of this level as *properties* and *propositions*. Notice that h-level n corresponds to the logical level $n-1$: the propositional logic (i.e., the propositional segment of our type theory) lives at h-level 1.

But what about $\emptyset$? It is known (see [1]) that there is a system ZF_1 of paraconsistent set theory that related to Church's version of Zermelo-Fraenkel set theory ZF_0 with a universal set as a da Costa paraconsistent first-order logic $C_1^=$ is related to the clasical first-order predicate calculus $C_0^=$. In essence, "ZF_1 should be 'partially' included in ZF_0, though the latter is is also to be contained, in a certain sense, in the former" [1, p. 170]. The basic set-theoretic concepts of ZF_1 are analogous to those of ZF_0, although the concepts involving negation give rise to two notions: one involving the weak negation ($\neg$) and the other the strong negation ($\neg*$). As a result we have, for instance, two empty sets: $\varnothing = \{x : x \neq x\}$ and $\varnothing^* = \{x : \neg * (x = x)\}$.

In this case (i.e. one will purport sets not from the class of ZF_0-models but from the class of ZF_1-models) the respective point from the definition above will be changed:

- There are exactly three types of h-level 1, pt, $\emptyset$ and $\varnothing^*$; i.e. types of h-level 1 are the *truth values*.

Of course, since it is known that each axiom scheme of ZF_0 generates two corresponding axiom schemes of ZF_1, one with the strong negation and another with the weak one, then we can say that ZF_1 simply includes ZF_0. Hence, it seems that our reformulation of the point above does not essentially distort the whole construction of usual homotopy theory. The only consequences is that the generality of this theory will be restricted.

But since ZF_1 should be considered as some extension of ZF_0 then one can assume that there are some (paraconsistent) sets which are outside the scope of usual set theory. Hence, our usual account of the category *Set* as the category of *all* sets will be incomplete and *Set* appears to be just a subcategory of some category *PSet* which should include paraconsistent sets too.

Anyway, will such kind of limitation be capable to affect the project of univalent foundations? The answer depends not only on the level of logical and mathematical tools exploited but also on the alternatives proposed.

13.4 Paraconsistent Categories and Types

Along with set-theoretical there are another aspects of restriction of generality of homotopy type theory and respectively univalent foundations. Analyzing this program A.Rodin writes: "in Voevodsky's univalent foundations homotopy types turn to be the elementary bricks for constructing the whole of the mathematical universe including its logic" [3, p. 224]. This inclusion having its effect in uniform hierarchical treatment of propositions simply as data of the specific type along with sets and categories. This uniformity plainly displayed in the semantic construction.

Syntactically a non-classical way of extending univalent foundations seems to be evident: one need to employ non-classical logical connectives and axioms in all formulations e.g. issue from typed λ—calculi with non-classical forming operations. But there are some deadends on this way—the open question of the general conception of negation forming operation (in many logical systems the negation is a primitive connective unlike the intuitionistic logic).

More challenging seems the semantic approach. To obtain a model with values in a category one need to construct a category with some additional structure, which is an object defned up to an equivalence. A technique for doing this which Voevodsky found very useful is based on the notion of a universe structure in a category [6, p. 4]. Let C be a category. By a universe structure on C we will mean a collection of data of the following form:

1. a final object pt in C,
2. a morphism $p : \tilde{U} \to U$,
3. for any morphism $f : X \to U$ a choice of a pull-back square

$$\begin{array}{ccc} (X, f) & \xrightarrow{Q(f)} & \tilde{U} \\ {\scriptstyle p(X,f)}\downarrow & & \downarrow{\scriptstyle p} \\ X & \xrightarrow[f]{} & U \end{array}$$

Usually then it is postulated that C is an *lccc*—a locally cartesian closed category. Doing this we obtain an opportunity to exploit a well known interpretation of logical connectives of intuitionistic logic as type forming operators since Martin-Lof's intuitionistic type theory has an *lccc* as a category-theoretic model. What will happen if we modify C?

Early in [4] the intepretation of da Costa paraconsistent logics in topos of functors was proposed. As the basic construction there have been implemented so-called CN-categories. Generalizing their definition (original construction was based on pre-order categories) we can shortly characterize such a category C as cartesian closed category for which the following conditions are fulfilled:

- for any object a of C there is an object Na such that we have arrows $NNa \to a$ and $a^o \to (Na)^o$ in C where $a^o = N\langle a, Na\rangle$ and for any arrow $d \to a$ there is an arrow $d \to Na$ in C;
- for any two objects a, b in C there is an arrow $a^o \to (b \Rightarrow a) \Rightarrow ((b \Rightarrow Na) \Rightarrow Nb)$ (here $\Rightarrow$ is an exponential);
- $1 \cong [a, Na]$ and $0 \cong \langle a^o, Na^o\rangle$.

Let us call such category a paraconsistent cartesian closed category—*pccc*. The main advantage of *pccc* is that there is the categorical interpretation of paraconsistent negation in it (along with da Costa paraconsistent logics). What should be done as one more step in our consideration it is an introduction of *lpccc*—a locally paraconsistent cartesian closed category. And if we succeed then we can try to transform our category C above in *lpccc*. Simultaneously one can ask the question: would it be right to speak in this case that we obtain in such a way a model of paraconsistent type theory? If yes then would we consider the situation from the point of view of non-classical univalent foundations project?

One more example would be obtained in a similar way by issuing from the categorical interpretation of relevant logic R in topos of functors (see [5]). Here the original subject were so-called RN-categories which generalization in nutshell should be defined as $\otimes$—cartesian cliosed categories with negation (again by generalizing the original construction based on pre-order categories). In detail one can defined them as the bicomplete categories endowed with a covariant bifunctor $\otimes : C \times C \to C$ and a contravariant functor $N : C \to C$ such that the following conditions are satisfied:

(i) for any objects a, b, c in C there are the following natural isomorphisms:

$$a \otimes [b, c] \cong [a \otimes b, a \otimes c],$$
$$[b, c] \otimes a \cong [b \otimes a, c \otimes a],$$

(ii) C allows exponentiation relative to $\otimes$, i.e. the following diagram commutes

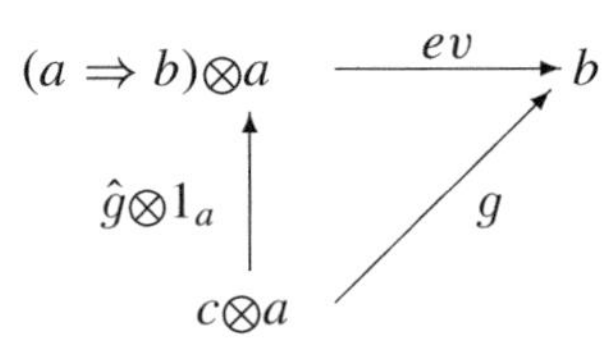

where $\Rightarrow$ is an exponential;

(iii) the following functorial equations are satisfied:

(a) $(g_1 f_1) \otimes (g_2 f_2) = (g_1 \otimes g_2)(f_1 \otimes f_2)$;
(b) $1_A \otimes 1_B = 1_{A \otimes B}$.

(iv) C has an object 1 such that $1 \otimes a \cong a$ and there is an arrow $a \to a \cong 1$ in C for all a in C;
(v) for any objects a, b, c in C, $a \otimes (b \otimes c) \cong (a \otimes b) \otimes c$.
(vi) for any objects a, b in C there is an arrow $a \otimes b \to b \otimes a$.
(vii) for any object a in C there is an arrow $a \to a \otimes a$.
(viii) $N^2 a \cong a$ for any a in C;
(ix) for any arrow $a \otimes b \to c$ there is an arrow $a \otimes Nc \to Nb$ in C.

If we denote such categories as relevant cartesian closed categories—*rccc*—and consider their more sophisticated version—a locally relevant cartesian closed categories *lrccc*—then one can conclude that we arrive at non-classical model of a hypothetical relevant type theory. Taking into account an internal paraconsistency of relevant logic R which is now mirroring in the *lrccc* due to the existence of an interpretation of R in topos of functors from RN-category to *Set* would it again be true to claim that there exists one more way to obtain a version of non-classical (paraconsistent) univalent foundations project?

13.5 Conclusion

The extremely non-classical point of view considered here seems to be too much marginal for taking him into account. But if the mathematical universe includes its logic in one or another way then it seems that logical pluralism inevitably will have direct influence on the mathematics. Hence, mathematical pluralism and emerging of non-classical mathematics are not accidental phenomena and these faces of plurality in mathematics are her real proper faces.

On the other hand, the implied lack of the uniqueness of univalent foundations project again make actual an issue of mathematical "paradise lost". It seems that we can turn out to be the witnesses of the process of paradigmal shifting (from set-theoretic to homotopy type paradigma). But is there one true mathematics still is an open question.

References

1. da Costa, N.C.A.: Paraconsistent mathematics. In: Batens, D., Mortensen, C., Priest, G., van Bendegem, J.-P. (eds.) Frontiers of Paraconsistent Logic, pp. 166–179. Research Studies Press Ltd, Baldock, Hartfordshire, England (2000)
2. Goldblatt, R.: Topoi. The Categorial Analysis of Logic. North-Holland, Amsterdam (1984)
3. Rodin, A.: Axiomatic Method and Category Theory. Springer, New York (2013)

4. Vasyukov, V.L.: Paraconsistency in categories. In: Batens, D., Mortensen, C., Priest, G., van Bendegem, J.-P. (eds.) Frontiers of Paraconsistent Logic, pp. 263–278. Research Studies Press Ltd, Baldock, Hartfordshire, England (2000)
5. Vasyukov, V.L.: Paraconsistency in categories: case of relevant logic. Stud. Logica **98**(3), 429–443 (2011)
6. Voevodsky, V.A.: Univalent Foundations Project (a modifed version of an NSF grant application), 1 Oct 2010 (2010a). http://www.math.ias.edu/~vladimir/Site3/Univalent_Foundations_files/univalent_foundations_project.pdf
7. Voevodsky, V.A.: Introduction to the Univalent Foundations of Mathematics (video recording of lecture given at princeton institute of advanced studies 10 Dec 2010), (2010b). http://www.math.ias.edu/node/2778

Chapter 14
A Method of Defining Paraconsistent Tableaus

Tomasz Jarmużek and Marcin Tkaczyk

Abstract The aim of this paper is to show how to simply define paraconsistent tableau systems by liberalization of construction of complete tableaus. The presented notions allow us to list all tableau inconsistencies that appear in a complete tableau. Then we can easily choose these inconsistencies that are effects of interactions between premises and a conclusion, simultaneously excluding other inconsistencies. A general technique we describe is presented here for the case of Propositional Logic, as the simplest one, but it can be easily extended to more complex cases. In other words, a kind of paraconsistent consequence relation is being studied here, and a simple tableau system is shown to exist that captures that consequence relation.

Keywords Blind rule · Paraconsistent consequence relation · Paraconsistent tableaus · t-inconsistency · Tableau rules · Tableaus

Mathematics Subject Classification (2000) Primary 99Z99 · Secondary 00A00

This work was completed with the support of Polish National Center of Science NCN UMO-2012/05/E/HS1/03542.

T. Jarmużek (✉)
Departament of Logic, Nicolaus Copernicus University in Toruń,
ul. Stanisława Moniuszki 16, 87-100 Toruń, Poland
e-mail: jarmuzek@umk.pl

M. Tkaczyk
Departament of Logic, The John Paul II Catholic University of Lublin,
Al. Racławickie 14, 20-950 Lublin, Poland
e-mail: tkaczyk@kul.pl

J.-Y. Beziau et al. (eds.), *New Directions in Paraconsistent Logic*,
Springer Proceedings in Mathematics & Statistics 152,
DOI 10.1007/978-81-322-2719-9_14

14.1 Introduction and Overview

In this article, we study some kind of paraconsistent consequence relation that is determined by tableau system. That is why we start with some remarks on tableau methods and a strategy we implement.

Usually, tableau methods are at the same time effective but rather intuitive and nonformal. One of the many problems of this is that when we develop a tableau, we can many times obtain the same expressions or apply rules to branches that contain inconsistent expressions. It is a reason why a formal approach to tableau methods that exclude such situations are studied.

Although we prepared a formal theory of tableaus [2] that prevents from a jeJune way of application of tableau rules—among others, they cannot be applied to branches that contain inconsistent expressions—here we defined rules as *blind*. It means that tableau inconsistencies that occur in tableaus do not stop developing of a given tableau. We do not stop a proof, till we decompose all expressions. It is because in the case of paraconsistent arguments we look for a special kind of inconsistency that follows from incompatibility of premises and a negated conclusion. In order to identify suitable inconsistencies in a tableau, we need to decompose all expressions to the level of literals in such a way that it would give an answer to the question whether there is a collision between premises and a negated conclusion or not.

In Sect. 14.3, we describe a mechanism of building such tableaus and choosing suitable inconsistencies. In further parts, we analyze some metatheoretical properties of this proposal and a paraconsistent consequence relation that it captures.

In the article, we consider the simplest case, the case of Propositional Logic, and its paraconsistent subrelation determined by described tableaus. This type of approach can be applied to other logics, being generalized as long as tableau rules are defined in the proposed style.

Finally, let us add that in our paper the tableau tools are treated as a fully syntactical method of checking whether arguments are correct and—as a counterpart of a consequence relation—defined semantically.

14.2 Basic Notions

In this part of paper, we remind some semantical notions and basic tableau notions we need to formulate and prove facts about paraconsistent tableaus that define a paraconsistent subrelation of the classical propositional consequence relation.

14.2.1 Classical Propositional Consequence Relation

Let For be the set of all formulas build over the following alphabet: $\mathsf{Var} \cup \{\neg, \wedge, \vee, \rightarrow, \leftrightarrow\}$. Let $V : \mathsf{For} \longrightarrow \{0, 1\}$ be a valuation of formulas, so for any $A, B \in \mathsf{For}$ the function V satisfies conditions:

- $V(\neg A) = 1$ iff $V(A) = 0$
- $V(A \wedge B) = 1$ iff $V(A) = 1$ and $V(B) = 1$
- $V(A \vee B) = 1$ iff $V(A) = 1$ or $V(B) = 1$
- $V(A \rightarrow B)$ iff $V(A) = 0$ or $V(B) = 1$
- $V(A \leftrightarrow B)$ iff $V(A) = V(B)$.

Let $\mathbf{V}$ be a set of all valuations of formulas. Having a set of formulas X and a valuation V we say that X is *true in* V (in short: $V(X) = 1$) iff for all $A \in X$, $V(A) = 1$. If a set of formulas X is not true in V, we say that is *false in* V and write $V(X) = 0$.

We define classical consequence relation $\models$ on $2^{\mathsf{For}} \times \mathsf{For}$ in a standard way. Hence, for any formula A and set of formulas X we say that A is *a consequence of* X (in short: $X \models A$) iff $\forall_{V \in \mathbf{V}}(V(X) = 1 \Longrightarrow V(A) = 1)$.

We say that a set of formulas X is *inconsistent* iff for any valuation $V(X) = 0$. Otherwise, we call X *consistent.*[1] Now, by definition of classical consequence relation $\models$, we have a conclusion that expresses a vulnerability of classical logic to inconsistent sets of premises.

Corollary 14.2.1 *Let X be an inconsistent set of formulas. Then $X \models A$, for all $A \in \mathsf{For}$.*

Of course, no paraconsistent logic should have the above property.

14.2.2 Tableau System for Propositional Logic

In work [3] and especially in [2], we presented a formal theory of tableau systems for a class of logics defined by some syntactical and semantical conditions.[2] Hence, we have precise tableau notions that incorporate standard, intuitive notions. The precise notions (of a tableau rule and various kinds of branches, tableaus etc.) with a notion of tableau system are necessary, when we generalize results, looking for some abstract properties of tableau methods.

---

[1] We use a word *inconsistent* instead—for example—*contradictory*, since it enables us a direct transition between semantical and tableau notions.

[2] We mean such logics that are logics of terms or propositions, and are two–valued.

However, here we dealt with tableaus for Boolean language, so we just use well-known intuitive tableau notions presented in many publications, for example in [1]. We remind them in turn giving a handful of nonformal but instructive definitions as usually authors do.

We assume that a set of formulas X are *t-inconsistent* iff for some $A \in \mathsf{For}$, X contains A and $\neg A$. A set of formulas X is *t-consistent* iff X are not *t-inconsistent*. By **R** we denote a set of all standard tableau rules for Boolean connectives. We have nine rules in **R**: four positive rules (for $\wedge$, $\vee$, $\rightarrow$, $\leftrightarrow$) and five negative (for $\neg\neg$, $\neg\wedge$, $\neg\vee$, $\neg\rightarrow$, $\neg\leftrightarrow$).[3]

A root is a set that contains premises and a negation of conclusion. A branch is a sequence of formulas that starts from a root. The rest of the branch contains results of applying of rules to former formulas. A branch is *complete* iff all applicable tableau rules were used.[4] A branch is *incomplete* iff it is not *complete*. Branches can be also closed or open. A branch is *closed* iff it contains t-incosistent set of formulas; it is *open* iff is not closed.

Tableau can be treated as sets of branches with the same roots. Tableaus that include all suitable and only complete branches are called *complete*. Complete tableaus can be closed or open. A tableau is *closed* iff it is complete and all branches that includes are closed; it is *open* iff is not closed.

Now, for any set of formulas X and a formula A, we define a tableau classical consequence relation $\rhd$, by putting:

Definition 14.2.2 $X \rhd A$ iff there exist a finite subset Y of X and a closed tableau with a root $Y \cup \{\neg A\}$.

Of course, the relation $\rhd$ is fully determined by tableau rules of **R**. Therefore, by $\langle \mathsf{For}, \rhd \rangle$ we understand a tableau system determined by classical tableau rules **R**. The tableau system defines Propositional Logic, since it is well-known that:

Fact 14.2.3 For all $X \subseteq \mathsf{For}$, $A \in \mathsf{For}$, $X \rhd A$ iff $X \models A$.

14.3 Paraconsistent Tableaus

As we said our aim is to define a paraconsistent tableau inference that defines some paraconsistent subrelation of classical propositional consequence relation. However, first we introduce some auxiliary notions.

---

[3]The rules modified a little in a manner that is good for our paraconsistent aim are presented in the Sect. 14.3.

[4]Generally, we divide complete branches into open and closed ones, since in our formal theory of tableau methods in [2] our aim is always to complete a branch, so a branch itself is just a technical concept. At the same time an occurrence of a t-inconsistency completes a branch. In the paper we change our point of view a bit: applying of rules is allowed as far as it is possible, ignoring any t-inconsistency—later we will come back to the idea, when explaining exactly what we mean by 'blind rules' (exactly in the Sect. 14.3).

Let $\mathbb{N}$ be the set of natural numbers—the set of indexes $\mathbb{I} = \mathbb{N} \cup \{0\}$. We distinguish an index zero: 0 for conclusions of a given tableau proof.

Next, we define a set of formulas indexed by superscripts: $\mathsf{For}' = \{A^i : A \in \mathsf{For}, i \in \mathbb{N}\}$. The expressions from For' will represent formulas from For in tableau proofs. Notice that no formula in For' has a superscript 0.

By the function $\bullet : \mathsf{For}' \cup \{A^0 : A \in \mathsf{For}\} \longrightarrow \mathbb{N} \cup \{0\}$, defined with the condition $\bullet(A^i) = i$, we can choose superscripts that occur in formulas For' and of those formulas that have a superscript 0.

Let X be a subset of For. By $X(x)$ we mean such a non-empty subset of powerset of For' that for all $Y \in X(x)$ and all $A, B \in \mathsf{For}$ the conditions are fulfilled:

1. $A \in X$ iff for some $i \in \mathbb{N}$, $A^i \in Y$
2. for any $i, j \in \mathbb{N}$, if $A^i, B^j \in Y$ then one of the below holds:
   (a) $A \neq B$ and $i \neq j$
   (b) $A = B$ and $i = j$.

Of course, for a set of formulas X there are usually many sets satisfying $X(x)$–conditions, so writing $Y \in X(x)$ we mean some arbitrary, but fixed set from $X(x)$ that we take into consideration.

Now we give a notion of a particular kind of t-inconsistency. We mean an inconsistency that is a result of expressions with some fixed indexes. Surely, this notion is based on a usual notion of inconsistency (defined here Sect. 14.2.2), so it is still about a set of formulas that contains A and $\neg A$, for some formula A, but additionally both inconsistent formulas should have indexes of some kind. Formally, let $i, j \in \mathbb{I}$ and $X \subseteq \mathsf{For}' \cup \{A^0 : A \in \mathsf{For}\}$. X is $t^{i,j}$*-inconsistent* iff for some A, B:

1. $\{A, B\} \subseteq X$
2. $\{A, B\}$ is a t-inconsistent set of formulas
3. $\{\bullet(A), \bullet(B)\} = \{i, j\}$.

It is a particular kind of t-inconsistency, because it refers to some superscripts omitting t-inconsistencies with other superscripts. Hence, a set Y of formulas with superscripts can be t-inconsistent, but not $t^{i,j}$-inconsistent, for some $i, j \in \mathbb{I}$, since no pair of t-inconsistent formulas in Y contains superscripts i, j. On the other hand, the opposite relationship holds: if a set is $t^{i,j}$-inconsistent, for some $i, j \in \mathbb{I}$, it is also just t-inconsistent.

Now, we reformulate the tableau rules of **R**. A new set of rules **R**$'$ is defined on $\mathsf{For}' \cup \{A^0 : A \in \mathsf{For}\}$. For all $i \in \mathbb{I}$ the schemas of new rules are as below:

$$R_\wedge : \frac{(A \wedge B)^i}{A^i, B^i} \qquad R_\vee : \frac{(A \vee B)^i}{A^i || B^i} \qquad R_\rightarrow : \frac{(A \rightarrow B)^i}{\neg A^i || B^i}$$

$$R_\leftrightarrow : \frac{(A \leftrightarrow B)^i}{A^i, B^i || \neg A^i, \neg B^i} \qquad R_{\neg\neg} : \frac{\neg\neg A^i}{A^i} \qquad R_{\neg\wedge} : \frac{\neg(A \wedge B)^i}{\neg A^i || \neg B^i}$$

$$R_{\neg\vee} : \frac{\neg(A \vee B)^i}{\neg A^i, \neg B^i} \qquad R_{\neg\rightarrow} : \frac{\neg(A \rightarrow B)^i}{A^i, \neg B^i} \qquad R_{\neg\leftrightarrow} : \frac{\neg(A \leftrightarrow B)^i}{\neg A^i, B^i || A^i, \neg B^i}$$

The tableau rules in $\mathbf{R}'$ have such a property that they preserve superscripts. For example, when we decompose a formula $\neg\neg p^1$ by rule for $\neg\neg$, we obtain p^1; if we decompose a formula $(p \to q)^0$ by rule for $\to$, we obtain on the left branch $\neg p^0$ and on the right branch q^0 etc. The technique allows us to trace a process of decomposition of formulas and find out the origin of t-inconsistencies.

As we already said, we resign here from internal mechanism nested in rules that blocks applying rules to branches including t-inconsistencies (it was one of distinguishing features of our last works [2, 3]). Here, we want to develop branches as long as it is possible in order to get all t-consistencies that a branch can generate. Now it is clear why we call these rules 'blind'—they just do not see that a branch is closed, which normally is a sufficient fact to stop applying rules.

Moreover, we assume all definitions for tableaus for Propositional Logic—obviously, now the notions depend on the new set of tableau rules $\mathbf{R}'$. However, we add one more definition for testing its properties.

Definition 14.3.1 Let $Y \in X(x)$, for some $X \subseteq \mathsf{For}$, and let $B \in \mathsf{For}$. A tableau T with a root $Y \cup \{\neg B^0\}$ is *paraconsistently closed* iff:

1. T is complete
2. for any branch b in T there is such index $i \in \bullet(Y \cup \{\neg B^0\})$ that a $t^{i,0}$-inconsistent set of formulas belongs to b.

Now, we explain the conditions in Definition 14.3.1 one by one. First, we have some set of formulas X and a formula B that is supposed to follow from X. We do not assume that X is a finite set, since by defining a suitable tableau consequence relation, we will impose a constraint that there must exist a finite set as a root for some complete and closed tableau (like in the case of classical tableau consequence relation Definition 14.2.2), so below we give examples only for finite cases.

We take a set $Y \in X(x)$, so Y has all and only formulas from X, each one with a different index. We build a complete tableau with the root $Y \cup \{\neg B^0\}$. Now, if on any branch there is a $t^{i,0}$-inconsistency, for some $i \in \bullet(Y \cup \{\neg B^0\})$, then the tableau is paraconsistently closed. If for some branch there is no $t^{i,0}$-inconsistency, for any $i \in \bullet(Y \cup \{\neg B^0\})$, then the tableau is not paraconsistently closed.

Now we present few simple examples of paraconsistently closed tableaus (according to our last Definition 14.3.1) for some key cases.

Example 14.3.1 Consider a set of premises $X = \{p \wedge \neg p\}$ and a possible conclusion q. We take a root $\{(p \wedge \neg p)^1, \neg q^0\}$ and draw a tableau.

$$
\begin{array}{c}
\{(p \wedge \neg p)^1, \neg q^0\} \\
| \\
p^1 \\
| \\
\neg p^1
\end{array}
$$

As we see it is not a paraconsistently closed tableau. In some branch (there is of course only one branch) there is no $t^{i,0}$-inconsistency, for any index i. Admittedly, we have a t-inconsistent set $\{p^1, \neg p^1\}$, but with no index 0. The example shows that a consequence relation completely determined by the notion of paraconsistently closed tableau Definition 14.3.1 is robust to unlimited *ex falso quodlibet*.

A positive point of the presented approach is also that we could sometimes infer a conclusion from a logically invalid formula, if a formula that is a conclusion follows from some part of it.

Example 14.3.2 Consider a set of premises $X = \{p \wedge \neg p\}$ and a possible conclusion p. We take a root $\{(p \wedge \neg p)^1, \neg p^0\}$ and draw a tableau.

$$\begin{array}{c} \{(p \wedge \neg p)^1, \neg p^0\} \\ | \\ p^1 \\ | \\ \neg p^1 \end{array}$$

As we see it is a paraconsistently closed tableau by Definition 14.3.1. In any branch (there is of course only one branch) there is $t^{i,0}$-inconsistency, for some index i. Admittedly, we have a t-inconsistent set $\{p^1, \neg p^0\}$. The example shows that a consequence relation completely determined by the notion of paraconsistently closed tableau Definition 14.3.1 enables to infer parts of a contradictory formula.

Another positive point of the presented approach is that we can of course infer any classical tautology.

Example 14.3.3 Consider a set of premises $X = \emptyset$ and a possible conclusion $p \vee \neg p$. We take a root $\{\neg(p \vee \neg p)^0\}$ and draw a tableau.

$$\begin{array}{c} \{\neg(p \vee \neg p)^0\} \\ | \\ \neg p^0 \\ | \\ \neg\neg p^0 \\ | \\ p^0 \end{array}$$

As we see it is a paraconsistently closed tableau by Definition 14.3.1. In any branch (there is of course only one branch) there is $t^{i,0}$-inconsistency, for some index i. Admittedly, we have a t-inconsistent set $\{p^0, \neg p^0\}$. The example shows that a consequence relation completely determined by the notion of paraconsistently closed tableau Definition 14.3.1 enables to infer parts of a contradictory formula.

However, we have some objections. First of all, we can accept that from $\{p, \neg p\}$ follows p and follows $\neg p$, since some part of information in the set $\{p, \neg p\}$ must be true (either $\{p\}$ or $\{\neg p\}$)—clearly, the inference also holds according to the Definition 14.3.1. But in the Example 14.3.2 we have an inference we do not accept, since a set $\{p \wedge \neg p\}$ cannot be true. One can say that sets $\{p \wedge \neg p\}$ and $\{p, \neg p\}$ are classically equivalent not because they are contradictory, but because in general sets $\{A \wedge B\}$ and $\{A, B\}$ are classically equivalent, for any formulas A, B. Our point of view is that the set $\{p \wedge \neg p\}$ is worthless. Contrary to the set $\{p, \neg p\}$ is not useless, if we do not know whether p or $\neg p$, then we can suspend for a moment one of the premises and use the classical consequence relation to a noncontradictory set $\{p\}$ or $\{\neg p\}$.

But the most striking fact and a fundamental weakness of that approach is pictured in the next example.

Example 14.3.4 Consider a set of premises $X = \{(p \wedge \neg p) \vee q)\}$ and a possible conclusion q. We take a root $\{((p \wedge \neg p) \vee q))^1, \neg q^0\}$ and draw a tableau. The tableau is complete, since all possible rules of decomposition were used.

$$\{((p \wedge \neg p) \vee q))^1, \neg q^0\}$$

$$(p \wedge \neg p)^1 \quad q^1$$

$$p^1$$

$$\neg p^1$$

It is a classically closed tableau, but—according to the Definition 14.3.1—it is not a paraconsistently closed tableau, since on the left branch we do not have $t^{1,0}$-inconsistency, and as a consequence the condition 2 of Definition 14.3.1 is not satisfied.

Since we cannot accept this situation, we propose a modification. This modification brings another additional benefit. Refusing inferences from inconsistent sets of premises, we can almost automatically define simple and intuitive semantics for the new tableaus. Obviously, one can say that a fact we refuse, for example, the inference from $\{p \wedge \neg p\}$ to p and simultaneously accept the inference from $\{p, \neg p\}$ to p is a cost we pay for natural semantics, but we also include such cases like the Example 14.3.4.

14.3.1 Paraconsistent Tableau Consequence Relation

Therefore we redefine the latter definition of paraconsistently closed tableau to capture some inferences we like, exclude some inferences we do not like, and at the same time have intuitive semantics.

Definition 14.3.2 Let $Y \in X(x)$, for some $X \subseteq \mathsf{For}$, and let $B \in \mathsf{For}$. A tableau T with a root $Y \cup \{\neg B^0\}$ is *paraconsistently closed* iff:

1. T is complete
2. for any branch b in T there are such indexes $i, j \in \bullet(Y \cup \{\neg B^0\})$ that $t^{i,j}$-inconsistent set of formulas belongs to b.
3. there is a branch b in T that for any pair of indexes $i, j \in \bullet(Y)$ no $t^{i,j}$-inconsistent set of formulas belongs to b.

Now, we explain the conditions in Definition 14.3.2 one by one. The first condition is identical to that in Definition 14.3.1. A novelty are the remaining two conditions.

In the second condition it is said that for some $i, j \in \bullet(Y \cup \{\neg B^0\})$ in all branches there must be $t^{i,j}$-inconsistent set of formulas. It means that at least on some branches a t-inconsistency may not contain index 0, so we capture cases like in the Example 14.3.4.

The third condition says that in a paraconsistently tableau at least on one branch there is no t-inconsistency generated on the ground of formulas from Y, which means that X is a consistent set of formulas itself. So although according to the former definition for $\{p \wedge \neg p, \neg p\}$ we have a paraconsistently closed tableau (Example 14.3.3), according to the latter one we do not have, which is the most convincing.

Again we have some simple examples.

Example 14.3.5 We come back to Example 14.3.2. Consider a set of premises $X = \{p \wedge \neg p\}$ and a possible conclusion p. We take a root $\{(p \wedge \neg p)^1, \neg p^0\}$ and draw a tableau.

$$\{(p \wedge \neg p)^1, \neg p^0\}$$
$$|$$
$$p^1$$
$$|$$
$$\neg p^1$$

As we see it is not a paraconsistently closed tableau according to Definition 14.3.2. On any branch (there is of course only one branch) there is $t^{i,0}$-inconsistency. Admittedly, we have a t-inconsistent set $\{p^1, \neg p^0\}$. But there is no branch on which for any pair of indexes $i, j \in \bullet(\{(p \wedge \neg p)^1\})$ no $t^{i,j}$-inconsistent set of formulas belongs to b (so the condition 3 is not satisfied), since on all branches we have $t^{1,1}$-inconsistency—$\{p^1, \neg p^1\}$.

Example 14.3.6 Consider a set of premises $X = \{p\}$ and a possible conclusion p. We take a root $\{p^1, \neg p^0\}$ and draw a tableau.

$$\{p^1, \neg p^0\}$$

As we see it is a paraconsistently closed tableau according to Definition 14.3.2. On any branch (there is of course only one branch) there is $t^{i,0}$-inconsistency. Admittedly, we have a t-inconsistent set $\{p^1, \neg p^0\}$. And there is a branch on which for any pair of indexes $i, j \in \{p^1\}$ no $t^{i,j}$-inconsistent set of formulas belongs to b (so the condition 3 is satisfied), since on the branch we do not have $t^{1,1}$-inconsistency.

At the end we present an example, where the mentioned liberalization of rules really works.

Example 14.3.7 Consider a set of premises $X = \{r, p \wedge \neg p\}$ and a possible conclusion r. We take a root $\{r^1, (p \wedge \neg p)^2, \neg r^0\}$ and draw a tableau.

$$\{r^1, (p \wedge \neg p)^2, \neg r^0\}$$

Classically, this is a closed and complete tableau, if we assume we cannot apply tableau rules to inconsistent sets of premisses. There is only one branch and we have $t^{1,0}$-inconsistency on it. Moreover, on some branches there is no $t^{i,j}$-inconsistency for $i, j \in \bullet(\{r^1, (p \wedge \neg p)^2)$. So there it would seem like a paraconsistently closed tableau. But it is not true, we can still make the branch longer and obtain some interesting formulas as below:

$$
\begin{array}{c}
\{r^1, (p \wedge \neg p)^2, \neg r^0\} \\
| \\
p^2 \\
| \\
\neg p^2
\end{array}
$$

As we see now it is not a paraconsistently closed tableau according to Definition 14.3.2, because on all branches we have $t^{i,j}$-inconsistency for $i, j \in \bullet(\{r^1, (p \wedge \neg p)^2)$ and the last condition of definition is not satisfied. The tableau we get, because we can apply tableau rules, even if we have t-inconsistencies. We should not worry about this, since as we have already said we shall define a paraconsistent tableau consequence relation in such a way that a formula A is a consequence of X iff for some finite subset Y of X we have a paraconsistently closed tableau. So although the example is not an example of paraconsistently closed tableau, from the premisses it follows the conclusion, because we can built a paraconsistently tableau with $\{r\}$—a finite subset of X.

$$\{r^1, \neg r^0\}$$

Now, we have a conclusion that expresses a connection between usual, classical tableaus, and paraconsistent tableaus.

Corollary 14.3.3 *Let $Y \in X(x)$, for some set of formulas X, and let B be a formula. A tableau T_1 with the root $Y \cup \{\neg B^0\}$ is paraconsistently closed iff*

1. *there is a complete and open tableau T_2 with a root X*
2. *there is a closed tableau T_3 with a root $X \cup \{\neg B\}$.*

Proof The proof is by conditions 2 and 3 of the Definition 14.3.2. □

It means that we could replace Definition 14.3.2 by the statements 1 and 2 of the Corollary 14.3.3 as definitional conditions. Theoretically, it would be simpler. However, practically it is difficult to choose a suitable subset of premises that generates complete and open tableau, but with a negated conclusion generates a closed tableau. In the presented approach, we consider all possible decompositions, tracking superscripts, and kinds of t-inconsistencies that appear, and finally we can choose a suitable and consistent set of premisses (if any exists) which on interaction with a negated conclusion generates some t-inconsistencies.

Now, we can define a paraconsistent tableau consequence relation $\rhd'$.

Definition 14.3.4 Let $X \subseteq \mathsf{For}$ and $A \in \mathsf{For}$. $X \rhd' A$ iff there exist a finite subset Y of X and a paraconsistently closed tableau with a root $Z \cup \{\neg A^0\}$, for some $Z \in Y(y)$.

One example of how to pass from a unsuccessful tableau to a paraconsistently closed one we give here.

Example 14.3.8 We consider a set of premisses $X = \{\neg p \vee q, r \wedge \neg r, \neg q, \}$ and a conclusion $\neg p$. The question is whether $X \rhd' \neg p$?

The set X is a finite subset of X and the tableau with a root $\{(\neg p \vee q)^1, (r \wedge \neg r)^2, \neg q^3, \neg\neg p^0\}$ is complete—all possible rules of decomposition were used.

$$\{(\neg p \vee q)^1, (r \wedge \neg r)^2, \neg q^3, \neg\neg p^0\}$$
$$p^0$$
$$r^2$$
$$\neg r^2$$
$$\neg p^1 \quad q^1$$

The condition 1 and 2 of the Definition 14.3.2 are satisfied—the tableau is complete and on all branches we have $t^{i,j}$-inconsistency, for some $i, j \in \{1, 2, 3, 0\}$. Unfortunately, the condition 3 of the Definition 14.3.2 is not satisfied, since on all branches we have $t^{i,j}$-inconsistency, for some $i, j \in \{1, 2, 3\}$.

Hence, it is not an example of a paraconsistently closed tableau, but it does not mean that it is not $X \rhd' \neg p$. When we take into account a subset of X, the subset $Y = \{\neg p \vee q, \neg q, \}$ we see that the conditions are fully satisfied. We can look at the latter tableau or draw another one.

$$\{(\neg p \vee q)^1, \neg q^3, \neg\neg p^0\}$$
$$p^0$$
$$\neg p^1 \quad q^1$$

The condition 1 and 2 of the Definition 14.3.2 are satisfied. Moreover at least on one branch—the left one—no $t^{i,j}$-inconsistency we have, for any $i, j \in \{1, 3\}$. Hence, it is an example of a closed tableau and simultaneously a paraconsistently closed tableau, and according to Definition 14.3.4 we have: $\{\neg p \vee q, r \wedge \neg r, \neg q, \} \rhd' \neg p$.

A demanded fact is that the paraconsistent, tableau consequence relation $\rhd'$ is a proper subrelation of a classical tableau consequence relation $\rhd$.

Corollary 14.3.5 $\rhd' \subset \rhd$.

Proof Let $X \rhd' A$, for some $X \subseteq \mathsf{For}$ and $A \in \mathsf{For}$. Then by Definition 14.3.4, there exist a finite subset Y of X and a paraconsistently closed tableau with a root $Z \cup \{\neg A^0\}$, for some $Z \in Y(y)$.

By Corollary 14.3.3 there exist a finite subset Y of X and a closed tableau with a root $Y \cup \{\neg A\}$. So, according to the Definition 14.2.2, $X \rhd A$, and $\rhd' \subseteq \rhd$.

On the other hand, we have an example of a closed tableau (Example 14.3.1), that is not paraconsistently closed. Hence, by Definitions 14.2.2 and 14.3.4, we get $\rhd \not\subseteq \rhd'$. □

Having the relation $\rhd'$, we straightforwardly determine a paraconsistent tableau system $\langle \mathsf{For}, \rhd' \rangle$ of a sublogic of Propositional Logic.

14.4 Semantics

As quick as there appears a question about semantics for $\langle \mathsf{For}, \rhd' \rangle$, we get a natural answer. A natural and commonsense approach to the problem of paraconsistency in Boolean language is to define a paraconsistent semantic relation of consequence by a set of valuations $\mathbf{V}$ as follows:

Definition 14.4.1 For all $X \subseteq \mathsf{For}$ and $A \in \mathsf{For}$, $X \models' A$ iff there is such $Y \subseteq X$ that Y is a consistent set of formulas and $Y \models A$.

Surely, the relation $\models'$ is identical to our relation $\rhd'$.

Theorem 14.4.2 $\models' = \rhd'$.

Proof We take any $X \subseteq \mathsf{For}$ and $A \in \mathsf{For}$.

First, we assume that $X \models' A$. Then, by Definition 14.4.1, there exists such $Y \subseteq X$ that Y is a consistent set of formulas, $Y \models A$ and Y is finite—by compactness of $\models$. By Fact 14.2.3 we have $Y \rhd A$, so there is a closed tableau with a root $Y \cup \{\neg A\}$. But because Y is a consistent set, so there is a complete and open tableau with a root Y. As a consequence, by Corollary 14.3.3, a tableau with a root $Z \cup \{\neg A^0\}$, for some $Z \in Y(y)$, is paraconsistently closed. Hence, by Definition 14.3.4, $X \rhd' A$.

Second, we assume that $X \rhd' A$. By Definition 14.3.4 there exist a finite subset Y of X and a paraconsistently closed tableau with a root $Z \cup \{\neg A^0\}$, for some $Z \in Y(y)$. By Corollary 14.3.3 Y is consistent, since there is a complete and open tableau with a root Y, and $Y \rhd' A$. Hence, $Y \rhd A$, by Corollary 14.3.5, and $Y \models A$, by Fact 14.2.3. As a consequence, since $Y \subseteq X$, Y is consistent and $Y \models A$, $X \models' A$. □

By Theorem 14.4.2 and Corollary 14.3.5 we have some final conclusion:

Corollary 14.4.3 $\rhd' = \rhd \cap \{\langle X, A \rangle : \langle X, A \rangle \subseteq 2^{For} \times For, \exists_Y \subseteq X, Y$ is consistent and $Y \models A\}$.

14.5 Further Applications

The presented mechanism can be used to other tableau systems/logics. Through a formal theory of tableau systems [2], we should aim at a general theorem:

if $\models = \rhd$, then $\models' = \rhd'$
where $\models$ and $\rhd$ are semantical and tableau consequence relations of a given logic, while $\models'$ and $\rhd'$ are their paraconsistent tableau counterparts.

Acknowledgments Many thanks to an anonymous reviewer for helpful comments and suggestions.

References

1. Priest, G.: An Introduction to Non-Classical Logic. Cambridge University Press, Cambridge (2001)
2. Jarmużek, T.: Formalizacja metod tablicowych dla logik zdań i logik nazw (Formalization of tableau methods for propositional logics and for logics of names). Wydawnictwo UMK, Toruń (2013)
3. Jarmużek, T.: Tableau Metatheorem for Modal Logics. In: Ciuni, R., Wansing, H., Willkommen, C. (eds.) Recent Trends in Philosophical Logic, pp. 105–123. Springer (2013)

Chapter 15
Some Adaptive Contributions to Logics of Formal Inconsistency

Diderik Batens

Abstract Some insights were gained from the study of inconsistency-adaptive logics. The aim of the present paper is to put some of these insights to work for the study of logics of formal inconsistency. The focus of attention is application contexts of the aforementioned logics and their theoretical properties in as far as they are relevant for applications. As the questions discussed are difficult but important, a serious attempt was made to make the paper concise but transparent.

Keywords Paraconsistent logic · Logics of formal inconsistency · Inconsistency-adaptive logic

Mathematics Subject Classification (2000) 03B53 · 03B60 · 03A05

15.1 Introduction

Logics of Formal Inconsistency, LFIs for short, exploit a typical property of Newton da Costa's $\mathbf{C}_n$-systems ($0 < n < \omega$), viz. that there is a connective that expresses consistency. The connective is explicitly definable within the $\mathbf{C}_n$-systems and the precise definition varies with n, but this matter need not concern us here. In [10], the consistency operator is studied in general, viz. in the context of a wide variety of paraconsistent logics. Many theorems are proved for classes of logics. The study is restricted to the propositional level; extending it to the predicative level involves some technical difficulties, which are studied in Sect. 15.5.

Within a paraconsistent context, consistency statements have a dramatic effect. Consider the logical symbols with their **CL** meaning, except that the negation may be paraconsistent. If A and $\neg A$ are true together, then $A \vee B$ and $\neg A$ may be true together while B is false. So Disjunctive Syllogism is invalid because the truth of the

D. Batens (✉)
Centre for Logic and Philosophy of Science,
Ghent University, Blandijnberg 2, B-9000 Gent, Belgium
e-mail: Diderik.Batens@UGent.be

J.-Y. Beziau et al. (eds.), *New Directions in Paraconsistent Logic*,
Springer Proceedings in Mathematics & Statistics 152,
DOI 10.1007/978-81-322-2719-9_15

premises $A \vee B$ and $\neg A$ may result from the inconsistent behaviour of A rather than from the truth of B and $\neg A$. Put differently, $A \vee B$ and $\neg A$ entail $(A \wedge \neg A) \vee (B \wedge \neg A)$ and if the first disjunct is true, then B may be false. A consistency statement $\circ A$ typically says that A is consistent, in other words that the first disjunct of $(A \wedge \neg A) \vee (B \wedge \neg A)$ is false, and hence that B is true if the disjunction is true. All this may be summarized by the comment that $A \vee B, \neg A \nvdash B$ but $A \vee B, \neg A, \circ A \vdash B$.

Every paraconsistent logic classifies a set of classically valid rules as invalid. Which rules are so classified depends on the logic. That there is a need for considering some inconsistencies as true does obviously not entail that all inconsistencies should be considered as true. So it is sensible to state that some formulas behave consistently. Whenever we state a formula to be consistent, some classical consequences of the premises are added to the paraconsistent consequence set. So we can separate formulas that behave consistently from those that might not so behave and in doing so we obtain a richer theory. This is what makes LFIs interesting.

Summarizing, the main idea behind LFIs is that a logic with a paraconsistent negation $\neg$ is extended with a connective, $\circ$, that expresses consistency with respect to $\neg$. Whenever $\circ A$ is derivable from the premises, A functions as it would function in classical logic. The effect may be phrased in two ways: $\circ$ offers a means to locally restore classicality and $\circ$ offers a means to locally give up paraconsistency. A consistency operator first occurred in da Costa's $\mathbf{C}_n$ logics, which were around at least from 1963 on [11]. In those days, the consistency operator did not receive much attention, partly because the idea of paraconsistency was so new for most logicians—their reaction was neither nice nor smart—partly because the consistency operator is defined in the $\mathbf{C}_n$ logics and actually not defined in a very elegant way.[1]

Several insights gained during the study of (the metatheory of) adaptive logics are useful for understanding and mastering LFIs. Presenting these insights was the initial aim of the present paper. On the road, a few further insights on LFIs were added. Moreover, it turned out useful to add a comparison, for some application contexts, between an approach in terms of LFIs and an adaptive approach. I shall not argue that one of the approaches is superior, but rather compare some of their properties.

As realistic applications of LFIs seem to be unavoidably predicative, the predicative case is included in the present paper. Some useful comments on predicative LFI are found in Sect. 15.5.

Next, I shall restrict my attention to LFI in which there is a (primitive or decidable) *consistency connective* $\circ$ that names a consistency operator. In some LFI, the role of the consistency operator is taken on by a set of formulas $\bigcirc(A)$ each member of which is built from logical symbols and occurrences of A. As we shall see, several consistency operators may occur within a LFI $\mathbf{L}$. Where $\circ$ is a consistency connective, $\circ A$ will be called a *consistency statement*; it states that A is consistent (or behaves consistently). Another restriction in focus is that I shall only consider paraconsistent logics that are not paracomplete.

[1] Being merely an abbreviation in $\mathbf{C}_n$ logics, the consistency operator adds nothing to the expressive power of the language. That the definiens, for example $\neg(A \wedge \neg A)$ in $\mathbf{C}_1$, expresses the consistency of A is somewhat awkward in the context of the $\mathbf{C}_n$ logics.

Finally, this paper is not phrased in the terminology from [10]. That terminology is precise as well as useful, but it would turn the present paper into one that is lengthy as well as difficult to read. This is especially so because the terminology would require modifications and extensions for the predicative case. I shall also depart from the terminology where an alternative is easier for present purposes.

It seems useful to state that the metalanguage will be classical—this actually holds for all my papers and it agrees with the Brazilian tradition. Also "true" and "false" will be used as excluding each other. So an inconsistent situation is one in which a formula A is true together with its negation $\neg A$ but such a situation cannot be described by saying that A is both true and false. This convention presupposes that paraconsistent logics can be consistently described, for example in that no formula is verified as well as falsified by a model, not both $M \Vdash A$ and $M \nVdash A$, and in that no formula is a semantic consequence as well as not a semantic consequence of any premise set, not both $\Gamma \vDash A$ and $\Gamma \nvDash A$. The two conventions are essential to interpret the theorems and other metatheoretic statements in the sequel of this paper.

Another useful warning is that this paper does not contain a decent survey of adaptive logics—other papers [5, 7] provide introductions. The essential dynamic proof theory is not even mentioned below. Yet, there will be sufficient information to make the present paper self-contained.

15.2 Preliminaries

Let $\mathcal{L}$ be a variable for languages, with $\mathcal{F}$ as its set of formulas and $\mathcal{W}$ as its set of closed formulas. The standard predicative language will be called $\mathcal{L}_s$, with $\mathcal{F}_s$ and $\mathcal{W}_s$ as expected. Where $\neg$ will be the standard negation, which will usually be paraconsistent, the symbol $\sim$ will always denote the classical negation—but you will be reminded. Let us impose a minimal requirement on negations.[2]

Definition 15.1 A unary connective ξ is a negation in a logic **L** iff there are Γ and A such that $\Gamma, A \nvdash_{\mathbf{L}} \xi A$ and $\Gamma, \xi A \nvdash_{\mathbf{L}} A$.

This very weak definition imposes requirements for being a negation that have no exceptions. A negation ξ is a unary connective. Moreover, ξA is neither logically stronger nor logically weaker than A. This property, which is expressed semantically by the first paragraph of Lemma 15.14 below, entails that for some A, ξA is not a logical truth. I suggest that Definition 15.1 holds for all connectives that were sensibly called negations in the literature, while stronger definitions do not—the "sensibly" obviously contains a conventional element.

---

[2] Some paraconsistent logicians defend a specific negation as the correct one. Priest [17], for example, seems to assign this role to the negation of **LP**. Other paraconsistent logicians, for example da Costa [12], consider a multiplicity of operators as paraconsistent negations, but sometimes impose certain conditions. Often a more general outlook is taken, as for example by Béziau [9].

As paracomplete negations are disregarded in this paper, every **L**-model will verify either A or ξA for every closed formula A and for every paraconsistent negation ξ. That a logic is not paracomplete can be defined as follows in syntactic terms.

Definition 15.2 A symbol ξ is a complementing negation in a logic **L** iff it is a negation in **L** and, for all Γ, A, and B, if $\Gamma, A \vdash_{\mathbf{L}} B$ and $\Gamma, \neg A \vdash_{\mathbf{L}} B$, then $\Gamma \vdash_{\mathbf{L}} B$.

The disjunction $\vee$ will be taken to be classical except where otherwise specified. Expressions like $A(x)$ and $A(a)$ will have their usual meaning. The existential closure of A, viz. the result of prefixing A with an existential quantifier over every variable free in A, will be denoted by $\exists A$. The universal closure of A will be denoted by $\forall A$. The **L**-consequence set of Γ is $Cn_{\mathbf{L}}(\Gamma) =_{df} \{A \mid A \in \mathcal{W}; \Gamma \vdash_{\mathbf{L}} A\}$.

An easy way to define what it means that a symbol is classical goes as follows. Every logic **L** defines a two-valued *inferential semantics*, obtained by turning every true inferential statement $A_1, \ldots, A_n \vdash_{\mathbf{L}} B$ $(n \geq 0)$ into the semantic clause "if $v_M(A_1) = \ldots = v_M(A_n) = 1$, then $v_M(B) = 1$".[3] Note that the usual **CL**-semantics contains a specific clause for every logical symbol of $\mathcal{L}_s$. The *classical clause* for disjunction, for example, reads "$v_M(A \vee B) = 1$ iff $v_M(A) = 1$ or $v_M(B) = 1$".

Definition 15.3 A logical symbol ξ is *classical* in a logic **L** iff extending the inferential semantics of **L** with the classical clause for ξ does not affect the semantic consequence relation.

Definition 15.4 A logic **L** is *explosive* with respect to a negation $\neg$ iff it holds that $\Gamma, A, \neg A \vdash_{\mathbf{L}} B$.[4]

Definition 15.5 A negation $\neg$ is paraconsistent in a logic **L** iff **L** is not explosive with respect to $\neg$.

As paracomplete negations are disregarded in this paper, a **L**-model that verifies the classical negation of A falsifies A, and hence verifies the paraconsistent negation of A.[5]

Fact 15.6 Where $\neg$ is a paraconsistent negation in **L** and $\sim$ is a classical negation in **L**, ${\sim}A \vdash_{\mathbf{L}} \neg A$.

Definition 15.7 A logic is *paraconsistent* iff one of its negations is paraconsistent.

In agreement with the announced restriction on $\bigcirc A$, the following definition is less general than the one in [10].

---

[3]The insight was Suszko's [18]. The resulting semantics may be ugly but is obviously adequate.

[4]The reference to Γ may be dropped for Tarski logics (reflexive, transitive, and monotonic logics).

[5]The syntactic justification refers to the complementing character of the non-paracomplete paraconsistent negation. $A, {\sim}A \vdash_{\mathbf{L}} \neg A$ by explosion for the classical $\sim$ and $\neg A, {\sim}A \vdash_{\mathbf{L}} \neg A$ by reflexivity. Both together entail ${\sim}A \vdash_{\mathbf{L}} \neg A$ in view of the is complementing character of $\neg$.

Definition 15.8 A logic **L** is *gently explosive* with respect to a negation $\neg$ iff there is a (primitive or defined) unary connective $\circ$ such that $\circ A, A \nvdash_{\mathbf{L}} B$ and $\circ A, \neg A \nvdash_{\mathbf{L}} B$ hold for some A and B, whereas $\Gamma, \circ A, A, \neg A \vdash_{\mathbf{L}} B$ hold for all Γ, A and B.[6]

Definition 15.9 **L** is a *Logic of Formal Inconsistency* iff there is a negation $\neg$ such that **L** is not explosive with respect to $\neg$ but is gently explosive with respect to $\neg$.

Let $\circ$ be a *consistency connective* for $\neg$ within a LFI iff it functions as described in Definition 15.8.

Fact 15.10 Where $\circ$ is a *consistency connective* for $\neg$ in a LFI **L**, $\nvdash_{\mathbf{L}} \circ A$ and, for some B, $\circ A \nvdash_{\mathbf{L}} B$.

Fact 15.11 Where $\circ$ is a *consistency connective* for $\neg$ in a LFI **L**, (i) $A, \neg A \vdash_{\mathbf{L}} \neg\circ A$, (ii) where $\sim$ is classical negation in **L**, $A, \neg A \vdash_{\mathbf{L}} {\sim}\circ A$, (iii) where $\wedge$ is a non-glutty conjunction in **L**, $\circ A \vdash_{\mathbf{L}} \neg(A \wedge \neg A)$, and (iv) where $\wedge$ is a non-glutty conjunction in **L** and $\sim$ is classical negation in **L**, $\circ A \vdash_{\mathbf{L}} {\sim}(A \wedge \neg A)$.

It is worth pointing out that a consistency connective $\circ$ of a LFI **L** need not be a truth-function in **L**. Every **L**-model that verifies $A \wedge \neg A$ falsifies $\circ A$, but some **L**-models may falsify both. Put differently, a **L**-model that verifies A may verify $\neg A$ and may also verify $\circ A$; it cannot verify both $\neg A$ and $\circ A$ but it can falsify both.[7] A consistency statement provides enough information to turn a specific inconsistency into triviality, but fulfils no further requirements.

Some people dislike this aspect of the approach. If $\neg$ is a paraconsistent negation, $\neg A$ is not a truth-function of A. One sensible way to understand the situation is this: if A is true, then whether $\neg A$ is true or false depends on a separate fact—separate in the sense that it is not part of the fact that causes A to be true. Note that this idea agrees with most of the two-valued semantics devised in Brazil or Belgium for paraconsistent logics. It seems more difficult, however, to argue for a notion of consistency such that $\circ A$ is not a truth-function of A and $\neg A$—whence $\neg A$ is not a truth-function of A and $\circ A$ either. If inconsistent situations are possible, then the truth-value of $\neg A$ depends on a fact independent of the one that determines A to be true. But which fact might determine the truth of $\circ A$ in case either A or $\neg A$ is false? Whether you like it or not, this is the way in which the people who devised LFI laid it out.[8] And there is nothing wrong with their decision to study, with respect to a negation $\neg$ and logic **L**, the behaviour of a connective $\circ$ that, however weak or strong, is such that $\circ A, A \nvdash_{\mathbf{L}} B$, $\circ A, \neg A \nvdash_{\mathbf{L}} B$, $\neg A, A \nvdash_{\mathbf{L}} B$, and $\circ A, A, \neg A \vdash_{\mathbf{L}} B$.

---

[6] Here too the reference to Γ may be dropped for Tarski logics.

[7] This is the reason why the converses of the inferences mentioned in Fact 15.11 do not hold for all consistency connectives.

[8] The situation may have been influenced by the somewhat odd behaviour of the (defined) consistency operator $A^{(n)}$ in da Costa's $\mathbf{C}_n$ systems with $n > 1$. Another relevant consideration might have been that the consistent behaviour of a formula A on a premise set Γ, viz. that the logic does not require Γ to entail A as well as $\neg A$, should not cause $\circ A$ to be derivable from Γ. However, this danger is nonexistent even in case $\circ A$ is the suitable truth-function of A and $\neg A$.

Needless to say some consistency connectives $\circ$ are such that $\circ A$ holds true just in case one of A and $\neg A$ is false. As I disregard paracomplete negations, this holds just in case $v_M(A) \neq v_M(\neg A)$.

Definition 15.12 The connective $\circ$ is a *complementing consistency connective* for $\neg$ within a LFI **L** iff extending the inferential semantics of **L** with the clause "$v_M(\circ A) = 1$ iff $v_M(A) \neq v_M(\neg A)$" does not affect the semantic consequence relation.

Given that I disregard paracomplete negations in the present paper, the clause may be replaced by "$v_M(\circ A) = 1$ iff ($v_M(A) = 0$ or $v_M(\neg A) = 0$)".

Fact 15.13 Where **L** is a paraconsistent logic, it is possible to extend the language of **L** with a connective $\circ$ and to devise a semantics for a LFI **L**$'$ by extending the **L**-semantics with a clause for $\circ$ in such a way that $\circ$ is a complementing consistency connective for $\neg$ in **L**$'$.

Whether **L**$'$ has a Hilbert axiomatization will depend on the logical symbols of **L**. However, it is possible to syntactically characterize **L**$'$ by extending the syntactic characterization of **L** with the rule $A, \neg A, \circ A/B$ and with two meta-rules: (i) if $\Gamma, \circ A \vdash B$ and $\Gamma, A \vdash B$, then $\Gamma \vdash B$, and (ii) if $\Gamma, \circ A \vdash B$ and $\Gamma, \neg A \vdash B$, then $\Gamma \vdash B$.

It is worthwhile to state the semantic equivalents of some of the definitions. The proof of the subsequent lemma is standard. Note that there is no need to refer to Γ in the lemma.[9]

Lemma 15.14 *A unary connective $\neg$ is a negation in a logic* **L** *iff there are* **L***-models M such that $M \Vdash A$ and $M \nVdash \neg A$ and there are* **L***-models M such that $M \Vdash \neg A$ and $M \nVdash A$.*

A negation $\neg$ is paraconsistent in **L** *iff there is a* **L***-model M such that $M \Vdash A$ and $M \Vdash \neg A$.*

Where $\neg$ is paraconsistent in **L***, a unary connective $\circ$ is a consistency connective for $\neg$ in* **L** *iff there is no non-trivial* **L***-model M such that $M \Vdash A$, $M \Vdash \neg A$ and $M \Vdash \circ A$.*

Where $\neg$ is paraconsistent in **L** *and $\circ$ is a consistency connective for $\neg$ in* **L***, $\circ$ is a complementing consistency connective for $\neg$ in* **L** *iff, for every* **L***-model M, the following holds: if $M \nVdash A$ or $M \nVdash \neg A$, then $M \Vdash \circ A$.*

The qualification "non-trivial" may obviously be dropped if the semantic clauses rule out the trivial model. Thus, the usual clause for the **CL**-negation rules out the trivial model and so does the clause mentioned in Definition 15.12. An alternative, which leads to an equally adequate semantics, is obtained by appending to that clause "or $v_M(B) = 1$ for all B".[10]

---

[9] A non-monotonic logic may assign to Γ a selection of models that verify all members of Γ. The lemma contains references to all **L**-models.

[10] Adding or removing the trivial model—the model verifying all closed formulas—to the set of models defined by a semantics may require that the semantic clauses are adjusted. In view of the definition of the semantic consequence relation, it is obvious that such addition or removal does not affect the consequence relation.

Lemma 15.15 *Where* $\neg$ *is paraconsistent in* **L**, $\circ$ *is a complementing consistency connective for* $\neg$ *in* **L**, *and* $\wedge$ *is classical or gappy in* **L**, $\sim A =_{df} \neg A \wedge \circ A$ *defines a classical negation.*

Proof It is easily seen that, if the antecedent is true, no **L**-model verifies A as well as $\neg A \wedge \circ A$ and every **L**-model verifies either A or $\neg A \wedge \circ A$. □

The following comments are meant to cause some reflection. In the presence of the complementing consistency connective $\circ$ for $\neg$ in **L**, every **L**-model agrees with one of three possibilities with respect to A, as represented in the top row of the following two tables. The LFI **L** is reduced to **CL** by the mapping that agrees with the following schema.

$\mathcal{L}_1$:	$A, \circ A$	$A, \neg A$	$\neg A, \circ A$
$\mathcal{L}_2$:	A		$\neg A$

It seems natural to read the so obtained version of **CL** as: either A is (consistently or inconsistently) true or else $\neg A$ is consistently true, but not both. However, the LFI **L** is also turned into **CL** by the mapping that agrees with the following schema.

$\mathcal{L}_1$:	$A, \circ A$	$A, \neg A$	$\neg A, \circ A$
$\mathcal{L}_2$:	A	$\neg A$	

This mapping gives us: either A is consistently true or else $\neg A$ is true, but not both. So even if the world is inconsistent, there are two ways to describe it in terms of **CL**. The first presupposes that consistent falsehood can be identified, the second that consistent truth can be located. In both cases, the transition to **CL** leads to a lack of expressive power—distinct situations are identified. If one wants to combine the paraconsistent view with the classical one in the same language, the first mapping merely requires that a new negation symbol is introduced, whereas the second mapping requires a consistently true symbol. Although most people will consider the second alternative as conceptually more difficult, both are perfectly symmetric.

An interesting insight in LFIs is that some $\circ$-free formulas establish logical relations between consistency statements. Let $\rightarrow$ be a detachable implication in a logic **L**—for all **L**-models M, if $M \Vdash A \rightarrow B$ and $M \Vdash A$, then $M \Vdash B$—but for which Modus Tollens does not hold.[11] Note that

$$A \rightarrow B, \neg B, \circ B \vdash_{\mathbf{L}} \neg A$$

holds. Indeed, no non-trivial models of the premises verify A. So all models of the premises verify $\neg A$. However,

$$A \rightarrow B, \neg B, \circ B \vdash_{\mathbf{L}} \circ A$$

[11] There is no reason to handle Modus Ponens and Modus Tollens on a par. The first states a property of the implication. The justification of Modus Tollens requires a reference to negation: if A is true, then B is true (by Modus Ponens); but $\neg B$ is true; so if inconsistencies are not true, then neither is A.

also holds for complementing consistency connectives $\circ$. Indeed, if a model of the premises would verify A, it would also verify B and hence would be trivial. So the model either falsifies A or is trivial. In both cases it verifies $\circ A$.

A final preliminary comment concerns weird logics. There are some paraconsistent logics and some LFI that we do not want to consider because they have exceptional properties and, as far as we can see at this moment, no one is interested in them. I shall call them irregular and now explain what I mean by that. Any decent semantics presupposes a complexity ordering $<$ which is such that if $A < B$, then all non-logical symbols that occur in A also occur in B. The valuation function defines the valuation value $v_M(A)$ in terms of the assignment function and in terms of valuation values $v_M(B_1), \ldots, v_M(B_n)$ such that $B_1 < A, \ldots, B_n < A$. Some paraconsistent models M verify both A and $\neg A$ while this is not determined by the truth values of formulas less complex than $\neg A$. There is nothing wrong with this, but if the logic is *regular* there should be a model M' such that M and M' verify the same formulas, except for $\neg A$ and formulas B such that $\neg A < B$.[12] Similarly, some LFI-models M falsify a member of $\{A, \neg A\}$ but also falsify $\circ A$. Again, this is all right, but if the LFI is *regular* there should be a LFI-model M' that verifies exactly the same formulas as M except for $\circ A$ and formulas B such that $\circ A < B$.[13]

15.3 Derivable Disjunctions of Contradictions

Let **L** be a LFI in which $\wedge$ and $\vee$ have their classical meaning and let us, for this section, restrict our attention to propositional premise sets. From some such sets, a set of contradictions is derivable. From others only a disjunction of contradictions is derivable. A common example of the latter is $\Gamma_1 = \{\neg p, \neg q, p \vee q, p \vee r, q \vee s, \neg t, u \vee t\}$. According to many paraconsistent logics, no contradiction is derivable from Γ_1, but a disjunction of contradictions is derivable from it, viz. $(p \wedge \neg p) \vee (q \wedge \neg q)$. Such disjunctions may count any (finite) number of disjuncts and infinitely many such disjunctions may be derivable from a premise set.

To save on terminology, I already introduce here some concepts from the metatheory of adaptive logics. As we shall see in Sect. 15.6, one of the elements of an adaptive logic is a 'set of abnormalities' Ω. Let us, for the present propositional discussion, identify abnormalities with contradictions, whence $\Omega = \{A \wedge \neg A \mid A \in \mathcal{W}\}$. A disjunction of members of Ω will be called a *Dab-formula* (a disjunction of abnormalities). In the expression $Dab(\Delta)$ (and in similar expressions), Δ is a finite subset of Ω and $Dab(\Delta)$ is the disjunction of the members of Δ. If $\Gamma \vdash_{\mathbf{L}} Dab(\Delta)$, we shall

---

[12]This regularity requirement is stronger than the requirement for being a negation in the sense of Lemma 15.14. If S contains all **CL**-models as well as the trivial model, $\neg$ is a negation but the regularity requirement is not fulfilled.

[13]Do not confuse the question whether a logic is regular with the question whether a specific semantics of this logics is deterministic. See Sect. 2 of another paper [7] in this volume for a method to turn an indeterministic semantics of a certain type into a deterministic one.

say that $Dab(\Delta)$ is a *Dab-consequence* of Γ.[14] Finally, consider a semantic notion. Where M is a **L**-model, define $Ab(M) = \{A \in \Omega \mid M \Vdash A\}$—the set of abnormalities verified by M, in some papers called the 'abnormal part' of M.

As the disjunction is classical, Addition holds. So if a premise set has a *Dab*-consequence, then it has infinitely many different *Dab*-consequences. In the sequel I shall need minimal *Dab*-consequences of a premise set. Where $Dab(\Delta)$ is a *Dab*-consequence of Γ, $Dab(\Delta)$ is a *minimal Dab-consequence of* Γ iff there is no $\Delta' \subset \Delta$ such that $Dab(\Delta')$ is a *Dab*-consequence of Γ.

If $Dab(\Delta)$ is a minimal *Dab*-consequence of Γ and $A \wedge \neg A \in \Delta$, then $Dab(\Delta - \{A \wedge \neg A\})$ is a minimal *Dab*-consequence of $\Gamma \cup \{\circ A\}$. Put differently, extending Γ with consistency statements may result in *Dab*-consequences that contain less disjuncts. The reader who frowns at the "may" should consider that extending Γ_1 with $\circ t$ does not have any effect on the minimal *Dab*-consequences of Γ_1.[15]

The previous paragraph hides an interesting insight. Instead of spelling it out here, I save it for Sect. 15.7 where its consequences can be highlighted.

15.4 A Logical Boundary

A theory may be seen (on the statement view) as a couple $T = \langle \Gamma, \mathbf{L} \rangle$ in which Γ is a set of non-logical axioms and **L** is a logic. Adding consistency statements to T only makes sense if at least one negation of **L** is paraconsistent and provided the consistency statements pertain to such a negation. The decision to add consistency statements to T is extra-logical. It is a decision to extend T with new non-logical theorems by strengthening a certain statement in a specific way. This is clearly extra-logical with respect to **L**. Strengthening A to $A \wedge \circ A$, or $\neg A$ to $\neg A \wedge \circ A$, may be justified by a general consistency presumption, but not if $A \wedge \neg A$ is a disjunct of a minimal *Dab*-consequence of Γ.

Although the decision is extralogical, there are logical constraints. If the non-logical axioms are $\Gamma_2 = \{p, q, \neg p \vee r, \neg q \vee s, \neg q\}$, then adding $\circ q$ causes triviality, whereas adding $\circ p$ does not. If the non-logical axioms are $\Gamma_1 = \{\neg p, \neg q, p \vee q, p \vee r, q \vee s, \neg t, u \vee t\}$, then neither adding $\circ p$ nor adding $\circ q$ causes triviality, but adding both does. In general there are, for every inconsistent theory T, sets of consistency statements such that adding all members of the set to T causes triviality but adding

[14]There is no need to add "with respect to **L**" as *Dab*-consequences of Γ will always be considered for a specific logic.

[15]The set of minimal *Dab*-consequences obviously depends on the logic. For some paraconsistent logics, like **CLuN** mentioned in a subsequent section, $(p \wedge \neg p) \vee (q \wedge \neg q)$ is the only *Dab*-consequence of Γ_1. Other paraconsistent logics assign infinitely many *Dab*-consequences to Γ_1. Still, I cannot picture any formal paraconsistent logic for which $\circ t$ has an effect on the minimal *Dab*-consequences of Γ_1. This is weaker than what is claimed in the text, but I shall buy you a beer if you show my imagination lacking at this point and that is stronger than what is said in the text.

all but one members does not.[16] Only a very limited number of sensible people will judge that extralogical reasons may outweigh reasons to avoid triviality. So in constructing LFI-theories, one should mind the *triviality danger*.

As soon as this is agreed upon, a further question surfaces: Given a paraconsistent theory, which are the maximal sets of consistency statements that can be added to it? This may be termed the *maximality question*. Reconsider the premise set Γ_1. It is possible that one has no good (extralogical) reason to prefer adding $\circ p$ to adding $\circ q$ and *vice versa*. In that case, opting for one of the extensions seems unjustifiable. Of course several alternatives are still open. One may simply not add either consistency statement. One may consider and study the two extended theories without choosing between them, for example in the hope that this may lead to a good reason to prefer one decision over the other. One may also extend the theory with the (classical or gappy) disjunction $\circ p \vee \circ q$. This will cause $r \vee s$ rather than one of its disjuncts to be a theorem.

It is not in general desirable that one tries to obtain a theory to which no further consistency statements can be added. After all, a person who devises a theory is free to organize it along his or her preferences. Theories are judged in view of what they state and in view of the way in which they 'react' to other knowledge. That an otherwise good theory does not contain a maximal set of consistency statements is at best a theoretical problem. Nevertheless, it is useful to solve the maximality question, viz. to study the maximal sets of consistency statements that extend a premise set without causing triviality. A set of consistency statements non-trivially extends the considered theory iff it is a subset of one of those maximal sets.

15.5 Predicative Consistency Statements

The transition from propositional LFI to predicative ones is not completely obvious and some definitions from Sect. 15.2 have to be adjusted, for example Definition 15.8. The matter is important in view of realistic applications.

The main technical difficulty concerns the typical predicative consistency statement. Indeed,

$$\exists(A \wedge \neg A) \vdash_{\mathbf{CL}} B \tag{15.1}$$

whereas, for many paraconsistent logics $\mathbf{L}$, there are A such that $\exists(A \wedge \neg A) \nvdash_{\mathbf{L}} B \wedge \neg B$ holds for all $B \in \mathcal{W}$—for example $\exists x(Px \wedge \neg Px) \nvdash_{\mathbf{CLuN}} B \wedge \neg B$.[17] In

[16] To be more precise, this is the case for some (not necessarily all) sets $\{A_1, \ldots, A_n\}$ such that $(A_1 \wedge \neg A_1) \vee \ldots \vee (A_n \wedge \neg A_n)$ is a minimal *Dab*-consequence of the non-logical axioms of the theory.

[17] The predicative logic **CLuN**, first introduced in [2], is the predicative extension of the propositional **PI** from [1]. The latter extended with a suitable axiom for a consistency operator is the LFI **mbC**—see Definition 42 of [10].

view of this situation, if a predicative LFI **L** has a suitable conjunction,[18] then it needs a formula X that functions as the consistency statement for A, viz.

$$X, \exists(A \wedge \neg A) \vdash_{\mathbf{L}} B \tag{15.2}$$

or perhaps

$$\exists(X \wedge A \wedge \neg A) \vdash_{\mathbf{L}} B\,. \tag{15.3}$$

In the presence of the **CL**-negation $\sim$, it obviously holds that

$$\sim\exists(A \wedge \neg A), \exists(A \wedge \neg A) \vdash_{\mathbf{L}} B\,, \tag{15.4}$$

equivalently

$$\forall(\sim A \vee \sim\neg A), \exists(A \wedge \neg A) \vdash_{\mathbf{L}} B\,, \tag{15.5}$$

and it also holds that

$$\exists((\sim A \vee \sim\neg A) \wedge A \wedge \neg A) \vdash_{\mathbf{L}} B\,. \tag{15.6}$$

Both $\sim\exists(A \wedge \neg A)$ and $\forall(\sim A \vee \sim\neg A)$ are complementing consistency operators for $\neg$. Similarly, for the option corresponding to (15.3), the 'internal' consistency operator $\sim A \vee \sim\neg A$ from (15.6) is complementing. However, in line with the way in which the consistency operator is introduced at the propositional level, one should also consider consistency operators that are not complementing. So we want to allow that the X in (15.2) is weaker than the first formula in (15.5) and that the X in (15.3) is a weaker than the open formula $\sim A \vee \sim\neg A$ in (15.6).

A little reflection readily reveals the road to be taken. Instead of explicitly defining a consistency operator by the definiens $\sim A \vee \sim\neg A$, we should replace this expression in (15.5) and (15.6) by $\circ A$ in which $\circ$ is any propositional consistency connective—remember the comment on Definition 15.8.

For the option corresponding to (15.2), this results in universally closed consistency statements,

$$\forall\circ A, \exists(A \wedge \neg A) \vdash_{\mathbf{L}} B\,, \tag{15.7}$$

supposing that the universal quantifier is classical. However weak the consistency connective, the consistency statement cannot warrant that $\exists(A \wedge \neg A)$ results in triviality unless $\circ A$ holds true independent of the way in which the free variables in A are mapped on the model's domain. Precisely this is warranted by $\forall\circ A$. If no

[18]Suitable are a classical conjunction or a gappy one. Glutty conjunctions have to be considered contextually because they allow for models that verify a conjunction and falsify one of the conjuncts. While such models are clearly abnormal with respect to **CL** and many other logics, it depends on further properties whether a consistency operator should handle this. See for example [8] on gluts and gaps of all kinds.

free variables occur in A, there are no quantifiers in (15.7), whence it reduces to the desired propositional property

$$\circ A, A, \neg A \vdash_{\mathbf{L}} B \tag{15.8}$$

provided the conjunction is not glutty.

For the option corresponding to (15.3) the possibly open formula $\circ A$ will do. However, this option does not seem very attractive, neither with respect to LFI properly nor with respect to adaptive LFI. Consider indeed $\Gamma_3 = \{\exists x Px, \forall(Qx \vee \neg Px)\}$ and let $\mathbf{L}$ be a logic in which conjunction, disjunction and the quantifiers behave classically, whence $\Gamma_3 \vdash_{\mathbf{L}} \exists x(Qx \vee (Px \wedge \neg Px))$. In order that $\exists x Qx$ be $\mathbf{L}$-derivable, it is obviously not sufficient to add $\exists x \circ Px$; we need to add at least $\exists x(Px \wedge \circ Px)$. In other words, we have to state that some object has property P and is consistent in this respect.

The situation is easily misleading. Indeed, $\exists x(Px \wedge \circ Px)$ cannot be seen as a consistency statement because it also contains the information $\exists x Px$, which is not part of the meaning of $\exists x \circ Px$. One might think that $\exists x(Px \wedge \circ Px)$ may be seen as a specification of the premise $\exists x Px$, as the addition 'under the quantifier' that the x which has property P has this property in a consistent way. This, however is mistaken. Consider indeed $\Gamma_4 = \{\exists x Px, \forall(Qx \vee \neg Px), \exists x(Px \wedge \neg Px)\}$. If we extend Γ_4 with $\exists x(Px \wedge \circ Px)$, then, just as in the case of Γ_3, $\exists x Qx$ is derivable. So it is quite obvious that $\exists x(Px \wedge \circ Px)$ cannot be seen as a specification of the premise $\exists x Px$ for the simple reason that, in the extended Γ_4, some x have property P in a consistent way whereas other x have it in an inconsistent way. Here is a different way of stating the matter: given that conjunction, disjunction and the quantifiers were presumed to be classical, the set of consequences of Γ_4 coincides with the set of consequences of $\Gamma_5 = \{\forall(Qx \vee \neg Px), \exists x(Px \wedge \neg Px)\}$. So $\exists x(Px \wedge \circ Px)$ is new information, viz. that an object has property P in a consistent way, and is not the specification that an object known to have property P has this property in a consistent way.

What precedes shows that formulas containing a consistency statement that is 'internal' in the sense of (15.3) introduces new information. It easily follows, however, that these are not consistency statements at all. The correct rendering of (15.3), implemented as $\exists(\circ A \wedge A \wedge \neg A) \vdash_{\mathbf{L}} B$ is

$$\exists(\circ A \wedge A \wedge \neg A), \exists(A \wedge \neg A) \vdash_{\mathbf{L}} B$$

because $\exists(\circ A \wedge A \wedge \neg A)$ constitutes new information and does not specify the statement $\exists(A \wedge \neg A)$. Note also that $\exists(\circ A \wedge A \wedge \neg A)$ itself is not a consistency operator. This is obvious from

$$\exists(\circ A \wedge A \wedge \neg A) \vdash_{\mathbf{L}} B$$

and Definition 15.8.

The preceding considerations, so you might think, show that the option corresponding to (15.3) leads to trouble and that the option corresponding to (15.2) is the right one. But that is wrong too. Extending Γ_4 (or Γ_5) with $\forall x \circ Px$ results in triviality. This, however, does not mean that the consistency connective does not allow one to extend Γ_4 or Γ_5 in such a way that $\exists x\, Qx$ is a consequence. Indeed, as we have seen, extending those premise sets with $\exists x(Px \wedge \circ Px)$ does the job.

Even more astonishing might be that the option corresponding to (15.2) is by no means exhausted by what was said before. Although $\forall \circ A$ seems to be the regular form of the predicative consistency statement, it sometimes also pays to add $\exists \circ A$. Consider indeed the premise set $\Gamma_6 = \{\forall x(Px \supset Qx), \exists x(Qx \wedge \neg Qx)\}$ and the LFI **mbC**. Extending Γ_6 with $\forall x \circ Qx$ results in triviality, but extending it with $\exists x \circ Qx$ does not. Moreover, it extends the **mbC**-consequence set with $\exists x((Qx \wedge \circ Qx) \vee (\neg Qx \wedge \circ Qx))$ and hence also with $\exists x((Qx \wedge \circ Qx) \vee (\neg Px \wedge \circ Px))$. By the same reasoning, if $\Gamma_7 = \{\forall x(Px \supset Qx), \forall x(Rx \supset \neg Qx), \exists x(Qx \wedge \neg Qx)\}$ is extended with $\exists x \circ Qx$, then its **mbC**-consequence set is extended with $\exists x((Qx \wedge \circ Qx) \vee (\neg Qx \wedge \circ Qx))$ and hence also with $\exists x((\neg Rx \wedge \circ Rx) \vee (\neg Px \wedge \circ Px))$.

Allow me to stress that what precedes is by no means a criticism of the LFI programme. I just want to point out that the transition from propositional LFIs to predicative LFIs is not obvious. Indeed, one of the oddities is that no logical form can function in general as *the* predicative consistency statement.

The only further noteworthy comment at the predicative level concerns decidability matters. Many propositional LFIs **L** assign a recursive consequence set $Cn_{\mathbf{L}}(\Gamma)$ to every finite premise set Γ. So it is decidable whether the **L**-consequence set of a finite propositional Γ is trivial. For infinite but decidable premise sets Γ, $Cn_{\mathbf{L}}(\Gamma)$ is only semi-decidable. By moving to the predicative level, $Cn_{\mathbf{L}}(\Gamma)$ is only semi-decidable even for most finite Γ. So it is in general only semi-decidable whether $Cn_{\mathbf{L}}(\Gamma)$ is trivial.

15.6 A Few Adaptive Basics

Adaptive logics are defined as triples consisting of (i) a lower limit logic **LLL**: a logic that has static proofs,[19] (ii) a set of abnormalities Ω: a set of closed formulas, characterized by a possibly restricted logical form,[20] and (iii) an adaptive strategy (as clarified below).

In this section, I consider the question what adaptive LFI should look like. By an adaptive LFI I mean an inconsistency-adaptive logic that has a LFI as lower limit and that enables one to derive consistency statements that, once derived, play the typical LFI-role.

The intuitive idea behind Ω is that it contains the formulas that are presumed to be false unless and until the premises require them to be true. The precise meaning of the latter expression depends on the strategy—only two strategies will be given some

[19] For present purposes, this may be identified with a compact Tarski logic.

[20] The set Ω may comprise formulas of the form $\exists(A \wedge \neg A)$. If A is any formula, the form is unrestricted; if A is required to be an atomic formula, the form is restricted.

attention in this paper—and on the (classical) disjunctions of abnormalities derivable by **LLL** from the premise set. As is the case for many Tarski logics, many LFI may be combined with different strategies and with different sets of abnormalities to obtain a multiplicity of adaptive logics. Where the lower limit logic **LLL** is a LFI, two hints help to avoid inadequate sets Ω. First, abnormalities should be **LLL**-contingent. Next, Ω should be such that the adaptive LFI maximally approaches the **CL**-consequence set without being trivial and without involving choices that are arbitrary from a logical point of view. This second hint requires some explanation.

Adding $\circ p$ rather than $\circ q$ to $\Gamma_1 = \{\neg p, \neg q, p \vee q, p \vee r, q \vee s, \neg t, u \vee t\}$ is an obvious example of a logically arbitrary choice. The choice may obviously be justified by non-logical preferences. So there is nothing wrong when a person applying a LFI chooses to add one consistency statement rather than another. Adaptive logics, however, whether their lower limit logic is a LFI or not, cannot make such choices. They may add consistency statements to premise sets, but only in a logically symmetric way—more detailed insights follow in this section. Adaptive LFIs interpret premise sets as consistently as possible in the following sense: if a consistency statement is not in the adaptive consequence set of Γ, then adding the statement either leads to the trivial consequence set or involves a logically arbitrary choice. In the specific case where Γ is a *normal* premise set, viz. one that has **CL**-models, the adaptive consequence set of Γ should be identical to its **CL**-consequence set. That **CL** is chosen as the upper limit logic[21] is a decision taken by the people who devised LFI. A neat comparison requires that a consistency operator is added to the language of **CL** and that it is explicitly or implicitly defined in such a way that $\circ A$ is a **CL**-theorem—an obvious choice is $\circ A =_{df} \neg(A \wedge \neg A)$.

In many inconsistency-adaptive logics, the set of abnormalities is

$$\Omega = \{\exists(A \wedge \neg A) \mid A \in \mathcal{X}\}$$

in which $\exists$ and $\wedge$ are classical[22] and $\mathcal{X}$ is the set of open and closed formulas of the standard predicative language or a subset of it, often the set of atomic formulas—note that the logical form is then restricted.[23] The need for the existential closure is obvious by the reasoning from Sect. 15.5.

In combination with a LFI as lower limit logic, Ω will only lead to an adaptive logic that maximally adds consistency statements if $\circ$ is complementing. A different choice, which will function well for any consistency connective $\circ$, is

$$\Omega = \{\exists\neg\circ A \mid A \in \mathcal{X}\} \tag{15.9}$$

[21]The *upper limit logic* **ULL** is obtained by extending **LLL** with a rule that causes all abnormalities to entail triviality.

[22]If one of those symbols would be glutty or gappy, the abnormalities would need to contain members that describe the gluts or gaps in the existential quantifier and the conjunction in order to handle the situation in an adequate way. See [8] for more information.

[23]The absence of the restriction may cause the adaptive logic to be a flip-flop, which means that the adaptive consequence set reduces to the lower limit consequence set whenever the premise set is abnormal.

in which $\exists$ and $\mathcal{X}$ are as before and, as agreed, the negation is not gappy.[24] The central desirable feature is that the falsehood of the abnormality should entail the truth of the corresponding consistency statement. And indeed, if the quantifiers are classical and the negation is not gappy, then $\forall \circ A$ is true whenever $\exists \neg \circ A$ is false.

For most paraconsistent logics **L** and premise sets Γ, it holds, first, that $Cn_{\mathbf{L}}(\Gamma)$ is inconsistent iff $Cn_{\mathbf{CL}}(\Gamma)$ is so and, next, that classical disjunctions of abnormalities are **L**-derivable from Γ while none of the disjuncts is. The second property is what interest us here. An obvious example is again $\Gamma_1 = \{\neg p, \neg q, p \vee q, p \vee r, q \vee s, \neg t, u \vee t\}$. When the logic is **CLuN**, or nearly any other sensible paraconsistent logic, $(p \wedge \neg p) \vee (q \wedge \neg q)$ is derivable from Γ_1 but neither disjunct is.

A few definitions were already hinted at in Sect. 15.3 but are repeated here in a more precise setting. By a *Dab-formula* I mean a classical disjunction of abnormalities, *including* the border case where there is only one disjunct. In expressions like $Dab(\Delta)$, Δ is a finite subset of Ω and $Dab(\Delta)$ is the classical disjunction of the members of Δ. $Dab(\Delta)$ is a *Dab*-consequence of Γ iff $\Gamma \vdash_{\mathbf{LLL}} Dab(\Delta)$. $Dab(\Delta)$ is a *minimal Dab-consequence of* Γ iff it is a *Dab*-consequence of Γ and no $\Delta' \subset \Delta$ is such that $Dab(\Delta')$ is a *Dab*-consequence of Γ. A *choice set* of $\Sigma = \{\Delta_1, \Delta_2, \ldots\}$ is a set that contains one element out of each member of Σ. A *minimal choice set* of Σ is a choice set of Σ of which no proper subset is a choice set of Σ.

Definition 15.16 Where $Dab(\Delta_1)$, $Dab(\Delta_2), \ldots$ are the minimal *Dab-consequences* of Γ, $\Phi(\Gamma)$ is the set of minimal choice sets of $\{\Delta_1, \Delta_2, \ldots\}$.[25]

Definition 15.17 Where M is a **LLL**-model, $Ab(M) = \{A \mid A \in \Omega;\ M \Vdash A\}$.

Let $\mathbf{AL}^m$ denote the adaptive logic defined by a given **LLL**, an Ω, and the Minimal Abnormality strategy. Let $\mathcal{M}_\Gamma^{\mathbf{LLL}}$ be the set of all **LLL**-models of Γ and let $\mathcal{M}_\Gamma^m$ be the set of all minimally abnormal models of Γ as defined below.

Definition 15.18 $M \in \mathcal{M}_\Gamma^m$ (M is a minimally abnormal model of Γ) iff $M \in \mathcal{M}_\Gamma^{\mathbf{LLL}}$ and no $M' \in \mathcal{M}_\Gamma^{\mathbf{LLL}}$ is such that $Ab(M') \subset Ab(M)$.

Definition 15.19 $\Gamma \vDash_{\mathbf{AL}^m} A$ iff $M \Vdash A$ for all $M \in \mathcal{M}_\Gamma^m$.

Note that there are no $\mathbf{AL}^m$-models, but only $\mathbf{AL}^m$-models *of* a set Γ. Theorem 15.20, proven as Lemma 4 in [5], establishes an important relation between the semantics and the syntactic level; it actually plays an essential role in the proof that the dynamic proof theory of $\mathbf{AL}^m$ is sound and complete with respect to the $\mathbf{AL}^m$-semantics.

Theorem 15.20 *If* Γ *has* **LLL**-*models, then* $\varphi \in \Phi(\Gamma)$ *iff* $\varphi = Ab(M)$ *for some* $M \in \mathcal{M}_\Gamma^m$.

---

[24]This Ω will also be adequate for some combinations of non-classical quantifiers, but that need not concern us in the present paper.

[25]If Γ has no *Dab*-consequences, $\Phi(\Gamma) = \{\emptyset\}$; if Γ has no **LLL**-models, $\Phi(\Gamma) = \{\Omega\}$; $\Phi(\Gamma) \neq \emptyset$ always holds.

The following corollary holds in view of Theorem 15.20 and Definitions 15.18 and 15.19 (and holds vacuously in case $\mathcal{M}_{\Gamma}^{\mathbf{LLL}} = \emptyset$):

Corollary 15.21 $\Gamma \vDash_{\mathbf{AL}^m} A$ *iff* $M \Vdash A$ *for all* $M \in \mathcal{M}_{\Gamma}^{\mathbf{LLL}}$ *for which* $Ab(M) \in \Phi(\Gamma)$.

While $\mathbf{AL}^m$ follows the Minimal Abnormality strategy, some adaptive logics follow the related Normal Selections strategy—this strategy was first invoked in [3]; the generic name of these adaptive logics is $\mathbf{AL}^n$.

Definition 15.22 $\Gamma \vDash_{\mathbf{AL}^n} A$ iff, for some $M \in \mathcal{M}_{\Gamma}^m$, $M' \Vdash A$ for all $M' \in \mathcal{M}_{\Gamma}^m$ for which $Ab(M') = Ab(M)$.

So $\Gamma \vDash_{\mathbf{AL}^n} A$ iff A is verified by all members of a set of minimally abnormal models of Γ that verify the same set of abnormalities.

If $M' \in \mathcal{M}_{\Gamma}^m$ and $Ab(M') = Ab(M)$ then M' verifies $\Gamma \cup Ab(M)$; so $M' \in \mathcal{M}_{\Gamma \cup Ab(M)}^{\mathbf{LLL}}$. If M' were not a minimally abnormal model of $\Gamma \cup Ab(M)$, then it would not be a minimally abnormal model of Γ in view of Definition 15.19. So $M' \in \mathcal{M}_{\Gamma \cup Ab(M)}^m$. In view of Theorem 15.20, this amounts to:

Corollary 15.23 $\Gamma \vDash_{\mathbf{AL}^n} A$ *iff, for some* $\varphi \in \Phi(\Gamma)$, $M \Vdash A$ *for all* $M \in \mathcal{M}_{\Gamma \cup \varphi}^m$.

The following theorem, proven as Theorem 5 in [5], is mentioned for future reference.

Theorem 15.24 *If* $M \in \mathcal{M}_{\Gamma}^{\mathbf{LLL}} - \mathcal{M}_{\Gamma}^m$*, then there is a* $M' \in \mathcal{M}_{\Gamma}^m$ *such that* $Ab(M') \subset Ab(M)$. *(Strong Reassurance for Minimal Abnormality.)*

The property fundamentally expresses that there are no infinite sequences of models of Γ such that every model is less abnormal than its predecessor. Other names for the property are Smoothness and Stopperedness.

15.7 Back to LFI

Information from the previous section will be put to use here. I shall only consider adaptive logics that have a LFI as **LLL** and Ω as in (15.9). Let $\Xi = \{\forall \circ A \mid A \in \mathcal{X}\}$—actually, for reasons that become clear later in this section, one may read $\mathcal{X}$ as $\mathcal{F}$.

Fact 15.25 Where $\Delta \subseteq \Xi$, $Cn_{\mathbf{LLL}}(\Gamma \cup \Delta)$ is not trivial iff some member of $\mathcal{M}_{\Gamma}^{\mathbf{LLL}}$ verifies all members of Δ.

As I have to make a decision anyway, I take $\forall \circ A$ to be the official predicative consistency statement and I take the notion of regularity from Sect. 15.2 upgraded accordingly.

Fact 15.26 $\Delta \subseteq \Xi$ is *maximal* with respect to Γ and **LLL** iff a member of $\mathcal{M}_\Gamma^{\mathbf{LLL}}$ verifies all members of Δ and no member of $\mathcal{M}_\Gamma^{\mathbf{LLL}}$ verifies all members of Δ and moreover some members of $\Xi - \Delta$.

Note that these facts are independent of the question whether $\circ$ is complementing or not. If M does not verify both A and $\neg A$ but nevertheless falsifies $\circ A$, then, by the regularity of **LLL**, a different **LLL**-model of the premises, say M', verifies the same formulas as M, except for $\circ A$ and formulas B such that $\circ A < B$.

Theorem 15.27 $\Delta \subseteq \Xi$ *is* maximal *with respect to* Γ *and* **LLL** *iff there is a* $M \in \mathcal{M}_\Gamma^m$ *such that (i) if* $A \in \Delta$, *then* $M \Vdash A$ *and (ii) if* $A \in \Xi - \Delta$, *then* $M \nVdash A$.

Proof $\Rightarrow$ Suppose that $\Delta \subseteq \Xi$ is maximal with respect to Γ and **LLL** but that there is no $M \in \mathcal{M}_\Gamma^m$ such that (i) and (ii) are fulfilled. In view of Fact 15.26, (1) some $M \in \mathcal{M}_\Gamma^{\mathbf{LLL}}$ verifies all members of Δ and no members of $\Xi - \Delta$, and (2) no $M \in \mathcal{M}_\Gamma^{\mathbf{LLL}}$ verifies all members of Δ as well as some members of $\Xi - \Delta$. (2) entails that (3) no $M \in \mathcal{M}_\Gamma^{\mathbf{LLL}}$ falsifies $\exists\neg\circ A$ whenever $\forall\circ A \in \Delta$ and moreover falsifies $\exists\neg\circ A$ for some $\forall\circ A \in \Xi - \Delta$.

By the regularity of **LLL**, (1) entails that there is a $M' \in \mathcal{M}_\Gamma^{\mathbf{LLL}}$ that verifies $\forall\circ A$ iff $\forall\circ A \in \Delta$ and that verifies $\exists\neg\circ A$ iff $\forall\circ A \in \Xi - \Delta$. So (i) and (ii) hold for M' and M' falsifies $\exists\neg\circ A$ iff $\forall\circ A \in \Delta$. But then $M' \in \mathcal{M}_\Gamma^m$ in view of (3).

$\Leftarrow$ Suppose that (1) (i) and (ii) hold for $M \in \mathcal{M}_\Gamma^m$, but that (2) $\Delta \subseteq \Xi$ is not maximal with respect to Γ and **LLL**. (1) entails that (3) $M \in \mathcal{M}_\Gamma^m$ and M falsifies all members of $\Xi - \Delta$ and hence verifies $\exists\neg\circ A$ whenever $\forall\circ A \in \Xi - \Delta$. By Fact 15.26, (2) entails that some $M' \in \mathcal{M}_\Gamma^{\mathbf{LLL}}$ verifies all members of Δ as well as some members of $\Xi - \Delta$. By the regularity of **LLL**, (4) some $M'' \in \mathcal{M}_\Gamma^{\mathbf{LLL}}$ verifies all members of Δ as well as some members of $\Xi - \Delta$, falsifies $\exists\neg\circ A$ whenever $\forall\circ A \in \Delta$ and moreover falsifies at least one $\exists\neg\circ A$ for which $\forall\circ A \in \Xi - \Delta$. But then $Ab(M'') \subset Ab(M)$, which is impossible (3).

Corollary 15.28 $\Delta \subseteq \Xi$ *is* maximal *with respect to* Γ *and* **LLL** *iff* $\{\exists\neg\circ A \mid \forall\circ A \in \Delta\} \in \Phi(\Gamma)$.

Theorem 15.27 relates minimally abnormal models to maximal sets of consistency statements. But what about maximal consistent models? Consider a LFI **LLL**, a premise set Γ, and a $\Delta \subseteq \Xi$ that is maximal with respect to Γ and **LLL**. Suppose that $M \in \mathcal{M}_\Gamma^{\mathbf{LLL}}$ verifies all members of Δ but that $M \notin \mathcal{M}_\Gamma^m$. In view of Theorem 15.27, this can only mean that there is a formula B such that M verifies $\forall\circ B \in \Delta$ but also verifies $\exists\neg\circ B$, whereas some $M' \in \mathcal{M}_\Gamma^{\mathbf{LLL}}$ falsifies $\exists\neg\circ A$ as well as $\neg\forall\circ A$ for all $\forall\circ A \in \Delta$, and hence falsifies $\exists\neg\circ B$. In view of Theorem 15.24, it follows that some $M'' \in \mathcal{M}_\Gamma^m$ falsifies $\exists\neg\circ A$ as well as $\neg\forall\circ A$ for all $\forall\circ A \in \Delta$, and hence falsifies $\exists\neg\circ B$ as well as $\neg\forall\circ B$. But this is impossible. Indeed, as $M'' \in \mathcal{M}_\Gamma^m$, it falsifies $\neg\circ\forall\circ B$ and hence verifies $\circ\forall\circ B$. But $\circ\forall\circ B \notin \Delta$, because M falsifies it and $\Delta \subseteq \Xi$ that is maximal with respect to Γ and **LLL**.

Corollary 15.29 $M \in \mathcal{M}_\Gamma^{\mathbf{LLL}}$ *verifies a* $\Delta \subseteq \Xi$ *that is maximal with respect to* Γ *and* **LLL** *iff* $M \in \mathcal{M}_\Gamma^m$.

Consider again the premise set $\Gamma_1 = \{\neg p, \neg q, p \vee q, p \vee r, q \vee s, \neg t, u \vee t\}$ and the LFI **mbC**—see footnote 17—in which conjunction and disjunction are classical. It is easily seen that $\Phi(\Gamma_1) = \{\{p\}, \{q\}\}$. Γ_1 may be extended with two different kinds of consistency statements. Every extension with a $\forall \circ A$ for which $\exists \neg \circ A \notin \bigcup \Phi(\Gamma_1)$ may be called a *consistency reclaim*. Such extensions are completely harmless. One may add as many consistency reclaims to Γ_1 as one desires and one may add them all together. Extending Γ_1 with a $\forall \circ A$ for which $\exists \neg \circ A \in \bigcup \Phi(\Gamma_1)$ may be called a *consistency decision*. Consistency decisions cannot always be combined—extending Γ_1 with both $\circ p$ and $\circ q$ results in triviality. Insights from adaptive logics teach us that consistency decisions may be combined iff the corresponding abnormalities belong to the same $\varphi \in \Phi(\Gamma)$. Needless to say, Γ_1 is an utterly simple toy example, but Corollary 15.28 shows that the matter holds generally.

Fact 15.30 If Γ is not **LLL**-trivial,[26] $\Delta \subseteq \Xi$, and $\exists \neg \circ A \notin \bigcup \Phi(\Gamma)$ whenever $\forall \circ A \in \Delta$, then a $\Gamma \cup \Delta$ is not **LLL**-trivial. (Consistency reclaims)

Fact 15.31 If Γ is not **LLL**-trivial, $\Delta \subseteq \Xi$, and $\exists \neg \circ A \in \bigcup \Phi(\Gamma)$ whenever $\forall \circ A \in \Delta$, then a $\Gamma \cup \Delta$ is not **LLL**-trivial iff there is a $\varphi \in \Phi(\Gamma)$ such that $\exists \neg \circ A \in \varphi$ whenever $\forall \circ A \in \Delta$. (Consistency decisions)

Theorem 15.32 *$Dab(\Delta) \in Cn_{\mathbf{AL}}(\Gamma)$ iff $Dab(\Delta) \in Cn_{\mathbf{LLL}}(\Gamma)$. (Immunity / **AL** is Dab-conservative with respect to **LLL**.)*

This theorem, proven as Theorem 10 in [5], shows that, if Γ is extended into Γ' by a (finite or infinite) set of consistency reclaims, then $\Phi(\Gamma') = \Phi(\Gamma)$. It follows that if a non-**LLL**-trivial Γ is extended by any set of consistency reclaims combined with any set of coherent consistence decisions—coherent in that they refer to the same $\varphi \in \Phi(\Gamma)$—then the resulting set is not **LLL**-trivial.

As we shall see in the next section, an inconsistency-adaptive logic restricts itself to extending a premise set with the full set of consistency reclaims—some extreme cases aside, this set is always infinite. Some consistency reclaims have an obvious effect—adding $\circ t$ to Γ_1 makes u derivable. For others, the gain is merely **CL**-theorems and combinations of them with already derivable formulas. Thus if $\circ v$ is added to Γ_1 and the LFI is **mbC**, then the following formulas are derivable among many others: $(v \wedge \neg v) \supset w$ and $(\neg t \wedge v) \vee (v \supset (w \wedge \neg w))$.

This is probably the best place to insert a brief comment on flip-flop adaptive logics. By minimizing abnormalities, adaptive logics interpret premise sets as much as possible in agreement with **ULL**, which is **CL** for most inconsistency-adaptive logics. However, some adaptive logics **AL** display the following odd behaviour. If the premise set Γ has no *Dab*-consequences, the **AL**-consequence set of Γ is identical to its **ULL**-consequence set. This is as it should be and as it is for all adaptive logics. However, if Γ has *Dab*-consequences, then the **AL**-consequence set of Γ is identical to its **LLL**-consequence set. This is obviously not all right. More correctly, it is nearly never what one wants. Usually one wants to isolate the unavoidable abnormalities and to consider at least some abnormalities as false.

---

[26]Where a logic **L** is defined over $\mathcal{L}$, a set Γ is **L**-trivial iff $Cn_{\mathbf{L}}(\Gamma) = \mathcal{W}$.

Some adaptive logics are flip-flops because the combination of a lower limit logic with a specific consequence set causes, in case Γ has *Dab*-consequences, all models of Γ to be minimally abnormal models of Γ. What this means in terms of consistency statements is that consistency reclaims are impossible while consistency decisions are possible. This looks odd from an adaptive point of view, but not from the viewpoint of LFIs. While the distinction between consistency reclaims and consistency decisions is heuristically and computationally interesting for a person applying a LFI, there is no reason for this person, or least for most applications, to restrict the addition of consistency statements to consistency reclaims. The target of inconsistency-adaptive logics is the set of formulas verified by every minimally abnormal model of the premises, the target of a person applying a LFI is the set of formulas verified by every model of the premises that falsifies a specific set of abnormalities.

15.8 Comparing Application Contexts

The last statement of the previous section identifies a central difference between LFIs and inconsistency-adaptive logics. Some more differences are worth being highlighted. One of them is that LFIs are Tarski logics (or very close to Tarski logics), are deductive logics, and have recursively enumerable consequence sets, whereas inconsistency-adaptive logics are defeasible, are formal characterizations of methods, and have very complex consequence sets—up to Π^1_1 for Minimal Abnormality—see [16, 19]. Another difference is that LFIs typically require ingenuity. The person who applies the logics should select consistency statements in order to extend the initial inconsistent theory T and to strengthen it with applications of **CL**-rules that are not generally valid. Inconsistency-adaptive logics do not depend on human decisions for their applications. They strengthen the initial T with the applications of the aforementioned **CL**-rules that are justifiable on logical grounds from the consistency presupposition—by way of comparison: consistency reclaims are so justified while consistency decisions are not. Some consequence sets that are too complex to be reached by human ingenuity can nevertheless be defined in adaptive terms.

The typical intended application context of LFIs is to phrase a theory $T = \langle \Gamma, \mathbf{L} \rangle$, in which **L** is a LFI and T is an inconsistent theory. Whether the theory was devised as inconsistent, or is the result of a failed attempt to formulate a consistent theory does not seem to be important. What is important is that the people devising T want it to be richer than a paraconsistent logic without definable consistency operator can define. They want T to be non-explosive but at the same time want T to contain the result of certain applications of **CL**-rules that are not validated by **L**. This is realized by adding consistency statements.

Given the LFI-theories that have been formulated in the da Costa tradition, I think it is fair to say that those theories basically came into being by starting from a T_0 that does not contain any consistency statements and stepwise extend it to T_1, T_2, etc. by adding a consistency statement at points where the available version T_i is judged to be too weak. Now and then, an extension will have turned out trivial, but then one may retract and remove a previously added consistency statement. This is perfectly all right; it is the way in which theories come into being in general. The only specific feature here is that T_0 is separated from the subsequent additions of consistency statements. Even if the addition and removal of consistency statements goes hand in hand with other additions and removals during the genesis of a theory, the steps that handle consistency statements are so specific and unusual that we may conceptually separate them from the other steps.

Are inconsistency-adaptive logics able to play the same role. Not quite. Given a premise set Γ, an adaptive LFI defines, all by itself, a consequence set of Γ that contains all **LLL**-consequences and moreover contains all consistency statements that are obtained by consistency reclaims. Incidentally, I write "all by itself" because the persons that apply the adaptive LFI do not need to add any consistency statements as premises. Comparing this to LFI for the typical intended application context of LFI, it seems that the adaptive approach does too much as well as not enough. It does too much by adding all consistency statements obtained from consistency reclaims. People applying the original LFI certainly do not do this and even cannot do this because the added set need not even be semi-recursive. Nevertheless, it is hard to see that anyone would object to consistency reclaims. If T can safely be extended with a consistency statement, if is can safely be so extended irrespective of the other consistency statements that are added to T[27] then what possible objection might one have to this extension? As announced, the adaptive approach does not add enough. Indeed, an adaptive LFI does not add any consistency statements obtained from consistency decisions.

One obviously might combine an adaptive LFI with a sequence of consistency decisions, just as in the application of the plain LFI. Another possibility is to apply a LFI and the Ω from (15.9) with the All Selections strategy, which is an obvious variant to the Normal Selections strategy. The resulting adaptive logic is somewhat unorthodox in that its consequence sets are sets of sets. Describing the approach here would take too much space. Moreover, the approach is somewhat arduous in that the adaptive logic defines all possible theories obtained by extending a premise set Γ with a $\Delta \subseteq \Xi$ that is maximal with respect to Γ and **LLL**.

There is, however, a further possibility. While inconsistency-adaptive logics handle inconsistency, other adaptive logics serve other purposes, for example defeasibly extend a set of statements (or formulas) with further entities, all on a par or in agreement with a preference ranking. One way to implement this is to add consistency statements preceded by a 'plausibility operator' $\Diamond$, which is governed by a modal logic, for example T. This lower limit logic may be combined, for example, with a set of abnormalities that (for the present application) have the form $\Diamond\forall\circ A \wedge \neg\forall\circ A$. Note that $\Diamond\forall\circ A$ entails $\forall\circ A \vee (\Diamond\forall\circ A \wedge \neg\forall\circ A)$. So if all minimally abnormal models of the premises falsify the abnormality $\Diamond\forall\circ A \wedge \neg\forall\circ A$, then $\forall\circ A$ is an adaptive consequence; otherwise it is not. The effect is that some plausible consistency statements function as actual consistency statements, whereas others remain merely

[27] If consistency decisions do not trivialise the theory, consistency reclaims do not either; if consistency decisions trivialise the theory, consistency reclaims cannot make that situation worse.

plausible. The effect may be enhanced by formulas that contain several diamonds. $\Diamond\forall\circ A$ expresses that $\forall\circ A$ has the highest plausibility, $\Diamond\Diamond\forall\circ B$ that $\forall\circ B$ has the next highest plausibility, and so on. I refer to [4] for a general description of this approach and for a related approach.

The logic that handles the plausibility-ordered consistency statements has some interesting features. The persons applying the logic have to decide to which premise set they apply it—so they have to fix which plausibility is attached to a consistency statement and the result functions as a premise. Once the premises are fixed, however, the adaptive logic defines the consequence set and does not demand ingenuity from the persons applying the logic. Moreover, adding consistency statements with a certain plausibility attached to them is a safe way to proceed. It never leads to triviality. If $\Diamond\forall\circ A$ is a premise, but adding $\forall\circ A$ would result in triviality, then $\forall\circ A$ will not be a member of the adaptive consequence set. As adaptive logics are reflexive, $\Diamond\forall\circ A$ will still be in the consequence set. But that is harmless anyway.

Allow me to repeat that it is not my intention to defend the adaptive approach or to attack the LFI approach. I am merely comparing. The computational problems are in principle similar for both approaches because they depend on the problem that has to be solved. Where a person applying a LFI is unable to choose the right set of consistency statements, a person following the adaptive approach will presumably be unable to figure out which consistency statements are in the adaptive consequence set. The consequence set is well-defined, which is clearly a very positive feature, but that does not make it available; its computational complexity may be just too high. Next, the fact that the adaptive approach eliminates the triviality danger should not be overestimated. The road through mistaken theories may very well be more efficient than the safe road, provided we reach the destination. Nevertheless, studying several approaches to the same problems may result in deeper insights and in improving one or both of the approaches.

The initial application context of inconsistency-adaptive logics was that a theory was intended as consistent and was given **CL** as its underlying logic, but later turned out to be inconsistent. Inconsistency-adaptive logics were devised with the aim to handle such situations by identifying and localizing the (minimal disjunctions of) inconsistencies present in a theory in the aforesaid situation. The idea was to devise a general means to 'interpret' such a theory as consistently as possible, viz. in such a way that, on the one hand, it is not trivial and, on the other hand, it maximally approaches the original intention (of those who devised the theory). The so obtained non-trivial and 'minimally inconsistent' theory was never meant as the ultimate goal. It is merely an intermediate goal on the road to consistency: once the inconsistencies in the theory are located and isolated, one may try to remove them. Forging of a consistent replacement, however, is not a logical matter. The central decisions require empirical considerations or conceptual considerations, and very often deep conceptual changes. Logics may guide this process, they may locate the interesting questions and their interrelations, they may dismiss proposals as inadequate, etc., but logics are unable to define the process.

Later inconsistency-adaptive logics turned out to have a second interesting application context. Especially in view of the twentieth century changes in the orthodoxy in mathematics, it turned out that inconsistency-adaptive and other adaptive mathematical theories have certain advantages over traditional **CL**-theories and, more generally, semi-recursive theories. Not too much was published until now [6, 20, 21], but even that seems to open interesting perspectives.

It seems to me that LFIs are not the right tools for any of the described application contexts of inconsistency-adaptive logics. For many a premise set or theory the set of consistency statements that need to be added is not only infinite but not even semi-recursive. Moreover, the work on adaptive theories in particular is mainly important from a theoretical point of view because it enables one to obtain sensible knowledge about well-defined but computationally complex sets. It might be hoped, however, that people committed towards LFIs would not be convinced by such arguments and would try to devise an approach for the typical application contexts of inconsistency-adaptive logics. Again, the interplay between competing approaches may lead to deeper insights as well as to new techniques.

15.9 Some Comments in Conclusion

I hope to have shown that the study of LFIs may benefit from insights gained in adaptive logics. The converse also holds but was not the topic of this paper. The apparently weak or less elegant features of other approaches allow one to discover weak or less elegant features of one's own approach. Given this and given that so much more can be said on the topic of this paper, I shall, by way of conclusion, mention some more results and insights from the adaptive side in the hope that LFI scholars will either locate flaws in my claims or will discover ways to profit from the insights and integrate the results. Before doing so, allow me to refer to the work on adaptive extensions of Jaśkowski's logics [13–15], which might provide new links between LFIs and Jaśkowski's logics.

Several techniques were developed to obtain criteria for final derivability within adaptive logics. Especially techniques in terms of prospective procedures seem to be transparent and promising and they are available in a single paper [22]. It seems likely that these techniques may be rephrased in terms of LFIs in order to cope with the triviality danger and the maximality question. Moreover, this transfer may lead to new insights and improved techniques.

In a similar vein, several arguments were developed in connection with adaptive logics in order to justify acting on the insights offered by a dynamic proof stage, even if one realizes that, given the defeasible character of the logics, these insights may be overruled in the future. Note that such arguments concern a specific form of acting under uncertainty. LFIs being deductive, they do not have to face difficulties related to defeasibility. Nevertheless they face related problems, summarized before as the triviality danger and the maximality question. Even if someone is not interested in sets of consistency statements that are maximal with respect to a given premise set and LFI, consistency reclaims are always selected from infinitely many possibilities and consistency decisions are not only selections, but may moreover cause triviality. At the predicative level, the set of minimal *Dab*-consequences of a decidable premise set Γ is only semi-recursive. What seems to be a consistency reclaim may later turn out to have been a consistency decision; disjuncts of a *Dab*-consequence may turn up again as disjuncts of later derived different *Dab*-consequences; *Dab*-consequences that count more than one disjunct may turn out not to be minimal in view of later derived *Dab*-consequences. So our present estimate of $\Phi(\Gamma)$—the estimate is defined in terms of a stage s of a dynamic proof and is called $\Phi_s(\Gamma)$—may be very different from $\Phi(\Gamma)$. But precisely our estimate of $\Phi(\Gamma)$ is our guide for consistency reclaims and consistency decisions; see also Facts 15.30 and 15.31. As far as I can see, it is our only guide.

Consider a paraconsistent logic **L** defined over a language $\mathcal{L}$ and suppose that disjunction and conjunction are classical and that no consistency connective is definable. Let $\mathcal{L}+$ be obtained from $\mathcal{L}$ by adding the symbol $\circ$ and let **L**+ result from extending **L** with the rule $\circ A, A, \neg A/B$. Next define $\mathbf{L}+^m$ by combining **L**+ with the Ω from (15.9) and the Minimal Abnormality strategy. The adaptive consequence set of a premise set Γ may contain consistency statements—all those that correspond to a consistency reclaim—and actually also disjunctions of consistency statements that are not themselves derivable. Moreover, these formulas will 'have an effect' on the adaptive consequences that belong to the initial language $\mathcal{L}$. Consider again the simplistic $\Gamma_1 = \{\neg p, \neg q, p \vee q, p \vee r, q \vee s, \neg t, u \vee t\}$ and let **L** be **CLuN** (or its propositional fragment **PI**). The $\mathbf{CLuN}+^m$-consequence set will contain $\circ p \vee \circ q$, as well as $\circ A$ for every sentential letter A different from p and from q.[28] Next, 'in line with' the presence of those consistency statements, the consequence will also contain u, $r \vee s$, $t \supset A$ for all formulas A, as well as infinitely more formulas from $\mathcal{L}$.

However, there is a little puzzle here. Suppose that we do not extend the language and logic, but proceed in terms of $\mathcal{L}$ and $\mathbf{L}^m$, the latter combining **L** with $\Omega = \{\exists(A \wedge \neg A) \mid A \in \mathcal{X}\}$.[29] There is a rather easy proof that, if $\Gamma \subseteq \mathcal{W}$, then $Cn_{\mathbf{L}+^m}(\Gamma) \cap \mathcal{W} = Cn_{\mathbf{L}^m}(\Gamma)$. So a consistency connective does not seem to have much use with respect to consistency statements that correspond to consistency reclaims. However, roughly the same holds true for consistency decisions. If these are handled in terms of plausibilities (as explained in the previous section), and the language is not extended with a consistency operator, one may still take formulas of the form $\Diamond\neg(A \wedge \neg A) \wedge (A \wedge \neg A)$ as abnormalities. I did not spell out the proof that this adaptive logic gives us the same formulas in $\mathcal{W}$ as the adaptive logic from the previous section. However, it seems extremely likely that there is such a proof. I phrased this point as a challenge in the hope that LFI scholars will show me wrong.

My final comment concerns the concentration on consistency. Remember the initial application context of inconsistency-adaptive logics: a theory was intended to be consistent but turned out inconsistent. My claim was that inconsistency-adaptive logics interpret such theory in a way that is maximally consistent and that the resulting

---

[28] Actually for all formulas not in the set $\{p, q\}$, but never mind.

[29] The reasons for the $\mathcal{X}$ is as in (15.9); it would be tiresome to make this more precise here.

adaptive theory may be taken as a starting point for devising a consistent replacement of the initial theory. It has turned out, however, that many theories may serve as the desired starting point. On the one hand, lower limit logics, sets of abnormalities, and strategies may be varied. But there is more. Inconsistencies may be seen as negation gluts. One may also consider negation gaps (both A and $\neg A$ false) and combinations of gluts and gaps. The same may be repeated for all other logical symbols. Moreover, non-logical symbols may be ambiguous. As was argued elsewhere [8], many inconsistent theories come out non-trivial if handled by logics that do not allow for negation gluts but allow for negation gaps, or for other types of gluts or gaps, or for ambiguities, or for combinations of the things mentioned. Next, gaps and gluts and ambiguities may be minimized, all at once or in a certain order. Each of these choices lead, for some inconsistent theories, to a desired starting point. All such starting points are in principle on a par. The idea that the only way out is minimizing inconsistency is just a prejudice.

All those gluts and gaps leave ample room for variants and combinations. Let me here just point to one such combination in connection with inconsistency [8, Sect. 4], the other logical symbols being kept classical. Instead of considering a complex inconsistency like $(p \vee q) \wedge \neg(p \vee q)$ as a single abnormality, one might consider three abnormalities instead: $p \wedge \neg(p \vee q)$, $q \wedge \neg(p \vee q)$, and $(p \vee q) \wedge \neg(p \vee q)$. In a sense, the first two offer a possible cause for the occurrence of the contradiction. The first and second abnormalities entail the third, but not vice versa. By minimizing all three abnormalities, one obtains a different selection of (for example **CLuN**-)models than when one minimizes only contradictions. This paragraph only sketches the vague idea in terms of an example, but a systematic approach was published.

What is common to all the cases just discussed is that the considered abnormalities are not matched by consistency statements. Take for example a conjunction glut—that $A \vee B$ is true while A is false or B is false. No consistency statement can eliminate it or make it to cause triviality. Similarly, some consistency statements reduce all three abnormalities from the previous paragraph to triviality, but if $(p \vee q) \wedge \neg(p \vee q)$ is true anyway and the logic is **mbC**, no consistency statement can rule out for example $p \wedge \neg(p \vee q)$ without causing triviality.[30] So the challenge to LFI scholars is to devise and study operators that eliminate gluts and gaps that are not inconsistencies. The fact that all gluts and gaps *surface* as inconsistencies shows that, in some cases, inconsistency may be merely the symptom rather than the actual disease.

The diversity of approaches within the paraconsistent community has been overwhelming from early on. We should not strive for unification. Actually, we will not strive for unification for most of us are pluralists. Yet we may continue to learn from each other—au choc des idées jaillit la lumière. This is why I hope that some comments from this paper may arouse interest of LFI scholars and of scholars interested in paraconsistency in general.

[30] If you frown here, realize that $\neg p$ is not a **mbC**-consequence of $\neg(p \vee q)$.

References

1. Batens, D.: Paraconsistent extensional propositional logics. Logique et Analyse 90–91, 195–234 (1980)
2. Batens, D.: Inconsistency-adaptive logics. In: Orłowska, E. (ed.) Logic at Work. Essays Dedicated to the Memory of Helena Rasiowa, pp. 445–472. Physica Verlag (Springer), Heidelberg (1999)
3. Batens, D.: Towards the unification of inconsistency handling mechanisms. Logic Log. Philos. **8**, 5–31 (2000). Appeared 2002
4. Batens, D.: Narrowing down suspicion in inconsistent premise sets. In: Malinowski, J., Pietruszczak, A. (eds.) Essays in Logic and Ontology. Poznań Studies in the Philosophy of the Sciences and the Humanities, vol. 91, pp. 185–209. Rodopi, Amsterdam (2006)
5. Batens, D.: A universal logic approach to adaptive logics. Log. Univers. **1**, 221–242 (2007)
6. Batens, D.: The consistency of Peano Arithmetic. A defeasible perspective. In: Allo, P., Van Kerkhove, B. (eds.) Modestly Radical or Radically Modest. Festschrift for Jean Paul van Bendegem on the Occasion of His 60th Birthday. Tributes, vols. 24, 22, pp. 11–59. College Publications, London (2014)
7. Batens, D.: Tutorial on inconsistency-adaptive logics. In: Béziau, J.-Y., Chakraborty,M., Dutta, S. (eds.) New Directions in Paraconsistent Logic. Springer (2015)
8. Batens, D.: Spoiled for choice? J. Logic Comput. (1913), in print. doi:10.1093/logcom/ext019
9. Béziau, J.-Y.: S5 is a paraconsistent logic and so is first-order classical logic. Log. Invest. **9**, 301–309 (2002)
10. Carnielli, W.A., Coniglio, M.E., Marcos, J.: Logics of formal inconsistency. In: Gabbay, D., Guenthner, F. (eds.) Handbook of Philosophical Logic, vol. 14, pp. 1–93. Springer (2007)
11. da Costa, N.C.A.: Calculs propositionnels pour les systèmes formels inconsistants. Comptes rendus de l'Académie des sciences de Paris **259**, 3790–3792 (1963)
12. da Costa, N.C.A.: On the theory of inconsistent formal systems. Notre Dame J. Form. Logic **15**, 497–510 (1974)
13. Nasieniewski, M.: Logiki adaptujące sprzeczność (Inconsistency adapting logics). Ph.d. thesis, Chair of Logic, N. Copernicus University, Toruń, Poland (2002)
14. Nasieniewski, M.: An adaptive logic based on Jaśkowski's logic $\mathbf{D}_2$. Logique et Analyse 185–188, 287–304 (2004). Appeared 2005
15. Nasieniewski, M.: Wprowadzenie do logik adaptywnych. Wydawnictwo Naukowe, Universytetu Mikołaja Kopernika, Toruń (2008)
16. Odintsov, S.P., Speranski, S.O.: Computability issues for adaptive logics in multi-consequence standard format. Stud. Logica **101**(6), 1237–1262 (2013). doi:10.1007/s11225-013-9531-2
17. Priest, G.: In Contradiction. A Study of the Transconsistent. Oxford University Press, Oxford (2006). Second expanded edition (first edition 1987)
18. Suszko, R.: The Fregean axiom and Polish mathematical logic in the 1920s. Stud. Logica **36**, 377–380 (1977)
19. Verdée, P.: Adaptive logics using the minimal abnormality strategy are Π^1_1-complex. Synthese **167**, 93–104 (2009)
20. Verdée, P.: Strong, universal and provably non-trivial set theory by means of adaptive logic. Logic J. IGPL **21**, 108–125 (2012)
21. Verdée, P.: Non-monotonic set theory as a pragmatic foundation of mathematics. Found. Sci. **18**, 655–680 (2013)
22. Verdée, P.: A proof procedure for adaptive logics. Logic J. IGPL **21**, 743–766 (2013). doi:10.1093/jigpal/jzs046

Chapter 16
Stipulation and Symmetrical Consequence

Bryson Brown

Abstract In this paper I lay some of the groundwork for a naturalistic, empirically oriented view of logic, attributing the special status of our knowledge of logic to the power of *stipulation* and expressing the stipulations that constitute the vocabulary of formal logic by *rules of inference*. The stipulation hypothesis does nothing *by itself* to explain the usefulness of logic. However, though I do not argue for it here, I believe the *selective adoption and application* of stipulations can. My concern here is with an issue that has already received a good bit of attention: it seems that we are free to make whatever stipulations we care to make, but we also know that logical stipulations must be carefully constrained, to avoid trivialization, as well as subtler impositions on the already established inferential practices to which we apply our logical vocabulary. I propose three increasingly stringent criteria that fully conservative extensions of a language should meet, and apply them to evaluate three symmetrical, multiple-conclusion logics. A new result, proven first for classical multiple-conclusion logics and then modified and extended to all reflexive, monotonic, and transitive consequence relations, undergirds the focus on proof-theoretic approaches to the consequence relation I adopt here.

Keywords Stipulation · Inference rule · Conservative extension · Multiple conclusions · Preservationism

Mathematics Subject Classification Primary 03A-02 · Secondary 03B-02

Andrew Tedder was my research assistant when the research presented here was begun. He provided both LaTeX expertise and helpful discussions during early stages of the project, and was coauthor of two conference presentations of that work at the Society for Exact Philosophy meetings of 2012 and 2013.

B. Brown (✉)
Department of Philosophy, University of Lethbridge,
4401 University Dr., Lethbridge, AB, Canada
e-mail: brown@uleth.ca

J.-Y. Beziau et al. (eds.), *New Directions in Paraconsistent Logic*,
Springer Proceedings in Mathematics & Statistics 152,
DOI 10.1007/978-81-322-2719-9_16

Empiricists have generally distinguished two kinds of knowledge, roughly following Hume's distinction between *relations of ideas* and *matters of fact*. For Hume, our knowledge of mathematics and logic was knowledge about relations of ideas, while matters of fact included individual experiences or observations and any true generalizations about such experiences or observations [8, p. 26f]. According to Hume, we can be certain about relations of ideas because we are directly aware of our own ideas in a way that makes their relations something we are also immediately aware of, whenever we think of the ideas in question. We can be similarly confident of individual matters of fact, but only when they are being experienced (or, with less vividness, remembered).

However, the special status of truths of mathematics and logic has occasioned some controversy amongst empiricists; Mill, in particular, argued for collapsing the distinction and subjecting all knowledge claims to the court of experience. In this paper I will defend an empiricist view, while dropping Hume's reliance on individual awareness of the relations of ideas. Instead I attribute the special status of our knowledge of mathematics and logic to the power of pragmatically constrained *stipulations*. Finally, I will express the stipulations that constitute the vocabulary of formal logic by *rules of inference*, rather than by appeal to some stipulated semantics.

Although the stipulation hypothesis does nothing *by itself* to explain the usefulness of logic and mathematics, the *selective adoption and application* of stipulations that help us express and systematize reliable inferential connections offers some insight into both why such systems are useful and what makes us so confident about them. But my concerns here begin with an issue that is already familiar. It is tempting to assume that we are free, like Tweedledee, to make whatever stipulations we care to make. But logical stipulations must be carefully constrained, if we wish to avoid disasters like A.N. Prior's 'tonk' [10]. In [1] Nuel Belnap presented a broad account of the kinds of constraint needed: the introduction of logical words with stipulated introduction and elimination rules must *respect* (i.e., it must not change) the inferential connections already in place in the language as it exists before the new word is added: "we are not defining our connective ab initio, but rather in terms of an antecedently given context of deducibility, concerning which we have some definite notions. By that I mean that before arriving at the problem of characterizing connectives, we have already made some assumptions about the nature of deducibility…(I)f we note that we already have some assumptions about the context of deducibility within which we are operating, it becomes apparent that by a too careless use of definitions, it is possible to create a situation in which we are forced to say things inconsistent with those assumptions" [1, p. 131]. Later, Belnap proposes a narrower, more specific view of the constraints on stipulation, invoking the standard definition of a conservative extension due to Post: "the extension must be conservative; i.e., although the extension may well have new deducibility statements, these new statements will all involve *plonk*" [1, p. 132]. In what follows I consider the criteria conservative extensions must meet, in the hope of capturing in more detail the broader account of the limits of stipulation that Belnap's paper begins with.

Following Belnap, I hold that we are free to stipulate whatever meanings we wish *in isolation*, the only limitations being those of *interest*: for example, a system of stipulated inference rules that trivializes is generally uninteresting, except perhaps as an illustration of how trivialization can arise. But when our aim is to apply some stipulations in already established linguistic contexts, the stipulations must respect commitments that are already in place. For example, if we wish to apply a formal theory of arithmetic to systematize an established practice of counting various kinds of things, comparing their quantities, and so on, the formal apparatus we add to the language of counting must be compatible with the practice of counting.[1] This amounts to a restatement of Belnap's initial, broader constraint, but in what follows we will identify some specific constraints that go beyond Belnap's narrower definition of conservative extension above, and that seem justified in the light of his broader view of the limits a satisfactorily conservative extension must observe.

The idea that logic is stipulated rather than discovered may seem peculiar. After all, natural languages do include words whose roles in inference correspond fairly well to those played by logical words in formal systems of logic. For instance, the English use of 'not,' 'and,' 'or,' 'all,' and 'some' come fairly close to the patterns of use prescribed by formal logic for the usual corresponding symbols (here we will use '$\neg$,' '$\vee$,' '$\wedge$,' '$\forall$,' and '$\exists$'), and of course we rely on these parallels in teaching our students both to understand those symbols and to translate from English into the language of formal logic and vice versa.

It would be surprising if this were not the case: reasoning and its presentation are important aspects of our use of natural language, and words that conveniently produce sentences bearing certain logical relations to other sentences often come in handy. But the correspondence between the use of such words in natural language and that of these symbols in a formal logic is not complete or regular enough to justify the assumption that these words in natural languages have the same *meanings* as their formal counterparts [9]. The conjunctive force of disjunction in some English contexts, where 'you may have A or B' is understood to imply both 'you may have A' and 'you may have B,' (though not 'you may have A and B') is one of many cases where a regular and well-understood pattern of use of 'or' does not match the formal $\vee$; the failed[2] attempt to capture an exclusive use 'or' by means of a binary connective expressing truth functional nonequivalence illustrates how easily we can fool ourselves into seeing a closer relation between natural language use and a formal stipulation than is really there.

Making a closely related point, Nuel Belnap[3] once asked how we can tell what logic a scientist is using: if we treat a logic as a theory of the meanings of words that

---

[1]This is not to say that some refinements of that practice may not result: as principles of arithmetic are applied, new and reliable ways of determining counts can emerge and become part of the practice. But the practice of counting predates the emergence of formal theories of arithmetic; when we add a theory of arithmetic to our rules for counting and comparing quantities of things, we want the stipulations that theory embodies to be conservative: it should not conflict with our established counting practices.

[2]Failed because iteration of such a connective to form n-ary 'disjunctions' gives a formula which is true iff the number of sentences 'disjoined' is odd, while a natural exclusive disjunction produces a sentence true iff exactly one of the disjoined sentences is true.

[3]In conversation.

are already part of the natural language, we saddle ourselves with difficult epistemic challenges. The imputed meanings will, at best, be imperfectly respected in natural language use, and what may seem to be 'errors' of logic on the part of natural language users can be explained in ways that reconcile actual use with different formal logics. Thus I believe it would be better to acknowledge that our formal logical vocabularies are extensions of natural usage. Such formal vocabularies have the advantage of being far more regular and systematic than the actual use of the words in natural languages that, at least on occasion, do similar work.[4] As already noted, this stipulative view of formal logic requires that, when applying formal logic to real languages, natural and scientific, we do so *conservatively*, i.e., in ways that do not impose on or make questionable assumptions about the languages we add them to.

16.1 Starting Points

When it comes to logical stipulations, I prefer multiple-conclusion logics for several reasons. They treat reasoning from assertions to further assertions and reasoning from denials to further denials in an elegant, symmetrical way, a symmetry whose preservation across a range of logical systems is one focus of this work. This allows such systems the flexibility to systematize a wider range of reasoning patterns. In particular, we do not always reason from assertions to further assertions: sometimes we reason from denials to further denials, as when, looking into the cupboard but failing to see either cups or saucers, we deny both 'I see cups in the cupboard' and 'I see saucers in the cupboard'.[5] Even if actual reasoning in some contexts does involve asymmetries between assertion and denial and in the use of certain logical words, a formal system that *allows* symmetrical inference patterns along with logical words that enable us to express them elegantly does not impose this *formal* symmetry on the actual practice we are extending with our proposed logical vocabulary and rules. So a multiple-conclusion logic can still be conservative with respect to such practices.[6] On the other hand, imposing an asymmetrical formal logic on a language that actually treats reasoning symmetrically would fail the test of conservatism.

In Sect. 16.2 we will develop a result about the relation between classical semantics and multiple-conclusion proofs. Section 16.3 draws on the Scott-Lindenbaum Lemma to extend the result (with some qualification) to a wider class of multiple-

[4]In legal and scientific contexts, the use of natural language is more regimented, and often highly redundant; I take this as evidence of efforts to achieve more uniform, shared understanding by means of regimentation and repetition.

[5]The inference here is from the *denial* of "I see cups or saucers" to the denial of "I see cups" and "I see saucers." In classical logic we can translate such reasoning into one that runs from assertions of *negations* as premises to a disjunction of negations that expresses our conclusion as an assertion. However, such translations are awkward, and may simply fail to work if we are dealing with a nonclassical negation.

[6]Similarly, the laws of classical physics are time-symmetric, but this symmetry allows for substantial matter-of-fact asymmetries in the course of actual events.

conclusion logics. Sequents of all these logics take the form $\Gamma \vdash \Delta$, where Γ and Δ are sets of sentences. A natural reading of such sequents takes them to show that a commitment to assert all $\gamma \in \Gamma$ is *incompatible* with a commitment to deny all $\delta \in \Delta$, i.e., that such commitments are *incoherent*. Restall [12] though other readings are possible, Restall's is a good place to start because it stays close to familiar applications of logic. The rest of the paper applies these results to argue that bi-intuitionistic logic [4] provides a good example of a symmetrical, conservative collection of logical words.

16.2 An Objection to Inferentialism in Logic

Before I present the result to be proven, there is some explaining and shoring up to do. We will be specifying the stipulated meanings of logical words by appeal to the inference rules that govern their use. But some have argued that this approach is inadequate, and that the semantic approach is more revealing. So we begin by defending the inferential approach to *classical logic* against one such critic. In [11] Panu Raatikainen draws on [5] to argue that the classical syntactic consequence relation does not adequately express the *meanings* of the classical logical words, because the inference rules for a standard classical system of natural deduction do not rule out the nonstandard valuations making all sentences true or making both A and B false while making $A \vee B$ true. These nonstandard valuations are compatible with the familiar single-conclusion natural deduction rules because these rules are regarded as "satisfactory," i.e., sound and complete, iff they capture *truth-preserving* character of $\models$, the semantic consequence relation, and the nonstandard valuations do not impose any further constraints on the inferences that preserve *truth*:

- Any inference preserves truth when all sentences are true!
- No truth-preserving inference takes us from $A \vee B$ to either A or B (of course this is why supervaluational semantics preserve the classical $\models$, despite allowing non-prime valuations).

Carnap's nonstandard valuations are compatible with (classical) *truth-preservation*, whether we characterize it in terms of either the syntactic $\vdash$ or the semantic $\models$. Raatikainen takes this fact to support a semantic understanding of the connectives because the rules for standard semantics rule out Carnap's nonstandard valuations:

- A is true(false) iff $\neg A$ is false(true);
- $A \vee B$ is false iff A is false and B is false.

However, this problem disappears if we adopt a multiple-conclusion proof system. Consider first the $\dashv$ relation, a falsehood-preserving 'proof' relation. We obtain the rules for $\dashv$ simply by forming the dual of the rules for $\vdash$, and it is immediately apparent that $\dashv$ imposes some constraints on acceptable valuations that $\vdash$ does not:

1. The all-true valuation assigns truth to all sentences, including classical contradictions. So the all-true valuation violates the *falsehood-preserving* inferences in $\dashv$, which derive these contradictions (showing they are correctly *deniable*) from no premises.
2. The classical falsehood-preserving $\dashv$ also includes an inference from denial of A, B to the denial of $A \vee B$, a mirror image of the truth-preserving inference from A and B to the truth of $A \wedge B$.

Of course $\dashv$ is *merely* falsehood-preserving. So it does not capture all the constraints of classical semantics either: the all-false valuation, along with valuations making $A \wedge B$ false but A and B both true are compatible with all the *falsehood-preserving* inferences. But combining $\vdash$ and $\dashv$ to produce a multiple-conclusion $\vdash$ relation that captures both truth-preserving inferences from left to right and falsehood-preserving inferences from right to left provides a simple response to Raatikainen's critique of inferentialism: for every tautology A, we have $\vdash A$; for every contradiction B we have $B \vdash$, for every disjunction we have $A \vee B \vdash A, B$, and for every conjunction, we have $A, B \vdash A \wedge B$. Carnap's nonstandard valuations and their duals violate these constraints, so multiple-conclusion systems provide a straightforward resolution of this challenge.[7]

This response to Raatikainen illustrates a useful point, which leads to the main lesson here: classical semantics is more elegantly linked to the inference rules of a (classical) multiple-conclusion proof system than it is to either a multiple premise, single-conclusion proof system or a single premise, multiple-conclusion proof system. That said, how best to *understand* this link needs further explication. Multiple-conclusion systems are rarely touched on in first or second logic courses, and when students are first introduced to multiple-conclusion logics they are often puzzled (even put off) by the long, indefinitely extendable conclusion sets that are declared to 'follow' from a given premise set. These conclusion sets contrast sharply with the simple, standard reading of what a single-conclusion consequence relation tells us about the *content* of a set of premises Γ: the *closure* of Γ under $\vdash$, i.e., the set of all sentences that follow from Γ, captures everything we are 'logically committed to' when we accept Γ. Nearby lies the standard definition of a *theory*, as a set of sentences closed under the consequence relation. There is no immediately obvious way to produce something similarly illuminating by 'gathering together' the conclusion sets following from some set of premises. Consider three simple candidates for this role:

[7]Raatikainen does recognize the potential of an appeal to multiple-conclusion logic here. His response is to raise concerns about how one could rule out the possibility of sentences being both true and false; we will not pursue his discussion any further here, though I do not see how this response helps to support a semantic as opposed to a proof-theoretic perspective on logic, since paraconsistent logics provide both proof theory and semantics tolerating such assignments.

1. The union of the conclusion sets includes all of L.
2. The set of all conclusion sets is a jumble, including every superset of every conclusion set.
3. The set of all *minimal* conclusion sets is better, but what it tells us about the 'contents' of the premise set is still obscure.

Happily, it turns out there is a more illuminating way to 'gather together' the family of sets that follow from a given premise set, Γ. Let $\mathfrak{D}$ be the family of sets that follow from a premise set, Γ. Then for consistent Γ, we define $\mathfrak{F}$ as the family of *least transverses*[8] of $\mathfrak{D}$, where a *transverse* of a family of sets $\mathfrak{S}$ is a set T that has a non-null intersection with every member of $\mathfrak{S}$ and a *least transverse* of $\mathfrak{S}$ is a transverse of S that has no proper subset that is also a transverse of S:

1. $\forall S \in \mathfrak{S}, S \cap T \neq \varnothing$
2. $T' \subset T \rightarrow \exists S \in \mathfrak{S}, S \cap T' = \varnothing$

Lemma
$A \in \mathfrak{F} \Leftrightarrow A$ is a maximal consistent extension of Γ.
$\Leftarrow$:
Let A be a maximal consistent extension of Γ.
RTS: A is a transverse of $\mathfrak{D}$
(i.e., A has a non-null intersection with every Δ such that $\Gamma \vdash \Delta$.)

Suppose $\Gamma \vdash \Delta^*$. Then, given compactness of $\vdash$, every maximal consistent extension (MCE) of Γ must include some member of Δ^*.

We need to show that if $\Gamma \vdash \Delta^*$, any consistent extension of Γ either includes a member of Δ^* or can be consistently extended by adding a member of Δ^*. Of course every maximal consistent extension of Γ, G, will include at least one disjunct from any disjunction G includes. But the disjunction of elements of Δ^*, $\bigvee(\Delta^*)$ is an element in every maximal consistent extension of Δ^*.

Therefore A includes an element of Δ^*.

$\Rightarrow$:

Suppose B is a transverse of $\{\Delta{:}\Gamma \vdash \Delta\}$.

Consider a set X including some sentence from each MCE of Γ which is not in B. By soundness and completeness of $\vdash$ each MCE of Γ includes all the sentences satisfied by some model of Γ, and every model of Γ satisfies some MCE of Γ. So every model of Γ satisfies at least one sentence in X. Thus by completeness and compactness, $\Gamma \vdash X'$, for some X' a finite subset of X. So B must include some member of X.

Therefore every transverse of $\Delta{:}\ \Gamma \vdash \Delta$ includes a maximal consistent extension of Γ.

Corollary
The symmetry of $\vdash$ implies that the converse also holds:

---

[8]This use of 'transverse' derives from the use of hypergraph transverses in the semantics of weakly aggregative logics [13, pp. 50–51].

For consistently deniable Δ, A is a transverse of $\mathfrak{G}$, the family of sets $\{\Gamma : \Gamma \vdash \Delta\}$ iff A includes a maximal *consistently deniable* extension of Γ.

This result provides a direct answer to Raatikainen's concerns by showing that a classical multiple-conclusion consequence relation determines the semantic $\vDash$ *along with* all the classical valuations on the language.

A related observation worth making here is that the symmetry of multiple-conclusion consequence allows classical multiple-conclusion systems to give a *proper* $\neg - intro$ rule, something not available in single-conclusion systems: suppose that $\Gamma, \alpha \vdash \Delta$. Then 'pushing' α from one side of the turnstile to the other while adding a negation *preserves* the incoherence of asserting all the members of Γ, α and denying all of Δ (of course the same goes for $\Gamma \vdash \alpha, \Delta$). So we can produce a consequence involving a negation without needing to have a negation in our premises. This example relies on the classical equivalence of asserting A and denying $\neg A$ (and of denying A and asserting $\neg A$), so we might wonder just how far these advantages of MC systems extend, and whether the result just presented generalizes. But in fact, the advantages extend quite widely, as does the result just proven. But before proceeding to show this, we consider an obvious objection.

Some might argue that I am making too much of this result. We can also express the 'aggregation' of finite conclusion sets here by replacing the ',' with '$\vee$': when and only when $\Gamma \vdash \Delta$ holds in a multiple-conclusion system, $\Gamma \vdash \bigvee(\Delta)$ also holds (where $\bigvee(\Delta)$ is the disjunction of all elements of Δ). So we can produce the maximal consistent extensions of Γ by forming the minimal transverses of every $\bigvee(\Delta)$ such that $\Gamma \vdash \bigvee(\Delta)$. There is no need to appeal to multiple conclusions after all!

But this line of argument proves too much. It is just as easy to aggregate our finite premise sets by forming their conjunctions. So we can discount the multiplicity of premise sets just as easily, reducing the usual consequence relation to a relation between individual sentences. This leaves us with a puzzle for those who object to multiple conclusions but are happy with multiple premises: why eliminate one but not the other, when the two are perfectly symmetrical?

Further, the syntactic expression of aggregation on the left and right can be carried out in many ways, and takes different forms in different logics, while a multiple-conclusion approach expresses whatever aggregation applies to premises and conclusions in a uniform way, simply by listing premises and conclusions: thus, in the classical case, syntactic aggregation can be expressed in many ways, since any way of producing sentences that are true when and only when all of some collection of sentences are true will do (similarly for aggregation on the right, we need to produce a sentence that is false when and only when all of some collection of sentences are false). As another example, in fixed-level n-forcing we standardly aggregate (on the left) by forming the disjunction of pairwise conjunctions amongst collections of n+1 sentences, but any non-n colourable hypergraph provides a 'template' for left-sided level n aggregation [3].[9]

[9]The template is applied on the left by conjoining the edges and then disjoining the conjunctions, and on the right by disjoining the edges and then conjoining the disjunctions.

16.3 Consequence and 1, 0 Semantics

Lemma 16.3.1 *Scott-Lindenbaum Lemma*

Every reflexive, monotonic, and transitive (RMT) consequence relation $\vdash$ *is such that, for some collection of* $1, 0$ *valuations, V on L:*

$\Gamma \vdash \Delta \Leftrightarrow \forall v \in V :, \Gamma v(\gamma) = 1 \rightarrow \exists \delta \in \Delta : v(\delta) = 1$ [14].

The Scott-Lindenbaum Lemma tells us that for $\vdash$ an RMT consequence relation, there is a set of allowed $1, 0$ valuations determining a $\models$ such that $\Gamma \vdash \Delta$ iff $\Gamma \models \Delta$. So just as in the classical case, the *minimal transverses* of $\{\Delta : \Gamma \vdash \Delta\}$ and of $\{\Gamma : \Gamma \vdash \Delta\}$ are closely connected to the allowed $1, 0$ valuations of the sentences of L. This makes the result above very general, since reflexivity, monotonicity, and transitivity are generally accepted as fundamental features of consequence relations in general.

Let V be a set $1, 0$ valuations on L that constitute a Scott-Lindenbaum semantics for $\vdash$, and w be any valuation in V. Then:

$\Gamma \models \Delta$ iff $\forall w$: if $\forall \gamma \in \Gamma$, $V_w(\gamma) = 1$, then $\exists \delta \in \Delta$, $V_w(\delta) = 1$

or (equivalently, given bivalence)

$\forall w$: if $\forall \delta \in \Delta$, $V_w(\delta) = 0$, then $\exists \gamma \in \Gamma$, $V_w(\gamma) = 0$.

- If some valuation assigns 1 to every member of Γ, A is a minimal transverse of $\{\Delta : \Gamma \vdash \Delta\}$ iff A includes all and only the sentences assigned 1 at some *1-minimal* valuation where Γ's members are assigned 1.
- If some valuation assigns 0 to every member of Δ, A is a minimal transverse of $\{\Gamma : \Gamma \vdash \Delta\}$ iff A includes all and only the sentences assigned 0 at some *0-minimal* valuation where Δ's members are assigned 0.

Note: by a 1 (0)-minimal valuation w we mean a complete assignment to the sentences of L such that there is no valuation w' that assigns 1 (0) to a proper subset of the sentences assigned 1 (0) at w.

As before we prove the first; the second follows by symmetry of $\vdash$.

$\Leftarrow$: Suppose A includes all the sentences assigned 1 at some w (i.e., in some allowed $1, 0$ valuation), i.e., $\exists w : \forall S \in \Gamma, V_w(S) = 1$ and $\{S : V_w(S) = 1\} \subseteq A$

RTS: A is a transverse of $\{\Delta : \Gamma \vdash \Delta\}$.

Suppose $\Gamma \vdash \Delta^*$.

Given soundness, if $\forall s \in \Gamma$, $V_w(s) = 1$, then $\exists s \in \Delta^* : V_w(s) = 1$.

So $\forall \gamma \in \Gamma$, $V_w \gamma = 1 \rightarrow \exists \delta \in \Delta^* : V_w \delta = 1$. But then $A \cap \Delta^* \neq \varnothing$, as required.

$\Rightarrow$: Suppose that A is a transverse of $\{\Delta; \Gamma \vdash \Delta\}$.

RTS: $\exists w : \forall \gamma \in Gamma, V_w(\gamma) = 1$ and $A \supseteq \{s : V_w(s) = 1\}$.

Suppose for reductio that no valuation w is such that $\forall \gamma \in \Gamma$, $V_w(\gamma) = 1$ and $A \supseteq \{s : V_w(s) = 1\}$. It follows that $\forall w : \gamma \in \Gamma : V_w(\gamma) = 1 \rightarrow \exists s : V_w(s) = 1$ & $s \notin A$.

Now consider the set of sentences produced by selecting some such s 'missing' from A from each valuation satisfying Γ:

$\mathbf{X}$: $s \in \mathbf{X} \Leftrightarrow \exists w, s : \forall\gamma \in \Gamma, V_w(\gamma) = 1 \;\&\; V_w(s) = 1 \;\&\; s \notin \mathfrak{F}$. Ex hypothesi, $\forall w[(\forall\gamma \in \Gamma : V_w(\gamma) = 1\,) \rightarrow (\exists s \in \mathbf{X}{:}V_w(s) = 1)]$.

That is, every model of Γ satisfies some sentence in $\mathbf{X}$. Therefore, given completeness and compactness, $\Gamma \vdash \mathbf{X}'$, for $\mathbf{X}'$ a finite subset of $\mathbf{X}$. But then A must include some member of $\mathbf{X}$ after all, contra our hypothesis.

16.4 A Preservationist Reading

It is natural for logicians to think of 1 and 0 here as expressing *truth-at-w* and *falsehood-at-w* for the sentences of L; more liberally, we can adopt Restall's readings, *correctly assertible at w* and *correctly deniable at w*. However, a still more modest (though also more abstract) reading is also available: we can take 1 and 0 in these valuations to indicate whether or not a sentence of L has some unspecified property, where what property that is depends on the consequence relation in question (and, for applied logics in which we take L to express/translate sentences in a specific language, on how it is applied in that language). As an offbeat example, consider the property of being among the sentences *asserted* by a given speaker of L. The history of each speaker S then determines a 1/0 valuation, and the resulting 'consequence' relation would be such that $\Gamma \vDash \Delta$ iff $\forall$ S: either S failed to assert some sentence in Γ or S asserted some sentence in Δ. Thus I take this result to be *preservationist* in spirit: if we call *classical* any property that *bipartitions* its domain, then Scott's result tells us that any RMT relation on L is determined by the preservation of some classical property of sentences.

In the introduction, following Belnap [1], I assumed that the logical words are introduced as stipulative extensions of a preexisting language. Such extensions are required to be conservative: they should not change how the language they are added to works. Like Belnap's first statement of this point, Dummett's [6, p. 221f] discussion of conservatism aims at broader results than the specific definition of conservatism subsequently provided by Belnap: it is not just that adding some new words with stipulated meanings to form $L*$ should not allow us to derive *new* consequences that involve only the vocabulary of the unextended language L. Dummett says the extension must not conflict with the understanding we already have of the *meanings* of sentences of L. In an effort to be more specific about what this broad notion of conservatism might require, and drawing on Dummett's insistence that we not assume every sentence in the underlying language is either correctly assertible or correctly deniable, I propose the following three 'degrees' of conservatism, raising two distinct kinds of concern about how adding new logical words might alter our understanding of the original language.

16.5 Three Degrees of Conservatism

- First degree (*minimal*) conservatism requires that the addition of new logical words to a language L, extending L to $L*$, does not allow proof or disproof of consequences in the 'base' language not already proven or disproven without them.[10]
- Second-degree conservatism imposes Dummett's requirement that the addition of our new logical words not imply that every sentence in the base language L is either correctly asserted or correctly denied.[11]
- Third degree conservatism requires the addition of our new logical words not imply that a consistently assertible (or deniable) set of sentences can be *extended* to a maximal consistently assertible (or deniable) set whose complement is a maximal, consistently deniable (assertible) set, unless the established *use* of L already assumes this.[12]

The first of these is, of course, the standard definition of conservatism adopted by Belnap, requiring that the original consequence relation of L be identical to the subrelation of the consequence relation of $L*$, the extended language, obtained by restricting the consequence relation of $L*$ to the vocabulary of L. In contrast, the second and third focus on the relation between the semantics of L and $L*$. Dummett's concern was that extending L to $L*$ may have nonconservative implications about what sort of meaning we take the sentences of L to have. Since consequence relations do not generally determine a fixed semantics [2], an inferentialist may take this concern with a grain of salt—but the close ties between the classical $\vdash$, *maximal consistent extensions* and their complementary duals, the maximal consistently deniable extensions of a conclusion set, suggest bivalence is a fundamental feature of the classical consequence relation, not just of its standard semantics. (The contrast here is with a consequence relation determined by $1/0$ valuations that cannot always be extended to valuations both assigning 1 to a maximal set, i.e., a set that is not a proper subset of sentences assigned 1 by any other allowed valuation *and* assigning 0 to a maximal set, i.e., a set that is not a proper subset of the sentences assigned 0 by any other allowed valuation.) So it seems that classical logic is conservative in only the first sense. Still, it is worth pausing to reflect on an alternative approach to

---

[10]This is the usual *formal* understanding of a conservative extension [1].

[11]Intuitively, this constraint rejects a bivalence condition; see again [6, p. 221f]. I propose to reconcile this condition with the bivalence of Scott-Lindenbaum semantics by suggesting that we take a sentence p to be 'unsettled' at a $1/0$ valuation v iff $v(p) = 0$ and there is a valuation $v*$ that *monotonically extends* the set of sentences assigned the value 1 at v such that $v*(p) = 1$.

[12]This condition is imposed because the consequence relation of L *may not* be determined by valuations of L that are *both* 1 and 0-maximal, where a valuation V is 1 (0) maximal iff there is no other valuation V' that gives the value 1 (0) to a proper superset of those assigned 1 (0) by V. For example, a language interpreted in epistemic terms might interpret 1 as *known* and 0 as *not known* for a non-omniscient being. Since the maximal 'known' assignments *overlap* all the maximal 'not-known' assignments, the consequence relation is not determined by what we might call the 'dual-maximal' assignments.

classical logic: if the classical consequence relation can be freed from the semantic assumption of complete valuations on L it may be more conservative than has been supposed.

16.6 Symmetrical Supervaluations

Supervaluations were proposed by Bas van Fraassen [15] to accommodate sentences including non-designating singular terms; their original logical function was to preserve the classical consequence relation while allowing sentences such as 'the present King of France is bald' to receive no truth value at all. They have been used in semantics for vagueness and in other philosophical applications where the principle that all sentences must receive a determinate truth value has seemed dubious. In general, a supervaluation is formed by quantifying across a collection, $\mathbf{V}$ of classical assignments to the sentences of a language L, and assigning 1 to all and only the sentences assigned 1 by *all* members of $\mathbf{V}$ and 0 to all and only the sentences assigned 0 by all members of $\mathbf{V}$. The set of sentences assigned 1 by all supervaluations assigning 1 to a set of premises Γ is the logical closure of Γ under the classical (single-conclusion) $\vdash$. But supervaluations do not produce a *bipartition* of L into truths and falsehoods, since some sentences receive no value at all. Thus supervaluations appear to be a promising candidate for 'second-degree' conservatism, since they do not assume all sentences have a determinate truth value. Of course we know already that the resulting logic is *conservative* in our first sense, since the consequence relation that results is just the classical $\vdash$.

But our aim here is to produce symmetrical multiple-conclusion consequence relations, treating premise and conclusion sets as duals of each other, so we still have some distance to go. We require that $\Gamma \vdash_S \Delta$ if and only if, for every supervaluation on L, if all of Γ receives the value 1, some $\delta \in \Delta$ also receives the value 1 (i.e., that no supervaluation assigns 1 to all members of Γ and 0 to all members of Δ). The resulting multiple-conclusion consequence relation is given by the familiar classical rules, and the use of supervaluations ensures that we can capture the familiar classical consequence relation without assuming that every sentence in L must be either correctly assertible (or more generally, have the property indicated by being assigned the value 1) or correctly deniable (have the property indicated by being assigned the value 0).

However, our supervaluations still assume complete valuations on L as a starting point, since they are produced by quantifying across a set of complete valuations, rather than by directly assigning some sentences 1, others 0, and leaving some sentences with neither value.[13] Since they depend on these complete classical valuations, supervaluations fail to achieve third degree conservatism. The close link between the

[13]This is why the familiar classical tautologies and contradictions always get the values 1 and 0, respectively: every extension of a partial assignment to a complete one winds up imposing 1 (0) as the value of a tautology (contradiction), even when some or all of its atoms wind up as 'gaps.'

classical multiple-conclusion consequence relation and maximal consistent/ consistently deniable extensions of the premise and conclusion sets adds to the conviction that complete valuations have not been exorcized. They are still there, in the complementary pairs of least transverses of consequences of the null set on the left and on the right: consider the family of sets $\Delta : \emptyset \vdash \Delta$: by the lemma above, a least transverse of all such sets Δ is a maximal consistent set of sentences, and the family of such transverses includes every maximal consistent set of sentences in L.

16.7 The Intuitionist Challenge to Symmetry

Thus far, we have noted some advantages of a symmetric, multiple-conclusion approach to the classical consequence relation and extended that approach to some other logics. But some logicians argue that the symmetries I regard as elegant and illuminating are just mistaken. If, as they argue, the roles of assertion and denial in reasoning are not really mirror images, then my attraction to symmetrical consequence relations is misguided. To answer some of these worries, we will turn to consider IL and related systems to see whether the concerns of their supporters can be accommodated within a symmetrical, multiple-conclusion approach.

On Negation

Classical multiple-conclusion rules ensure a symmetrical treatment of negation very simply:

$$\frac{\Gamma, A \vdash \Delta}{\Gamma \vdash \neg A, \Delta} \qquad\qquad \frac{\Gamma \vdash A, \Delta}{\Gamma, \neg A \vdash \Delta}$$

Applying these rules to a given classical MC consequence, we can arrive a set whose assertion is incoherent, and another whose denial is incoherent.

Intuitionistic Negation

Intuitionists have questioned two assumptions that underpin this classical multiple-conclusion approach:

1. The assumption that every atom is either correctly assertible or correctly deniable.
2. The assumption that the practice underlying correct assertion and denial treats these two attitudes symmetrically.

Dispensing with the first assumption requires at least second-degree conservatism. Defending the second requires a modification of IL that restores symmetry without undermining its conservatism. Intuitionistic negation restricts the left rule, applying it only when $\Delta = \emptyset$. This blocks the classical proof,

$$\frac{A \vdash A}{\vdash \neg A, A}$$

Thus the intuitionistic negation does not produce a *subcontrary* of the sentence it is applied to. But it does preserve the other half of the classical reading, producing a *weakest contrary* of A. That is,

- $A, \neg A \vdash_{IL}$ (from $A \vdash A$, by $\neg\vdash$)
- If $A, B \vdash_{IL}$, then $B \vdash_{IL} \neg A$ (by $\vdash\neg$) (Note the right-hand side of the premise sequent is empty here.)

To remedy this asymmetry, we propose adding a mirror image 'negation', $\llcorner$.

$$\frac{\Gamma \vdash A, \Delta}{\Gamma, \llcorner A \vdash \Delta} \text{ iff } \Gamma = \varnothing \qquad\qquad \frac{\Gamma, A \vdash \Delta}{\Gamma \vdash \llcorner A, \Delta}$$

We retain the other Kleene rules for classical logic, including thinning on left and right. The result is a symmetrical, multiple-conclusion logic.

As intended, $\llcorner$ is a *strongest subcontrary* forming operator:

$$\frac{A \vdash A}{\vdash A, \llcorner A} \qquad\qquad \frac{\vdash A, B}{\llcorner A \vdash B}$$

The traditional intuitionist objection to $\vdash A \vee \neg A$ has been that a proper proof of a disjunction must prove at least one disjunct. Perhaps an intuitionist might insist on this basis that we should not be able to prove $\vdash A, \llcorner A$ without having a proof of either A or $\llcorner A$. In reply, a 'tu quoque' comes to mind: when it comes to *denial*, how can the intuitionist justify her own negation's role in $A, \neg A \vdash$? From our point of view this amounts to *denying* $A, \neg A$ even though she has no grounds to deny either of $A, \neg A$ individually! It may be that for some linguistic practices, assertions require a kind of justification not demanded of denials—but a formal logic intended to provide a tool kit for systematizing our inferences and simplifying their expression should not assume this. A symmetrical treatment of assertion and denial will provide the same kind of 'flexibility' to both reasonings concerning what to assert (as embodied, for instance, in a practice of 'proofs') and reasonings concerning what to deny (as embodied in a practice of 'disproofs'): the monotonic progress of proofs envisioned by intuitionists may suit some practices (for example, that of an *idealized* mathematics), but a broader examination of forms of inquiry reveals both advances and retreats in what we take to be correctly asserted or denied. Further, reading $\llcorner A$ as 'A is not proven at a point less or equal to this point in the frame' makes it hard to resist the conclusion that $(A \vee \llcorner A)$ will hold at every point, just as $(A \wedge \neg A)$ fails at every point. I would also like to emphasize here that the symmetrical/directional reading of Kripke frames being suggested here requires that we do not take either 'proofs' or 'disproofs' to be *in general* the last word; we can describe sentences receiving the value 1 as *accepted* and sentences receiving the value 0 as *rejected*, but neither status is assumed to be *permanent*, and epistemic 'progress' can involve accepting the heretofore rejected or rejecting the heretofore accepted.

More directly, we have explicitly introduced '$\llcorner$' with rules that ensure it produces a strongest subcontrary, just as '$\neg$' produces a weakest contrary. The language may have lacked such sentences before, but it has them now. So long as this act of introduction is conservative with respect to the consequence relation of the sublanguage lacking $\llcorner$ *and* does not force us to say that every sentence is either correctly asserted or correctly denied, or that it is, in principle, always possible to extend a consistent partial assignment the values 1 and 0 to some subset of L to a correct bipartition of the language into a maximal consistently assertable set of sentences and a maximal consistently deniable set of sentences, the addition is conservative in all three of our senses.

Of course the points of a Kripke frame always do partition the sentences of L, assigning 1 to some and 0 to the rest. But we need not read the partition of L at a given point as *settled* either with respect to sentences receiving 1 or those receiving 0: instead, we can adopt a symmetrical epistemic perspective in which, looking 'up' along the frame relation we find points where some as yet rejected sentences come to be accepted, while looking down along the relation we find points where some currently accepted sentences come to be rejected. In both directions the changes are monotonic, expanding the set of sentences assigned the value 1 or expanding the set of sentences assigned the value 0, and transitions in either direction can be epistemic improvements on our current position. A key point here which distinguishes this extended intuitionistic logic from classical logic is that changing the value of an atom from 0 to 1 in intuitionistic logic does not, in general, require that any other sentences change value from 1 to 0: *monotonic* increases in the set of sentences receiving the value 1 occur in the 'upward' direction along the Kripke frame relation, while monotonic increases in the set of sentences receiving the value 0 occur in the downward direction. This feature makes our logic *symmetrically* conservative in all of our three senses.

16.8 Janus-Faced Negation

Classical logic fails to be conservative in the second or third sense because it assumes that the operator forming a weakest contrary and the operator forming a strongest subcontrary are one and the same. This *Janus-faced* negation makes the sets of sentences, $\{A, \neg A\}$ trivial on both sides of $\vdash$ and forces A, $\neg A$ to have opposite values in any $1, 0$ semantics. Once we assume that these values stand for (correct) *assertibility* we are forced to conclude that, since one of A or $\neg A$ must receive the value 1, and one must receive the value 0, the assertion of one (and the denial of the other) must be correct. Further, however, we interpret the property of sentences that is preserved (from left to right) by our consequence relation, from the point of view adopted here, the assumption that every sentence is such that either it or its weakest contrary must have the property remains a substantial assumption about the language to which we propose to add these logical words.

Bi-intuitionistic logic [4] avoids both the Janus-faced negation of classical logic and the asymmetrical negation of intuitionistic logic. We have a symmetrical consequence relation while avoiding the assumption that every sentence will receive a *settled* value: instead, the logic places us in *medias res*: our current commitments are treated as tentative and an assignment of either the value 1 or the value 0 to a given atom may be withdrawn. The familiar transitive, reflexive (Kripke) frame provides an intuitively appealing semantics for bi-intuitionistic logic. At each point atoms are evaluated into $1, 0$ where 1 is heritable and 0 is *ancestral* for the atoms. But the interpretation of 1 and 0 is quite different from the standard reading of Kripke semantics for intuitionistic or dual-intuitionistic logic: neither represents a 'fixed' commitment, because we do not think of inquiry *in general* as proceeding via the monotonic accumulation of proven or disproven sentences. Instead, we think of the values $1, 0$ as representing *in principle changeable* commitments, with new assertions arising along the direction of the frame relation, and new denials arising in the opposite direction.

Of course the semantics for $\neg$ and $\llcorner$ are dual:

- $\neg : v_s(\neg\phi) = 1 \text{ iff } \forall s' : Rss' \implies v_{s'}(\phi) = 0$
- $v_s(\neg\phi) = 0 \text{ iff } \exists s' : Rss' \wedge v_{s'}(\phi) = 1$
- $\llcorner : v_s(\llcorner\phi) = 1 \text{ iff } \exists s' : Rs's \wedge v_{s'}(\phi) = 0$
- $v_s(\llcorner\phi) = 0 \text{ iff } \forall s' : Rs's \implies v_{s'}(\phi) = 1$

As in intuitionistic logic, the usual extensional semantics applies to $\wedge$ and $\vee$:

- $\wedge$: $v_s(\phi \wedge \psi) = 1$ only if $v_s(\phi) = v_s(\psi) = 1$, else $v_s(\phi \wedge \psi) = 0$
- $\vee$: $v_s(\phi \vee \psi) = 1$ only if $v_s(\phi) = 1$ or $v_s(\psi) = 1$, else $v_s(\phi \vee \psi) = 0$

Kripke-style intuitionist and dual-intuitionist semantics are contained here as sublogics, each of which involves only one of our negations along with its corresponding conditional: these two systems, along with our system dividing the classical negation into two distinct negations while ignoring conditionals, are sublogics of bi-intuitionistic logic. The key constructivist results which hold for each are preserved in the full logic (including the collapse of series of negations beyond two).[14] By not assuming that maximal consistent extensions of consistent premise sets are always available, bi-intuitionistic logic achieves the third form of conservatism, while the symmetry of the logic allows room for a *fallibilistic* epistemology, allowing epistemic changes including both denial of sentences once tentatively endorsed, and endorsement of sentences once tentatively denied. I believe the conservative thing to do is allow for both—if there are indeed any certainties, we can recognize them individually, in due course, in the context of the particular language that underwrites them.

As emphasized above, Scott's result shows that any reflexive, monotonic, and transitive multiple-conclusion logic can be given a semantics based on $1, 0$ valuations of the sentences of the language. The close link we have noted between minimal transverses of the family of conclusion sets following from a given premise set and the valuations assigning 1 to sentences in the premise set, and between families of premise sets from which a given conclusion set follows and the valuations assigning 0 to the conclusion set, helps to establish that such semantics are not more *fundamental* to specifying a consequence relation than the rules of inference are. Bi-intuitionistic logic shows that the elegant symmetries of multiple-conclusion logic can be had without giving up the logical conservatism of intuitionistic logic. But our interpretation of the frame needs to be subtler now that we have shifted to a bi-intuitionistic logic. Intuitionistic logic and its dual involve a single-minded 'forward look' along

---

[14] A cut-free proof theory for bi-intuitionistic logic can be found in [4].

a frame relation in which either more and more sentences are proven or more and more sentences are disproven. Combining the two gives us what might be called a *Hempelian* [7] logical perspective, allowing both that some accepted sentences may later be rejected, and some rejected sentences may later be accepted. This is a logic for fallible inquirers, not infallible provers.

Of course the adoption of two separate negations, each capturing half of the Janus-faced classical negation, brings with it a straightforwardly paraconsistent negation. Some have objected to the paraconsistent bona-fides of such negations, but this divide-and-conquer approach to negation remains a tempting path to capturing both a robust (weakest) contrary-forming operator and a robust (strongest) subcontrary-forming operator without committing ourselves to classical negation. Since the stipulation of both connectives is fully conservative, this approach offers no barrier to a form of paraconsistency grounded in the dual-intuitionistic negation.

16.9 Conclusion

My aim here has been to arrive at a better understanding of what is required for a logical system to be a truly conservative extension of any language we might add it to, and therefore admissible as a *stipulation* extending a preexisting linguistic practice. One element in this effort has been to develop at a more general understanding of the relation between a syntactic consequence relation and a semantics for that relation. For the first aim, I have argued that bi-intuitionistic logic is a step in the right direction. It provides a symmetrical logical framework which is conservative in the usual sense of not imposing new consequences in the language it is added to, and also conservative in the stronger senses that it does not assume every sentence is either correctly assertible or correctly deniable, and does not assume that every valuation can be extended to a valuation, that is, both 1-maximal and 0-maximal. This latter property seems particularly significant in the light of the incompleteness of arithmetic: the density of the order of provability in arithmetic rules out the possibility of arriving at a 1-maximal theory of arithmetic. For the second aim, transverses of conclusion sets provide a new perspective on the relation between multiple-conclusion logics and $1/0$ semantics. In classical logic (and in its supervaluational variant) they are the maximal consistently assertible extensions of the premise set, while the transverses of the premise sets from which a given conclusion set follows are the maximal consistently deniable extensions of the conclusion set. But in bi-intuitionistic logic, transverses of conclusion sets are the 1-minimal extensions of valuations assigning 1 to all members of a given premise set, while transverses of the premise sets from which a conclusion set follows are the 0-minimal extensions of valuations assigning 0 to all members of the conclusion set. These sets are not maximal (as they are in the case of classical and supervaluational logic), that is, they are not such that for each right-transverse C there is a left-transverse P such that $C \cup P = L$. Thus bi-intuitionistic logic truly avoids the assumption that every sentence must either receive the value 1 or the value 0; it also avoids the dual-intuitionist assumption that every disjunction of a sentence with its negation must be trivial on the right, and the intuitionistic assumption that every conjunction of a sentence with its own negation is trivial on the left, by distinguishing two negations, each of which plays just one of these roles.

References

1. Belnap, N.D.: Tonk, Plonk and Plink. Analysis **22**(6), 130–134 (1962)
2. Brown, B.: Refining preservation. In: Tanaka, K., Berto, F., Mares, E., Paoli, F. (eds.) Paraconsistency: Logic and Applications, pp. 123–140. Springer, Dordrecht (2012)
3. Brown, B., Schotch, P.K.: Logic and Aggregation. J. Philos. Log. **28**, 265–287 (1999)
4. Buisman, L., Goré, R.: A cut-free sequent calculus for bi-intuitionistic logic (2007). arXiv:0704.1707v2
5. Carnap, R.: Formalization of Logic. Harvard University Press, Cambridge (1943)
6. Dummett, M.: Truth and Other Enigmas. Harvard University Press, Cambridge (1978)
7. Hempel, C.G.: Empiricist criteria of cognitive significance: problems and changes. In: Hempe, l1965a, pp. 101–119 (1965)
8. Hume, D.: A treatise of human nature. In: Selby-Bigge, L.A. (ed.). Clarendon Press, Oxford (1896)
9. Jennings, R.E.: The Genealogy of Disjunction. Oxford University Press, New York (1994)
10. Prior, A.: The runabout inference ticket. Analysis **21**, 38–39 (1960)
11. Raatikainen, P.: On rules of inference and the meanings of logical constants. Analysis **68**, 282–287 (2008)
12. Restall, G.: Multiple conclusions. In: Hajek, et al. (eds.), pp. 189–206 (2005)
13. Schotch, P., Brown, B., Jennings, R.: On Preserving: Essays on Preservationism and Paraconsistent Logic. University of Toronto Press, Toronto (2009)
14. Scott, D.: Completeness and axiomatizability in many-valued logic. In: Proceedings of the Tarski Symposium (Proceedings of Symposia in Pure Mathematics, vol. 25, University of California, Berkeley California, 1971) Providence, R.I., pp. 411–435 (1974)
15. van Fraassen, B.: Presuppositions, supervaluations, and free logic. In: Lambert (ed.), pp. 67–91 (1969)

Chapter 17
Logic—The Big Picture

Ross T. Brady

Abstract The big picture is my big picture, as I see it, based on a lifetime of research into logic. We will cover a reasonably wide range of topics, with some level of author focus. The paper builds on my earlier work [9, 11, 19] (with Rush). Indeed, it can also be seen as an update of the approach to logic taken in [9]. We start with the issue of what logic is about, identifying two inference concepts, one of meaning containment (a connective) and one of deductive argument in general (a rule). Examining the other connectives, we point out the difference between disjunction, as understood in proof-theoretic systems, as opposed to that understood in standard semantics, and show why distribution is not an instance of meaning containment. Negation is judged as being incompletely captured, due to the non-recursive nature of deductive systems in general, but with Boolean negation being the intended concept. We then focus on the logic MC of meaning containment, setting out its axiomatization, content semantics and metavaluation. Quantification is added in a standard way, based on the connectives. We finally deal with applications, focusing on set theory and arithmetic.

Keywords General logic · Disjunction · Negation · Meaning containment · Applications

Mathematics Subject Classification (2000) 03A05

R.T. Brady (✉)
Philosophy Program, La Trobe University, Melbourne VIC 3086, Australia
e-mail: Ross.Brady@latrobe.edu.au

J.-Y. Beziau et al. (eds.), *New Directions in Paraconsistent Logic*,
Springer Proceedings in Mathematics & Statistics 152,
DOI 10.1007/978-81-322-2719-9_17

17.1 Introduction

The first thing to note is that the big picture is my big picture, as I see it, based on a lifetime of research into logic.[1] The title should probably be "A Big Picture" but the aim is to present it as a picture of logic that follows from a series of rational arguments, to be presented in this paper. The big picture will attempt to cover a reasonably wide range of topics, though not necessarily every topic the reader will think of. There will be some level of author focus. Also, in order to provide such a wide coverage, the depth and referencing will have to suffer a little as a result, but it is hoped the reader will follow-up some of the given references to add clarity.

An earlier paper [11] covered some similar ground, by first presenting 23 concerns about classical logic and then showing that a logic such as the author's entailment system MC deals with all of these. Also, Brady and Rush "Four Basic Logical Issues" [19] deals with the following issues, which are also of relevance:

(i) The choice of logic between classical and non-classical logic,
(ii) The determination of the particular non-classical logic,
(iii) Classical deduction versus relevant deduction, and
(iv) Classical versus non-classical meta-logic.

These two papers provide useful background for the current enterprise and indeed some of their arguments will find their way into the current paper. As in [11], we will consider a wide range of issues, but update our thinking and make use of more recent results. Further, this can also be seen as an update of the approach to logic taken in [9].

17.2 What Is Logic About?

A natural starting point is the issue of what logic is. In textbooks, logic is split into deductive and inductive reasoning. For deductive reasoning to be valid, *the conclusion must follow as a matter of certainty, given the premises*. On the other hand, for good inductive reasoning, the conclusion is not certain, but remains of a high probability, given the premises. Uncertainty pervades inductive logic, not just because that is what it deals with but also, I believe, in providing an acceptable theory of it. Hence, our primary concern here is with deductive logic; we assume that logic is deductive, unless otherwise indicated.

The scope of application of (deductive) logic should be restricted only by its ability to deduce conclusions from premises, given what deductive logic is. As in

---

[1]This paper was presented to the Logic Seminar at the University of Melbourne, on 30th November, 2012. I would like to thank those present for their useful comments and discussion, in particular Greg Restall, Lloyd Humberstone and Rohan French. I also wish to thank Che-Ping Su for his help in getting the paper ready for presentation.

the textbooks, valid deductive arguments, in pure logic anyway, adhere to patterns of argument with the use of formula schemes. Some texts go further and put nonsense terms into such patterns, such as follows:

All bodkins are claptens.
All claptens are goobies.
Hence, all bodkins are goobies.

This, I believe, goes too far, as the terms must have meanings in order to be applied. So, meanings provide a natural outer boundary of deductive argument. Without meanings, how can an argument be applicable?

The main other semantic concept of relevance to logic is that of truth. The role of truth in deductive arguments is standardly encapsulated in the following definition of a valid deductive argument:

A deductive argument is valid iff, whenever the premises are all true then the conclusion is true.

However, there are three concerns that can be raised here:

(1) What if a premise is false? We can just as satisfactorily argue about Middle Earth as about Earth. The context is built up in a fictional novel, in a similar way to the context that one might be aware of in the real world, i.e. through a body of statements about characters, their discussions and their surrounds. It does not appear to matter whether the premises are true or false in reality.
(2) Further, what if a premise is inconsistent? One would hope that the premises, together with the entire context, are consistent. Also, there are paraconsistent (or indeed relevant) concerns about deriving any formula from a contradiction.
(3) This definition is usually interpreted semantically with a classical meta-logic, which seems to divorce it from deduction (and deductive logic). That is, a valid deductive argument is determined by sheer truth-preservation (presumably in a model of some sort, based on truth-conditions), rather than on the passage of argument from premises to conclusion.

What such truth-preservation does do is to provide a criterion for assessing a deductive argument. Indeed, every valid deductive argument should preserve truth. The converse is what is the problem. Just because an argument preserves truth does not mean to say that there is a piece of derivation getting one from the premises to the conclusion, as the respective truths can be unrelated, i.e. irrelevant.[2] Truth-preservation is usually used, however, as a criterion for rules to satisfy, where what is derivable from false premises is determined according to what is derivable from true instantiations into the premise schemes of the rule. In a sense, we can say that the false premises are *assumed to be true* for the sake of the deduction. To clarify this, take the rule Modus Ponens, $A, A \to B \Rightarrow B$, for example. It basically says

[2]However, Stephen Read, in his book [25], claims that relevance is already contained in truth-preservation, rather than an added extra. Here, we follow the standard approach of separating the two concepts.

that, for any true entailment $A \rightarrow B$, truth is preserved from A to B. Although based on truth, this rule continues to be applied at any point in a logical deduction, even when there are premises involved which may turn out to be false. Thus, we say that such premises are assumed to be true for the sake of the deduction.

How does this latter definition relate to the earlier one where the conclusion is certain, given the premises? The earlier one seems to have a different interpretation. The "giving" of the premises means that one is assuming the premises, whether true or not. The certainty seems to be obtained from somewhere, presumably from the meanings of the premises, unless the conclusion is already certain. The "obtaining" seems to involve some derivation process, based on these meanings. In the case of the conclusion being already certain, its derivation would have already been done for it as a theorem, whilst the premises would be irrelevant. This case, I think, should be allowed, even though there is no likely derivation process involved. The reason is that the definition of a valid deductive argument is conclusion-focussed. In contrast, I do not think $A, {\sim}A \Rightarrow B$ follows, unless the system is consistent, in which case A and ${\sim}A$ cannot both be derived. In the inconsistent case, since instances of B are not certain, B would have to be derived from A and ${\sim}A$, which is not in general possible in a non-trivial system.

Moreover, there is what might be called in-between cases where the conclusion is certain, except for some piece of incidental information. In such a case, the meanings of the premises might relate to only part of the conclusion. In this category, I would put the substitution of identity, $x = y \Rightarrow A(x) \leftrightarrow A(y)$, and the substitution of equivalence, $A \leftrightarrow B \Rightarrow C(A) \leftrightarrow C(B)$, arguments, where both the substitutions are single-place. I would also put restricted quantification arguments of the form $\exists x A(x) \Rightarrow (\forall x A(x))B(x)$, representing 'All As are Bs', where $A(x)$ provides the non-empty restricted domain. Where the whole conclusion is meaning-dependent on the meanings of the premises, we have what I have called meaning containment of the conclusion in the premises. (We will continue this discussion in the next section when we distinguish between rules and meaning containments.)

Thus, in the debate between the proof-theoretic and semantic approaches to logic, we clearly favour the former, and our overall conclusion being that deductive logic is about proof of conclusions from premises, making the exception for conclusions already proved and thus taken as certain. The semantics comes afterwards when trying to make sense of the deductive system.

17.3 The Two Inference Concepts

We take up the discussion from Sect. 17.2, where we distinguished between

(i) Conclusions whose whole meaning is dependent on that of the premises,
(ii) Conclusions that only in part are meaning-dependent on the premises, and
(iii) Conclusions that are certain, independent of the premises.

Each of these cases is taken to satisfy the definition of a valid deductive argument, but is there a case for an inference connective for case (i)?

There are reasons for an inference connective here, just as there are reasons for the material '$\supset$' which is essentially truth-preservation, though with classical negation features applying to its antecedent. Truth and meaning are the two main semantic concepts and, though there are some problems with truth-preservation as a determinant for deductive inference (as above), meaning containment seems to be a better alternative in providing a yardstick to determine the validity of the core deductive arguments under (i). And, meaning is, at least prima facie, prior to truth, in which one needs to be sure of the meanings of terms, or at least their salient parts, in order to work out the truth of sentences, through examination of the real world and/or a comparison of concepts. So, meaning containment would then be a tighter concept than truth-preservation and so a meaning containment connective is justified if that is so for truth-preservation. And, the classical negation encapsulated in '$\supset$' is not appropriate for meaning containment, as will be seen in Sect. 17.5.

It is important to note that there are two inference concepts developing here: meaning containment, covered by (i) above, represented by a connective '$\rightarrow$' and valid deductive argument in general, covered by (i), (ii) and (iii), represented by a broader rule '$\Rightarrow$'. The latter is understood in meta-theoretic terms as it represents broad-scale deduction in a logical system, as opposed to the more intimate meaning containment of the former, which just relates antecedent and consequent as opposed to the whole deductive system. We will see in Sect. 17.6 that such a logic of meaning containment will be a weak relevant entailment logic, which we call MC.

We make two further points. It is hard to construct an axiomatic logic based on meanings from the philosophical accounts of meaning as they are very general. I do not think the philosophical accounts are precise enough to do the job and hence further specification needs to be added. (More detail on this in Sects. 17.6. and 17.7., especially on "intensional set-theoretic containment".)

Meanings may work well for well-defined concepts, as occur in arithmetic and set theory, but how do we deal with meanings in more vague contexts? Considering the vast literature on vagueness, we do not want to get locked into the very many discussions there. Instead, we make a quick response that is related to the discussion in hand. As such, we would have to rely on the premises of an argument to determine sufficient of the meanings of their components to make the respective valid deductions. Carnap called such premises "meaning postulates", which function as axioms for the concepts concerned. (See [20].) Nevertheless, logic carries on quite nicely without total meanings of all the concepts involved and not all questions need be answered in the process. The same applies to vagueness, where there is a lack of definition or specificity of concepts and it should be treated in the same way as far as logical deduction is concerned. So, no special logic should be needed for vagueness, as it fits in with normal logical deduction.

17.4 Conjunction and Disjunction

We take conjunction as simply "both of" and disjunction as "at least one of". Though straightforward, there are some subtleties that need to be drawn out concerning disjunction and the distribution laws.

Disjunction, as understood in natural deduction, does not require that one of the disjuncts be proved in order for the disjunction to be proved, primarily within a subproof. (Here, we take natural deduction as being the best representation of human reasoning, though what is said can apply to Hilbert-style axiomatic systems as well.) This can be seen from the $\vee E$ rule, where a formula C is proved from both disjuncts A and B, taken hypothetically, meaning that it does not matter whether a specific disjunct has been proved or not, but it is assumed that at least one has, for the purposes of understanding why the $\vee E$ rule works. Note here that $A \vee B$ can be introduced by hypothesis, rather than through applying $\vee I$.

However, in standard model theory, in order for $A \vee B$ to be true, a specific formula A or B must first be shown to be true, which can be called a disjunctive "witness", analogously to the case of the existential quantifier. This is because the evaluation structure is based on formula induction, as occurs in the truth-tables. Here, the induction procedure starts with the atomic formulae and introduces each connective, including $\vee$, one by one, until the whole formula is built. Thus, either A or B is required in order to build $A \vee B$. So, model theory does not represent disjunction, as it occurs in proof-theoretic settings. For the sake of clarifying this point, let us consider worlds in model theory as corresponding to subproofs in natural deduction systems, where the worlds are inductively set up, whilst the subproofs can have disjunctive hypotheses. (Also, see [13] for a witness-free semantics, which is natural deduction based.)

One can see this in the Henkin-style proofs of completeness for classically based logics upon which the Routley–Meyer proofs for relevant logics are based. (See [21] and Chap. 4 of [26].) Whereas Henkin was concerned about obtaining negation-completeness for logics with the Law of Excluded Middle (LEM), $A \vee {\sim}A$, Routley and Meyer were concerned about disjunctive completeness or Priming: if $A \vee B \in a$ then $A \in a$ or $B \in a$, for theories a. (This is more general than Henkin, as logics with the LEM would then have to satisfy $A \in a$ or ${\sim}A \in a$, in particular.) Further, the Henkin-style proof involves an extension obtained by adding witnesses to create consistent theories, or, in the Routley–Meyer case, witnesses to satisfy priming. This process shows that the starting point, usually obtained deductively, is expanded upon to create the truth-tables for negation and disjunction within the canonical model being built. Thus, this can separate deduction from the semantic extension. However, the soundness theorem is fine, as this theorem extends to applications of logics as well, but the completeness theorem is a result for the pure logic all right, but not necessarily for applied logics such as Peano Arithmetic, and, as such, it is generally oversold, I believe. (See Sect. 17.10. for further discussion.)

But, let us have a look at theorems, i.e. formulae proved without premises. If $A \vee B$ is proved in a main proof, i.e. without hypothesis, one would expect there to be a proof of A or a proof of B, as otherwise how is $A \vee B$ to be proved? $\vee I$ is the obvious mode of proof, at least at a simplistic level, given that there is no LEM. (More on the LEM and reductio arguments in Sect. 17.5., and constructivity in Sects. 17.6 and 17.8.) This proof-theoretic witness does then align with the model theory, at the level of theorems.

We now look into the distribution laws to see in what forms they should hold, keeping in mind the distinction between the rule '$\Rightarrow$' and the meaning containment connective '$\rightarrow$'. Let us start with $A \mathbin{\&} (B \vee C) \rightarrow (A \mathbin{\&} B) \vee (A \mathbin{\&} C)$, asking the question is this distribution law an instance of meaning containment?

Let us have a look at Schroeder-Heister's work on his Proof-Theoretic Semantics, introduced in his [27]. In his tutorial [28], he talked about the uniqueness of logical connectives for classical logic. In particular, upon proving $A \supset B \vdash A \supset' B$, $A \supset' B \vdash A \supset B$ and hence $\vdash (A \supset' B) \equiv (A \supset B)$, with both '$\supset$' and '$\supset'$' having the same introduction and elimination rules in natural deduction, he declared '$\supset$' to be unique. Indeed, '$A \supset B$' and '$A \supset' B$' are inter-substitutable in all contexts within the logic. Moreover, this also applies to conjunction and to disjunction in classical, intuitionist and relevant logic settings. So, the introduction and elimination rules, $\& \ I$, $\vee I$, $\& \ E$ and $\vee E$, suffice to uniquely characterize conjunction and disjunction, but these rules can be seen to be independent of distribution. To prove $A \mathbin{\&} (B \vee C) \rightarrow (A \mathbin{\&} B) \vee (A \mathbin{\&} C)$, one would need some structural rule or some such consideration, over and above the standard introduction and elimination rules. In Fitch-style natural deduction, one needs to conjoin A in a subproof α with B and also with C in respective subproofs of α. This manoeuvre is not allowed in relevant logic as it leads to irrelevance, as $A \rightarrow .B \rightarrow A \mathbin{\&} B$ and hence $A \rightarrow .B \rightarrow A$ would be derivable. So, given that the meanings of conjunction and disjunction are given by their introduction and elimination rules, $(A \mathbin{\&} B) \vee (A \mathbin{\&} C)$ does not follow by meaning from $A \mathbin{\&} (B \vee C)$, and hence $A \mathbin{\&} (B \vee C) \rightarrow (A \mathbin{\&} B) \vee (A \mathbin{\&} C)$ is not valid as a meaning containment. (See [17] for a detailed account.)

What now about the rule-form, $A \mathbin{\&} (B \vee C) \Rightarrow (A \mathbin{\&} B) \vee (A \mathbin{\&} C)$? This, however, seems fine as the disjunctive meta-rule, if $A, B \Rightarrow C$ then $D \vee A, D \vee B \Rightarrow D \vee C$, suffices to prove it, and this is seen as an obvious extension to two premises of the single premise disjunctive meta-rule, if $A \Rightarrow B$ then $C \vee A \Rightarrow C \vee B$, which was the previous meta-rule of a pre-tweaked version of MC.

What these meta-rules rely on, at base, is that for a disjunction $B \vee C$ to be a theorem then either B is a theorem or C is a theorem. This also applies to the premises which are assumed to be the case for the sake of the derivation. (See Sect. 17.2.) This is what gives the certainty to $(A \mathbin{\&} B) \vee (A \mathbin{\&} C)$, given $A \mathbin{\&} (B \vee C)$, as required for deductive validity. This also provides a way to give some restricted Hilbert-style representation to the Priming Principle: $\vdash A \vee B \Rightarrow \vdash A$ or $\vdash B$, without adding a structural connective for the meta-linguistic disjunction. (See for [17] further discussion of distribution.)

17.5 Negation

Classically, negation is taken as just the fall-back for truth, i.e. $\sim A$ holds precisely when A fails to be the case. Thus, one has just the two values T and F, with negation effectively changing the value, as per the truth-tables. This is the most immediate and recognizable conception of negation, which is often called Boolean negation.

The problem occurs when one puts negation into a deductive context, and one cannot avoid doing this as deduction can be applied to any meaningful sentences. Then, one gets what Penelope Rush called the *four deductive outcomes* for a formula A within a deductive system (see [12]):

(i) A, without $\sim A$,
(ii) $\sim A$, without A,
(iii) neither A nor $\sim A$, and
(iv) both A and $\sim A$.

This applies for both classical and non-classical logic, as indicated in [11], since case (iii) applies for the Godel sentence G in Peano Arithmetic and case (iv) applies to the Russell set R in naive set theory, both based on classical logic. So, the negation of the truth-tables does not always align with what is provable, and there is a conflict between the immediate Boolean concept and what is achievable in proof theory. Further, the negation occurring in these four deductive outcomes could be virtually any operator at all, as there is no relationship between the occurrence of A and that of $\sim A$.

Let us consider the Russell set R, where $x \in R$ iff $x \notin x$, for all x. In almost all cases one would be likely to consider $x \in R$ will be classically true since relatively few x would satisfy $x \in x$, i.e. $x \in x$ is false and $x \notin x$ is true for almost all of the likely cases of x. There would be relatively few cases where $x \in x$ is true and $x \notin x$ false—also classical. Of course, the familiar $R \in R$ behaves non-classically, but this case seems to be the only one, or at least rare and unusual, except perhaps amongst logicians. The large bulk of cases is still classically behaved. This leads us to believe that this negation is intended to be classical, but becomes non-classical in this deductive context, due to the presence or absence of the LEM for $R \in R$. [$A \to \sim A \to \sim A$ would alternately suffice for outcome (iv).] That is, the LEM yields outcome (iv), since $R \in R \leftrightarrow R \notin R$ is deductively equivalent to $R \in R \vee R \notin R \leftrightarrow R \in R \vee R \notin R$ (given weak logical assumptions), and its absence will yield outcome (iii), since the absence of $R \in R \vee R \notin R$ will mean that neither $R \in R$ nor $R \notin R$ is present.

So, I believe the classical Boolean negation, which is the clear negation concept, is intended, but the deductive systems may not bear it out because of the way they work. This intention is due to the sheer simplicity and obvious usage of the Boolean concept, in contrast with the lack of concept based on the four deductive outcomes. Why is this so, though? The classical deductive outcomes (i) and (ii) are meta-theoretic. That is, one needs to examine the whole deductive system to determine whether $\sim A$ or A is absent in (i) or (ii), respectively, rather than just examining meanings of components within it. This is similar to what goes on in practice, usually over a finite domain. The negation 'I am not in the corridor' is determined by examining the domain consisting of the corridor and finding me not in it. Because of finiteness, one concludes by negation-completeness that "I am not in the corridor" is true. So, we are treating the deductive context much like the corridor, as it determines a domain to be examined. However, in general, in deductive systems, this domain can be infinite and checking whether some formula is absent can be non-recursive. This also applies to outcome (iii), though outcome (iv) can be determined proof-theoretically and thus by recursively enumerable means. Further, as seen in naive set theory, there can be a non-classically determined domain of self-membered sets, as happens for the Russell set R.

The forthcoming logic MC has De Morgan negation rather than Boolean negation. However, De Morgan negation is characterized by the principles of double negation, $\sim\sim A \leftrightarrow A$, and contraposition, $A \leftrightarrow B \leftrightarrow .\sim B \leftrightarrow \sim A$. From these two principles, the De Morgan's Laws for & and $\vee$, $\sim(A \,\&\, B) \leftrightarrow \sim A \vee \sim B$ and $\sim(A \vee B) \leftrightarrow \sim A \,\&\, \sim B$, can easily be derived, rounding off the characterization. Thus, this concept of negation specifically applies to the connectives, $\sim$, & and $\vee$, and not to atomic sentences. (See [12] for further discussion.) So, one really needs some atoms, like $x \notin \emptyset$ in set theory or $0 \neq x$ in Peano arithmetic, to kick start negations at an atomic level, so that further negative sentences can be proved from the De Morgan principles. Conceptually, De Morgan negation, by itself, is deficient and acts like a shell for negation whilst the Boolean concept is better in this respect as it applies to all sentences, atomic as well. Further, Boolean negation is applicable to atomic sentences when there are no truth-value gaps or gluts, i.e. no instances of outcomes (iii) or (iv) above.

In conclusion, since outcomes (i), (ii) and (iii) are meta-theoretic, as explained above, negation, of all the connectives, does not have a complete concept. However, it does have the (intended) Boolean concept, but this is restricted to classical contexts, which are determinate, either due to finitude or recursion. More can be said about negation, but we have focused on its concept and on its logical laws.

One last point to mention regarding negation concerns the use of reductio arguments. There is a distinction between those that are really contrapositions, like $A \rightarrow B, \sim B \Rightarrow \sim A$, and those that involve the derivation of a contradiction, like $A \rightarrow B \,\&\, \sim B \Rightarrow \sim A$. The former are easily accommodated in the logic via contraposition, but the latter involves the use of the LEM, i.e. $\sim(B \,\&\, \sim B) \rightarrow \sim A \Rightarrow \sim A$, where $\sim(B \,\&\, \sim B) \leftrightarrow B \vee \sim B$. Further, reductio proofs of form: If $A \Rightarrow B \leftrightarrow \sim B$ then $\vdash \sim A$ also use the LEM, in that $B \leftrightarrow \sim B$ is deductively equivalent to $B \vee \sim B \rightarrow B \,\&\, \sim B$, and the rule $A \Rightarrow B \,\&\, \sim B$ can be contraposed to the form $\sim(B \,\&\, \sim B) \Rightarrow \sim A$, subject to the LEM holding for A and the disjunctive syllogism (DS), holding for $B \,\&\, \sim B$, with help from the meta-rule MR1 of MC below. (Here, the DS takes the form: $\sim(B \,\&\, \sim B), (B \,\&\, \sim B) \vee \sim A \Rightarrow \sim A$.) Thus, these two latter forms of reductio involve a certain level of classicality, which would have to be independently argued for.

17.6 The Logic MC of Meaning Containment

This is an attempt to formally capture a logic based on the meaning containment '$\to$' with the more general use of the deductive rule '$\Rightarrow$' as discussed in Sect. 17.3. For logical purposes, the meaning containment is thought of as an *intensional set-theoretic containment*. The idea is that logical contents represent the meanings that are contained in one another, and that these contents are best represented as sets of formulae, as appears in the canonical model of the content semantics initially presented in [5], and also appearing in [9] in a slightly more complex form. (See Sect. 17.7. for treatment of contents.) And, the set-theoretic containment needs to be intensional rather than the usual extensional containment, as it is meanings that are to be related after all. Note that [5, 9] contains distribution, but it was later removed in [17] in favour of the rule-form, as discussed in 4. So, the following axiomatization of the logic MC attempts to capture such a logic.

One should also note the difference between the logic MC and Anderson and Belnap's logic E of entailment in their [1]. MC is conceptualized independently of necessitated implication, despite the '$\to$' of MC being a reasonable entailment in itself in that necessity and analyticity are closely related. The main difference here is the lack of an implication which, when necessitated, would yield such an entailment.

MC.

Primitives:

$\sim, \&, \vee, \to.$

Axioms:

1. $A \to A.$
2. $A \mathbin{\&} B \to A.$
3. $A \mathbin{\&} B \to B.$
4. $(A \to B) \mathbin{\&} (A \to C) \to .A \to B \mathbin{\&} C.$
5. $A \to A \vee B.$
6. $B \to A \vee B.$
7. $(A \to B) \mathbin{\&} (B \to C) \to .A \vee B \to C.$
8. $\sim\sim A \to A.$
9. $A \to \sim B \to .B \to \sim A.$
10. $(A \to B) \mathbin{\&} (B \to C) \to .A \to C.$

Rules:

1. $A, A \to B \Rightarrow B.$
2. $A, B \Rightarrow A \mathbin{\&} B.$
3. $A \to B, C \to D \Rightarrow B \to C \to .A \to D.$

Meta-rule:

1. If $A, B \Rightarrow C$ then $D \vee A, D \vee C \Rightarrow D \vee C.$

We can say that A is a classical formula when the LEM and the DS hold for A. Note that A is classical iff the classical deduction theorem, $A \Rightarrow B$ iff $\vdash \sim A \vee B$, holds. Also, if both A and B are classical then $\sim$, $A \,\&\, B$ and $A \vee B$ are classical. So, classical logic can be built up from a set of classical formulae, using conjunctive normal forms with distribution in rule-form.

The logic MC is paraconsistent as the DS fails in general, although it can hold in particular cases, as it does for classical formulae. This logic has been used to prove the non-triviality of (inconsistent) naive set theory with the addition of the LEM, in [4, 9] by three-valued modelling and in [14] by metavaluations. (Metavaluations are introduced in Sect. 17.8.)

In the following sections, we introduce some other logics, which we axiomatize below in relation to MC, as well as the system E, already mentioned in Sect. 17.6.

Additional Axioms:

11. $A \,\&\, (B \vee C) \to (A \,\&\, B) \vee (A \,\&\, C)$.
12. $A \to B \to .B \to C \to .A \to C$.
13. $A \to B \to .C \to A \to .C \to B$.
14. $A \to .A \to B \to B$.
15. $A \to .B \to A$.
16. $(A \to .A \to B) \to .A \to B$.
17. $A \to \sim A \to \sim A$.

Additional Rules:

4. $A \Rightarrow \sim(A \to \sim A)$.
5. $A \Rightarrow A \to B \to B$.

Other Systems.

(We assume these systems (except E) all have the meta-rule MR1, for the sake of uniformity, though most earlier versions did not include it, or included a one-premise version.)

DW = MC + A11 − A10.
TJK = MC + A11 + A12 + A13 + A15.
RW = DW + A12 + A13 + A14.
RWK = RW + A15.
E = RW − A14 + A16 + A17 + R5 − MR1.

17.7 Content Semantics

What sort of semantics is appropriate for a logic based on meaning containment? One needs to distinguish between the technical needs of a logic and its appropriate semantics. Standard model theory tries to do both and certainly succeeds in the technical sphere, e.g. with non-provability and conservative extension results. I do not think it succeeds in providing a "real" semantics, though. (See Sect. 17.4. regarding disjunction and Sect. 17.9. regarding existential quantification.)

A "real" semantics ought to capture the essence of what the logic represents and as such it ought to distinguish what axioms and rules should be included in the logic and those that should not. It should not be a general semantics for a wide range of logics. It should encapsulate the "one true logic" with appropriate distinguishing features, which will be shown below. It may or may not be of much technical value. Indeed, as it turns out, showing the non-provability of formulae will require an understanding of what the contents are, rather than a knock-down technical argument as occurs in standard model theory.[3]

The following content semantics for the logic MC aims to be such a "real" semantics. We take this from [5, 9], though a more general content semantics was introduced earlier in two parts in [3, 4]. This general semantics covered a wide range of logics from Lavers' BB right through to classical logic (see [22] for BB). However, we do wish to focus the semantics on just the one logic MC, as appears below. Furthermore, we will exclude the distribution property for content containments, following the tweaking of the logic MC in [17].

Contents, as introduced in [5, 9], are analytic closures of a sentence or set of sentences, whereby the meanings of the sentences are analysed and entailments made until no further meaning can be drawn out from the sentences. Initially, in this section, we focus on the formulae of MC to capture the semantics at the sentential logical level, and then reach out to quantifiers and various application areas later. Since contents are sets of formulae, content containment, which interprets the '$\rightarrow$' of MC, can then be understood as *intensional set-theoretic containment*. This is what determines the semantic postulates below and thereby provides specificity for entailment itself.

A content model structure (c.m.s.) consists of the five concepts $\langle T, C, \overline{\cup}, *, c\rangle$, where C is a set of sets (called contents), $T \neq \emptyset$, $T \subseteq C$ (the non-empty set T of all true contents), '$\overline{\cup}$' is a 2-place function on C (the closed union of contents), '$*$' is a 1-place function on C (the '$*$' function on contents), and 'c' is a 1-place function from containment sentences, $c_1 \supseteq c_2$ between contents c_1 and c_2 of C, to members of C, subject to the semantic postulates p1–p15 below. Unlike other algebraic-style semantics, the concepts '$\cap$', '$=$'and '$\supseteq$', are taken from the background set theory, '$\cap$' being a 2-place function on C (the intersection of contents), '$=$' being a 2-place relation on C (identity) and '$\supseteq$' being a 2-place relation on C (content containment). This gives the semantics some specific contact with "reality".

Whilst the intersection of two contents is always a content, this does not apply to the union of contents. In this case, we form the closed union $c_1 \overline{\cup} c_2$ of contents c_1 and c_2, which is the analytic closure of their set-theoretic union. In [9], the '$*$' function on contents takes contents to ranges, a dual concept to that of contents embracing the De Morgan properties.

[3] I wish to thank Greg Restall for making the point that content semantics does not behave like model theory in this respect and that this specialized content semantics requires an understanding of what these contents are.

Let c_1, c_2 and c_3 be contents.

p1. $c_1 \overline{\cup} c_2 \supseteq c_1, c_1 \overline{\cup} c_2 \supseteq c_2$.
p2. If $c_1 \supseteq c_2$ and $c_1 \supseteq c_3$ then $c_1 \supseteq c_2 \overline{\cup} c_3$.
p3. $c_1 \supseteq c_1 \cap c_2, c_1 \supseteq c_1 \cap c_2$.
p4. If $c_1 \supseteq c_3$ and $c_2 \supseteq c_3$ then $c_1 \cap c_2 \supseteq c_3$.
p5. $c_1^{**} = c_1$.
p6. If $c_1 \supseteq c_2$ then $c_2^* \supseteq c_1^*$.

We add the *set T of all true contents*, in order to define validity. We regard a content as true when all of its elements are true. The following are straightforward, given the conjunctive interpretation of content arising from its definition as an analytic closure:

p7. If $c_1 \supseteq c_2$ and $c_1 \in T$ then $c_2 \in T$.
p8. If $c_1 \in T$ and $c_2 \in T$ then $c_1 \overline{\cup} c_2 \in T$.
p9. If $c_1 \cap c_2 \in T$ then $c_1 \in T$ or $c_2 \in T$.

(For p9, let $c_1 \notin T$ and $c_2 \notin T$. Then, there are sentences p in c_1 and q in c_2 which are not true. In which case, $p \vee q$ in $c_1 \cap c_2$ is not true and $c_1 \cap c_2 \notin T$.)

So, T is prime truth filter.

p10. $c(c_1 \supseteq c_2) \overline{\cup} c(c_2 \supseteq c_3) \supseteq c(c_1 \supseteq c_3)$.
p11. $c(c_1 \supseteq c_2) \overline{\cup} c(c_1 \supseteq c_3) \supseteq c(c_1 \supseteq c_2 \overline{\cup} c_3)$.
p12. $c(c_1 \supseteq c_3) \overline{\cup} c(c_2 \supseteq c_3) \supseteq c(c_1 \cap c_2 \supseteq c_3)$.
p13. $c(c_1 \supseteq c_2) \supseteq c(c_2^* \supseteq c_1^*)$.
p14. $c(c_1 \supseteq c_2) \in T$ iff $c_1 \supseteq c_2$.
p15. If $c_1 \supseteq c_2$ then $c(c_3 \supseteq c_1) \supseteq c(c_3 \supseteq c_2)$ and $c(c_2 \supseteq c_3) \supseteq c(c_1 \supseteq c_3)$.

(p15 follows due to the semi-substitution of meaning containment into consequent and antecedent positions.)

An interpretation I on a c.m.s. is an assignment, to each sentential variable, of an element of C.

Interpretations I are extended to all formulae, inductively as follows:

(i) $I(\sim A) = I(A)^*$.
(ii) $I(A \,\&\, B) = I(A) \overline{\cup} I(B)$.
(iii) $I(A \vee B) = I(A) \cap I(B)$.
(iv) $I(A \rightarrow B) = c(I(A) \supseteq I(B))$.

A formula A is true under an interpretation I on a c.m.s. M iff $I(A) \in T$.

A formula A is valid in a c.m.s. M iff A is true under all interpretations I on M.

A formula A is valid in the content semantics iff A is valid in all c.m.s.

This can also be extended to an *argument* $A_1, A_2, \ldots, A_n \Rightarrow B$, which *preserves truth under interpretation I on a c.m.s.* M iff $I(B) \in T$ whenever $I(A_1) \in T$, $I(A_2) \in T, \ldots,$ and $I(A_n) \in T$.

MC is sound and complete with respect to the above content semantics. Also, the rules preserve truth under I on c.m.s. M, for all I and all M. Note that, in the Lindenbaum-style completeness proof, instead of equivalence classes, canonical contents $[A]$ consist of the set of all formulae entailed by A, including itself.

Importantly, key non-theorems can be rejected by the semantics through our understanding of contents, thus pinning down the logic MC, e.g. $I(A)\overline{\cup}c(I(A) \supseteq I(B)) \supseteq I(B))$, and hence $I(A \,\&\, (A \to B) \to B) \in T$, is rejected on account of the content $c(B)$ not being contained in the closed union of $c(A)$ and $c(A \to B)$. The point here is that $c(B)$ can be any content whilst $c(A \to B)$ is a content containment statement, which is a specific type of content, thus creating a mismatch. The same problem happens when trying to relate $c(A)$ with $c(A \to B)$. Thus, $I(A \,\&\, (A \to B) \to B) \notin T$. One can compare this with $c(I(A) \supseteq I(B))\overline{\cup}c(I(B) \supseteq I(C)) \supseteq c(I(A) \supseteq I(C))$, where the content containment statements do properly interact (p10), yielding $I((A \to B) \,\&\, (B \to C) \to .A \to C) \in T$.

The content semantics is doing an appropriate semantical job, picking out the logic MC. However, being an algebraic semantics (without the algebra), there is not a simple knock-down method of rejecting non-theorems. The normalized natural deduction system would be best, as the Routley-Meyer semantics includes the distribution axiom. (See [10] for the neighbouring logic DW, and [13] for DW without the distribution axiom A11.) Gentzen systems are very useful too, especially without the distribution axiom. (See [6, 7] in two parts.) Note the usefulness of proof theory generally over the standard model theory here.

17.8 Metavaluations

Without the Routley-Meyer semantics, we need to search around for other technical systems to prove results about the logic MC. This brief presentation will be a general account of metavaluations and their applications, as an alternative to standard model-theoretic approaches. These metavaluations are basically proof-theoretic, even though they are set up in semantical style. Importantly, they are formula-inductive, like model theory, and so they have some of the benefits of model theory, but without having to leave the proof theory. So, some proof-theoretic results like priming, 'if $\vdash A \vee B$ then $\vdash A$ or $\vdash B$', come out easily. (See [23] for the introduction of metavaluations.)[4]

Metavaluations generally work for metacomplete logics which are roughly the contraction-less relevant logics, with the possible additions of Conjunctive Syllogism, $(A \to B) \,\&\, (B \to C) \to .A \to C$, and the irrelevant, $A \to .B \to A$. (See [29, 30] for the introduction of metacomplete logics with negation, including the M1- and M2-logics distinguished below.) Importantly, the logic MC is included, as well as surrounding systems, which can be generally characterized as entailment-focussed logics. The entailment focus can be seen in the following inductive specification of a metavaluation v and its star valuation v*.

[4]One could try to base content semantics on metavaluations with a tight correspondence between the semantic postulates of the content semantics and each of the metavaluations, substituting the latter for the standard Hilbert-style axioms and rules. This metavaluation-driven content semantics, however, has never been tried, to the author's knowledge.

We understand $v(A) = T$ as 'A is provable' and $v^*(A) = T$ as '$\sim A$ is unprovable', as can be seen from the soundness and completeness theorems to follow.

$v(p) = F$; $v^*(p) = T$, for sentential variables p.
$v(A \,\&\, B) = T$ iff $v(A) = T$ and $v(B) = T$.
$v^*(A \,\&\, B) = T$ iff $v^*(A) = T$ and $v^*(B) = T$.
$v(A \vee B) = T$ iff $v(A) = T$ or $v(B) = T$.
$v^*(A \vee B) = T$ iff $v^*(A) = T$ or $v^*(B) = T$.
$v(\sim A) = T$ iff $v^*(A) = F$.
$v^*(\sim A) = T$ iff $v(A) = F$.
$v(A \to B) = T$ iff $\vdash A \to B$ and if $v(A) = T$ then $v(B) = T$, and if $v^*(A) = T$ then $v^*(B) = T$.
$v^*(A \to B) = T$, for M1-logics.
$v^*(A \to B) = T$ iff, if $v(A) = T$ then $v^*(B) = T$, for M2-logics.

Completeness: if $v(A) = T$ then $\vdash A$, for all formulae A, and hence if $v^*(A) = F$ then $\vdash \sim A$.

Consistency: if $v(A) = T$ then $v^*(A) = T$.

Soundness: if $\vdash A$ then $v(A) = T$, and hence if $\vdash \sim A$ then $v^*(A) = F$.

Metacompleteness: $\vdash A$ iff $v(A) = T$, and hence $\vdash \sim A$ iff $v^*(A) = F$.

Priming Property: if $\vdash A \vee B$ then $\vdash A$ or $\vdash B$.

Negated Entailment Property: $\nvdash \sim(A \to B)$ (M1-logics); $\vdash \sim(A \to B)$ iff $\vdash A$ and $\vdash \sim B$ (M2-logics).

All logics from MC up to TJK are metacomplete M1-logics, and all logics from MC + $A \Rightarrow \sim(A \to \sim A)$ up to RWK are metacomplete M2-logics.

We can use metavaluations to prove the simple consistency of naive set theory and arithmetic (see Sect. 17.10.). We can also produce counter valuations for unprovable formulae through substitution, as can be seen from these two examples:

(1) Put A as p and B as $\sim(A \to A)$ in $A \,\&\, (A \to B) \to B$. This suffices to show $v(A \,\&\, (A \to B) \to B) = F$ in M1- and M2-logics.
(2) Put A as p in $A \vee \sim A$. Thus, $v(A \vee \sim A) = F$, again in both types of logics.

17.9 Quantifiers

We have left the quantifiers until now to simplify the foregoing. There are some parallels with conjunction and disjunction, the universal quantifier being a complication of conjunction, and the existential quantifier having similar concerns to that of disjunction. Starting with the universal quantifier, there are three methods of establishing $\forall x A(x)$, brought about by the prospect of infinite domains:

(1) $A(x)$ holds over a finite domain.
(2) $A(x)$ holds by recursion over a denumerable domain, or by transfinite induction over an infinite well-ordered domain.
(3) $A(x)$ holds by universal introduction $\forall I$.

Method (1) is extensional, being a simple expansion of conjunction. Method (2) is a constructive extension of extensionality, involving recursion with two steps or transfinite induction with three steps. Method (3) is intensional, as it requires x to be an unconstrained free variable, without a mention of the domain, usually within a natural deduction system. It would apply to any domain, whether recursive or not. Here, $A(x)$ would have to hold for all x to start with, prior to introducing or re-introducing the universal $\forall$. Further, method (3) is the form of universal quantifier introduction used in soundness arguments, as it is completely general.

As with disjunction, there is a difference between the proof-theoretic interpretation of the existential quantifier and the model-theoretic interpretation. The natural deduction rule $\exists E$ does not require a witness, in subproofs anyway. For the main proof, however, given metacompleteness (see below), the witness would be previously proved, even if it is a variable. For model theory there is always a witness, due to the inductive procedure on formulae and this is followed up in the Henkin-style completeness proofs, as with disjunction in Routley-Meyer completeness proofs. Also, $\forall I$ does not use a domain, whilst a domain is required for model theory, which again differentiates proof theory and model theory.

We next examine the distribution laws. Clearly, $\forall x(A \vee B) \rightarrow .A \vee \forall x B$ fails due to the failure of its sentential version over a two-element domain, and similarly for $A \mathbin{\&} \exists x B \rightarrow \exists x(A \mathbin{\&} B)$. The rule $\forall x(A \vee B) \Rightarrow .A \vee \forall x B$ also fails as it does in intuitionist logic, since the '$\forall$' and '$\vee$' in MC and intuitionist logic are essentially the same concepts, the differences between the two logics being in '$\sim$' and '$\rightarrow$'. We now make the required quantificational additions.

MCQ.

Primitives:

$\forall$, $\exists$.

$x, y, z,$ ... bound variables.

$a, b, c,$ ... free variables.

$s, t, u,$ ... terms, i.e. free variables, or constants that might be added in applications.

Axioms:

1. $\forall x A \rightarrow At/x$
2. $\forall x(A \rightarrow B) \rightarrow .A \rightarrow \forall x B$
3. $At/x \rightarrow \exists x A$
4. $\forall x(A \rightarrow B) \rightarrow .\exists x A \rightarrow B$

Rule.

1. $Aa/x \Rightarrow \forall x A$, where a is not in A.

Meta-Rule.

1. If $A, Ba/x \Rightarrow Ca/x$ then $A, \exists x B \Rightarrow \exists x C$, where R1 is not used to generalize on any free variables occurring in the A nor in the B of the rule $A, Ba/x \Rightarrow Ca/x$. This also applies to the rule $A, B \Rightarrow C$ of MR1 for the sentential component.

Note that the existential distribution rule, $A \mathbin{\&} \exists x B \Rightarrow \exists x(A \mathbin{\&} B)$, follows from R2 and QMR1. This form does hold in intuitionist logic, as well.

The quantificational additions for the content semantics are quite complex (See [9] for details.). We proceed with the additional metavaluations:

v($\forall x A$) = T iff v(At/x) = T, for all terms t.
v*($\forall x A$) = T iff v*(At/x) = T, for all terms t.
v($\exists x A$) = T iff v(At/x) = T, for some term t.
v*($\exists x A$) = T iff v*(At/x) = T, for some term t.

Completeness, consistency, soundness and metacompleteness follow as for the sentential logic, with the addition of the quantificational axioms, rule and meta-rule above. Thus, by metacompleteness:

if $\vdash \exists x A$ then $\vdash At/x$, for some term t. (The Existential Property)

Finally, we consider restricted quantification. Two attempts have been made here, in [2, 8]. However, after talking with Sam Butchart, who raised the question, and Hartry Field, who enforced the solution, I am currently of the view that the classical restricting connectives of '$\supset$'and '&' suffice for the two respective quantifiers '$\forall$' and '$\exists$' as they do for classical logic. This is because the restricting predicate needs to be classical, i.e. satisfy the LEM and DS. This generally holds for the various universes of discourse and, as with their non-emptiness, should extend to the restricted domains as well. (See Sect. 17.2. for the formal setting out of this.) This classicality also prevents fuzziness in the specification of the domain, which would then spread to the restricted quantification.

17.10 Applications

Before focusing on set theory and arithmetic, we should consider broader-based applications. Application in general depends on the premises that one chooses to elucidate the concepts, together with all relevant background information. It is then that one deductively derives conclusions. As pointed out at the beginning, we are not doing inductive reasoning, based on the high probability of a conclusion holding. Most studies of conditionals do not specify their assumptions fully, but rely on *ceteris paribus* (other things being equal), which introduces a lack of certainty of conclusions drawn. Thus, I think the general study of conditionals should belong to inductive reasoning, though model theory can play a role in their analysis, as neighbourhood semantics in particular has done.

Even if the concepts are somewhat vague, the premises are what drive a deductive inference, however well they might capture the intended concepts. Indeed, in the case of vague concepts, certain positive and certain negative statements can be derived, but there will be some statements that are unresolved either way, due to the vagueness. So, derivation itself can be used to provide a dividing line between, say, the red objects and the others, and between the non-red objects and the others.

Generally, Carnap's idea of meaning postulates seems like a good way of providing suitable premises or axioms for concepts, but it is in mathematics and computer science where the high degree of specificity lies. In mathematics, the concepts and objects are quite abstract, so much so that there can be little room for disagreement and deductive reasoning can be used at its best. In computer science, it is what is stored in the knowledge base that provides the premises and what is not explicit cannot be used.

We now turn to set theory and arithmetic, which are the more logically accessible parts of mathematics and where the most work is done. There are a number of concepts of set, but the naive concept is the most immediate, the others involving some retreat from paradox and this would include the classical sub-theory of [9]. The naive concept is based on the generation of sets from predicates, i.e. the set y of all x such that ...x.... It is axiomatically represented by the Comprehension Axiom, $\exists y \forall x(x \in y \leftrightarrow A)$, where y is not free in A, and the extensionality rule, $\forall z(z \in x \leftrightarrow z \in y) \Rightarrow x \in w \leftrightarrow y \in w$. Since the set y is contextually defined, the equivalence involved should be a meaning equivalence, which is appropriately represented by the '$\leftrightarrow$' connective of MCQ. This naive set-theoretic concept is simply consistent based on the logic MCQ, proved in [9] using models and in [16], using metavaluations. The classical sub-theory in [9] needs more work to capture the constructive approach adopted here.

Nevertheless, in reference to this sub-theory, in [18], it was argued that Cantor's diagonal argument fails to go through for constructive (or, indeed, recursive) functions f from the set of natural numbers $\mathbb{N}$ to its power set $\wp(\mathbb{N})$. The Cantor argument takes the shape: $1-1C \Rightarrow k \in f(k) \leftrightarrow k \notin f(k)$, subject to a definition of $k \in \mathbb{N}$, where $1-1C$ stands for 'f is a one-to-one correspondence between $\mathbb{N}$ and $\wp(\mathbb{N})$'. Again, '$\leftrightarrow$' is appropriate here to capture the effect of the definition. In order to ensure that the classical Cantor argument goes through, we can make some classicality assumptions along the way. We can use the LEM for $k \in f(k)$ to derive $k \in f(k)$ & $k \notin f(k)$ and, to contrapose the rule, we use the LEM for $1-1C$ and the DS for $k \in f(k)$ & $k \in f(k)$, arriving at the contraposed rule:$\sim(k \in f(k)$ & $k \notin f(k)) \Rightarrow \sim(1-1C)$. Given the LEM for $k \in f(k)$, $\sim(1-1C)$ follows, as required. However, what we dispute here is the LEM for $k \in f(k)$, used twice in the above argument, given that this classical argument requires all functions f, rather than constructively generated functions such as recursive functions. And, in order to justify the LEM for $k \in f(k)$, $k \in f(k) \vee k \notin f(k)$ would have to be proved by some constructive proof, and that would normally mean that either $k \in f(k)$ or $k \notin f(k)$ would be constructively proven using MCQ. This leaves Cantor's theorem as an open question. (See [18] for discussion on this.) However, since MCQ is weaker than classical logic, what we can say is that a one-to-one correspondence between $\mathbb{N}$ and $\wp(\mathbb{N})$ is not provable using MCQ if it is not provable classically.

A similar concern occurs in Peano Arithmetic, based on a slight weakening of the quantificational extension of MC, and proved simply consistent in [15] using finitary methods. Here, the LEM needs to be proved, usually by mathematical induction, for all instances where it is used. Generally, the Peano axioms are replaced by rules and, in order to apply mathematical induction, one needs to convert these rules into single formulae. To do this, the LEM is used, since, if $A \Rightarrow B$ then, by MR1, $\sim A \vee A \Rightarrow \sim A \vee B$, yielding $A \supset B$, as the single formula. The method of proving simple consistency is that of metavaluations, with use of the priming property: if $\vdash A \vee B$ then $\vdash A$ or $\vdash B$. Thus, for the Godel sentence G, if $\vdash G \vee \sim G$ then $\vdash G$ or $\vdash \sim G$. Since the logic is weaker than classical, Godel's classically stated incompleteness result still applies, as it concerns unprovability which is then projected down to the weaker logic, assuming classical consistency. So, $\nvdash G \vee \sim G$, and the LEM fails for G. However, it can be seen in [15] that the LEM does hold for primitive recursive functions, but general recursion is left to be completed.

The naive truth concept can be dealt with in a similar way to that of naive set theory, as it also involves the use of a definition and a general schema. There, the Liar sentence p is defined as 'p is not-true' and formalized as $p \leftrightarrow \sim Tp$. Then, since $Tp \leftrightarrow p$, $p \leftrightarrow \sim p$. (See Chap. 8 of [9] for examples, proofs and discussion of semantic paradoxes.)

Meta-theory is just another application of the logic and, as such, need not be classical. There is quite some classical gain over the object theory as the whole deductive system becomes an object with reasonably clear definition, and as a result the LEM and the DS hold for many meta-theoretic concepts. Assignments to formulae can be T or F, but not both, but the classicality of provability would require decidability of the logic, given our constructive approach.

17.11 Conclusion

Please accept my apologies for the absence of modal logic and other intensional logics, as these have not been in my core areas of research. Otherwise, we have covered a lot of ground, even though relatively briefly.

The central thread is that logic is about deduction of conclusions from premises and this induces a constructive proof-theoretic approach to logic, which in turn requires a re-examination of many of the main proofs. Of particular interest are those that use reductio arguments from the negation of the desired conclusion to a conclusion which is of the form $A \,\&\, \sim A$, or indeed $A \leftrightarrow \sim A$. This argument structure occurs in a number of contexts: Cantor's diagonal argument, the proof of Godel's first theorem, and the undecidability of the Turing halting problem. (For the last, see Chap. 3 of [24].) We have discussed the first two of these, but the last one remains to be dealt with in detail. We have seen in the first two that a classically proved negative meta-theoretic result can be transmitted to a weaker logic such as MCQ, but this does not apply to the last one, as decidability is "two-sided" as it affects both theoremhood and non-theoremhood. For example, the sentential logic RW is decidable, whilst R is undecidable (i.e. classically proved as such) and classical logic is decidable.

Whilst the paradoxes occurring in the object-language are solved using MCQ, this does leave the questions regarding the extended paradoxes and semantic closure. Briefly, the extended paradoxes involve some item which is in both the object- and meta-languages, and a solution consists simply in separating these two languages as we normally do in the course of logical study. Nevertheless, there is a philosophical preference for semantic closure which requires that a statement and the statement of its truth occur in the same language. As argued above, the LEM and the DS hold very widely in the meta-language, and this is due to the consideration of the object-language as a whole, from outside of itself. Thus, the object-language has sufficient specificity for many of its meta-theoretic statements to be classical. So, these two languages cannot be meshed together. In response, one might wish to achieve semantic closure at a limit point of an infinite sequence of meta-languages. However, here, there must be some self-reference between languages to achieve this closure, whereupon the LEM would fail, given the lack of separation of languages. This would still apply if all the languages are somehow compacted into one.

It remains for us to briefly consider the scope of classical logic, as much of logical investigation has focused on it and its use. In order for classical logic to be the only logic used, the LEM and the DS must hold for every atomic statement. So, every predicate, when applied to any object, must have such clarity of application so that it must either apply or not, without overlap. The main examples of this would be with a finite domain, where every predicate is specified as such. In the case of infinite domains, the logic in its application would need to be decidable and also for each sub-domain determined by a predicate the logic would also need to be decidable. Each such sub-domain, if non-empty, would be usable for restricted quantification. Thus, this is very restrictive.

Non-classical use of the logic MCQ would be required not only when there is any vagueness or lack of specificity of concepts but also in the case of conceptual clash resulting in overspecification rather than underspecification. In the former case, we would have neither A nor $\sim A$ in the system, for some A, but, in the latter case, we would have both A and $\sim A$, for some A. And, these would commonly occur. Thus, the use of the title "Universal Logic" in [9] is vindicated, in which it constitutes a widening out of the application of classical logic to meaningful sentences. However, such meaning need not be full meaning, as partial meaning may suffice to drive an argument from premises to conclusion, as occurs in vague contexts.

References

1. Anderson, A.R., Belnap, N.D.: Entailment, The Logic of Relevance and Necessity, vol. 1. Princeton University Press (1975)
2. Beall, J.C., Brady, R.T., Hazen, A., Priest, G.G., Restall, G.: Relevant restricted quantification. J. Philos. Logic **35**, 587–598 (2006)
3. Brady, R.T.: A content semantics for quantified relevant logics I. Studia Logica **47**, 111–127 (1988)
4. Brady, R.T.: A content semantics for quantified relevant logics II. Studia Logica **48**, 243–257 (1989)
5. Brady, R.T.: Relevant implication and the case for a weaker logic. J. Philos. Logic **25**, 151–183 (1996)

6. Brady, R.T.: Gentzenizations of relevant logics without distribution—I. J. Symbolic Logic **61**, 353–378 (1996)
7. Brady, R.T.: Gentzenizations of relevant logics without distribution—II. J. Symbolic Logic **61**, 379–401 (1996)
8. Brady, R.T.: Relevant Logics and Their Rivals, Vol 2: A Continuation of the Work of R. Sylvan, R.K. Meyer, V. Plumwood, R.T. Brady, ed. by R.T. Brady with contributions by M. Bunder, A. Fuhrmann, A. Loparic, E. Mares, C. Mortensen, A. Urquhart, Ashgate, Aldershot (2003)
9. Brady, R.T.: Universal Logic. CSLI Publs, Stanford (2006)
10. Brady, R.T.: Normalized natural deduction systems for some relevant logics I: The logic DW. J. Symbolic Logic **71**, 35–66 (2006)
11. Brady, R.T.: Entailment: a blueprint. In: Beziau, J.-Y., Carnielli, W., Gabbay, D. (eds.) Handbook of Paraconsistent Logic, College Publications, pp. 127–151. King's College, London (2007)
12. Brady, R.T.: Negation in metacomplete relevant logics. Logique et Anal. **51**, 331–354 (2008)
13. Brady, R.T.: Free semantics. J. Philos. Logic **39**, 511–529 (2010)
14. Brady, R.T.: Metavaluations, naive set theory and inconsistency. In: Beziau, J.-Y. Coniglio, M.E. (eds.) Logic without Frontiers, Festschrift for Walter Alexandre Carnielli on the occasion of his 60th Birthday, pp. 339-360. College Publications, London (2011)
15. Brady, R.T.: The consistency of arithmetic. Based Logic Meaning Containment, Logique et Anal. **55**, 353–383 (2012)
16. Brady, R.T.: The simple consistency of naive set theory using metavaluations. J. Philos. Logic **43**, 261–281 (2014)
17. Brady, R.T., Meinander, A.: Distribution in the logic of meaning containment and in quantum mechanics. In: Tanaka, K., Berto, F., Mares, E., Paoli, F. (eds.) Paraconsistency: Logic and Applications, pp. 223–255. Springer, Dordrecht (2013)
18. Brady, R.T., Rush, P.A.: What Is wrong with cantor's Diagonal argument? Logique et Anal. **51**, 185–219 (2008)
19. Brady, R.T., Rush, P.A.: Four basic logical issues. Rev. Symbolic Logic **2**, 488–508 (2009)
20. Carnap, R.: Meaning And necessity, 2nd edn. University of Chicago Press (1956)
21. Henkin, L.: The completeness of the first order functional calculus. J. Symbolic Logic **14**, 159–166 (1949)
22. Lavers, P.S.: Generating intensional logics, M.A. Thesis, University of Adelaide (1989)
23. Meyer, R.K.: Metacompleteness. Notre Dame J. Formal Logic **17**, 501–516 (1976)
24. Papadimitriou, C.H.: Computational Complexity. Addison Wesley Longman, Reading, Massachusetts (1994)
25. Read, S.: Relevant Logic. Blackwell, Oxford (1988)
26. Routley, R., Meyer, R.K., Plumwood, V., Brady, R.T.: Relevant Logics and their Rivals, vol. 1. The Basic Philosophical and Semantic Theory. Ridgeview, Atascadero, California (1982)
27. Schroeder-Heister, P.: Uniform proof-theoretic semantics for logical constants (abstract). J. Symbolic Logic **56**, 1142 (1991)
28. Schroeder-Heister, P: Proof-Theoretic Semantics, Tutorial given to the World Universal Logic Conference, Xian, China (2007)
29. Slaney, J.K.: A metacompleteness theorem for contraction-free relevant logics. Studia Logica **43**, 159–168 (1984)
30. Slaney, J.K.: Reduced models for relevant logics without WI. Notre Dame J. Formal Logic **28**, 395–407 (1987)

Chapter 18
The Evil Twin: The Basics of Complement-Toposes

Luis Estrada-González

For Christian Edward Mortensen, long-distance mentor, in his 70th birthday.

Abstract In this paper I describe how several notions and constructions in topos logic can be dualized, giving rise to complement-toposes with their paraconsistent internal logic, instead of the usual standard toposes with their intuitionistic logic.

Keywords Standard topos · Complement-topos · Internal logic

Mathematics Subject Classification (2000) 03A05 · 03B53 · 03G30 · 18B25

This paper is conceptually the first one of a tetralogy gathering up my logico-philosophical investigations on topos logic, deeply motivated by the philosophy behind Universal Logic: The other papers are [12, 15, 16], in that order. However, insofar as each paper is self-contained for it can be read independently, the reader might find several similarities between them in their introductions, speaking of the general motivation, or their presentations of the basics of topos logic, but each deals with special, specific problems. This paper has been written under the support from the CONACyT project CCB 2011 166502 "Aspectos filosóficos de la modalidad," as well as from the PAPIIT project IA401015 "Tras las consecuencias. Una visión universalista de la lógica (I)." I thank Axel Barceló-Aspeitia, Jean-Yves Béziau, Carlos César Jiménez, Chris Mortensen, Zbigniew Oziewicz, Ivonne Pallares-Vega and now countless referees for their comments, suggestions or encouragement over the years I have spent working on this topic. Diagrams were drawn using Paul Taylor's diagrams package v. 3.94.

L. Estrada-González (✉)
Instituto de Investigaciones Filosóficas, Universidad Nacional Autónoma de México,
Circuito Maestro Mario de la Cueva s/n, Ciudad universitaria, C.P. 04510 Coyoacán,
México D.F., México
e-mail: loisayaxsegrob@gmail.com

J.-Y. Beziau et al. (eds.), *New Directions in Paraconsistent Logic*,
Springer Proceedings in Mathematics & Statistics 152,
DOI 10.1007/978-81-322-2719-9_18

18.1 Introduction

A category can be thought of as a universe of *objects* and their transformations or connections, called *morphisms*, subject to some very general conditions. An example of a category is **Set**, whose objects are sets and its morphisms are functions between sets. In **Set** there is a special kind of objects, namely objects with two elements. As objects with two elements, all these objects are isomorphic to each, and each of them has all and only those mathematical properties (as expressible in categorial terms) as any other, so the sign '$\mathbf{2_{Set}}$' can be used to denote any of them and speak as if there were only one of them. We will say that an object with the property of having exactly two elements is unique up to isomorphism. $\mathbf{2_{Set}}$ act as truth values object in **Set** in the sense that suitable compositions with codomain $\mathbf{2_{Set}}$ serve to expresses that certain sets are part of others. Hence, the two elements of $\mathbf{2_{Set}}$ are conveniently called $true_{\mathbf{Set}}$ and $false_{\mathbf{Set}}$.

Zero- and higher-order connectives can also be defined in **Set**. It can be proved that the right logic to study the objects in **Set**, its *internal logic*, is that induced by the algebra formed by $\mathbf{2_{Set}}$ and the connectives, which turns out to be classical. This logic is called *internal* because it is formulated exclusively in terms of the objects and morphisms of the topos in question and it is the right to reason about the topos in question because it is determined by the definition of its objects and morphisms in a way that using a different logic for that purpose would alter the definitory properties of those objects and morphisms and thus it would not be a logic for the intended objects and morphisms; it cannot be a canon imposed "externally" to reason about the topos.

As in usual axiomatic membership-based set theories like ZFC, most of mathematics can be interpreted and carried out in **Set**. However, a set theory developed from a category-theoretic point of view is not based on the notion of membership, but rather on those of function and composition (of functions).

There are other **Set**-like categories, called *elementary toposes* or simply *toposes*. In a topos $\mathcal{E}$ there are objects which play the role of $\mathbf{2_{Set}}$ in the particular case of **Set**, i.e., they serve to express that certain objects are part of others via suitable compositions of morphisms. An object that plays such a role in a topos is also unique up to isomorphism and any of them can be denoted by the sign '$\Omega_{\mathcal{E}}$' and speak as if there were only one of them. Logical notions like truth values and zero- and higher-order connectives can also be defined in a topos. However, in general, $\Omega_{\mathcal{E}}$ has more than two elements and, since $\Omega_{\mathcal{E}}$ has all the same universal properties as $\mathbf{2_{Set}}$ and the latter can be considered a truth values object, so can the former. In addition, the logic appropriate for dealing with the objects and morphisms in a topos, its internal logic, is in general intuitionistic, not classical. This is precisely a logic arising from objects and morphisms themselves, not from our devices to reason about them. Like **Set**, toposes also allow for the interpretation of set-theoretical notions and hence of significant parts of mathematics, but the reconstruction of mathematics carried out in a topos corresponds to mathematics as done in an intuitionistic set theory. If toposes can be considered universes of sets and, given that at least parts of mathematics can be reconstructed in a set theory, toposes also allow for the reconstruction of those parts of mathematics, then the universal laws of mathematics are those valid across all universes of sets, viz. the laws of intuitionistic logic.

This beautiful picture of logic in a topos can be summarized in the following slogans[1]:

(S1) $\Omega_{\mathcal{E}}$ is (or at least can be seen as) a truth values object. (Common categorial wisdom, see for example [18, 29–31].)
(S2) The internal logic of a topos is in general many-valued. (Common categorial wisdom, but see [3–5, 18, 29, 30, 37].)
(S3) The internal logic of a topos is in general (with a few provisos) intuitionistic. (This also is common categorial wisdom, just to name but two important texts where this is asserted see [18, 31].)
(S4) Intuitionistic logic is the objective logic of variable sets. (A powerful metaphor widely accepted, see [24, 25, 38].)
(S5) The universal, invariant laws of mathematics are those of intuitionistic logic. (Cf. again [3–5].)

With the exception of (S5),[2] which is a claim specifically due to Bell, these slogans are theses so widely endorsed by topos-theorists as accurate readings of some definitions, results and constructions in topos theory that it is hardly worth documenting, but I have done it just to show that they appear in several major texts by leading category theorists.

Slogans (S1) and (S2) can be challenged by invoking a family of results initiated by Roman Suszko which state that every logic satisfying certain conditions—a Tarskian structural logic in the case of Suszko, i.e., a logic whose consequence relation is reflexive, transitive, monotoni, and structural—has a bivalent semantics. However, I do not want to discuss (S1) and (S2) here, which is left for another paper (see [15]).

Let me use '(T)' to denote a well-known theorem in topos theory, which is read as stating as in one of the slogans:
(S3) The internal logic of a topos is (with a few provisos) intuitionistic.
Recent research on so-called complement-toposes (cf. [39, 40] or [13]) suggests that (T) should not be paraphrased laxly as in (S3) nor as in (S4). Neither the rather philosophical reading in (S5) proclaiming the intuitionistic character of the mathematical universal laws would be justified. Mortensen and Lavers' view implies that certain (equalities expressed in) diagrams entail very abstract truth conditions which do not determine a unique kind of logic. Whereas in standard topos theory, the categorial reconstruction of logic starts by naming "*true*" one of the elements Ω, in complement-toposes it is called "*false*." The authors stress the fact that the very categorial structure of toposes supports different names for some crucial morphisms, so

---

[1] I use the word 'slogan' here pretty much in the sense of van Inwagen: "a vague phrase of ordinary English whose use is by no means dictated by the mathematically formulated speculations it is supposed to summarize" [51, p. 163], "but that looks as if it was," I would add.

[2] And maybe also of (S4), due mostly to the appearance of Hegelian terminology ("objective"), very frequent in Lawvere but not in other topos-theorists. Omitting that, one can add [2, 18] as supporters of this slogan.

the internal logic of a topos could be equally described as a kind of dual-intuitionistic, paraconsistent logic. The difference is imperceptible in **Set**, where there are only two truth values, but gives rise to different logics in more general cases. That bare structure of toposes which I referred to above seems to support different logics, so one can even ask whether there is such a thing like *the* internal logic of a topos at all. Moreover, if toposes are regarded as universes of sets where mathematics can be reconstructed, the above question amounts to asking whether there is something like the universal, invariant laws of mathematics at all.

Mortensen [40] talks about "a considerable Public Relations Exercise" done on behalf of intuitionistic logic in topos theory leading to the denial of the latter interpretation, but as I see it it is derived from a cognitive bias, the "prejudice towards truth." I take the expression "prejudice towards truth" from Marcos [34]. He there rightly asks why theoremhood is going to be preferred over inference, truth over falsity, single-conclusion over multiple-conclusion or acceptance over rejection. It may be that the slant towards certain members of those pairs is due to some important psychological, cognitive, historical, evolutive, and whatnot reasons, but it seems that as far as logic is concerned both members of each pair deserve equal attention. Thus, this is not changing one preference for another, but for the sake of symmetry: Asymmetric considerations and preferences permeate from the definition of a topos to the definition of logical consequence in the internal logic and then to theorems and proofs, which finally leads to distorted philosophical claims based on those results, as in (S1)–(S5).

This paper is, yes, an exposition of the basics of complement-toposes, but in the end is thus a philosophically motivated logical discussion of topos logic. My objective is threefold. First, to bring a clearer understanding of the philosophical foundations of the common categorial wisdom regarding topos logic. Second, to show that the notion of complement-topos allows to question certain philosophical associated to topos theory, and especially to the theory of its internal logic—the aforementioned slogans. Third, to show that, and how, the notion of complement-topos motivate further steps into abstraction in topos-theoretic notions and constructions, and in particular in the logic therein, mostly from a logical point of view but exploiting the power of the conceptual clarification provided by category theory, to gain a deeper conceptual understanding of what an internal logic is. More shortly, and in Lawverean terms, it seems to me that complement-toposes show that there is still a lot of "substance" in topos theory, and deeper "invariant forms" wait to emerge, and that is what I want to show in this paper.

I have divided the paper into four main sections. Next section, Sect. 18.2, serves to introduce all the required elements about the standard topos-theoretical analysis of logic and the means to nailing down the aforementioned slogans, in a way that philosophers familiar with first-order (classical) logic could be able to follow the discussion. In Sect. 18.3 I describe complement-toposes. This is the core of the paper, since the notion of complement-topos gives rise to doubts about the standard description of the internal logic of a topos, it calls (S3), (S4), and (S5) into question and opens the door for further abstraction. I show that complement-toposes can be described as pretending that standard toposes do not exist, i.e., giving directly a complement reading of the categorial structure of toposes, as Mortensen and Lavers did. Then I provide a proof theory for the internal logic of complement-toposes. Finally, in Sect. 18.4 I analyze some objections against complement-toposes. One of them says that complement-toposes are not categorically distinct from standard toposes, which is true but does not imply that those alternative labelings give rise to the same internal logic. Another objection says that complement-toposes contradict some well-established results of topos theory, but those theorems presuppose one or another feature of standard toposes, so they beg the question. The third objection is a more general criticism against inconsistency in mathematics, which is answered appealing basically to the free and pluralist nature of mathematics.

The present work has some limitations which deserve to be made explicit. Perhaps, the most important of them is that, although the new notions and constructions introduced here do apply to first- or higher-order topos logic, for simplicity most of my examples use only zero-order topos logic, so the reader will have to figure out the higher-order cases. Another important omission is at least a sketch of a paraconsistent set theory allowed by the notion of complement-topos, and thus a full discussion of (S5). Most paraconsistent set theories thus far have been motivated by the idea of doing justice to naïve set theory (in addition to being still membership-based). It would be interesting comparing motivations and results, but I will leave the investigation on this field for further work.[3]

Finally, the reader is assumed to know classical logic and to understand first-order languages and naïve set-theoretic notation. Those are the prerequisites. A fluent reading presupposes the knowledge of some category theory, order theory, and algebra. In most cases, I will give definitions in a nice format and prove theorems only when they are original contributions from this work. When presenting theorems already proved I only state them and refer the reader to a text where they can find a proof (even if not necessarily the original one). There is a convention to keep on mind: I use the adjective "categorial" exclusively used as shorthand for "category theoretic," but note that this convention has not been applied to quotations.

18.2 Standard Toposes and Their Internal Logic

18.2.1 The Comprehension Axiom

For our convenience, think of an object O of a topos as a type, collection of things, or generalized set—the O's.[4] Thus an object O is the objects of o's, in the same way that a product is the object of pairs $\langle x, y\rangle$ such that x is in X and y is in Y. The basic means of getting logic in a topos will be by a generalized notion of comprehension

[3] A discussion of other topics in philosophy of logic, like the issue of meaning variance or the discussion of the connections between degenerate categories and trivialism, is certainly worth, but that is material for separate work.

[4] This elucidation of toposes in logical terms follows closely [1].

of subobjects by “properties.” There are two things one needs to know about such properties:

Properties are local: A property is always a property of o’s of some O, thus every property has a fixed domain of significance.

Properties are variable propositions: If φ is a property with domain of significance O, and a is a constant element of type O, then $\varphi(a)$ is a proposition.[5]

So, in a topos, a property with domain of significance X will be called a *propositional function on O*. Every morphism must have a codomain, so a topos will include an Ω of propositions or algebraic truth values. Its elements (if any) $p:\mathbf{1}\longrightarrow\Omega$ are propositions, and its generalized elements $\varphi:X\longrightarrow\Omega$ are variable propositions, hence propositional functions. If the proposition p factors as $p = \varphi(a) : \mathbf{1} \longrightarrow O \longrightarrow \Omega$, then p results from evaluating the propositional function φ for the element a of O.

It is be assumed that there is a proposition $true:\mathbf{1}\longrightarrow\Omega$ satisfying a certain comprehension axiom. Note that there are two very important assumptions here, one categorial, “formal” or “structural,” merely concerning the existence of a certain morphism

(i) There is a morphism $\nu:\mathbf{1}\longrightarrow\Omega$ and the other “material” or more substantive concerning a very loaded conceptualization of such a morphism:
(ii) it is better thought of as ‘$true:\mathbf{1}\longrightarrow\Omega$’ provided a plausible conceptualization of the properties it satisfies.

Let me unpack this idea. A morphism $\nu:\mathbf{1}\longrightarrow\Omega$, called *(bare) subobject classifier*, has the following property:

(Bare) Comprehension axiom. For each $\varphi:O\longrightarrow\Omega$ there is an equalizer of φ and ν_O, and each monic $m:M\rightarrowtail O$ is such an equalizer for a unique φ. In diagrams, ν is such that for every φ and every object T and morphism $o:T\longrightarrow O$, if $m\circ\varphi = m\circ\nu_O$ and $x\circ\varphi = x\circ\nu_O$, there is a unique $h:X\longrightarrow M$ that makes the diagram below commutative:

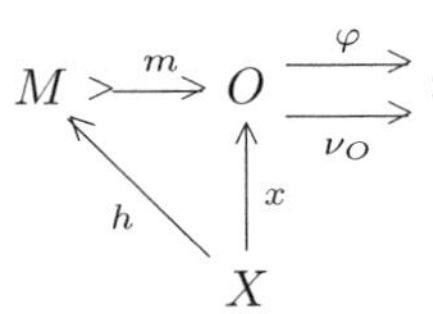

The connection of these definitions with traditional logical notions is less mysterious than it appears at first sight.[6] Consider the diagram for the **(Bare) Comprehension axiom** (which is a special case of the diagram corresponding to the definition of equalizer):

[5] As Awodey has noted, this is Russell’s notion of propositional function, for example in *The Principles of Mathematics* Sect. 22 or *Principia Mathematica*, pp. 14 and 161.

[6] Note by the way that, unlike many authors, I prefer the equalizers presentation of logic, not the pullbacks one.

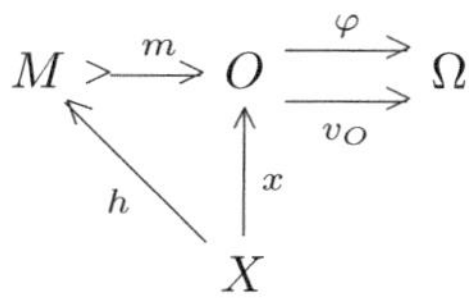

The only morphism from X to Ω which makes the diagram commutative is ν_X:

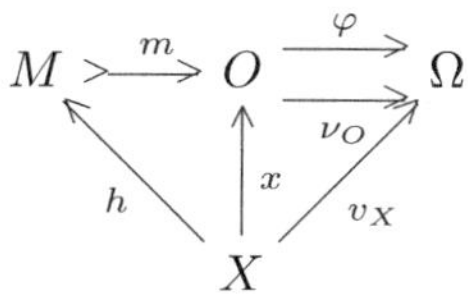

and the following diagram is obtained:

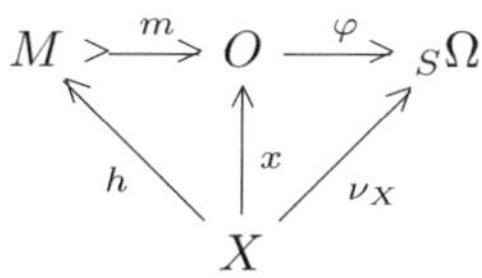

Note that, according to the definition of equalizer, h must be the only morphism that, among other things, $x = m \circ h$. But this satisfies the definition of $x \in m$. Thus, what the **(Bare) Comprehension axiom** says is that $\varphi(x) = \upsilon_X$ (by the right commutative triangle) if and only if $x \in m$ (by the left commutative triangle). This clearly invites the reading of ν as $true$, if m is thought of as the *extension* of the property φ, as is natural to think. In formal terms, this conceptualization is a particular Skolemization of the (equational) formula describing the (bare) subobject classifier.

Assumption (ii), the "standard Skolemization" of ν, obligates certain names for other categorial ingredients but let me state this more formally. Let '$\ulcorner\urcorner$' denote an instantiation device, such that '$\ulcorner x \urcorner$' denotes a constant which is the replacement of x and thus '${}_S\ulcorner x \urcorner$' denotes the standard instantiation of x. Thus,

$({}_S true)$ ${}_S\ulcorner \nu : \mathbf{1} \longrightarrow \Omega \urcorner = {}_S true : \mathbf{1} \longrightarrow {}_S\Omega$

According to this, '${}_S\Omega$' denotes that

$({}_S\Omega)$ for every $f : X \longrightarrow \Omega$ in a given topos $\mathcal{E}$, $\ulcorner f : X \longrightarrow \Omega \urcorner$ is standard according to the initial Skolemization for $\nu : \mathbf{1} \longrightarrow \Omega$.

'${}_S\mathcal{E}$' denotes something similar to '${}_S\Omega$,' but emphasizing the ambient topos '$\mathcal{E}$':

$({}_S\mathcal{E})$ in a given topos $\mathcal{E}$, for every $f : X \longrightarrow \Omega$, $\ulcorner f : X \longrightarrow \Omega \urcorner$ is standard according to the initial Skolemization for $\nu : \mathbf{1} \longrightarrow \Omega$.

'${}_S f$' denotes quite the same as the two symbols above but emphasizing the morphism f:

$({}_S f)$ for the morphism $f : X \longrightarrow \Omega$ in a given topos $\mathcal{E}$, $\ulcorner f : X \longrightarrow \Omega \urcorner$ is standard according to the initial Skolemization for $\nu : \mathbf{1} \longrightarrow \Omega$.

I must confess I do not know how to rinse the phrase '$\ulcorner f : X \longrightarrow \Omega \urcorner$ is standard according to the initial Skolemization for $\nu : \mathbf{1} \longrightarrow \Omega$' otherwise than by saying that the $\ulcorner f : X \longrightarrow \Omega \urcorner$'s correspond with some prior knowledge or conception of logical notions which is coherent with the initial choice of name for $\nu : \mathbf{1} \longrightarrow \Omega$. Consider the (partial) truth condition $p\#q = \nu$ if and only if $p = \nu$ and $q = \nu$: If one has chosen the name '*true*' for ν then the best name for # is 'conjunction,' not 'disjunction,' or some other.

So, a *(standard) topos* is a category ${}_S\mathcal{E}$ with equalizers, (binary) products, coequalizers, coproducts, exponentials, and a morphism ${}_Strue : \mathbf{1} \longrightarrow {}_S\Omega$, called *(standard) subobject classifier*, which has the following property:

(Standard) Comprehension axiom. For each ${}_S\varphi : O \longrightarrow {}_S\Omega$ there is an equalizer of ${}_S\varphi$ and ${}_Strue_O$, and each monic $m : M \rightarrowtail O$ is such an equalizer for a unique ${}_S\varphi$. In diagrams, ${}_Strue$ is such that for every ${}_S\varphi$ and every object T and morphism $o : T \longrightarrow O$, if $m \circ_S \varphi = m \circ {}_Strue_O$ and $x \circ_S \varphi = x \circ_S true_O$, then there is a unique $h : X \longrightarrow M$ that makes the diagram below commutative:

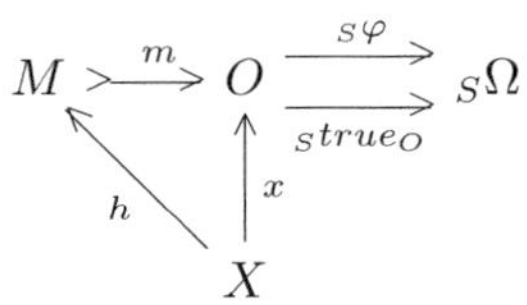

The propositional function ${}_S\varphi$ is also called "the (standard) characteristic (or classifying) morphism of m," denoted ${}_S\varphi_m$ for more convenience. A subobject classifier is unique up to isomorphism and so is ${}_S\varphi_m$.

Then, for any object O in a topos, the composite ${}_Strue \circ !_O : O \longrightarrow \mathbf{1} \longrightarrow {}_S\Omega$ denotes a constant, ${}_S$true-valued propositional function on O, abbreviated to ${}_Strue_O$. Propositional functions specify subobjects as follows. Given a propositional function $\varphi : O \longrightarrow {}_S\Omega$, one gets the part of the o's of which ${}_S\varphi$ is true, if any, as an equalizer $m : M \rightarrowtail O$ of ${}_S\varphi$ and ${}_Strue_O$. This subobject will be named accordingly the *extension* of the propositional function ${}_S\varphi$.

Let me make explicit the following two very important properties of ${}_S\Omega$ and ${}_Strue : \mathbf{1} \longrightarrow {}_S\Omega$:

1. If X is an object of ${}_S\mathcal{E}$ and $m : M \rightarrowtail X$ is a subobject of X, then there is exactly one morphism ${}_S\varphi_m : X \longrightarrow {}_S\Omega$ such that for every $x : \mathbf{1} \longrightarrow X$, $x \in m$ if and only if ${}_S\varphi_m \circ x = {}_Strue$.
   (Succinctly, a subobject classifier says, for every object X and every subobject m of X, what elements of X are included in the subobject m.)
2. If a morphism f has ${}_S\Omega$ as codomain, then it is the characteristic morphism of some other morphism g such that its codomain is the domain of f.
   (A morphism to ${}_S\Omega$ is fully determined by the part of its domain that it takes to ${}_Strue$, that is, by the subobject of its domain that it classifies.)

All the above supports the following slogan:

(S1) Ω is (or at least can be regarded) a truth values object.

It is well-known that the subobjects of a given object in a category form a partial order. In particular, the elements of ${}_S\Omega$ form a partial order, which means that propositions form a partial order, i.e., for every propositions p, q, and r:

- $p \leq p$,
- If $p \leq q$ and $q \leq r$ then $p \leq r$,
- If $p \leq q$ and $q \leq p$ then $p = q$.

If $\leq$ is interpreted as a deducibility relation, $\vdash$, the properties above say that deducibility is reflexive, transitive, and that interdeducible propositions are equivalent.

Given the notion of subobject classifier, one can define also $_{S}false : \mathbf{1} \longrightarrow {}_{S}\Omega$ as the character of 0_1, the only morphism from an initial object to a terminal one:

$$\mathbf{0} \xrightarrow{0_1} \mathbf{1} \underset{strue_1}{\overset{sfalse =_{def.}\ s\varphi_{0_1}}{\rightrightarrows}} {}_{S}\Omega$$

The full commutative diagram for this equalizer looks like this:

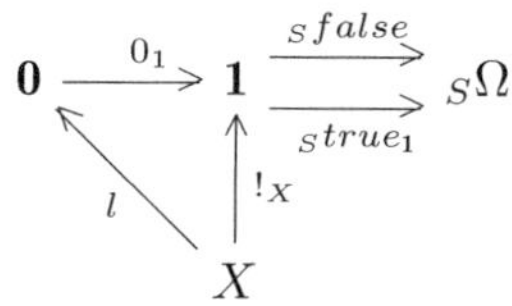

The only morphism from X to ${}_{S}\Omega$ that makes the diagram above commutative is $_{S}true_X$:

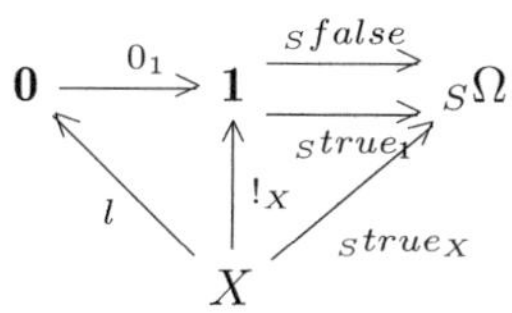

Thus, the following diagram is obtained:

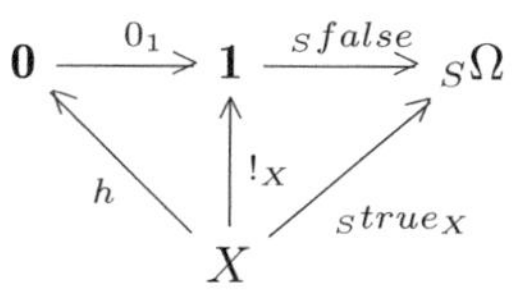

What the diagram above expresses is that $_{S}false = {}_{S}true_X$ (because of the right commutative triangle) if and only if $\mathbf{0}$ has a generalized element (because of the left commutative diagram). Note that this would obtain at least once, namely considering the maximal element of $\mathbf{0} : \mathbf{0}$ itself. But this does not count yet as a case where there is a proper element of $\mathbf{0}$ such that it equates $_{S}false$ and $_{S}true_X$. Is there any such case?

Suppose that X is a terminal object, $\mathbf{1}$:

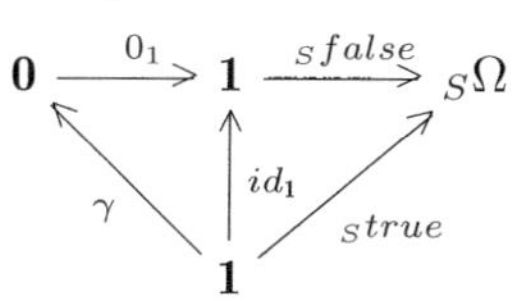

Given that this diagram commutes, what it expresses is that ${}_{S}false = {}_{S}true$ (because of the right commutative triangle) if and only if $\mathbf{0}$ has at least one element (because of the left commutative triangle). But if $\mathbf{0}$ has an element other than the generalized element that is the identity morphism, then all the objects of the category are isomorphic, so for any mathematical purpose is as if there were only one object (and only one morphism) in the category. Only in that case, a "degenerate category," one obtains ${}_{S}false = {}_{S}true$.[7] Now, if $\mathbf{0}$ has no elements, not all objects of the category are isomorphic and ${}_{S}false \neq {}_{S}true$. In that case, either ${}_{S}false < {}_{S}true$, or ${}_{S}true < {}_{S}false$ or they are incomparable. They are not incomparable since ${}_{S}true$ is the greatest element in the order formed by propositions. This also excludes ${}_{S}true < {}_{S}false$. Then the only option for a nondegenerate category is ${}_{S}false < {}_{S}true$.

Example 18.2.1 Let **Set** be the (standard) category of (abstract constant) sets as objects and functions as morphisms. ${}_{S}\Omega_{\mathbf{Set}}$ has only two elements with the order ${}_{S}false_{\mathbf{Set}} < {}_{S}true_{\mathbf{Set}}$. Hence, in this category ${}_{S}\Omega_{\mathbf{Set}} = \mathbf{2}_{{}_{S}\mathbf{Set}}$. Thus, for every element t of O, $t : \mathbf{1} \longrightarrow O$, $t \in O$ if and only if ${}_{S}\varphi \circ t = {}_{S}true_{\mathbf{Set}}$, and $t \notin O$ if and only if ${}_{S}\varphi_m \circ t = {}_{S}false_{\mathbf{Set}}$, since ${}_{S}false_{\mathbf{Set}}$ is the only morphism distinct from ${}_{S}true_{\mathbf{Set}}$. According to the aforementioned convention, I will use '${}_{S}\mathbf{Set}$' to denote that $\Omega_{\mathbf{Set}}$ is ${}_{S}\Omega_{\mathbf{Set}}$. A similar convention will be used for the categories below.

Example 18.2.2 ${}_{S}\mathbf{Set}^{\rightarrow}$ is the standard category of functions. A terminal object in this category, $\mathbf{1}_{{}_{S}\mathbf{Set}^{\rightarrow}}$, is the identity function from $\mathbf{1}_{{}_{S}\mathbf{Set}}$ to $\mathbf{1}_{{}_{S}\mathbf{Set}}$.

Consider two objects of $\mathbf{Set}^{\rightarrow}$, $f : A \longrightarrow B$ and $g : C \longrightarrow D$. If f is a subobject of g, then $A \subseteq C$, $B \subseteq D$ and f is the restriction of g, that is, $f(x) = g(x)$ for $x \in A$. To the question "Is a given element x of C also an element of B?" there are only two possible answers: Either it is or it is not, so the codomain of a function playing the role of a subobject classifier can be ${}_{S}\Omega_{{}_{S}\mathbf{Set}}$. But before giving that definite answer, one must compute whether x is in A or not. One has then three options:

(i) Either $x \in A$, so the final answer to original question is "Yes," because $g(x) \in B$; or
(ii) $x \notin A$, but the final answer to the original question will be "Yes," because $g(x) \in B$ after all; or
(iii) $x \notin A$, but the final answer will be "No" because $x \notin B$ too.

Then, the domain of a function playing the role of a subobject classifier will be any three-element set to represent these three options. Let me use '1,' '$\frac{1}{2}$,' and '0' to denote each of those elements, respectively. So ${}_{S}\Omega_{\mathbf{Set}^{\rightarrow}}$ looks like this:

$$t : \mathbf{3}_{{}_{S}\mathbf{Set}} \longrightarrow {}_{S}\Omega_{\mathbf{Set}}$$

with $t(0) = {}_{S}false_{\mathbf{Set}}$ and $t(\frac{1}{2}) = t(1) = {}_{S}true_{\mathbf{Set}}$.

Thus, a subobject classifier in this category is ${}_{S}true_{\mathbf{Set}^{\rightarrow}} : \mathbf{1}_{\mathbf{Set}^{\rightarrow}} \longrightarrow {}_{S}\Omega_{\mathbf{Set}^{\rightarrow}}$, i.e., a pair of morphisms $\langle t'_{\mathbf{Set}}, true_{{}_{S}\mathbf{Set}} \rangle$ from $id_{\mathbf{1}_{\mathbf{Set}}} : \mathbf{1}_{\mathbf{Set}} \longrightarrow \mathbf{1}_{\mathbf{Set}}$ to ${}_{S}\Omega_{\mathbf{Set}^{\rightarrow}}$. There are

[7] This is not as odd as it might seem at first sight; see [14].

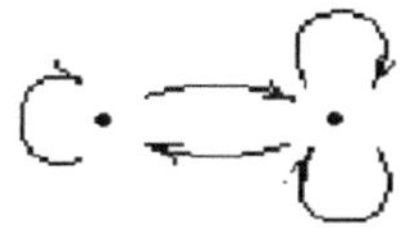

Fig. 18.1 Truth values object in $S^{S\downarrow\downarrow}$

only two truth values in this category. The calculation is straightforward and can be left to the reader (Hint: There seems to be an additional value, let us denote it $_S\alpha_{\mathbf{Set}^{\rightarrow}} = \langle t''_{\mathbf{Set}}, true_{_S\mathbf{Set}}\rangle$. Note that although $t' \neq t''$, $_S\alpha_{\mathbf{Set}^{\rightarrow}} = {_S}true_{\mathbf{Set}^{\rightarrow}}$).

Example 18.2.3 $_S\mathbf{S}^{\downarrow\downarrow}$ is the category of (standard irreflexive directed multi-) *graphs* and graph structure preserving maps.[8] An object of $_S\mathbf{S}^{\downarrow\downarrow}$ is any pair of sets equipped with a parallel pair of maps $A \underset{t}{\overset{s}{\rightrightarrows}} V$ where A is called the set *arrows* and V is the set of *dots* (or *nodes* or *vertices*). If a is an element of A (an arrow), then $s(a)$ is called the *source* of a, and $t(a)$ is called the *target* of a.

Morphisms of $_S\mathbf{S}^{\downarrow\downarrow}$ are also defined so as to respect the graph structure. That is, a morphism $f:(A \underset{t}{\overset{s}{\rightrightarrows}} V) \longrightarrow (E \underset{t'}{\overset{s'}{\rightrightarrows}} P)$ in $_S\mathbf{S}^{\downarrow\downarrow}$ is defined to be any *pair* of morphisms of **Set** $f_a : A \longrightarrow V$, $f_v : E \longrightarrow P$ for which both equations

$$f_v \circ s = s' \circ f_a$$

$$f_v \circ t = t' \circ f_a$$

are valid in $_S\mathbf{Set}$. It is said that f preserves the structure of the graphs if it preserves the source and target relations.

A terminal object in this category, $\mathbf{1}_{_S\mathbf{S}^{\downarrow\downarrow}}$, is any arrow such that its source and target coincide.

This topos provides a simple yet good example of a truth values object with more than two elements. $_S\Omega_{\mathbf{S}^{\downarrow\downarrow}}$ has the form of a graph like that in Fig. 18.1. There are exactly three morphisms $\mathbf{1}_{_S\mathbf{S}^{\downarrow\downarrow}} \longrightarrow {_S\Omega_{\mathbf{S}^{\downarrow\downarrow}}}$ in this category, which means that $_S\Omega_{\mathbf{S}^{\downarrow\downarrow}}$ has three truth values with the order $_S false_{\mathbf{S}^{\downarrow\downarrow}} < {_S}\binom{s}{t}_{\mathbf{S}^{\downarrow\downarrow}} < {_S}true_{\mathbf{S}^{\downarrow\downarrow}}$.

Now, these examples show that, in general, $_S\Omega$ has more than two elements, and it is thus that one comes with the slogan

(S2) The internal logic of a topos is in general many-valued.

18.2.2 The Connectives

Before studying the internal logic of standard toposes let me introduce some notions that will prove very helpful in what follows.

---

[8]Nice introductions to this category can be found in [30, 53].

Let $i : X^Y \longrightarrow X$, $o : X^Y \longrightarrow Z$, $t : W \longrightarrow X$ be morphisms in a category $\mathbf{C}$. i, o and t are called *operations in* X, *on* X, and *to* X, respectively. An *operation of* X is any of these kinds of operations.

If X is an object of a standard topos, I shall write $\Delta_X : X \longrightarrow X \times X$ for the *diagonal of* X, i.e., the unique morphism whose composite with both projections is the identity on X. In particular, from $p_i \circ \Delta_X = 1_X$ one can deduce that Δ_X is a monomorphism. The (standard) "equality" on an object X of a topos is the characteristic morphism ${}_S =_X : X \times X \longrightarrow {}_S\Omega$ of the diagonal $\Delta_X : X \longrightarrow X \times X$.

The *product/exponential adjunction* states that to every morphism $f : X \times Y \longrightarrow Z$ corresponds a morphism $\lambda_x . f : X \longrightarrow Z^Y$. In particular, to $=_X : X \times X \longrightarrow {}_S\Omega$ corresponds a morphism $\lambda_x . {}_S =_X : X \longrightarrow {}_S\Omega^X$. This latter morphism is called the *(standard) singleton on* X.

For an object X of a topos, the *(standard) membership* relation ${}_S \in_X$ on X is the subobject ${}_S \in_X \rightarrowtail X \times {}_S\Omega^X$ whose characteristic morphism, still written ${}_S \in_X \ : {}_S\Omega^X \times X \longrightarrow {}_S\Omega$, corresponds to the identity on ${}_S\Omega^X$ by the product/exponential adjunction.

The internal logic of a standard topos will be the algebra of operations of ${}_S\Omega$, that is, the algebra of operations of its object of propositions. Variable propositions or propositional functions have been defined as morphisms $\varphi : X \longrightarrow {}_S\Omega$, i.e., as *operations to* ${}_S\Omega$. One defines next *operations in* ${}_S\Omega$ or, more commonly, *(standard) connectives*.[9]

In a standard topos, a morphism $k : ({}_S\Omega \times \cdots \times {}_S\Omega)^{{}_S\Omega^{\cdot^{\cdot^{{}_S\Omega^X}}}} \longrightarrow {}_S\Omega$ (with ${}_S\Omega \times \cdots \times {}_S\Omega$ n times and ${}_S\Omega^{\cdot^{\cdot^{{}_S\Omega}}}$ t times, $n, t \geq 0$), abbreviated to $k : {}_S\Omega^{n_m} \longrightarrow {}_S\Omega$, will be said to be a *(standard) n-ary connective* of order m, where $m = 0$ if and only if ${}_S\Omega^{\cdot^{\cdot^{{}_S\Omega^X}}} \cong \mathbf{1}$ and $m = (t + 1)$ otherwise.[10]

This enables us to define the more usual standard connectives, three binary ($\wedge$, conjunction; $\vee$, disjunction; $\supset$, conditional) and three unary ($\neg$, negation; $\forall$, universal quantifier; $\exists$, particular quantifier). Propositions, i.e., morphisms $\mathbf{1} \longrightarrow {}_S\Omega$, can be thus considered 0-ary connectives (and can be of any order) with $({}_S\Omega \times \cdots \times {}_S\Omega) \cong \mathbf{1}$. The above-mentioned standard connectives can be defined as follows.[11]

Negation. Let be ${}_S true : \mathbf{1} \longrightarrow {}_S\Omega$. Then $\neg : {}_S\Omega \longrightarrow {}_S\Omega$ is the characteristic morphism of ${}_S false$:

$$\mathbf{1} \xrightarrow{\ s false\ } {}_S\Omega \underset{strue_{{}_S\Omega}}{\overset{\neg =_{def.}\ S\varphi_{S false}}{\rightrightarrows}} {}_S\Omega$$

[9] I will omit *operations on* ${}_S\Omega$ for simplicity, since it does not constitute a fundamental missing in the theme of the internal logic.

[10] For brevity, I often will talk only of "connectives," since their standard character can be obviated in this chapter, and their arity and order will be made explicit only when needed.

[11] See for example [18, Sect. 6.6].

The full commutative diagram for this equalizer is as follows:

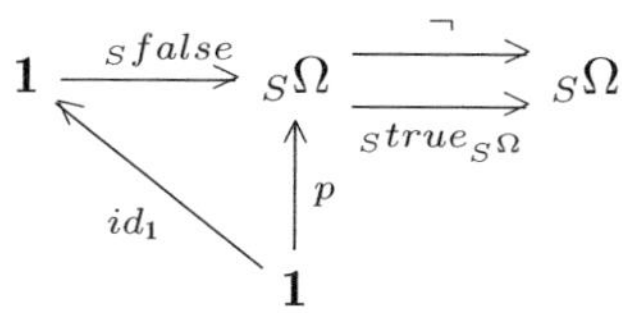

The only morphism from **1** to $_S\Omega$ that makes the diagram above commutative is $_Strue$:

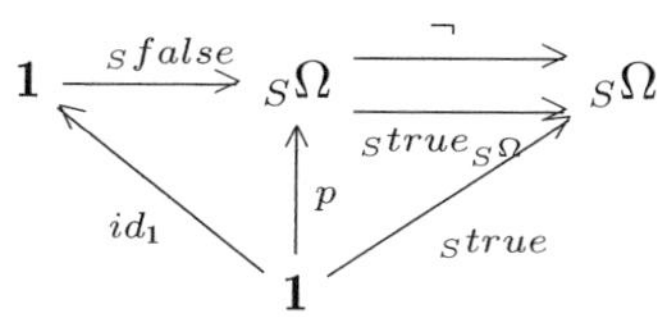

Thus, the following diagram is obtained:

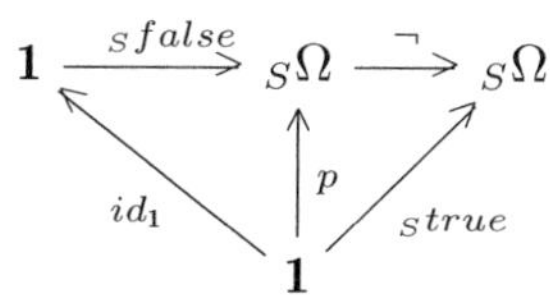

Given that this diagram commutes, what it expresses is that $\neg \circ p = {}_Strue$ (because of the right commutative triangle) if and only if $p = {}_Sfalse$ (because of the left commutative triangle). In fact, the full truth condition implied by this definition is that $\neg \circ p = {}_Strue$ if and only if $p = {}_Sfalse$ and $\neg \circ p = {}_Sfalse$ otherwise.

Conjunction. Conjunction $\wedge : {}_S\Omega \times {}_S\Omega \longrightarrow {}_S\Omega$ is defined as the characteristic morphism of $\langle {}_Strue, {}_Strue \rangle : \mathbf{1} \longrightarrow {}_S\Omega \times {}_S\Omega$:

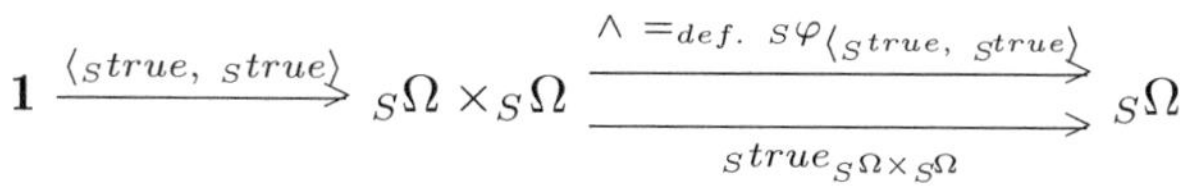

The full commutative diagram for this equalizer is as follows:

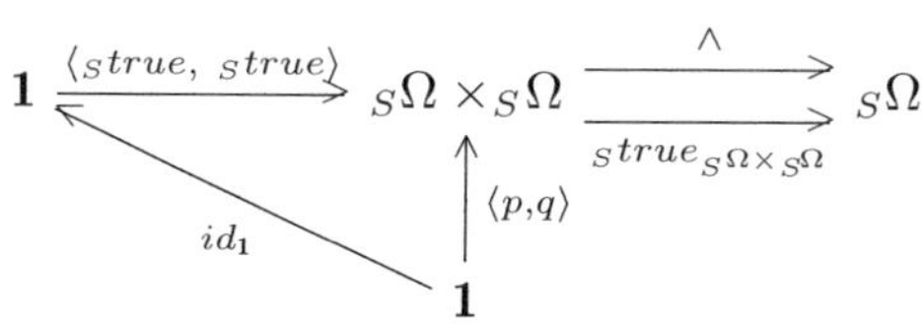

The only morphism from **1** to ${}_S\Omega$ that makes the diagram above commutative is ${}_Strue$:

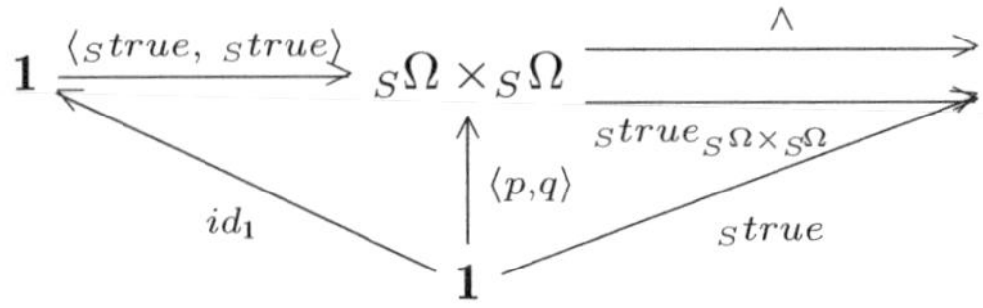

Thus, the following diagram is obtained:

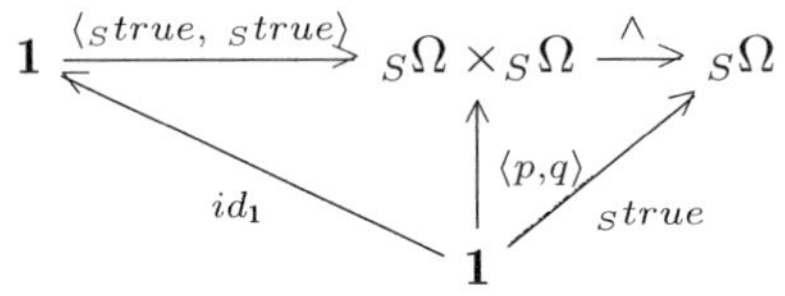

Given that this diagram commutes, what it expresses is that $\wedge \circ \langle p, q\rangle = {}_Strue$ (because of the right commutative triangle) if and only if $\langle p, q\rangle = \langle {}_Strue, {}_Strue\rangle$, and hence $p = {}_Strue$ and $q = {}_Strue$ (because of the left commutative triangle). $\wedge \circ \langle p, q\rangle$ is more commonly written '$p \wedge q$' and I will do so throughout the rest of this work. In fact, the complete truth condition implied by this definition is that $p \wedge q = \inf(p, q)$ (with respect to the partial order formed by the elements of ${}_S\Omega$).

Disjunction. Disjunction $\vee : {}_S\Omega \times {}_S\Omega \longrightarrow {}_S\Omega$ is defined as the characteristic morphism of the image of $[\langle {}_Strue,\ id_{{}_S\Omega}\rangle,\ \langle id_{{}_S\Omega}, {}_Strue\rangle]$:

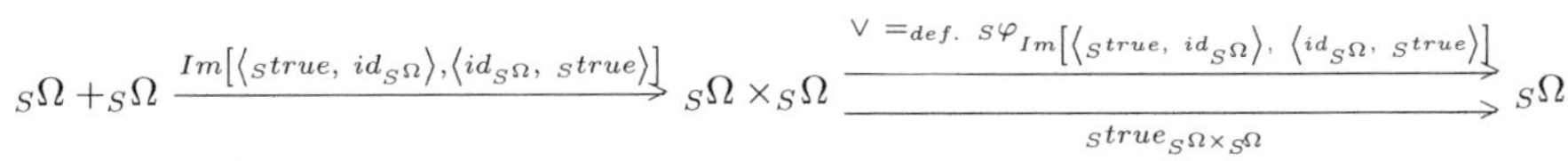

The full development of the corresponding equalizer for this and the remaining connectives can be left to the reader. The truth condition implied by this definition is that $p \vee q = \sup(p, q)$ (again with respect to the partial order formed by the elements of ${}_S\Omega$).

Conditional. Conditional $\supset : {}_S\Omega \times {}_S\Omega \longrightarrow {}_S\Omega$ is defined as the characteristic morphism of $e : {}_S\leq \longrightarrow {}_S\Omega \times {}_S\Omega$, where e is the equalizer of $\wedge : {}_S\Omega \times {}_S\Omega \longrightarrow {}_S\Omega$ and the first projection, p:

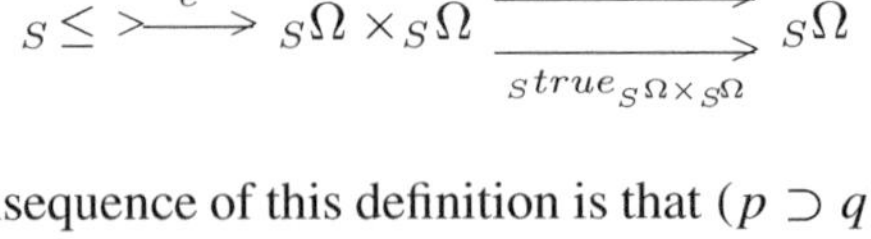

An immediate consequence of this definition is that $(p \supset q) = {}_Strue$ if and only if $(p \wedge q) = p$ (as can be noted, the equalizer e expresses the condition on the right: It equals conjunction and the first projection). The complete truth condition is that $(p \supset q) = {}_Strue$ if and only if $p \leq q$ and $(p \supset q) = q$ otherwise.

Recall that the product/exponential adjunction states that to every morphism $f : X \times Y \longrightarrow Z$ corresponds a morphism $\lambda_x.f : X \longrightarrow Z^Y$. As a particular case,

$\lambda_x.f:\mathbf{1} \longrightarrow Y^X$ is the element corresponding to $f:X \longrightarrow Y$ under such "exponential transposition." Thus, each element $x:\mathbf{1} \longrightarrow {}_S\Omega^X$ corresponds uniquely to a propositional function ${}_S\varphi:X \longrightarrow {}_S\Omega$, and hence to a subobject $m:M \rightarrowtail X$. Accordingly, $\lambda_x.\varphi:\mathbf{1} \longrightarrow \Omega^X$ is the element corresponding to the propositional function $\varphi:X \longrightarrow \Omega$. Now the universal quantifier can be defined as follows:

Universal quantifier. Universal quantifier $\forall_X: {}_S\Omega^X \longrightarrow {}_S\Omega$ is defined as the characteristic morphism of $\lambda_x.{}_Strue_X$, that is, of the exponential transposition of ${}_Strue_X \circ pr_X : \mathbf{1} \times X \longrightarrow X \longrightarrow {}_S\Omega$:

$$\mathbf{1} \xrightarrow{\lambda_x.{}_Strue_X} {}_S\Omega^X \underset{{}_Strue_{{}_S\Omega^X}}{\overset{\forall_X =_{def.}\ {}_S\varphi_{\lambda_x.{}_Strue X}}{\rightrightarrows}} {}_S\Omega$$

The universal quantifier has the property that $\forall_X\varphi(x) = {}_Strue$ if and only if $\varphi(x) = {}_Strue$, for all x. The exact truth condition implied by the definition above is $\forall_X\varphi(x) = \inf(\varphi(x))$.

Particular quantifier. Particular quantifier $\exists_X : {}_S\Omega^X \longrightarrow {}_S\Omega$ is defined as the characteristic morphism of the image of the composite ${}_Spx_X \circ {}_S\in_X (\in): \in \rightarrowtail {}_S\Omega^X \times X \longrightarrow {}_S\Omega^X$ (where ${}_Spx_X$ is the first projection and ${}_S\in_X$ is the subobject of ${}_S\Omega^X \times X$ whose character is the evaluation morphism $e_X: {}_S\Omega^X \times X \longrightarrow {}_S\Omega$):

$${}_Spx_X \circ \in_X (\in) \xrightarrow{Im({}_Spx_X \circ\ {}_S\in_X)} {}_S\Omega^X \underset{{}_Strue_{{}_S\Omega^X}}{\overset{\exists_X =_{def.}\ {}_S\varphi_{Im({}_Spx_X \circ {}_S\in_X)}}{\rightrightarrows}} {}_S\Omega$$

The particular quantifier has the property that $\exists_X\varphi(x) = {}_Strue$ if and only if $\varphi(x) = {}_Strue$, for some x. The exact truth condition implied by the definition above is $\exists_X\varphi(x) = \sup(\varphi(x))$.

Given that ${}_S\mathcal{E}$ is a category with exponentials, one has ${}_S\Omega^X$, ${}_S\Omega^{{}_S\Omega^X}$, etc. for any X in ${}_S\mathcal{E}$, which may be regarded as representing collections of properties, properties of properties, etc. defined over X, so one can also have higher-order propositions.

18.2.3 The Internal Logic of a Standard Topos

There is a correspondence between classical logic and Boolean algebra, to wit, such a logic can be represented as a two-element Boolean algebra in which operations satisfy certain conditions. Roughly, those two elements are truth values, operations are the connectives, and the conditions they must satisfy are determined by the well-known truth conditions. Taking into account the categorial generalization over algebraic truth values, one can define the internal logic of a topos as follows: Given a standard topos ${}_S\mathcal{E}$, its internal logic is the algebra induced by its object of propositions or algebraic truth values, ${}_S\Omega_\mathcal{E}$, and the connectives. This logic is called *internal* because (i) it is formulated exclusively in terms of the objects and morphisms of the topos in question and (ii) it is the right to reason about the topos in question, since it is determined by the definition of its objects and morphisms. It is not a canon imposed "externally" to reason about the topos: Using a different logic for that purpose would alter the definitory properties of those objects and morphisms and thus it would not be a logic for the intended objects and morphisms. Below I will offer an example of how a logic distinct of the internal would alter those definitory properties.

Logical consequence is defined in the usual, "Tarskian" way: Let ${}_Sp \models {}_{S\mathcal{E}}\, {}_Sq$ denote that whenever the morphism ${}_Sp$ is the same morphism as ${}_Strue$ in ${}_S\mathcal{E}$, so is ${}_Sq$, and we will say that ${}_Sq$ is a *(Tarskian) logical consequence of* ${}_Sp$ ($\models_{fS\mathcal{E}}\, {}_Sp$ means that ${}_Sp$ is the same morphism as ${}_Strue$ in ${}_S\mathcal{E}$).

There is a theorem establishing necessary and sufficient conditions for a proposition ${}_Sp$ being the same morphism as ${}_Strue$ in a given standard topos ${}_S\mathcal{E}$. Let '$\models_I$' indicate that logical consequence gives the results as in intuitionistic logic. Then the following theorem holds:

Theorem 18.2.4 *For every proposition* ${}_Sp$, $\models_{S\mathcal{E}}\, {}_Sp$ *for every topos* ${}_S\mathcal{E}$ *if and only if* $\models_I {}_Sp$.

i.e., ${}_S\Omega$ is a Heyting algebra.[12] Hence, the slogan

(S3) The internal logic of a topos is, in general, intuitionistic.

Summarizing, the standard categorial analysis of logic implies the following:

(IL1) Propositions form a partial order, i.e., for every propositions p, q, and r:
(IL1a) $p \leq p$,
(IL1b) If $p \leq q$ and $q \leq p$ then $p = q$,
(IL1c) If $p \leq q$ and $q \leq r$ then $p \leq r$.
(IL2) There is a truth value called ${}_Strue$ with the following property:

$$\text{For every proposition } p,\ p \leq {}_Strue$$

(IL3) One can define a truth value called ${}_Sfalse$ that has the following property:

$$ {}_Sfalse \leq {}_Strue$$

and

$$\text{For every proposition } p,\ {}_Sfalse \leq p$$

(IL4) Connectives obey the following truth conditions:
$\neg p = {}_Strue$ if and only if $p = {}_Sfalse$, otherwise $\neg p = {}_Sfalse$,
$(p \wedge q) = \inf(p, q)$,
$(p \vee q) = \sup(p, q)$,
$(p \supset q) = {}_Strue$ if and only if $p \leq q$, otherwise $(p \supset q) = q$,
$\forall x \varphi(x) = \inf(\varphi(x))$,
$\exists x \varphi(x) = \sup(\varphi(x))$.

---

[12] I have made a little abuse of notation, for I used '${}_Sp$' in both $\models_{S\mathcal{E}}$ and $\models_I$. In rigor, ${}_Sp$ is a morphism which corresponds to a formula $({}_Sp)^*$ in a possibly different language, but there is no harm if one identifies them. A proof can be found in [18, see Sect. 8.3 for the soundness part and Sect. 10.6 for the completeness part].

(IL5) The categorial analysis of logic does not imply, but rather assume, the traditional, "Tarskian," notion of logical consequence:
Let '$p \models_{s\mathcal{E}} q$' denote that q is a logical consequence of p in a standard topos $\mathcal{E}$, i.e., that whenever p is the same morphism as $_{S}true$ in $_{S}\mathcal{E}$, so is q. Equivalently, if q is not the same morphism as $_{S}true$, p neither is. $\models_{s\mathcal{E}} p$ means that p is the same morphism as $_{S}true$ in $_{S}\mathcal{E}$.
(IL6) From (IL1)–(IL5), the internal logic of a standard topos is in general intuitionistic.

Example 18.2.5 The internal logic of $_{S}\mathbf{Set}$ is classical. For example, in $_{S}\mathbf{Set}$, every proposition p is the same as one and only one of $_{S}true_{\mathbf{Set}}$ and $_{S}false_{\mathbf{Set}}$. $\neg \circ_S true_{\mathbf{Set}} = {}_{S}false_{\mathbf{Set}}$ and $\neg \circ_S false_{\mathbf{Set}} = {}_{S}true_{\mathbf{Set}}$. Hence, for any p, $\neg\neg p = p$. Also, for any p $(p \vee \neg p) = \vee \circ \langle p, \neg p \rangle = \sup(p, \neg p) = {}_{S}true_{\mathbf{Set}}$.

Example 18.2.6 Even though it is many-valued, the internal logic of $_{S}\mathbf{Set}^2$ is classical: $_{S}\Omega\mathbf{Set}^2$ is a Boolean algebra with four elements, which in turn is the Cartesian product of a two-element Boolean algebra with universe $\{_{S}true_{\Omega_{\mathbf{Set}}}, false_{\Omega_{\mathbf{Set}}}\}$ with itself (i.e., operations act coordinatewise). For example, negation gives

$$\neg_{S}true_{\mathbf{Set}^2} = \langle \neg true_{_{S}\mathbf{Set}}, \neg true_{_{S}\mathbf{Set}} \rangle = \langle false_{_{S}\mathbf{Set}}, false_{_{S}\mathbf{Set}} \rangle = {}_{S}true_{\mathbf{Set}^2}$$

$$\neg_{S}\alpha_{\mathbf{Set}^2} = \langle \neg true_{_{S}\mathbf{Set}}, \neg false_{_{S}\mathbf{Set}} \rangle = \langle false_{_{S}\mathbf{Set}}, true_{_{S}\mathbf{Set}} \rangle =_S \beta_{\mathbf{Set}^2}$$

The cases of $_{S}\alpha_{\mathbf{Set}^2}$ and $_{S}false_{\mathbf{Set}^2}$ are left to the reader. It is easy to verify that for every p in $_{S}\mathbf{Set}^2$, $\neg\neg p = p$ and that $(p \vee \neg p) = {}_{S}true_{\mathbf{Set}^2}$.

Example 18.2.7 As I have mentioned, $_{S}\Omega_{S^{\downarrow\downarrow}}$ has three truth values with the order $_{S}false_{S^{\downarrow\downarrow}} < {}_{S}\binom{s}{t}_{S^{\downarrow\downarrow}} < {}_{S}true_{S^{\downarrow\downarrow}}$. Negation gives the following identities of morphisms:

$$\neg_{S}true_{S^{\downarrow\downarrow}} = {}_{S}false_{S^{\downarrow\downarrow}}, \quad \neg_{S}\binom{s}{t}_{S^{\downarrow\downarrow}} = false_{S^{\downarrow\downarrow}}, \quad \neg_{S}false_{S^{\downarrow\downarrow}} = {}_{S}true_{S^{\downarrow\downarrow}}$$

Since $(p \supset q) = {}_{S}true$ if and only if $(p \wedge q) = p$, in general $(\neg\neg p \supset p) \neq {}_{S}true$ in $S^{\downarrow\downarrow}$ because even though $(\neg\neg p \supset p) = {}_{S}true_{S^{\downarrow\downarrow}}$ either when $p = {}_{S}true_{S^{\downarrow\downarrow}}$ or when $p = {}_{S}false_{S^{\downarrow\downarrow}}$, $(\neg\neg p \wedge p) \neq \neg\neg p$ when $p =_S \binom{s}{t}_{S^{\downarrow\downarrow}}$. Given that $(\neg\neg p \supset p) \neq {}_{S}true_{S^{\downarrow\downarrow}}$ but there is no formula Φ such that $\Phi = true$ in classical logic and $\Phi = false$ in intuitionistic logic, $(\neg\neg p \supset p) = {}_{S}\binom{s}{t}_{S^{\downarrow\downarrow}}$ when $p = {}_{S}\binom{s}{t}_{S^{\downarrow\downarrow}}$. Moreover, $p \vee \neg p$ fails to be the same morphism as $_{S}true_{S^{\downarrow\downarrow}}$ since $(p \vee q) = {}_{S}true$ if and only if either $p = {}_{S}true$ or $q = {}_{S}true$. If $p = {}_{S}\binom{s}{t}_{S^{\downarrow\downarrow}}$, $\neg p = {}_{S}false_{S^{\downarrow\downarrow}}$, so neither $p = {}_{S}true_{S^{\downarrow\downarrow}}$ nor $\neg p = {}_{S}true_{S^{\downarrow\downarrow}}$ and hence $(p \vee \neg p) \neq {}_{S}true_{S^{\downarrow\downarrow}}$.

18.2.4 *Internal Logic and the Algebras of Subobjects and of Operations of $_S\Omega$*

In the literature there is a somewhat erratic usage of the expression 'internal logic.' Something similar happens with 'subobject classifier': Some authors use it to name the object $_S\Omega$, whereas others use it to name the morphism $_Strue : \mathbf{1} \longrightarrow {_S\Omega}$ and reserve the name 'truth values object' for $_S\Omega$.

Intuitively, an internal logic is a logic which is internal to the category in question in the following two senses:

(ILC 1) It is exclusively formulated in terms of the objects and morphisms of the category in question, and
(ILC 2) it is the logic adequate to reasoning about that category.

But at first sight at least two things could be called 'internal logic':

- The algebra of subobjects of the objects in a category (AS);
- The algebra of operations of Ω (AΩ).

The basic ideas of the internal logic induced by a given category **C** via the algebra of subobjects are these:

- An object X of **C** is regarded as a collection of things of type X;
- morphisms $X \to Y$ are regarded as terms of type Y containing a free variable of type X;
- a subobject $p : P \rightarrowtail X$ is regarded as a proposition, i.e., as indicating a certain part P of the things of type X for which a property is true;
- the maximal subobject of X (the identity morphism of X) is the proposition always true (of X), $\top_X$;
- the minimal subobject of X ($\mathbf{0} \longrightarrow X$) is the proposition always false (of X), $\bot_X$;
- one proposition implies another (with regard to X) if $p \leq q$ in the poset of subobjects of X;
- other connectives are obtained by means of constructions on subobjects: The conjunction of p and q is their product, their disjunction is their coproduct, and so on.[13]

Against the second interpretation, (AΩ), and in favor of (AS) is the fact that not all categories possess such an Ω but still have something like an internal logic. This is expressed most naturally in the "hierarchy of doctrines": One starts with categories with very simple features and then enriches them, so that certain logical notions and properties arise as the enrichment progresses. For example, if one starts with categories with products and equalizers, one will have the constant $\top$ and

[13] It must be said that some categories might not have enough structure to support some of the logical notions mentioned here. For example, an object in a category might have no minimal subobject distinct from the maximal one (so no proposition always false is distinct from always true) or might have no coproducts (and therefore, would lack disjunctions), etc.

conjunctions for all p and q; if one also allows that coequalizers exist under certain conditions and that epimorphisms be the coequalizers of certain kind of morphisms, the particular quantifier arises, and so on, all this without a subobject classifier. The logic adequate to reason about those categories is the logic so obtained, satisfying thus the requisite (ILC 2). But, as nice as all this appears, the logic in this sense is not completely internal as required by (ILC 1): The collection of subobjects of an object X might not be an actual object of the category in question.[14]

However, in the case of toposes there is a rather simple argument for (AΩ) and against (AS). Lawvere (see [27]) and Reyes and his collaborators (see [44–46]) have shown that the algebra of subobjects of a given object (in, say, a standard topos) can be described as a Heyting algebra (and thus supports intuitionistic logic) but also as a Brouwerian algebra (an algebra dual to Heyting algebra and thus supports a dual-intuitionistic logic).[15] But this is not true for the algebra of operations of ${}_S\Omega$. They have proved that if certain Brouwerian laws hold in the algebra of operations of ${}_S\Omega$ then it would be a Boolean algebra and thus classical reasoning would be allowed. But reasoning classically in certain categories, like some standard toposes, would alter the identity of their objects. In that case, it would fail to be an internal logic in the sense (ILC 2). Moreover, an internal logic based on (A${}_S\Omega$) is fully internal in the sense (ILC 1), unlike (AS): The collection of subobjects of an object constitutes an actual object of the category due to the presence of exponentials. Then (A${}_S\Omega$) and not (AS) would be the internal logic of a topos.

Let me say a word on what does it mean to say that reasoning about a category with a logic different to the internal one would alter the identity of the objects and morphisms of the category. Suppose that the internal logic of ${}_S S^{\downarrow\downarrow}$ was other than an intuitionistic logic, say, classical logic. If $Fx \vee \neg Fx$ were to hold for all x (and F) in ${}_S S^{\downarrow\downarrow}$, then there would be some x such that would not be an object of ${}_S S^{\downarrow\downarrow}$, and so this law would not be a law for the objects of ${}_S S^{\downarrow\downarrow}$. Let me make this clearer by means of an example. Consider a graph G and a part $S \rightarrowtail G$ of it, the S's. There is another part $N \rightarrowtail G$ such that its (generalized) elements are those which are completely out $S \rightarrowtail G$, the *not* S's. Consider a (generalized) element E of G consisting of an arrow and its two vertexes, such that one of its vertexes (a graph on its own) is in $S \rightarrowtail G$ and the other is in $N \rightarrowtail G$. Requiring that the S's and the *not* S's be exhaustive with an empty intersection would lead to "break up" E, since having the arrow either in $S \rightarrowtail G$ or in $N \rightarrowtail G$ would alter the identity of E: No matter where the arrow stays at the end, there would be something in those subobjects that it is not an object of ${}_S S^{\downarrow\downarrow}$, namely an arrow without one of its vertexes. Thus, $Fx \vee \neg Fx$ does not hold in general for the objects of ${}_S S^{\downarrow\downarrow}$.

[14]The difference is clearly explained in [18, Sect. 7.3].

[15]For more on this, see Sect. 18.3.

18.2.5 Intuitionistic Logic and Variable Sets

Colin McLarty has addressed the question of why topos logic is so close to intuitionistic logic (cf. [36–38]). His favorite answer is that of Lawvere, namely that if toposes are regarded as universes of variable sets then it is not surprising that intuitionistic logic is their internal logic, for Kripke models for intuitionistic logic can be adapted to toposes in a way that its objects can be considered set varying over a preorder (see [24, 25]). Moreover, for Lawvere such logic is "objective," in the sense that it reflects how variable sets are and thus establishes how to reason about them. We have then another slogan grounded on some mathematical results:

(S4) Intuitionistic logic is the objective logic of variable sets.

This is by far the most difficult slogan in terms of exposition, so I have decided to set up the necessary machinery to motivate it, although not for explaining it in all exactness, by employing a strategy halfway between the use of powerful metaphors (very common among those who reproduce the slogan) and enough technical detailing (albeit by no means exhaustive) of the Kripke semantics to ground those metaphors. First, I quickly expound what Lawvere understands by 'objective logic' and 'variable set' in "intuitionistic logic is the objective logic of variable sets." Then I study a special (standard) unary connective called Lawvere–Tierney topology, which is central to several topos-theoretical constructions and has important consequences for the study of intuitionistic logic and some of its modal extensions that are expounded after that.

Lawvere distinguishes (more or less following Hegel) between (i) the connections between (conceptual) universes, their objects (concepts) and their parts, on one hand, and (ii) the (inference) relations between statements (see [28, p. 43f]). According to Lawvere, the former are "structures" and "invariant mathematical content" which are "objective," whereas the latter are "presentations" which "reflect" that content and may vary. (i) would be the "objective logic" (of the connections between universes, objects, and parts) and (ii) the "subjective logic" (of the reflex of those connections).[16]

[16]In Hegel's *Science of Logic*, logic is divided into two parts, one of which is the logic of Being, the objective logic. Being is thought of here as an outer world beyond any particular subjective mind but still conceptually informed (and thus with a logic): It is "objectified spirit." The other part is the logic of thinking, the subjective logic (Hegel calls it thus because thinking requires a thinking subject, more or less like a subjective right is the right of a subject). Hegel's subjective logic is what today (and also in Hegel's times) is commonly called 'logic.' An excellent treatment of the connections between Lawvere's and Hegel's views on logic can be found in [47]. Further clarification of the term 'objective' as applied to logic by Lawvere comes also from Hegel, but not from his work in logic but from his work on ethics, especially in *Philosophy of Right* (intended to be read with his *Science of Logic* as background), which is essentially a developed version of the section "Objective Spirit" in the *Encyclopaedia of the Philosophical Sciences*' *Philosophy of Spirit*. According to Hegel, the objective spirit consists of collective, social practices, whereas the subjective spirit is the individual. Part of Lawvere's "historical and dialectical realism" is that mathematical entities exist objectively, like being in the *Logic*, but that existence is determined in mathematical experience as a whole, including its collective practice (like the objective spirit)—not in a Platonic realm independent of any subject, nor in merely subjective, individual, experience.

Sometimes, presentations introduce noninvariant, nonobjective content in the sense that mathematical practice strongly suggests that such content does not belong to the given universes, their objects, and their parts. For example, taking membership as primitive in the study of sets and functions introduces some oddities like debates over whether the members of the natural number 5 are 0, 1, 2, 3, 4, or not. One may wonder whether categorial mathematics does not fail too in studying merely mathematical content. It might be so but, as in the case of foundations, neither categorial theories nor methods are committed to fixedness on this issue and in a sense they foresee their own improvement.

On its side, the term 'variable set' is sometimes explained in a way which might be not clear enough for those unfamiliar with topos theory, namely in terms of sheaves. One of the best and most concise characterizations for those acquainted with the notions of this paper is the following one:

> (…) while non-standard analysis, the forcing method in set theory, and Kripke semantics all involved (…) sets varying along a poset **X**, it was [the work of several algebraic geometers] who, by developing topos theory, made the qualitative leap (…) to consideration of sets varying along a small category **X** and at the same time emphasized that the fundamental object of study is the whole category of sets so varying. Those insisting on formal definitions may thus (…) consider that "variable set" simply means an object in some (elementary) topos (…). [25, p. 102]

A neat example of what Lawvere means is the (standard) category of functions (between sets) ${}_S\mathbf{Set}^{\rightarrow}$ from Example 18.2.2 above. In an object of this category, $f : S_t \longrightarrow S_n$, S_t can be considered the "state" of the variable set f at stage t and S_n as its state at a "later" stage n. The set may be thought of having undergone, via f, a change from what it was at t to what it is at n. Any element x of S_t, that is, of f at t, becomes the element $f(x)$ of S_t at n. Pursuing this informal description, and because f is a morphism, two elements at t may become one at n (unless f is monic), or a new element may arise at n, but no element at t can split into two or more at n or vanish altogether. These latter options would be allowed and variation would be more complex if f were a relation. Let make more precise the connection between intuitionistic logic and variation.

Let $\mathbf{P} = (P, R)$ be a poset. A set $A \subseteq P$ is *hereditary* in $\mathbf{P}$ if for all x, y, if $x \in A$ and xRy then $y \in A$, i.e., if it is closed "upwards" under R. Let $\mathbf{P}^h$ be the collection of hereditary subsets of $\mathbf{P}$. A $\mathbf{P}$-*valuation* is a function $\mathfrak{V} : F \longrightarrow \mathbf{P}^h$, where F stands for a collection of formulas on a usual zero-order language, such that to each p_i is assigned an hereditary subset $\mathfrak{V}(p_i) \subseteq P$.

A *model based on* $\mathbf{P}$ is a triple $\mathfrak{M} = \langle P, R, \mathfrak{V} \rangle$, where $\mathfrak{V}$ is a $\mathbf{P}$-valuation. The idea that A is *true at* x, in symbols $\mathfrak{M} \Vdash_x A$, is defined recursively as follows[17]:

For every $x \in P$:

$\mathfrak{M}, x \Vdash_t A$ if and only if $x \in \mathfrak{V}(p_i)$;
$\mathfrak{M}, x \Vdash_t \neg A$ if and only if for all y with xRy, $\mathfrak{M}, y \nVdash_x A$;
$\mathfrak{M}, x \Vdash_t A \wedge B$ if and only if $\mathfrak{M}, x \Vdash_t A$ and $\mathfrak{M}, x \Vdash_t B$;

[17] For simplicity I will omit quantifiers here and in the following section.

$\mathfrak{M}, x \Vdash_t A \vee B$ if and only if either $\mathfrak{M}, x \Vdash_t A$ or $\mathfrak{M}, x \Vdash_t B$;
$\mathfrak{M}, x \Vdash_t A \supset B$ if and only if for all y with xRy, if $\mathfrak{M}, y \Vdash_t A$ then $\mathfrak{M}, y \Vdash_t B$.

A is *true in a model* $\mathfrak{M}$, denoted $\mathfrak{M} \Vdash A$, if A is true at every $x \in P$ in $\mathfrak{M}$. A formula A is *valid on* **P**, denoted $\mathbf{P} \Vdash A$, when it is true in every model based on **P**.

Informally, **P** represents a collection of states of, say, possible mathematical knowledge. The truth value of a proposition depends on the knowledge established at a certain state, so their truth value is relative to a certain state. However, once a proposition is true at a state, it is true at all later states. This conveys the idea that once a proposition is established as true in a given evidential situation then it remains true. One state comes after, or is later than, another, but the order is not necessarily linear because the states are not merely of actual but of possible mathematical knowledge, so immediately later than a state of knowledge may come several other states, each with a different value for a certain proposition: Due to the persistence of truth, if a proposition is true at a state later than a certain state and false at another later state, these later states cannot be connected. The logic valid for this variation on states happens to be intuitionistic logic.

For example, take $\mathbf{P} = (\{0, 1\}, \leq)$ (with $0 \leq 1$ as usual) and $\mathfrak{V}(p) = \{1\}$, which is hereditary. Then with $\mathfrak{M} = \langle \{0, 1\}, \leq, \mathfrak{V} \rangle$ one has that $\mathfrak{M}, 0 \nVdash_t p$. But $\mathfrak{M}, 1 \Vdash_t p$ and $0 \leq 1$, so $\mathfrak{M}, 0 \nVdash_t \neg p$. Thus, $\mathfrak{M}, 0 \nVdash_t p \vee \neg p$, so $p \vee \neg p$ is not valid on this **P**.

Operations of ${}_S\Omega_{\mathcal{E}}$ with the order ${}_{S\mathcal{E}} \leq$ between them constitute a frame to provide a categorial version of the above.[18] What the **Comprehension axiom** says is that ${}_S\varphi(x) = {}_Strue_X$ if and only if $x \in m$, i.e., it specifies when a generalized element x belongs to a part m of O (that part of the O's which satisfy φ, described as $\{y \mid {}_S\varphi(y)\}$). As above, these specifications depend on the way the formula ${}_S\varphi(x)$ is built up from connectives and are formulated in terms of the relation "X makes ${}_S\varphi(x)$ true," or also "X forces $\varphi(x)$," written $X \Vdash_{ST} \varphi(x)$ and defined as follows:

For every $x : X \longrightarrow O$ with its image $Imx \in_S \Omega_{\mathcal{E}}^X$:
$X \Vdash_{ST} {}_S\varphi(x)$ if and only if $x \in m = \{y \mid \varphi(y)\}$

that is
$X \Vdash_{ST} {}_S\varphi(x)$ if and only if ${}_S\varphi(x) = {}_Strue_{\mathcal{E}}$

For brevity, 'p_X' will denote ${}_S\varphi(x)$, i.e., a (generalized) proposition about a certain (generalized) element x of an object O; 'q_X' will be then ${}_S\psi(x)$, i.e., another (generalized) proposition about x. Accordingly, 'p_{XY}' and 'q_{XY}' stand for $\varphi\,(x \circ y) : Y \longrightarrow X \longrightarrow O \longrightarrow {}_S\Omega_{\mathcal{E}}$ and $\psi\,(x \circ y) : Y \longrightarrow X \longrightarrow O \longrightarrow {}_S\Omega_{\mathcal{E}}$, respectively.

$X \Vdash_{ST} \neg p_X$ if and only if, for every morphism $y : Y \longrightarrow X$ such that $Y \Vdash_{ST} p_{XY}$, $Y \cong \mathbf{0}$;
$X \Vdash_{ST} (p_X \wedge q_X)$ if and only if $X \Vdash_{ST} p_X$ and $X \Vdash_{ST} q_X$;
$X \Vdash_{ST} (p_X \vee q_X)$ if and only if there are morphisms $y : Y \longrightarrow X$ and $w : W \longrightarrow X$ such that $y + w : Y + W \twoheadrightarrow X$ is an epimorphism while both $Y \Vdash_{ST} p_{XY}$ and $W \Vdash_{ST} q_{XW}$;
$X \Vdash_{ST} (p_X \supset q_X)$ if and only if, for every morphism $y : Y \longrightarrow X$ such that $Y \Vdash_{ST} p_{XY}$, also $Y \Vdash_{ST} q_{XY}$.

[18]For more details on what follows, see [31].

A is *true in a (standard) topos* ${}_S\mathcal{E}$, denoted $\Vdash_{{}_S\mathcal{E}} A$, if $A =_S true_{\mathcal{E}}$ at every x in ${}_S\mathcal{E}$. A formula A is *(standardly) valid*, denoted $\Vdash_{ST} A$, when it is true in every (standard) topos. As I mentioned before, the theorems of intuitionistic logic are those standardly valid propositions; it would be then the logic holding across all the variations on objects, properties of objects and elements of objects.

Let me conclude this survey of standard topos logic Kripke-style describing a Lawvere–Tierney topology. Define a *standard local operator* on an algebra $\mathfrak{A}$ with meets ($\cap$) as any operation j that is
multiplicative: $j(x \cap y) = jx \cap jy$
idempotent: $j(jx) = jx$
inflationary: $x \leq jx$
and call the pair $(\mathfrak{A}, j)$ a *standard local algebra*. In the present context $\mathfrak{A}$ can be restricted to Heyting algebras.

Macnab [32, 33] showed that standard local operators on a Heyting algebra can be alternatively defined using relative pseudocomplement $\rightarrow$ by the single equation

$$(x \rightarrow jy) = (jx \rightarrow jy)$$

A *(standard) Lawvere–Tierney topology* is a morphism $j :_S \Omega \longrightarrow {}_S\Omega$ in a standard topos ${}_S\mathcal{E}$ such that
$(j1)$ $j \circ \wedge = \wedge \circ (j \times j)$;
$(j2)$ $j \circ j = j$;
$(j3)$ $j \circ {}_S true = {}_S true$;
$(j1)$ and $(j2)$ say that j is multiplicative and idempotent. It can be proved that
$(j4)$ $p_S \leq j \circ p$
so j is also inflationary and hence a standard Lawvere–Tierney topology is a standard local operator. Following Macnab's result, j can be defined using the conditional by the single equation

$$\supset \circ \langle p, j \circ q \rangle = \supset \circ \langle j \circ p, j \circ q \rangle;$$

Let $J \rightarrowtail {}_S\Omega$ be the subobject classified by j. Every subobject $f : P \rightarrowtail X$ of any object X has a *j-closure*, defined as the subobject classified by $j \circ_S \varphi_f$ (where ${}_S\varphi_f$ classifies f), denoted '$\bar{f} : \bar{P} \rightarrowtail X$.'

A subobject $f : P \rightarrowtail X$ is called *j-dense* if its j-closure is all of X, that is, if $\bar{f} = id_X$.

For any standard topos ${}_S\mathcal{E}$ and standard Lawvere–Tierney topology on it, an object S of ${}_S\mathcal{E}$ is a *sheaf* for j if and only if for every object X and j-dense subobject $f : P \rightarrowtail X$ and every $g : P \longrightarrow S$, there is a unique $h : X \longrightarrow S$ with $g = h \circ f$. Intuitively, since f is j-dense, it is j-true that f is all of X and a sheaf for j, S, is an object that "sees" f as all of X, and hence that a morphism from P to S completely determines a morphism from X to S. The pair $({}_S\mathcal{E}, j)$ is called a *(standard) elementary site*.

Putting $j = \neg \circ \neg$ defines a topology on any standard topos ${}_S\mathcal{E}$ for which the associated standard topos ${}_S\mathcal{E}_{\neg\neg}$ of sheaves is a model of classical logic.[19]

Let ${}_S\mathcal{E}$ be $\mathbf{Set}^{\mathbf{X}}$ ($\mathbf{X} = (X, R)$). In the case that $\mathbf{X}$ is an appropriate set of "forcing conditions," the topos ${}_S\mathcal{E}_{\neg\neg}$ of "double-negation sheaves" becomes a model showing that the continuum hypothesis (for example) is independent of the axioms for topos theory including classical logic (see [49]).

If $j : {}_S\Omega \longrightarrow {}_S\Omega$ is a standard Lawvere–Tierney topology on a standard topos ${}_S\mathcal{E}$, then the site $({}_S\mathcal{E}, j)$ can be used to interpret modal formulas as truth values $\mathbf{1} \longrightarrow {}_S\Omega$ in ${}_S\mathcal{E}$. The morphism j induces a local operator on the Heyting algebra ${}_S\Omega$ of truth values in ${}_S\mathcal{E}$. If a formula is satisfied by the resulting local algebra then it is said to be valid in the site $({}_S\mathcal{E}, j)$. The modal propositions that are valid in all (standard) sites correspond to intuitionistic logic supplemented with the corresponding axioms for the modality j (see [17]), for example intuitionistic logic plus $(p \supset \nabla q) \Leftrightarrow (\nabla p \supset \nabla q)$ (where ∇ is the counterpart of j outside toposes).

Lawvere [23] has suggested reading the topology j as a geometric modality saying "it is cofinally the case that." Given two subsets X and Y of a partially ordered set, Y is *cofinal with* X if and only if for all $x \in X$ there is a $y \in Y$ such that $x \leq y$. Actually, the Kripke modeling sketched above implies that $\mathfrak{M}, x \Vdash_t \neg\neg A$ if and only if A is cofinal with $\{y : x \leq y\}$ and remember that the double negation is a case of a Lawvere–Tierney topology j.

18.2.6 Proof Theory for Topos Logic

Sound rules of inference can be given to characterize topos logic. A *sequent* is an expression $\Gamma : p$, where Γ is a finite (possibly empty) set of formulas and p is a formula. A sequent is *true* if and only if Γ does imply p. When a sequent $\Gamma : p$ is true we write

$$\Gamma \vdash p$$

(for example, one has $\neg\neg p \vdash p$ in classical logic, but not in intuitionistic logic, and one has $p, \neg p \vdash q$ in both classical and intuitionistic logic, but not in an inconsistency-tolerant logic).

First are the *structural rules*. From the sequent(s) above the line one can infer the one below. An asterisk shows that the sequent below follows from an empty set of assumptions:

$$\text{Trivial sequent:} \quad \frac{*}{p : p}$$

[19]These constructions are mathematical manifestations of the double-negation translation of classical zero-order logic into intuitionistic zero-order logic which works by inserting $\neg\neg$ in relevant places. See the first section of [9] for more about these translations.

True and false: $\dfrac{*}{:\ true}$ $\dfrac{*}{false\ :\ p}$

Thinning: $\dfrac{\Gamma\ :\ p}{\Gamma,\ q\ :\ p}$

Cut: $\dfrac{\Gamma,\ q\ :p\ \text{ and }\ \Gamma\ :q}{\Gamma\ :\ p}$

There is one reversible *connective rule* for each connective. From the sequent(s) above the double line one can infer the sequent below, and from the one below one can infer the either of the two above:

$$\frac{\Gamma, p\ :\ false}{\Gamma\ :\ \neg p}$$

$$\frac{\Gamma\ :p\ \text{ and }\ \Gamma\ :q}{\Gamma\ :\ p \wedge q}$$

$$\frac{\Gamma, p\ :\theta\ \text{ and }\ \Gamma, q\ :\theta}{\Gamma,\ p \vee q\ :\ \theta}$$

$$\frac{\Gamma, p\ :\ q}{\Gamma\ :\ p \supset q}$$

and if the variable x is not free in Γ or q:

$$\frac{\Gamma\ :\ p}{\Gamma\ :\ (\forall x)p} \qquad \frac{\Gamma, p\ :\ q}{\Gamma, (\exists x)p\ :\ q}$$

It is worth noting that the structural rule for $false$ and the connective rules for negation and disjunction are derived, just as the corresponding morphisms $false$, $\neg$ and $\vee$ needed the other ones in order to be defined. The proof is straightforward; see [37, Chap. 15].

Colin McLarty has rightly pointed out that the internal logic coincides with no intuitionistic logic studied before toposes (cf. [37, p. vii] and [36, pp. 153ff]). The internal logic strikingly resembles intuitionistic logic, indeed there is no difference at the zero-order level. Differences lie at the higher-order level, where traditional intuitionistic principles like the existence property ($\exists x Fx$ is accepted only if for some constant c Fc is accepted) or the disjunction property (accept a disjunctive statement $p \vee q$ only if either p or q is accepted) do not hold.

18.3 Complement-Toposes

18.3.1 A Categorial Approach to Inconsistency: Bi-Heyting Toposes

A *bi-Heyting algebra* is a distributive lattice which is both a Heyting algebra and a Brouwerian algebra (the dual of a Heyting algebra, also called 'co-Heyting algebra'). Clearly, a Boolean algebra is a bi-Heyting algebra. Let $c()$ be the operation of Boolean complement. Define then $a \rightarrow b = c(a) \bigvee b$ and $a - b = a \bigwedge c(b)$. In this case $-a = \rceil a = c(a)$.

A *bi-Heyting topos* is a standard topos for which the algebra of subobjects of any object is a Brouwerian algebra. Since the algebra of subobjects of any object in a standard topos is a Heyting algebra, a bi-Heyting topos can be defined as a standard topos for which the algebra of subobjects of any object is bi-Heyting.

There might be objects in a standard topos whose algebra of subobjects is a co-Heyting algebra. In [46], following the work in [27], some examples in the category ${}_S S^{\downarrow\downarrow}$ are given. This is the closest one which will get paraconsistency in standard toposes, though. The internal logic of a bi-Heyting topos is never dual to an intuitionistic or superintuitionistic logic. Remember that the internal logic of a topos is determined by the algebra of ${}_S\Omega$ and the connectives, not by the algebra of its subobjects, and it is a co-Heyting algebra only if it is a Boolean algebra. This is assured by the following theorems:

Theorem 18.3.1 *Let* $\delta : {}_S\Omega \longrightarrow {}_S\Omega$ *a morphism such that* $\delta \leq id_{{}_S\Omega}$ *and* $\delta \circ {}_S true = {}_S true$. *Then* $\delta = id_{{}_S\Omega}$.

(This is Corollary 1.12 in [44] or Proposition 4.1 in [46], where a proof is given.)

Theorem 18.3.2 *In any topos* ${}_S\mathcal{E}$ *the following conditions are equivalent:*

(a) ${}_S\mathcal{E}$ *is Boolean.*
(b) $\neg \circ \neg = id_{{}_S\Omega}$.

(This is proved as Theorem 7.3.1 in [18].)

Theorem 18.3.3 *If* ${}_S\mathcal{E}$ *is Boolean, then its internal logic is classical.*

(This is proved as Theorem 7.4.1 in [18].)

However, these results rely heavily on the *standard* character of a topos, i.e., on a particular description of its categorial structure. In what follows I will show that the same categorial structure can be described in an alternative, coherent way, such that the internal logic of a topos can also be described as dual intuitionistic or paraconsistent.

18.3.2 Introducing Complement-Toposes

Mortensen's argument for developing an inconsistency-tolerant approach to category theory is that every topological space gives a topos (the category of pre-sheaves on the space), mathematically

> (...) specifying a topological space by its closed sets is as natural as specifying it by its open sets. So it would seem odd that topos theory should be associated with open sets rather than closed sets. Yet this is what would be the case if open set logic were the natural propositional logic of toposes. At any rate, there should be a simple 'topological' transformation of the theory of toposes, which stands to closed sets and their logic [i.e., inconsistency-tolerant], as topos theory does to open sets and intuitionism. [39, p. 102]

If the duality between intuitionistic logic and CSL is as deep as topological, then a representation of CSL as the internal logic of a topos should be equally natural. So Mortensen's remark amounts to this: The same categorial structure described as supporting intuitionistic logic should also be describable as supporting inconsistency-tolerance. Note that the crucial motivation is topological, and does not turn on paraconsistent ideology (even though Mortensen subscribes to the latter). In what follows I expound Mortensen and Lavers's dualization of logical connectives in a topos.[20]

Think of the objects of complement-toposes as the objects of standard toposes in Sect. 18.2 and retain the definition of propositional functions. It will be assumed that there is a proposition $false\colon \mathbf{1} \longrightarrow \Omega$. This assumption will obligate certain names for other morphisms. "${}_D\Omega$" will denote this initial assumption about the name of a certain morphism with codomain Ω ("$false$" in this case) and from now on, ${}_Df$ will denote that there is a monomorphism from ${}_D\Omega$ to the codomain of f and ${}_D\mathcal{E}$ that the morphisms with codomain the object of propositions of $\mathcal{E}$ receive their names according to this initial assumption, which is an alternative Skolemization for the purely equational structure of a topos.

---

[20]It is important to set their individual contributions. Of the ten diagrams in [39, Chap. 11], Mortensen drew the first one and the final five, while Lavers drew the remaining four. The diagram for the dual-conditional never was explicitly drawn, but it was discussed in [39, p. 109]. The full story, as told by Mortensen in personal communication, is as follows. Mortensen gave a talk at the Australian National University (Canberra) in late 1986, on paraconsistent topos logic, arguing the topological motivation for closed set logic. He defined a complement-topos, drew the first three diagrams from *Inconsistent Mathematics*, Chap. 11, that is including the complement versions of $strue$ and paraconsistent negation, and criticized Goodman's views on the conditional. But it was not seen clearly at that stage how the logic would turn out. Peter Lavers was present (also Richard Routley, Robert K. Meyer, Michael A. McRobbie, Chris Brink and others). For a couple of days in Canberra, Mortensen and Lavers tried without success to thrash it out. Mortensen returned home to Adelaide and two weeks later Lavers' letter arrived in Adelaide, in which he stressed that inverting the order is the key insight to understanding the problem, drew the diagrams for conjunction and disjunction, and pointed out that subtraction is the right topological dual for the conditional. Mortensen then responded with the four diagrams for the $S5$ conditional, and one for quantification (last five diagrams in *Inconsistent Mathematics*, Chap. 11). A few months later (1987) Mortensen wrote the first paper, with Lavers as co-author, and sent it to Saunders Mac Lane and William Lawvere (also Routley, Meyer, Priest). Mac Lane replied but Lawvere did not. A later version of that paper became the eleventh chapter of *Inconsistent Mathematics*. I thank Prof. Mortensen for providing me this information.

Then, for any object X in a complement-topos, the composite ${}_Dfalse \circ !_X : X \longrightarrow \mathbf{1} \longrightarrow {}_D\Omega$ denotes a constant, ${}_D$false-valued propositional function on X, abbreviated to ${}_Dfalse_X$. Propositional functions will specify subobjects as follows. Given a propositional function ${}_D\varphi : X \longrightarrow {}_D\Omega$, one gets the part of the x's of which ${}_D\varphi$ is false, if any, as an equalizer $m : M \rightarrowtail X$ of ${}_D\varphi$ and ${}_Dfalse_X$. This subobject will be named the *anti-extension* of the propositional function ${}_D\varphi$. A morphism ${}_Dfalse : \mathbf{1} \longrightarrow {}_D\Omega$, called *dual classifier*, has the following property:

Anti-comprehension axiom. For each ${}_D\varphi : O \longrightarrow {}_D\Omega$ there is an equalizer of ${}_D\varphi$ and ${}_Dfalse_O$, and each monic $m : M \rightarrowtail O$ is such an equalizer for a unique ${}_D\varphi$. In diagrams, ${}_Dfalse$ is such that for every ${}_D\varphi$ and every object T and morphism $o : T \longrightarrow O$, if $m \circ {}_D\varphi = m \circ {}_Dfalse_O$ and $x \circ_D \varphi = x \circ {}_Dfalse_O$, then there is a unique $h : X \longrightarrow M$ that makes the diagram below commutative:

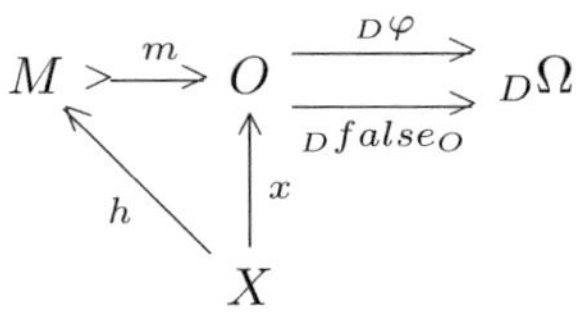

The propositional function ${}_D\varphi$ is also called "the dual characteristic (or classifying) morphism of m," denoted ${}_D\varphi_m$ for more convenience. A dual classifier is unique up to isomorphism and so is ${}_D\varphi_m$. Now a complement-topos can be defined more precisely: A category ${}_D\mathcal{E}$ with equalizers, (binary) products, coequalizers, coproducts, exponentials, and a dual classifier is called *elementary complement-topos*.[21]

The connection of these definitions with more traditional logical notions is much less mysterious than it might appear at first sight. Consider the diagram in the definition of an equalizer:

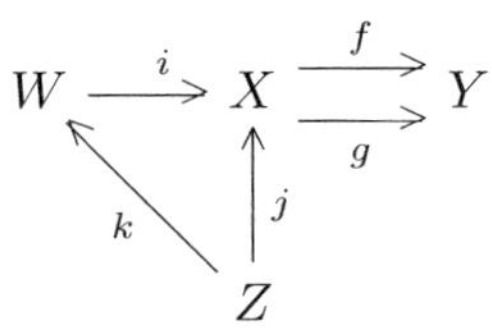

As a particular case for the **Anti-comprehension axiom** one has:

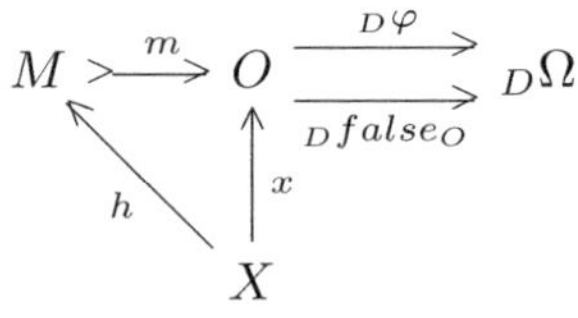

[21]Mortensen and Lavers use the names *complement-classifier* and *complement-topos*, which are now the names set in the literature (cf. [13, 39, 40, 52]). Although the use of 'dual topos' would be appropriate here but misleading given the usual understanding of 'dual category,' I think there is no similar problem for the classifier of a complement-topos.

The only morphism from X to ${}_D\Omega$ such that makes the diagram above commutative is ${}_D false_X$:

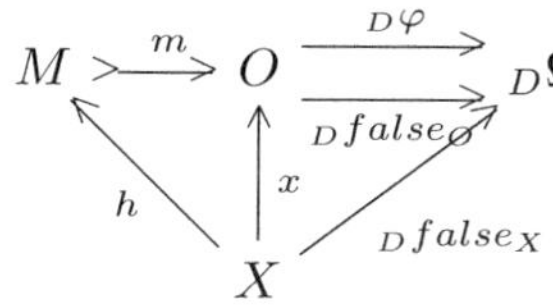

Thus, the following diagram is obtained:

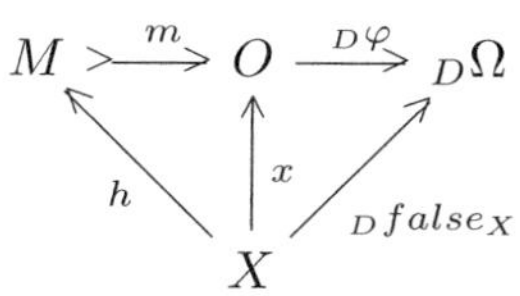

Note that, according to the definition of an equalizer, h must be the only morphism that, among other properties, $x = m \circ h$. But this suffices to satisfy the definition of $x \in m$. Hence, what **Anti-comprehension** states is that ${}_D\varphi(x) = {}_D false_X$ (because of the right commutative triangle) if and only if $x \in m$ (because of the left commutative diagram). It is nice and expectable that ${}_D\varphi(x) = {}_D false_X$ if and only if $x \in m$, since M is (thought of as) the anti-extension of ${}_D\varphi$ and the elements belonging to M are those which would make ${}_D\varphi$ false.

Let me make explicit the following two very important properties of ${}_D\Omega$ and ${}_D false : \mathbf{1} \longrightarrow {}_D\Omega$:

1. If X is an object of ${}_D\mathcal{E}$ and $m : M \rightarrowtail X$ is a subobject of X, then there is exactly one morphism ${}_D\varphi_m : X \longrightarrow {}_D\Omega$ such that for every $x : \mathbf{1} \longrightarrow X$, $x \in m$ if and only if ${}_D\varphi_m \circ x = {}_D false$.
   (Succinctly, a dual classifier says, for every object X and every subobject m of X, what elements of X are included in the subobject m.)
2. If a morphism f has ${}_D\Omega$ as codomain, then it is the dual characteristic morphism of some other morphism g such that its codomain is the domain of f.
   (A morphism to ${}_D\Omega$ is fully determined by the part of its domain that it takes to ${}_D false$, that is, by the subobject of its domain that it classifies dually.)

Given the notion of dual classifier, one can define also ${}_D true : \mathbf{1} \longrightarrow {}_D\Omega$ as the dual character of 0_1, the only morphism from an initial object to a terminal one:

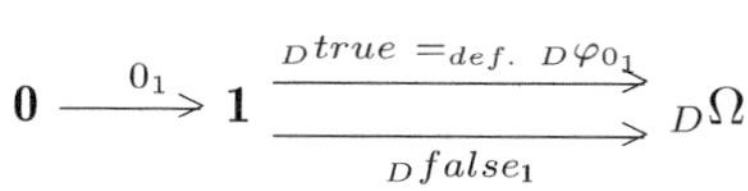

The diagram above implies that ${}_D true = {}_D false_X$ (because of the right commutative triangle) if and only if $\mathbf{0}$ has a generalized element (because of the left commutative diagram). $\mathbf{0}$ has non-generalized elements only in a degenerate category, so only then one obtains ${}_D true = {}_D false$.[22] The situation for nondegenerate categories can be easily dualized.

Finally, it is well-known that the subobjects of a given object form a partial order. In particular, the elements of ${}_D\Omega$ form a partial order, which means that propositions form a partial order, i.e., for every propositions p, q and r:

- $p \geq p$
- If $p \geq q$ and $q \geq r$ then $p \geq r$
- If $p \geq q$ and $q \geq p$ then $p = q$

Mutatis mutandis, one can use $\leq$ instead of $\geq$. Such order relation can be interpreted as a deducibility relation, $\vdash$, and the properties above say that deducibility is reflexive, transitive and that interdeducible propositions are equivalent.

n-ary connectives of order m are defined as in Sect. 18.2.2. [23] This enables us to define connectives dual to the standard ones, namely three binary ($\curlyvee$, disjunction; $\curlywedge$, conjunction; $-$, subtraction) and three unary ($\sim$, negation; E, particular quantifier; A, universal quantifier). The definitions go as follows.

Negation. Let be ${}_D false : \mathbf{1} \longrightarrow {}_D\Omega$. Then $\sim : {}_D\Omega \longrightarrow {}_D\Omega$ is the dual characteristic morphism of ${}_D false$:

$$\mathbf{1} \xrightarrow{{}_D true} {}_D\Omega \underset{{}_D false_{{}_D\Omega}}{\overset{\sim \,=_{def.}\, {}_D\varphi_{{}_D false}}{\rightrightarrows}} {}_D\Omega$$

The full truth condition implied by this definition is that $\sim \circ p = {}_D false$ if and only if $p = {}_D true$ and $\sim \circ p = {}_D true$ otherwise.

Disjunction. Disjunction $\curlyvee : {}_D\Omega \times {}_D\Omega \longrightarrow {}_D\Omega$ is defined as the dual characteristic morphism of $\langle {}_D false, {}_D false \rangle : \mathbf{1} \longrightarrow {}_D\Omega \times {}_D\Omega$:

$$\mathbf{1} \xrightarrow{\langle {}_D false,\, {}_D false \rangle} {}_D\Omega \times {}_D\Omega \underset{{}_D false_{{}_D\Omega \times {}_D\Omega}}{\overset{\curlyvee \,=_{def.}\, {}_D\varphi_{\langle {}_D false,\, {}_D false \rangle}}{\rightrightarrows}} {}_D\Omega$$

The complete truth condition implied by this definition is that $p \curlyvee q = \sup(p, q)$ (with respect to the partial order formed by the elements of ${}_D\Omega$).

Conjunction. Conjunction $\curlywedge : {}_D\Omega \times {}_D\Omega \longrightarrow {}_D\Omega$ is defined as the dual characteristic morphism of the image of $[\langle {}_D false,\ id_{{}_D\Omega} \rangle,\ \langle id_{{}_D\Omega}, {}_D false \rangle]$:

$${}_D\Omega + {}_D\Omega \xrightarrow{Im[\langle {}_D false,\, id_{{}_D\Omega} \rangle, \langle id_{{}_D\Omega},\, {}_D false \rangle]} {}_D\Omega \times {}_D\Omega \underset{{}_D false_{{}_D\Omega \times {}_D\Omega}}{\overset{\curlywedge \,=_{def.}\, {}_D\varphi_{Im[\langle {}_D false,\, id_{{}_D\Omega} \rangle,\, \langle id_{{}_D\Omega},\, {}_D false \rangle]}}{\rightrightarrows}} {}_D\Omega$$

The truth condition implied by this definition is that $p \curlywedge q = \inf(p, q)$ (again with respect to the partial order formed by the elements of ${}_D\Omega$).

---

[22] The same remark on note 7 in Sect. 18.2 applies here.

[23] Again, for brevity I often will talk only of "connectives," since their dual character can be obviated, and their arity and order will be made explicit only when needed.

Dual-conditional, *subtraction* or *pseudo-difference*. Subtraction $- : {}_D\Omega \times {}_D\Omega \longrightarrow {}_D\Omega$ is defined as the dual characteristic morphism of $\bar{e} : {}_D\geq \longrightarrow {}_D\Omega \times {}_D\Omega$, where $\bar{e}$ is the equalizer of $\curlyvee : {}_D\Omega \times {}_D\Omega \longrightarrow {}_D\Omega$ and the first projection ${}_D p_1$, so it makes the following diagram a pullback:

$$ {}_D\geq \overset{\bar{e}}{\rightarrowtail} {}_D\Omega \times {}_D\Omega \underset{{}_D false_{{}_D\Omega \times {}_D\Omega}}{\overset{- =_{def.}\ {}_D\varphi_{\bar{e}}}{\rightrightarrows}} {}_D\Omega $$

Following Mortensen and Lavers' method of dualization, this is the right complement-topos connective corresponding to the conditional in standard toposes. Remember that $(p \supset q) = {}_S true$ if and only if $(p \wedge q) = p$, and it is dualized $(q - p) = {}_D false$ if and only if $(p \curlyvee q) = p$, which is expressed by the equalizer above.

An immediate consequence of this definition is that $(q - p) =_D false$ if and only if $(p \curlyvee q) = p$ (as can be noted, the equalizer $\bar{e}$ expresses the condition on the right: It equals disjunction and the first projection). The complete truth condition is that $(q - p) = {}_D false$ if and only if $p_D \geq q$ and $(q - p) = q$ otherwise.

The dualization of the conditional is a delicate matter, though. To begin with, the conditional in usual topos theory may be defined in several ways, for example by considering it the characteristic morphism of $e' : \leq \longrightarrow \Omega \times \Omega$, the equalizer of disjunction and the second projection, which would lead to several different dualizations. Goodman [19] proved that in CSL no connective definable in terms of the connectives $\curlyvee$, $\curlywedge$, $\sim$, $-$ has $\top$ as semantic assignment if and only if the assignment of its antecedent is less or equal than the assignment of its consequent. It can be argued that a connective like $-$ cannot be regarded as a conditional at all, since $(p - p) = false$ for every p and does not satisfy *modus ponens*, that CSL lacks of a reasonable conditional and therefore it is not a serious logic and much less a logic strong enough for developing some mathematics based on it. Certainly, $-$ might not be regarded as a conditional, in the same way that the dual of conjunction is not even a kind of conjunction. $-$ should be regarded rather as an "anti-implication," as Popper once suggested (cf. [41]).

Mortensen has argued against this alleged deficiency of CSL. He points out that it is not clear how much of mathematics depends on an object-language conditional. What mathematics needs, he says, is a deducibility relation, but that is provided by ordering and an adequate proof theory; Goodman himself proved that derivability in CSL respects the natural semantic ordering of set inclusion.[24] Moreover, nothing in the above rules out the possibility of defining a reasonable conditional in CSL or in complement-topos theory. That a conditional cannot be defined in terms of the other connectives is not a strong argument; after all connectives in, e.g., intuitionistic logic, are not interdefinable and it is not thought of as defective. Mortensen proposed

[24]This is a controversial point. Mortensen thinks that functionality is mathematically prior to, and a more important matter than some logical notions. Someone might object to this by saying that ordinary math books use conditional constantly, and for example use definitions stated as conditionals so that one constantly quantifies into conditional contexts, and every time one proves that some object does not have a defined property one is negating a conditional. Seemingly, this cannot all be pushed into the metalanguage without severe contortions.

a conditional for complement-toposes which, however, should not be regarded necessarily as a dualization of usual conditional, but a more general case. I will discuss it in the following sections.

Particular quantifier. Particular quantifier $E_X : {}_D\Omega^X \longrightarrow {}_D\Omega$ is defined as the dual characteristic morphism of $\lambda_x.{}_Dfalse_X$, the exponential transposition of ${}_Dfalse_X \circ pr_X : \mathbf{1} \times X \longrightarrow X \longrightarrow {}_D\Omega$:

$$\mathbf{1} \xrightarrow{\lambda_x.{}_Dfalse_X} {}_D\Omega^X \underset{{}_Dfalse_{{}_D\Omega^X}}{\overset{E_X =_{def.} {}_D\varphi_{\lambda_x.{}_DfalseX}}{\rightrightarrows}} {}_D\Omega$$

The particular quantifier has the property that $E_X\varphi(x) = {}_Dfalse$ if and only if $\varphi(x) = {}_Dfalse$, for all x. The exact truth condition implied by the definition above is $E_X\varphi(x) = \sup(\varphi(x))$.

Universal quantifier. Universal quantifier $A_X : {}_D\Omega^X \longrightarrow {}_D\Omega$ is defined as the dual characteristic morphism of the image of the composite ${}_Dpx_X \circ_D \in_X (\in) : \in \rightarrowtail {}_D\Omega^X \times X \longrightarrow {}_D\Omega^X$ (where ${}_Dp_X$ is the first projection and ${}_D\in_X$ is the subobject of $\Omega^X \times X$ whose dual character is the evaluation morphism $\bar{e}_X : {}_D\Omega^X \times X \longrightarrow {}_D\Omega$) so it makes the following diagram a pullback:

$${}_Dp_X \circ \in_X (\in) \xrightarrow{Im({}_Dp_X \circ\, {}_D\in_X)} {}_D\Omega^X \underset{{}_Dfalse_{{}_D\Omega^X}}{\overset{A_X =_{def.} {}_D\varphi_{Im({}_Dp_X \circ {}_D\in_X)}}{\rightrightarrows}} {}_D\Omega$$

The universal quantifier has the property that $A_X\varphi(x) = {}_Dfalse$ if and only if $\varphi(x) = {}_Dfalse$, for some x. The exact truth condition implied by the definition above is $A_X\varphi(x) = \inf(\varphi(x))$.

Given that ${}_D\mathcal{E}$ is a category with exponentials, one has ${}_D\Omega^X$, ${}_D\Omega^{{}_D\Omega^X}$, etc. for any X in ${}_D\mathcal{E}$, which may be regarded as representing collections of properties, properties of properties, etc. defined over X, so one can also have higher-order dual propositions.

If ${}_S\mathcal{E}$ is a standard topos and ${}_D\mathcal{E}$ is the category obtained by assuming not the name ${}_Strue$, but ${}_Dfalse$ for a given morphism with codomain Ω and making the corresponding suitable choice of names for connectives, then ${}_D\mathcal{E}$ and ${}_S\mathcal{E}$ are categorically indistinguishable since equalizers, (binary) products, coequalizers, coproducts and exponentials are notions and constructions prior to the characterization of classifiers and connectives. Moreover, Mortensen proved the following.

Theorem 18.3.4 (Duality Theorem) *Let S be a statement about ${}_D\mathcal{E}$ obtained by the above relabeling method from a statement S' about ${}_S\mathcal{E}$. Then S' is true of ${}_S\mathcal{E}$ if and only if S is true of ${}_D\mathcal{E}$.*

A proof can be found in [39, p. 106].

Clearly, Heyting algebras and Brouwerian algebras, on one hand, and the logics they give rise to, on the other, are dual. Nonetheless, toposes ${}_S\mathcal{E}$ and ${}_D\mathcal{E}$ are not dual in the traditional categorial sense, so this other kind of duality has to be studied. A categorial characterization of the "duality" between standard toposes and complement-toposes would be most welcome, but for now I will describe in more detail the internal logic of complement-toposes.[25]

The *internal logic* of a complement-topos ${}_D\mathcal{E}$ is the algebra induced by the object of propositions or algebraic truth values, ${}_D\Omega$, and the connectives ($\sim$, $\curlywedge$, etc.). Consequence is defined as usual: Let ${}_Dp \models_{{}_D\mathcal{E}} {}_Dq$ denote that whenever the morphism ${}_Dp$ is the same morphism as ${}_Dtrue$ in ${}_D\mathcal{E}$, so is ${}_Dq$ ($\models_{{}_D\mathcal{E}} {}_Dp$ means that ${}_Dp$ is the same morphism as ${}_Dtrue$ in ${}_D\mathcal{E}$).

There is a theorem establishing necessary and sufficient conditions for a proposition ${}_Dp$ being the same morphism as ${}_Dtrue$ in a given complement-topos ${}_D\mathcal{E}$. Let $\models_{CSL}$ be the consequence relation of closed set logic. Then the following theorem holds:

Theorem 18.3.5 *For every topos ${}_D\mathcal{E}$ and proposition ${}_Dp$, $\models_{{}_D\mathcal{E}} {}_Dp$ if and only if $\models_{CSL} {}_Dp$.*

i.e., ${}_D\Omega$ is a Brouwerian algebra (by THEOREM 18.2.4 and the DUALITY THEOREM above).[26]

Summarizing, the complement-categorial analysis of logic implies the following[27]:

(DIL1) Propositions form a partial order, i.e., for every propositions p, q and r:

(DIL1a) $p \geq p$;

(DIL1b) If $p \geq q$ and $q \geq p$ then $p = q$;

(DIL1b) If $p \geq q$ and $q \geq r$ then $p \geq r$.

(DIL2) There is a truth value called ${}_Dfalse$ with the following property:

$$\text{For every proposition } p,\ p \geq {}_Dfalse.$$

(DIL3) One can define a truth value called ${}_Dtrue$ that has the following property:

$$ {}_Dtrue \geq {}_Dfalse.$$

and

$$\text{For every proposition } p,\ {}_Dtrue \geq p.$$

(DIL4) Connectives obey the following truth conditions:

$\sim p = {}_Dfalse$ if and only if $p = {}_Dtrue$, otherwise $\sim p = {}_Dtrue$,

$(p \curlyvee q) = \sup(p, q)$,

$(p \curlywedge q) = \inf(p, q)$,

$(q - p) = {}_Sfalse$ if and only if $q_D \leq p$, otherwise $(q - p) = q$,

$E_X\varphi(x) = \sup(\varphi(x))$,

$A_X\varphi(x) = \inf(\varphi(x))$.

---

[25] I have attempted such a categorial description of this kind of duality in [12].

[26] Again, I have made a little abuse of notation, for I used '${}_Dp$' in both $\models_{{}_D\mathcal{E}}$ and $\models_I$. In rigor, ${}_Dp$ is a morphism which corresponds to a formula $({}_Dp)^*$ in a possibly different language.

[27] By abuse of notation but to simplify reading I will not indicate that the order here is dual to that in standard toposes, unless there is risk of confusion.

(DIL5) The categorial analysis of logic in complement-toposes assumes the Tarskian notion of logical consequence too:

Let '$p \models_{D\mathcal{E}} q$' denote that q is a logical consequence of p in a complement-topos ${}_D\mathcal{E}$, i.e., that whenever p is the same morphism as ${}_D true$ in ${}_D\mathcal{E}$, so is q. Equivalently, if q is not the same morphism as ${}_D true$, p neither is. $\models_{S\mathcal{E}} p$ means that p is the same morphism as ${}_D true$ in ${}_D\mathcal{E}$.

(DIL6) From (DIL1)–(DIL5), in the internal logic of a complement-topos hold at least the laws of dual-intuitionistic, as shown in Theorem 18.3.5.[28]

Example 18.3.6 Since classical logic is its own dual (just as a Boolean algebra is its own dual), the internal logic of **Set** is not modified by the renaming and thus complement-**Set** (${}_D$**Set**) is the same as **Set**.[29]

Example 18.3.7 Complement-$S^{\downarrow\downarrow}$ or ${}_DS^{\downarrow\downarrow}$ has, mutatis mutandis, the same three truth values with its original order,[30] but negation gives now the following identities of morphisms:

$$\sim {}_D false_{S^{\downarrow\downarrow}} = {}_D true_{S^{\downarrow\downarrow}}, \quad \sim {}_D\binom{s}{t}_{S^{\downarrow\downarrow}} = {}_D true_{S^{\downarrow\downarrow}}, \quad \sim {}_D true_{S^{\downarrow\downarrow}} = {}_D false_{S^{\downarrow\downarrow}}$$

In $S^{\downarrow\downarrow}$ one has $(p \vee \neg p) \neq {}_S true_{S^{\downarrow\downarrow}}$, and in the alternative labeling one obtains $(p \curlywedge \sim p) \neq {}_D false_{S^{\downarrow\downarrow}}$. Remember that in a complement-topos $(p \curlywedge q) = {}_D false$ if and only if either $p = {}_D false$ or $q = {}_D false$. If $p = {}_D\binom{s}{t}_{S^{\downarrow\downarrow}}$ then $\sim p = {}_D true_{S^{\downarrow\downarrow}}$, so neither $\sim p = {}_D false_{S^{\downarrow\downarrow}}$ nor $p \neq {}_D false_{S^{\downarrow\downarrow}}$ and hence $(p \curlywedge \sim p) \neq {}_D false_{S^{\downarrow\downarrow}}$. Besides, in a Heyting algebra (like the algebra ${}_S\Omega$) in general it is not the case that $q \leq (p \vee \neg p)$, which in the alternative labeling corresponds to the fact that in a Brouwerian algebra (like ${}_D\Omega$) in general it is not the case that $(p \curlywedge \sim p) \leq q$. So, the internal logic of complement-$S^{\downarrow\downarrow}$ is not classical (nor intuitionistic!), but inconsistency-tolerant. Moreover, in complement-$S^{\downarrow\downarrow}$ both $p \curlyvee \sim p$ and $\sim (p \curlywedge \sim p)$ are the same morphism as ${}_D true_{S^{\downarrow\downarrow}}$, unlike their standard counterparts. In standard $S^{\downarrow\downarrow}(p \wedge \neg p) = {}_S false_{S^{\downarrow\downarrow}}$, which in the alternative labeling gives $(p \curlyvee \sim p) = {}_D true_{S^{\downarrow\downarrow}}$. In standard $S^{\downarrow\downarrow} \neg(p \vee \neg p) = {}_S false_{S^{\downarrow\downarrow}}$ (for in intuitionistic logic the negation of a classical theorem is always false), and the alternative labeling gives $\sim (p \curlywedge \sim p) = {}_D true_{S^{\downarrow\downarrow}}$.

---

[28] Inconsistency-tolerant categorial structures are studied further in [39, Chap. 12, written by William James] and in [22].

[29] Thus, as Vasyukov [52, p. 292] points out: "(…) in **Set** we always have paraconsistency because of the presence of both types of subobject classifiers (…)" just as we always have in it (at least) intuitionistic logic. The presence of paraconsistency within classical logic is not news. See for example [7], where some paraconsistent negations in S5 and classical first-order logic are defined.

[30] It is easy to verify that after making all the necessary changes, i.e., changing ${}_S true_{S^{\downarrow\downarrow}}$ for ${}_D false_{S^{\downarrow\downarrow}}$, etc., the names are ordered in the same way as they are in ${}_SS^{\downarrow\downarrow}$.

18.3.3 Complement-Toposes and Variable Sets

Bell says, maybe with a wink to the Marxist Lawvere (and perhaps also to his former student friend of contradictions Graham Priest):

> Certain philosophers —notably Hegel and Marx— believed that achieving a true understanding of the phenomenon of change would require the fashioning of a dialectical logic or "logic of contradiction," in which the law of noncontradiction —"no statement can be both true and false"— is repudiated. It is a striking fact that, so far at least, rather more light has been thrown on the problem of variation by challenging the law of excluded middle than by questioning the law of noncontradiction. [6, p. 179]

But complement-toposes, with their paraconsistent internal logic, also challenge Lawvere's idea that intuitionistic logic is the objective logic of variable sets. Intuitionistic logic may be the objective logic of some variable sets, but not of all of them. Furthermore, variable sets taken together might be so variable that no logic is preserved under such variation, but I am not going to press that point. My aim here is just to indicate the objects of a topos can be described as varying in a way such that they are also variable sets yet their internal logic is not intuitionistic.

Let me present first the formulation of Kripke semantics for CSL. Let $\mathbf{P} = (P, R)$ be a poset. A set $A \subseteq P$ is *anti-hereditary* or *rooting* in $\mathbf{P}$ if for all x, y, if $x \in A$ and xRy then $y \in A$, i.e., if it is closed "downwards" under R. Let $\mathbf{P}^r$ be the collection of anti-hereditary subsets of $\mathbf{P}$. A $\mathbf{P}$-*valuation* is a function $\mathfrak{V}: F \longrightarrow \mathbf{P}^r$, where F stands for a collection of formulas on a usual zero-order language, such that to each p_i is assigned an anti-hereditary subset $\mathfrak{V}(p_i) \subseteq P$.

A *model based on* $\mathbf{P}$ is a triple $\mathfrak{M} = \langle P, R, \mathfrak{V}\rangle$, where $\mathfrak{V}$ is a $\mathbf{P}$-valuation. The idea that A is *false at* x, in symbols $\mathfrak{M}, x \Vdash_f A$, is defined recursively as follows: For every $x \in P$:

$\mathfrak{M}, x \Vdash_f A$ if and only if $x \in \mathfrak{V}(p_i)$,
$\mathfrak{M}, x \Vdash_f \sim A$ if and only if for all y with xRy, $\mathfrak{M}, y \nVdash_f A$,
$\mathfrak{M}, x \Vdash_f A \curlyvee B$ if and only if $\mathfrak{M}, x \Vdash_f A$ and $\mathfrak{M}, x \Vdash_f B$,
$\mathfrak{M}, x \Vdash_f A \curlywedge B$ if and only if either $\mathfrak{M}, x \Vdash_f A$ or $\mathfrak{M}, x \Vdash_f B$,
$\mathfrak{M}, x \Vdash_f B - A$ if and only if for all y with xRy, if $\mathfrak{M}, y \Vdash_f A$ then $\mathfrak{M}, y \Vdash_f B$.

A is *false in a model* $\mathfrak{M}$, denoted $\mathfrak{M} \Vdash_f A$, if A is false at every $x \in P$ in $\mathfrak{M}$. A formula A is *invalid on* $\mathbf{P}$, denoted $\mathbf{P} \Vdash_f A$, when it is false in every model based on $\mathbf{P}$.

Take $\mathbf{P} = (\{1, 0\}, \geq)$ (with $1 \geq 0$ as usual) and $\mathfrak{V}(p) = \{0\}$, which is anti-hereditary. Then with $\mathfrak{M} = \langle\{1, 0\}, \geq, \mathfrak{V}\rangle$ one has that $\mathfrak{M}, 1 \nVdash_f p$. But $\mathfrak{M}, 0 \Vdash_f p$ and $1 \geq 0$, so $\mathfrak{M}, 1 \nVdash_f \sim p$. Thus, $\mathfrak{M}, 1 \nVdash_f p \curlywedge \sim p$, so contradictions are not automatically refuted in this model.

Informally, $\mathbf{P}$ represents a collection of states, say, of possible mathematical knowledge. The truth value of a proposition depends on the knowledge established at a certain state, so their truth value is relative to a certain state. However, once a proposition is false at a state, it is false at all earlier states. One state comes before, or is earlier than, another, but the order is not necessarily linear because the states are not merely of actual but of possible mathematical knowledge, so immediately earlier than a state of knowledge may be several other states, each with a different value for a certain proposition: Due to the rooting of falsity, if a proposition is false at a state earlier than a certain state and true at another later state, these later states cannot be connected. This conveys the idea that if it is established that p is false in a given evidential situation then it has been false always in the past of that situation. Goodman [19] thought that dualization of "any formula, once true, remains true" should be "Popperian," i.e., preservative of falsity: "any formula, once false, remains false." But Goodman gives no further details about the semantics for his logic besides this short remark. I have opted here for systematically inverting the order, so the dualization of the intuitionistic preservation of truth would be the rooting of falsity: "any formula, once false, was always false."[31]

Operations of ${}_D\Omega$ with the order ${}_D\leq$ between them constitute a frame to provide a categorial version of the Kripke semantics above in terms of complement-toposes. What the **Anti-comprehension axiom** says is that ${}_D\varphi(x) = {}_D\mathit{false}_X$ if and only if $x \in m$, i.e., it specifies when a generalized element x belongs to a part m of O (the o's for which φ is false). These specifications depend on the way the formula $\varphi(x)$ is built up from connectives, but this time formulated in terms of the relation "X makes $\varphi(a)$ false," or also "X refutes $\varphi(a)$," written $X \Vdash_{DF} \varphi(x)$ and defined as follows:

For every $x : X \longrightarrow O$ with its image $Imx \in_D \Omega^X$:

$X \Vdash_{DF} {}_D\varphi(x)$ if and only if $x \in m$

that is

$X \Vdash_{DF} {}_D\varphi(x)$ if and only if ${}_D\varphi(x) = {}_D\mathit{false}$.

In what follows 'p_X,' 'q_X,' 'p_{XY},' and 'q_{XY}' (just changing '${}_S\Omega$' by ${}_D\Omega$) are as in 4.4:

$X \Vdash_{DF} {\sim}\, p_X$ if and only if, for every morphism $y : Y \longrightarrow X$ such that $Y \Vdash_{DF} \varphi$ $(a \circ v)$, $Y \cong \mathbf{0}$

$X \Vdash_{DF} (p_X \curlyvee q_X)$ if and only if $X \Vdash_{DF} p_X$ and $X \Vdash_{DF} q_X$

$X \Vdash_{DF} (p_X \curlywedge q_X)$ if and only if there are morphisms $y : Y \longrightarrow X$ and $w : W \longrightarrow X$ such that $y + w : Y + W \twoheadrightarrow X$ is an epimorphism while both $Y \Vdash_{DF} \varphi(a \circ y)$ and $W \Vdash_{DF} \psi(a \circ w)$

$X \Vdash_{DF} (p_X - q_X)$ if and only if, for every morphism $y : Y \longrightarrow X$ such that $Y \Vdash_{DF}$ $\varphi(a \circ v)$, also $V \Vdash_{DF} \psi(a \circ v)$.

Now, applying the Mortensen–Lavers dualization and using the DUALITY THEOREM, it is possible to make explicit some categorial constructions in complement-toposes and some facts about CSL and some of its modal extensions. In particular, dualization also extends to topologies. Note that one of the conditions of a standard Lawvere–Tierney topology is typical of an interior operator (that it is multiplicative), other is typical of a closure operator (that it is inflationary), and the remaining one

[31] A dualization of Kripke semantics for intuitionistic logic similar to that presented here was studied in [48]. Shramko maintains the Popperian reading to provide a "logic of refutation"; the crucial distinction lies in the condition for $\sim$. Moreover, Shramko, like Goodman, also omits the discussion of subtraction. Goré [20, p. 252] even says that "[Goodman] annoyingly fails to give the crucial clause for satisfiability for his 'pseudo-difference' connective."

is both (that it is idempotent).[32] Then, following the Mortensen–Lavers method, the dualization proceeds by changing the typical interior condition by a closure one, the typical closure condition by an interior one, and changing the idempotent condition of an interior operator from the idempotent condition of a closure operator (that is, leaving it unchanged).

Thus, *dual Lawvere–Tierney topology* is then a morphism $\vartheta :_D\Omega \longrightarrow {}_D\Omega$ in a complement-topos ${}_D\mathcal{E}$ such that
(ϑ1) $\vartheta \circ \Upsilon = \Upsilon \circ (\vartheta + \vartheta)$,
(ϑ2) $\vartheta \circ \vartheta = \vartheta$,
(ϑ3) $\vartheta \circ_D false = {}_D false$,
(ϑ1) and (ϑ2) say that ϑ is additive and idempotent. It can be proved that
(ϑ4) $p_D \geq \vartheta p$,
so ϑ is also deflationary.

Dualizing Macnab's proof, it can be proved that ϑ can be defined using subtraction by the single equation

$$- \circ \langle \vartheta \circ q, p \rangle = - \circ \langle \vartheta \circ q, \vartheta \circ p \rangle.$$

There are a number of topics which resemble situations in intuitionistic logic and the modality j; let me give just two examples on how to push forward the study of ϑ. As I have said, the (standard) topos ${}_S\mathcal{E}_{\neg\neg}$ of double-negation sheaves is a model showing that, for example, the continuum hypothesis is independent of the axioms for topos theory including classical logic. What does the complement-topos ${}_D\mathcal{E}_{\sim\sim}$ show, say, about (models of set theory based on) complement-toposes, classical logic and the continuum hypothesis? On the other hand, Gödel used a double-negation translation to make some connections between classical and intuitionistic theories of natural numbers. Maybe there is another double-negation translation giving hints about classical and paraconsistent theories of natural numbers. All this requires further separate work.

Finally, note that if j is a geometric modality meaning "it is cofinally the case that," ϑ is also a geometric modality but meaning "it is coinitially the case that" (dualization is left to the reader) which, incidentally, happens to be conceptually closer to forcing through the notion of density.

18.3.4 Proof Theory for Complement-Topos Logic

When studying dual-intuitionistic logic from the topos-theoretical perspective, Mortensen and Lavers did not present a sequent calculus. In this section, I will present a sequent calculus for complement-topos logic, which completes the one sketched in [13]. Just as the dualization of connectives in a topos, the rules for

---

[32] An *interior operator* is an operator which is multiplicative, idempotent, and deflationary, i.e., $x \geq jx$. A *closure operator* is additive ($j(x \cup y) = jx \cup jy$), idempotent and inflationary.

complement-toposes logic also mirror the "topologico-algebraic" dualization. That is, any occurrence of $\wedge$, $\vee$, $true$, $false$ on a formula must be replaced by $\vee$, $\wedge$, $false$, $true$, respectively. To dualize a given formula $a \supset b$, replace the antecedent by the consequent and *vice versa* and then replace $\supset$ by $-$. Since $a \leq b$ can be interpreted as a sequent $a : b$ and the dual of $a \leq b$ is $b \leq a$ (or $a \geq b$), the dual of $a : b$ is $b : a$.[33] Thus, the corresponding rules for complement-toposes logic are the following ones:

Structural Rules:

Trivial sequent: $$\frac{*}{p \ : \ p}$$

False and true: $$\frac{*}{false \ :} \qquad \frac{*}{p \ : \ true}$$

Thinning: $$\frac{p \ : \ \Gamma}{p \ : \ \Gamma, \ q}$$

Cut: $$\frac{p \ : \Gamma, \ q \quad \text{and} \quad q \ : \Gamma}{p \ : \ \Gamma}$$

if every variable free in q is free in Γ or in p. The restriction is due to the fact that if q has free variables over empty objects then the upper sequents are trivially true even if the lower one is false.

Substitution: $$\frac{p \ : \ \Gamma}{p(x/s) \ : \ \Gamma(x/s)}$$

for any term s free for x in all the formulas.

Connective rules:

$$\frac{true \ : \ \Gamma, p}{\neg p \ : \ \Gamma}$$ (double line)

$$\frac{p \ : \Gamma \quad \text{and} \quad q \ : \Gamma}{p \vee q \ : \ \Gamma}$$ (double line)

$$\frac{\theta \ : \Gamma, p \quad \text{and} \quad \theta \ : \Gamma, q}{\theta \ : \ \Gamma, p \wedge q}$$ (double line)

$$\frac{q \ : \ \Gamma, p}{q - p \ : \ \Gamma}$$ (double line)

[33]Such dualization is (the zero-order part of) the mapping $\star$ discussed in [50, p. 444], which builds upon one described in [10].

and if the variable x is not free in Γ or q:

$$\frac{p \ : \ \Gamma}{\overline{(Ax)p \ : \ \Gamma}} \qquad \frac{q \ : \ \Gamma, p}{\overline{q \ : \ \Gamma, (Ex)p}}$$

Dually to the case of toposes, the structural rule for $true$ and the connective rules for negation and conjunction are derived, just as the corresponding morphisms $true$, $\neg$ and $\wedge$ needed $false$ and $\vee$ in order to be defined. A soundness proof for these rules can be adapted from the original soundness proof for the rules for standard topos logic as presented, for example, in [37, 15.2], and the remarks there on completeness also apply here.

18.4 Some Objections to the Legitimacy of Complement-Toposes

In this section, I address three potential objections to the mathematical legitimacy of complement-toposes. The first one, "The *Just Definitional Variants* Objection," says that the definition of a complement-topos does not differ from the usual definition of a topos and thus the relabelings involved in complement-toposes are uninteresting and pointless. The second one, "The Theorems Objection," says that the alternative reading of the categorial structure of toposes giving rise to complement-toposes conflicts certain well-known theorems in topos theory. The third one, "The Working Mathematician Objection," asks for the importance of complement-toposes in particular, and of inconsistency-tolerances and paraconsistency in general, for the working mathematician. I will show that the first two objections do not take the dualization seriously, and thus judge complement-toposes from the pre-dualization point of view. About the third objection I will make the point that there is more to mathematics than current, actual practice and mainstream research lines, and even thus the study of inconsistency-tolerance has found its place among leading mathematicians.

18.4.1 The Just Definitional Variants *Objection*

Given that complement-toposes have been practically unnoticed, there are no public statements of the first objection I am going to discuss here.[34] A topos- or category-

[34] As I have said before, Mac Lane knew about complement-toposes via Mortensen. His stance was that they were just the old toposes, but he gave no argument. I have been trying to figure out why did Mac Lane thought that and this is the best I can imagine. I think this would be part of what a mainstream topos- or category-theorist would say at first glance on complement-toposes (and in my experience, this is what they invariably say).

theorist might think of complement-toposes as just definitional variants of the already well-known toposes. Given that they share all the categorial ingredients, toposes and complement-toposes could not be different (kinds of) categories; in particular, they do not have different internal logics.

The mainstream topos-theorist can correctly insist on the categorial indistinguishableness between standard toposes and complement-toposes, but this amounts rather to a proof that both kinds of toposes equally deserve the name "topos," since for all mathematical purposes they have the same constituents independently of Skolemization for the morphisms whose codomain is Ω. However, the internal logic induced is in fact different in each case, true, not by differences in the categorial structure, but in the way that categorial structure is described.

Although it sounds repetitive, it must be emphasized that the claim is not that complement-toposes are categorically different from toposes, nor to say that standard connectives acquire further categorial properties *qua* morphisms after the being renamed, but rather to stress the fact that the same categorial stuff, which is essentially equational with variables, can be described in at least two different ways. The categorial reconstruction of logic in a standard topos starts with a certain object O *thought of* as the **extension** of a propositional function φ and that a certain element belongs to O is *thought of* as making φ $true$. Hence, in a standard topos the basic proposition is ${}_{S}true$. What complement-toposes say is that one can start the categorial reconstruction of logic with the same categorial data but interpreted in a different way. A certain object O is *thought of* as the **anti-extension** of a propositional function φ and that certain element belongs to O is accordingly *thought of* as making φ $false$. Hence, in a complement-topos the basic proposition is ${}_{D}false$. Neither of the labelings is imposed by the categorial structure of toposes itself so, in its current and mainstream form, there is more than just categorial structure in the study of toposes, ex. gr. there are particular Skolemizations of it.

One could also start with complement-toposes and then obtain standard toposes by proposing an alternative description of the underlying of the equational structure. This means that, even if at first glance, the categorial structure invites to be conceptualized in certain ways, and it does not force them. All this helps to solve the initial perplexity: If ${}_{S}\mathcal{E}$ and ${}_{D}\mathcal{E}$ should be indistinguishable because they are categorically indistinguishable, how can one in fact distinguish between them, as one does by noting their different internal logics? The answer is this: To date, there is more than categorial structure in the study of toposes, to wit, special, intuitive names conceptually laden for some of the morphisms, invited, yes, but not necessitated, by the categorial structure. Neither of the names is imposed by the categorial structure of toposes itself so, in its current form, there is more than just categorial structure in the study of toposes.

Another reason to deny the difference in the internal logic might be that topos-theorists think that one is using the traditional modeling of intuitionistic logic using Brouwerian logic with the least element as designated (as in [35]). But it is not the case. It is clear that Ω can support the structure of a Brouwerian algebra (otherwise, the usual semantics for intuitionistic logic using them would not be possible) and, in order to use not the least element but the top one as designated one has to introduce some conceptual changes, but not in the categorial structure. That is why the internal logic of a complement-topos is different.

This is not a mere play with labels and, even though the underlying dualities between Heyting algebras and Brouwerian algebras are well-known, the choice of names affects what we are considering as the internal logic of a topos because the names are conceptually laden. Even if from a mathematical point of view all this might be regarded as uninteresting (which is not, for it invites us to rethink an important theorem), preferring one way of naming above the other may have (and has had) important philosophical consequences. As I have said, complement-toposes bring in question, for example, Lawvere's "Intuitionistic logic is the objective logic of variable sets" (slogan (S4)), as well as John Lane Bell's claims that "The universal, invariant laws of mathematics are intuitionistic" (slogan (S5)) or that "Intuitionistic logic sheds more light on the issue of mathematical variation than paraconsistent logics." The categorial structure of toposes support paraconsistent logics as well, so any philosophical claim aiming to emphasize the mathematical supremacy of intuitionistic logic has to take into account this fact.

18.4.2 The Theorems Objection

Now, there might be a worry concerning a potential conflict between the alternative names for certain parts of the structure of toposes giving rise to complement-toposes and certain well-known theorems in topos theory. In particular, those theorems imply that, if Ω were a Brouwerian algebra as it is in complement-toposes, then their internal logic should be classical. But the internal logic of a complement-topos in general is not classical, so this tension should be explained away.

In his doctoral dissertation, Giovanni da Silva de Queiroz [11] argues against the possibility of toposes whose internal logic is dual to the intuitionistic one. He begins his Chap. 5, "Categories, *topos* and duality" in this way:

> It is known that for every elementary *topos* ξ, the object Ω (subobject classifier) has the structure of a Heyting algebra. It can be guessed that, by duality, to every *topos* can be associated a dual *topos* such that Ω has also the structure of a Brouwer algebra. This is not true, though. We are going to show that (...) a "Brouwerian" negation, distinct from Boolean negation, cannot be defined.

and in the introduction he had said "We also show that a negation defined in a Brouwer algebra cannot be "captured" by the truth values object. If that happened, the internal logic of a topos comes to be the Boolean logic (henceforth, classical)" [11, p. 5].

Reyes and Zolfaghari [45, 46] have given the conditions under which the subobject algebra of a topos is a Brouwerian algebra. Nonetheless, most authors agree that Ω cannot be but a Heyting algebra. de Queiroz makes a case based on the following theorems already mentioned at the end of Sect. 18.3.1[35]:

---

[35]Although de Queiroz does not make explicit his use of the last of them. As a side remark, the converse of that theorem does not hold.

Theorem 18.4.1 Let $\delta : {}_S\Omega \longrightarrow {}_S\Omega$ be a morphism such that $\delta \leq id_{{}_S\Omega}$ and $\delta \circ {}_S true = {}_S true$. Then $\delta = id_{{}_S\Omega}$.

Theorem 18.4.2 In any topos ${}_S\mathcal{E}$ the following conditions are equivalent:

(a) ${}_S\mathcal{E}$ is Boolean.
(b) $\neg \circ \neg = id_{{}_S\Omega}$.

Theorem 18.4.3 If ${}_S\mathcal{E}$ is Boolean, then its internal logic is classical.

But in Brouwerian algebras $\sim\sim p \leq p$ and $\sim\sim true = true$, and hence, by the theorems above, Brouwerian negation would collapse into classical negation. So, the internal logic of a topos with a Brouwerian negation would not be of a new kind, but classical. This is precisely de Queiroz conclusion: There cannot be a topos whose internal logic is dual to intuitionistic logic, unless its logic is classical, which is not an interesting case.

If the alternative reading is possible and the internal logic of a topos can be described as a paraconsistent one, as showed in the preceding sections, what about de Queiroz's argument? The answer is that the proofs of these theorems presuppose the standard names for certain morphisms with codomain Ω. Even though it is true that in a Brouwerian algebra both $\sim\sim p \leq p$ and $\sim\sim true = true$ hold, it does not make it to collapse into Boolean algebra, in the same way that a Heyting algebra does not collapse into Boolean algebra just because in it $p \leq \neg\neg p$ and $\neg\neg false = false$ hold. A Brouwerian algebra would collapse into a Boolean one if it has added to it that $p \leq \sim\sim p$, and a Heyting algebra would collapse if it is added to it that $\neg\neg p \leq p$, i.e., each would become into a Boolean algebra if their respective double negations where the same as the identity. What the argument by de Queiroz shows is that once *true*, *false* and *negation* are defined in the usual, standard way, Brouwerian negation cannot be added unless the internal logic is classical. However, the argument does not work once these notions have been defined in the way Mortensen and Lavers have suggested, and what one has then is the following theorem (by the DUALITY Theorem 18.3.4 above):

Theorem 18.4.4 *Let* ${}_D\delta : {}_D\Omega \longrightarrow {}_D\Omega$ *be a morphism such that* $id_{{}_D\Omega} \leq {}_D\delta$ *and* ${}_D\delta \circ {}_D false = {}_D false$. *Then* ${}_D\delta = id_{{}_D\Omega}$.

Then, by Theorems 18.4.4, 18.3.2 and 18.3.3, the only complement-toposes with intuitionistic features ($p \leq \sim\sim p$ and $\sim\sim false = false$) would be those whose internal logic is classical.[36]

So, the categorial structure of a topos supports *separately* both kinds of connectives. What de Queiroz's arguments prove (or at least imply) is that the internal logic of an arbitrary topos, be it standard or complement, and under the given characterization of logical consequence, cannot be H-B logic, intuitionistic logic plus two

[36] There is another problem in de Queiroz's analysis of Brouwerian connectives in a topos. For example, the following diagram

connectives satisfying the dual conditions for the conditional and negation.[37] It is clear since H-B algebras[38] have both negations without being necessarily Boolean, while in topos theory the only toposes in which one can have both intuitionistic and paraconsistent features are the Boolean ones, with the shortcoming that they do not have two negations, like H-B logic, but just one. Summarizing, Theorems 18.3.1, 18.3.2 and 18.3.3 say that if Ω is a Heyting algebra, it cannot be also a Brouwerian algebra unless Ω is Boolean (and therefore, the internal logic cannot have paraconsistent features unless it is classical). Theorems 18.4.4, 18.3.3 and the dual of 18.3.2 (for that just put $\sim$ instead of $\neg$) say that if Ω is a Brouwerian algebra, it cannot be also a Heyting algebra unless it is Boolean (and therefore, the internal logic cannot have intuitionistic features unless it is classical). Together, they say that Ω cannot be a B-H algebra unless it is Boolean (and therefore, the internal logic cannot have H-B logic features unless it is classical).

A similar diagnosis applies to a criticism based on the following theorem:

Theorem 18.4.5 *For every object X in a topos, the power object $\mathcal{P}X$ is an internal Heyting algebra in the topos. When $X = \mathbf{1}$, $\mathcal{P}\mathbf{1}$ is Ω. So Ω is an internal Heyting algebra in the topos.*

(A proof of this can be found in [31, IV.8, Theorem 1 (internalversion)], or in [8, Proposition 6.2.1]).

A careful examination of the standard proofs of this theorem reveals that what the topos structure implies is not enough to determine a Heyting algebra. Let me

---

(Footnote 36 continued)

$$\begin{array}{ccc} \leq & \xrightarrow{\ e_1\ } & \Omega \times \Omega \\ {\scriptstyle !}\downarrow & & \downarrow {\scriptstyle - =_{def.}\ \chi_{e_1}} \\ \mathbf{1} & \xrightarrow[\ true\ ]{} & \Omega \end{array}$$

given as definition of Brouwerian subtraction in [11, p. 131] makes no sense. de Queiroz's definition implies that the equalizer of $p \vee q$ and the first projection p is $e_1 : \leq \rightarrowtail \Omega \times \Omega$, but it is not correct for when one dualizes

$$(p \supset q) = true \text{ if and only if } (p \wedge q) = p$$

one obtains

$$(q - p) = false \text{ if and only if } (p \curlyvee q) = p,$$

and not

$$(p - q) = true \text{ if and only if } (p \vee q) = p.$$

Briefly, de Queiroz's definition gives us that Brouwerian subtraction is true in crucial cases where it should be false. Note that one cannot save de Queiroz's proposal by reverting $p - q$ as $q - p$ in his suggested truth condition: In that case the truth conditions of $\supset$ and $-$ would coincide and no connective different from $\supset$, let alone its dual, would have been defined.

[37]This logic and the corresponding algebra were defined in [42] and extensively investigated in [43, 46] and now in practically every text on "dual intuitionistic logic" (although it is not properly a dual-intuitionistic logic). A gentle introduction can be found in [21].

[38]Or *bi-Heyting* algebras, as has been widely called recently mainly by influence of Reyes and Zolfaghari. See for example [45].

analyze how McLarty [38] explains the theorem to see more clearly what is wrong with its alleged consequences. The topos structure implies that for any object O in a topos $\mathcal{E}$ there is a *top* element id_O, a *bottom* element $\mathbf{0} \longrightarrow O$, and that for any two subobjects $i : S \longrightarrow O$ and $j : T \longrightarrow O$ of O there is an intersection $S \cap T$ (the largest subobject of O contained in both S and T)

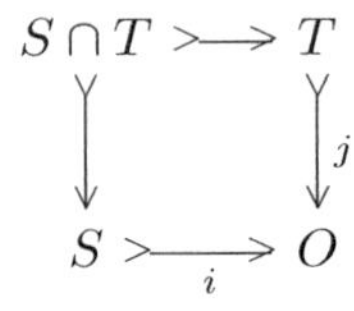

and a union $S \cup T$ (the smallest subobject of O containing both S and T)[39]

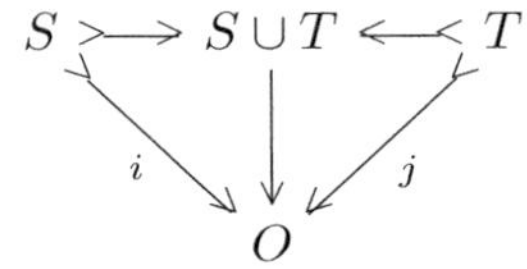

Now, if the negation of a subobject i, $\neg i$, is defined as the largest subobject of O disjoint from i, any subobject i of an object O has a negation:

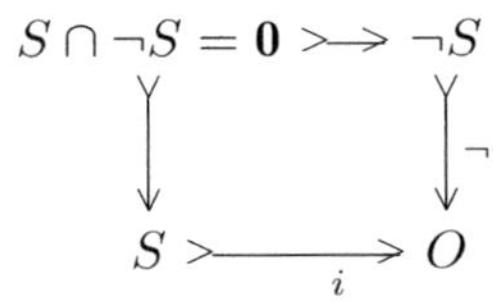

A consequence of this is that $\neg\neg S \subseteq S$ and that the union of $\neg i$ and i is not necessarily equal to O.

But such a definition of negation is not forced by the topos structure! It is intuitive indeed, but not forced. If one defines the negation of i, $\sim i$, as the smallest subobject of O such that joint with i it is the whole of O

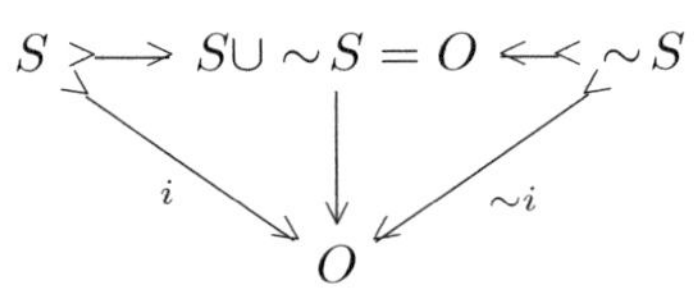

[39] A proof that every pair of subobjects in a topos has a union can be found in [8, Proposition 5.10.1]. Here is a sketch. Consider any two subobjects $i : S \longrightarrow O$ and $j : T \longrightarrow O$ of O in a topos $\mathcal{E}$. Take their coproduct $S + T$ and the morphism $f : S + T \longrightarrow O$. As every morphism in a topos, f factors uniquely into an epimorphism followed by a monomorphism (see [8, Corollary 5.9.4]) $f = k \circ p : S + T \twoheadrightarrow I \rightarrowtail O$. Morphisms from S and T to I, composed with i and j, imply that these are included in k (I, k is thus the smallest subobject of O which contains both S, i and T, j.) A similar reasoning can be applied to show that every pair of subobjects in a topos has an intersection using products.

what fails then is $S \subseteq \sim\sim S$ and the intersection of $\sim i$ and i is not necessarily empty.[40]

Making a few moves more in the proofs, one has the consequence that there must be a conjunction morphism classifying intersections such that it is the (same morphism as the) infimum of its conjuncts, and a disjunction classifying unions such that it is the (same morphism as the) supremum of its disjuncts. But one has $\wedge$ and $\vee$ in standard toposes and $\curlywedge$ and $\curlyvee$ in complement-toposes doing exactly that. In order to obtain a Heyting algebra for Ω, i.e., intuitionistic logic, what follows in the standard proofs are definitional matters. One defines $\supset$ and $\neg$, and all these data already induce a Heyting algebra. Nonetheless, neither $\supset$ nor $\neg$ are implied by the topos structure: They *can be* defined, i.e., they are merely allowed. Of course, once they are defined, nothing else altering the Heyting algebra structure can be added or defined. This heavily relies on the assumed morphism $\mathbf{1} \longrightarrow \Omega$. If one thought of it as $true$, the alternative definitions do not run, but nothing in the categorial structure prevents interpreting it as $false$ and then, given top and bottom elements, as well as meets and joins, one can define a subtraction $-$ instead of $\supset$ and a negation $\sim$ instead of $\neg$ to obtain a Brouwer algebra.

Shortly, what Theorem 18.4.2 asserts is that Ω has the structure of a Heyting algebra, but not that it is the only structure it can have. A more accurate reading of the theorem is that the power object of X can have the structure of a Heyting algebra, or certain data induce a Heyting algebra over the power object of X. From this does not follow that the power object of X must have the structure of a Heyting algebra, nor that no other data can induce a different algebraic structure over the power object X.[41]

There is nothing wrong with filling in it with certain names so as to produce a Heyting algebra. What is wrong is forgetting that those names have been chosen by hand, so to speak, and that different names can be chosen. Again, names or labels are not determined by the categorial structure itself.

I have already dealt with the theorem stating that the internal logic of a topos is intuitionistic. Just to recall it, the theorem holds when the standard definitions are given, and not when the categorial structure is described in the alternative way. This must suffice to show that certain theorems of topos theory using standard names cannot be used against complement-toposes.

---

[40] The next objection can be seen as trying to defend the original definition of negation against the suggested by complement-toposes, but this time using extra-categorial considerations.

[41] Yet in [22] it is argued, with [27], that inconsistency-tolerant constructions in sheaf categories in general are not categorically natural (are not preserved by pullbacks along morphisms). A similar point was behind de Queiroz's objection and the theorems by Reyes et al. I do not know whether there are still hidden "standard" assumptions behind the results in [22]; I will leave the investigation of that for future work.

18.4.3 The Working Mathematician Objection

The working mathematician might say at this point: "OK. You can provide an alternative description of the internal logic of a topos such that it turns out to be dual-intuitionistic or paraconsistent rather than intuitionistic. All that is formally right. However, there is no guarantee that the alternative reading returns something meaningful. In particular, a paraconsistent logic cannot be regarded as a logic for mathematics at all, as would be required for the reconstruction of mathematics in a complement-topos. Inconsistency has no place in mathematics."

I think this worry is misguided, although I am not going to go very deep into this issue. Inconsistent mathematics is the study of mathematical objects, like sets, numbers, functions, etc. where some contradictions are allowed. Tools from formal logic are used to make sure that any contradictions are contained and that the overall theories remain coherent. Although there are several examples of what can be seen as inconsistencies tolerated in mathematics in spite of classical logic (pre-Weierstrass calculus, Bohr theory of atom, delta functions, identification in quotient structures, etc.), inconsistent mathematics began in its actual form as a response to the set-theoretic and semantic paradoxes such as Russell's and the Liar—the response being that the premises of those reasonings seem true, the proofs seem valid, and the conclusions seem true is because they are so—and has so far been of interest primarily to logicians and philosophers. More recently, though, the techniques of inconsistent mathematics have been extended into wider mathematical fields, such as analysis, vector spaces, and topology, to study inconsistent structures for their own sake.

How can it be? Mainstream research lines and current practice do not necessarily dictate what is mathematically legitimate. These are different questions. Even if a mathematical theory were not worth of study because of its lack of profound results or applications, it could still be a mathematical theory. A mathematical theory is a collection of formulas, the theorems, which are obtained through logical proofs. A contradiction is a formula together with its negation, and a theory is inconsistent if it includes a contradiction. Inconsistent mathematics considers inconsistent mathematical theories and requires then careful attention to logic. There is not only one logical way to obtain a result, for one may use, say, the full force of classical or opt for an intuitionistically valid proof. In many logics, such as classical and intuitionistic, a contradiction implies every other formula. A theory containing every formula is called trivial, and is considered absurd. Classical and intuitionistic logics therefore make nonsense of inconsistency and are inappropriate for inconsistent mathematics. A paraconsistent logic guides proofs so that inconsistencies do not necessarily lead to triviality. With a paraconsistent logic, mathematical theories can be both inconsistent and interesting. So inconsistency, or at least inconsistency-tolerance, has place in mathematics. It is there. True, it is not mainstream nor widely known, but that does not make inconsistent mathematics unintelligible. As Mortensen says, the idea is not necessarily to deny or replace the obviously excellent corpus of traditional mathematics, but to expand it. That the categorial structure of toposes supports a paraconsistent logic as readily as it supports intuitionistic logic should not be viewed as an objection to intuitionism, however, so much as an argument that inconsistent theories are equally reasonable as items of mathematical study.

Finally, I think that ultimately I have conceded too much to the working mathematician. Leading mainstream category theorists have devoted several pages to the issue of inconsistency-tolerance in topos theory, like the aforementioned Reyes, Zolfaghari and Zawadowski, as well as Lawvere, who has advocated its study in order to broaden our insights into the connections between logic and geometry to recapture, for example, the geometric notion of boundary (cf. [26, 27]).

18.5 Conclusions

I have expounded some important topos-theoretical notions and results, and have showed how they seem to support some philosophical claims. I first described a subobject classifier and how its codomain, Ω, can be thought of as a truth values object. In general, Ω has more than two elements, and thus mathematical statements are essentially many-valued. The notion of a subobject classifier allows to define zero-, first-, and higher-order connectives. The standard definitions imply certain truth conditions that make the internal logic of a topos intuitionistic. Given that ordinary mathematics can be interpreted in an arbitrary topos just as it is interpreted in **Set** and usual membership-based set theories, the intuitionistic laws are the invariant or universal laws of mathematics.

The usual theory of toposes leads to set the following slogans:

(S1) Ω is (or at least can be thought of as) a truth values object.
(S2) In general, the internal logic of a topos is many-valued.
(S3) In general, the internal logic of a topos is (with some provisos) intuitionistic.
(S4) Intuitionistic logic is the objective logic of variable sets.
(S5) The universal, invariant laws of mathematics are intuitionistic.

Mortensen and Lavers' notion of complement-topos constitute an audacious approach to topos logic. They give a different but sound interpretation of certain basic categorial situation in a topos, and they show how to describe the internal logic of a topos as dual-intuitionistic or paraconsistent. I have presented the essential notions, dualizations, techniques, constructions—including a sequent calculus for complement-topos logic—of the theory of complement-toposes, and showed that complement-toposes have the same categorial structure of standard toposes, but it is given different names for some crucial morphisms and constructions. It is worth noting that nothing in the categorial structure of toposes forces one labeling above the other. Furthermore, I have discussed three objections against the legitimacy of complement-toposes, all based on subtle misunderstandings and hidden assumptions.

The theory of complement-toposes implies that current topos theory gives just part of the concept of topos, that some common theorems on topos logic tell just part of the relevant story, and that in a further, more abstract development the slogans above have just limited application. Thus, a topos, and especially its internal logic, is a truly

Protean categorial creature which can accommodate the most diverse descriptions and support an enormous variety of logics besides that mentioned in the slogans. The main obstacle to any attempt of thinking differently about topos logic seems to be Theorem 18.2.4. Like most mathematical theorems, Theorem 18.2.4 is the case only under certain conditions $c_1, \ldots, c_n$, among them the name of the initial morphism or the presupposed notion of logical consequence. However, $c_1, \ldots, c_n$ are just one way among many others to describe certain parts of the categorial structure of toposes, and when philosophical consequences are drawn from a mathematical theorem, these assumptions do matter, especially when are so deeply internalized within the working mathematician as to be forgotten as important assumptions.

However, I think the philosophical significance of complement-toposes is much broader than I have been able to show here. In Lawverean terms, complement-toposes show that there is still a lot of "substance" in topos theory, and deeper "invariant forms" wait to emerge. Complement- toposes bring in question, for example, John Lane Bell's (S4) "The universal, invariant laws of mathematics are intuitionistic" but also his idea of "local mathematics" and his claim "Intuitionistic logic sheds more light on the issue of mathematical variation than paraconsistent logics," as well as Lawvere's (S5) "Intuitionistic logic is the objective logic of variable sets." I have not given a study of paraconsistent set theories based on complement-toposes but I hope to deal with such issues in further work.

Thus, the notion of complement-topos goes against (S3)–(S5), but I think it is possible to advance further. The main morals of Mortensen and Lavers' study of complement-toposes seem to be the following:

(a) There is a "bare" or "abstract" categorial structure of toposes that can filled in at least two ways (the standard way and the way suggested by Mortensen and Lavers).
(b) The theorem stating the intuitionistic character of the internal logic should be read rather as follows: Under certain conditions $c_1, \ldots, c_n$, most of them extra-categorial, (S1) is the case.
(c) The universal, invariant laws of mathematics are not those of intuitionistic logic. They seem to be so only when $c_1, \ldots, c_n$ are adopted.

This leaves some questions open, though. Among them I can mention the following ones:

- Does the categorial structure of toposes support other logics as readily as it supports intuitionistic logic and its topologico-algebraic dual?
- What are all those conditions $c_1, \ldots, c_n$? The names of certain morphisms are clearly among them, but perhaps the underlying notion of logical consequence should be also taken into account.
- Does the categorial structure of toposes supports a minimal logic, stable under all changes of conditions $c_1, \ldots, c_n$?
- If intuitionistic laws cannot be regarded as the universal laws of mathematics, what are these, if any?

References

1. Awodey, S.: Structure in mathematics and logic: a categorical perspective. Philos. Math. **4**(3), 209–237 (1996)
2. Awodey, S.: Continuity and logical completeness: an application of sheaf theory and topoi. In: Johan van Benthem, M.R., Heinzmann, G., Visser, H. (eds.) The Age of Alternative Logics. Assesing Philosophy of Logic and Mathematics Today, pp. 139–149. Springer, The Netherlands (2006)
3. Bell, J.L.: Lectures on the foundations of mathematics. http://publish.uwo.ca/~jbell/foundations%20of%20mathematics.pdf
4. Bell, J.L.: From absolute to local mathematics. Synthese **69**, 409–426 (1986)
5. Bell, J.L.: Toposes and Local Set Theories: An Introduction. Oxford Clarendon Press (1988)
6. Bell, J.L.: Oppositions and paradoxes in mathematics and philosophy. Axiomathes **15**(2), 165–180 (2005)
7. Béziau, J.-Y.: S5 is a paraconsistent logic and so is first-order classical logic. Logical Investigations **9**, 301–309 (2002)
8. Borceux, F.: Handbook of Categorical Algebra. Volume 3: Categories of Sheaves.In: Encyclopedia of Mathematics and its Applications, vol. 50. Cambridge University Press (1994)
9. Carnielli, W.A., Coniglio, M.E., D'Ottaviano, I.M.L.: New dimensions on translations between logics. Log. Univers. **3**(1), 1–18 (2009)
10. Czermak, J.: A remark on Gentzen's calculus of sequents. Notre Dame J. Form. Logic **18**(3), 471–474 (1977)
11. de Queiroz, G.d.S.: On the duality between intuitionism and paraconsistency. Ph.D. thesis, Universidade Estadual de Campinas (1998). In Portuguese
12. Estrada-González, L.: Quaternality in a topos-theoretical setting. Unpublished typescript, 201X
13. Estrada-González, L.: Complement-topoi and dual intuitionistic logic. Australas. J. Logic **9**, 26–44 (2010)
14. Estrada-González, L.: Models of possibilism and trivialism. Logic Log. Philos. **22**(4), 175–205 (2012)
15. Estrada-González, L.: Disputing the many-valuedness of topos logic. In: Béziau, J.-Y., D'Ottaviano, I.M.L., Costa-Leite, A. (eds.) Aftermath of the Logical Paradise, pp. 193–213. CLE-Universidade Estadual de Campinas, Campinas (2015)
16. Estrada-González, L.: From (paraconsistent) topos logic to Universal (topos) logic. In: Koslow, A., Buchsbaum, A.R.d.V. (eds.) The Road to Universal Logic. Festschrift for Jean-Yves Béziau, vol. 2, pp. 263–295. Birkhäuser (2015)
17. Goldblatt, R.: Grothendieck topology as geometric modality. Math. Logic Q. **27**(31–35), 495–529 (1981)
18. Goldblatt, R.: Topoi: The Categorial Analysis of Logic. Studies in Logic and the Foundations of Mathematics, vol. 98. North Holland Publishing Co., Amsterdam (1984). Revised edition
19. Goodman, N.: The logic of contradiction. Zeitschrift für Mathematische Logik und Grundlagen der Mathematik **27**(8–10), 119–126 (2002)
20. Goré, R.P.: Dual intuitionistic logic revisited. In: Proceedings of the International Conference on Automated Reasoning with Analytic Tableaux and Related Methods. Lecture Notes In Computer Science, vol. 1847, pp. 252–267. Springer, London (2000)
21. Iturrioz, L.: Les algèbres de Heyting-Brouwer et de Łukasiewicz trivalentes. Notre Dame J. Form. Logic **17**(1), 119–126 (1981)
22. James, W., Mortensen, C.: Categories, sheaves, and paraconsistent logic. Unpublished typescript (1997)
23. Lawvere, F.W: Quantifiers and sheaves, In: Berger, M., Dieudonné, J., Leray, J., Lions, J.L., Malliavin, P., and Serre, J.P. (eds.) Actes du Congrés International des Mathematiciens, 329–334, Nice, Gauthier-Villars (1970)
24. Lawvere, F.W.: Continuously variable sets: algebraic geometry = geometric logic. In: Rose, H.E., Shepherdson, J.C. (eds.) Logic Colloquium '73 (Bristol, 1973), pp. 135–156. North Holland, Amsterdam (1975)

25. Lawvere, F.W.: Variable quantities and variable structures in topoi. In: Heller, A., Tierney, M. (eds.) Algebra, Topology, and Category Theory. A Collection of Papers in Honor of Samuel Eilenberg, pp. 101–131. Academic Press (1976)
26. Lawvere, F.W.: Introduction. In: Lawvere, F.W., Schanuel, S.H. (eds.) Categories in Continuum Physics. Springer (1982)
27. Lawvere, F.W.: Intrinsic co-Heyting boundaries and the Leibniz rule in certain toposes. In: Carboni, A., Pedicchio, M.C., Rosolini, G. (eds.) Proceedings of the 1990 Meeting on Category Theory held in Como, Italy. Lecture Notes in Mathematics, vol. 1488, pp. 279–281. Springer (1991)
28. Lawvere, F.W.: Tools for the advancement of objective logic: closed categories and toposes. In: Macnamara, J., Reyes, G.E. (eds.) The Logical Foundations of Cognition, pp. 43–56. Oxford University Press (1994)
29. Lawvere, F.W., Rosebrugh, R.: Sets for Mathematics. Cambridge University Press, Cambridge (2003)
30. Lawvere, F.W., Schanuel, S.H.: Conceptual Mathematics. A First Introduction to Categories. Cambridge University Press (2000). Second reprint
31. Mac Lane, S., Moerdijk, I.: Sheaves in Geometry and Logic: A First Introduction to Topos Theory. Springer (1992)
32. Macnab, D.S.: An Algebraic Study of Modal Operators on Heyting Algebras, with Applications to Topology and Sheafification. Ph.D. thesis, University of Aberdeen (1976)
33. Macnab, D.S.: Modal operators on Heyting algebras. Algebra Univers. **12**(1), 5–29 (1981)
34. Marcos, J.a.: Ineffable inconsistencies. In: Béziau, J.-Y., Carnielli, W.A. (eds.) Paraconsistency with No Frontiers, pp. 301–311. Elsevier, Amsterdam (2006)
35. McKinsey, J.C.C., Tarski, A.: Some theorems about the sentential calculi of Lewis and Heyting. J. Symb. Logic **13**(1), 1–15 (1948)
36. McLarty, C.: Book review: John Bell. Introduction to toposes and local set theory. Notre Dame J. Form. Logic **31**(1), 150–161 (1989)
37. McLarty, C.: Elementary Categories. Elementary Toposes. Oxford Clarendon Press, Toronto (1995)
38. McLarty, C.: Two constructivist aspects of category theory. Philosophia Scientiæ **27**(Cahier spécial 6), 95–114 (2006)
39. Mortensen, C.: Inconsistent Mathematics. Kluwer Mathematics and Its Applications Series. Kluwer Academic Publishers (1995)
40. Mortensen, C.: Closed set logic. In: Brady, R.T. (ed.) Relevant Logics and Their Rivals, vol. II, pp. 254–262. Ashgate Publishing, Aldershot (2003)
41. Popper, K.: On the theory of deduction, parts I and II. Indagationes Mathematicae **10**, 173–183 and 322–331 (1948)
42. Rauszer, C.: Semi-Boolean algebras and their applications to intuitionistic logic with dual operations. Fundam. Math. **83**, 219–249 (1974)
43. Rauszer, C.: An algebraic and Kripke-style approach to a certain extension of intuitionistic logic. Diss. Math. **167**, 1–62 (1980)
44. Reyes, G., Zawadowski, M.: Formal systems for modal operators on locales. Stud. Logica **52**(4), 595–613 (1993)
45. Reyes, G., Zolfaghari, H.: Topos-theoretic approaches to modality. In: Carboni, A., Pedicchio, M.C., Rosolini, G. (eds.) Proceedings of the 1990 Meeting on Category Theory held in Como, Italy. Lecture Notes in Mathematics, vol. 1488, pp. 359–378. Springer (1991)
46. Reyes, G., Zolfaghari, H.: Bi-Heyting algebras, toposes and modalities. J. Philos. Logic **25**(1), 25–43 (1996)
47. Rodin, A.: Axiomatic Method and Category Theory. http://arxiv.org/pdf/1210.1478v1.pdf
48. Shramko, Y.: Dual intuitionistic logic and a variety of negations: the logic of scientific research. Stud. Logica **80**(2–3), 347–367 (2005)
49. Tierney, M.: Sheaf theory and the continuum hypothesis. In: Lawvere, F.W. (ed.) Toposes, Algebraic Geometry and Logic. Lecture Notes in Mathematics, vol. 274, pp. 13–42. Springer (1972)

50. Urbas, I.: Dual intuitionistic logic. Notre Dame J. Form. Logic **37**(3), 440–451 (1996)
51. van Inwagen, P.: Metaphysics. Westview Press, Boulder, Colorado (2008). Third Edition
52. Vasyukov, V.: Structuring the universe of universal logic. Log. Univers. **1**(2), 277–294 (2007)
53. Vigna, S.: A guided tour in the topos of graphs. Technical Report 199–7, Università di Milano, Dipartimento di Scienze dell'Informazione (1997). http://vigna.dsi.unimi.it/papers. ToposGraphs.pdf

Chapter 19
Topological Semantics for da Costa Paraconsistent Logics C_ω and C_ω^*

Can Başkent

Abstract In this work, we consider a well-known and well-studied system of paraconsistent logic which is due to Newton da Costa, and present a topological semantics for it.

Keywords Paraconsistent logic · Topological semantics · da Costa logics

Mathematics Subject Classification (2000) 03B53 · 03B45

19.1 Introduction

In a paraconsistent logic, contradictions do not entail everything. Namely, in a paraconsistent logic, there are some formula φ, ψ such that $\{\varphi, \neg\varphi\} \nvdash \psi$ for a logical consequence relation $\vdash$. In this work, we will focus on a well-known and well-studied paraconsistent logic, which is due to Newton da Costa, and present a topological semantics for it.

Da Costa's hierarchical systems C_n and C_n^* are one of the earliest examples of paraconsistent logic [7]. Da Costa systems C_n where $n < \omega$ are consistent and finitely trivializable. Yet, for the limit ordinal ω, it is possible to obtain a logic C_ω which is not finitely trivializable [8]. In this work, we focus on C_ω and its first-order cousin C_ω^* both of which are paraconsistent.

Da Costa systems are not unfamiliar. As Priest remarked, the logic C_ω can be thought of as the positive intuitionistic logic with dualized negation to give truth value gluts [21]. We define C_ω with the following postulates [7, 8].

C. Başkent (✉)
Department of Computer Science, University of Bath, Bath BA2 7AY, UK
e-mail: c.baskent@bath.ac.uk

J.-Y. Beziau et al. (eds.), *New Directions in Paraconsistent Logic*,
Springer Proceedings in Mathematics & Statistics 152,
DOI 10.1007/978-81-322-2719-9_19

1. $\varphi \to (\psi \to \varphi)$
2. $(\varphi \to \psi) \to ((\varphi \to (\psi \to \chi)) \to (\varphi \to \chi))$
3. $\varphi \wedge \psi \to \varphi$
4. $\varphi \wedge \psi \to \psi$
5. $\varphi \to (\psi \to \varphi \wedge \psi)$
6. $\varphi \to \varphi \vee \psi$
7. $\psi \to \varphi \vee \psi$
8. $(\varphi \to \chi) \to ((\psi \to \chi) \to (\varphi \vee \psi \to \chi))$
9. $\varphi \vee \neg\varphi$
10. $\neg\neg\varphi \to \varphi$

The rule of inference that we need is modus ponens: $\varphi, \varphi \to \psi \therefore \psi$.

Based on this axiomatization, Baaz gave a Kripke-type semantics for C_ω [3]. Baaz's C_ω-Kripke model is a tuple $M = \langle W, \leq, V, T \rangle$ where W is a nonempty set, $\leq$ is a partial order on W, and V is a valuation that returns a subset of W for every propositional variable in the language. We will call the members of W as *possible worlds* or *states*. By *accessible states from* $w \in W$, we will mean the set $\{v : wRv\}$. The additional component T is a function defined from possible worlds to the sets of negated propositional forms. The imposed condition on T is the monotonicity: $w \leq v$ implies $T(w) \subseteq T(v)$. Monotonicity condition resembles the hereditary condition of intuitionistic logic. The valuation respects the monotonicity and is assumed to return upsets. In this context, an upset U is a subset of W such that if $w \in U$ and $w \leq v$, then $v \in U$.

Also note that the relation $\leq$ is a partial order rendering the frame $\langle W, \leq \rangle$, an **S4**-frame. The fact that the frame of Baaz's model is **S4** and will be central in our topological investigations later.

One of the most interesting properties of da Costa systems is the principle of nonsubstitution for negated formulas. For instance, even if p and $p \wedge p$ are logically equivalent, i.e., $p \equiv p \wedge p$, we do not necessarily have that $\neg p \equiv \neg(p \wedge p)$ in da Costa systems, where $\equiv$ denotes logical equivalence. In Baaz's construction, the function T returns a set of formulas which are negated at that possible world. Yet, for a possible world w, the set $T(w)$ is not necessarily a theory as it need not be closed under logical equivalence. In short, at w, we can have $\neg p \in T(w)$, but this does *not* imply that $\neg(p \wedge p) \in T(w)$. Monotonicity of T, on the other hand, reflects the *intuitionistic* side of da Costa systems. In the partially ordered Kripkean frame for C_ω, children nodes have the same formulas as their parents and possibly more under T.

Baaz gave a Kripkean semantics for C_ω as follows [3]. But first, let us set up some notation. We put $\neg^0\varphi \equiv \varphi$ and $\neg^{n+1}\varphi \equiv \neg(\neg^n\varphi)$ for a formula φ which does not include a negation sign in the front.

$w \models p$	iff for all v such that $w \leq v$, $v \models p$ for atomic p
$w \models \varphi \wedge \psi$	iff $w \models \varphi$ and $w \models \psi$
$w \models \varphi \vee \psi$	iff $w \models \varphi$ or $w \models \psi$
$w \models \varphi \supset \psi$	iff for all v such that $w \leq v$, $v \models \varphi$ implies $v \models \psi$
$w \models \neg^1 \varphi$	iff $\neg^1 \varphi \in T(w)$ or $\exists v.v \leq w$ and $v \not\models \varphi$
$w \models \neg^{n+2} \varphi$	iff $\neg^{n+2} \varphi \in T(w)$ and $w \models \neg^n \varphi$, or $\exists v.v \leq w$ and $v \not\models \neg^{n+1} \varphi$
$w \models \varphi_1, \ldots, \varphi_n \rightarrow \psi_1, \ldots, \psi_n$	iff $\forall v.w \leq v$, $w \models \varphi_1, \ldots, w \models \varphi_n$ imply $v \models \psi_1$ or ...or $v \models \psi_n$

Let us now briefly comment on the semantics. First of all, the above semantics admits the hereditary condition for propositional variables. The truth of propositional variables persists throughout the accessible states. This is an interesting property that resembles what is commonly known as the hereditary condition of intuitionistic logic. Another similarity between da Costa systems and the intuitionistic logic is the way the semantics of implication is defined. Perhaps the most interesting and distinguishing part of the above semantics is the semantics of negation. A negated formula is true at a state w if it is in $T(w)$ or there is a predecessor state at which the formula does not hold. It is important to underline that the function T renders the negation as a nonfunctional operator. This is another way of saying that substitution principle for negated formulas does not hold in da Costa systems. Finally, in the syntax of the operator $\rightarrow$, we can very well have the empty set as the antecedent. The statement $\emptyset \rightarrow p, q$ will be shortened as $\rightarrow p, q$.

Using the proof theory of (propositional) intuitionistic logic and Gentzen style calculus, Baaz showed the soundness, completeness, and decidability of this system [3]. We will henceforth denote this system as KC_ω.

19.2 Topological Models TC_ω

In this section, we give a topological semantics for da Costa's system C_ω, and call our formalism as TC_ω. The topological semantics precedes the Kripke semantics, and was first presented in early 1920s [12]. The major developments in the field of topological semantics for (modal) logics have been initiated by J.C.C. McKinsey and Alfred Tarski in 1940s in a series of papers [14–16].

A topology σ is a collection of subsets of a set S that satisfies the following conditions. The empty set and S are in σ, and σ is closed under arbitrary unions and finite intersections. The elements of σ are called *opens*. Complement of an open set is called a *closed set*. A *topological space* is defined as the tuple (S, σ). For a given set U, the interior operator Int returns the largest open set contained in U whereas the closure operator Clo returns the smallest closed set that contains U. For a set U, we define the boundary of U as $\mathsf{Clo}(U) - \mathsf{Int}(U)$, and denote it as $\partial(U)$. Therefore, by definition, closed sets include their boundary whereas open sets do not.

In the classical modal case, McKinsey and Tarski associated the modal operator $\Box$ with the topological interior operator (and, dually $\Diamond$ with the closure operator), and observed that the interior and closure operators behave as **S4** modalities (normal, reflexive, and transitive). The well-known McKinsey–Tarski result showed that **S4** is the modal logic of topological spaces, in fact, of any metric, separable, dense-in-itself space. This result has been extended to various other nonclassical logics, and the topological semantics for intuitionistic and some paraconsistent logics have also been given [4, 13, 17].

In this section, we first give a topological semantics for C_ω based on the Kripkean semantics, and then we discuss various topological notions that relate topological spaces and TC_ω.

19.2.1 Topological Semantics

The language of TC_ω is the language of propositional logic with the usual Boolean conjunction, disjunction, and implication, and we will allow iterated negations. We denote the closure of a set X by $\mathsf{Clo}(X)$. If a set $\{x\}$ is a singleton, we write $\mathsf{Clo}(x)$ instead of $\mathsf{Clo}(\{x\})$ provided no confusion arises. Also note that in this case $\mathsf{Clo}(x)$ is the intersection of all closed sets containing x.

The language of TC_ω is built using a countable set of propositional variables which we denote by $\mathbf{P}$. Now, we start with defining TC_ω models.

Definition 19.2.1 A TC_ω model M is a tuple $M = \langle S, \sigma, V, N \rangle$ where S is a non-empty set, σ is an Alexandroff topology on S, $V : \mathbf{P} \to \wp(S)$ is a valuation function, and N is a (full) function which takes possible worlds $s \in S$ as inputs and returns sets of negated propositional forms (possibly empty) in such a way that $w \in \mathsf{Clo}(v)$ implies $N(w) \subseteq N(v)$.

Here, we resort to the standard translation between topological models and Kripke frames. Given a topological space, we put $w \leq v$ for $w \in \mathsf{Clo}(v)$ to obtain a partially ordered tree, which produces the Kripke frame. Conversely, given a Kripke frame with a partial order, we consider the upward closed (or dually, downward closed) branches of the tree as open sets (dually, closed sets) to construct a topology. The topology we obtain from a given Kripke frame is an Alexandroff topology which is closed under arbitrary intersections. In other words, since the Baaz's frames are already **S4**, the topology we obtain (after translating the given **S4** frame) is an Alexandroff topology. We refer the reader to [23] for a detailed treatment of the subject from a modal logical perspective.

Interestingly, the fact that we obtain Alexandroff spaces in TC_ω raises the question of handling non-Alexandroff spaces in the topological models of C_ω. This is a very interesting question in-self, and can help us identify a variety of formalisms that are weaker than C_ω. In order not to digress from our current focus, we leave it for a future work.

Now, we give the semantics of TC_ω as follows. We abbreviate $\neg^0\varphi := \varphi$, and $\neg^{n+1}\varphi := \neg(\neg^n\varphi)$ for a φ which does not include a negation sign in the front. Similarly, we assume that the valuation function V returns closed sets [4].

$w \models p$	iff $\forall v.w \in \mathsf{Clo}(v)$, $v \models p$ for atomic p
	iff $w \in V(p)$
$w \models \varphi \wedge \psi$	iff $w \models \varphi$ and $w \models \psi$
$w \models \varphi \vee \psi$	iff $w \models \varphi$ or $w \models \psi$
$w \models \varphi \supset \psi$	iff $\forall v.w \in \mathsf{Clo}(v)$, $v \models \varphi$ implies $v \models \psi$
$w \models \neg^1 \varphi$	iff $\neg^1 \varphi \in N(w)$ or $\exists v.v \in \mathsf{Clo}(w)$ and $v \not\models \varphi$
$w \models \neg^{n+2} \varphi$	iff $\neg^{n+2}\varphi \in N(w)$ and $w \models \neg^n \varphi$ or $\exists v.v \in \mathsf{Clo}(w)$ and $v \not\models \neg^{n+1}\varphi$
$w \models \varphi_1, \ldots, \varphi_n \to \psi_1, \ldots, \psi_n$	iff $\forall v.w \in \mathsf{Clo}(v)$, $v \models \varphi_1, \ldots, v \models \varphi_n$ imply $v \models \psi_1$ or ...or $v \models \psi_n$

Following the usual representation, we denote the extension of a formula φ in a model M by $[\varphi]^M$, and define it as follows $[\varphi]^M := \{w : M, w \models \varphi\}$.

Now we can discuss the satisfiability problem (SAT) and its complexity in logic C_ω and KC_ω. First of all, note that the complexity of SAT for basic modal logic is known to be PSPACE-complete. In Kripkean frames, searching for a satisfying assignment may not be efficient timewise, but it uses the space efficiently yielding a PSPACE-complete complexity. This procedure can be thought of as searching the branches of a Kripke model (which is a tree or a forest) starting from the root. Once you are done with one branch, you do not need to remember it, thus you can reuse the same space. And, the extent of the tree you need to search, i.e., the depth, solely depends on the length of the formula. Therefore, the given formula determines the space you need to check. In KC_ω, the only issue is checking the satisfiability for negation. However, a careful examination shows that it has a rather immediate solution. The case for $\neg^1$ requires two operations: check whether a given $\neg^1$ is in the image set of T at the given state, and check if there exists a state that is accessible from the current state with the desired condition. The latter part is PSPACE considering the standard modal argument for SAT. The prior part is also polynomial—it is a sequential check for membership. Moreover, one can easily construct a polynomial transformation from modal SAT with topological semantics to KC_ω satisfiability yielding the fact that SAT for KC_ω is also PSPACE. Considering $\neg^n$ as a nested (intuitionistic) modality, one can come up with the obvious translation giving the PSPACE-completeness of the satisfiability problem for KC_ω. In order to show the complexity of TC_ω, we need to reduce it to KC_ω. Yet, we already saw how to obtain a topological model given a Kripke model, and this translation reduces KC_ω to TC_ω.

Now, based on the above-mentioned argument, and the efficient model transformations between topological spaces and Kripke frames which we discussed earlier, it is immediate to observe that SAT for TC_ω is also PSPACE-complete.

Theorem 19.2.2 *The satisfiability problem for both KC_ω and TC_ω is PSPACE-complete.*

Corollary 19.2.3 *Both KC_ω and TC_ω are decidable.*

In his work, Baaz gave several results using Kripke semantics [3]. Here, we observe that they hold in TC_ω as well. Our aim is to clarify the use of topological concepts in TC_ω, and make sure that the function N works as expected. The following results will also exemplify the behavior of negation in TC_ω.

Proposition 19.2.4 *$w \models \varphi$ iff for all v such that $w \in \mathsf{Clo}(v)$, we have $v \models \varphi$.*

Proof The proof is by induction on the length of the formula. The only interesting case is the negation. We assume $\varphi \equiv \neg^1\psi$. Then, let us suppose $w \models \neg^1\psi$. By definition, either $\neg^1\psi \in N(w)$ or there exists a x such that $x \in \mathsf{Clo}(w)$ and $x \not\models \psi$. Now, let v be such that $w \in \mathsf{Clo}(v)$. Then, by the definition of N, we observe $N(w) \subseteq N(v)$. Thus, $\neg^1\psi \in N(v)$. On the other hand, $w \in \mathsf{Clo}(v)$ implies that $\mathsf{Clo}(w) \subseteq \mathsf{Clo}(v)$. Therefore, $w \in \mathsf{Clo}(w) \subseteq \mathsf{Clo}(v)$ with $v \not\models \psi$. Then, we have either $\neg^1\psi \in N(w)$ or there exists x such that $x \in \mathsf{Clo}(v)$ with $x \not\models \psi$. Thus, $v \models \neg^1\psi$.

The cases for $\neg^{n+1}$ are similar using the induction hypothesis. □

Proposition 19.2.5 *$w \not\models \varphi$ implies that there is no $v \in \mathsf{Clo}(w)$ such that $v \not\models \neg\varphi$.*

Proof Let $w \not\models \varphi$. Toward a contradiction, we assume that there is a $v \in \mathsf{Clo}(w)$ with $v \not\models \neg\varphi$. On the other hand, by Proposition 19.2.4, $v \not\models \neg\varphi$ means that for all w such that $v \in \mathsf{Clo}(w)$, we have $w \not\models \neg\varphi$. Thus, we conclude $w \not\models \varphi$ and $w \not\models \neg\varphi$. Contradiction. □

Proposition 19.2.6 *$w \models \neg\neg\varphi \to \varphi$.*

Proof We will show that $w \not\models \varphi$ implies $w \not\models \neg\neg\varphi$. Let $w \not\models \varphi$. Then, by Proposition 19.2.5, there is no $v \in \mathsf{Clo}(w)$ with $v \not\models \neg\varphi$. Then, by definition of $\neg^2$, we conclude that $w \not\models \neg\neg\varphi$. □

Substitution principle for negated formulas $\neg\varphi \leftrightarrow \neg(\varphi \wedge \varphi)$ does not hold in KC_ω. Next, we observe that it does not hold in TC_ω as well.

Proposition 19.2.7 *$\neg\varphi \leftrightarrow \neg(\varphi \wedge \varphi)$ is not valid.*

Proof Let us take a state w such that $\mathsf{Clo}(w) \subseteq [\varphi]$ and $\neg\varphi \in N(w)$. Thus, $w \models \neg\varphi$. We now stipulate further that $\neg(\varphi \wedge \varphi) \notin N(w)$ to get a countermodel. □

Proposition 19.2.8 *$w \models\to \varphi, \neg\varphi$.*

Proof We recall that $\to \varphi_1, \ldots, \varphi_n$ means that $\varphi_1 \vee \cdots \vee \varphi_n$ holds. Then, the result follows from the axiomatization of C_ω. □

For the completeness of our arguments in this work, we now present the semantical counterpart of cut elimination. The proof is a straightforward manipulation of formulas, hence, we skip it.

Proposition 19.2.9 *$w \models \Pi \to \Gamma, \varphi$ and $w \models \varphi, \Delta \to \Lambda$ imply $w \models \Pi, \Delta \to \Gamma, \Lambda$.*

We now state the soundness theorem without a proof.

Theorem 19.2.10 *$\vdash \varphi \to \psi$ implies $\models \varphi \to \psi$.*

Baaz used Gentzen style sequent calculus to show the completeness of his system. He then concluded that if $\Pi \rightarrow \Gamma$ is not provable without cuts, there is a KC_ω-Kripke model $M = \langle W, \leq, v, T \rangle$ such that $0 \in W$ and $0 \not\models \Pi' \rightarrow \Gamma'$ where $\Pi' \equiv \Pi, \Delta$ and $\Gamma' \equiv \Gamma, \Psi$. Namely, $M \not\models \Pi \rightarrow \Gamma$. Here, 0 is the lowest top sequent in the reduction tree of $\Pi \rightarrow \Gamma$.

Now, in order to show the completeness of our system TC_ω, we will resort to the model translation which we mentioned earlier. Given a KC_ω model $M = \langle W, \leq\ , v, T \rangle$, we can construct a TC_ω model $M' = \langle S, \sigma, V, N \rangle$ as follows. Let $S := W$, and $V := v$. Now, we need to define the topology σ, and the open and closed sets in σ. Define closed sets as the upsets, and observe that $v \in \mathsf{Clo}(w)$ whenever $v \leq w$. For a tree model, it is easy to observe that the closed sets we defined produces an Alexandroff topology, as we already observed. Furthermore, we put $N(w) := T(w)$. Therefore, given a TC_ω model, we can effectively convert it to KC_ω which is known to be complete. This is the immediate method to show the completeness of TC_ω. Alternatively, we can start with the logic TC_ω, and give a topological completeness proof. This is what we achieve next.

For the completeness of TC_ω, we use *maximal nontrivial sets of formulas*. A set X is called *trivial* if every formula in the language is deducible from X, otherwise it is called nontrivial. A nontrivial set X is called a *maximal nontrivial set* if $\varphi \notin X$, then $X \cup \{\varphi\}$ is trivial, for an arbitrary formula φ.

We start by observing the following:

Proposition 19.2.11 *If Γ is a maximal nontrivial set of formulas, then we have $\Gamma \vdash \varphi$ iff $\varphi \in \Gamma$.*

Using canonical sets, we construct the canonical TC_ω model $\langle S', \sigma', V', N' \rangle$. Let us first start with the canonical topological space. The canonical topological space is the pair $\langle S', \sigma' \rangle$ where S' is the set of all maximal nontrivial sets, and σ' is the set generated by the basis $B = \{\widehat{\neg\varphi} :$ any formula $\varphi\}$ where we define $\widehat{\varphi} := \{s' \in S' : \varphi \in s'\}$. Here, our construction is very similar to the classical case: instead of (classical) modal formula, we use negated formulas in the construction of the canonical model (and its topology). The reason for this choice is the fact that in TC_ω negated formulas resort to the closure operator—similar to the modal operators in the classical case.

In order to show that B is a basis for the topology σ', we need to show that

1. For any $U, U' \in B$ and any $x \in U \cap U'$, there is $U_x \in B$ such that $x \in U_x \subseteq U \cap U'$,
2. For any $x \in S'$, there is $U \in B$ with $x \in U$.

For the first item, we observe that $\widehat{\neg\varphi \wedge \chi} = \widehat{\neg\varphi} \wedge \widehat{\neg\chi}$. Therefore, $U \cap U' \in B$ which argues for finite intersection.

For the second item, we observe that $\neg\bot \in x$ for any maximal consistent set x in the canonical TC_ω. Therefore, for any $x \in S'$, there is a $\widehat{\neg\bot} \in B$ that includes x.

This argument shows that B is a basis for the topology of the canonical model.

Now, the valuation V' is defined in the standard way: $V'(p) := \{s' \in S' : p \in s'\}$. Similarly, we define N' from S' to sets of formulas, and put, $N'(s') \subseteq N'(t')$ if

$s' \in \mathsf{Clo}(t')$ for $s', t' \in S'$. Additionally, we impose that $N'(s') \subseteq s'$ to handle the negated formula correctly. Another way of looking at it is to include $N'(s')$ into s' in the construction of the maximal nontrivial set s'. Therefore, we close maximally consistent sets under logical connectives and also under the N' function. This simply reflects how the negation is defined in TC_ω.

The truth of classical Booleans are defined as usual in the canonical models. For negation, we put the following.

$$s' \models \neg^1\varphi \text{ iff } \neg^1\varphi \in N'(s') \text{ or } \exists t' \in \mathsf{Clo}(s') \text{ such that } t' \not\models \varphi$$

For the truth lemma, we only need to observe that, $s' \models \varphi$ if and only if $\varphi \in s'$.

The standard Boolean cases are immediate. So, let us take $\varphi = \neg^1\psi$ for some ψ. For "truth to membership" direction, if $\neg^1\psi \in N'(s')$, then we are done as $N'(s') \subseteq s'$. Otherwise, we need to find a t' in $\mathsf{Clo}(s')$ which does not satisfy ψ. Since the topology σ is constructed using a basis with opens, we can select t' from the boundary $\partial(s')$ which is not in the interior of the extension, but in the closure of the extension by definition.

For instance, if the space is discrete and the boundary is empty, then we can take any point from s' as each subset of the space is clopen (both closed and open) so that $\mathsf{Clo}(s') = s' = \mathsf{Int}(s')$. Therefore, let us here argue assuming that the boundary is not empty (if it is, we still know what to do as described above).

Take such a $t' \in \partial(s')$ such that $t' \not\models \psi$. Then, by the induction hypothesis, $\psi \notin t'$. The set t' is maximal and nontrivial, so $\neg^1\psi \in t'$. Recall that $t' \in \mathsf{Clo}(s')$, thus $\neg^1\psi \in \mathsf{Clo}(s')$.

This was the direction from "truth to membership". The direction from "membership to truth" is similar using some properties of closure operators, so we skip it. Similarly, we leave the case $\varphi = \neg^{n+2}\psi$ to the reader as it only requires an inductive proof.

After establishing the truth lemma, we have the following completeness result.

Theorem 19.2.12 *For any set of formulas Σ in TC_ω, if $\Sigma \models \varphi$ then, $\Sigma \vdash \varphi$.*

Proof Assume $\Sigma \nvdash \varphi$. Then, $\Sigma \cup \{\neg\varphi\}$ is nontrivial, and can be extended to a maximal nontrivial set Σ'. By the truth lemma, $\Sigma' \models \neg\varphi$ yielding $\Sigma' \not\models \varphi$. This is the countermodel we were looking for. □

So far, we have showed that Baaz's results in KC_ω can be carried over to TC_ω without much difficulty. This is achieved relatively easily as a consequence of the immediate and effective translation between KC_ω and TC_ω, and the similarity between the classical modalities and the da Costa negation operator. Such similarities between classical modalities and paraconsistent operators were also addressed in some other work [5, 6].

19.2.2 Further Results

In this section, we reconsider TC_ω models in various topological spaces, and investigate how topological properties and TC_ω models interact. Here, our focus will be separation axioms, regular spaces, and connected spaces. The main motivation behind choosing these structures is the fact that the semantics of the negation operator in TC_ω deals with the closure (and then indirectly, with the boundary) of the sets. Thus, topological notions that are relevant to the boundary become our main subject in this section.

We also remind the reader that our treatment is by no mean exhaustive. Various other topological, mereotopological, and geometrical notions can further be investigated within the framework of da Costa logics or paraconsistent logics in general. Nevertheless, in this work, we confine ourselves to the aforementioned issues, and leave the rest for future work.

19.2.2.1 Separation Axioms

Let us first recall some of the well-known separation axioms for topological spaces. Two points are called *topologically indistinguishable* if both have the same neighborhoods. They are topologically distinguishable if they are not topologically indistinguishable. Indiscrete space (trivial topology) is perhaps the simplest example where any two points are topologically indistinguishable. Moreover, two points are *separated* if each of the points has a neighborhood which is distinct from the other's neighborhoods. Two points x, y are *distinct* if $x \neq y$.

Separation axioms present an interesting perspective to analyze paraconsistent models. Traditionally, paraconsistent logics are known as the logics with truth value *gluts* as opposed to intuitionistic logics which have truth value *gaps*. Theory of truth value gluts suggests that some propositions can have multiple (including inconsistent) truth values. Topological models then identifies the superimposed truth values with the intersection of sets that denote the truth and falsity of logical formulas. Separation axioms become relevant when we want to *separate* the superimposed truth values in order to render the model and the formulas consistent.

Let us now define the separation axioms that we need. A topological space is called

- $\mathbf{T_0}$ if any two distinct points in it are topologically distinguishable,
- $\mathbf{T_1}$ if any two distinct points in it are separated,
- $\mathbf{R_0}$ if any two topologically distinguishable points are separable,
- $\mathbf{T_2}$ if any two distinct points in it are separated by neighborhoods,
- $\mathbf{T_{2^1/_2}}$ if any two distinct points in it are separated by closed neighborhoods.

While discussing the semantics of TC_ω above, we made use of the relation $w \in \mathsf{Clo}(v)$ quite often. This relation is called *the specialization order*: $w \leq v$ if and only if $w \in \mathsf{Clo}(v)$. It is a partial order if and only if the space is $\mathbf{T_0}$. In this case, if the relation $\leq$ is symmetric, then the space we obtain is $\mathbf{R_0}$. Throughout the paper, we will call a model a $\mathbf{T_x}$-model if its topological space is a $\mathbf{T_x}$ space for $x \in \{0, 1, 2, 2^1\!/2\}$.

We do not force TC_ω models to be $\mathbf{T_1}$ models or even $\mathbf{R_0}$ models. Then the natural question is the following: Can we have TC_ω models which are not even $\mathbf{T_0}$ or $\mathbf{T_1}$?

Proposition 19.2.13 *Given a KC_ω model M, the TC_ω model M' obtained from M is $\mathbf{T_0}$.*

Proof Given a KC_ω model M, the specialization order that we defined above, generates a TC_ω model M'. In this case, the topology we obtain in M' is an Alexandroff topology as the specialization order of the Alexandroff topology is precisely the partial order that comes from the Kripke model. Therefore, since the specialization order is a preorder, the space we obtain is $\mathbf{T_0}$, so is M'. □

In Proposition 19.2.13, the model M' is proved to be $\mathbf{T_0}$. Therefore, it is worthwhile to note that M' is not necessarily $\mathbf{T_1}$. Alexandroff spaces are $\mathbf{T_1}$ if only if they are discrete—each s having a neighborhood of $\{s\}$ only [1].

Now, we focus on $\mathbf{T_{2^1/2}}$ spaces as the closed sets and closure operator play a central role in paraconsistent semantics. Our main theorem is the following:

Theorem 19.2.14 *Let $M = \langle S, \sigma, V, N\rangle$ be a $\mathbf{T_{2^1/2}}$ TC_ω model which admits true contradictions, then N cannot be empty.*

Proof In TC_ω (and similarly in KC_ω) models, N (or T) function tracks the negated formulas in an *ad hoc* way. In this fashion, nonemptiness of N means that the model cannot have superimposition of truth values which can produce inconsistencies. Intuitively, this is because of the assumption of the separation axiom. Let us now see the proof.

Let $M = \langle S, \sigma, V, N\rangle$ be a $\mathbf{T_{2^1/2}}$ TC_ω model. Assume N is empty. Let w be a state where we have a true contradiction $\varphi \wedge \neg\varphi$ for some φ. Thus, $w \models \varphi$, and moreover, since N is empty, there is $v \in \mathsf{Clo}(w)$ such that $v \not\models \varphi$. Since we are in a $\mathbf{T_{2^1/2}}$ space, w and v must be separable. However, since $v \in \mathsf{Clo}(w)$, it means that v is in the intersection of all closed sets in σ containing w. Thus, they are not separable by closed neighborhoods. Contradiction. Thus, N cannot be empty, and such a point v cannot exist in a $\mathbf{T_{2^1/2}}$ space that admit true contradictions. □

The contrapositive of Theorem 19.2.14 can also be useful, let us specify it here.

Proposition 19.2.15 *Let $M = \langle S, \sigma, V, N\rangle$ be a TC_ω space with true contradictions. If N is empty, then M cannot be $\mathbf{T_{2^1/2}}$.*

In order to see the correctness of the above proposition in an example, we will construct the following model. Now, under the assumption that N is empty, let us consider a formula φ and its negation $\neg\varphi$. Then, we choose w, w' in a way that $w \in [\varphi]$ and $w' \in [\neg\varphi]$, and also that the only closed sets around w and w' will be $[\varphi]$ and $[\neg\varphi]$, respectively. Let $S = \{1, 2, 3\}$, and $\sigma = \{\emptyset, S, \{1, 2\}, \{2, 3\}, \{2\}\}$. Let $[\varphi] = \{1, 2\}$, and $[\neg\varphi] = \{2, 3\}$. (Consider the formula $\varphi \wedge \neg\varphi$ at 2.) Then, observe that the points 1 and 3 are not separable by closed sets. Thus, this model cannot be $\mathbf{T}_{2^1/_2}$. However, if N was not empty, in an ad hoc way, we would have defined the truth of negated formula $\neg\varphi$ in a way to overcome this issue by letting $N(2) = \{\neg\varphi\}$.

Mortensen, in an earlier paper, investigated the connection between similar separation axioms and paraconsistent theories where he made several observations about discrete spaces, and $\mathbf{T}_1$ and $\mathbf{T}_2$ spaces [19].

Moreover, similar connections can be made between paraconsistent logics, topological semantics, and the topological properties of connectedness and continuity. We refer the reader to [4] where such properties are studied in detail.

19.2.2.2 Regular Spaces

Regular (open) sets are the sets which are equal to the interior of their closure. They play an important role not only in topology but also in mereotopology where the relationship between parts and the whole is investigated [20].

Even if we do not dwell on it further in this paper, it is important to underline that the algebra of closed sets and the topological models for paraconsistent logic do have the same algebraic structure, they both are co-Heyting algebras. Co-Heyting algebras are duals of Heyting algebras which were first proposed as the algebraic counterpart of intuitionistic logics. Some region-based logics, on the other hand, utilize both Heyting and co-Heyting algebras [18, 22]. From an algebraic perspective, we observe that regular sets play an important role in paraconsistency. Now we will consider the matter from a model theoretical perspective, and focus on TC_ω. We start with definitions.

Definition 19.2.16 Let $\langle S, \sigma\rangle$ be a topology. A subset $X \subseteq S$ is called a *regular open* set if X is equal to the interior of its closure, namely if $X = \mathsf{Int}(\mathsf{Clo}(X))$. Similarly, a subset $Y \subseteq S$ is called a *regular closed* set if Y is equal to the closure of its interior, namely if $Y = \mathsf{Clo}(\mathsf{Int}(Y))$. We call a space *regular open (closed)* if all the open sets (or dually, closed sets) are regular. A model is *regular open (closed)* if its topological space is regular open (closed).

For example, regular open sets in the standard topology of $\mathbb{R}^2$ are the open sets with no "holes" or "cracks". Also note that the complement of a regular open is a regular closed and vice versa.

We now observe the following:

Proposition 19.2.17 *Let $M = \langle S, \sigma, V, N\rangle$ be a TC_ω model with discrete topology σ. If $N = \emptyset$, then we have $w \models \neg\varphi$ if and only if $w \not\models \varphi$, for all $w \in S$ and for all φ.*

Proof It is a well-known fact that in a discrete topology every subset is closed (or open dually). In this proof, similar to our earlier remarks for the same issue (such as in the proof of Theorem 19.2.14), we assume that N is empty. Let us see the proof now.

First, we assume that N is empty. Then, let us suppose, for an arbitrary $w \in S$, an arbitrary formula φ, we have $w \models \neg\varphi$. Then, by definition, considering the discrete topology and the emptiness of N, we have $w \not\models \varphi$. Converse direction is also similar, and we leave it to the reader. □

Clearly, the converse of the above statement is not necessarily true, as it is very much possible to add "redundant" elements to N to make it nonempty.

19.2.2.3 Connectedness

A topological space is called *connected* if it cannot be written as the disjoint union of two open sets. We define *connected component* as the maximal connected subset of a given space. Moreover, in a connected topological space $\langle S, \sigma \rangle$, the only subsets with empty boundary are S and $\emptyset$. This fact, together with the semantics of the negation, plays an important role in TC_ω.

Proposition 19.2.18 *Let $M = \langle S, \sigma, V, N \rangle$ be a TC_ω model that admits a true contradiction whose extension is in the topology. If the space is disconnected and $|M| > 1$, then N cannot be empty.*

Proof Proof follows from the fact that in disconnected spaces, there are sets with empty boundary other than S itself and the empty set. So, we briefly mention the proof idea here. Let a contradiction $\varphi \wedge \neg\varphi$ satisfied in the model. Then, in this case, the positive φ and negative φ conjuncts of the contradiction will lie in the different connected components. However, if N is empty, it means that the extensions of each conjunct is connected via the boundary—which creates the contradiction as the space is assumed to be disconnected. □

Again, the contrapositive of the above theorem can help clarify the matter.

Proposition 19.2.19 *If N is empty, and M admits true contradictions whose extensions are in the topology, then M cannot be disconnected.*

19.3 Topological First-Order Models TC_ω^*

The logic C_ω can be extended to first-order level by introducing quantifiers, and the resulting first-order da Costa logic is called C_ω^* [7]. In his work, Baaz considered only the propositional case for KC_ω, and did not take the next step to introduce a Kripke semantics for C_ω^*. Priest, later on presented a Kripke semantics and tableaux style completeness for first-order da Costa logic [21]. Here, we introduce a topological semantics for C_ω^*, and call our system TC_ω^*.

First, let us set a piece of notation. For a formula φ, we abbreviate $\varphi^\circ := \neg(\varphi \wedge \neg\varphi)$. Moreover, we let, $\varphi^{(1)} := \varphi^\circ$, $\varphi^{(n)} := \varphi^{(n-1)} \wedge (\varphi^{(n-1)})^\circ$ for $2 \leq n \leq \omega$. We will often abuse the notation, and write φ^n instead of $\varphi^{(n)}$ for easy reading.

Let us now start with introducing the axioms for C_ω^*. The axioms of C_ω^* are the axioms of C_ω together with the following additional axioms [7].

1. $\forall x F(x) \to F(y)$.
2. $F(y) \to \exists x F(x)$.
3. $\forall x (F(x))^{(n)} \to (\forall x F(x))^{(n)}$ for $n \leq \omega$.
4. $\forall x (F(x))^{(n)} \to (\exists x F(x))^{(n)}$ for $n \leq \omega$.
5. Given F and F', if either one is obtained from the other by replacing bound variables or by suppressing vacuous quantification (without confusion of variables), then $F \leftrightarrow F'$ is an axiom.

The rules of inference are modus ponens, $\varphi \to F(x) \therefore \varphi \to \forall x F(x)$, and $F(x) \to \varphi \therefore \exists x F(x) \to \varphi$. Based on the given axiomatization, C_n^* is finitely trivializable for $n < \omega$ while C_ω^* is not. Also, it is important to note that C_0^* is the classical first-order logic.

Our goal now is to give a topological semantics for C_ω^*. In order to achieve this, we will make use of *denotational semantics* akin to Awodey and Kishida's work on topological first-order classical modal logic. In their work, they used sheaves to express the quantification domain of predicated modal formulas [2]. Their semantics is elegant, and simply explains how we should read predication in a natural way in the case of topological modal models. The use of denotational semantics will also be helpful for TC_ω^* as it presents a quite natural way to handle the non-truth functional behavior of the negation.

We start by introducing TC_ω^* models, and the related denotational interpretation function.

Definition 19.3.1 A first-order topological da Costa model TC_ω^* is given as the tuple $\langle S, D, |\cdot|, N^*, \sigma \rangle$ where S is a nonempty set with topology σ on it, $\emptyset \neq D \subseteq S$ is called the domain of individuals, $|\cdot|$ is a denotational interpretation function that assigns denotations in S to formulas, and N^* is the extension of the propositional negation function N to the first-order case defined over S.

Let us now give a brief explanation of TC_ω^* models here. The denotational interpretation function $|\cdot|$ takes formulas (with or without free variables), and returns individuals from S. Domain D, on the other hand, is given to precise the quantification. Similar to first-order classical modal logic, we use the domain set in the definition of the semantics of the quantifiers [11]. Here, we take D as a subset of S, so that we can make use of the topology σ defined on S for the objects in the domain. Alternatively, domain D and the topological space S can be taken as disjoint, and there can be defined a homeomorphic map from D to S [2]. Nevertheless, for simplicity, we choose the former. Finally, the function N^* is similar in purpose to the propositional N, and makes the semantics for negation non-truth functional, which we need in da Costa systems.

For variables $x_1, \ldots, x_n$ of appropriate arity n in the formula F, the vector $\overline{x}$ is the function that maps all free variables in F to some objects. We denote the denotational interpretation of F with $\overline{x}$ by $|\overline{x}; F|$, which is a tuple in S^n. For the formulas with different arity for free variables, we simply adjust the arity of the function $\overline{x}$ for each of its occurrence. The complement of $|\overline{x};\, F|$ will be denoted by $|\overline{x};\, F|^c$. By a slight abuse of the notation, $|\overline{x};\, N^*|$ will denote set of denotations of the formulas returned by N^*. Therefore, $|\overline{x};\, N^*| := \bigcup_{F \in N^*} |\overline{x};\, F|$.

The variable assignment is denoted by v. The function v assigns objects of the model to the variables present in a logical term, and this construction is a familiar one from first-order logic matching individual atoms with objects in the model. For a formula F, by a slight abuse of notation, $v(F)$ will denote the objects assigned to the variables of F. Moreover, we also define *terms* following the standard construction in first-order logic.

As we have remarked earlier, da Costa negation, in both propositional and first-order cases, is not truth functional. Note that there are, however, some paraconsistent logics with topological semantics where negation behaves truth functionally [13]. In such systems, the extension of each and every propositional variable is associated with a closed set while this condition is not a requirement in the topological semantics for classical modal logics. The reason for this is that in classical modal logics, only modal formulas are forced to have open or closed extensions. Propositional formulas do not necessarily have such extensions in classical case. Then, the negation in paraconsistent logics with topological semantics is defined as the *closure of the complement* [13]. The reason for this is quite immediate. While attempting to take the negation of a given formula, the usual way is to consider the set theoretical complement of the extension of the given formula. However, the complement of a closed set (which is the extension of the given formula) may not be closed, thus, may not be in the topology since the topology in question is a closed set topology. Therefore, in order to maintain the closed set topological structure, negation needs to be defined in that way to produce a closed set.

This idea, however, does not work in da Costa logics. For instance, assume that we endorse the aforementioned definition of negation for TC_ω^*. Namely, consider the following definition for the denotational interpretation of the negated formula $\neg F$ with respect to variables $\overline{x}$: $|\overline{x};\, \neg F| = \mathsf{Clo}(S^n - |\overline{x};\, F|)$.

A closer inspection immediately reveals that the above semantics for negation is indeed truth functional. In order to see the failure of this definition within the context of TC_ω^*, consider the logically equivalent formulas $\neg p$ and $\neg(p \wedge p)$. Based on the proposed semantics, the denotations of $\neg p$ and $\neg(p \wedge p)$ are necessarily the same. However, in da Costa systems, recall that the extensions of both $\neg p$ and $\neg(p \wedge p)$ are not necessarily identical. Therefore, the proposed (standard) topological semantics for paraconsistency does not work for da Costa systems.

Here, we suggest a working topological semantics for C_ω^*.

- $|\overline{x};\, c| \in S$ for a constant c,
- $|\overline{x};\, F| \subseteq S^n$ for a n-place predicate F,
  In particular, take an atomic sentence $F(t_1, \ldots, t_n)$ with terms t_i for $1 \leq i \leq n$. If $d_1, \ldots, d_n$ are the evaluation of the terms $t_1, \ldots, t_n$ under the variable assignment v, then we have the following in S: $|\overline{t};\, F(t_1, \ldots, t_n)| = v(F)(v(t_1), \ldots v(t_n))$,
- $|\overline{x};\, F \wedge G| = |\overline{x};\, F| \cap |\overline{x};\, G|$,

- $|\overline{x};\, F \vee G| = |\overline{x};\, F| \cup |\overline{x};\, G|$,
- $|\overline{x};\, \neg F| = |\overline{x};\, N^*| \cup \mathsf{Clo}(|\overline{x};\, F|^c)$,
- $|\overline{x};\, \exists y F| = \bigcup_{d\in D} |\overline{d}, d;\, F|$ where $\overline{d} \in D^n$,
- $|\overline{x};\, \forall y F| = \bigcap_{d\in D} |\overline{d}, d;\, F|$ where $\overline{d} \in D^n$.

We can furthermore define the truth in a TC_ω^* model M. We say that a formula $F(\overline{x})$ is true in the denotational interpretation $|\cdot|$, if $|\overline{x};\, F| = S$.

Let us now explicate the given semantics a bit further. The denotational semantics for the negation ensures that the negated denotation is among the formulas determined by N^* function. So, $|\overline{x};\, N^*|$ can be thought of as the collection of the denotations of the formulas returned by N^*. The closure operator Clo in the definition functions as the *classical* (or standard) part of the semantics. Similarly, the denotational semantics for the quantifier varies over the objects in the domain even though the denotation of the formula in question will eventually be in S.

As an illustration, let us consider the denotational semantics of the formula $\exists y(\neg F \wedge F)$ with a variable x.

$$
\begin{aligned}
|x;\, \exists y(\neg F \wedge F)| &= \bigcup_{d\in D} |d, d';\, (\neg F \wedge F)| \\
&= \bigcup_{d\in D} \{|d, d';\, \neg F| \cap |d, d';\, F|\} \\
&= \bigcup_{d\in D} \{(|d, d';\, N^*| \cup \mathsf{Clo}(|d, d';\, F|^c)) \cap |d, d';\, F|\} \\
&= \bigcup_{d\in D} \{(|d, d';\, N^*| \cap |d, d';\, F|) \cup \partial(|d, d';\, F|)\}
\end{aligned}
$$

where $\partial(\cdot)$ is the topological boundary operator. In this example, the individuals $d \in D$ which exist and satisfy the contradictory formula $F \wedge \neg F$ lie in the boundary of the denotation of F, or in the intersection of the denotation of F and the denotation of the formulas returned by N^*.

Also, it is worth noting that quantified De Morgan's laws are not valid in da Costa systems—even if the set theoretical De Morgan's laws hold [9]. As an illustration, we consider the following classical first-order logical equality $\forall x Fx \leftrightarrow \neg\exists x \neg Fx$. Let us first see the denotation of $\neg\exists x \neg Fx$.

$$
\begin{aligned}
|x;\, \neg\exists x \neg Fx| &= |d;\, N^*| \cup \mathsf{Clo}(|d;\, \exists x \neg Fx|^c) \\
&= |d;\, N^*| \cup \mathsf{Clo}((\cup_{d\in D} |d;\, \neg F|)^c) \\
&= |d;\, N^*| \cup \mathsf{Clo}((\cup_{d\in D} (|d;\, N^*| \cup \mathsf{Clo}(|d;\, F|)^c))^c)
\end{aligned}
$$

Therefore, if $|d;\, N^*|$ is not empty, we cannot generally obtain $\bigcap_{d\in D} |d;\, F|$, which is the denotation of $\forall x Fx$. Other quantified De Morgan laws can be given similar arguments [10].

Soundness of the axioms of TC^*_ω with respect to the given semantics above is a straightforward symbolic manipulation. However, we will still consider some of the axioms which are unique to da Costa systems, and show their soundness.

Now, as the first case, we take the following formula as an instantiation of the axiom scheme (3) with $n = 1$.

$$\forall x F^1 x \rightarrow (\forall x F x)^1$$

In order to have an idea of what to expect, let us first note the following logical equalities.

$$\forall x F^1 x = \forall x F^\circ x = \forall x \neg(Fx \wedge \neg Fx)$$

and

$$(\forall x F x)^1 = (\forall x F x)^\circ = \neg(\forall x F x \wedge \neg\forall x F x).$$

So let us now assume, $\forall x F^1 x$, which is equivalent to $\forall x.\neg(Fx \wedge \neg Fx)$. Then, we have the following:

$$
\begin{aligned}
|x; \forall x.F^1 x| &= |x; \forall x.\neg(Fx \wedge \neg Fx)| \\
&= \bigcap_{d \in D} |d; \neg(Fx \wedge \neg Fx)| \\
&= \bigcap_{d \in D} \{|d; N^*| \cup \mathsf{Clo}(|d; Fx \wedge \neg Fx|^c)\} \\
&= \bigcap_{d \in D} \{|d; N^*| \cup \mathsf{Clo}((|d; Fx| \cap |d; \neg Fx|)^c)\} \\
&= \bigcap_{d \in D} \{|d; N^*| \cup \mathsf{Clo}((|d; Fx| \cap (|d; N^*| \cup \mathsf{Clo}(|d; Fx|^c)))^c)\} \\
&= \bigcap_{d \in D} \{|d; N^*| \cup \mathsf{Clo}((|d; Fx| \cap |d; N^*|) \cup (|d; Fx| \cap \mathsf{Clo}(|d; Fx|^c)))^c)\}
\end{aligned}
$$

(as intersection operation commutes with closure operator)

$$
\begin{aligned}
&= \bigcap_{d \in D} |d; N^*| \cup \cap_{d\in D}\mathsf{Clo}(|d; Fx| \cap |d; N^*|)^c \cup (|d; Fx| \cap \mathsf{Clo}(|d; Fx|^c))^c)) \\
&= \bigcap_{d \in D} |d; N^*| \cup \mathsf{Clo}(\cap_{d\in D}|d; Fx| \cap |d; N^*|)^c \cup \cap_{d\in D}(|d; Fx| \cap \mathsf{Clo}(|d; Fx|^c))^c))
\end{aligned}
$$

(as the interior of a set is its subset)

$$
\begin{aligned}
&\subseteq |d; N^*| \cup \mathsf{Clo}(\cap_{d\in D}|d; Fx|)^c \cup (\cap_{d\in D}(|d; Fx| \cap \mathsf{Clo}(|d; Fx|^c))^c)) \\
&\subseteq |d; N^*| \cup \mathsf{Clo}(\cap_{d\in D}|d; Fx|)^c \cup ((|d; Fx| \cap \mathsf{Clo} \cap_{d\in D} (|d; Fx|^c))^c)) \\
&\subseteq |d; N^*| \cup \mathsf{Clo}(\cap_{d\in D}|d; Fx| \cap ((|d; Fx| \cap \mathsf{Clo} \cap_{d\in D} (|d; Fx|^c))) \\
&\subseteq \neg(\forall x F x \wedge \neg\forall x F x) \\
&\subseteq (\forall x F x)^1
\end{aligned}
$$

Thus, we obtain $\forall x F^1 x \rightarrow (\forall x F x)^1$.

As the second case, let us take the axiom scheme (4) instantiated with $n = 1$. Thus, we consider the following implication,

$$\forall x F^1 x \rightarrow (\exists x Fx)^1$$

Now, below is what follows from the above statement.

$$
\begin{aligned}
|x; \forall x.F^1 x| &= |x; \forall x.\neg(Fx \wedge \neg Fx)| \\
&= \bigcap_{d\in D} \{|d; N^*| \cup \mathsf{Clo}(|d; Fx \wedge \neg Fx|^c)\} \\
&= \bigcap_{d\in D} \{|d; N^*| \cup \mathsf{Clo}(|d; Fx| \cap |d; \neg Fx|)^c\} \\
&= \bigcap_{d\in D} \{|d; N^*| \cup \mathsf{Clo}(|d; Fx|^c \cup |d; \neg Fx|^c)\} \\
&= \bigcap_{d\in D} \{|d; N^*| \cup \mathsf{Clo}(|d; Fx|^c \cup (|d; N^*| \cup \mathsf{Clo}(|d; Fx|^c))^c))\} \\
&= \bigcap_{d\in D} \{|d; N^*| \cup \mathsf{Clo}(|d; Fx|^c \cup (|d; N^*|^c \cap \mathsf{Int}(|d; Fx|)))\} \\
&\subseteq \bigcap_{d\in D} |d; N^*| \cup \mathsf{Clo}(\cap_{d\in D}|d; Fx|^c \cup (\cap_{d\in D}|d; N^*|^c \cap \mathsf{Int}(\cap_{d\in D}|d; Fx|))) \\
&\subseteq |d; N^*| \cup \mathsf{Clo}(\cap_{d\in D}|d; Fx|^c \cup (|d; N^*|^c \cap \mathsf{Int}(\cup_{d\in D}|d; F|))) \\
&\quad \text{by set theoretical De Morgan's Laws} \\
&\subseteq |d; N^*| \cup \mathsf{Clo}(\cup_{d\in D}|d; Fx| \cap (|d; N^*| \cup \mathsf{Clo}(\cup_{d\in D}|d; F|^c)^c)) \\
&\subseteq |d; N^*| \cup \mathsf{Clo}((\exists x Fx \wedge \neg\exists x Fx)^c) \\
&\subseteq \neg(\exists x Fx \wedge \neg\exists x Fx) \\
&\subseteq (\exists x Fx)^1
\end{aligned}
$$

Finally, we obtain $\forall x F^1 x \rightarrow (\exists x Fx)^1$.

The remaining axioms can also be given rather straightforward arguments for their soundness, thus we leave them to the reader.

*

This was soundness. However, we still do not have a completeness result (or lack thereof) for TC_ω^*. We leave it for further work.

Acknowledgments This paper was partially written while I was visiting the Arché Philosophical Research Centre for Logic, Language, Metaphysics and Epistemology at the University of St Andrews in the summer of 2011, and I am grateful for the help and encouragement of Graham Priest. The welcoming environment of St. Andrews facilitated the production of this paper. I am also very thankful for the tremendous help of the referee.

References

1. Arenas, F.G.: Alexandroff spaces. Acta Mathematica Universitatis Comenianae **68**(1), 17–25 (1999)
2. Awodey, S., Kishida, K.: Topology and modality: the topological interpretation of first-order modal logic. Rev. Symb. Logic **1**(2), 146–166 (2008)
3. Baaz, M.: Kripke-type semantics for da costa's paraconsistent logic C_ω. Notre Dame J. Form. Logic **27**(4), 523–527 (1986)
4. Başkent, C.: Some topological properties of paraconsistent models. Synthese **190**(18), 4023–4040 (2013)
5. Béziau, J.-Y.: S5 is a paraconsistent logic and so is first-order classical logic. Log. Stud. **9**(1) (2002)
6. Béziau, J.-Y.: Paraconsistent logic from a modal viewpoint. J. Appl. Logic **3**(1), 7–14 (2005)
7. da Costa, N.C.A.: On the theory of inconsistent formal systems. Notre Dame J. Form. Logic **15**(4), 497–510 (1974)
8. da Costa, N.C.A., Alves, E.H.: A semantical analysis of the calculi c_n. Notre Dame J. Form. Logic **18**(4), 621–630 (1977)
9. da Costa, N.C.A., Krause, D., Bueno, O.: Paraconsistent logics and paraconsistency. In: Jacquette, D. (ed.) Philosophy of Logic, vol. 5, pp. 655–781. Elsevier (2007)
10. Ferguson, T.M.: Notes on the model theory of demorgan logics. Notre Dame J. Form. Logic **53**(1), 113–132 (2012)
11. Fitting, M., Mendelsohn, R.: First-Order Modal Logic. Kluwer (1998)
12. Goldblatt, R.: Mathematical modal logic: a view of its evolution. In: Gabbay, D.M., Woods, J. (eds.) Handbook of History of Logic, vol. 6. Elsevier (2006)
13. Goodman, N.D.: The logic of contradiction. Zeitschrift für Mathematische Logik und Grundlagen der Mathematik **27**(8–10), 119–126 (1981)
14. McKinsey, J.C.C.: On the syntactical construction of systems of modal logic. J. Symb. Logic **10**(3), 83–94 (1945)
15. McKinsey, J.C.C., Tarski, A.: The algebra of topology. Ann. Math. **45**(1), 141–191 (1944)
16. McKinsey, J.C.C., Tarski, A.: On closed elements in closure algebras. Ann. Math. **47**(1), 122–162 (1946)
17. Mints, G.: A Short Introduction to Intuitionistic Logic. Kluwer (2000)
18. Mormann, T.: Heyting mereology as a framework for spatial reasoning. Axiomathes (2012)
19. Mortensen, C.: Topological seperation principles and logical theories. Synthese **125**(1–2), 169–178 (2000)
20. Pratt-Hartman, I.E.: First-order mereotopology. In: Aiello, M., Pratt-Hartman, I.E., van Benthem, J. (eds.) Handbook of Spatial Logics, pp. 13–97. Springer (2007)
21. Priest, G.: First-order da costa logic. Stud. Logica **97**, 183–198 (2011)
22. Stell, J.G., Worboys, M.F.: The algebraic structure of sets of regions. In: Hirtle, S.C., Frank, A.U. (eds.) Spatial Information Theory. Lecture Notes in Computer Science, vol. 1329, pp. 163–174. COSIT 97. Springer (1997)
23. van Benthem, J., Bezhanishvili, G.: Modal logics of space. In: Aiello, M., Pratt-Hartman,I.E., van Benthem, J. (eds.) Handbook of Spatial Logics. Springer (2007)

Part V
Philosophical Aspects and Applications of Paraconsistent Logic

Chapter 20
Perceiving and Modelling Brightness Contradictions Through the Study of Brightness Illusions

Ashish Bakshi and Kuntal Ghosh

Abstract In this paper, we argue in the light of visual perceptual experience in favour of the fact that true contradictions are very much perceivable and happen to be an inherent part of the real world. We first describe the phenomenon of perception of two contrary types of brightness illusions, termed as the brightness–contrast and the brightness assimilation type illusions. Next, we present a model of brightness induction which can envisage the above-mentioned contradictions in visual brightness perception. The proposed model, called DDOG (Difference of Difference of Gaussians) is based on two aspects. First, two Difference of Gaussians (DOG) functions acting in opposition in two complementary channels, Magno & Parvo, in the central visual pathway and second, a two-pass model of attentive vision. Although the Oriented Difference of Gaussian (ODOG) model of Blakeslee et al. (Vis Res 45:607–615, 2005) can already account for most of these types of illusions, our model is significantly simpler, more consistent than ODOG and biologically more plausible as a neurocomputational model for explaining brightness contradictions in the brain.

Keywords Brightness contradiction · Brightness perception · Computational model · Vision · Optical illusions

Mathematics Subject Classification (2000) Primary 68T45 · Secondary 68T27 · 68T10 · 68T05

A. Bakshi (✉)
Machine Intelligence Unit, Indian Statistical Institute,
203 B.T. Road, Kolkata 700108, India
e-mail: ashishbakshi@outlook.com

K. Ghosh
Machine Intelligence Unit and Center for Soft Computing Research,
Indian Statistical Institute, 203 B.T. Road, Kolkata 700108, India
e-mail: kuntalghos@gmail.com

J.-Y. Beziau et al. (eds.), *New Directions in Paraconsistent Logic*,
Springer Proceedings in Mathematics & Statistics 152,
DOI 10.1007/978-81-322-2719-9_20

20.1 Introduction

One of the most intriguing questions raised by sections of the paraconsistency research community, is whether contradictions are at all perceivable through our sense organs, or in other words, whether true contradictions do actually exist in the observable world [1, 2, 16]. The present work attempts to address this issue in the light of what it claims to be one such class of perceivable contradictions in visual perception. This may be referred to as the occurrence of brightness contradictions. Like the example from Graham Priest, in which by wearing filters of the opposing colours, viz. red and green on a special pair of glasses, a colour contradiction becomes perceivable, where red may be seen as green and green as red [1]—here the same pair of objects that appear dark and bright, respectively, because of their surroundings, may be sensed oppositely, in apparently unaffected surroundings, thus giving rise to brightness contradiction.

Graham Priest [16] has argued that the observable world is consistent, but for the odd visual illusion. Here we attempt to demonstrate that on the contrary, the existence of the visual illusions actually provide the indication that the observable world is both consistent and inconsistent, or in other words it is neither consistent nor inconsistent; the world is what it is. The eye–brain system is one sense organ system through which we perceive the world and build up an opinion of it. Accordingly, like any other system, it is endowed with its own contradictions which are expressed through such phenomenon as brightness contradictions which may be looked upon as an example paraconsistency too. There are other well-known visual contradictions leading to paraconsistent statements like for instance, looking at a 170 cm woman one concludes that she is both tall and not tall, a purplish blue object is both blue and not blue to the observer and so on [2]. The very observation that the observable real world do not display true contradictions is itself contradictory. Let us take the case of subatomic world. Is it observable and perceivable? The answer will be no if we consider only our five sense organs, but the answer is yes if these sense organs or one or some of them take the help of some sophisticated instruments from experimental physics, and in that world under the observation through such instruments, one shall come across contradictory experiences when compared with those perceived and observed by the five sense organs only. This is the well-known contradiction between classical and quantum physics. Numerous such examples can be given that historically gave rise to such crises in physics like the experimental observations of Rutherford, Becquerel, Roentgen and so on. But none of these should actually be considered as crises, since it is the experience of such contradictions only that give birth to new theories which in turn again lead to new contradictions in observations with the passage of time, seeking further development of theory. Hence contradictions are an inherent part of the world. The present work on brightness contradictions also demonstrates the need to develop new theories in visual psychophysics compared to the present ones. The paper tries to propose one such neurobiologically plausible theory for explaining the contradictions in brightness perception.

Brightness contradictions are found to occur in a type of visual illusions in which various surfaces of perfectly equal luminance have different apparent brightness [9, 12]. It so seems that, what apparent brightness those surfaces will have, depends on the nature of the rest of the field of view. A few examples of such illusions are shown in Fig. 20.1: the Simultaneous brightness–contrast (SBC) illusion, the White effect, the Checkerboard illusion and the Grating induction (GI) illusion. In all these illusions the grey regions have the same real intensity but apparently look differently bright, depending on their surroundings and on the background.

There are two basic and contrary types of brightness illusions. These are the brightness–contrast and brightness assimilation types. In the brightness–contrast type, the apparent brightness of a region changes in the opposite direction to the brightness of its surrounding regions. This increases the apparent contrast of the region with respect to its surroundings. Examples of this include the SBC illusion (Fig. 20.1a) and the GI illusion (Fig. 20.1d). In the SBC stimulus, the left-hand side grey patch, being on a brighter background, looks darker, compared to the right-hand side grey patch which has a darker surrounding, though both the grey patches are of same luminance. Similarly, in the GI illusion, the uniform grey test patch in the foreground looks brighter over dark stripes and darker over white stripes. Therefore, the grating in the background appears to have been inducted in opposite phase into the uniform foreground patch. On the other hand, in the brightness assimilation type of illusions, the apparent brightness of a region changes in the same direction as

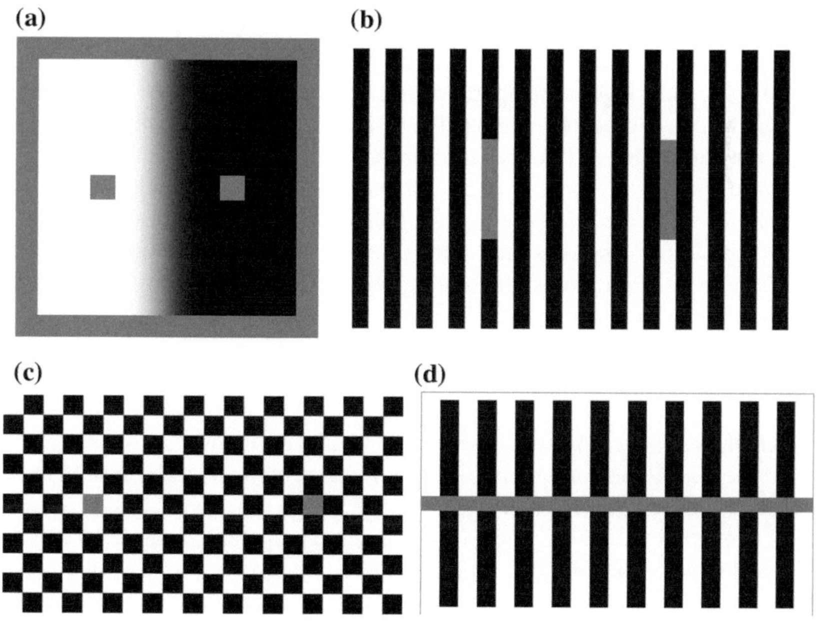

Fig. 20.1 Some brightness induction illusions. **a** Simultaneous brightness–contrast illusion. **b** White effect. **c** Checkerboard illusion. **d** Grating induction illusion

its surroundings. This reduces the apparent contrast of the region with respect to its surroundings. Examples of this include the White effect [21] (Fig. 20.1b) and the Checkerboard illusion (Fig. 20.1c). In the White effect, the test patch on the left, shares a brighter neighbourhood to a greater extent than that on the right, and yet appears brighter than the latter, which is contrary to the previously mentioned brightness–contrast phenomenon. Same is the case for the two grey patches in the Checkerboard illusion. In both Fig. 20.1b, c the grey patch on the left appears brighter than the grey patch on the right.

Why does the human brain perceive brightness illusions of such opposing natures and what factors influence the type of illusion perceived, is still unknown. It is also possible to gradually transform the input stimulus in a continuous fashion so that the illusion changes from one type to the other. Howe's stimulus represents an interim position in one such type of smooth transition in which the White effect can be smoothly converted into SBC [4, 11].

20.2 Modelling Brightness Induction Illusions

Traditionally, some of these illusions have been explained using a spatial filtering function applied on the input stimulus. Such models have been supported by the experimental observation of lateral inhibition within the retina of the eye and the LGN [19]. Lateral inhibition is the phenomenon in which the nerve response of a spot of light falling on the retina is inhibited by a neighbouring spot of light falling within a zone known as the receptive field of the original nerve cell. Lateral inhibition can be modelled using the convolution operation in signal processing. The image falling on the eye can be considered as the input signal. The response produced by the receptor nerves is the output signal. Using a convolution function, which takes positive values at the centre and negative values at the surroundings, we can simulate the effect of lateral inhibition on the response signal. Thus a strong signal on the surrounding parts will inhibit the response signal at the centre. A typical spatial filtering function, expressed here in one dimension for simplicity, that has been frequently used to model lateral inhibition is the DOG function defined as:

$$DOG(x; A_1, A_2, \sigma_1, \sigma_2) = A_1 exp(-x^2/\sigma_1^2) - A_2 exp(-x^2/\sigma_2^2) \tag{20.1}$$

The parameters σ_1 , σ_2 represent spatially the widths of the two Gaussian functions, and A_1, A_2 represent the maximum responses in the central and the surround regions, respectively. When $\sigma_2 > \sigma_1$ and $A_1 > A_2$, the DOG function looks as shown in Fig. 20.2. This function has a positive central region (spatially limited by σ_1) and a negative value in the surrounding regions (limited by σ_2).

Although lateral inhibition can partially explain some brightness–contrast illusions such as SBC, it cannot explain brightness assimilation illusions such as the White effect. In fact, any simple linear spatial filtering algorithm has not yet explained the opposing natures of various brightness illusions simultaneously. The most

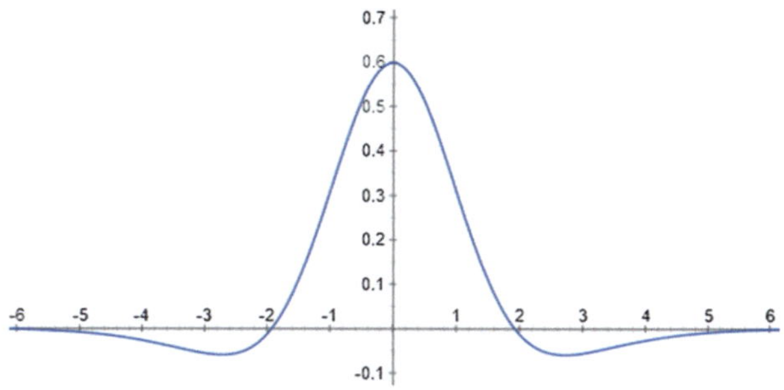

Fig. 20.2 Typical shape of the DOG function. The x-axis is the spatial distance from a given point. The y-axis is the response produced at the origin by a spot of light falling at a distance x from the origin

successful attempt in the direction of explaining brightness contradictions in visual perception has been the Oriented DOG (ODOG) model of [4]. The ODOG model consists of a set of 42 anisotropic DOG functions with seven different length scales and six different orientation directions. The outputs from the ODOG filters are then nonlinearly combined to produce the final output. But the ODOG algorithm has its own weaknesses and is computationally complex and also has no known neural correlate, i.e. biological analogue in the brain [12]. We describe below our attempt to model brightness induction phenomena, by the use of what we have termed as the DDOG function (Difference of Difference of Gaussians). Hence we may call our model the DDOG model.

20.3 Description of DDOG Model

We have combined the DOG function, as expressed in Eq. 20.1, with another DOG function each having different spatial widths. This we term as the Difference of DOG function (DDOG).

The DDOG function is a difference of two DOG functions:

$$\begin{aligned} &DDOG(x, y; A_1, A_2, A_3, A_4, \sigma_1, \sigma_2, \sigma_3, \sigma_4) \\ &\qquad = DOG(x, y; A_1, A_2, \sigma_1, \sigma_2) - DOG(x, y; A_3, A_4, \sigma_3, \sigma_4) \end{aligned} \quad (20.2)$$

We shall use two different versions of the DDOG function to explain the above-mentioned contrary behaviour of brightness–contrast and brightness induction. Each of the two versions has a different set of parameter values A_1 , A_2 , A_3 , A_4, etc.

The two sets of DDOG functions are called the P-channel and M-channel, respectively, (explained later) and differ only by the values of the coefficients A_1, A_2, A_3, A_4 and the spatial sampling interval used to implement the digital filters. The parameter values used are summarized in Table 20.1.

We take $\sigma_1 = 0.7$, $\sigma_2 = 3\sigma_1$, $\sigma_3 = 3\sigma_1$, $\sigma_4 = 9.3\sigma_1$ for both filters. These parameter values have been arrived at through trial and error based on several neurobiological inputs related to low-level vision [6] that justifies the combination of a DOG

Table 20.1 Parameter values of DDOG filter

	A_1	A_2	A_3	A_4	Sampling interval
M-channel	10	0.5	0.5	0.07	0.5
P-channel	10	0.25	0.25	0.01	0.25

representing a very localized receptive field (which is the first DOG on the right-hand side of Eq. (20.2), above) with another DOG that represents a comparatively wider receptive field (the other DOG in Eq. (20.2)). Some other neurophysiological inspirations, like the values of sampling interval, will be elaborated later in this section. These values mostly work only for spatial widths between 10 and 30 pixels in size, demonstrations of which have been shown in this paper. For very large or very small sizes these parameter values will have to be changed appropriately. This has also been explained later in this paper through appropriate plots.

The M and P filters are graphically shown and compared in Fig. 20.3. It is to be noticed that although the M/P filters in Fig. 20.3a roughly look just like the DOG function (Fig. 20.2), there are small but crucial differences, too. The M/P filter functions are found to cut the x-axis at four points whereas DOG cuts the x-axis only at two points. This is clearly seen in Fig. 20.3b, which represents magnified views of the x-axis region of Fig. 20.3a. The M/P filters have five local extrema unlike the DOG,

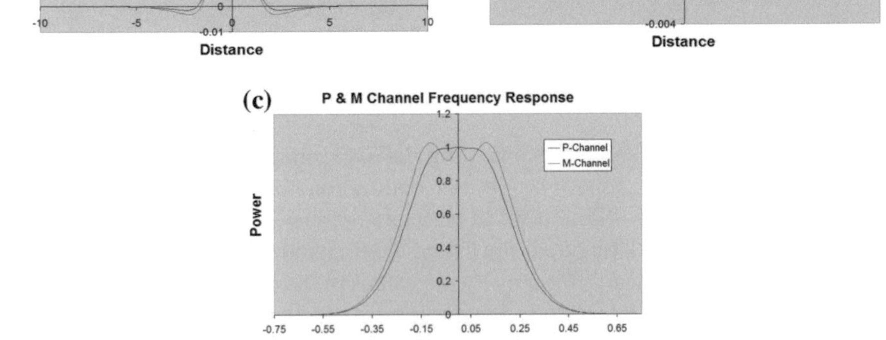

Fig. 20.3 **a** Graphs of P- & M-channel filter functions (superimposed for comparison). **b** Magnified view of same graphs. **c** Frequency domain graphs

which has only three local extrema, although the local maxima farthest from the origin are so small that they are not visible in Fig. 20.3a, and almost remains so for P filter, even in Fig. 20.3b. Figure 20.3c displays the frequency response curves of M & P channels respectively. Even though the frequency response curves largely resemble a sort of combination, of a low-pass and a band-pass filters but the frequency response of M-channel shows an extra depression at very low frequencies. These differences turn out to be important for our model to work properly in explaining brightness contradictions.

It has previously been observed by [6–8] that some brightness illusions could be explained by a linear combination of Gaussians (called the ECRF filter) with certain parameter values while some other brightness illusions could be explained by different parameter values. But the decision of when to apply which filter has to be made by the user in the ECRF model. Our new model, here, builds upon the above-mentioned works by Ghosh et al. by dynamically combining the M-channel and P-channel functions in a proportion which itself is a function of the input image. In this way our algorithm is effectively nonlinear in nature. With our model we have been able to explain both the SBC and White effect illusions through an automated algorithm. Our algorithm also uses isotropic filters, so that if the input stimulus is rotated by certain angle then the output also rotates by the same angle. The simplicity of our model also makes it a much more plausible model for the neural networks involved with low-level vision. The method of combination of the above-mentioned two sets of filters is inspired by a two-pass model of attentive vision according to which the visual process is divided into two stages [5, 10], based on the anatomy of the parallel pathways in the central visual system [15]. In the light of this proposed model, in the first stage called vision at a glance, the brain first interprets the contents of the magnocellular pathway. If it can find sufficient detail in this stage then it mostly ignores the contents of the parvocellular pathway, which is called vision with scrutiny. If it cannot find sufficient detail then it gives more importance to this second stage. In this stage the brain examines the contents of the parvocellular pathway to find further details in those regions of the Magno output where sufficient details were not found. It is well known that the parvocellular pathway carries much more detail than the magnocellular pathway. But the magnocellular pathway can carry information much faster than the parvocellular pathway and therefore it is processed in the first stage of the process. In our model we implement the magnocellular and parvocellular pathway using the M and P filters as described in Table 20.1 above. These two channels vary only slightly in the values of their coefficients, but a major difference between them is in the sampling interval used to implement the filters digitally. The M-channel has a larger sampling interval reflecting the fact that the magnocellular pathway has lower spatial resolution whereas the P-channel has much finer spatial resolution just as in the biological visual system [15]. So the above-mentioned two-stage process can be described as follows:

If

(the initial M-channel output identifies that the background around the test patch is uniform)

Then

the brightness percept is formed mostly by contribution from P.

Else

M-channel contributes maximally to produce the final percept.

The condition of background uniformity under the If clause can be implemented in a variety of ways. We take recourse to a very simple way of evaluating background uniformity based on Marr–Hildreth's operator [14] that is supposed to produce the raw primal sketch in the primary visual cortex. First, we calculate the Laplacian at every pixel of the M-channel output image and take its squared value. This gives us a positive number at every pixel. We then compute its average value per pixel. This gives us a single positive number λ for every image. If this number λ is very low then the background must be very uniform and therefore the P-channel must play the major role to determine the output. On the other hand if λ is large then the background must be highly non-uniform. So the M-channel must be more important to form the output. Instead of choosing any threshold value of λ to simply switch from P- to M-channel, we thus choose to have a more gradual transition from P to M by linearly combining both the M-output and the P-output in some proportion $f(\lambda)$ depending on λ. $f(\lambda)$ must be such that if λ is small the proportion of M is small whereas if λ is high then the proportion of M is high. After trial and error we therefore chose the following form of $f(\lambda)$ (Fig. 20.4):

$$f(\lambda) = \lambda/(\lambda + 4) \tag{20.3}$$

Therefore, the final output is determined by the following equation:

$$\text{FINAL OUTPUT} = f(\lambda) * (\text{M-OUTPUT}) + (1 - f(\lambda)) * (\text{P-OUTPUT}) \tag{20.4}$$

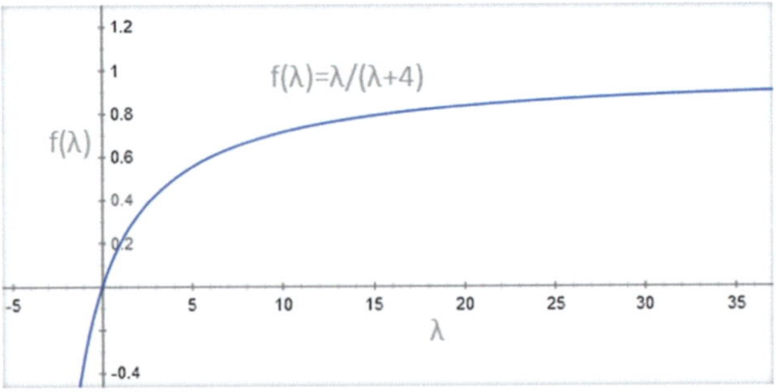

Fig. 20.4 Graph of the function $f(\lambda) = \lambda/(\lambda + 4)$

20.4 Results

The results shown in this section were produced using the space constant values, $\sigma_1 = 0.7, \sigma_2 = 3\sigma_1, \sigma_3 = 3\sigma_1, \sigma_4 = 9.3\sigma_1$. We tried various values of σ_1 before settling at the value of 0.7, as this value seems to work, more or less, for all illusions. If the length scale (i.e. strip width for White effect, patch length for SBC, grating period for the grating illusion, etc.) of the input stimulus is changed then σ_1 also needs to be changed appropriately. Other space constants, i.e. σ_2, σ_3 and σ_4 are kept in constant proportion to σ_1. For the purpose of illustration, we show below how the results vary with typical length scale, in the cases of White's illusion and SBC.

The DDOG algorithm was implemented in the language C whereas for ODOG we have used Alan Robinson's MATLAB implementation [17, 18]. The same input images were fed into the two algorithms to produce the results shown in this section. Below we show some input stimuli and the corresponding brightness profiles of both input and outputs.

20.4.1 White Effect

The image in Fig. 20.5b shows the output profile of our model when applied to the stimulus in Fig. 20.5a, which shows the White effect. The input profile has also been shown for comparison. It is well known and can also be seen in Fig. 20.5a that in White's illusion the grey patch over a black stripe appears brighter than the grey patch over a white stripe. This is correctly predicted by the DDOG filter, as shown in Fig. 20.5b. We can also observe that, as expected, the brightness of the patches has shifted in the same direction as the stripes on either sides of the patches, i.e. brightness has been assimilated. Figure 20.5c shows the ODOG output profile for the image in Fig. 20.5a. The ODOG filter also correctly predicts the direction of brightness change for the two grey patches. The output brightness levels of the grey patches, although in the correct direction, are less pronounced in the ODOG model than in the DDOG model (Fig. 20.5c).

As already stated, the result shown in Fig. 20.5b correspond to the space constant values of $\sigma_1 = 0.7$, $\sigma_2 = 3\sigma_1$, $\sigma_3 = 3\sigma_1$, $\sigma_4 = 9.3\sigma_1$. The length scale of the input stimulus in this case happens to be 16 pixels. If a different length scale input is used we do not always get our desired output. For example, for the values of the space constants as given above, if we plot the grey patch intensities of the output as a function of the length scale, then we obtain a graph as shown in Fig. 20.5d. It can be seen from this graph that the brightness assimilation effect is more pronounced at some length scales while being very small at other length scales. This implies that our algorithm gives good results only for a certain range of length scales. This fact is even clearer for the SBC illusion.

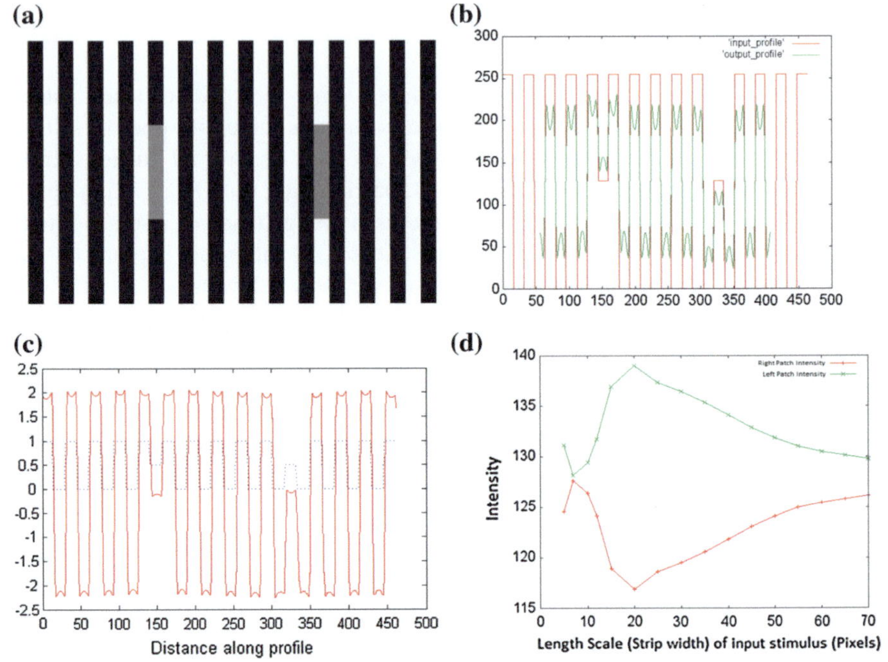

Fig. 20.5 **a** Input stimulus for the White effect. **b** Output profile of our model (*green*) and the input profile (*red*). **c** Output profile of ODOG model (*red*) and input profile (*dotted blue*). **d** Graph of the grey patch intensities with our proposed model as a function of input length scale (strip width). *Left* grey patch is *green* while the *right* grey patch is *red*

20.4.2 SBC

The image in Fig. 20.6b shows the output result of our model for the SBC stimulus in Fig. 20.6a. As expected we get a brightness–contrast effect, i.e. the patch with a brighter surrounding looks darker while the patch with a darker surrounding looks brighter. This is also reflected in the DDOG output profile shown in Fig. 20.6b where the predicted brightness level for the grey patch on bright background is lower than the other grey patch. The ODOG output shown in Fig. 20.6c also shows the same brightness–contrast effect, i.e. the brightness level of the grey patch on black background is higher than its counterpart. Hence in the case of SBC illusion, both ODOG and DDOG produce equally good results.

Similarly as in Fig. 20.5d above, Fig. 20.6d shows the variance of patch intensities as a function of length scale, for the space constant values given above. It can be seen that we get desirable results (i.e. in accordance with brightness–contrast) only for length scales between 10 and 30 pixels. Beyond this range the space constants need to be changed, since the direction of brightness induction is not found to change

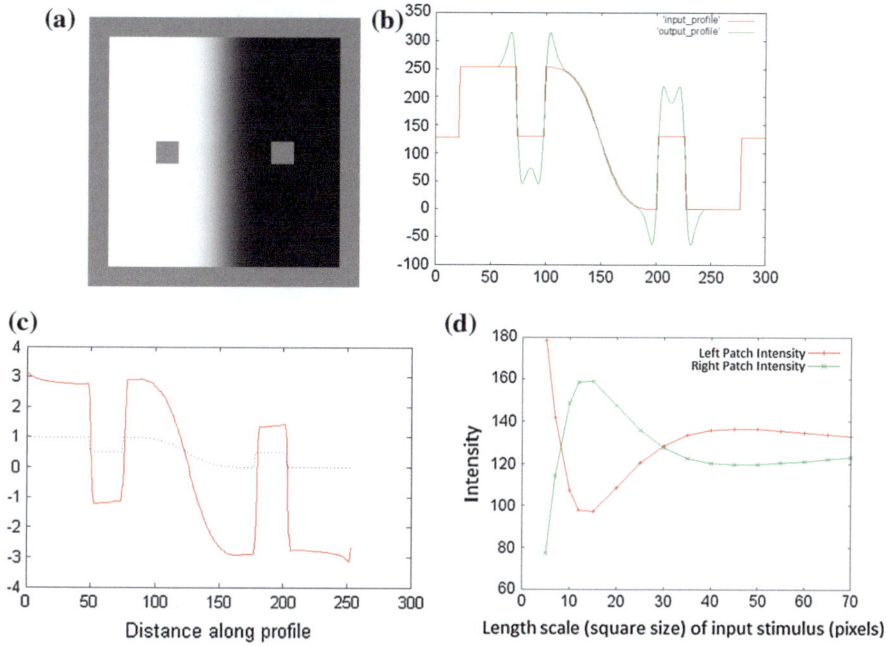

Fig. 20.6 **a** Input stimulus of the SBC illusion. **b** Output profile of our model (*green*) and the input profile (*red*). **c** Output profile of ODOG model (*red*) and input profile (*dotted blue*). **d** Graph of the grey patch intensities as a function of input length scale (square size). *Left* grey patch is *red* while the *right* grey patch is *green*

beyond this range, as is seen in the graph. It is to be noted that this range has already been found to work perfect for the contrary type of illusion also, i.e. White effect as evident from Fig. 20.5d.

20.4.3 Checkerboard

The Checkerboard illusion shows a brightness-assimilation effect as seen in Fig. 20.7a. The Checkerboard illusion is known to be a brightness assimilation illusion, i.e. the grey patch surrounded by white squares appears brighter than the grey square surrounded by black squares. The output of the DDOG model, as shown in Fig. 20.7b, shows an overall shift in patch brightness in the same direction as the surroundings, although the amount of shift is quite small. In the case of the Checkerboard illusion the ODOG output shows a shift in the brightness levels of the grey patches in the opposite direction as their surroundings, although the shift is small in magnitude. Thus the ODOG here clearly fails to explain the Checkerboard illusion,

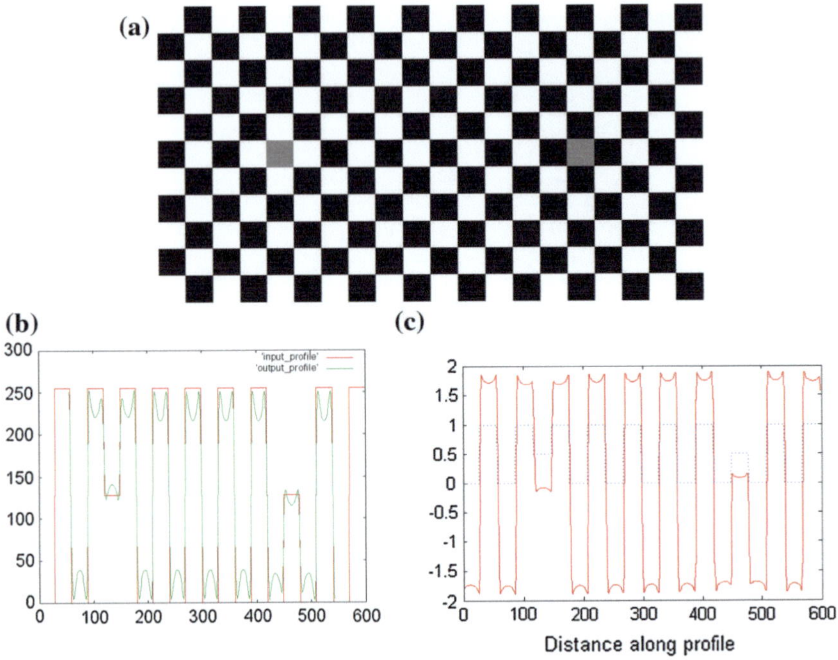

Fig. 20.7 **a** Input stimulus of the Checkerboard illusion. **b** Output profile of our model (*green*) and the input profile (*red*). **c** Output profile of ODOG model (*red*) and input profile (*dotted blue*)

and incorrecly predicts a brightness–contrast effect. The proposed DDOG model therefore, for the case of the Checkerboard illusion, performs better than the ODOG model.

20.4.4 Sine Grating Induction

The Sine grating stimulus as shown in Fig. 20.8a consists of a sinusoidally varying background over which a thin uniformly grey strip is placed. The grey patch shows an apparent brightness that varies in the opposite direction as the background brightness. Therefore, this is a brightness–contrast type of illusion. The output profile of our model as shown in Fig. 20.8b correctly shows the background sinusoid peaks coinciding with the output troughs and vice versa. The ODOG model also successfully explains this illusion, as is evident from Fig. 20.8c, where the peaks of the output profile coincide with the troughs of the input profile.

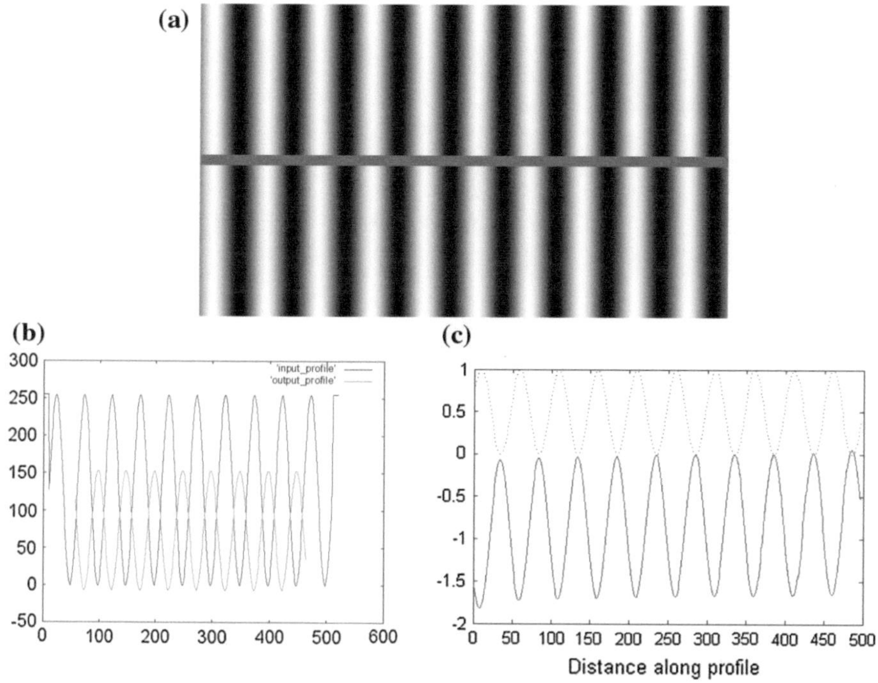

Fig. 20.8 **a** Input stimulus of the Sine grating illusion. **b** Output profile of our model (*green*) and the background sinusoid profile (*red*). **c** Output profile of ODOG model (*red*) and background profile (*dotted blue*)

20.4.5 Howe's Illusion

Howe's illusion [11] represents an interim stage in a gradual transition of the White effect into the SBC illusion [4] (Fig. 20.9c). As the stimulus gradually changes from White's illusion to SBC (Fig. 20.9a–g), the illusory effect also changes gradually from brightness assimilation to brightness–contrast. In particular it must be noted that during the transition the left and the right edges of both the grey patches, which form the major portion of the edge boundary, do not change any colour. Only the top and bottom edges, which form a minor portion of the edge boundary, change their colour. Yet the apparent brightness of the two grey patches gets inverted. Fig. 20.9h plots the DDOG predicted brightness of the two grey patches with respect to transition stage. The DDOG predicted that intensity curves reflect the observed crossover in brightness of the two grey patches. This can also be seen in the ODOG predicted intensity plots (Fig. 20.9i). Therefore for this illusion also, the DDOG model is just as good as the ODOG model.

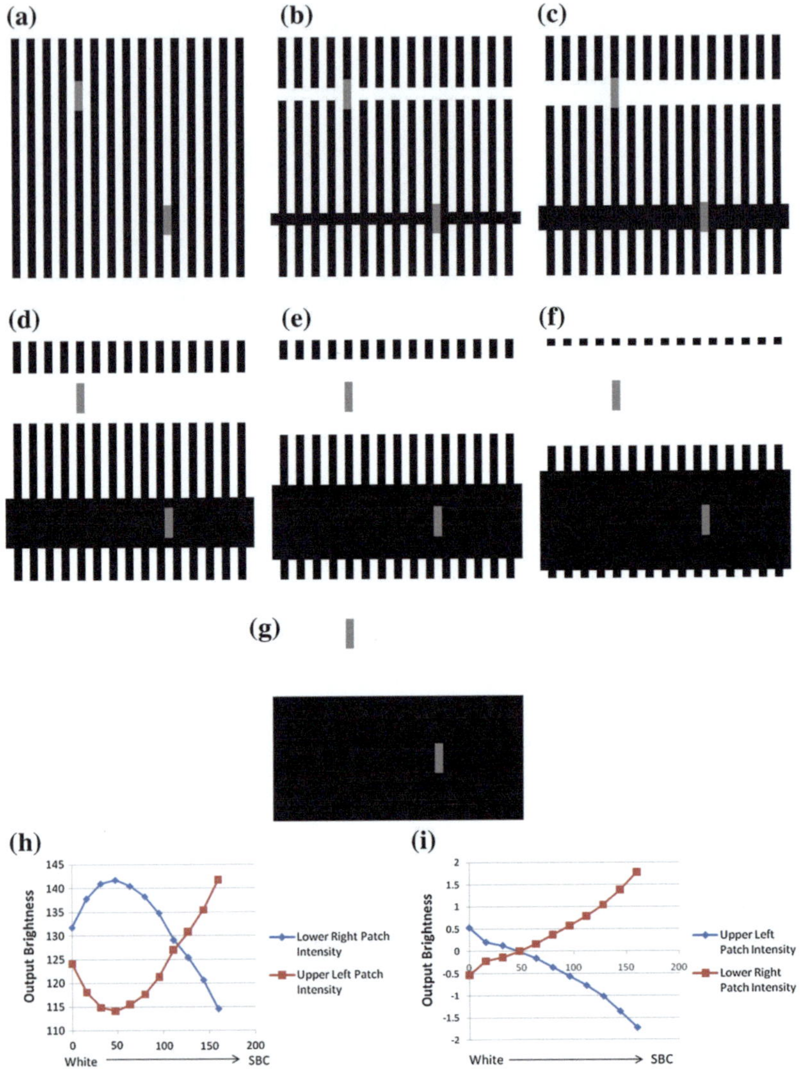

Fig. 20.9 **a–g** Howe's transition from White effect to SBC. **h** Plot of output intensities for our model. **i** Plot of output intensities for ODOG model

20.5 Discussion and Conclusions

In this work, we have simultaneously dealt with two distinct but highly correlated domains, viz. visual psychophysics and neurobiophysics in general, and correspondingly, brightness perception and receptive field structure in particular. While thus proposing an alternative neurophysiologically plausible model of perceiving brightness contradictions, we have actually expounded the concept that, not only are contradictions real, but they actually represent the mode and mechanism by which we perceive the visual world. Let us consider the phenomenon of lateral inhibition in receptive field for instance. It consists of two oppositely acting concentric regions the excitatory centre and the inhibitory surround, that has been modelled by the physiologists by two Gaussian functions in opposite phase. Such a DOG model, we have seen, though capable of explaining brightness–contrast, cannot predict the brightness assimilation. We have here proposed the existence of a higher level of contradiction in the visual system to provide a unified explanation to these two contrary types of brightness phenomena. The proposed model envisages two DOGs acting in opposition to each other, which we call the DDOG model. In fact, neurobiologically, these two opposing DOGs may be looked upon as the neural correlates of the two opposite types of neurons called the on-centre cell and the off-centre cells actually existing in different layers of the eye–brain system, with the former having an excitatory centre and inhibitory surround that is vice versa for the later. Neuroscientists have pointed out several times that such on–off type of contradictions are present throughout in visual computational process having different other manifestations [3]. One such manifestation has been mentioned in the present paper too. These are the magnocellular (M) and parvocellular (P) channels which represent two complimentary pathways in visual signal processing. The former comprises of parasol neurons with large receptive fields, high temporal and low spatial and colour sensitivity. The latter comprising what are called the midget neurons represent just the opposite characteristics. Some of these facts and also the fact that the former channel (M) is having a much higher signal conduction velocity compared to the later (P), has actually been utilized in our proposed model in fixing the role of M-channel as the candidate for the initial vision at a glance in the 2-pass mechanism.

The ODOG model also cannot account for the Checkerboard illusion, as shown in a previous section, where ODOG predicts brightness–contrast instead of brightness assimilation. Although many parameter values of our model are yet to be substantiated strongly from neurophysiological data, we have been able to define a scale range, within which, the DDOG model, works at least as good as ODOG in most cases, and better than ODOG in some cases such as the Checkerboard illusion.

The ODOG filter uses 42 DOGs of varying widths and orientations. The fact that ODOG has filters involving seven different length scales suggests that it will require several layers of ganglia to implement ODOG with neurons and therefore is unlikely to be a model of low-level vision. Indeed as pointed out by [12] there is no known physiological analogue, i.e. neural correlate of the ODOG filter. This is especially true for the contrast normalization step of the ODOG filter, which is primarily responsible for producing the response to brightness assimilation illusions (e.g. White effect). On the other hand, compared to the ODOG filter, the DDOG filter has the distinct advantage of being much simpler with far fewer number of DOG filters. DDOG only uses 4 DOGs, 2 each for the M- and P-channels. Also, the DOG-like centre–surround response profiles have long been known to be present in neural responses [19]. So a difference of two DOGs can easily be performed within low-level vision. Even the background uniformity detection step can easily be performed with DOG functions since the LOG-DOG equivalence has already been shown by [14] and others [6]. The DDOG model is therefore not only neurobiologically plausible, it substantiates the fact pointed out by several neuroscientists, like Rudolfo Llinas and Mriganka Sur that, computations in nature in general, and also the brain in particular, are actually carried out through a series of contradictions only [13], and that there is nothing 'intrinsically visual about visual thalamus and cortex' in the brain [20]. The two DOGs in the DDOG model, represent a unity of opposites, with variable degree of opposition (as indicated by the parameters in Table 20.1), such variability being responsible for giving birth yet again to two opposite mechanisms, viz. the M- and P-channels in the central visual pathway and so on. Thus, through a continuous process of contradiction, builds up our perception of the world which itself is always in a state of change and hence always in contradiction.

Acknowledgments We express our sincere gratitude to Alan Robinson for providing us with the code of his implementation of the ODOG filter [17, 18] whose outputs have been shown in this article, and Mr. Sourya Roy, Mr. Arijit Mallick and Mr. Debjyoti Bhattacharjee for helping us with experiments and discussions. One of the authors, Ashish Bakshi, would like to thank the Council of Scientific and Industrial Research (CSIR) for providing financial support to him that enabled him to perform the present work.

References

1. Albergo, G.: Is it possible to perceive contradictions. In: Beziau, J.Y. et al. (eds.) Handbook of the 5th World Congress of Paraconsistency, pp. 33–35. Kolkata (2014)
2. Beall, J.C., Collivan, M.: Looking for contradictions. Australas. J. Philos. **79**, 564–569 (2001)
3. Beziau, J.Y. et al. (eds.) Round table, brain contradictions and computability. In: Handbook of the 5th World Congress of Paraconsistency. Kolkata (2014). ISBN: 978-93-83373-02-4
4. Blakeslee, B., Pasieka, W., McCourt, M.E.: Oriented multiscale spatial filtering and contrast normalization: a parsimonious model of brightness induction in a continuum of stimuli including White, Howe and simultaneous brightness contrast. Vis. Res. **45**, 607–615 (2005)
5. Bullier, J.: Integrated model of visual processing. Brain Res. Rev. **36**, 96–107 (2001)
6. Ghosh, K., Sarkar, S., Bhaumik, K.: A possible mechanism of zero-crossing detection using the concept of extended classical receptive field model of retinal Ganglion cells. Biol. Cybern. **93**, 1–5 (2005)
7. Ghosh, K., Sarkar, S., Bhaumik, K.: A possible explanation of the low-level brightness contrast illusions in the light of an extended classical receptive field model of retinal ganglion cells. Biol. Cybern. **94**, 89–96 (2006)
8. Ghosh, K.K., Pal, S.K.: Explaining and classifying brightness perception illusions by a new receptive field approach. Perception **36**(s), 54–55 (2007)
9. Ghosh, K., Bhaumik, K.: Complexity in the human perception of brightness: a historical review on the evolution of the philosophy of visual perception. Online J. Biol. Sci. **10**, 17–35 (2010)
10. Hochstein, S., Ahissar, M.: View from the top: hierarchies and reverse hierarchies in the visual system. Neuron **36**, 791–804 (2002)
11. Howe, P.D.L.: A comment on the Anderson (1997), the Todorovic (1997), and the Ross and Pessoa (2000) explanations of White's effect. Perception **30**, 1023–1026 (2001)
12. Kingdom, F.A.A.: Lightness, brightness and transparency: a quarter century of new ideas, captivating demonstrations and unrelenting controversy. Vis. Res. **51**, 652–673 (2011)
13. Llins, R.: I of the vortex: From Neurons to Self. MIT Press, Cambridge (2001)

14. Marr, D., Hildreth, E.: Theory of edge detection. Proc. R. Soc. Lond. B **207**, 187–217 (1980)
15. Merigan, W.H., Maunsell, J.R.H.: How parallel are the primate visual pathways? Ann. Rev. Neurosci. **16**, 369–402 (1993)
16. Priest, G.: Perceiving contradictions. Australas. J. Philos. **77**, 439–446 (1999)
17. Robinson, A.E., Hammon, P.S., de Sa, V.R.: Explaining brightness illusions using spatial filtering and local response normalization. Vis. Res. **47**, 1631–1644 (2007a)
18. Robinson, A.E., Hammon, P.S., de Sa, V.R.: A filtering model of brightness perception using frequency-specific locally-normalized oriented difference-of-gaussians (FLODOG). Poster at: Vision Sciences Society Meeting (VSS2007). Sarasota, FL (2007)
19. Rodieck, R.W., Stone, J.: Analysis of receptive fields of cat retinal ganglion cells. J. Neurophysiol. **28**, 833–849 (1965)
20. Sur, M., Garraghty, P.E., Roe, A.W.: Experimentally induced visual projections into auditory thalamus and cortex. Science **242**, 1437–1441 (1988)
21. White, M.: A new effect of pattern on perceived lightness. Perception **8**, 413–416 (1979)

Chapter 21
Truth, Trivialism, and Perceptual Illusions

Otávio Bueno

Abstract Dialetheism is the view according to which some contradictions (i.e., statements of the form, *A* and not-*A*) are true. In this paper, I discuss three strategies to block dialetheism: (i) Contradictions cannot be true because some theories of truth preclude them from emerging. (ii) Contradictions cannot be true because we cannot see what it is like to perceive them. Although that does not undercut the possibility that there are true contradictions that we cannot perceive, it makes their introduction a genuine cost. (iii) Contradictions cannot be true because if they were, we would end up sliding down into believing that everything is true (trivialism). Even if the dialetheist is not committed to that slippery slope, it is crucial that the dialetheist establishes that trivialism is unacceptable; but it is not clear how that could be done successfully. Graham Priest has considered these strategies (in his *Doubt Truth to be a Liar*), but I argue that none of his responses successfully block them.

Keywords Dialetheism · Truth · Trivialism · Perceptual illusions · Graham Priest

Mathematics Subject Classification (2000) Primary 03B53 · Secondary 03A05

21.1 Introduction

According to dialetheism, some contradictions that is, statements of the form *A* and not-*A* are true (Priest [14, 15]). Few would be naturally disposed to agree with the view, but Graham Priest has defended it with great ingenuity and care. And Priest is certainly right in emphasizing the significant role that inconsistency plays in our understanding of logic, rationality, and various methodological issues. Even if you are not prepared to follow Priest all the way down, and believe that there are true contradictions, the encounter with the dialetheist is bound to make you rethink some of your deepest held assumptions.

O. Bueno (✉)
Department of Philosophy, University of Miami, Coral Gables, FL 33124-4670, USA
e-mail: otaviobueno@mac.com

J.-Y. Beziau et al. (eds.), *New Directions in Paraconsistent Logic*,
Springer Proceedings in Mathematics & Statistics 152,
DOI 10.1007/978-81-322-2719-9_21

It is not easy to argue with dialetheists, and some, such as David Lewis, have noted that one cannot argue against them, since there is no foothold on uncontested ground (Lewis [12]). Whether this is right or not is not an issue I need to address here (see, e.g., Bueno and Colyvan [5]). What I would like to explore are some strategies in terms of which one could resist dialetheism. Priest has considered them in *Doubt Truth to Be a Liar* (Priest [14]), but it is not clear to me that he has adequately blocked them. In particular, I will discuss three such strategies:

(a) *First strategy*: Contradictions cannot be true because some theories of truth preclude them from emerging. A serious commitment to these theories rules out the commitment to dialetheism.
(b) *Second strategy*: Contradictions cannot be true because we cannot see what is like to perceive them. Although that does not undercut the possibility that there are true contradictions that we cannot perceive, it makes their introduction a genuine cost.
(c) *Third strategy*: Contradictions cannot be true because if they were, we would end up sliding down into believing that everything is true (trivialism). Even if the dialetheist is not committed to that slippery slope, it is crucial that the dialetheist establishes that trivialism is unacceptable. But it is not clear how that could be done successfully.

Let me consider each of these strategies in turn.

21.2 First Strategy: Resisting True Contradictions from Some Theories of Truth

Here is a line of argument that someone who is committed to certain theories of truth could invoke in order to resist dialetheism. Consider the traditional coherence theorist who believes that truth is ultimately a matter of having a coherent belief system. There is, of course, the familiar issue as to whether this is indeed an account of truth or rather an account of justification. However the issue is resolved, the traditional coherence theorist would insist that there are independent reasons to think that consistency is a necessary component in a coherent system. On the traditional coherence picture, without consistency, we obtain triviality, and a trivial system—in which everything is true—would not be coherent, since it would both have and lack every conceivable property. This is a good reason for the traditional coherence theorist to keep consistency as a requirement. In other words, given that the traditional coherence theorist has no independent reason to believe that a coherent belief system can be inconsistent, or that we can be justified in believing in inconsistent systems, the coherence theorist has no reason to entertain the possibility of inconsistent coherent systems. This provides the resources for the traditional coherence theorist to avoid being committed to dialetheism.

Of course, the traditional coherence theorist did not have any knowledge of paraconsistent logics. Such logics would only be explicitly developed much later. Consider then an enlightened coherence theorist, who is now aware that inconsistency and triviality should be distinguished. With a paraconsistent logic in place, an inconsistent system need not be trivial. There is no need to keep consistency as a necessary requirement for a coherent system if what we really want to avoid is triviality. Is the enlightened coherence theorist a dialetheist?

Not necessarily. After all, even though an inconsistent coherent system is entertained, the enlightened coherence theorist need not be committed to the conclusion that there are true contradictions. Strictly speaking, in a coherence theory, truth is a property of overall systems rather than individual statements. Thus, even though an inconsistent coherent system may be true, on the coherence theory it does not follow that there are true contradictions.

The idea is that the enlightened coherence theorist can resist dialetheism without begging the question. At no point has the enlightened coherence theorist assumed that contradictions cannot be true. In fact, given that a paraconsistent logic has been invoked, the central logical resource for dialetheism is in place. So, the dialetheist can no longer complain that the position has been illegitimately excluded from consideration. Rather, the enlightened coherence theorist has an argument to resist dialetheism, since on her conception truth (or justification) is a property of whole systems, not of individual statements.

It may be objected that this response is not adequate. After all, what is at issue now is whether according to the enlightened coherence theorist there are inconsistent but true coherent systems. So the issue has moved to the level of whole systems rather than particular statements. But are there reasons to think such systems exist in the first place? Nothing from the enlightened coherence account—*as an account of truth (or justification)*—settles the matter. The world would have to be such that it allowed for true but inconsistent coherent systems. But the argument now turns on how the world is rather than on what is required from a theory of truth. So a very different kind of argument than the one provided by the dialetheist needs to be offered.

But is this not precisely the dialetheist's argument, namely, that nothing in the enlightened coherence theory rules out dialetheism? To answer this question, it is important to be clear about what we expect from a theory of truth. Clearly, theories of truth have been formulated quite independently of dialetheism. So why should they suddenly be required to rule out this particular philosophical view? Consider an analogous argument. A coherence theory of truth—as a theory of truth—does not rule out Aristotelian physics, Newtonian mechanics, alchemy, or a number of other false theories. If these theories are ruled out, it is because they are ultimately false; but that is the outcome of the relation these theories bear to the world. It is not a feature of the coherence theory alone. So just because the coherence theory does not rule out alchemy, that does not mean, of course, that we should now all be alchemists, any more than Aristotelian natural philosophers, Newtonian physicists, or dialetheists. In the end, it is unclear what exactly the dialetheist gains by insisting that certain theories of truth do not rule out dialetheism. Why should anyone expect that dialetheism—or any other philosophical or scientific theory—be ruled out on the basis of a theory of truth alone?

Clearly, this general response is similarly open to the traditional coherence theorist. If coherence is taken seriously—and it is by coherence theorists—then contradictions cannot be true. Since contradictions are individual statements, and not features of overall systems, they are not the kinds of things that, according to the coherence theorist, truth can be suitably assigned to. And whether there are true inconsistent whole systems turns on how the world ultimately is rather than on particular commitments emerging from one's theory of truth—and this is as it should be. (Of course, a paraconsistent logic would be needed to accommodate such inconsistencies at the level of entire systems without triviality.)

It should be noted that, on this coherentist conception (enlightened or not), the statement "snow is white" is similarly not truth-apt, even though it is part of a coherent system. Some may argue that this consequence shows that the version of the coherence theory I am considering is just inadequate, since it flies in the face of ordinary practice and common use of the truth predicate. But I take it that, at least in a context where dialetheism is at issue, flying in the face of ordinary practice is not an objection that has much force. What counts is the overall explanatory balance of the resulting view, and, in this particular case whether there are suitably formulated versions of the coherence theory of truth that block dialetheism.

I insist that this move does not beg the question against the dialetheist. Dialetheism was not even a possibility when the traditional coherence theories were first formulated. One would need to wait until the development of paraconsistent logics before dialetheism could be seriously entertained. Without a paraconsistent logic explicitly in place, we would immediately obtain triviality if we were to be committed to an inconsistent system.

It might be objected that all that the dialetheist needs is an implicit paraconsistent logic, in the way in which the Aristotelian syllogistic system is paraconsistent. There is no need for an explicit, fully articulated paraconsistent logic to be developed in order for dialetheism to get going. Consider, for instance, the syllogism:

($P1$) All men are mortal.
($P2$) Some men are not mortal.
Therefore, all men are blue.

This is, of course, an invalid argument according to Aristotelian logic. The premises are contradictory, but not everything follows from them. Explosion (the principle according to which everything follows from a contradiction) is then blocked. Since blocking explosion is considered a central feature for a logic to be deemed paraconsistent, and given that Aristotelian logic has that feature, it is indeed paraconsistent (see da Costa and Bueno [7], and Priest [13]). Despite being minimally paraconsistent, Aristotelian logic was not developed as a way of providing resources to handle reasoning involving inconsistencies. In this sense, such a logic—whatever its paraconsistent status—would not be of much help to the dialetheist. It is not surprising that dialetheism only emerged when the suitable resources of paraconsistent logic were in place.

The point here is that, by taking into closer account the particular features of the coherence theory of truth, this proposal—both in the traditional and enlightened forms—has the resources to block dialetheism. The dialetheist may complain that the coherence theory is inadequate for other reasons. But this means changing the argumentative strategy. The argument no longer can be: the coherence theory of truth does not block dialetheism, and so, if you are a coherence theorist, there is no reason, based on your theory of truth alone, for you not to be a dialetheist. The argument needs to be: the dialetheist rejects some features of the coherence theory of truth (such as, the fact that it relies on consistency as a requirement for coherence), and by rejecting these features, the resulting theory no longer blocks dialetheism.

This raises the issues as to whether the resulting coherence theory, without the consistency requirement on coherence, would still count as a coherence theory, and whether the changes envisaged by the dialetheist are independently motivated. These are points that the dialetheist would still need to argue for. But one may wonder what would be gained from this exercise. After all, it would not be philosophically surprising that a reformulation of the coherence theory that is offered so that it becomes compatible with dialetheism turns out to be so compatible!

A similar style of argument applies to other theories of truth. For example, according to Priest, the semantic conception of truth does not block dialetheism (see Priest [14, pp. 45–47]). In order to avoid the semantic paradoxes, Tarski introduced a hierarchy of languages in which the truth predicate of the object language could not be defined in that language. As a result, the liar sentence is not expressible in any language of the hierarchy, and the paradox is blocked.

On Priest's view, such a hierarchy is not essential to the semantic conception of truth, and should be rejected (see Priest [14, p. 46], and [15, Chap. 1]). Of course, by rejecting the hierarchy of languages, the semantic conception is unable to block the semantic paradoxes. It is no longer surprising that dialetheism could not be resisted anymore.

The problem here should now be familiar. Similarly to what happened in the last point made in the discussion of the coherence theory, Priest has rejected the feature of the semantic conception of truth that allowed the theory to resist the paradox. By substantially weakening that theory, it is not surprising that dialetheism can no longer be avoided. But anyone who takes the semantic conception of truth seriously, as Tarski did, is unlikely to recognize the weakened version of the theory as still a candidate for a semantic conception. For a distinctive feature of that conception, at least as Tarski developed it, is the acknowledgement that semantically closed languages, such as English, are ultimately inconsistent. In order to ensure that the formal languages under consideration are not semantically closed, and thus are not open to semantic paradoxes, Tarski formulated the now familiar hierarchy of languages. As a result, such a hierarchy is indeed constitutive of the semantic conception, which would thereby be entirely disfigured without it. So, the conclusion that the semantic conception of truth does not avoid dialetheism is not warranted. And it should be expected that if one reformulated that conception by excluding the hierarchy of languages, it would become compatible with dialetheism. But little would be gained by this exercise. It is a much more controversial point to claim that one should reject the hierarchy of languages—a point that the dialetheist does make— but which the defender of the semantic conception is unlikely to concede, for the reason just noted.

I think that the same point applies to all of the other theories of truth that Priest considers in Chap. 2 of *Doubt Truth to Be a Liar* (Priest [15]). Properly developed, and perhaps a little more sympathetically presented, each of these theories has the resources to resist dialetheism. The only exception is the deflationist theory. This theory, as presented, for example, by Horwich [11], since it does not impose any constraints whatsoever on the disquotational schema, does seem to invite dialetheism. It is not surprising then that those who are sympathetic to deflationism about truth end up working so hard to develop a well-motivated account of how the semantic paradoxes can be resisted (see, e.g., Field [8]).

21.3 Second Strategy: Can We Perceive Contradictions?

A second strategy to resist dialetheism insists that contradictions cannot be true because we cannot see what is like to perceive them. Although that does not undercut the possibility that there are true contradictions that we cannot perceive, it makes their introduction significantly more costly. After all, given that we have no access to what contradictions are like, it is just expected that some would try to resist being committed to their existence.

The dialetheist, however, argues that we can perceive contradictions—or, at least, we can know what is like to perceive them (Priest [14, pp. 57–61]). Various kinds of perceptual illusions illustrate that. Consider, for instance, the Penrose figure. These are ascending stairs, and by starting at any point in the figure and moving upward anti-clockwise, you return to the same spot where you started. Thus, you end up at a point that is higher than itself (and that is also not higher than itself). At this moment, you have perceived a contradiction.

The Penrose figure should be distinguished from the Schuster figure. According to Priest ([14, p. 59]), the latter is not a case of perceiving a contradiction, since it does not depict an inconsistent situation. Rather, the picture is constituted by two different perfectly consistent drawings (a three-legged object and a two-legged one) pasted together. We cannot visually parse the whole drawing, but we can clearly see each of its two parts. Priest takes the situation with the Penrose figure to be different, given that the whole object can be perceived: we perceive here a truly inconsistent object.

It is not clear, however, that the two cases are different. In both examples, rules of perspective are violated. In the Schuster figure, convention codes about how to draw a three-legged object and a two-legged one are violated. A central principle that governs the representation of an object under perspective is that one should not draw what cannot be seen from the point of view that is adopted when picturing that object (see Gombrich [10]). This principle is clearly violated in the Schuster figure, in which two different perspectives are simultaneously adopted in order to produce the image. It is not surprising that we have trouble parsing it. The same point applies to the Penrose figure, since it also adopts two distinct perspectives: one according to which the point you started ascending the stairs is lower than the other points in the stairs, and one according to which that point is higher. The trick, once again, emerges from the simultaneous combination, in a perfectly symmetric way, of two different perspectives. The figure can be perfectly divided into two planes: a right one and a left one, crossing vertically the stair that is closest to the viewer. The careful symmetry used in the composition of the image gives the sense that it is an ordinary stair.

Note that the point here is not to state that, in the Schuster figure, there is an object that has two legs from one perspective and three from another. I have no reason to believe that any such object exists. Rather, the point is that the Schuster figure involves two distinct perspectives that are brought together in order to produce the image—quite independently of the issue as to whether there is any object that the figure represents. (Priest presumably will not disagree with this point.)

Can we say that the Penrose figure is an inconsistent object? It is not clear to me that we can. The figure violates a central principle of perspective, since two different perspectives are used simultaneously to produce the image. But within each plane of perspective, the rules of perspective are thoroughly followed. This explains the perfect sense of familiarity that the image initially has. It is in virtue of this symmetry that the image initially just seems to be an unremarkable arrangement of stairs. In this sense, the Penrose figure yields a slightly different phenomenology than the Schuster figure. But in the end it also generates some dissonance—a sense of puzzlement—when after always walking down a set of stairs one suddenly reaches the highest point of those stairs! Now, clearly, the fact that an image is composed in terms of two different perspectives does not entail that the object that is being depicted is inconsistent. Picasso's drawings of Dora Maar violate the same principle of perspective that the Penrose image does. But clearly, this gives us no reason to believe that Picasso's lover was an inconsistent object. At best, a different convention code is invoked to produce the image.

Now, what is the significance of realizing that the Penrose figure does not depict an inconsistent situation? If we are not really looking at an inconsistent object, then the argument to the effect that we know what is like to perceive a contradiction no longer seems persuasive. Remember that the dialetheist is trying to offer us an account of what is like to perceive an inconsistent situation (a contradiction). If we can find that out without effectively looking at an inconsistent object, one may wonder how reliable the answer is.

But perhaps it is enough for the dialetheist's purposes simply to address the phenomenology of the perception of an inconsistent situation—whether that situation is indeed inconsistent or not. To the question: "What would it be like to perceive a contradiction?" the dialetheist replies: It would be like getting trapped in the situation depicted in the Penrose figure, where one finds oneself both at a point higher and lower than itself. Even if the object depicted in the figure is not really inconsistent, we know what is like to perceive a contradiction by examining carefully what happens when we look at the Penrose image.

I do not think this response succeeds, though. The difficulty here is that how can we know that contradictions do look the way the dialetheist reports them as looking—unless we already have reason to believe that (a) the object that we are looking at is indeed inconsistent, and (b) perception works reliably in the presence of inconsistent objects? With regard to (a), if the object in question is not inconsistent, what grounds would we have to think that the phenomenology described corresponds to the relevant sort of object? Being trapped in the situation depicted by the Penrose figure is not phenomenologically unlike perceiving an anamorphosis (a distorted drawing that appears correctly only when viewed from a particular point). Both cases involve careful, perverse manipulation of the rules of perspective. And both fail to guarantee that the objects in question are inconsistent. With regard to (b), what grounds do we have to believe that perception works reliably in the presence of inconsistent objects? If the observable world is consistent, as Priest argues it is (Priest [14, pp. 62–64]), it does not seem that we have even grounds to determine whether perception functions reliably in an inconsistent context. We lack the opportunity to do that. So, more needs to be said before we can be reassured that we know what is like to perceive an inconsistent situation.

JC Beall and Mark Colyvan [3] argue that we do have good reason to believe that the world—including, in particular, its observable parts—is inconsistent, and that contradictions in fact abound. Their considerations are not based on drawings of presumably impossible objects—they explicitly mention, and disregard, Escher's drawings—but on the pervasive vagueness of language and the assumption that a paraconsistent approach to vagueness is the only one that does not fail (Beall and Colyvan [3, p. 565]). On their view:

> You might think that some Escher's drawings apparently represent inconsistent objects but that these drawings do not give us reason to believe that the world is inconsistent. There's an important difference, though, between the Escher-like figures and our case: it's hardly plausible that Escher's drawings are the best representations of the world. (Indeed, most people don't think they represent at all.) On the other hand, the language of our best scientific theories is supposed to not only represent, but accurately represent. Thus, if the language of our best scientific theories (indispensably) involves vague predicates, then as naturalistic philosophers we have good reason to believe that this vagueness is a feature of the world. [...] This is not intended to be an argument for vagueness-in-the-world, we merely wish to show that considerations of vagueness provide a significantly better case for observable contradictions than Escher-like drawings (Beall and Colyvan [3, p. 568]).

So whenever we experience a Sorites-like patch of colors ranging from blue to purple, "the purplish-blue object is both blue and not blue" (Beall and Colyvan [3, p. 565]). Contradictions, Beall and Colyvan insist, do abound!

Does this argument go through? I do not think it does. It relies on two assumptions that turn out to be highly problematic. The first is that a paraconsistent approach to vagueness is the only that is not ruled out. But what reasons do we have to believe that this assumption is true? Let us return to Beall and Colyvan's purplish-blue object. The phenomenology of seeing such an object has none of the puzzlements we experience when we look at allegedly inconsistent objects, such as the Penrose or the Schuster figures. In contrast with experiencing these figures, there is no difficulty to parse a purplish-blue object: no cognitive dissonance is involved there. (A related puzzlement emerges when we reason through a contradiction, such as the one involved in the liar paradox: we experience conflicting conclusions that we derive—*prima facie*, the liar sentence is true and it is not true.) The fact that experiencing the purplish-blue object is not accompanied by any such corresponding cognitive phenomenology suggests that rather than exemplifying a contradictory vague entity, we have more reason to consider it as a consistent vague object: it is not determined that the object is blue and it is not determined that the object is not blue. That is all. No contradiction is in fact needed—or invoked—here. To assume a paraconsistent approach of vagueness in this case just seems unmotivated.

The second assumption that Beall and Colyvan rely on is the indispensability argument. According to this argument, we ought to be ontologically committed to all and only the entities that are indispensable to our best theories of the world (Colyvan [6]). And, they continue, since our best scientific theories invoke vague predicates, we should conclude that there are vague objects in the world. But clearly we quantify over all kinds of things—even indispensably so—we have no reason to believe exist: fictional objects, average moms, and perfectly frictionless planes are obvious examples. Quantification and ontological commitment are, thus, best kept apart (Azzouni [1]). To mark ontological commitment, one needs an existence predicate. Quantification only indicates the range of the objects that one is considering—whether they exist or not is a separate matter (Bueno [4]).

As a result, even if we granted, for the sake of argument, that a paraconsistent approach to vagueness were the only workable solution to the problem of vagueness, we cannot conclude that vagueness is a feature of the world. Thus, the intended conclusion that there are observable contradictions in the world is, once again, blocked.

21.4 Third Strategy: Sliding Down into Trivialism

A source of worry against dialetheism—although definitely not a good one—is that if we accept that some contradictions are true, we will end up having to believe that everything is true. Of course, this would only be the case if we adopted classical logic. Since the dialetheist does not do that, but adopt instead a paraconsistent logic, this worry does not get off the ground. But once it is claimed that *some* contradictions are true, it is important for the dialetheist to distance the view from the claim that *all* contradictions are true. After all, under very reasonable assumptions, this is equivalent to the claim that everything is true. This is trivialism. To make sure that there is no slippery slope from dialetheism to trivialism, it is important for the dialetheist to resist that view.

Given that the trivialist does not discriminate anything, since he or she takes everything to be true, trivialism is obviously not very plausible. But how can it be refuted? It turns out to be something far from trivial. One would need to show that something is untrue. But the trivialist would agree with that, since "something is untrue" is also true, given that everything is true!

To refute the trivialist, Priest provides a transcendental argument, according to which the conditions of possibility of choosing something is that one believes that certain actions will have certain effects. This means believing certain things and not believing others. And this is precisely what the trivialist is unable to do, given that he or she takes everything to be true. As Priest notes:

> To choose how to act is to have a purpose: to (try to) to bring about *this* rather than *that*. [. . .] Choosing is an irredeemably goal-directed activity. And [. . .] such an action is incompatible with believing everything. It follows that I cannot but reject trivialism. Phenomenologically, it is not an option for me. This does not show that trivialism is untrue. As far as the above considerations go, it is quite possible that everything is the case; but not for me—or for any other person. (Priest [14, p. 70])

This is an intriguing argument. It concedes that one cannot refute trivialism. But it tries instead to undermine the existence of trivialists. It is not clear, however, that the argument succeeds. The phenomenology of the trivialist may actually not be all that different from the dialetheist's. The dialetheist believes that some contradictions are true, but decides to act on some other beliefs: the consistent beliefs, not the contradictory ones. Of course, the dialetheist has a story to offer here. Since the observable world—in particular, the parts of the world in which we act—is consistent, the dialetheist will typically not face a situation in which a course of action requires commitment to an inconsistent state of affairs.

But precisely the same argument also applies to the trivialist. The trivialist also selects the beliefs he or she will act on. Some will seem more appropriate than others for certain tasks. The trivialist can even tell a story—exactly the same story as the dialetheist's—about the consistency of the observable world. This would explain the phenomenology of the trivialist's choice, in exactly the same way as the dialetheist explains it.

One may resist this conclusion by insisting that Priest is not constrained at all to act only on consistent beliefs, since actions depend on beliefs about more than just what is directly observable. Even if we grant this last point, the objection still would not go through. For whatever the sources of one's actions—whether they are grounded on beliefs about the observable or not—given that these actions can only be realized in the observable part of the world (which is the only part we inhabit), and since that part, according to the dialetheist, is consistent, the dialetheist will never face a course of action that requires the realization of an inconsistency. And precisely the same point goes for the trivialist.

It might be argued that this cannot be right. After all, the trivialist will also believe in the negation of every aspect of the story just told. That is right. But, of course, that does not undermine the trivialist's account! In an interesting way, believing everything is remarkably similar to not believing anything, which is often taken to be a skeptical position.[1] Interestingly enough, the phenomenology of choice in Pyrrhonian skepticism won't be much different from the descriptions given so far.

[1]Priest in fact considers the skeptic in this context (see Priest [14, p. 69]).

The Pyrrhonist acts without beliefs in anything having to do with the ultimate nature of things (since the skeptic suspends judgment about these matters). But this does not prevent the Pyrrhonist from acting, since similarly to the dialetheist and the trivialist, it seems to the Pyrrhonist that certain courses of action are better than others. This does not require the skeptic to be committed to any substantive belief about the way things are. It is enough that things just seem to be in a certain way.

Note also that the dialetheist's response to the trivialist, which is content with just describing the dialetheist's own phenomenology, would be welcomed by the Pyrrhonist, who similarly just describes, in a non-dogmatic way, the content of his or her own experience. Trivialism, although obviously not a plausible view, has still to be undermined. Unless the dialetheist is able to do that, the plausibility of dialetheism can always be questioned.

21.5 Conclusion

I discussed three strategies that one could invoke to resist dialetheism: one based on the theory of truth, another on perceptual illusions, and a third on trivialism. In each case, the dialetheist's arguments were found wanting. In the end, despite all of the alleged benefits brought by the view, dialetheism can still be resisted.

Acknowledgments My thanks go to Jeremy Morris, Shane Oakley, and Scott Shalkowski for very helpful discussions. Thanks are also due to an anonymous reviewer for the comments on an earlier version of this work.

References

1. Azzouni, J.: Deflating Existential Consequence. Oxford University Press, New York (2004)
2. Beall, J.C., Armour-Garb, B. (eds.): Deflationism and Paradox. Clarendon Press, Oxford (2005)
3. Beall, J.C., Colyvan, M.: Looking for contradictions. Australas. J. Philos. **79**, 564–569 (2001)
4. Bueno, O.: Dirac and the dispensability of mathematics. Stud. Hist. Philos. Mod. Phys. **36**, 465–490 (2005)
5. Bueno, O., Colyvan, M.: Logical non-apriorism and the 'Law' of non-contradiction. In: Priest, G., Beall, J.C., Armour-Garb, B. (eds.), pp. 156–175 (2004)
6. Colyvan, M.: The Indispensability Argument. Oxford University Press, New York (2001)
7. da Costa, N.C.A., Bueno, O.: Paraconsistency: towards a tentative interpretation. Theoria **16**, 119–145 (2001)
8. Field, H.: Is the liar sentence both true and false? In: Beall, J.C., Armour-Garb, B. (eds.) pp. 23–40 (2005)
9. Gabbay, D., Guenthner, F. (eds.): Handbook of Philosophical Logic. Kluwer Academic Publishers, Dordrecht (2002)
10. Gombrich, E.: The Image and the Eye: Further Studies in the Psychology of Pictorial Representation. Phaidon, London (1982)
11. Horwich, P.: Truth, 2nd edn. Clarendon Press, Oxford (1998)

12. Lewis, D.: Logic for equivocators. Noûs **16**, 431–441 (1982)
13. Priest, G.: Paraconsistent logic. In: Gabbay, D., Guenthner, F. (eds.), vol. 6, pp. 287–393 (2002)
14. Priest, G.: Doubt Truth to Be a Liar. Clarendon Press, Oxford (2006)
15. Priest, G.: In Contradiction, 2nd edn. Oxford, Clarendon Press (2006)
16. Priest, G., Beall, J.C., Armour-Garb. B. (eds.): The Law of Non-Contradiction: New Philosophical Essays. Oxford, Clarendon Press (2004)

Chapter 22
Being Permitted, Inconsistencies, and Question Raising

Andrzej Wiśniewski

Abstract A semantic relation of being permitted by a set of possible worlds is defined and analysed. We call it "permittance". The domain of permittance comprises declarative sentences/formulas. A paraconsistent consequence relation which is both permittance-preserving and truth-preserving is characterized. An application of the introduced concepts in the analysis of question raising is presented.

Keywords Being permitted · Consequence · Paraconsistency · Logic of questions

Mathematics Subject Classification (2000) Primary 03B60 · Secondary 03B20, 03B53

22.1 Introduction

We are often confronted with a number of alternative accounts of how things are, yet without knowing which of the accounts, if any, is the right one. These accounts disagree on some issues and agree on others. Despite discrepancies, however, some facts still remain known, some states of affairs are considered impossible, and some statements are *permitted* while other are not.

In this paper we define the relation "a declarative sentence is permitted by a set of possible worlds" and we analyse its basic properties. The possible worlds in question are supposed to represent alternative accounts of how things are. We dub the relation

---

Work on this paper was supported by funds of the National Science Centre, Poland (DEC-2012/04/A/HS1/00715). I am indebted to Dorota Leszczyńska-Jasion, Mariusz Urbański and Tomasz Skura for their comments. Some remarks of Ivano Ciardelli were also of help.

A. Wiśniewski (✉)
Department of Logic and Cognitive Science, Institute of Psychology,
Adam Mickiewicz University, Poznań, Poland
e-mail: Andrzej.Wisniewski@amu.edu.pl

J.-Y. Beziau et al. (eds.), *New Directions in Paraconsistent Logic*,
Springer Proceedings in Mathematics & Statistics 152,
DOI 10.1007/978-81-322-2719-9_22

"permittance". The definition proposed is an explication of the corresponding intuitive notion of permitting, taken in one of its meanings. Our intuitions are presented in Sect. 22.1.2. For clarity, we start with a short description of the basic logical tools used throughout the paper.

22.1.1 Logical Preliminaries

We remain at the propositional level only. We consider a non-modal propositional language, L. The vocabulary of L includes a non-empty set $\mathcal{P} = \{p, q, r, \ldots\}$ of propositional variables, the propositional constant $\bot$ (*falsum*), and the connectives $\neg$, $\vee, \wedge, \rightarrow$. Well-formed formulas (wffs for short) of L are defined in the usual manner. We shall use the letters $A, B, C, \ldots,$ with subscripts if needed, as metalanguage variables for wffs of L. The letters $X, Y, \ldots$ are metalanguage variables for sets of wffs of L.

The connectives, as well as $\bot$, are understood, at the truth-functional level, as in Classical Propositional Logic. By an *L-model* we mean an ordered pair $\mathbf{M} = \langle \mathcal{W}, \mathcal{V} \rangle$, where $\mathcal{W} \neq \emptyset$ and $\mathcal{V} : \mathcal{P} \times \mathcal{W} \longmapsto \{\mathbf{1}, \mathbf{0}\}$ is a valuation of propositional variables in $\mathcal{P}$ w.r.t. elements of $\mathcal{W}$. As usual, the elements of the domain, $\mathcal{W}$, are called possible worlds. The concept of truth of a wff A in a world $w \in \mathcal{W}$ of $\mathbf{M}$, in symbols $\mathbf{M}, w \models A$, is defined in the standard manner. The inscription $\mathbf{M} \models A$ means "A is true in $\mathbf{M}$", that is, A is true in each world of the domain of $\mathbf{M}$.

Elements of domains of L-models, the possible worlds, will be intuitively thought of here as *alternative accounts of how things are*. This has no impact on the formalism, however. As long as we remain at the propositional level, the only condition imposed on $\mathcal{W}$ is non-emptiness. It follows that the domain of an L-model need not contain all the relevant alternatives.

By a *state* we will mean a non-empty set of possible worlds. In view of the intuitive interpretation of possible worlds adopted above, a non-singleton state comprises a number of alternative accounts of how things are.

Let $\mathbf{M} = \langle \mathcal{W}, \mathcal{V} \rangle$ be an L-model.

Definition 22.1 (*Truth set of a wff in an L-model*) $|A|_{\mathbf{M}} = \{w \in \mathcal{W} : \mathbf{M}, w \models A\}$.

Of course, $|\bot|_{\mathbf{M}} = \emptyset$.

Definition 22.2 ($\mathbf{M}$-*state*) An $\mathbf{M}$-state is a non-empty subset of $\mathcal{W}$.

Note that $\mathcal{W}$ is (also) an $\mathbf{M}$-state, and that, for each $w \in \mathcal{W}$, the singleton set $\{w\}$ is an $\mathbf{M}$-state.

22.1.2 *Intuitions*

Our basic intuition concerning the analysed concept of being permitted is:

(I) *A declarative sentence/wff* γ *is permitted by a state* σ *iff it is not the case that* σ *rules out* γ.

However, what "rules out" means depends on the form of γ.

γ can be *positive* that is not of the form $\neg\zeta$ (where $\neg$ stands for sentential negation and ζ is a declarative sentence/wff). It is natural to postulate:

(II) *Let* γ *be positive. State* σ *rules out* γ *iff* γ *is false in each world of* σ.

For example, "Andrew is a bachelor" is ruled out by a state which comprises (only) possible worlds in which Andrew is married.

γ can be *negative* that is of the form $\neg\xi$, where ξ is positive.[1] We seem justified in saying:

(III) *Let* γ *be negative and* $\gamma = \neg\xi$. *State* σ *rules out* γ *iff* ξ *is true in some world of* σ.

For instance, a state that contains a possible world in which Andrew is a bachelor rules out the sentence "It is not the case that Andrew is a bachelor".

Assuming bivalence, by (I) and (II) we get:

(II*) *A positive,* γ, *is permitted by a state* σ *iff* γ *is true in some world of* σ.

By (I) and (III), in turn, we get:

(III*) *A negative,* γ, *is permitted by a state* σ *iff* γ *is true in each world of* σ.

An analogy can be of help. A civil servant is permitted to issue a positive decision if there is a rule that entitles him/her to do so, and is permitted to decide to the negative if the disputed activity is forbidden by each rule that is applicable to the case. Similarly, a negative is permitted by a state if there is no world of the state that makes the negated sentence true, while for a positive being permitted by a state amounts to the existence of a world of the state which makes it true. Our usage of "being permitted" is thus akin to that of its deontic cousin. Yet, we do not aim at analysing "being permitted" deontically construed. Permittance in our sense is a relation between a declarative sentence/wff on the one hand, and a state on the other. What is (or is not) permitted is a declarative sentence/wff, and what permits it (or does not permit) is a set of possible worlds, where possible worlds are intuitively thought of as alternative accounts of how things are.[2]

---

[1]Observe that $\neg\neg\xi$ is neither negative nor positive. We will come back to this issue later on.

[2]Looking from a formal point of view, permittance belongs to the same category as *support* analysed in Inquisitive Semantics (see, e.g., [1, 2, 7]). However, the underlying intuitions are different. Moreover, Inquisitive Semantics conceives states/sets of possible worlds as information states.

The paper is organized as follows. In Sect. 22.2, we define the concept of permittance, characterize its basic properties, and show how knowledge and epistemic possibility can be modelled in our framework. Section 22.3 is devoted to permittance of inconsistencies. In Sect. 22.4 we analyse a paraconsistent consequence relation of transmission of permittance. Section 22.5 addresses the issue of question raising, in particular the problem of question raising by inconsistencies.

22.2 Permittance

22.2.1 Definition and Basic Properties

Let $\mathbf{M} = \langle \mathcal{W}, \mathcal{V} \rangle$ be an arbitrary but fixed L-model. "$\sigma \looparrowright A$" reads: "wff A is permitted by an $\mathbf{M}$-state σ". "$\looparrowright$" is thus the sign of the permittance relation.

Given the considerations presented above, the following definition comes with no surprise.

Definition 22.3 (*Permittance*)

1. $\sigma \looparrowright \mathrm{p}$ iff $|\mathrm{p}|_{\mathbf{M}} \cap \sigma \neq \emptyset$, for any propositional variable p;
2. $\sigma \looparrowright \neg A$ iff $\sigma \not\looparrowright A$;
3. $\sigma \looparrowright (A \vee B)$ iff $|(A \vee B)|_{\mathbf{M}} \cap \sigma \neq \emptyset$;
4. $\sigma \looparrowright (A \wedge B)$ iff $|(A \wedge B)|_{\mathbf{M}} \cap \sigma \neq \emptyset$;
5. $\sigma \looparrowright (A \rightarrow B)$ iff $|(A \rightarrow B)|_{\mathbf{M}} \cap \sigma \neq \emptyset$;
6. $\sigma \looparrowright \bot$ iff $|\bot|_{\mathbf{M}} \cap \sigma \neq \emptyset$.

Observe that permittance *is not* defined inductively. This is intended.

For positive wffs, being permitted by a state amounts to being true in some world(s) of the state. To be more precise, as an immediate consequence of Definition 22.3 we get:

Corollary 22.1 *Let σ be an $\mathbf{M}$-state and let A be a positive wff. Then $\sigma \looparrowright A$ iff $\mathbf{M}, w \models A$ for some $w \in \sigma$.*

However, the case of negative wffs is different. By Corollary 22.1 and clause (2) of Definition 22.3 we have:

Corollary 22.2 *Let σ be an $\mathbf{M}$-state. Let D be a wff of any of the forms: p, $\bot$, $(B \vee C)$, $(B \wedge C)$, $(B \rightarrow C)$. Then $\sigma \looparrowright \neg D$ iff $\mathbf{M}, w \not\models D$ for each $w \in \sigma$.*

Hence:

Corollary 22.3 *Let σ be an $\mathbf{M}$-state and let A be a negative wff. Then $\sigma \looparrowright A$ iff $\mathbf{M}, w \models A$ for each $w \in \sigma$.*

Corollary 22.3 shows that negatives behave in the context of permittance as it has been required in Sect. 22.1.2.

But what about wffs which are neither positive nor negative? As for L, there is only one kind of such wffs, namely wffs falling under the general schema:

$$\neg \ldots \neg D \tag{22.1}$$

where D is positive and the number of negations preceding D is greater than 1. If the number is even, we say that (22.1) is a $\neg_e$-wff; otherwise (22.1) is a $\neg_o$-wff. By D_A we designate the positive wff which occurs in a $\neg_e$-wff A or in a $\neg_o$-wff A after the string of negations.[3]

One can prove:

Corollary 22.4 *$\sigma \looparrowright \neg\neg A$ iff $\sigma \looparrowright A$.*

Proof By the clause (2) of Definition 22.3 we have:

$$\sigma \looparrowright \neg\neg A \text{ iff } \sigma \not\looparrowright \neg A$$
$$\sigma \not\looparrowright \neg A \text{ iff } \sigma \looparrowright A$$

and hence $\sigma \looparrowright \neg\neg A$ iff $\sigma \looparrowright A$. □

Thus, taking into account Corollaries 22.1, 22.2 and 22.4 we get:

Corollary 22.5 *1. Let A be a $\neg_e$-wff. Then $\sigma \looparrowright A$ iff $\mathbf{M}, w \models D_A$ for some $w \in \sigma$ iff $\mathbf{M}, w \models A$ for some $w \in \sigma$.*
2. Let A be a $\neg_o$-wff. Then $\sigma \looparrowright A$ iff $\mathbf{M}, w \not\models D_A$ for each $w \in \sigma$ iff $\mathbf{M}, w \models A$ for each $w \in \sigma$.

For brevity, let us introduce:

Definition 22.4 (*p-wffs and n-wffs*)

1. A p-wff is a positive wff or a $\neg_e$-wff.
2. A n-wff is a negative wff or a $\neg_o$-wff.

As we have shown, the categories of p-wffs and n-wffs are semantically homogeneous: a p-wff is permitted by a state iff it is true in at least one world of the state, while a n-wff is permitted by a state iff it is true in each world of the state. Permittance could had been concisely defined in terms of p-wffs and n-wffs. However, doing this would require an ad hoc acceptance of the claim of Corollary 22.4.

22.2.1.1 Remarks

Remark 22.1 For a singleton state permittance amounts to truth in the only world of the state. As an immediate consequence of the above corollaries we get:

[3] When A is neither positive nor negative, D_A is in the scope of the rightmost negation of the string.

Corollary 22.6 *Let* $\mathbf{M} = \langle \mathcal{W}, \mathcal{V} \rangle$ *be an* L*-model and* $\{w\}$ *be a (singleton)* $\mathbf{M}$*-state. Then* $\{w\} \Vdash A$ *iff* $\mathbf{M}, w \models A$.

Remark 22.2 Permittance becomes intensional when non-singleton states enter the picture. It happens that wffs which have equal truth sets (i.e. are classically equivalent) are not simultaneously permitted by a state. For example, we have:

$$|\neg(p \to q)|_{\mathbf{M}} = |p \wedge \neg q|_{\mathbf{M}}$$

Now take an L-model and its state $\{w_1, w_2\}$ such that:

- $\mathcal{V}(p, w_1) = \mathbf{1}$ and $\mathcal{V}(q, w_1) = \mathbf{0}$,
- $\mathcal{V}(p, w_2) = \mathbf{0}$ and $\mathcal{V}(q, w_2) = \mathbf{0}$.

We get:

$$\{w_1, w_2\} \nVdash \neg(p \to q)$$

$$\{w_1, w_2\} \Vdash p \wedge \neg q$$

Remark 22.3 Note that wffs of the forms:

$$\neg A \tag{22.2}$$

$$A \to \bot \tag{22.3}$$

do no differ as to their truth conditions in a world, but can differ with respect to permittance by states. When A is a p-wff, (22.2) is permitted only by a state in which A is false in each world of the state, whereas (22.3) can be permitted by a state in which A is false only in some, but not all worlds. This does not mean, however, that the negation connective $\neg$ has a non-classical meaning in L. Its meaning is determined by the standard truth condition. But $\neg$ behaves in a somewhat non-standard way in the context of permittance.

Remark 22.4 Observe that for any wff A, any L-model $\mathbf{M}$, and any $\mathbf{M}$-state σ we have:

$$\sigma \Vdash (\neg A \to \bot) \text{ iff } \sigma \Vdash A \wedge A \tag{22.4}$$

Hence we are able to express in terms of permittance, and without using $\bot$, that a n-wff, B, is true in at least one, but not necessarily all worlds of a state σ; this holds just in case the wff $B \wedge B$ is permitted by σ.

Note also that in general permittance is neither downward closed (if A is a p-wff, permittance of A by σ need not yield permittance of A by a proper subset of σ) nor upward closed (a n-wff permitted by a state need not be permitted by an extension of the state). However, permittance is upward closed for p-wffs and downward closed in the case of n-wffs.

22.2.2 Modalization

Let us now augment the initial language L with the modalities $\Box$ (necessity) and $\Diamond$ (possibility). Wffs of the enriched language are defined in the standard manner. We label the new language as $\mathcal{L}$. We use $\phi, \psi, \ldots$ as metalanguage variables for wffs of $\mathcal{L}$, and $\Phi, \Psi, \ldots$ as metalanguage variables for sets of wffs of the language. Whenever $\Box$ or $\Diamond$ precedes a metalanguage expression referring to wffs of L, it is understood that the wff in the scope of a modality belongs to L (i.e. is a wff of $\mathcal{L}$ in which no modality occurs).

Definition 22.5 (**S5**-*model*) An **S5**-model is a structure:

$$\langle \mathcal{W}, \mathcal{R}, \mathcal{V} \rangle$$

where $\mathcal{W} \neq \emptyset$, $\mathcal{V}$ is a valuation of $\mathcal{P}$ w.r.t. elements of $\mathcal{W}$, and $\mathcal{R} = \mathcal{W} \times \mathcal{W}$.

Thus by **S5**-models we will mean here only these relational models in which the accessibility relation $\mathcal{R}$ is universal. In the case of **S5**-models we have:

$$\mathcal{M}, w \models \Box\phi \textit{ iff } \mathcal{M}, w \models \phi \textit{ for each } w \in \mathcal{W}, \tag{22.5}$$

$$\mathcal{M}, w \models \Diamond\phi \textit{ iff } \mathcal{M}, w \models \phi \textit{ for some } w \in \mathcal{W}. \tag{22.6}$$

It is well known that **S5** is sound and complete w.r.t. the class of models of the above kind.

Definition 22.6 (*Accompanied* **S5**-*model*) Let $\mathbf{M} = \langle \mathcal{W}, \mathcal{V} \rangle$ be an L-model, σ be an $\mathbf{M}$-state, and $\mathcal{R} = \mathcal{W} \times \mathcal{W}$. Let $\mathcal{M}_\sigma$ be an **S5**-model such that:

$$\mathcal{M}_\sigma = \langle \sigma, \mathcal{R}|\sigma, \mathcal{V}|\sigma \rangle$$

$\mathcal{M}_\sigma$ is called the **S5**-model accompanied with $\mathbf{M}$ w.r.t. state σ.

It is obvious that for each L-model $\mathbf{M}$ and each state of the model there exists exactly one **S5**-model accompanied with $\mathbf{M}$ w.r.t. the state. For each wff A of L we have:

Corollary 22.7 *Let* $\mathbf{M}$ *be an* L*-model,* σ *be an* $\mathbf{M}$*-state, and* $w \in \sigma$*. Then* $\mathbf{M}, w \models A$ *iff* $\mathcal{M}_\sigma, w \models A$.

The following is true as well:

Lemma 22.1 *For each* $\mathbf{M}$*-state* σ*:*

1. *if* A *is a p-wff, then:* $\sigma \nrightarrow A$ *iff* $\mathcal{M}_\sigma \models \Diamond A$,
2. *if* A *is a n-wff, then:* $\sigma \nrightarrow A$ *iff* $\mathcal{M}_\sigma \models \Box A$.

Proof As for (1), it suffices to recall that for a p-wff A we have $\sigma \looparrowright A$ iff A is true in at least one world of σ. On the other hand, the accessibility relation in $\mathcal{M}_\sigma$ is universal and thus $\mathcal{M}_\sigma \models \Diamond A$ iff $\mathcal{M}_\sigma, w \models A$ for at least one $w \in \sigma$.

Clause (2) is an immediate consequence of Corollary 22.3 and the fact that $\mathcal{R}|\sigma$ is universal in σ. □

Let us also prove:

Lemma 22.2 *Let A be a wff of L. For each* **M**-*state* σ:

1. $\sigma \looparrowright \neg(A \rightarrow \bot)$ *iff* $\mathcal{M}_\sigma \models \Box A$,
2. $\sigma \looparrowright (A \rightarrow \bot)$ *iff* $\mathcal{M}_\sigma \models \Diamond\neg A$,
3. $\sigma \looparrowright (\neg A \rightarrow \bot)$ *iff* $\mathcal{M}_\sigma \models \Diamond A$.

Proof As for (1), $\neg(A \rightarrow \bot)$ is a n-wff and hence, by Corollary 22.3, $\sigma \looparrowright \neg(A \rightarrow \bot)$ iff for each $w \in \sigma$: $\mathbf{M}, w \not\models (A \rightarrow \bot)$, that is, $\mathcal{M}_\sigma, w \models A$ for any $w \in \sigma$, which, due to the universality of $\mathcal{R}|\sigma$ gives $\mathcal{M}_\sigma \models \Box A$.

Concerning (2): $\sigma \looparrowright (A \rightarrow \bot)$ iff $|(A \rightarrow \bot)|_\mathbf{M} \cap \sigma \neq \emptyset$ iff for some $w \in \sigma$: $\mathbf{M}, w \models \neg A$ iff $\mathcal{M}_\sigma \models \Diamond\neg A$.

(3) is a direct consequence of (2). □

22.2.3 *Epistemization*

As it is well known, **S5** can be interpreted as an epistemic logic, where the box, $\Box$, represents the knowledge operator, and the diamond, $\Diamond$, represents, generally speaking, epistemic possibility. This suggests a kind of purely epistemic readings of some *metalanguage* expressions of the form "$\sigma \looparrowright A$".

Consider:

$$\sigma \looparrowright \neg(A \rightarrow \bot) \tag{22.7}$$

Due to clause (1) of Lemma 22.2, this can be read:

$$A \textit{ is known in state } \sigma \tag{22.8}$$

where "*A is known in state* σ" means:

$$\mathcal{M}_\sigma \models \Box A \tag{22.9}$$

that is, $\Box A$ is true in an **S5**-model whose domain is σ (more precisely: $\Box A$ is true in the **S5**-model whose domain is the **M**-state σ and which agrees with **M** on the values of propositional variables w.r.t. worlds in σ).

Observe that, by Lemma 22.2, "being known in σ" does not differentiate between n-wffs and p-wffs.

Example 22.1 Let us consider the case of implication. In our setting $(A \rightarrow B)$ is said to be known in a state σ iff:

$$\sigma \looparrowright \neg((A \rightarrow B) \rightarrow \bot) \tag{22.10}$$

It follows that:

$$\mathcal{M}_\sigma \models \Box(A \rightarrow B) \tag{22.11}$$

and:

$$\textit{for each } w \in \sigma : \mathbf{M}, w \models (A \rightarrow B) \tag{22.12}$$

Thus an implication constitutes an item of knowledge in a state if, and only if it is true in each world of the state. Or, to put it differently, an implication is known in a state just in case it is a strict implication w.r.t. the state.

Example 22.2 Now consider the case in which the negation of an implication, i.e. $\neg(A \rightarrow B)$, is known in state σ. This means:

$$\sigma \looparrowright \neg(\neg(A \rightarrow B) \rightarrow \bot) \tag{22.13}$$

which gives:

$$\mathcal{M}_\sigma \models \Box\neg(A \rightarrow B) \tag{22.14}$$

Hence:

$$\textit{for each } w \in \sigma : \mathbf{M}, w \not\models (A \rightarrow B) \tag{22.15}$$

So a negated implication is an item of knowledge in a state just in case the implication itself is false in each world of the state. It follows that a negated implication is known in a state if, and only if it is permitted by the state.

Let us consider expressions of the form:

$$\sigma \looparrowright (A \rightarrow \bot) \tag{22.16}$$

By Lemma 22.2 we have:

$$\sigma \looparrowright (A \rightarrow \bot) \textit{ iff } \mathcal{M}_\sigma \models \Diamond\neg A \tag{22.17}$$

Hence an expression of the form (22.16) can be read:

$$\neg A \textit{ is epistemically possible in } \sigma \tag{22.18}$$

again uniformly for all the wffs of L.

Now let us consider:

$$\sigma \Vdash (\neg A \rightarrow \bot) \tag{22.19}$$

By Lemma 22.2 we get:

$$\sigma \Vdash (\neg A \rightarrow \bot) \textit{ iff } \mathcal{M}_\sigma \models \Diamond A \tag{22.20}$$

Thus one can read (22.19) as:

$$A \textit{ is epistemically possible in } \sigma \tag{22.21}$$

For convenience, we introduce:

Definition 22.7 (*"Epistemic" modalities*)

1. $\boxplus A =_{df} \neg(A \rightarrow \bot)$
2. $\oplus A =_{df} (\neg A \rightarrow \bot)$
3. $\ominus A =_{df} (A \rightarrow \bot)$

Observe that $\boxplus$ *is not* the **S5** necessity/knowledge operator. Let A be a wff of L and let $\mathcal{M} = \langle \mathcal{W}, \mathcal{R}, \mathcal{V} \rangle$ be an **S5**-model. Consider an arbitrary but fixed world $w \in \mathcal{W}$. Clearly, $\boxplus A$ is true in w iff A is true in w, while $\Box A$ is true in w just in case A is true in each world of $\mathcal{W}$. Thus $\boxplus A$ and $\Box A$ have different truth conditions in worlds.[4] But $\boxplus$ behaves similarly as the **S5** knowledge operator. One can easily prove:

Corollary 22.8 *Let A, B be wffs of L.*

1. *The following:*

$$\boxplus (A \rightarrow B) \rightarrow (\boxplus A \rightarrow \boxplus B) \tag{22.22}$$

$$\boxplus A \rightarrow A \tag{22.23}$$

$$\neg \boxplus A \rightarrow \boxplus \neg \boxplus A \tag{22.24}$$

   are true in each L-model.
2. *If A is true in each L-model, then $\boxplus A$ is true in each L-model.*

However, $\sigma \Vdash \boxplus A$ only means $\mathcal{M}_\sigma \models \Box A$ (or equivalently: $\langle \sigma, \mathcal{V}|\sigma \rangle \models \neg(A \rightarrow \bot)$). Thus, a wff known in a state σ of an L-model $\langle \mathcal{W}, \mathcal{V} \rangle$ must be true in each world of the state σ, but not necessarily in each world of the whole model. In other words, knowledge in a state is factive w.r.t. worlds of the state, but need not be factive with regard to all worlds of the model. Yet, when one considers a singleton state, it is impossible that a wff A is known in the state (in the sense of $\boxplus$) when A is false in the (only) world of the state.

[4]However, since $\mathcal{R}$ is supposed to be universal, $\boxplus A$ in true in an L-model or in an **S5**-model iff $\Box A$ is true in the model(s).

22.2.3.1 A Philosophical Comment

The standard philosophical concept of knowledge conceives it as a *true* justified belief about the actual world. In the framework of an epistemic logic supplemented with a relational semantics "being known in a world w of a model" is explicated by "being true in each world w^* of the model such that w^* is accessible from w". When **S5** is used as an epistemic logic, this amounts to being true in each world of the model. Since we usually assume that the actual world is among the possible worlds considered (or is represented by a certain possible world of a model), the truth of $\Box A$ in a model yields the truth of A in the actual world, and $\Box A$ is true in the actual world only if A is true in the world.

Knowledge in a state behaves differently. If A is known in a state σ, it is true in each world of the state and thus also in the actual world *if* the actual world "is" in σ. This, however, need not be the case.

22.3 Permittance and Inconsistency

As above, we assume that $\mathbf{M} = \langle \mathcal{W}, \mathcal{V} \rangle$ is an arbitrary but fixed L-model. The **M**-*permittance class* of a wff A of L, in symbols: $\|A\|_{\mathbf{M}}$, comprises all the **M**-states that permit A. The **M**-permittance class of a set of wffs X, $\|X\|_{\mathbf{M}}$, in turn, is the intersection of **M**-permittance classes of elements of X. More formally:

Definition 22.8 (*Permittance class*)

1. $\|A\|_{\mathbf{M}} = \{\sigma \subseteq \mathcal{W} : \sigma \neq \emptyset \text{ and } \sigma \looparrowright A\}$
2. $\|X\|_{\mathbf{M}} = \{\sigma \subseteq \mathcal{W} : \sigma \looparrowright B \text{ for each } B \in X\}$.

Definition 22.9 X has a non-empty permittance class iff there exists an L-model **M** such that $\|X\|_{\mathbf{M}} \neq \emptyset$.

When $\{A\}$ has a non-empty permittance class, we will be saying briefly: "A has a non-empty permittance class".

One can show that some inconsistent sets of wffs have non-empty permittance classes. For clarity, let us first introduce:

Definition 22.10 (*Inconsistent sets*) A set of wffs X of L is:

1. inconsistent iff $\bigcap_{B \in X} |B|_{\mathbf{M}} = \emptyset$ for each L-model **M**;
2. plainly inconsistent iff:
   (a) for some wff A, both $A \in X$ and $\ulcorner \neg A \urcorner \in X$, or
   (b) for some wff $A \in X$, $\{A\}$ is inconsistent.

Clearly, permittance classes of plainly inconsistent sets are always empty. However, the situation is different in the case of some sets of wffs which are inconsistent, but not plainly inconsistent.

For example, $\{A, A \to \bot\}$ is inconsistent. But the following holds:

Corollary 22.9 *Let A be a p-wff of L such that $\ulcorner \Diamond A \to \Box A \urcorner \notin$ **S5**. Then there exists an L-model **M** such that $\|\{A, A \to \bot\}\|_{\mathbf{M}} \neq \emptyset$.*

Proof When $\ulcorner \Diamond A \to \Box A \urcorner \notin$ **S5**, there exists a **S5**-model $\mathcal{M} = \langle \mathcal{W}, \mathcal{R}, \mathcal{V} \rangle$ and a world $w \in \mathcal{W}$ such that $\mathcal{M}, w \models \Diamond A$ and $\mathcal{M}, w \models \Diamond \neg A$. So for some $w_1 \in \mathcal{W} : \mathcal{M}, w_1 \models A$, and for some $w_2 \in \mathcal{W} : \mathcal{M}, w_2 \models \neg A$. Consider the following L-model **M**:

$$\langle \{w_1, w_2\}, \mathcal{V}|\{w_1, w_2\} \rangle$$

As both A and $(A \to \bot)$ are p-wffs, it is easily seen that for the state $\{w_1, w_2\}$ of the model we have:

$$\{w_1, w_2\} \nrightarrow A$$

$$\{w_1, w_2\} \nrightarrow (A \to \bot)$$

Hence $\|\{A, (A \to \bot)\}\|_{\mathbf{M}} \neq \emptyset$. □

In particular, the permittance class of $\{p, p \to \bot\}$ is non-empty.

Thus the following is true:

Corollary 22.10 *There exist: inconsistent sets of wffs of L and L-models such that the sets have non-empty permittance classes in the models.*

Here is another example of an inconsistent set which has a non-empty permittance class.

Example 22.3 The set $\{p \to q, p, \neg q\}$ is inconsistent, but not plainly inconsistent. Let $\mathbf{M} = \langle \mathcal{W}, \mathcal{V} \rangle$ be an L-model such that for some $w_1, w_2 \in \mathcal{W}$:

- $\mathcal{V}(p, w_1) = \mathbf{0}$,
- $\mathcal{V}(q, w_1) = \mathbf{0}$,
- $\mathcal{V}(p, w_2) = \mathbf{1}$,
- $\mathcal{V}(q, w_2) = \mathbf{0}$.

Clearly we have:

- $\mathbf{M}, w_2 \models p$ and hence $\{w_1, w_2\} \nrightarrow p$,
- $\mathbf{M}, w_1 \models (p \to q)$ and thus $\{w_1, w_2\} \nrightarrow (p \to q)$,
- $\mathbf{M}, w_1 \models \neg q$ as well as $\mathbf{M}, w_2 \models \neg q$; therefore $\{w_1, w_2\} \nrightarrow \neg q$.

Thus $\|\{p \to q, p, \neg q\}\|_{\mathbf{M}} \neq \emptyset$.

22.4 Transmission of Permittance

22.4.1 Definition and Basic Properties

Let us now introduce:

Definition 22.11 (*Transmission of permittance*) $X \hookrightarrow_L A$ iff for each L-model $\mathbf{M}$ and each $\mathbf{M}$-state σ:

$$\text{if } \sigma \in \|X\|_{\mathbf{M}}, \text{ then } \sigma \in \|A\|_{\mathbf{M}}.$$

The intuitive content of the above concept is: if each element of X is permitted by a state, then A is permitted by the state. This condition is supposed to hold for each L-model and each state of the model.

Let "$\sigma \Vdash X$" abbreviate "for each $B \in X : \sigma \Vdash B$".

Corollary 22.11 $X \hookrightarrow_L A$ *iff the following condition:*

$$\text{if } \sigma \Vdash X, \text{ then } \sigma \Vdash A \tag{22.25}$$

is fulfilled by each state σ of any L-model.

$\hookrightarrow_L$ is a consequence relation. One can easily prove:

Corollary 22.12 $\hookrightarrow_L$ *has the following properties:*

(Overlap): If $A \in X$, then $X \hookrightarrow_L A$.
(Dilution): If $X \hookrightarrow_L A$ and $X \subseteq Y$, then $Y \hookrightarrow_L A$.
(Cut for sets): If $X \cup Y \hookrightarrow_L A$ and $X \hookrightarrow_L B$ for every $B \in Y$, then $X \hookrightarrow_L A$.

$\hookrightarrow_L$ is not structural, however. The following examples illustrate this[5]:

Example 22.4

$$\{\neg(p \wedge \neg q), p\} \hookrightarrow_L q \tag{22.26}$$

To prove (22.26) suppose that for some state σ of an L-model $\mathbf{M}$ it holds that:

(1) $\sigma \Vdash \neg(p \wedge \neg q)$, and
(2) $\sigma \Vdash p$.

By (2) there exists $w \in \sigma$, say, w_1, such that $\mathbf{M}, w_1 \models p$. But since (1) holds as well, we have $\mathbf{M}, w_1 \models \neg(p \wedge \neg q)$ and hence $\mathbf{M}, w_1 \models q$. Thus $\sigma \Vdash q$.

Example 22.5

$$\{\neg(p \wedge \neg\neg q), p\} \not\hookrightarrow_L \neg q \tag{22.27}$$

To see this it suffices to consider an L-model $\mathbf{M} = \langle\{w_1, w_2\}, \mathcal{V}\rangle$ in which $\mathcal{V}(p, w_1) = \mathbf{1}$, $\mathcal{V}(q, w_1) = \mathbf{0}$, $\mathcal{V}(p, w_2) = \mathbf{0}$, and $\mathcal{V}(q, w_2) = \mathbf{1}$. We get:

---

[5]For brevity, we use, here and below, object-level language expressions instead of their metalinguistic names.

- $\mathbf{M}, w_1 \models p$,
- $\mathbf{M}, w_1 \models \neg(p \wedge \neg\neg q))$,
- $\mathbf{M}, w_2 \models \neg(p \wedge \neg\neg q))$

Thus $\{w_1, w_2\} \looparrowright \{\neg(p \wedge \neg\neg q), p\}$. On the other hand, since $\mathbf{M}, w_1 \not\models \neg q$, we have $\{w_1, w_2\} \not\looparrowright \neg q$.

Generally speaking, $\hookrightarrow_L$ is not structural because substitution can change the categories of wffs, that is, can turn p-wffs into n-wffs, or n-wffs into p-wffs.[6]

22.4.2 Transmission of Permittance Versus Entailment

Entailment in L, $\models_L$, can be defined by:

Definition 22.12 (*Entailment in L*) $X \models_L A$ iff for each L-model $\mathbf{M}$:

$$\bigcap_{B \in X} |B|_{\mathbf{M}} \subseteq |A|_{\mathbf{M}}$$

Entailment in L amounts to entailment determined by Classical Propositional Logic.

Transmission of permittance is a special case of entailment. By Corollary 22.6 we get:

Corollary 22.13 *If* $X \hookrightarrow_L A$, *then* $X \models_L A$.

Hence $\hookrightarrow_L$ is a *truth- preserving* consequence relation.

The converse of Corollary 22.13 does not hold. The following examples illustrate this:

Example 22.6

$$\neg p \vee \neg q \not\hookrightarrow_L \neg(p \wedge q) \tag{22.28}$$

For, consider an L-model $\mathbf{M} = \langle\{w_1, w_2\}, \mathcal{V}\rangle$ such that $\mathcal{V}(p, w_1) = \mathbf{0}, \mathcal{V}(p, w_2) = \mathbf{1}$, and $\mathcal{V}(q, w_2) = \mathbf{1}$. Since $\neg p \vee \neg q$ is a p-wff, $\{w_1, w_2\} \looparrowright \neg p \vee \neg q$. On the other hand, $\neg(p \wedge q)$ is a n-wff and we have $\{w_1, w_2\} \not\looparrowright \neg(p \wedge q)$ because $\mathbf{M}, w_2 \models (p \wedge q)$.

Example 22.7

$$\{p \to q, \neg q\} \not\hookrightarrow_L \neg p \tag{22.29}$$

To see this it suffices to consider an L-model $\mathbf{M} = \langle\{w_1, w_2\}, \mathcal{V}\rangle$ in which $\mathcal{V}(p, w_1) = \mathbf{0}$, $\mathcal{V}(q, w_1) = \mathbf{0}$, $\mathcal{V}(p, w_2) = \mathbf{1}$, and $\mathcal{V}(q, w_2) = \mathbf{0}$. Since $\mathbf{M}, w_1 \models (p \to q)$, we get $\{w_1, w_2\} \looparrowright (p \to q)$. Clearly, $\{w_1, w_2\} \looparrowright \neg q$. But $\{w_1, w_2\} \not\looparrowright \neg p$ because $\mathcal{V}(p, w_2) = \mathbf{1}$.

[6]This can happen when the wff being substituted is a propositional variable or has the form $\neg \ldots \neg p$, where p is a propositional variable.

22.4.3 *Paraconsistency*

As we have shown in Sect. 22.3, some inconsistent sets have non-empty permittance classes. It follows that $\hookrightarrow_L$ is *paraconsistent* in the following sense of the word: it is not the case that for every inconsistent set X and every wff B it holds that $X \hookrightarrow_L B$.

Example 22.8 The set $\{p \rightarrow q, p, \neg q\}$ has a non-empty permittance class (see Example 22.3). Hence, in particular:

$$\{p \rightarrow q, p, \neg q\} \not\hookrightarrow_L r \tag{22.30}$$

Example 22.9 The set $\{p, p \rightarrow \bot\}$ is inconsistent, but has a non-empty permittance class. One can easily show that:

$$\{p, p \rightarrow \bot\} \not\hookrightarrow_L q \tag{22.31}$$

Observe, however, that we still have:

$$\{p, \neg p\} \hookrightarrow_L q \tag{22.32}$$

22.4.4 *Translation* ( )*

The operation $(\ )^*$ assigns to a wff of L the corresponding wff of $\mathcal{L}$. It is defined as follows:

Definition 22.13 1. If A is a p-wff, then $(A)^* = \Diamond A$.
2. If A is a n-wff, then $(A)^* = \Box A$.

Let us stress that A in $\Diamond A$ or in $\Box A$ represents a wff of L. The operation $(\ )^*$ is performed on A only once; the subformulas of A remain unaffected. In other words, $(\ )^*$ is a kind of "surface translation" of wffs of L into wffs of $\mathcal{L}$.[7]

For convenience, we put:

$$(X)^* =_{df} \{(A)^* : A \in X\}$$

Let us now prove:

Lemma 22.3 *If* $\mathcal{M} = \langle \mathcal{W}, \mathcal{R}, \mathcal{V} \rangle$ *is a* **S5***-model such that* $\mathcal{M} \models (X)^*$*, then* $\mathcal{W} \nrightarrow X$.

[7]The idea of using translations into **S5** in constructing paraconsistent logics goes back to Jaśkowski (cf. [3], or [4] for an English translation). However, Jaśkowski's translation is defined recursively and enables an introduction of "discussive" connectives. The operation ( )* behaves differently.

Proof First observe that $\mathcal{M}$ is the **S5**-model accompanied with an L-model $\mathbf{M} = \langle \mathcal{W}, \mathcal{V} \rangle$ w.r.t. the **M**-state $\mathcal{W}$.

The elements of $(X)^*$ are either of the form $\Diamond B$ or of the form $\Box B$, where $B \in X$. If $\Diamond B \in (X)^*$, then, by Lemma 22.1, $\mathcal{M} \models \Diamond B$ yields $\mathcal{W} \nrightarrow B$.

The case in which $\Box B \in (X)^*$ is analogous. □

The following holds:

Theorem 22.1 *X has a non-empty permittance class iff there exists a* **S5**-*model* $\mathcal{M}$ *such that* $\mathcal{M} \models (X)^*$.

Proof (⇒). Let **M** be an L-model for which $\|X\|_{\mathbf{M}} \neq \emptyset$. Let $\sigma \in \|X\|_{\mathbf{M}}$. We consider the **S5**-model $\mathcal{M}_\sigma$ accompanied with **M** w.r.t. σ, and we apply Lemma 22.1.

(⇐). By Lemma 22.3. □

Example 22.10 As we have shown (see Example 22.3), the inconsistent set $\{p \to q, p, \neg q\}$ has a non-empty permittance class. The following takes place on the modal side:

$$\mathcal{M}_{\{w_1,w_2\}} \models \{\Diamond(p \to q), \Diamond p, \Box\neg q\} \tag{22.33}$$

where $\mathcal{M}_{\{w_1,w_2\}}$ is the **S5**-model accompanied (w.r.t. state $\{w_1, w_2\}$) with the L-model considered in Example 22.3.

However, the following holds:

Corollary 22.14 *If X is inconsistent and each element of X is a n-wff, then the permittance class of X is empty.*

Proof Suppose that the permittance class of X is non-empty. Then, by Theorem 22.1, for some **S5**-model $\mathcal{M}$ we have $\mathcal{M} \models (X)^*$. But the elements of $(X)^*$ are of the form $\Box A$, where $A \in X$. Since $\mathcal{W}$ is non-empty, there exists a world w of $\mathcal{M}$ such that $\mathcal{M}, w \models X$. It follows that X is consistent. □

The situation can be different when X contains some p-wffs.

22.4.5 *Transmission of Permittance Versus Global* S5-*Entailment*

Recall that Φ stands for a set of wffs of $\mathcal{L}$ (i.e. the modal extension of L), and ϕ is a metalanguage variable for wffs of $\mathcal{L}$.

Let us introduce:

Definition 22.14 (*Global* **S5**-*entailment*) $\Phi \models_{\mathbf{S5}} \phi$ iff for each **S5**-model $\mathcal{M}$: if $\mathcal{M} \models \Phi$, then $\mathcal{M} \models \phi$.

We will now prove:

Theorem 22.2 (Reduction modulo ( )*) $X \hookrightarrow_L A$ *iff* $(X)^* \models_{\mathbf{S5}} (A)^*$

Proof Suppose that $X \hookrightarrow_L A$, but $(X)^* \not\models_{\mathbf{S5}} (A)^*$. Thus for some **S5**-model $\mathcal{M} = \langle \mathcal{W}, \mathcal{R}, \mathcal{V} \rangle$ we have $\mathcal{M} \models (X)^*$ and $\mathcal{M} \not\models (A)^*$. But $\mathcal{M}$ is accompanied with the L-model $\mathbf{M} = \langle \mathcal{W}, \mathcal{V} \rangle$ w.r.t. $\mathcal{W}$, that is, $\mathcal{M} = \mathcal{M}_{\mathcal{W}}$. By Lemma 22.3 we get $\mathcal{W} \looparrowright X$ and hence, due to the transmission of permittance, $\mathcal{W} \looparrowright A$. If A is a p-wff, then, by Lemma 22.1, $\mathcal{M} \models \Diamond A$, that is, $\mathcal{M} \models (A)^*$. A contradiction. Similarly, if A is a n-wff, by Lemma 22.1 we get $\mathcal{M} \models \Box A$, i.e. $\mathcal{M} \models (A)^*$. A contradiction again.

Now suppose that $(X)^* \models_{\mathbf{S5}} (A)^*$, but $X \not\hookrightarrow_L A$. Then there exists a state σ of a certain L-model $\mathbf{M}$ such that $\sigma \looparrowright X$ and $\sigma \not\looparrowright A$. We consider the **S5**-model $\mathcal{M}_\sigma$ accompanied with $\mathbf{M}$ w.r.t. σ. By Lemma 22.1 we get $\mathcal{M}_\sigma \models (X)^*$ and $\mathcal{M}_\sigma \not\models (A)^*$. A contradiction. □

According to Theorem 22.2, transmission of permittance amounts to (global) **S5**-entailment among the relevant *-wffs. This does not mean that transmission of permittance can be *identified with* global **S5**-entailment. Recall that the *-wffs are either of the form $\Box A$ or of the form $\Diamond A$, where A is a wff of the non-modal language L (and thus does not involve modal operators).

Remark 22.5 Necessity and possibility are, in a sense, expressible in L (cf. Sect. 22.2.2). But when we have $\phi \models_{\mathbf{S5}} \psi$ for $\mathcal{L}$-wffs ϕ, ψ which are of neither of the forms: $\Box A$, $\Diamond A$, the systematic replacement in ϕ and ψ of $\Box A$ by $\neg(A \to \bot)$ as well as of $\Diamond A$ by $(\neg A \to \bot)$ need not turn **S5**-entailment between ϕ and ψ into the transmission of permittance between the resultant wffs of L. For example, we have:

$$\neg\Box p \models_{\mathbf{S5}} \Box\neg\Box p \tag{22.34}$$

By the systematic replacement we get:

$$\neg\neg(p \to \bot) \hookrightarrow_L \neg(\neg\neg(p \to \bot) \to \bot) \tag{22.35}$$

(22.35) *does not* hold, however. To see this let us take an L-model $\mathbf{M}^* = \langle \{w_1, w_2\}, \mathcal{V} \rangle$ such that $\mathcal{V}(p, w_1) = \mathbf{0}$ and $\mathcal{V}(p, w_2) = \mathbf{1}$. Clearly, we have:

$$\{w_1, w_2\} \looparrowright \neg\neg(p \to \bot) \tag{22.36}$$

since $\mathbf{M}^*, w_1 \models \neg\neg(p \to \bot)$. At the same time we have:

$$\{w_1, w_2\} \not\looparrowright \neg(\neg\neg(p \to \bot) \to \bot) \tag{22.37}$$

because $\mathbf{M}^*, w_2 \not\models \neg(\neg\neg(p \to \bot) \to \bot)$.

To sum up: Theorem 22.2 does not reduce the "logic of permittance" to **S5**, but shows that one can "calculate" transmission of permittance by well-known means.

22.4.6 What is Retained and What is Lost

22.4.6.1 The Case of Single Wffs

Let us first prove:

Lemma 22.4 *If* $B \models_L A$ *and (a)* B *and* A *are p-wffs, or (b)* B *and* A *are n-wffs, or (c)* B *is a n-wff and* A *is a p-wff, then* $B \hookrightarrow_L A$.

Proof If $B \models_L A$, then $\models_L (B \to A)$ and hence $\ulcorner\Box(B \to A)\urcorner \in \mathbf{S5}$.

Assume that B and A are p-wffs. Suppose that $\Diamond B \not\models_{\mathbf{S5}} \Diamond A$. So there exists an **S5**-model $\mathcal{M} = \langle \mathcal{W}, \mathcal{R}, \mathcal{V} \rangle$ such that $\mathcal{M} \models \Diamond B$ and $\mathcal{M} \not\models \Diamond A$. Hence $\mathcal{M}, w \not\models A$ for each $w \in \mathcal{W}$, and $\mathcal{M}, w \models B$ for some $w \in \mathcal{W}$. It follows that for some $w \in \mathcal{W}$ we have $\mathcal{M}, w \not\models (B \to A)$ and therefore $\ulcorner\Box(B \to A)\urcorner \notin \mathbf{S5}$. A contradiction. Thus $\Diamond B \models_{\mathbf{S5}} \Diamond A$ and hence, by Theorem 22.2, $B \hookrightarrow_L A$.

Assume that B and A are n-wffs. Suppose that $\Box B \not\models_{\mathbf{S5}} \Box A$. So for some **S5**-model $\mathcal{M} = \langle \mathcal{W}, \mathcal{R}, \mathcal{V} \rangle$ we get: $\mathcal{M}, w \models B$ for any $w \in \mathcal{W}$, and $\mathcal{M}, w \not\models A$ for some $w \in \mathcal{W}$. Thus $\ulcorner\Box(B \to A)\urcorner \notin \mathbf{S5}$. A contradiction. Therefore, by Theorem 22.2, $B \hookrightarrow_L A$.

Finally, assume that B is a n-wff and A is a p-wff. Suppose that $\Box B \not\models_{\mathbf{S5}} \Diamond A$. Thus, for some **S5**-model $\mathcal{M} = \langle \mathcal{W}, \mathcal{R}, \mathcal{V} \rangle$, $\mathcal{M}, w \models B$ for any $w \in \mathcal{W}$, and $\mathcal{M}, w \not\models A$ for each $w \in \mathcal{W}$. Hence $\ulcorner\Box(B \to A)\urcorner \notin \mathbf{S5}$. A contradiction again. Therefore, by Theorem 22.2, $B \hookrightarrow_L A$. □

Thus, for instance, the following hold:

$$p \hookrightarrow_L \neg\neg p \tag{22.38}$$

$$\neg\neg p \hookrightarrow_L p \tag{22.39}$$

$$(p \to q) \hookrightarrow_L (\neg q \to \neg p) \tag{22.40}$$

$$(\neg q \to \neg p) \hookrightarrow_L (p \to q) \tag{22.41}$$

$$p \hookrightarrow_L (q \to p) \tag{22.42}$$

$$(p \to q) \wedge p \hookrightarrow_L q \tag{22.43}$$

$$(p \vee q) \wedge \neg q \hookrightarrow_L p \tag{22.44}$$

$$(p \vee \neg q) \wedge q \hookrightarrow_L p \tag{22.45}$$

$$(p \to (q \to r)) \hookrightarrow_L ((p \to q) \to (p \to r)) \tag{22.46}$$

$$(p \to (q \to r)) \hookrightarrow_L (p \wedge q \to r) \tag{22.47}$$

$$(p \wedge q \rightarrow r) \hookrightarrow_L (p \rightarrow (q \rightarrow r)) \tag{22.48}$$

$$(p \rightarrow (q \rightarrow r)) \hookrightarrow_L (q \rightarrow (p \rightarrow r)) \tag{22.49}$$

$$((p \rightarrow q) \wedge (q \rightarrow r)) \hookrightarrow_L (p \rightarrow r) \tag{22.50}$$

$$\neg(p \wedge q) \hookrightarrow_L (\neg p \vee \neg q) \tag{22.51}$$

$$\neg(p \vee q) \hookrightarrow_L (\neg p \wedge \neg q) \tag{22.52}$$

$$\neg(p \wedge \neg q) \hookrightarrow_L (p \rightarrow q) \tag{22.53}$$

$$\neg(p \rightarrow q) \hookrightarrow_L (p \wedge \neg q) \tag{22.54}$$

Observe, however, that the converses of (22.51)–(22.54) do not hold. The counterpart of *Modus Tollendo Tollens* does not hold either, i.e.:

$$((p \rightarrow q) \wedge \neg q) \not\hookrightarrow_L \neg p \tag{22.55}$$

because:

$$\Diamond((p \rightarrow q) \wedge \neg q) \not\models_{\mathbf{S5}} \Box\neg p \tag{22.56}$$

Hence:

Corollary 22.15 *There are cases in which: B is a p-wff, A is a n-wff, $B \models_L A$, and $B \not\hookrightarrow_L A$.*

Yet, the following holds:

$$((p \rightarrow q) \wedge \neg q) \hookrightarrow_L \oplus\neg p \tag{22.57}$$

(Recall that $\oplus\neg p$ claims that $\neg p$ is epistemically possible in a state.) This can be generalized.

Corollary 22.16 *If $B \models_L A$, B is a p-wff and A is a n-wff, then $B \hookrightarrow_L \oplus A$.*

Proof If $B \models_L A$, then $\ulcorner\Box(B \rightarrow A)\urcorner \in$ **S5**. Suppose that $\Diamond B \not\models_{\mathbf{S5}} \Diamond \oplus A$. So for some **S5**-model $\mathcal{M} = \langle \mathcal{W}, \mathcal{R}, \mathcal{V} \rangle$ there exists $w_1 \in \mathcal{W}$ such that $\mathcal{M}, w_1 \models B$ and, at the same time, $\mathcal{M}, w \not\models \oplus A$ for any $w \in \mathcal{W}$. Recall that $\oplus A =_{df} (\neg A \rightarrow \bot)$. Hence for each $w \in \mathcal{W}$ we have $\mathcal{M}, w \not\models A$. Therefore $\ulcorner\Box(B \rightarrow A)\urcorner \notin$ **S5**. A contradiction. $\Box$

22.4.6.2 The Case of Sets of Wffs

The direct counterpart of *Modus Ponens* holds for $\hookrightarrow_L$ (cf. 22.43). But we have[8]:

$$\{p \to q, p\} \not\hookrightarrow_L q \tag{22.58}$$

So conjunction behaves is a non-standard way in the context of $\hookrightarrow_L$: $A_1 \wedge \ldots \wedge A_n \hookrightarrow_L B$ need not be tantamount to $\{A_1, \ldots, A_n\} \hookrightarrow_L B$. The reason is that a permittance class of a set of wffs need not be equal with the permittance class of a conjunction of all the wffs in the set.[9]
Yet, the following is true:

$$\{p \to q, \boxplus p\} \hookrightarrow_L q \tag{22.59}$$

Recall that $\boxplus p$ can be read "p is known in a state in question".

Here are further "negative" examples:

$$\{p, q\} \not\hookrightarrow_L (p \wedge q) \tag{22.60}$$

$$\{p, p \to \bot\} \not\hookrightarrow_L (p \wedge \neg p) \tag{22.61}$$

$$\{p \vee \neg q, q\} \not\hookrightarrow_L p \tag{22.62}$$

$$\{p \to q, q \to r\} \not\hookrightarrow_L (p \to r) \tag{22.63}$$

Observe, however, that the following hold:

$$\{\boxplus p, q\} \hookrightarrow_L (p \wedge q) \tag{22.64}$$

and similarly for q,

$$\{\boxplus p, \boxplus q\} \hookrightarrow_L \boxplus(p \wedge q) \tag{22.65}$$

$$\{p \vee \neg q, \boxplus q\} \hookrightarrow_L p \tag{22.66}$$

$$\{\neg(\neg p \wedge q), q\} \hookrightarrow_L p \tag{22.67}$$

---

[8]Since $\{\Diamond(p \to q), \Diamond p\} \not\models_{\mathbf{S5}} \Diamond q$. (22.43) holds because $\Diamond((p \to q) \wedge p) \models_{\mathbf{S5}} \Diamond q$.

[9]For example, take an L-model $\mathbf{M} = \langle\{w_1, w_2\}, \mathcal{V}\rangle$ such that $\mathcal{V}(p, w_1) = \mathbf{0}$, $\mathcal{V}(q, w_1) = \mathbf{0}$, $\mathcal{V}(p, w_2) = \mathbf{1}$, and $\mathcal{V}(q, w_2) = \mathbf{0}$. Clearly, $\{w_1, w_2\} \in \|(p \to q), p\|_{\mathbf{M}}$, but $\{w_1, w_2\} \notin \|(p \to q) \wedge p\|_{\mathbf{M}}$. In general, a conjunction of p-wffs carries information that the conjuncts are simultaneously true in some world(s) of a state, while the information carried by the set of conjuncts amounts to the claim that each conjunct is true in a certain world of the state. When we have a "mixed" conjunction (that is, involving both p-wff and n-wffs), the information carried by n-wffs "weakens": the consecutive conjuncts, n-wffs included, are supposed to simultaneously hold in a certain world of a state.

$$\{\boxplus(p \to q), \boxplus(q \to r)\} \hookrightarrow_L \boxplus(p \to r) \tag{22.68}$$

$$\{\neg(p \wedge \neg q), \neg(q \wedge \neg r)\} \hookrightarrow_L \neg(p \wedge \neg r) \tag{22.69}$$

It happens that conjunction behaves in the "standard" way in the context of $\hookrightarrow_L$ although the conjuncts belong to diverse categories, as in:

$$\{p \to q, \neg(q \wedge \neg r)\} \hookrightarrow_L (p \to r) \tag{22.70}$$

$$\{\neg p \to q, \neg p\} \hookrightarrow_L q \tag{22.71}$$

$$\{p \vee q, \neg q\} \hookrightarrow_L p \tag{22.72}$$

$$\{\neg p \to q, \neg q\} \hookrightarrow_L p \tag{22.73}$$

Let us now turn to inconsistent sets. As we have shown, $\hookrightarrow_L$ is paraconsistent. But, for instance, we still have:

$$\{p \to q, p, \neg q\} \hookrightarrow_L (\neg p \vee q) \tag{22.74}$$

$$\{p \to q, p, \neg q\} \hookrightarrow_L (\neg(p \to q) \vee \neg p) \tag{22.75}$$

$$\{p \to q, p, \neg q\} \hookrightarrow_L ((\neg(p \to q) \vee \neg p) \vee q) \tag{22.76}$$

$$\{r, s, (r \to p), (s \to \neg p)\} \hookrightarrow_L (p \vee \neg p) \tag{22.77}$$

$$\{r, s, \boxplus(r \to p), \boxplus(s \to \neg p)\} \hookrightarrow_L (\boxplus r \vee \boxplus s) \tag{22.78}$$

22.5 Question Raising

22.5.1 *Questions*

Let us now augment the language L with *questions*. In order to achieve this we enrich the vocabulary of L with the following signs: {, }, ?, and the comma. The new language is labelled as $L^?$. *Declarative well-formed formulas* of $L^?$ are simply the wffs of L. Questions of $L^?$ are expressions of the language falling under the schema:

$$?\ \{A_1, \ldots, A_n\} \tag{22.79}$$

where $n > 1$ and $A_1, \ldots, A_n$ are nonequiform, i.e. pairwise syntactically distinct, wffs of L. An expression of the form (22.79) can be read:

$$\textit{Is it the case that } A_1, \textit{ or } \ldots, \textit{ or is it the case that } A_n? \tag{22.80}$$

If $?\,\{A_1, \ldots, A_n\}$ is a question, then each of the wffs $A_1, \ldots, A_n$ is called a *direct answer* to the question, and these are the only direct answers to the question. A direct answer is a *possible* answer. Moreover, it constitutes a *sufficient* answer: a direct answer is supposed to provide neither less nor more information than it is requested by the corresponding question. It is *not* assumed that a direct answer must be true.[10]

We shall use $Q, Q_1, \ldots$ as metalanguage variables for questions. The set of direct answers to a question Q will be denoted by $\mathbf{d}Q$.

22.5.1.1 Soundness of a Question

We do not assign truth or falsity to questions. However, we introduce the concepts of *soundness* of a question in a world of an L-model and in a state of an L-model.[11]

Definition 22.15 (*Soundness of a question*)

1. A question Q is sound in a world w of an L-model iff at least one direct answer to Q is true in w.
2. A question Q is sound in a state σ of an L-model iff Q is sound in at least one world of the state σ.

Clearly, there are questions which are not sound in some worlds of certain L-models. Similarly, there are questions which are not sound in any states of some L-models. For example, $?\,\{p, q\}$ is not sound in any state of an L-model in which for each world w of the model it holds that $\mathcal{V}(p, w) = \mathcal{V}(q, w) = \mathbf{0}$. On the other hand, $?\,\{p, \neg p\}$ is sound in each state of any L-model, and in each world of the model.

22.5.2 *From Permittance to Soundness: Proto-raising*

Let us now define the following relation between sets of declarative formulas of $L^?$ (i.e. wffs of L) and questions of $L^?$.

Definition 22.16 (*Proto-raising*) A set of wffs X proto-raises a question Q (in symbols: $\mathbf{R_P}(X, Q)$) iff for each L-model $\mathbf{M}$ and each $\mathbf{M}$-state σ:

---

[10] For details of this approach to propositional questions of formal languages see, e.g. [11], Chap. 2.

[11] Cf. [9].

$(\bullet)$ if $\sigma \in \|X\|_{\mathbf{M}}$, then $\mathbf{M}, w \models A$ for some $w \in \sigma$ and $A \in \mathbf{d}Q$.

The underlying intuition is: if all the wffs in X are *permitted* by a state, then Q is sound in the state, that is, at least one direct answer to Q is *true* in at least one world of the state.

We have:

Lemma 22.5 *Let* $n > 1$. $\mathbf{R_P}(X, ?\ \{A_1, \ldots, A_n\})$ *iff* $X \hookrightarrow_L A_1 \vee \ldots \vee A_n$.

Proof ($\Rightarrow$). Suppose that $X \not\hookrightarrow_L A_1 \vee \ldots \vee A_n$. So there exist an L-model $\mathbf{M}$ and an $\mathbf{M}$-state σ such that $\sigma \looparrowright X$ and $\sigma \not\looparrowright A_1 \vee \ldots \vee A_n$. Since $n > 1$, $A_1 \vee \ldots \vee A_n$ is a p-wff. Thus there is no $w \in \sigma$ such that $\mathbf{M}, w \models A_i$, where $1 \leq i \leq n$. Hence it is not the case that $\mathbf{R_P}(X, ?\ \{A_1, \ldots, A_n\})$.

($\Leftarrow$) Take an L-model $\mathbf{M} = \langle \mathcal{W}, \mathcal{V} \rangle$ and an $\mathbf{M}$-state σ. Let $\sigma \looparrowright X$. Then $\sigma \looparrowright A_1 \vee \ldots \vee A_n$. Since $n > 1$, $A_1 \vee \ldots \vee A_n$ is a p-wff. Thus $\mathcal{M}_\sigma \models \Diamond(A_1 \vee \ldots \vee A_n)$. Hence there exists $w \in \sigma$ such that $\mathcal{M}_\sigma, w \models A_1 \vee \ldots \vee A_n$. Therefore $\mathcal{M}_\sigma, w \models A_i$ for some $1 \leq i \leq n$ and thus $\mathbf{M}, w \models A_i$ for some $1 \leq i \leq n$. It follows that $\mathbf{R_P}(X, ?\ \{A_1, \ldots, A_n\})$. □

Lemma 22.5 together with Corollary 22.13 yield:

Corollary 22.17 *If* $\mathbf{R_P}(X, ?\ \{A_1, \ldots, A_n\})$, *then* $X \models_L A_1 \vee \ldots \vee A_n$.

Thus, if X proto-raises Q, a disjunction of all the direct answers to Q is (classically) entailed by X. Hence:

Corollary 22.18 *Let w be a world of an L-model. If* $\mathbf{R_P}(X, Q)$ *and all the wffs of X are true in w, then Q is sound in w.*

In other words, proto-raising secures the transmission of truth into soundness (w.r.t. worlds), which, in turn, constitutes the basic criterion of adequacy of an explication of the intuitive notion "a question Q arises from a set of declaratives X" (cf. [9–11]). However, Definition 22.16 cannot be regarded as providing an adequate explication of the concept. Proto-raising allows for a situation which seems forbidden in view of the intuitive notion: the permittance of all the elements of X in a state transforms into knowledge of some direct answer(s) to Q in the state. But if Q arises from X, the permittance of (all the elements of) X by a state is insufficient for knowing a direct answer to Q in the state; otherwise X would resolve Q and thus Q would not arise from X.

22.5.3 Giving Rise

The following definition can be regarded as providing an explication of the intuitive notion "a question arises from a set of declaratives". Again, we assume that X is a set of declarative formulas of $L^?$, and Q is a question of the language.

Definition 22.17 (*Giving rise*) A set of wffs X gives rise to a question Q (in symbols: $\mathbf{R}(X, Q)$) iff $\mathbf{R_P}(X, Q)$ and for each $A \in \mathbf{d}Q : X \not\hookrightarrow_L A$.

Thus X gives rise to Q just in case X proto-raises Q, but there is no transmission of permittance between X and direct answers to Q.

We have:

Corollary 22.19 *If* $X \not\hookrightarrow_L A$, *then* $X \not\hookrightarrow_L \boxplus A$.

Thus the permittance of all the elements of X by a state does not yield the knowledge of any direct answer to Q in the state. To be more precise, there is no transmission of permittance between X and formulas which express that direct answers to Q are known.

By Lemma 22.5 and Definition 22.17 we get:

Corollary 22.20 *Let* $Q = ?\,\{A_1, \ldots, A_n\}$. *Then* $\mathbf{R}(X, Q)$ *iff*

1. $X \hookrightarrow_L A_1 \vee \ldots \vee A_n$, *and*
2. $X \not\hookrightarrow_L A_i$ *for* $i = 1, \ldots, n$.

Therefore, by the Reduction Theorem (i.e. Theorem 22.2):

Corollary 22.21 $\mathbf{R}(X, ?\,\{A_1, \ldots, A_n\})$ *iff*

1. $(X)^* \models_{\mathbf{S5}} \Diamond(A_1 \vee \ldots \vee A_n)$ *and*
2. $(X)^* \not\models_{\mathbf{S5}} (A_i)^*$ *for* $i = 1, \ldots, n$.

Hence, the following examples come with no surprise[12]:

$$\mathbf{R}(p, q, ?\,\{p \wedge q, \neg(p \wedge q)\}) \tag{22.81}$$

$$\mathbf{R}(\neg p \vee \neg q, ?\,\{\neg(p \wedge q), \neg(p \vee q)\}) \tag{22.82}$$

$$\mathbf{R}(\neg p \wedge \neg q, ?\,\{\neg(p \vee q), p \wedge q\}) \tag{22.83}$$

$$\mathbf{R}(p \to q, \neg q, ?\,\{p, \neg p\}) \tag{22.84}$$

Observe that the raised questions have direct answers which are classically entailed by the raising sets. This is not a general rule, however.

$$\mathbf{R}(p \vee \neg p, ?\,\{p, \neg p\}) \tag{22.85}$$

$$\mathbf{R}(p \vee q, ?\,\{p, \neg p\}) \tag{22.86}$$

$$\mathbf{R}(p \vee q, ?\,\{p, q\}) \tag{22.87}$$

[12]For brevity, we simply list the elements of sets of wffs.

$$\mathbf{R}(p \vee q, ?\,\{p \wedge q, \neg(p \wedge q)\}) \tag{22.88}$$

$$\mathbf{R}(p \vee q, ?\,\{p \wedge q, p \wedge \neg q, \neg p \wedge q\}) \tag{22.89}$$

$$\mathbf{R}(p \to q \vee r, ?\,\{p \to q, p \to r\}) \tag{22.90}$$

$$\mathbf{R}(p \to q \vee r, p, ?\,\{q, r\}) \tag{22.91}$$

$$\mathbf{R}(\neg(q \wedge r), ?\,\{\neg q, \neg r\}) \tag{22.92}$$

$$\mathbf{R}(p \wedge q \to r, ?\,\{p \to r, q \to r\}) \tag{22.93}$$

$$\mathbf{R}(p \wedge q \to r, \neg r, ?\,\{\neg p, \neg q\}) \tag{22.94}$$

$$\mathbf{R}((p \vee q) \vee r, ?\,\{p, q \vee r\}) \tag{22.95}$$

$$\mathbf{R}(p, ?\,\{\oplus\neg p, \boxplus p\}) \tag{22.96}$$

$$\mathbf{R}(p \to \bot, ?\,\{\oplus p, \boxplus\neg p\}) \tag{22.97}$$

22.5.4 Question Raising by Inconsistencies

Questions often arise from inconsistencies. The presented account of question raising does justice to that. To be more precise, we are able to model the case in which questions arise from inconsistent sets with non-empty permittance classes. The following holds:

Corollary 22.22 *If* $\mathbf{R}(X, Q)$*, then the permittance class of* X *is non-empty.*

Proof By assumption, $\mathbf{d}Q \neq \emptyset$. Let $A \in \mathbf{d}Q$. If $\mathbf{R}(X, Q)$, then $X \not\twoheadrightarrow_L A$, so there exist an L-model $\mathbf{M}$ and an $\mathbf{M}$-state σ such that $\sigma \looparrowright X$ as well as $\sigma \not\looparrowright A$. Therefore X has a non-empty permittance class. □

Hence plainly inconsistent sets do not give rise to (in the sense of Definition 22.17) any questions. Similarly, by Corollary 22.14, inconsistent sets which comprise only n-wffs do not give rise to questions. The case of inconsistent sets having non-empty permittance classes in different, however.

Let us start with examples. The following hold:

$$\mathbf{R}(p \to q, p, \neg q, ?\,\{\neg(p \to q), \neg p, q\}) \tag{22.98}$$

$$\mathbf{R}(p \to q, p, \neg q, ?\,\{\neg(p \to q), \neg p\}) \tag{22.99}$$

$$\mathbf{R}(p \to q, p, \neg q, ?\,\{\neg(p \to q), q\}) \tag{22.100}$$

$$\mathbf{R}(p \rightarrow q, p, \neg q, ?\,\{\neg p, q\}) \tag{22.101}$$

$$\mathbf{R}(r, s, r \rightarrow p, s \rightarrow \neg p, ?\,\{p, \neg p\}) \tag{22.102}$$

$$\mathbf{R}(r, s, \boxplus(r \rightarrow p), \boxplus(s \rightarrow \neg p), ?\,\{\boxplus r, \boxplus s\}) \tag{22.103}$$

Let us now introduce:

Definition 22.18 (*Complement*)

1. If A is of the form $\neg C$, then $\overline{A}$ is C.
2. If A is not of the form $\neg C$, then $\overline{A}$ is $\neg A$.

Recall that, by Theorem 22.1, X has a non-empty permittance class iff $(X)^*$ has a **S5**-model.

Theorem 22.3 *If* $\{A_1, \ldots, A_n\}$, *where* $n > 1$, *is inconsistent, but has a non-empty permittance class, then* $\mathbf{R}(\{A_1, \ldots, A_n\}, ?\,\{\overline{A_1}, \ldots, \overline{A_n}\})$.

Proof If $\{A_1, \ldots, A_n\}$ is inconsistent, then for each L-model $\mathbf{M} = \langle \mathcal{W}, \mathcal{V} \rangle$ and each $w \in \mathcal{W}$ we have $\mathbf{M}, w \not\models \{A_1, \ldots, A_n\}$ and hence $\mathbf{M}, w \models \overline{A_1} \vee \ldots \vee \overline{A_n}$. Suppose that $\mathbf{R_P}(\{A_1, \ldots, A_n\}, ?\,\{\overline{A_1}, \ldots, \overline{A_n}\})$ does not hold. Thus $\{A_1, \ldots, A_n\} \not\hookrightarrow_L \overline{A_1} \vee \ldots \vee \overline{A_n}$ and hence $(\{A_1, \ldots, A_n\})^* \not\models_{\mathbf{S5}} \Diamond(\overline{A_1} \vee \ldots \vee \overline{A_n})$. So there exists a world w of an **S5**-model $\mathcal{M}$ such that $\mathcal{M}, w \not\models \Diamond(\overline{A_1} \vee \ldots \vee \overline{A_n})$. Thus the argument of $\Diamond$ is false in each $w \in \mathcal{W}$. Therefore $\mathcal{M}, w \models \neg(\overline{A_1} \vee \ldots \vee \overline{A_n})$, that is, $\mathcal{M}, w \models \neg\overline{A_1} \wedge \ldots \wedge \neg\overline{A_n}$. But $\mathcal{M}, w \models \neg\overline{A_i}$ iff $\mathcal{M}, w \models A_i$ for $1 \leq i \leq n$. It follows that there exists a L-model $\mathbf{M}$ for which it holds that $\mathbf{M}, w \models \{A_1, \ldots, A_n\}$ and thus the analysed set is consistent. A contradiction.

Since $\{A_1, \ldots, A_n\}$ has a non-empty permittance class, there exist: an L-model $\mathbf{M}' = \langle \mathcal{W}', \mathcal{V}' \rangle$ and an $\mathbf{M}'$-state σ such that $\sigma \nrightarrow A_i$ for $1 \leq i \leq n$. Suppose that $\{A_1, \ldots, A_n\} \hookrightarrow_L \overline{A_i}$ for some $1 \leq i \leq n$. Thus $\sigma \nrightarrow \overline{A_i}$ and hence $\sigma \not\nrightarrow A_i$. A contradiction. □

According to Theorem 22.3, an at least two-element finite inconsistent set of wffs gives rise to a question whose direct answers are complements of the wffs in the set—provided that the set has a non-empty permittance class. For instance:

$$\mathbf{R}(p \vee q \rightarrow r, p \wedge \neg r, ?\,\{\neg(p \vee q \rightarrow r), \neg(p \wedge \neg r)\}) \tag{22.104}$$

$$\mathbf{R}(p, p \rightarrow \bot, ?\,\{\neg p, \neg(p \rightarrow \bot)\}) \tag{22.105}$$

$$\mathbf{R}(r, s, r \rightarrow p, s \rightarrow \neg p, ?\,\{\neg r, \neg s, \neg(r \rightarrow p), \neg(s \rightarrow \neg p)\}) \tag{22.106}$$

The "complement" question is also raised by the empty set. In order to show this we need an auxiliary concept and two lemmas.

Definition 22.19 Let Q be a question and C be a wff. By Q_C we designate a question such that $\mathbf{d}Q_C = \mathbf{d}Q \cup \{C\}$.

When $C \notin \mathbf{d}Q$, any Q_C may be called an *extension* of Q by C.[13]

Lemma 22.6 *If* $\mathbf{R_P}(X \cup \{B\}, Q)$, *then* $\mathbf{R_P}(X, Q_{\overline{B}})$.

Proof Let $Q = ?\,\{A_1, \ldots, A_n\}$ and $Q_{\overline{B}} = ?\,\{A_1, \ldots, A_n, \overline{B}\}$.

Suppose that $\mathbf{R_P}(X, ?\,\{A_1, \ldots, A_n, \overline{B}\})$ does not hold. Then, by Lemma 22.5, $X \not\twoheadrightarrow_L A_1 \vee \ldots \vee A_n \vee \overline{B}$. Hence for some L-model $\mathbf{M} = \langle \mathcal{W}, \mathcal{V} \rangle$ and some $\mathbf{M}$-state σ we have: $\sigma \Vdash X$ and $\sigma \nVdash A_1 \vee \ldots \vee A_n \vee \overline{B}$. It follows that $\mathbf{M}, w \models \overline{A_1} \wedge \ldots \wedge \overline{A_n} \wedge B$ for each $w \in \sigma$. Therefore $\mathcal{M}_\sigma \models \Box(\overline{A_1} \wedge \ldots \wedge \overline{A_n} \wedge B)$ and hence $\mathcal{M}_\sigma \not\models \Diamond(A_1 \vee \ldots \vee A_n)$ as well as $\mathcal{M}_\sigma \models \Box B$. Thus $\mathcal{M}_\sigma \models (B)^*$ regardless of whether B is a p-wff or a n-wff, and $\mathcal{M}_\sigma \not\models (A_1 \vee \ldots \vee A_n)^*$. Since $\sigma \Vdash X$, we have $\mathcal{M}_\sigma \models (X)^*$ and hence $\mathcal{M}_\sigma \models (X \cup \{B\})^*$. Therefore $(X \cup \{B\})^* \not\models_{\mathbf{S5}} (A_1 \vee \ldots \vee A_n)^*$. Thus, by Theorem 22.2, $X \cup \{B\} \not\twoheadrightarrow_L A_1 \vee \ldots \vee A_n$. Since $?\,\{A_1, \ldots, A_n\}$ is a question, $n > 1$ and thus Lemma 22.5 applies. Hence $\mathbf{R_P}(X \cup \{B\}, ?\,\{A_1, \ldots, A_n\})$ does not hold as well.

We have assumed that $\overline{B}$ is the "last" direct answer to $Q_{\overline{B}}$. Yet, nothing essential changes when we place $\overline{B}$ at some other position. □

Lemma 22.7 (Deduction) *If* $\mathbf{R}(X \cup \{B\}, Q)$, *then* $\mathbf{R}(X, Q_{\overline{B}})$.

Proof If $\mathbf{R}(X \cup \{B\}, Q)$, then $X \cup \{B\}$ has a non-empty permittance class. Let σ be an element of the class. Since $\sigma \Vdash B$, we get $\sigma \nVdash \overline{B}$ and hence $X \not\twoheadrightarrow_L \overline{B}$. Clearly $X \not\twoheadrightarrow_L A$ for any $A \in \mathbf{d}Q$. On the other hand, by Definition 22.17 and Lemma 22.6, $\mathbf{R}(X \cup \{B\}, Q)$ yields $\mathbf{R_P}(X, Q_{\overline{B}})$. □

Lemma 22.7 enables us to derive new examples from already established ones. Thus, for instance, from (22.101) we get:

$$\mathbf{R}(p \to q, p, ?\,\{\neg p, q\}) \tag{22.107}$$

while (22.99) gives:

$$\mathbf{R}(p \to q, \neg q, ?\,\{\neg(p \to q), \neg p\}) \tag{22.108}$$

From (22.100) we get:

$$\mathbf{R}(p \to q, p, ?\,\{\neg(p \to q), q\}) \tag{22.109}$$

However, the most important consequence of Lemma 22.7 is:

[13] Despite of their form, questions of $L^?$ are not sets of direct answers, but object-language expressions. Thus, for example, $?\,\{p, q\} \neq ?\,\{q, p\}$, although $\{p, q\} = \{q, p\}$. Hence Q_C denotes a class of expressions.

Theorem 22.4 *Let* $\{A_1, \ldots, A_n\}$, *where* $n > 1$, *be an inconsistent set which has a non-empty permittance class. Then* $\mathbf{R}(\emptyset, ?\ \{\overline{A_1}, \ldots, \overline{A_n}\})$.

Proof By Theorem 22.3 and Lemma 22.7 (since $\{\overline{A_1}, \ldots, \overline{A_n}\} \cup \{\overline{A_i}\} = \{\overline{A_1}, \ldots, \overline{A_n}\}$). □

Thus, for instance:

$$\mathbf{R}(\emptyset, ?\ \{\neg(p \to q), \neg p, q\}) \tag{22.110}$$

$$\mathbf{R}(\emptyset, ?\ \{\neg p, \neg(p \to \bot)\}) \tag{22.111}$$

By the way, the following holds as well:

$$\mathbf{R}(\emptyset, ?\ \{\boxplus p, \boxplus \neg p\}) \tag{22.112}$$

because we have:

$$\mathbf{R}(p \to \bot, \neg p \to \bot, ?\ \{\neg(p \to \bot), \neg(\neg p \to \bot)\}) \tag{22.113}$$

22.5.5 Some Comparisons

The intuitive notion "a question arises from a set of declaratives" is explicated in Inferential Erotetic Logic[14] by the concept "a set of declaratives evokes a question". Leaving aside the general schema of definition of evocation,[15] in the case of the language L evocation can be defined as follows:

Definition 22.20 (*Evocation of questions*) A set of wffs X evokes a question Q iff X entails a disjunction of all the direct answers to Q, but does not entail any single direct answer to Q.

By "entails" we mean "entails in L"; cf. Definition 22.12. We write $\mathbf{E}(X, Q)$ for "X evokes Q".

Clearly we have:

Corollary 22.23 *Let* $Q = ?\ \{A_1, \ldots, A_n\}$. *Then* $\mathbf{E}(X, Q)$ *holds iff*

---

[14]Generally speaking, Inferential Erotetic Logic (IEL for short) is a logic that analyses inferences in which questions play the role of conclusions and proposes criteria of validity for these inferences. For IEL see, e.g. [9–11].

[15]Formulated in terms of multiple-conclusion entailment (mc-entailment for short): a set of wffs X evokes a question Q iff X mc-entails the set of direct answers to Q, but does not mc-entail any singleton set whose element is a direct answer to Q. The concept of mc-entailment generalizes the concept of entailment. Mc-entailment is a relation between *sets* of wffs. Roughly, X mc-entails Y iff the truth of all the wffs in X warrants the existence of a true wff in Y. For mc-entailment see [8].

1. $X \models_L A_1 \vee \ldots \vee A_n$, *and*
2. $X \not\models_L A_i$ *for* $i = 1, \ldots, n$.

For examples of evocation see, e.g. [9–11].

Since no direct answer to an evoked question is (classically) entailed by the evoking set, we get:

Corollary 22.24 *If* $\mathbf{E}(X, Q)$, *then* X *is consistent.*

So evocation behaves differently than giving rise understood in the sense of Definition 22.17; as we have shown, some inconsistent sets give rise to questions.

However, evocation can be defined in terms of giving rise. Let us introduce:

Definition 22.21 $\boxplus X =_{df} \{\boxplus A : A \in X\}$

Recall that $\boxplus A$ abbreviates $\neg(A \rightarrow \bot)$ and thus can be read "A is known".

Theorem 22.5 $\mathbf{E}(X, ?\,\{A_1, \ldots, A_n\})$ *iff* $\mathbf{R}(\boxplus X, ?\,\{A_1, \ldots, A_n\})$.

Proof For conciseness, let us write "$A_1 \vee \ldots \vee A_n$" as "$\bigvee A_{1,n}$".
($\Rightarrow$) If $\mathbf{E}(X, ?\,\{A_1, \ldots, A_n\})$, then $X \models_L \bigvee A_{1,n}$.
Suppose that $\mathbf{R_P}(\boxplus X, ?\,\{A_1, \ldots, A_n\})$ is not the case. Thus $\boxplus X \not\hookrightarrow_L \bigvee A_{1,n}$. So there exists an L-model $\mathbf{M} = \langle \mathcal{W}, \mathcal{V} \rangle$ such that for some $\mathbf{M}$-state σ: $\sigma \looparrowright \boxplus X$ and $\sigma \not\looparrowright \bigvee A_{1,n}$. Since $\bigvee A_{1,n}$ is a p-wff, it follows that $\mathbf{M}, w \not\models \bigvee A_{1,n}$ for any $w \in \mathcal{W}$. The elements of $\boxplus X$ are n-wffs of the form $\neg(B \rightarrow \bot)$. Hence $\mathbf{M}, w \models X$ for any $w \in \sigma$. Thus $X \not\models_L \bigvee A_{1,n}$. A contradiction.

Since $\mathbf{E}(X, ?\,\{A_1, \ldots, A_n\})$, then $X \not\models A_i$ for $1 \leq i \leq n$. Thus for each i, where $1 \leq i \leq n$, there exists an L-model $\mathbf{M} = \langle \mathcal{W}, \mathcal{V} \rangle$ such that for some $w \in \mathcal{W}$: $\mathbf{M}, w \models X$ and $\mathbf{M}, w \not\models A_i$. Hence $\{w\} \looparrowright \boxplus X$ and $\{w\} \not\looparrowright A_i$, that is, $\boxplus X \not\hookrightarrow_L A_i$.

Therefore $\mathbf{R}(\boxplus X, ?\,\{A_1, \ldots, A_n\})$.

($\Leftarrow$). Assume that $\mathbf{R}(\boxplus X, ?\,\{A_1, \ldots, A_n\})$. Hence $\boxplus X \hookrightarrow_L \bigvee A_{1,n}$. Suppose that $X \not\models_L \bigvee A_{1,n}$. So there exists a singleton state, $\{w\}$ of a certain L-model such that $\{w\} \looparrowright \boxplus X$ and $\{w\} \not\looparrowright \bigvee A_{1,n}$. Thus $\boxplus X \not\hookrightarrow_L \bigvee A_{1,n}$. A contradiction.

Since $\mathbf{R}(\boxplus X, ?\,\{A_1, \ldots, A_n\})$, we have $\boxplus X \not\hookrightarrow_L A_i$ for $1 \leq i \leq n$. Thus for each A_i, where $1 \leq i \leq n$, there exists a state σ_i of an L-model $\mathbf{M}$ such that $\sigma_i \looparrowright \boxplus X$ and $\sigma_i \not\looparrowright A_i$. Recall that states are, by definition, non-empty sets. Suppose that A_i is a p-wff. Hence for any $w \in \sigma_i$ we have $\mathbf{M}, w \models X$ and, at the same time, $\mathbf{M}, w \not\models A_i$. Thus $X \not\models_L A_i$ for $1 \leq i \leq n$. Now suppose that A_i is a n-wff. Since $\sigma_i \not\looparrowright A_i$, we get $\sigma_i \looparrowright \neg A_i$, where $\neg A_i$ is a p-wff. Hence for some $w \in \sigma_i$ we have $\mathbf{M}, w \not\models A_i$ and thus $X \not\models_L A_i$.

Therefore $\mathbf{E}(X, ?\,\{A_1, \ldots, A_n\})$. □

Thus, generally speaking, evocation is just giving rise by premises supposed to be known. This explains why Corollary 22.24 holds.

22.6 Final Remarks

Since *Ex Falso Quodlibet* holds in Classical Logic, in order to model the phenomenon of the arising of questions from inconsistencies we have to use some non-classical tools. The concept of permittance analysed in this paper is useful in this respect, although the solution offered is not fully general. An advantage of the solution lies in staying closer to the standard logical format than the alternative solutions proposed within the adaptive logic programme (see [5, 6]).

Besides its applicability in the area of questions, the concept of permittance seems interesting on its own. As we pointed out in Sect. 22.2.3, one can express the fact that A is known in a state directly in a non-modal language. The relativization to states, in turn, seems to resolve the old philosophical problem: one can legitimately claim that A is an item of knowledge in some initial state and ceases to constitute knowledge as the initial state is enriched with a new possible world/ a new account of how things are in which A is not true anymore.[16] Moreover, let us consider the case of conflicting hypotheses being general statements of the form $\forall x_i \mathcal{A}$. Assuming that they are treated semantically as we have treated p-wffs, conflicting hypotheses can be simultaneously permitted by a state and this is not tantamount to falling into a contradiction. A hypothesis of this kind constitutes an item of knowledge in a state if it is true in *each* world of the state, and extending the state with a new world in which the claim of the hypothesis does not hold only changes its epistemic status, but does not require the rejection of the hypothesis: it remains an item of knowledge in the "old" state and becomes (only) permitted in the "new" state. Permitted counterparts of n-wffs, in turn, perform the role of *state-constraints*, since in their case permittance by a state equals being true in each world of the state.

Last but not least: $\hookrightarrow_L$ seems to be an interesting truth-preserving paraconsistent consequence relation and the logic determined by it is worth further study.

References

1. Ciardelli, I., Roelofsen, F.: Inquisitive logic. J. Philos. Log. **40**, 55–94 (2011)
2. Ciardelli, I., Groenendijk, J., Roelofsen, F.: On the semantics and logic of declaratives and interrogatives. Synthese. **192**, 1689–1728 (2015)
3. Jaśkowski, S.: Rachunek zdań dla systemów dedukcyjnych sprzecznych, Studia Societatis Scientiarum Toruniensis. Sect. A. **I**, 55–77 (1948)
4. Jaśkowski, S.: A propositional calculus for inconsistent deductive systems. Log. Log. Philos. **7**, 35–56 (1999)
5. Meheus, J.: Erotetic arguments from inconsistent premises. Log. Anal. **165–167**, 49–80 (1999)
6. Meheus, J.: Adaptive logic for question evocation. Log. Anal. **173–175**, 135–164 (2001)
7. Roelofsen, F.: Algebraic foundations for the semantic treatment of inquisitive content. Synthese **190**, 79–102 (2013)
8. Shoesmith, D.J., Smiley, T.J.: Multiple-Conclusion Logic. Cambridge University Press, Cambridge (1978)

[16]More precisely, A ceases to constitute knowledge with respect to the "new" state.

9. Wiśniewski, A.: The Posing of Questions: Logical Foundations of Erotetic Inferences. Kluwer, Dordrecht (1995)
10. Wiśniewski, A.: Questions and inferences. Log. Anal. **173–175**, 5–43 (2001)
11. Wiśniewski, A.: Questions, Inferences, and Scenarios. College Publications, London (2013)

On the Type-Free Paracoherent Foundation of Mathematics with the Sedate Extension of Classical Logic by the Librationist Set Theory £, and Specifically on Why £ Is Neither Inconsistent nor Contradictory nor Paraconsistent

Frode Bjørdal

To my godson Jon Lervåg Walderhaug (born March 11 2012), and to Jon Vegard Lervåg (February 17 1979–July 22 2011).

Abstract The main purpose of this article is to discuss and clarify philosophical issues in connection with the librationst set theory £. We take technical results in and upon £ for granted though make references to them as that is useful. We defend the view that £ is neither an inconsistent nor a contradictory system and point out that it is neither paraconsistent nor dialetheist; in contrast we consider £ a *bialethic* and *paracoherent* theory.

Keywords Bialethism · Complementarity · Contrasistency · Foundations of mathematics · Librationism · Mathematicalism · Mathematical optimism · Nominism · Paracoherency · Paradoxes · Parasistency · Set theory

Mathematics Subject Classification (2000) Primary 03A99 · Secondary 03E70

The librationist foundational system, now denoted by £ ("libra"), is published most completely in [5], but the arxived e-print [4] gives more details and improvements with consequences for fundamental mathematical and philosophical matters; we send the reader to these for expositions of and results about the system.

£ gives a novel comprehensive and fully type-free account on how to deal with the paradoxes and a foundation for mathematics and semantics without compromising any classical logical theorems. Bjørdal [5] established that £ by itself is slightly stronger than the "Big Five" of the Reverse Mathematics Program, so £ is fully

---

F. Bjørdal (✉)
Universidade Federal do Rio Grande do Norte, Natal, Brazil
e-mail: bjordal.frode@gmail.com

F. Bjørdal
Universitetet i Oslo, Oslo, Norway

J.-Y. Beziau et al. (eds.), *New Directions in Paraconsistent Logic*,
Springer Proceedings in Mathematics & Statistics 152,
DOI 10.1007/978-81-322-2719-9_23

impredicative. Subsequent work shows that £ in combination with further arguably plausible assumptions is very much stronger in its interpretative power.

In [1] we showed that if $ZF\Omega^- = ZF$ *minus extensionality plus 'there are omega inaccessible cardinals'* is consistent then £ has an interpretation of $ZF\Omega^-$ which £ believes is a standard (i.e., well founded) model of $ZF\Omega^-$. As a consequence £ has an interpretation of *ZF* with extensionality given theorem 1 of [8], which shows that a system *S* slightly weaker than *ZF* minus extensionality—with collection in lieu of replacement—has an interpretation of *ZF* with extensionality.

In [4] we prove that £ plus the *Skolem-Fraenkel Postulation* (*SFP*) interprets *ZFC* by an extension of Friedman's interpretation of *ZF* just referred to. Notice well that £ does not by itself commit to the consistency of something as strong as *ZFC*, and it may be that we should rest with £ as a much weaker system in the spirit of Feferman's attitude that classical set theory is as problematic as medieval theology. On the other hand, £ + SFP^X, where SFP^X is an arguably plausible extension of *SFP*, may even interpret *ZFC* + '*there are X inaccessible cardinals*' (if such systems are consistent) in an essential countable framework—on the last point compare [2] and superseded publications dating back to 2004 as well as [12]; recent results show that *X* may be replaced by arbitrary Mahlo-cardinals.

In [3] we showed how to isolate the definable real numbers in £. It is noteworthy that this cannot be done in classical set theories, so £ is in some important respects stronger than them; regarding issues of strength, compare its negjunction completeness discussed below. The isolation of the definable real numbers becomes possible inter alia because of the presence of a nontypical truth predicate in £, and we can have such a truth predicate in £ because £ importantly resists Cantor's conclusion that there are uncountable infinities (for this, see again also [2]). Note well that we do not claim that Cantor's arguments are invalid, as they are eminently and undeniably logally valid arguments. Nevertheless, Cantor's reductio arguments involve properly paradoxical objects as premises when adapted in the librationist setting, and so the arguments turn out to not be sound according to the librationist points of view.

We adopt a limitative attitude and go a step further with postulating a bijection € from a set of finite von Neumann ordinals to a universal set of £; notice the use of indefinite articles on account of the essential and pervasive non-extensionality of sets in £ (see [5], pp. 345–346). Given € we adopt a *nominality interpretation* of what are standardly taken as free variables and take them as *noemata* which serve as standard names of sets. Given the noemata and that there are only countably many objects we define a parametrized coding of the language so that the truth predicate can serve its proper nontypical role. We remark that the isolation of *definable* real numbers mentioned above becomes possible also on account of the availability in £ of a peculiar fixed point construction going back to [6], p. 76 with a crucial precedent in [13], which we call *manifestation points*.

Incidentally, as this question has been raised in conversation, we point out that £ has choice-principles akin to the axiom of choice for kinds (non-paradoxical sets) on account of €. Given Gödel's result that any model of *ZF* has a model of *ZFC*, this state of affairs is not so important as one would otherwise assume; the fact that £ + *SFP* has an interpretation of *ZFC* is also of relevance here. It is of interest to consider the possible role of principles akin to the Axiom of Determinacy (*AD*) in the context of £ as the refutation of *AD* as carried out in *ZFC* cannot be transferred to £.

In the following we concentrate on how £ deals with the paradoxes in a novel manner, and we present an external way of thinking about the situation which cannot be matched by just gleaning upon theses of £. This external viewpoint involves the definition of a series of novel concepts and we introduce corresponding neologicisms.

Distinguish between theorems *about* a system and theorems *in* a system by adopting "thesis" for the latter usage and maintaining the previous usage. As becomes clear below we use slightly deviant names for connectives in order to forestall irrelevant philosophical objections appealing to something like their one and only true meaning; in particular we use the novel term "negjunction" in lieu of "negation". Formula A is an *antithesis* of a system iff the negjunction $\neg A$ is a thesis of it, and A is a *nonthesis* of a system iff it is not a thesis of it. S is an extension of T iff all theses of T are theses of S and some thesis of S is a nonthesis of T. S is a *sedation* of T iff no thesis of S is an antithesis of T. S is a *sedate extension* of T iff it is a sedation and an extension of T. *Sedationism* is the view that we should only accept sedate extensions of classical logic. £ is a sedate extension of classical logic.

The austere alphabet of £ is | and ., and *symbols* of £ just those strings of these that count as powers of two if | is taken as '1' and . as '0' in the binary numerical system; combining the Peirce arrow (dualized later by Sheffer) and Łukasiewicz' notation strategy we understand the formulas £ posits by the appropriate strings of symbols (as by the definitions and formation rules and elementary number theoretic concatenation definition) of £ to **be** the finite von Neumann ordinals of $\mathbf{L}_\varsigma$ so denoted where ς is the level of Gödel's constructible hierarchy needed for our semantic construction à la [6] (Bjørdal [5] and descendents merit comparison); it is sufficient for ς to be a Σ_3-admissible ordinal. £ thinks it has the language of set theory minus identity plus truth predicate T and enumerator sign €. The latter pins down a bijection from the finite von Neumann ordinals of £ to any universal set (*denumerabilism*), and the former connects to a semantic predicate; these both use the *internal* Gödel coding $\ulcorner\urcorner$ which mirrors the *external* coding invoked above in a manner so that all sets have names (*expressionism*).

A is a *maxim* of £ iff A is a thesis of £ and $\neg A$ is a nonthesis of £, and a *minor* iff a thesis of £ and an antithesis of £. £ has novel *regulations* and *prescriptions*; unlike classical inference rules the regulations of £ are sensitive to whether antecedent theses are maxims or minors, and prescriptions are similarly unlike traditional axioms and axiom schemas. We consider £ a *super-formal* and *contentual* system; the former is on account of its countenance of crucial infinitary regulations and the latter on account of its equally crucial expressionistic treatment of noemata as names.

We underline here with this as background that £ is negjunction complete, so that for any formula A of the language of £ either A is a thesis of £ or $\neg A$ is a thesis of £. On this, see [5], p. 338 and [4], p. 12; it is also with this in mind that it must be stressed that £ is not a traditional formal system and our account here is semantical.

The librationist treatment of the Curry paradoxes (see [5], p. 356f) virtually proves that such a semantical approach is needed in order to account for the paradoxes quite liberally such as in £.

We point out that £ supports a *mathematicalist* point of view to the effect that mathematics is more fundamental than logic. In particular, from the point of view of $\mathbf{L}_\varsigma$ formulas of £ simply **are** natural numbers of $\mathbf{L}_\varsigma$ and the prescriptions of £ are regarded as functions from real numbers (sets of natural numbers) to real numbers. Moreover, £ cannot be understood as a logic X plus comprehension principles Y. It is rather so that the adherence of £ to classical logic, in as far as £ extends it sedately, can be read off as a superficial phenomenon. Librationist comprehension is the sum total of principles which entail existence statements in £, and this sum cannot be stated effectively. Importantly, the points of view of £ are different from that of $\mathbf{L}_\varsigma$.

£ amazes in connection with paradoxical phenomena, and we consider this a forte of the librationist framework as paradoxical phenomena *are* amazing. Let Russell's set r be $\{x : x \in x\}$ and Russel's sentence R be $r \in r$. Although £ has R as a thesis and also as an antithesis, £ is neither inconsistent nor contradictory.

The valency of a sentence are the set of ordinals where it holds in the Herzbergerian style semi-inductive semantics (cfr. [10] and the related work [9]) with the *librationist twist* that formulas unbounded under the closure ordinal ϙ (archaic Greek Koppa) are the ones taken as designated, and not only those formulas stably in as from some ordinal below the closure ordinal. The *valor* of a sentence is the least upper bound of its valency. The *contravalence* of a sentence is the closure ordinal ϙ minus the valency of that sentence, and the *ambovalence* of two sentences is the intersection of their valencies. Induced set theoretic definitions introduce the concepts of *velvalence*, *subvalence of ... under _* and *homovalence* for veljunction, subjunction (material conditional) and equijunction (material biconditional), respectively. A sentence is true iff its valor is the closure ordinal ϙ, and a sentence is false iff its negjunction is true.

Connectives are *valency functional*: The valency of $\neg A$ is the contravalency of A, the valency of the adjunction $A \wedge B$ is the ambovalenc of A and B, the valency of the veljunction $A \vee B$ is the velvalence of A and B, the valency of the subjunction $A \rightarrow B$ is the subvalence of A under B and the valency of the equijunction $A \leftrightarrow B$ is the homovalency of A and B. In the special and preferable case of maxims (non-paradoxical formulas) the valency functionality of connectives induce their truth functionality.

A sentence dictates its valor, and its valency is the way the valor is dictated. Two sentences contradict each other iff they are contravalent and do not dictate the same. Two sentences are complementary iff they are contravalent and not contradictory. Two complementary sentences consequently dictate the same, and thence they dictate the closure ordinal. If R be Russell's sentence as above, R and its negjunction $\neg R$ dictate ϙ in complementary, or opposite, ways.

By these definitions two sentences A and B which are contradictory or complementary are contravalent and *incompatible* in that the adjunction $A \wedge B$ does not hold. Let us agree that a theory is *contrasistent* iff a sentence A is both a thesis and an antithesis of the theory. We take a theory to be inconsistent iff it has a thesis of the form $B \wedge \neg B$ and to be trivial iff all its formulas are theses. Trivial systems and inconsistent theories with simplification (adjunction elimination) are contrasistent. £ is contrasistent, but neither trivial nor inconsistent. We retain the term "parasistent" of [5] for the somewhat different idea that £ lets us *stand beyond* and shift between perspectives.

We say that two formulas A and B of a theory T *cohere* iff A and B are theses of T only if also $A \wedge B$ is a thesis of T. Russell's sentence R and its negjunction $\neg R$ are theses of £ which are *incoherent* with each other. We say that theories that have relatively incoherent and incompatible theses as these are *paracoherent* theories. Some paraconsistent logics, such as the ones following the approach by Jaskowski, are *non-adjunctive*; but such logics do not in and of themselves have incoherent theses, though extensions of such logics with appropriate comprehension principles may be paracoherent.

We have underlined that £ is a super (semi) formal system, and that it is not recursively axiomatisable. Nevertheless, a lot of highly informative prescriptions ("axiom schemas") and prescripts ("axioms") and regulations ("inference rules") are isolated. Importantly, modus ponens is not an unexceptional regulation or modus ponens as classically understood is appropriately interpreted by the regulation *modus maximus* in £ and the other regulations in it are appropriately considered *super classical*; arguably it is in the novelty of regulations that £ most deviates from and supersedes classical approaches. We take the super formal system £ to be a *contentual* system as it is categorical with respect to content.

There is the question as to whether £ should be considered a property theory or a set theory which we dealt with in [5], p. 324, where I inter alia on the basis of the opening lines of [11] decided to settle for the term "sort theory." I uphold the view that the term "property theory" is unfortunate and that "sort theory" is a better term when one wants to include, e.g., *individuals* (Urelemente) and *crowds* (properties) and *queues* (tuples) thereof besides pure sets. But I am at this point no longer satisfied that the idea of the cumulative hierarchy should be allowed to monopolize the significance of the term "set" as, e.g., suggested in [7], Chap. 1, Sect. 4. This is not only because librationism holds that the power set operation is paradoxical, as shown in [5], but also because we in more recent literature encounter a wide variety of proposals that are suggested as set theories without abiding by the strictures of being iterative or extensional or non-paradoxical. Thence I now hold that £ indeed is a set theory as now presented, but certain extensions of £ that deal with extra-mathematical objects should be studied under the name "sort theory."

We have above pointed out that £ is negjunction complete and that arguably plausible extensions of £ interpret *ZFC* and stronger systems. It is important to be aware, then, that according to such librationist interpretations of *ZFC* all statements are settled in the interpretation. As a consequence of Gödel's second incompleteness theorem we cannot settle all such questions with resources that are weaker, and so strength must be imported in the arguably plausible extensions of £ and in the methods used to show that the interpretation of *ZFC* settles the question in the determinate manner. Importantly, librationism on account of the foregoing supports a *mathematical optimism* according to which there are no **absolutely** unsolvable mathematical problems.

We take *nominism* to be the view that all mathematical objects have a name while it, as opposed to nominalism, nonetheless upholds the platonist view that mathematical objects are abstract. Nominism is supported by librationism in the latter's avoidance of Cantor's conclusion that there are uncountable infinities and insistence instead that there are only denumerably many obects; as there, according to £, are only a denumerable infinity of objects, we have enough names to name all mathematical objects and we have invoked a nominality policy with that as objective in our semantics.

Notice that the term "platonism" in the context of set theory is sometimes taken to stand for the *cantorianist* view that the endless hierarchy of ever larger alephs exists in a nonrelative and absolute sense. Here we abide by what we take as a more plausible usage of the term in the philosophy of mathematics where it denotes the view that takes mathematical objects to be abstract objects. Nominism rejects cantorianism.

Arguably, a defining feature of paraconsistent systems is that they do not include the scheme of *ex contradictione quodlibet*, but £ extends classical logic sedately so it is not paraconsistent according to such standards. £ does not articulate a *dialetheist* point of view for the latter is canonicaly characterized as the view that some contradictions are true; but quite on the contrary, £ does not condone the assertion of any contradictions. We adopt the name *bialethism* for the peculiar way of dealing with truth supported by the super formal, super classical, contentual, contrasistent, complementary, paracoherent, librationist set theory £.

Acknowledgments Many thanks to the interlocutors at recent invited presentations I had on librationist matters at the seminaries of logic at the Mathematics Department of the University of Oslo on May 22, 2014, the Steklov Institute of Mathematics in Moscow on September 4, 2014, and the Mathematics Department of Universidade Federal da Bahia in Salvador, Brazil, on October 9, 2014.

References

1. Bjørdal, F.: A Librationist Interpretation of ZF. Lecture for the Non-Classical Mathematics Workshop at the 4th World Congress and School on Universal Logic in Rio de Janeiro, Brasil, 29 Mar–7 April 2013
2. Bjørdal, F.: Considerations contra cantorianism. In: Pelis, M., Puncochar, V. (eds.) The LOGICA Yearbook 2010, pp. 43–52. College Publications (2011)
3. Bjørdal, F.: Definable Real Numbers with Librationist Capture and Domination. Lecture for the Encontro Brasileiro de Lógica in Petrópolis, Brazil, 8 April 2014
4. Bjørdal, F.: Elements of Librationism. http://arxiv.org/abs/1407.3877Arxiv e-reprint 1407.3877 (2014)
5. Bjørdal, F.: Librationist Closures of the Paradoxes. Logic Logical Philos. **21**(4), 323–361 (2012)
6. Cantini, A.: Logical Frameworks for Truth and Abstraction. Elsevier (1996)
7. Drake, F.R.: Set Theory: An Introduction to Large Cardinals. North Holland (1974)
8. Friedman, H.: The consistency of classical set theory relative to a set theory with intuitionistic logic. J. Symbolic Logic **38**(2), 315–319 (1973)

9. Gupta, A.: Truth and paradox. J. Philos. Logic **11**(1), 1–60 (1982)
10. Herzberger, H.: Notes on naive semantics. J. Philos. Logic **11**(1), 61–102 (1982)
11. Myhill, J.: Paradoxes. Synthese **60**, 129–143 (1984)
12. Skolem, T.: Axiomatized set theory. In: van Heijenort, J. (ed.) From Frege to Gödel: A Source Book in Mathematical Logic, 1879–1931, pp. 291–301. Harvard University Press (1967)
13. Visser, A.: Semantics and the liar paradox. In: Gabbay, D., Guenthner, F. (eds.) Handbook of Philosophical Logic, vol. 4, pp. 617–706. Dordrecht Reidel (1989)

Chapter 24
None of the Above: The Catuṣkoṭi in Indian Buddhist Logic

Graham Priest

Abstract The catuṣkoṭi (*Greek*: tetralemma; *English*: four corners) is a venerable principle of Indian logic, which has been central to important aspects of reasoning in the Buddhist tradition. What, exactly, it is, and how it is applied, are, however, moot—though one thing that does seem clear is that it has been applied in different ways at different times and by different people. Of course, Indian logicians did not incorporate the various interpretations of the principle in anything like a theory of validity in the modern Western sense; but the tools of modern non-classical logic show exactly how to do this. The tools are those of the paraconsistent logic of First Degree Entailment and some of its modifications.

Keywords Catuṣkoṭi · Buddhism · Nāgārjuna · Ineffability · First Degree Entailment · Plurivalent logic

Mathematics Subject Classification (2000) Primary 03B53 · Secondary 03B50

24.1 Introduction

The catuṣkoṭi (*Greek*: tetralemma; *English*: four corners) is a venerable principle of Indian logic, which has been central to important aspects of reasoning in the Buddhist tradition. What, exactly, it is, and how it is applied, are, however, moot—though one thing that does seem clear is that it has been applied in different ways at different times and by different people. Of course, Indian logicians did not incorporate the various interpretations of the principle in anything like a theory of validity in the modern Western sense; but the tools of modern non-classical logic show exactly how

---

G. Priest (✉)
Department of Philosophy, The Graduate Center, City University of New York, New York, USA
e-mail: priest.graham@gmail.com

Department of Philosophy, University of Melbourne, Melbourne, Australia

J.-Y. Beziau et al. (eds.), *New Directions in Paraconsistent Logic*, Springer Proceedings in Mathematics & Statistics 152,
DOI 10.1007/978-81-322-2719-9_24

to do this. The tools are those of the paraconsistent logic of First Degree Entailment (FDE), and some of its modifications.[1] We will approach the matter chronologically, interlacing philosophical and technical material, as appropriate.[2] The point of the exercise is to show how the history of philosophy and the techniques of contemporary non-classical logic can profitably inform each other. Positions which one might have taken to be unintelligible can be shown to be perfectly coherent with the aid of these techniques; conversely, the positions may themselves suggest the development of interesting new logical techniques.

24.2 Back to the Beginning

So let us go back to the earliest applications of the catuṣkoṭi.

The four koṭis (corners) of the catuṣkoṭi are four options that one might take on a question. Given any question, there are four possibilities, *yes*, *no*, *both*, and *neither*. Who first formulated this thought would appear to be lost in the mists of time, but it seems to be fairly orthodox in the intellectual circles of Siddhārtha Gautama (*Pali*: Gotama), the historical Buddha (c. 6c BCE). Thus, canonical Buddhist texts often set up issues in terms of these four possibilities. For example, in the *Mijjhima Nikāya,* when the Buddha is asked about one of the profound metaphysical issues, the text reads as follows[3]:

> 'How is it, Gotama? Does Gotama believe that the saint exists after death, and that this view alone is true, and every other false'?
>
> 'Nay, Vacca. I do not hold that the saint exists after death, and that this view alone is true, and every other false'.
>
> 'How is it, Gotama? Does Gotama believe that the saint does not exist after death, and that this view alone is true, and every other false'?
>
> 'Nay, Vacca. I do not hold that the saint does not exist after death, and that this view alone is true, and every other false'.
>
> 'How is it, Gotama? Does Gotama believe that the saint both exists and does not exist after death, and that this view alone is true, and every other false'?
>
> 'Nay, Vacca. I do not hold that the saint both exists and does not exist after death, and that this view alone is true, and every other false'.
>
> 'How is it, Gotama? Does Gotama believe that the saint neither exists nor does not exist after death, and that this view alone is true, and every other false'?
>
> 'Nay, Vacca. I do not hold that the saint neither exists nor does not exist after death, and that this view alone is true, and every other false'.

---

[1]For FDE, see Priest [10], Chap. 8.

[2]I note right at the start there are some Buddhist logicians in whose thinking the catuṣkoṭi played no role. This is true, in particular, of the school of Dignāga and Dharmakīrti. Like the Nyāyā, this school of logic endorsed both the Principles of Non-Contradiction and Excluded Middle. See Scherbatsky [18], pt. 4, Chap. 2.

[3]Radhakrishnan and Moore [14], p. 289 f. The word 'saint' is a rather poor translation. It refers to someone who has attained enlightenment, a Buddha (Tathāgata).

It seems clear from the dialogue that the Buddha's interlocutor thinks of himself as offering an exclusive and exhaustive disjunction from which the Buddha is to choose. That there are four such possibilities, was the standard view.[4]

Later Buddhists echoed the thought. Thus, in the *Mūlamadhyamakakārikā* (hereafter, MMK) Nāgārjuna frequently addresses an issue by considering these four cases. Thus, in Chap. XXV, he considers nirvāṇa. First, he considers the possibility that it exists (vv. 4–6); then that it does not exist (vv. 7–8); then that it both exists and does not exist (vv. 11–14); and finally, that it neither exists nor does not (vv. 14–15). As Āryadeva, Nāgārjuna's disciple, was to put it[5]:

> Being, non-being, [both] being and non-being, neither being [nor] non-being: such is the method that the wise should always use with regard to identity and all other [theses].

Thus, it would seem, originally, the catuṣkoṭi functioned as something like a Principle of the Excluded Fifth. Aristotle held a principle of the Excluded Third: any statement must be either true or false; there is no third possibility; moreover, these two are exclusive. In a similar but more generous way, the catuṣkoṭi gives us an exhaustive and mutually exclusive set of four possibilities.

24.3 First Degree Entailment

How to formulate such a simple idea has bemused many commentators, and resulted in many dud ideas.[6] However, to anyone familiar with the rudiments of contemporary non-classical logic, there is an obvious way. First Degree Entailment (FDE) is a system of logic that can be set up in many ways, but one of these is as a four-valued logic whose values are t (true only), f (false only), b (both), and n (neither). The values are standardly depicted by the following Hasse diagram:

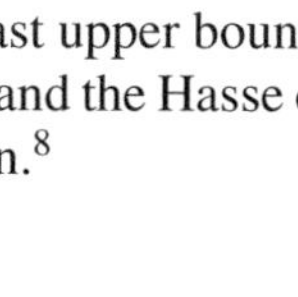

Negation maps t to f, vice versa, n to itself, and b to itself. Conjunction is greatest lower bound, and disjunction is least upper bound. The set of designated values, D, is $\{t, b\}$.[7] The four corners of truth and the Hasse diagram seem like a marriage made for each other in a Buddhist heaven.[8]

[4]See Ruegg [17], p. 1.

[5]Tillemans [20], p. 189.

[6]For a survey, see Ruegg [17], p. 39ff. And for a critique, see Priest [12], 2.2.

[7]See Priest [10], Chap. 8.

[8]As observed in Garfield and Priest [4].

FDE can be characterised by the following sound and complete rule system. (A double line indicates a two-way rule, and overlining indicates discharging an assumption.)[9]

$$\frac{A, B}{A \wedge B} \qquad \frac{A \wedge B}{A\ (B)}$$

$$\frac{A\ (B)}{A \vee B} \qquad \frac{A \vee B \quad \begin{matrix} \overline{A} \\ \vdots \\ C \end{matrix} \quad \begin{matrix} \overline{B} \\ \vdots \\ C \end{matrix}}{C}$$

$$\begin{array}{c} \neg(A \wedge B) \\ \hline\hline \neg A \vee \neg B \end{array} \qquad \begin{array}{c} \neg(A \vee B) \\ \hline\hline \neg A \vee \neg B \end{array} \qquad \begin{array}{c} \neg\neg A \\ \hline\hline A \end{array}$$

24.4 Denying All the Koṭis

So far so good. Returning to the question of the Tathāgata after death, the Buddha, as we observed, refused to endorse any of the koṭis on this matter—and on a number of similar "unanswerable" metaphysical questions. In some sūtras it appears that this is because speculation over the matter is simply a waste of time.[10] Thus, in the *Cula-Malunkyovada Sutta*, we read[11]:

> It is just as if a man were wounded with an arrow thickly smeared with poison. His friends and companions, kinsmen, and relatives would provide him with a surgeon, and the man would say, 'I won't have this arrow removed until I know whether the man who wounded me was a noble warrior, a priest, a merchant, or a worker' He would say, 'I won't have this arrow removed until I know the given name and clan name of the man who wounded me... until I know whether he was tall, medium, or short... until I know whether he was dark, ruddy-brown, or golden-colored... until I know his home village, town, or city... '. The man would die and those things would still remain unknown to him.
>
> In the same way, if anyone were to say, 'I won't live the holy life under the Blessed One as long as he does not declare to me that 'The cosmos is eternal',... or that 'After death a Tathagata neither exists nor does not exist', the man would die and those things would still remain undeclared by the Tathagata....
>
> So, Malunkyaputta, remember what is undeclared by me as undeclared, and what is declared by me as declared. And what is undeclared by me? 'The cosmos is eternal', is undeclared by me. 'The cosmos is not eternal', is undeclared by me. ... 'After death a Tathagata exists'... 'After death a Tathagata does not exist'... 'After death a Tathagata both exists and does not exist'... 'After death a Tathagata neither exists nor does not exist', is undeclared by me.

[9] See Priest [7], 4.6.

[10] See Ruegg [17], pp. 1, 2.

[11] Thanissaro [19].

And why are they undeclared by me? Because they are not connected with the goal, are not fundamental to the holy life. They do not lead to disenchantment, dispassion, cessation, calming, direct knowledge, self-awakening, unbinding. That is why they are undeclared by me.

However, in some of the sūtras there is a hint of something else going on. The Buddha seems to explicitly *reject* all the options, the suggestion being that all the answers have a common and false presupposition. Thus, in the *Mijjhima Nikāya*, the Buddha says that none of the four kotis 'fits the case' in such issues. When questioned how this is possible, he says[12]:

> But Vacca, if the fire in front of you were to become extinct, would you be aware that the fire in front of you had become extinct?
>
> Gotama, if the fire in front of me were to become extinct, I would be aware that the fire in front of me had become extinct.
>
> But, Vacca, if someone were to ask you, 'In which direction has the fire gone,—east, or west, or north, or south'? what would you say O Vacca?
>
> The question would not fit the case, Gotama. For the fire which depended on fuel of grass and wood, when all that fuel has gone, and it can get no other, being thus without nutriment, is said to be extinct.

The thought seems to be that if fires or Tathāgatas have gone out of existence, one can say nothing about them.

We find Nāgārjuna and some of his Madhyamaka successors appearing to deny all the koṭi sometimes too. For example, as part of an argument that the Tathāgata has no self-being (svabhāva), MMK XXII: 11, 12 says[13]:

> 'Empty' should not be asserted.
> 'Non-empty; should not be asserted.
> Neither both nor neither should be asserted.
> These are used only nominally.
>
> How can the tetralemma of permanent and impermanent, etc.
> Be true of the peaceful?
> How can the tetralemma of finite, infinite, etc.
> Be true of the peaceful?

and in the course of an argument for the same conclusion about nirvāṇa, MMK XXV: 17 says:

> Having passed into nirvāṇa, the Victorious Conqueror
> Is neither said to be existent
> Nor said to be nonexistent.
> Neither both nor neither are said.

---

[12] Radhakrishnan and Moore [14], p. 290.

[13] All translations from the MMK are from Garfield [2].

Rejecting all four koṭis in this way is sometimes, and for obvious reasons, called the 'four-cornered negation'. And just to confuse matters, the word 'catuṣkoṭi' is sometimes taken to refer to this.

24.5 A 5-Valued Logic

How can one understand the rejection of all four koṭis in terns of modern logic? The fact that none of the four koṭis sometimes holds would seem to imply that there is a fifth possibility: (e) none of the above. Technically, the obvious thought is to add a new value, e, to our existing four (t, f, b and n), expressing this new status.

Since e is the status of claims such that neither they nor their negations should be accepted, it should obviously not be designated. Thus, we still have that $D = \{t, b\}$. How are the connectives to behave with respect to e? Both e and n are the values of things that are, in some sense, neither true nor false, but they had better behave differently if the two are to represent distinct alternatives. The simplest suggestion is to take e to be such that whenever any input has the value e, so does the output: e-in/e-out.

The logic that results by modifying FDE in this way is obviously a sub-logic of it. It is a proper sub-logic. It is not difficult to check that all the rules of FDE are designation-preserving except the rule for disjunction-introduction, which is not, as an obvious counter-model shows. However, replace this with the rules:

$$\frac{\varphi(A) \quad C}{A \vee C} \qquad \frac{\varphi(A) \quad C}{\neg A \vee C} \qquad \frac{\varphi(A) \quad \psi(B) \quad C}{(A \wedge B) \vee C}$$

where $\varphi(A)$ and $\psi(B)$ are any sentences containing A and B.[14] Call these the φ *Rules*, and call this system FDE_φ. FDE_φ is sound and complete with respect to the semantics.[15]

24.6 *e* and Ineffability

Whether or not *Nāgārjuna himself is best interpreted as really denying all the* koṭis is a question of interpretation that I won't go into here. There is no doubt that later philosophers did.[16] This is particularly the case when the Yogacārā influence came to be felt in subsequent developments. According to this, there is an ultimate reality. Our

[14]Instead of $\varphi(A)$ (etc.), one could have any sentence that contained all the propositional parameters in A.

[15]For the proof, see the technical appendix of Priest [12].

[16]The Buddhists tadition was not alone in appearing to reject all four of the koṭis sometimes. See Raju [15].

conventional (lived) reality is produced by the imposition of a conceptual/linguistic structure onto this. What is this ultimate reality like? One cannot say. To do so would require the employment of linguistic and conceptual categories; and the ultimate reality is what remains after all such categories have been "peeled off". It is a simple *thatness* (tathāta), often referred to as *emptiness*. One may have a direct perception of it under appropriate circumstances, but describe it one cannot. It is ineffable. In some Buddhist philosophers, the fifth status given by denying the four standard values of the catuṣkoṭi is the value of the ineffable.

The interpretation of the catuṣkoṭi and fourfold negation which takes ineffability on board is spelled out perhaps most clearly and explicitly by the Tibetan philosopher Gorampa. He says in his *Synopsis of Madhyamaka*, 75[17]:

> The scriptures which negate proliferations of the four extremes refer to ultimate truth but not to the conventional, because the ultimate is devoid of conceptual proliferations, and the conventional is endowed with them.

The fifth value, e, then, is the value of the ineffable.

Care is needed here over the word 'truth' in this quotation. It is a translation of the Tibetan *bden-pa* (*Sanskrit*: satya). This can mean either *truth* or *reality*. In the quote from Gorampa, it clearly means 'reality'. Now, it is states of affairs which are effable or ineffable, not sentences. This requires us to rethink our formal language and its interpretation.

We must now think of the bearers of the truth values as states of affairs. Connectives generate complex states of affairs. Thus, if A and B are states of affairs, then $A \wedge B$, $A \vee B$ and $\neg A$ are the related conjunctive, disjunctive and negative state of affairs. As for the truth values themselves: a state of affairs that receives the value t exists and its negation does not. A state of affairs that receives the value b is such that both it and its negation exists. Similarly for f and n. And a state of affairs that receives the value e is ineffable.

24.7 Talking of the Ineffable

Matters are still more complex, though. Ultimate reality is, on this understanding, ineffable. Yet Gorampa himself talks about it. Thus, as I just quoted him as saying, 'the ultimate is devoid of conceptual proliferations'. This explains why, indeed, it is ineffable; but it is also says something about it. Some things about the ineffable *can* be expressed.[18]

One might react to this in various ways. One is to write off the whole project as misconceived. Obviously, this was not Gorampa's reaction. Indeed, nor is this

---

[17]The translation is taken from Kassor [5].

[18]It is not just Gorampa who finds himself in this position. Any theory according to which there is something ineffable and which explains why it *is* ineffable is going to be in the same situation. There are many such theories, East and West. See Priest [9].

obviously required in a context where the possibility of contradictions is clearly allowed for in the shape of one of the koṭis.

Gorampa's own response to the situation is to draw a distinction. Kassor (2103) describes matters thus (her italics):

> In the *Synopsis*, Gorampa divides ultimate truth into two: the nominal ultimate (*don dam rnam grags pa*) and the ultimate truth (*don dam bden pa*). While the ultimate truth ... is free from conceptual proliferations, existing beyond the limits of thought, the nominal ultimate is simply a conceptual description of what the ultimate is *like*. Whenever ordinary persons talk about of conceptualize the ultimate, Gorampa argues that they are actually referring to the nominal ultimate. We cannot think or talk about the *actual* ultimate truth because it is beyond thoughts and language; any statement or thought about the ultimate is necessarily conceptual, and is, therefore, the nominal ultimate.

It does not take long to see that this hardly avoids contradiction. If all talk of the ultimate is about the nominal ultimate, then Gorampa's own talk of the ultimate is this. And the nominal ultimate is clearly effable. Hence, Gorampa's own claim that the ultimate is devoid of conceptual proliferations is just false.

A similar situation was to arise about 500 years later and a few miles to the West. In the *Critique of Pure Reason* Kant explains that there are noumenal objects about which one cannot talk/think. For talk/thought constitutes *phenomenal* objects. Realising the bind he is in here, Kant drew a distinction between an illegitimate positive notion of a noumenon and a legitimate negative, or limiting, notion. This does not help: according to Kant, the negative notion is there to place a limit on the area in which we can apply thought/language. But to say that there is an area to which we cannot apply thought/language is clearly to say something about this area, and so apply thought/language to it.[19]

Indeed, the Gorampa/Kant predicament is inevitable. If one wishes to explain why something is ineffable, one *must* refer to it and say something about it. To refer to something *else*, which one can talk about, is just to change the subject.

24.8 Accepting More Than One Koṭi

The honest thing to do, then, is to admit that the situation is a contradictory one. We have here a contradiction at the limits of thought, of a kind to which certain Buddhist views are committed. Nor is this irrational. Given those views, and the fact that the contradictions can be controlled, this is exactly the rational position to hold.[20]

Given this, we must allow for things to be (truly) sayable and ineffable as well—that is, to take more than one semantic value. In fact, there is some precedent for this in Nāgārjuna as well. Thus, MMK XVIII: 8 says:

---

[19] See Priest [8], 5.5.

[20] See Garfield and Priest [3], and Deguchi et al. [1]. The contradiction we are dealing with here is closely related to Nāgārjuna's paradox that the ultimate truth is that there is no ultimate truth. (See Garfield and Priest [3], Sect. 5.) One can say nothing true about ultimate reality—either because there is no such thing, or because it is ineffable. But either way, that is itself an ultimate truth.

Everything is real and is not real.
Both real and not real,
Neither real nor not real.
This is Lord Buddha's teaching.

Exactly how to interpret this passage from Nāgārjuna is moot. But whatever the truth of that matter, in Gorampa, at least, we seem to be stuck with the idea that something can be true and ineffable, and so inhabit more than one of our five values.

24.9 Relational Semantics

But how to make sense of this technically? There is, in fact, an easy way to do so.

In classical logic, evaluations are functions which map sentences to one of the values 1 and 0. In one semantics for FDE, evaluations are thought of, not as functions, but as relations, which relate sentences to some number of these values. This gives the four possibilities represented by the four values of our many-valued semantics.[21]

We may do the same with the values t, b, n, f and e themselves. So if P is the set of propositional parameters (or atomic states of affairs), and $V = \{t, b, n, f, e\}$, an evaluation is a relation, $\rhd$, between P and V. We insist that every formula has at least one of these values. That is, the values are exhaustive:

Exh: for all $p \in P$, there is some $v \in V$, such that $p \rhd v$.

However, there is no reason why $\rhd$ cannot relate a sentence/state of affairs to *more* than one value. Thus, p may relate to both *true* (t) and *ineffable* (e).

How might the connectives behave in this context? If we denote the many-valued truth functions corresponding to the connectives $\neg$, $\vee$, and $\wedge$ in FDE_φ, by $f_\neg$, $f_\vee$, and $f_\wedge$, then the most obvious extension of $\rhd$ to all formulas is given by the point-wise clauses:

- $\neg A \rhd v$ iff for some x such that $A \rhd x$, $v = f_\neg(x)$
- $A \vee B \rhd v$ iff for some x, y, such that $A \rhd x$ and $B \rhd y$, $v = f_\vee(x, y)$
- $A \wedge B \rhd v$ iff for some x, y, such that $A \rhd x$ and $B \rhd y$, $v = f_\wedge(x, y)$

One can show, by a simple induction, that for every A there is some $v \in V$ such that $A \rhd v$. I leave the details as an exercise.

Where, as before, $D = \{t, b\}$, we may simply define validity as follows: $\Sigma \models A$ iff for all evaluations, $\rhd$:

- if for every $B \in \Sigma$, there is a $v \in D$ such that $B \rhd v$, then there is a $v \in D$ such that $A \rhd v$

That is, an inference is valid if it preserves the property of relating to *some* designated value.

---

[21] See [10], 8.2.

A moment's reflection will show that if we insist that every parameter takes *exactly* one of the five values, the same is true for *all* formulas. These semantics are, then, just a variation of the functional semantics for FDE_φ which we have already employed. Let us call them the *single-valued relational semantics*.

But what is this logic which allows multiple values? In fact, it is FDE_φ. Let is write the single-valued consequence relation as $\models_s$ and the many-valued consequence relation as $\models_m$. Any single-valued interpretation is a many-valued interpretation. Hence if $\Sigma \not\models_s A$ then $\Sigma \not\models_m A$; so if $\Sigma \models_m A$ then $\Sigma \models_s A$. Conversely, suppose that $\Sigma \models_s A$. Then by the completeness result mentioned in Sect. 24.5, the inference is delivered by the rules for FDE_φ. But it is easy to check that each of these rules is sound with respect to the many-valued semantics. Hence, $\Sigma \models_m A$.

A final technical comment. One can turn a relational semantics into an equivalent functional semantics by taking the functional values to be *sets* of the many values ($\{t, e\}$, etc.). In this way, it is possible to iterate the construction to higher orders, taking sets of values, sets of sets of values, etc. For the case, where we start with the simple classical truth values, 1 and 0, this is done in Priest [6]. Again, there, applying the construction (after the first iteration) does not destabilise the consequence relation.[22]

24.10 Coda: Jaina Logic

In conclusion, it is worth nothing the similarity of the view we have just been looking at with that which is to be found in another Indian logical tradition: Jainism.[23] In this, there are three basic "truth values", *true*, *false*, and a third truth value. The precise meaning of this third value is somewhat moot, since different writers gloss it in different ways: *both true and false*, *neither true nor false*, *ineffable*, *non-assertable*.[24] Sentences can take any number of these values, as long as this is at least one, giving seven possibilities in all ($2^3 - 1$)—rather than the 31 ($2^5 - 1$) we have in the Buddhist case. One can turn this trilogy into a relational logic in exactly the way we have done in the Buddhist case.[25]

The Jains endorsed a metaphysical view about the nature of reality, according to which is it "multi-faceted". One can then think of each of the values, v, as one of the basic values of the sentence if it has that value at some facet. Perhaps the most natural way to develop this picture in terms of modern logic is to take each facet to be something like a possible world. Each world is many-valued, but the resulting logic is not a many-valued one, but a modal one. One can do exactly the same with our

[22]For a fuller discussion of the construction described in this section, see Priest [13].

[23]For details of what follows, see Priest [11].

[24]See Priest [11], Sect. 5.

[25]In [11], this is formulated not as a relational semantics but equivalently as a functional semantics, where the functional values are sets of truth values. The possibility of applying this construction to the Buddhist four (or five) values, as we hae done here, is noted there in footnote 15.

four or five Buddhist values instead of the three Jaina ones. I leave the details as a relatively straightforward exercise. The fact that the Buddhists do not subscribe to a similar metaphysical doctrine concerning the many-faceted nature of reality makes this sort of logical development much less natural.

References

1. Deguchi, Y., Garfield, J., Priest, G.: The way of the dialetheist: contradictions in buddhism. Philos. East West **58**, 395–402 (2008)
2. Garfield, J.: The Fundamental Principles of the Middle Way: Nāgārjuna's Mūlamadhyamikakārikā. Oxford University Press, New York (1995)
3. Garfield, J., Priest, G.: Nāgārjuna and the limits of thought. Philos. East West **53**, 1–21 (2003) (Reprinted as ch. 5 of Garfield's, Empty Words. Oxford University Press, New York (2002) and as ch. 16 Priest (2002))
4. Garfield, J., Priest, G.: Mountains are Just Mountains (2009) (ch. 7 of D'Amato, M., Garfield, J., Tillemans, T. (eds.): Pointing at the Moon: Buddhism, Logic, Analytic Philosophy. Oxford University Press, New York)
5. Kassor, C.: Is Gorampa's "Freedom from Conceptual Proliferations" dialetheist? A response to Garfield, Priest, and Tillemans. Philos. East West **63**, 399–410 (2013)
6. Priest, G.: Hypercontradictions. Logique et Analyse **107**, 237–243 (1984)
7. Priest, G.: Paraconsistent Logic. In: Gabbay, D., Guenther, D. (eds.) Handbook of Philosophical Logic, 2nd edn, Vol. 6, pp. 287–393. Kluwer Academic Publishers, Dordrecht (2002)
8. Priest, G.: Beyond the Limits of Thought, 2nd edn. Oxford Unversity Press, Oxford (2002)
9. Priest, G.: The Limits of Language. In: Brown, K. (ed.) Encyclopedia of Language and Linguistics, 2nd edn, Vol. 7, pp. 156–159. Elsevier, Amsterdam (2005)
10. Priest, G.: Introduction to Non-Classical Logic: From If to Is. Cambridge University Press, Cambridge (2008)
11. Priest, G.: Jaina logic: a contemporary perspective. History Philos. Logic **29**, 263–278 (2008)
12. Priest, G.: The logic of the Catuṣkoṭi. Comp. Philos. **1**, 32–54 (2010)
13. Priest, G.: Plurivalent logic. Australas. J. Logic **11** (2014) article 1. http://ojs.victoria.ac.nz/ajl/article/view/1830
14. Radhakrishnan, S., Moore, C. (eds.): A Source Book in Indian Philosophy. Princeton University Press, Princeton (1957)
15. Raju, P.: The principle of four-cornered negation in Indian philosophy. Rev. Metaphys. **7**, 694–713 (1953)
16. Robinson, R.: Some logical aspects of Nāgārjuna's system. Philos. East West **6**, 291–308 (1956)
17. Ruegg, D.: The uses of the four positions of the Catuṣkoṭi and the problem of the description of reality in Mahāyāna buddhism. J. Indian Philos. **5**, 1–71 (1977)
18. Scherbatsky, Th: Buddhist Logic, vol. 1. Matilal Banarsidass, Delhi (1993)
19. Thanissaro: Cula-Malunkyovada Sutta: the shorter instructions to Malunkya. http://www.accesstoinsight.org/tipitaka/mn/mn.063.than.html. Accessed Mar 2010
20. Tillemans, T.: Is buddhist logic non-classical or deviant? Chap. 9 of Scripture, Logic, Language: Essays on Dharmakīrti and his Tibetan Successors. Wisdom Publications, Boston (1999)

Chapter 25
Eastern Proto-Logics

Fabien Schang

Abstract An alternative semantic framework is proposed in the following to reconstruct and make sense of "Eastern logics": a Question-Answer Semantics (thereafter: **QAS**), including a set of questions-answers and a finite number of ensuing non-Fregean logical values. Thus, meaning is provided by yes-no answers to corresponding questions about relevant properties. These logical values help to show that the *saptabhaṅgī* (and its dual, viz., the Buddhist Mādhyamaka *catuṣkoṭi*) is not a many-valued paraconsistent logic but, rather, a one-valued proto-logic: a constructive machinery that serves as a formal theory of judgment, rather than a Tarskian-like theory of consequence. Such an explanatory model of contradiction assumes a deep redefinition of logical values.

Keywords Avaktavyam · Bhaṅgī · *Catuṣkoṭi* · Logical values · Negation · Syād · Question-Answer Semantics · *Saptabhaṅgī*

Mathematics Subject Classification(2000) Primary 03B53 · Secondary 03B50

25.1 Paraconsistency

Two philosophical schools are put into focus in the following, namely: on the one hand, the Jains' *nayavāda* or theory of standpoints and the corresponding *saptabhaṅgī* or theory of sevenfold predication (thereafter: SB); on the other hand, the

This research was supported by the Indo-European Research Training Network in Logic (IERT-NiL) funded by the Institute of Mathematical Sciences of Chennai, the Institute for Logic, Language and Computation of the University of Amsterdam, and the Faculty of Mathematics, Computer Science and Natural Sciences of the University of Hamburg. This work is an output of a research project implemented as part of the Basic Research Program at the National Research University Higher School of Economics (HSE).

F. Schang (✉)
National Research University - HSE, Moscow, Russia
e-mail: schang.fabien@voila.fr

J.-Y. Beziau et al. (eds.), *New Directions in Paraconsistent Logic*,
Springer Proceedings in Mathematics & Statistics 152,
DOI 10.1007/978-81-322-2719-9_25

Mādhyamaka's *catuṣkoṭi*), or theory of Four-Cornered Negation (thereafter: 4CN).[1] Let us consider these two peculiar "logics" or, better, sets of statements similar to the Middle Age *disputatio*. A number of questions is to be put about their nature.

As is well known, paraconsistency is seen as a common property of a set of logical systems that infringe the so-called principle of *explosion*, according to which everything follows from a contradiction in a given logical system $\mathbb{L}$. In semantic terms, it means that the acceptance of a sentence (α, say) and its negation ($\sim\alpha$, say) need not entail the acceptance of any other sentence (ψ, say) in $\mathbb{L}$: $\alpha, \sim\alpha \not\models \psi$.

Is SB a paraconsistent logic? Our answer to this question will be negative, for a number of reasons.

First of all, SB is not an *inferential* system, crucially including the relation of consequence between sets of sentences. Indeed, [9] acknowledged the very special status of Indian "logics" in these terms:

> This chapter is somewhat tangential to the main thread in the book. The reader whose main interest is in the development of the notion of an inference-warranting relation and associated concepts may wish to skip it and move to Chap. 7.[2]

Secondly, SB is not a formal language endowed with a set of sentential connectives. It is another reason to qualify its logical status, since it hardly make sense to refer to a "logical" system without any logical constant in it. This is why [9] tampers his investigation on Indian "logics" anew:

> My reference to the *non-bivalence* or *paraconsistent* logic, in connection with Jainism, should not be overemphasized. I have already noted that Jaina logicians did not develop, unlike the modern logicians, truth matrices for Negation, Conjunction, and so on. It would be difficult, if not impossible, to find intuitive interpretations of such matrices, if one were to develop them in any case.[3]

Third, although SB has been currently mentioned as a pioneer of paraconsistency it wrongly seems to accept contradictory sentences: the reference to metalogical rules or *paribhāṣā* in ancient texts has been equally made to by all the commentators who plainly subscribed to Aristotle's Principle of Non-contradiction and refused to assimilate SB with an inconsistency-friendly system.

Fourthly, SB often relates to paraconsistency by dealing with other logical topics like many-valuedness, modalities, and sentential negation. In this respect, subscribing to many-valuedness could be a way to block explosion by introducing nonclassical logics and redefining the logical constant of negation. Does it really? Just as it has been noted that no logical constant clearly appears in SB, it is not clear whether the latter can be seen seriously as a grandfather of modal or nonclassical systems.

Therefore, let us check if many-valuedness and SB are really on a par before rejecting its paraconsistent nature once for all.

---

[1] About SB, see especially Vādiveda Sūri: "Pramāṇa-naya-tattvālokālaṃkāra"; about 4CN, see Nāgārjuna: "Mūlamadhyamaka-kārikā."

[2] Matilal [9], p. 127.

[3] Matilal [9], p. 139.

25.2 Many-Valuedness

A logical system is said to many-valued whenever it includes more than the two classical values of truth and falsity. Despite the failure of explicit logical constants in SB and 4CN, their interpretations by means of nonclassical values largely contributed to their logical flavor.

In SB, a good deal of analyses led to a clear lack of consensus among them as to the number of truth-values, all the more than one and the same commentator may argue for several different options: either three [3, 5, 6, 9], four [1], seven [5, 11, 14], eight [1], twelve [1], or fifteen values [16, 17].

The same trouble arises in the other theory 4CN, where the list of values goes decreasingly from five [11] to the two classical ones [4] and even until a single one [2, 15]. Not only is there no reference to logical constants in 4CN either, or maybe with the exception of negation; but also, the occurrence of only one truth-value should definitely condemn the logical status because no difference can be made therefore between valid and invalid arguments. To motivate this surprising assumption of a single value, [2] argued that a blatant confusion has been made (in SB, at the least) between truth-values and the property of a statement:

> All constructivist interpretations in terms of *many-valued* logic seem to tacitly assume that at least some *bhaṅgī* can be hierarchically ordered with respect to their truth-value, ranging from false and indeterminate to true. The matter of fact is, however, that all seven statements are true (. . .) It is not the case that each member of this septuplet has a different truth-value; what each of these figures actually expresses is a different property![4]

The main reason why **QAS** has been advocated here is the dialectical nature of Indian theories, whether logical or merely argumentative ones. Thus SB proceeds as an interrogative model of yes-no questions concerning very special topics:

> A *metaphysical* thesis was usually expressed in the canonical literature of Buddhism and Jainism in the form of a question, "Is A B?" or "Is everything F?"—to which an answer was demanded, either *yes* or *no*. If yes, the thesis was put forward as an assertion, that is, the proposed assertion "A is B" or "Everything is F" was claimed to be true. If *no*, it was denied, that is, it was claimed to be false. Therefore, yes and no were substitutes for the truth-values, true and false. The Buddhist canons describe such questions as *ekāṃsā-akaraṇīya*, those that can be answered by a direct *yes* or *no*.[5]

This accounts for the following set of seven statements, these expressing yes-answers with assertions and no-answers with denials about one and the same initial sentence at hand.

---

[4]Balcerowicz [2], p. 13.

[5]Matilal [9], p. 128.

(1) *syād asty eva*
arguably, it exists
(assertion)
(2) *syān nāsty eva*
arguably, it does not exist
(denial)
(3) *syād asty eva, syān nāsty eva*
arguably, it exists; arguably, it does not exist
(successive assertion and denial)
(4) *syād avaktavyam eva*
arguably, it is unspeakable
(simultaneous assertion and denial)
(5) *syād asty eva*; *syād avaktavyam eva*
arguably, it exists; arguably, it is unspeakable
(assertion ans simultaneous assertion and denial)
(6) *syān nāsty eva*; *syād avaktavyam eva*
arguably, it does not exist; arguably, it is unspeakable
(denial and simultaneous assertion and denial)
(7) *syād asty eva*; *syān nāstyeva syād avaktavyam eva*
arguably, it exists; arguably, it does not exist; arguably, it is unspeakable
(assertion and denial and simultaneous assertion and denial)

To make the connection with many-valuedness, the structure of SB is like a combination of $2^3 - 1 = 7$ predications about speaker's attitudes toward a given sentence. These seven predications, or *bhaṅgī*, are formed on the basis of three basic predications or *mulabhaṅgī*. The meaning of the first two ones is obviously truth T and falsity F, where the alleged truth-values are embedded into the speech-acts of assertion and denial. The real difficulty stems from the third basic predication, i.e., "avaktavyam," whose ambiguity will be symbolized by #. Thus SB is said to include three basic values for its three basic statements, namely: T for "syād asty eva," where the sentence is claimed to be (arguably) true; F for "syān nāsty eva," where the sentence is claimed to be (arguably) false; and finally, # for "syād avaktavyam," where the sentence is claimed to be (arguably) unsayable or unspeakable. Each of the above seven statements can be rendered by the following combination of several basic predications: $(1) = \{T\}$, $(2) = \{F\}$, $(3) = \{T, F\} = B$, $(4) = \{\#\}$, $(5) = \{T, \#\}$, $(6) = \{F, \#\}$, $(7) = \{B, \#\}$, whilst $(8) = \{\varnothing\} = N$ is an impossible value that never gives any yes-answer to the *exhaustive* (but not exclusive) set of three basic questions.

Our thesis is that these seven predications do *not* amount to seven proper truth-values, thus doing justice to our ultimate refusal of a seven-valued logical system to depict SB.

As to 4CN, it consists of a combination of $2^2 = 4$ predications, all equally *denied* in the form of no-answers. That is:

(i) Does a being come out itself?	No.
(ii) Does a being come out the other?	No.
(iii) Does a being come out of both itself and the other?	No.
(iv) Does a being come out neither?	No.

In a nutshell, three basic predications occur in SB, whereas only two are used in 4CN, i.e., the classical copula "is" for assertion and its negation "is not" for denial. We will note later on, however, that two sorts of denial are to be distinguished from each other in order to make sense of Nāgārjuna's purely negative stance (i′)–(iv′) toward affirmative and negative sentences.

The numerical distinction between SB and 4CN may relie upon the different speech-acts manifested in them. Does the third predication "*avaktavyam*" not occur in 4CN? Does it correspond rather to the constant no-answer in 4CN? Or does it correspond to the forbidden value (8) in SB, i.e., the case where only no-answers are given to the three basic questions? Be this as it may, many-valuedness and paraconsistency seem to be on a par in 4CN once one notes the blatantly inconsistent set of answers (i)–(iv) from a classical, bivalent point of view. For let us formalize this dialectical game by a classical logic form, where the no-answer is parsed into the classical negation $\sim$. Then we have the following result:

(i′) $\sim(\alpha)$

(ii′) $\sim(\sim\alpha) \Rightarrow$ (ii″) α
By (ii′), double negation

(iii′) $\sim(\alpha \wedge \sim\alpha) \Rightarrow$ (iii″) $\sim\alpha \vee \alpha$
By (iii′), De Morgan, double negation

(iv′) $\sim(\sim(\alpha \vee \sim\alpha)) \Rightarrow$ (iv″) $\sim\alpha \vee \alpha$
By (iv′), De Morgan, double negation, commutativity

It turns out that (i′) contradicts (ii′), and (iv′) is redundant with (iii′). Again, our thesis is that another use of modern logical tools is in position to maintain the classical metalogical principles of Indian logicians (or dialecticians, if you please) without entering into the realm of paraconsistency by any means. This alternative reconstruction of ancient texts requires a preliminary distinction between a *sentence* and a *statement*, given the way in which an information is conveyed by means of a common question-answer game.

25.3 Sentences and Statements

What is talked about in both SB and 4CN? The former predications are statements given by an answerer and are about sentences, or propositions. By a *proposition* it is meant that extralinguistic entity with a constant property, following the so-called "Frege's axiom" to the effect that each proposition refers to a (unique) truth-value among the True (T) and the False (F).

At the same time, *sentences* are linguistic, context-dependent vehicles of communication expressing propositions. No further distinction will be made throughout the paper between a proposition and a sentence: although several sentences may express one and the same proposition, it will be assumed in the following that a sentence relates a proposition without ambiguity and can be said to be true or false in an elliptic way.

Now can a sentence be said to be neither true nor false, or even both? The use of statements in the Indian tradition seems to concern speech-acts with a sentential content, so that both assertions and denials are cases in point when we deal with SB and 4CN. Accordingly, the seven *bhaṅgī* are statements upon one given sentential content, and the crucial combination of different standpoints may explain how one and the same "sentence" (a statement, actually) can be said to be true and false at once: not from the same standpoint, as Aristotle stated in his Principle of Non-contradiction. We will return, however, to the fourth predication of SB later on, since it has been viewed as a genuine case of asserted contradiction by some commentators.

25.4 Modalities

Due to the central role of standpoints in the corresponding "nayavāda," SB seems to float between many-valued and modal logic by putting statements into the scope of the relative notion of "*syād*." Indeed, this can be translated as "arguably" and naturally alludes to the modal concept of *possibility*. A way to corroborate this reading is to assess judgments in the Jain epistemology, depending upon whether these are taken to be complete (and thereby lead to truth) or incomplete. For instance, it is argued that

> Mallisena distinguishes a *pramāna* from a *durnaya* and a *naya*. According to him, a *pramāna* is always true and for which we assign the truth-value T, but a *durnaya* is always false for which we assign the truth-value F. The truth-value of a *naya* (incomplete judgment) is different from the truth-value T or the truth-value F; hence it is intermediate between these truth-values. This gives rise to a third *intermediate* truth-value I.[6]

While these commentators see in the intermediate stage between plain truth and plain falsity something similar with the third truth-value (neither true nor false), the correlated distinction between different grades of truth (from absolutely to absolutely not) is also reminiscent of Hugh MacColl's "symbolic logic" (see [13]).

According to [8], sentences can be true, false, certain, impossible, or variable; these semantic predicates appear as five truth-values that can be equally rendered in a bivalent modal logic of necessity, contingency, and impossibility. In a sense, this many-valued or modal system includes SB insofar as every sentence α cannot be but variable, i.e., *contingent* from the Jain perspective. To put it in formal terms, each Jain statement α is such that there is both some standpoint w_i at which α is

---

[6]Bharucha [3], p. 182.

arguably true: $v(w_i, \alpha) = \text{T}$, and some standpoint w_j at which α is arguably false: $v(w_j, \alpha) = \text{F}$.

More importantly, the common analogy between *modality* and *quantification* brings more light upon the Jain valuation of statements: SB is not a modal logic in the sense of iterated modalities, like in S4 or S5; rather, it is a modal system that assesses the complex value of every object by ranging over a set of viewpoints.

First of all, let us emphasize that the Jain model proceeds as a "supermodel," that is, a set of heterogeneous models in which the contradictory speech-acts of assertion and denial are not separate from each other but, on the contrary, combined to form a list of inconsistent but complementary viewpoints. Such a semantic frame relies upon an epistemology, viz., a theory of justification that can accept an indefinite number of particular standpoints among seven main sorts of *naya*: *naigama-naya* (non-distinguished standpoint), *saṃgraha-naya* (collective standpoint), *vyavahāra-naya* (particular standpoint), *ṛju-sūtra-naya* (momentary standpoint), *śabda-naya* (synonym standpoint), *samabhirūḍha-naya* (etymological standpoint), *evaṃbhūta-naya* (momentary etymological standpoint). Taking again the Jain model as a supermodel W, i.e., a set of heterogeneous sets of worlds $\{w_1, \ldots, w_n\} \in W$ with different assessments, W can be viewed as a set of various sorts of standpoints: ontic (w_1), epistemic collective (w_2), epistemic individual (w_3), temporal (w_4), grammatical (w_5, w_6), and grammatical temporal (w_7). By this way, SB somehow leads to a multimodal logic by conflating ontic, epistemic, temporal, topological (and grammatical) standpoints or "worlds" assessing a (non-indexical) sentence at once.

Another reason to promote the use of **QAS** in the present context is that the bitstrings are logical values that help take account of the holistic sense of *not-being*. Unlike the mainstream view of nonexistence as an absolute absence of the corresponding object, the following explanation of the Jain conflated standpoints puts into emphasis the meaning of existence as a relative not-being-as (some given property) that accommodates with some other existent properties.

> The existence of an entity such as a pot, depends upon its being a particular *substance* (an earth-substance), upon its being located in a particular *space*, upon its being in a particular *time*, and also upon its having some particular (say, dark) *feature*. With respect to a water substance, it would be nonexistent, and the same with respect to another spatial location, another time (when and where it was nonexistent), and another (say, red) feature. It seems to me that the *indexicality* of the determinants of existence is being emphasized here.[7]

While echoing the *Nyaya* school, according to which negative qualities are located into objects by equating the absence of a quality (F, say) with the presence of its negative counterpart (not-F, say), this perspectivist view of existence amounts to say that, if an object x is F, then there is some property G such that not-G is in x. This naturally argues for the logical values of **QAS**, i.e., these bitstrings assigning a set of present or absent properties to every object α in L. In other words, "being" and "not-being" are treated as object determiners such that, for every object α, $\exists\, \mathbf{a}_i \exists\, \mathbf{a}_j$ $(\mathbf{a}_i(\alpha) = 1 \;\&\; \mathbf{a}_j(\alpha) = 0)$.

---

[7]Matilal [9], p. 132.

Moreover, this *holistic* ontology also entails that every object α (whether a concept, sentence, or whatever meaningful in a language) is a determinate set of properties that stands between absoluteness, or *pramāna*, and nothingness, or *durnaya*. In **QAS**, absoluteness means that some property is assigned to a given object in every context: $\forall\, \mathbf{a}_i\ (\mathbf{a}_j\ (\alpha) = 1)$, while nothingness is the contrary case in which it is assigned in no context: $\forall\, \mathbf{a}_i\ (\mathbf{a}_j\ (\alpha) = 0)$. All the objects stand between these polar unconditional judgments, so that, for every object α in SB, the logical value of is given by a set of different property assignments from the seven distinctive standpoints.

$$\begin{array}{ll}
\mathbf{A}(\alpha) = \langle \mathbf{a}_1(\alpha), \ldots, \mathbf{a}_n(\alpha) \rangle & w_1 \in W \\
\mathbf{A}(\alpha) = \langle \mathbf{a}_1(\alpha), \ldots, \mathbf{a}_n(\alpha) \rangle & w_2 \in W \\
\vdots & \vdots \\
\mathbf{A}(\alpha) = \langle \mathbf{a}_1(\alpha), \ldots, \mathbf{a}_n(\alpha) \rangle & w_n \in W
\end{array}$$

Assuming that every object α is defined not only by what it is but, also, by what it is not, this complex valuation leads to a *many-sorted* set of answers $\mathbf{a}_i^j$, whereby i stands for the ith defining property assigned to α (successfully or not) and j for the ontological status (presence or absence, respectively) of α in a given context.

A way to account for *bivalence* as a *paribhāṣā* of SB is to rephrase it in terms of questions-answers: every sentence is either affirmed or denied from a single standpoint. Ontologically speaking, the duality between presence and absence entails that affirming the presence of a property G in F is equivalent with not affirming, i.e., denying its absence or, equivalently, denying the presence of its negative counterpart not-G (from one single standpoint). In symbols:

$$\mathbf{a}_i^1(\alpha) = 1 \Leftrightarrow \mathbf{a}_i^2(\sim \alpha) = 0$$

It also follows from this that α refers to an ordered join of bits in $\mathrm{W} = \{w_1, \ldots, w_n\}$. The general form of this semantic analysis in SB can be depicted as follows.

What is α in a given set W of complementary standpoints?
In symbols: $\mathbf{A}(\alpha, W) = ?$ (with $W = \{w_1, w_2, w_3, w_4, w_5, w_6, w_7\}$)

The ith property (A, say) is *present* in α iff, in *some* standpoint w_i, α is A in w_i.
In symbols: $\mathbf{a}_i^1(\alpha) = 1$ iff $\exists w_i\ (\mathbf{a}_i^1(\alpha, w_i) = 1)$.

The ith property (A, say) is *absent* in α or, alternatively, the negative property not-A is present in α, iff in *some* standpoint w_i α is not A or, alternatively, α is not-A in w_i.
In symbols: $\mathbf{a}_i^2(\alpha) = 1$ iff $\exists w_i\ (\mathbf{a}_i^2(\alpha, w_i) = 1)$.

Note that there is no difference in SB between *external* and *internal* negation: not to be F and to be not-F mean the same, whereas a clear difference will be made afterwards in 4CN in this respect. Likewise, the call for different standpoints entails that the presence of A in α need not amount to negate the absence of A in the object α: two affirmative answers can be actually given, when A is present in α in a given context and absent in α in another one.

In the light of this multiple valuation, every object α of SB is such that it can be said to be somehow *everything* by sharing any of the properties at hand from the different complementary standpoints. Recalling the *disjunctive* combination of standpoints in SB, it results in the following sample that accounts for the Jain non-one-sided theory to truth (*anekantavada*):

	α is A	α is B	α is C	α is D	α is E	
$\mathbf{A}(\alpha)=$	1	1	1	1	1	
	1	0	0	1	0	w_1
	0	1	1	0	0	w_2
	0	0	0	1	1	w_3
	1	1	0	0	0	w_4
	0	0	1	1	0	w_5
	0	1	1	0	1	w_6
	1	0	1	1	1	w_7

A corollary of this valuation is that any two objects of $\mathbb{L}$ are *subcontrary* (compatible) to each other by having at least one common property in contexts: everything is watered, firing, smooth, hard, and so on, from similar or different standpoints.

The converse result can be observed with the Mādhyamaka valuation in 4CN, within the common background of **QAS**. Unlike the disjunctive and complementary set of viewpoints in SB, the following states that only one context must be assigned to every object α: it is A or not so in an absolute, one-sided way. This leads to an alternative conjunctive reading of standpoints, such that a property cannot be assigned to an object once the least evidence (standpoint) may count against it. Without entering into the ontological foundations that underlie the Buddhist logic, the upshot of such a skeptic-minded assessment is the following one, where absence expresses a commitment for the speaker that relevantly differs from non-presence.

What is α in a given set W of complementary standpoints?
In symbols: $\mathbf{A}(\alpha, W)=?$ (with $W = \{w_1, w_2, w_3, w_4, w_5, w_6, w_7\}$)

The ith property (A, say) is *present* in α iff in *every* standpoint w_i, α is A in w_i.
In symbols: $\mathbf{a}_i^1(\alpha)=1$ iff $\forall w_i\ (\mathbf{a}_i^1(\alpha, w_i)=1)$.

The ith property (A, say) is *absent* in α or, alternatively, the negative property not-A is present in α, iff in *every* standpoint w_i, α is not A or, alternatively, α is not-A in w_i.
In symbols: $\mathbf{a}_i^2(\alpha)=1$ iff $\exists w_i\ (\mathbf{a}_i^2(\alpha, w_i)=1)$.

The ith property (A, say) is *not present* in α or, alternatively, the negative property not-A is present in α, iff in *some* standpoint w_i, α is not A in w_i.
In symbols: $\mathbf{a}_i^1(\alpha)=0$ iff $\exists w_i\ (\mathbf{a}_i^1(\alpha, w_i)=0)$.

The Mādhyamaka epistemology is such that no object can be absolutely said to be present or absent in this more stringent perspective. The multidimensional valuation in 4CN is such that α is said to be "nothing," but not as an asserted absence of any property from any standpoint. Rather, it should be said that α is not *something* for want of any definite property to be assigned to it. Here is below the resulting valuation for a denial of present properties, instead of an assertion of absent properties. It means that any two objects are *contrary* (incompatible) to each other, insofar as none of these are said to have some property in common.

	α is A	α is B	α is C	α is D	α is E	
$\mathbf{A}(\alpha)=$	0	0	0	0	0	
	1	0	0	1	0	w_1
	0	1	1	0	0	w_2
	0	0	0	1	1	w_3
	1	1	0	0	0	w_4
	0	0	1	1	0	w_5
	0	1	1	0	1	w_6
	1	0	1	1	1	w_7

The multiple valuation of **QAS** in the form of bitstrings attempted to show how the Indian judgments can be streamlined in a quite consistent and bivalent way, once the context-dependence of statements is made explicit in the metalanguage. This clearly questions the paraconsistent aspect of SB and 4CN, especially in their treatment of the so-called Principle of Contradiction (thereafter: PNC).

25.5 Contradictoriness

Is the PNC to be taken as one of the basic *paribhāṣā* of Indian thought, unlike the paraconsistent reading of the so-called "Eastern thought"?

A first task is to recall what this Aristotelian principle actually means in symbols. A "strong" version of paraconsistency is to the effect that *every* object has a property P and does not have it at once, thereby meaning that a given property is both present and absent in it. Not only did Aristotle wrongly assume this interpretation to reduce the Heraclitean position *ad absurdum*, thus claiming that nothing can be said any more under such a position. Indeed, a proper classical denial of PNC should lead to the denial of the following logical form:

(PNC$_1$) $\forall x \forall \mathrm{P} \sim(\mathrm{P}(x) \wedge \sim\mathrm{P}(x))$

where x corresponds to any object α in **QAS**. Accordingly, any infringement of this principle means that *not every* property *cannot* be both present and absent in an object, i.e., some property can be both present and absent in some object. In symbols:

(PNC$_2$) $\sim(\forall x \forall \mathrm{P} \sim(\mathrm{P}(x) \wedge \sim\mathrm{P}(x))) \equiv \exists x \exists \mathrm{P}\, (\mathrm{P}(x) \wedge \sim\mathrm{P}(x))$

This obviously differs from the much stronger view that every property cannot not be present and absent in an object, i.e., *every* property *must* be both present and absent in an object. In symbols:

(PNC_3) $\forall x \forall P\, (P(x) \wedge \sim P(x))$

The point is that (PNC_3) is a contrary of (PNC_1), while (PNC_2) is its mere contradictory. Now any infringement to PNC amounts to contradict its universal application and accept some exception with a mere counterexample, thereby sustaining its validity from a metalinguistic point of view (as Aristotle himself argued rightly in the Book 3 of his *Metaphysics*, but from an unduly strong interpretation of the Heraclitean stance).

A second task is to outweigh the role of *contexts* with respect to PNC. Aristotle stated that no object can have and not have some property "from the same respect," i.e., from the *same* standpoint. SB does not refute PNC as it stands, accordingly, so long as its acceptance of inconsistent statements relies upon the acceptance of different standpoints at once. Now the real difficulty remains with the meaning of the third basic predication, namely: "avaktavyam." Does the latter mean that some property is both present and absent in an object *from the same standpoint*? If it does, then dialetheism gives a right analysis of it; if it does not, then dialetheism is nothing but misleading.

In order to settle this problem, let us consider the theoretical background in which the Jain theory has been developed. On the one hand, it consists in a discussion about metaphysical statements or related subjects. On the other hand, it should be noticed that an overall opposition prevailed between Indian scholars about what there is. To quote some of these ancient schools, "Everything is Brahman (the ultimate reality)" for the Advaita Vedantin whereas "Everything is empty by its own nature" for the Mādhyamika school of Buddhism. "Everything is a vijñapti (awareness maker)" for the Yogācāri, whereas "Everything is non-spiritual, impermanent, and painful" for the Buddhists. "Everything is knowable and nameable" for the Nyāya-Vaiséṣika, whereas "Everything is non-one-sided" for the Jains. Useless to say these schools somehow overlap each other by epistemological or ontological assumptions such as realism or antirealism.

At any rate, our main point is that each of their statements is a categorical statement that includes a universal quantification with "everything." It is clear that an assessment of a particular sentence concerning empirical data cannot be the same as one about a universal sentence concerning metaphysical entities. Therefore, is the logical form of these statements the crucial source of the trouble in SB and its fourth predication "*avaktavyam*"?

Let be a sentence of the form "F is G," in accordance to the Indian concern with general terms like F or G. Then two sorts of inconsistency can be then distinguished from a set-theoretical approach, assuming that the copula "is" has the sense of *membership* in the Indian above statements. In the first case, a universal predication of G is

(U_1) "Everything is G"

which is inconclusive because *irrelevant* from a relevant sense of information. In the second case, a universal predication of G is

(U_2) "G is everything"

which is inconclusive because *self-contradictory* and leads to the well-known paradox of self-reference.

A comparison between first-order logic and set theory may help give a better view of the two different levels of inconsistency. On the one hand, (U_1) means that G is a property of any object, whether F or not F. That is,

(U_1) $(F \subseteq G \cap \bar{F} \subseteq G)$
$\forall x\ ((Fx \supset Gx) \wedge (\sim Fx \supset Gx))$

This means that the sentence is expressible but irrelevant, given that nothing can be denied of the object F. This universal object does make sense, however, so that we hardly believe why it should be said to be "unsayable" as the Sanskrit word "avaktavyam" is supposed to mean in SB. On the other hand, (U_2) means that G is a property that includes every other one, *including itself* (as the subject of the sentence). Thus:

(U_2) $(F \subseteq G \cap F \nsubseteq G)$
$\forall x\ ((Fx \supset Gx) \wedge \sim(Fx \supset Gx))$

The latter version of inconsistency seems to be a more appropriate transcription of what the Jains meant by something "inexpressible" or "unsayable," provided that PNC is seen by them as a universal *paribhāṣā*. Now the fundamental motivation of dialetheism is an acceptance of (U_2), thereby accepting its paradoxical self-contradiction from the outset. Is this what is also meant by "avaktavyam" in SB?

In **QAS**, an interpretation of (U_2) is such that $\mathbf{A}(G) = \top$, that is, for every $\mathbf{a}_i(G) \in \mathbf{A}(G)$, $\mathbf{a}_i(G) = 1$. Now let us recall the open-minded or non-one-sided import of the Jain philosophy, in order to settle the problem of meaning with "avaktavyam." Starting from the latter and the question whether it amounts to infringe the Principle of Explosion, an *inferentialist* account of meaning would be such that no object can make sense once it does not correspond to a sequence of determinate properties. In this sense, a sentence is "avaktavyam" whenever it is about a metaphysical subject that behaves as an ultimate class. Let G for "Brahman," or "ātman," and F for any predicate in $\mathbb{L}$. Then not only is F both G and not G, because of its being universal. But also, F is neither (only) G nor (only) not G by being an ultimate class that goes beyond the realm of subsumed entities (i.e., objects that fall under the scope of other concepts). The result is a "hypercontradictory" statement about the sentence α, such that the object G is both true-and-false of F and neither-true-nor-false of F! While being reminiscent of what has been devised elsewhere by [10, 17] under the heading of "impossible truth-values," we hardly believe that the Jain combination of one single speech-act (assertion or denial, i.e., negative assertion) in several viewpoints could be reversed into a combination of several speech-acts into one single standpoint. Rather, the interpretation we favor here is that an object F (which is equally a property of other objects, according to our holistic theory of meaning) is said to be "unsayable" once nothing determinate can be predicated of it. If so, then F is *impredicable*: nothing meaningful can be said about F because of its self-contradictoriness. Although dialetheism is a technically feasible alternative logic, a large number of comments about the Jain philosophy opt for the opposite view that PNC is a *paribhāṣā* that lies behind "avaktavyam" and cannot accept self-contradictory sentences.

To sum up, two mainstream options have been proposed thus far to make sense of the fourth predication in SB.

Either we opt for the glutty option of a "*simultaneous* assertion and denial," stating that a given sentence is both true and false from the same standpoint. This yields a double yes-answer to whether α is true (i.e., G is present in F) or false (i.e., G is absent in F, or not-G is present in F). In symbols:

$$\mathbf{a}_i^1(\alpha) = \mathbf{a}_i^2(\alpha) = 1$$

Or we opt for the gappy option of an "unassertable" sentence, meaning that the latter is neither true nor false (simultaneous non-truth and non-falsity) from the same standpoint. This yields a double no-answer about the sentence's truth and falsity. In symbols:

$$\mathbf{a}_i^1(\alpha) = \mathbf{a}_i^2(\alpha) = 0$$

Above these two usual interpretations in the literature about SB, our own one is an inferentialist option of "inexpressibility" or mere *silence* (a non-answer in **QAS**). Indeed, we just inferred from the statement about the universal class G that such a metaphysical sentence results in a double yes-no answer: it is both true-and-false and neither-true-nor-false and, therefore, nothing as it stands, leading to a plain nonsense. This constitutes a lack of logical value for such a sentence, rather than an inflationist valuation combining nonclassical values toward an increasingly indefinite expansion of combined speech-acts. In symbols:

$$\mathbf{a}_i^1(\alpha) = \mathbf{a}_i^2(\alpha) = \{1, 0\} = \varnothing$$

By virtue of this threefold scenario to account for "avaktavyam," the misunderstanding about PNC can be disentangled in at least three views of inconsistency.The first reading is a "light" version of inconsistency that is harmless for PNC, where a given statement accepts a sentence α (that, e.g., F is G) and its negation $\sim\alpha$ (that F is not G) from different standpoints (i and j, say). In symbols: $\mathbf{a}_i^1(\alpha)=\mathbf{a}_j^2(\alpha)=1$. The second version is a "mild" one where both α and $\sim\alpha$ are accepted from one and the same standpoint (i, say). In symbols: $\mathbf{a}_i^1(\alpha)= \mathbf{a}_i^2(\alpha)=1$. And the third is a "strong" version, such that the single sentence α is both accepted and denied from a single standpoint. In symbols: $\mathbf{a}_i^1(\alpha)=\mathbf{a}_i^2(\alpha)= \{1,0\}$.

Actually, it turns out by (EQ) that mild and strong contradiction are nothing but one single form of inconsistency: whoever asserts α and its negation $\sim\alpha$ from the standpoint does also assert and deny α at once, so that there is no grade of inconsistent involvement but only one expression of PNC, i.e., the allegedly "mild" one which departs from the undue form of "light" inconsistency. No genuine inconsistency stands between a sentence and its negation from *different* standpoints, just as no inconsistency occurs between a sentence and its negation from two different models. The "supermodel" generated by SB may have created some misunderstanding in this respect; indeed, the Principle of Explosion can be dismissed under the proviso that a sentence and its negation belong to a unique model including different standpoints, as the case is with our reconstruction of SB. However, this conflation of models into a single big one does not imply any infringement of PNC since no sentence can be said to be true and false at once. It is well known that not every paraconsistent system violates PNC, unlike Priest's dialetheism or da Costa's hierarchy of inconsistent systems. In this respect, such a distinction between explosion and contradiction does justice to those who equate SB with paraconsistency. Our objection mostly concerns the formal character of Jain "logic," given that no set of logical constants is assigned to its ancient texts.

Again, we defend a conservative, consistent-preserving reading of SB at the meta-level of speech-acts, where the non-Fregean logical values of **QAS** still obey bivalence. This treatment can be also applied to 4CN, with the difference that the latter set of statements is not about a single sentence but, rather, an extension of an initial atomic sentence into its negated conjuncts or disjuncts. While tampering our preceding assumption that no logical constant is mentioned in the ancient Indian "proto-logics" (i.e., theories of judgment without proper formal languages), the point is that falsity and negation seem to be on a par so long as the latter constant connotes something like exclusion and separation (between two sentences or speech-acts about them, i.e., corresponding statements). Let us see if the so-called "Indian negations" are really consistency-preserving, and if nothing differs between the use of negation in SB and 4CN.

25.6 Negation

What of negation if SB contains no connectives, and can SB and 4CN be said to be *dual* (proto-)logics? Once again, we said that SB is a set of judgments about one single, atomic sentence, whereas 4CN is depicted as a set of four atomic and molecular statements.

However, the dialetheist Graham Priest denied the latter reconstruction by replacing the molecular sentences (iii′)–(iv′) (see Sect. 25.2, p. 3) with single valuations. He claims that 4CN requires a jump into a 5-valued logic, both reminiscent of MacColl's modal logic whilst departing from it in the interpretation of its logical values (as shown in [15]):

> The most obvious way to proceed is now to take this possibility as a fifth semantic value, and construct a 5-valued logic. Thus, we add a new value, E, to our existing four (T, B, F, and N).[8]

[8] Priest [12], p. 15.

This many-valued interpretation of a single atomic sentence α can be rendered with the following formalization, which extends Belnap's 4-valued logic FDE and its combined bivalent values (truth and falsity) with an additional element (i.e., the empty set).

4CN	Semantic reading
(i′)	$v(\alpha) \neq$ T
(ii′)	$v(\alpha) \neq$ F
(iii′)	$v(\alpha) \neq$ B
(iv′)	$v(\alpha) \neq$ N

For every sentence α, let $v(\alpha)$ be a sentential valuation function that maps α into $V_5 = \{$T, B, F, N, E$\}$. Assuming that the domain of values V_5 is both *exclusive* and *exhaustive*, the set of denials (i′)–(iv′) entails that $v(\alpha) =$ E. Now it should be noted that the above syntactic reading presents the repeated denial as a *sentential* negation but proceeds semantically as a *metalinguistic* negation that applies to values rather than sentences.

While being a further reminiscence of MacColl's use of denial, the advantage of Priest's 5-valuedness is that no more connectives are needed in the predications of 4CN. At the same time, this many-valued approach presents at least two disadvantages.

First, seven or three "truth-values" have been proposed in [11] as a logical analysis of SB, whereas five values are advocated for 4CN. This entails that *duality* is lost between SB and 4CN, contrary to our view according to which, as suggested by the title of [1], 4CN is like a dual of SB by reverting its (unique) logical value for every statement.

Second, bivalence is equally lost by favoring many-valuedness and assigning nonclassical truth-values to sentences.

Alternatively, **QAS** aims at restoring the duality of SB and 4CN with respect to their respective valuations, focusing on their dialectic nature by means of a question-answer game. It also purports to clarify the status of *silence* as a failure of logical value, rather than an extra value which has been exemplified by Priest's 5 truth-value. By making a clear-cut distinction between the Fregean truth-value of a sentence and the non-Fregean logical value of a statement, let us see now to what extent the Mādhyamaka system of 4CN can be rendered as a counterpart of the Greek (Pyrrhonian) *ou mallon*.

To begin with, there are two readings of negation in the light of **QAS**.

The first negation occurs in SB and corresponds to the mainstream sentential, *locutionary* negation that attaches to the content of a speech-act. By analogy with Frege's unary theory of judgment (only one judgment prevails for the latter, i.e., assertion as a process of truth acknowledgment), the Sanskrit concept of *paryudāsa pratiṣedha* is translated by the speech-act of a negative assertion and amounts to the assertion of a negative sentence. That is: in the answer value $\mathbf{a}_i(\alpha)$, negation belongs to the sentential content of α and is symbolized by the unary functor $\sim$.

The second negation pertains to 4CN and corresponds to an *illocutionary* negation, related to the speech-act of denial and expressed by a no-answer. A Sanskrit counterpart of it is the concept of *prasajya pratiṣedha* and amounts to a metalinguistic negation, or nonassertion. In the valuation $\mathbf{a}_i(\alpha) = 0$, denial expresses a negative answer to the question whether the sentence α is either true or false and recalls the skeptic suspension of judgment. This does not imply that denial is not a judgment at all, however: suspension is a proper decision made by the Mādhyamaka, for want of any access to the mundane reality; by doing so, his speech-acts of denial is like a second-order commitment toward his noncommitment in the truth-value of a given sentence, contrary to the Jain who always makes first-order commitments about the truth-value of sentences.

Borrowing from the symbolism used in Searle's *Speech-Acts*, the first negation of a sentence α is an assertion of the negated sentence $\sim\alpha$. This positive speech-act is symbolized by Frege's turnstile as follows: $\vdash\sim\alpha$, by opposition to the negative speech-act applied to a sentence (whether affirmative or negative): $\dashv \alpha$. These distinctive negations of SB and 4CN make the two judicative systems appear as clear mutual *mirror images* of each other ($\vdash$ versus $\dashv$), but the properties of the first sentential negation remain to be clarified.

For this purpose, we propose to construct a general logical system within the conceptual background of **QAS**. This system includes both SB and 4CN as two one-valued subsystems, and sentential negation is viewed in both as a unique logical constant.

First of all, the meaning of a sentence α is a pair of two answers to corresponding questions about its bivalent truth-value (i.e., true or false):

$$\mathbf{A}(\alpha) = \langle \mathbf{a}_i^1(\alpha), \mathbf{a}_i^2(\alpha) \rangle$$

This twofold questioning includes the first two basic predications of SB (and 4CN) while omitting the third one, since "avaktavyam" means a lack of value rather than a third one.

Then the logical constant of sentential negation proceeds as a *permutation* operator upon the ordered answers. By extension, the following definition of sentential negation holds irrespective of the number m of questions about the properties of α. For every sentence α such that $\mathbf{a}_i(\alpha) = \langle\, \mathbf{a}_i^1(\alpha), \ldots, \mathbf{a}_i^m(\alpha) \rangle$:

$$\mathbf{a}_i(\sim \alpha) = \langle \mathbf{a}_i^m(\alpha), \ldots, \mathbf{a}_i^1(\alpha) \rangle$$

25.7 Logical Values

The general logical system at hand, namely: $\mathbf{AR}_4$, is a logic of acceptance and rejection with four logical values. It is a common logic for both Buddhist and Jain schools as well as the Aristotelian, classical logic with its two exclusive logical values. A comparison can be made between this non-Fregean system and Belnap's 4-valued logic FDE, while noticing that only one logical value occurs separately in SB and 4CN.

$\mathbf{AR}_4$	FDE	Interpretation
$\mathbf{A}(\alpha)=\langle 1{,}0\rangle$	$v(\alpha)=\mathrm{T}$	α is true from some standpoint false from no standpoint
$\mathbf{A}(\alpha)=\langle 1{,}1\rangle$	$v(\alpha)=\mathrm{B}$	α is true from some standpoint false from some standpoint
$\mathbf{A}(\alpha)=\langle 0{,}1\rangle$	$v(\alpha)=\mathrm{F}$	α is true from no standpoint false from some standpoint
$\mathbf{A}(\alpha)=\langle 0{,}0\rangle$	$v(\alpha)=\mathrm{N}$	α is true from no standpoint false from no standpoint

Indeed, this valuation results in 2 *dual* one-valued logics where the logical values are answers to questions concerning sentential truth-values. Bivalence is clearly defended as a common *paribhāṣā* of the Indian proto-logics, given that, for every α in $\mathbf{AR}_4$:

For every $\mathbf{A}_i(\alpha) = \langle\, \mathbf{a}_i^1(\alpha), \mathbf{a}_i^2(\alpha)\rangle$, $\mathbf{a}_i(\alpha)=1$ or $\mathbf{a}_i(\alpha)=0$
(where $1=$yes and $0=$no)
$\mathbf{a}_i^1(\alpha)=1$ iff $v(\alpha)=\mathrm{T}$
$\mathbf{a}_i^2(\alpha)=1$ iff $v(\alpha)=\mathrm{F}$

By interpolation, we can also devise a number of definitions for the sentential connectives of conjunction and disjunction in $\mathbf{AR}_4$, in the light of its classical metalanguage.[9] That is:

$\mathbf{a}_i^1(\alpha \wedge \psi)=1$ iff $v(\alpha \wedge \psi)=\mathrm{T}$, i.e., $v(\alpha)=v(\psi)=\mathrm{T}$
$\mathbf{a}_i^2(\alpha \wedge \psi)=1$ iff $v(\alpha \wedge \psi)=\mathrm{F}$, i.e., $v(\alpha)=\mathrm{F}$ or $v(\psi)=\mathrm{F}$

$\mathbf{a}_i^1(\alpha \vee \psi)=1$ iff $v(\alpha \vee \psi)=\mathrm{T}$, i.e., $v(\alpha)=\mathrm{T}$ or $v(\psi)=\mathrm{T}$
$\mathbf{a}_i^2(\alpha \vee \psi)=1$ iff $v(\alpha \vee \psi)=\mathrm{F}$, i.e., $v(\alpha)=v(\psi)=\mathrm{F}$

The motivation behind this all-embracing system relies upon the different epistemologies advocated by SB, 4CN, and the traditional Aristotelian logic. This has already been explained elsewhere (e.g., in [17]), but we have to return to this point in order to justify our non-Fregean valuation.

As [5] put it, ancient logics did not share the same criteria of truth-assignment to a sentence. This is due to their metaphysical discrepancies about what makes a sentence true, which means that not only one constraint was put on their truth-ascriptions. On the one hand, [5] presented the Aristotelian logic as a realist "doctrinalism" (or dogmatism) where every sentence can be said to be true or false without any evidence at hand:

[9]We omit here the definition of conditional in $\mathbf{AR}_4$, because we do not need it for our present concerns. However, a strengthened definition of it in $\mathbf{AR}_4$ will be given in a forthcoming paper: "'If', and only 'if,'" such that $(p \rightarrow q) \neq_{df} ({\sim}p \vee q)$.

> it is always possible, in principle, to discover which of two inconsistent sentences is true, and which is false.[10]

It results in a standard valuation where only one relevant question is to be asked about the property of a sentence, i.e., whether it is true. As to the second question imposed by $\mathbf{AR}_4$, it is irrelevant insofar as the truth of a sentence implies the falsity of its sentential negation (and vice versa). This explains why only two logical values of $\mathbf{AR}_4$ can be included into the Aristotelian logic. Let us depict the Aristotelian question-answer game as follows, where the sentence is about, e.g., α's being B or not:

$$\mathbf{q}_i(\alpha) = \langle \mathbf{q}_i^1(\alpha), \mathbf{q}_i^2(\alpha) \rangle$$

The *exclusive* answers can be easily understood by the interpretation of the corresponding questions $\mathbf{q}_i^1$ (α): "Is α B?" and $\mathbf{q}_i^2$ (α): "Is α not B?". For if it is true that α is B, then it is not true (i.e., false) that α is not B, and conversely. Bivalence is thereby imposed on these two logical values by a restriction upon the possible answers. Thus for every α:

$$\mathbf{a}_i^1(\alpha) = 1 \text{ iff } \mathbf{a}_i^2(\alpha) = 0$$

in a subdomain of $\mathbf{AR}_4$ of the Aristotelian valuation $V_A = \{\langle 1{,}0 \rangle; \langle 0{,}1 \rangle\}$.

Turning again to the general valuation of **QAS**, where the objects α are not only sentences but whatever meaningful, an application to Aristotle's standard logic amounts to say that every sentence is true or false from a *unique* standpoint j which is the real world as it is. Thus, for every object α in $\mathbb{L}$:

$$\mathbf{A}(\alpha) = \langle \mathbf{a}_1^j(\alpha), \ldots, \mathbf{a}_n^j(\alpha) \rangle, \text{ with } W = j = \{w_1\}$$

so that every property of α refers to a single bit in w_0 (the "real world").

To take an example, let us take an arbitrary object α in **QAS**, e.g., a general term that can be individuated by $n = 5$ relevant properties A, B, C, D, E. Each of the ensuing predications yields a definite sentence about what the object α is, and the logical form of α is thus given by the 5-tuple of answers $\mathbf{A}(\alpha) = \langle \mathbf{a}_1(\alpha), \mathbf{a}_2(\alpha), \mathbf{a}_3(\alpha), \mathbf{a}_4(\alpha), \mathbf{a}_5(\alpha) \rangle$. Let $\mathbf{A}(\alpha) = 11010$, where $\mathbf{a}_2(\alpha_1) = 1$, $\mathbf{a}_1(\alpha, w_1) = \mathbf{a}_2(\alpha, w_1) = \mathbf{a}_4(\alpha, w_1) = 1$, and $\mathbf{a}_3(\alpha, w_1) = \mathbf{a}_5(\alpha, w_1) = 0$. Then it can be stated that only one Boolean value can be assigned to each element of the whole 5-bitstring, since only one standpoint is available for every compound predication.

This is not so, with the Indian epistemologies underlying SB and 4CN.

Against Aristotle's dogmatism, the Jain condition for truth-ascription is more tolerant and can be compared to the Greek traditions of eclecticism, or even sophism. Ganeri [5] calls by "perspectivism" this weaker two-sided truth-ascription theory. According to this conventional view of truth, to be expressed by the Sanskrit word "*saṃvṛti-satya*," the precondition for truth is

---

[10] [5],p. 268.

> to find some way conditionally to assent to each of the sentences, by recognizing that the justification of a sentence is internal to a standpoint.[11]

The difference with Aristotle's theory of judgment can be expressed by a nonstandard, twofold questioning about single properties. The "supermodel" attached to SB entails that, unlike the previous case, a same sentence can be affirmed *and* negated at once (from distinctive standpoints, however). More precisely, the Jain questioning in 4CN is still of the form

$$\mathbf{q}_i(\alpha) = \langle \mathbf{q}_i^1(\alpha), \mathbf{q}_i^2(\alpha) \rangle,$$

the real difference coming from the interpretation of the two corresponding questions. To account for the crucial role of the concept of evidence ("*syād*", i.e., "arguably") in the Jain tolerant truth ascription, we take $\mathbf{q}_2^1(\alpha)$ to mean "Is α arguably B?" and $\mathbf{q}_2^2(\alpha)$, "Is α arguably not B?" (where $i = 2$ denotes the second predication about B). The occurrence of at least one affirmative standpoint for every statement about A leads to another, non-Aristotelian subdomain of $\mathbf{AR}_4$ to characterize the Jain valuation, namely: $V_J = \{\langle 1,1 \rangle\}$. Taking again the set of five properties that define **A**: $\mathbf{A}(\alpha) = \langle \mathbf{a}_1(\alpha), \mathbf{a}_2(\alpha), \mathbf{a}_3(\alpha), \mathbf{a}_4(\alpha), \mathbf{a}_5(\alpha) \rangle$, each of the Jain predications (or "bhaṅga") also results from a questioning of the form $\mathbf{q}_i(\alpha) =$ "Is α so?".

Furthermore, each of the seven Jain predications can be reconstructed in the light of **QAS** as a number of either absolute ("pramana": α is always so, or "durnaya": α is always not-so) or relative judgments about what α is (α is sometimes so, sometimes not-so). Due to the essential relativity of standpoints in the Jain philosophy, we claim that some of the seven predications cannot be endorsed by a Jain speaker, i.e., those which amount to absolute (not relative) judgments. We make the difference in the following exemplification of a Jain valuation, by declaring as "available" these possible predications about whether α is so: $\mathbf{a}_i^1(\alpha)$, as opposed to the cases of unavailability from the Jain perspectivist stance. The corresponding Fregean truth-value is associated under each of these predications, together with its non-Fregean sentential counterpart in $\mathbf{AR}_4$.

Bhaṅga (1): *syad asty eva* (unavailable: α is so from *every* standpoint)
$\forall w_i \in W, \mathrm{a}_i^1(\alpha, w_i) = 1$
Fregean valuation: $v(\alpha) = \{\mathrm{T}\}$
Non-Fregean valuation in $\mathbf{AR}_4$: $\mathbf{a}_i^1(\alpha) = 1$, $\mathbf{a}_i^1(\alpha) = 0$, hence $\mathbf{a}_2(\alpha) = \langle 1,0 \rangle$
Bhaṅga (2): *syan nasty eva* (unavailable: α is so from *no* standpoint)
$\forall w_i \in W, \mathrm{a}_i^1(\alpha, w_i) = 0$
Fregean valuation: $v(\alpha) = \{\mathrm{F}\}$
Non-Fregean valuation in $\mathbf{AR}_4$: $\mathbf{a}_i^1(\alpha) = 0$, $\mathbf{a}_i^1(\alpha) = 1$, hence $\mathbf{a}_2(\alpha) = \langle 0,1 \rangle$

Bhaṅga (3): *syad asty eva syan nasty eva* (available: α is so from *some* (*but not every*) standpoint)
$\exists w_i \exists w_j \in W, \mathrm{a}_i^1(\alpha, w_i) = \mathrm{a}_i^1(\alpha, w_j) = 1$
Fregean valuation: $v(\alpha) = \{\mathrm{T,F}\} = \{\mathrm{B}\}$

[11] [5], p. 268.

Non-Fregean valuation in $\mathbf{AR}_4$: $\mathbf{a}_i^1(\alpha) = \mathbf{a}_i^2(\alpha) = 1$, hence $\mathbf{a}_2(\alpha) = \langle 1,1 \rangle$

Bhaṅga (4): *syad avaktavyam eva* (unavailable: α is so from *no* standpoint)
For every $w_i \in W$, $\mathrm{a}_i^1(\alpha, w_i) = \varnothing$
Fregean valuation: $\{\varnothing\}$
Non-Fregean valuation in $\mathbf{AR}_4$: $\mathbf{a}_i^1(\alpha) = \mathbf{a}_i^1(\alpha) = \varnothing$, hence $\mathbf{a}_2(\alpha) = \varnothing$

Bhaṅga (5): *syad asty eva syad avaktavyam eva* (unavailable: α is not so from *no* standpoint)
$\exists w_i \exists w_j \in W$, $a_i^1(\alpha, w_i) = 1$ and $a_i^1(\alpha, w_j) = \varnothing$
Fregean valuation: $\{T,\varnothing\}$
Non-Fregean valuation in $\mathbf{AR}_4$: $\mathbf{a}_i^1(\alpha) = 1$, $\mathbf{a}_i^1(\alpha) = \varnothing$, hence $\mathbf{a}_2(\alpha) = \langle 1,0 \rangle$

Bhaṅga (6): *syan nasty eva syad avaktavyam eva* (unavailable: α is so from *no* standpoint)
$\exists w_i \exists w_j \in W$, $a_i^1(\alpha, w_i) = 0$ and $a_i^1(\alpha, w_j) = \varnothing$
Fregean valuation: $\{F,\varnothing\}$
Non-Fregean valuation in $\mathbf{AR}_4$: $\mathbf{a}_i^1(\alpha) = 1$, $\mathbf{a}_i^1(\alpha) = \varnothing$, hence $\mathbf{a}_2(\alpha) = \langle 0,1 \rangle$

Bhaṅga (7): *syad asty eva syan nasty eva syad avaktavyam eva* (available: α is so from *some* (*but not every*) standpoint)
$\exists w_i \exists w_j \exists_k \in W$, $a_i^1(\alpha, w_i) = 1$, $a_i^1(\alpha, w_i) = 0$, and $a_i^1(\alpha, w_j) = \varnothing$
Fregean valuation: $\{T,F,\varnothing\}$
Non-Fregean valuation in $\mathbf{AR}_4$: $\mathbf{a}_i^1(\alpha) = 1$, $\mathbf{a}_i^1(\alpha) = 1$, hence $\mathbf{a}_2(\alpha) = \langle 1,1 \rangle$

Finally, 4CN embeds a skeptic-minded or Pyrrhonist epistemology, such that a strong two-sided truth-ascription theory is required by its absolute approach to truth (in Sanskrit: "*param ārtha-satya*"). Ganeri [5] depicts it as the view that

> the existence both of a reason to assert and a reason to reject a sentence itself constitutes a reason to deny that we can justifiably either assert or deny the sentence.[12]

The Mādhyamaka questioning is not as demanding as the Jain perspectivism, because of the unclear logical form of the denied Tetralemma: is the set of four denied statements about one atomic sentence α, as suggested by Priest's many-valued analysis, or a number of molecular sentences related to the initial α?

Let us first suppose an *atomic* interpretation of 4CN. Then it requires a nonstandard, fourfold questioning outside the domain of $\mathbf{AR}_4$, namely: $\mathbf{q}_i(\alpha) = \langle\, \mathbf{q}_i^1(\alpha)$, $\mathbf{q}_i^2(\alpha)$, $\mathbf{q}_i^3(\alpha)$, $\mathbf{q}_i^4(\alpha)\rangle$, with $\mathbf{q}_i^1(\alpha)$: "Is α definitely so?", $\mathbf{q}_i^2(\alpha)(\alpha)$: "Is α definitely not-so?", $\mathbf{q}_i^3(\alpha)(\alpha)$: "Is α definitely so and not-so?", and $\mathbf{q}_i^4(\alpha)(\alpha)$: "Is α definitely neither so nor not-so?". Correspondingly, the systematic denial of the four statements should lead to a unique logical value in $V_M = \{\langle 0, 0, 0, 0 \rangle\}$ where α cannot be made true (or true-and-false) and cannot be made false (or neither-true-nor-false) by every standpoint. At the same time, Priest's 5-valuedness is replaced here by 1-valuedness in the form of a fourfold no-answer. That is:

[12] Ganeri [5], p. 268.

4CN	Semantic reading
(i)	$v(\alpha) \neq \langle 1,0,0,0\rangle$
(ii)	$v(\alpha) \neq \langle 0,1,0,0\rangle$
(iii)	$v(\alpha) \neq \langle 0,0,1,0\rangle$
(iv)	$v(\alpha) \neq \langle 0,0,0,1\rangle$

Some problems arise with this interpretation, however.

First, the third question $\mathbf{q}_i^3(\alpha)$ violates PNC (a *paribhāṣā*, again) and is not independent from the two first ones. Indeed, α is definitely so and α is definitely not-so whenever α is definitely so and not-so.

Second, this non-Fregean valuation is not complete because there still remains eleven other combinations among the $2^4 = 16$ bitstrings.

In order to avoid these troubles, let us admit by now a *molecular* interpretation of 4CN. This yields another nonstandard, twofold questioning about single properties: $\mathbf{q}_i(\alpha) = \langle \mathbf{q}_i^1(\alpha), \mathbf{q}_i^2(\alpha)\rangle$, with $\mathbf{q}_i^1(\alpha)$: "Is α definitely so?" and $\mathbf{q}_i^2(\alpha)$: "Is α definitely not-so?". The result is a unique logical value inside the domain of $\mathbf{AR}_4$, $V_M = \{\langle 0, 0\rangle\}$. This valuation means that α cannot be made true and cannot made false by every standpoint, in such a way that Priest's 5-valuedness is replaced by one-valuedness in the form of a twofold no-answer. We can reconstruct this result as follows, rendering the systematic denial by the speech-act of no-answer rather than the sentential negation.

4CN	Syntactic reading	Semantic reading
(a)	$\dashv \alpha$	$\mathbf{a}_i^1(\alpha) = 0$
(b)	$\dashv {\sim}\alpha$	$\mathbf{a}_i^2(\alpha) = 0$
(c)	$\dashv (\alpha \wedge {\sim}\alpha)$	$\mathbf{a}_i^1(\alpha) = 0$
(d)	$\dashv ({\sim} (\alpha \vee {\sim} \alpha))$	$\mathbf{a}_i^1(\alpha) = 0$

The valuation of 4CN is such that, for every sentence α, $\mathbf{a}_i^j(\alpha) = 0$, thereby meaning that the set of denied tetralemma (a)–(d) entails that $\mathbf{a}_i(\alpha) = \langle 0, 0\rangle$.

General statements can be made in $\mathbf{AR}_4$ about the previous free epistemologies, whether in the form of committed assertions or merely cautious denials.

According to Aristotle's dogmatism, the world of independent substances is such that no thing is everything and every thing is something. In symbols: $\vdash \forall\alpha(\mathbf{A}(\alpha) \neq \top \ \& \ \mathbf{A}(\alpha) \neq \bot)$.

According to the Jain perspectivism, the participation of the things to common properties from different standpoints entails that every thing is somehow everything and no thing is nothing, due to the existence of at least one positive statement for every object. In symbols: $\vdash \forall\alpha(\mathbf{A}(\alpha) = \top)$.

As for the skeptic trend of the Mādhyamaka, the impermanence of the world leads to the paradoxical negative statement that not something is something. In symbols: $\dashv \exists\alpha\ (\mathbf{A}(\alpha) \neq \bot)$. But again, the latter does not lead to a nihilist assertion about everything: $\vdash \forall\alpha(\mathbf{A}(\alpha) = \bot)$ does not hold for him.

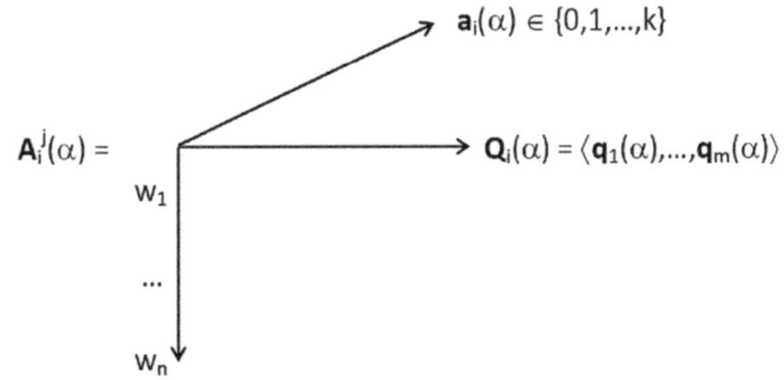

Fig. 25.1 A three-dimensional view of logical values

25.8 Conclusion: Logic and Logic

A general reflection has been proposed in the present paper upon a number of theories, in order to make sense of what is currently presented as "paraconsistent Indian logics."

Concerning the *theory of valuation*, a distinction has been made in the way to establish the meaning of different entities like objects, concepts, or sentences. Truth-values have been replaced by some "marked" values, in the vein of Belnap's combination of bivalent truth-values. By doing so, we took valuation as an increasing process of *dichotomy* from a primary class (the True) to a range of values obtained by partition.

Following the Principle of Bivalence, valuation is a class of two elements that serve as the common proper names of sentences. The corresponding domain of values V_2 occurs in all the bivalent, *classical Tarskian logics*.

Once many-valuedness is allowed, the logical values do not proceed as proper names and are more like definite descriptions. The set of logics sharing such a domain of values $V_{>2}$ are said to be *nonclassical Tarskian logics* and admit of more than two classes of elements to characterize the sentences.

As a last step of this dichotomous process, our non-Fregean valuation applies to any sort of meaningful information (beyond the sole domain of sentences) and is such that every valued object is a *unique* proper name: no other object shares the same logical value and appears as the singleton of a unique class, insofar as a logical value helps individuate every given information in the dialectical context of a question-answer game. The corresponding domain of values V_m^n has been described in **QAS** and leads to a number of *non-Tarskian logics*.

A general representation of this logical value results in the following many-dimensional object (see Fig. 25.1), to be characterized by three main features: a *predicative* set of m questions; a *many-sorted* set of k corresponding sorts of answer ("yes," "no," or some further ones); and a *quantified* set of standpoints where the question-answer game is played.

Concerning the *theory of logic*, its main feature is consequence but another one has been sketched throughout the present paper. In the case of Indian "proto-logics," these can be considered to be "logics" provided that the general theory of logic is questioned. Otherwise, the Indian systems SB and 4CN would be nothing but trivial set of formulas. For if every sentence α is said to be true in SB, then the logic of SB is (maximally) trivial. In symbols: $\varnothing \models_{SB} \alpha$. If no sentence α is said to be true in 4CN, then the logic of 4CN is (minimally) trivial. In symbols: $\alpha \models_{4CN} \varnothing$. Actually, our view is that SB and 4CN serve as two proper logics with a different *language-game*: any logic of *consequence* relies upon rules for *truth*-preservation between sets of formulas, whereas a logic of *opposition* relies upon rules for *difference*-preservation between set of formulas. An overview of this alternative way of doing logic has been described by [7] as follows:

> On the standard view, logic is concerned with reasoning, more in particular with fixing criteria for the soundness and validity of arguments. (...) Reasoning is just one particular language-game. And if we think of our daily conversations, it does not have the same central position it has in logic. Cooperative information exchange seems a more prevailing linguistic activity.[13]

In other words, our main point is that SB and 4CN are *difference*-friendly theories whose final aim of the game differs from the majority of formal systems with scientific purposes.

This alternative way of doing logic also includes a reply to the objection of monovalence: what is the point of a one-valued logical system, assuming that any logic requires at least two values to make a distinction between valid and invalid arguments? Again, our reference to two different language-games opposes two sorts of logics: Tarskian logics that are scientific theories upon (nothing but) *truth*, on the one hand; non-Tarskian logics like SB and 4CN, which are soteriological theories centered upon *openness*. If so, the latter Indian "logics" are not just alternative logics with a standard Tarskian structure ⟨L,Cn⟩ but, rather, alternative views of Logic with a deviant structure ⟨L,Op⟩ (where Op is an abstract relation of opposition). After all, should any logic serve as a decidable, problem-*solving* machinery? Alternatively, SB and 4CN seem to proceed as problem-*posing* machineries whose final aim does consist in their resolution. Here is a way to reinforce the affinity between such nonstandard activities of logic and the medieval games of *disputatio*.

References

1. Bahm, A.J.: Does seven-fold predication equal four-cornered negation reversed? Philos. East West **7**, 127–130
2. Balcerowicz, P.: Do attempts to formalize the syād-vāda make sense? In: Paper presented at the 11th Jaina Studies Workshop: Jaina Scriptures and Philosophy, pp. 1–51. SOAS, London
3. Bharucha, F., Kamat, R.: Syadvada theory of Jainism in terms of deviant logic. Indian Philos. Q. **9**, 181–187
4. Chakravarti, S.S.: The mādhyamika catuṣkoṭi or tetralemma. J. Indian Philos. **8**, 303–306
5. Ganeri, J.: Jaina logic and the philosophical basis of pluralism. Hist. Philos. Log. **23**, 267–281
6. Gokhale, P.: The logical structure of Syādvāda. J. Indian Counc. Philos. Res. **8**, 73–81

[13][7], p. 109.

7. Groenendijk, J.: The logic of interrogation. In: Matthews, T., Strolovitch, D. (eds.) SALT IX, pp. 109–126. Cornell University, Ithaca (1999)
8. MacColl, H.: Symbolic Logic and its Applications. Longmans, London (1906)
9. Matilal, B.K.: The Jaina contribution to logic. In: Ganeri, J., Tiwari, H.H. (eds.) The Character of Logic in India, pp. 127–139. State University of New Press (1998)
10. Priest, G.: Hypercontradiction. Log. Anal. **107**, 237–243
11. Priest, G.: Jaina logic: a contemporary perspective. Hist. Philos. Log. **29**, 263–278
12. Priest, G.: Four corners - east and west. In: Banerjee, M., Seth, A. (eds.) Logic and its Applications (4th Indian Conference, ICLA 2011), pp. 12–18. Springer, Delhi (2011)
13. Schang, F.: MacColl's modes of modalities. Philosophia Scientiae **15(1)**, 163–188
14. Schang, F.: A plea for epistemic truth: Jaina logic from a many-valued perspective. In: Schumann, A. (ed.) Logic in Religious Discourse, pp. 54–83. Ontos Verlag, Frankfurt (2009)
15. Schang, F.: A one-valued logic for non-one-sidedness. Int. J. Jaina Stud. **9(1)**, 1–25
16. Sylvan, R.: A generous Jainist interpretation of core relevant logics. Bull. Sect. Log. **16**, 58–66
17. Shramko, Y., Wansing, H.: Hyper-contradiction, generalized truth-values and logics of truth and falsehood. J. Log. Lang. Inf. **15**, 403–424

MIX
Papier aus verantwortungsvollen Quellen
Paper from responsible sources
FSC® C105338

If you have any concerns about our products,
you can contact us on
ProductSafety@springernature.com

In case Publisher is established outside the EU,
the EU authorized representative is:
Springer Nature Customer Service Center GmbH
Europaplatz 3, 69115 Heidelberg, Germany

Printed by Libri Plureos GmbH
in Hamburg, Germany